ISBN 978-0-656-11617-1
PIBN 10019873

A SYSTEM

OF

INORGANIC CHEMISTRY

BY

WILLIAM RAMSAY, Ph.D., F.R.S.

PROFESSOR OF CHEMISTRY IN UNIVERSITY COLLEGE, LONDON.

LONDON
J. & A. CHURCHILL
11, NEW BURLINGTON STREET
1891

PREFACE.

For more than twenty years, the compounds of carbon have been classified in a rational manner; and the relations between the different groups of compounds and between the individual members of the same groups have been placed in a clear light. It is, doubtless, owing to the brilliant originators of this method of classification—Kekulé, Hofmann, Wurtz, Frankland, and others too numerous to mention, but whose names occupy a prominent place in the history of our science—that the domain of organic chemistry has been so systematically and successfully enlarged, and that it presents an aspect of orderly arrangement which can scarcely be surpassed.

This has unfortunately not been the fate of the chemistry of the other elements. Nearly twenty-five years have elapsed since the discovery by Newlands, Mendeléeff, and Meyer of the periodic arrangement of the elements; and, in spite of the obvious guide to a similar classification which it furnishes, no systematic text-book has been written in English with the periodic arrangement of the elements as a basis.

The reasons for this neglect have probably been that the ancient and arbitrary line of demarcation between the non-metals and the metals has been adhered to; that too great importance (from the standpoint of pure chemistry) has been assigned to the distinction between acid hydroxides and basic hydroxides (acids and bases), which has tended to obscure the fact that they belong essentially to the same class of compounds, viz., the hydroxides; and that the chemistry of text-books has almost always been influenced by commercial considerations. The first of these reasons has often led, among other anomalies, to the separation of such closely allied elements as boron and aluminium, antimony and bismuth, silicon and tin; the second reason has often led to the ignoring of the double halides, except in a few special instances, and to the neglect of compounds such as double oxides of the sesquioxides of the iron group, in which these oxides play an acidic part; while, for the third reason, those methods of preparing compounds which are of commercial importance are usually given, while other methods, as important from a scientific point of view, are often ignored; the borides, nitrides, &c., have been almost completely neglected since the time of Berzelius; and the less easily obtained elements and compounds have been dismissed with scant notice because of their rarity; whereas they should obviously be considered as important as the commoner ones in any treatise on *scientific* chemistry.

The methods of classification adopted in this book are, as nearly as the difference of subject will permit, those which have led to the systematic arrangement of the carbon compounds. After a short historical preface, the elements are considered in their order; next their compounds with the halogens, including the double halides; the oxides, sulphides, selenides, and tellurides follow next, double oxides, such as sulphates, for example, being considered among the compounds of the simple oxides with the oxides of other elements; a few chapters are then occupied with the borides, carbides, and silicides, and the nitrides, phosphides, arsenides, and antimonides; and in these the organo-metallic compounds, the double compounds of ammonia, and the cyanides are considered; while a short account is given of alloys and amalgams. The chemistry of the rare earths, which must at present be relegated to a suspense account, is treated along with spectrum analysis in a special chapter; and the systematic portion of the book concludes with an account of the periodic table.

The periodic arrangement has been departed from in two instances: the elements chromium, iron, manganese, cobalt, and nickel have been taken after those of the aluminium group; and the elements copper, silver, gold, and mercury have been grouped together and considered after the other elements. It appeared to me that the analogies of these elements would have been obscured, had the periodic arrangement been strictly adhered to.

It has been thought desirable, instead of treating of processes of manufacture under the heading of the respective elements or compounds, to defer a description of them to the end of the book, and to group them under special headings, those compounds being considered together which are generally manufactured under one roof. In describing manufactures, chemical principles have been considered, rather than the apparatus by means of which the manufactures are carried on. The student, having acquired the requisite acquaintance with facts, is now better able to appreciate these principles.

The physical aspects of chemistry have generally been kept in the background, and introduced only when necessary to explain modern theories. I hold that a student should have a fair knowledge of a wide range of facts before he proceeds to the study of physical chemistry, which, indeed, is a science in itself. But short tables of the more important physical properties of elements, and of the simpler compounds, have been introduced for purposes of reference.

It may be asked if such a system is easily grasped by the student, and if it is convenient for the teacher. To this question I can reply that, having used it for four years, I am perfectly satisfied with the results. For the student, memory work is lightened; for the teacher, the long tedious description of metals and their salts is avoided; and I have found that the student's interest is retained, owing to the fact that all the "fire-works" are not displayed at the beginning of the course, but are distributed pretty evenly throughout.

It need hardly be mentioned that the teacher is not required to teach, nor the student to remember, all the facts as they are here set forth. It is necessary to make a judicious selection. But it is of advantage to have the list fairly complete for purposes of reference. It should be stated that, in the case of compounds of questionable existence, they have received the benefit of the doubt. It is at least well that they should be known, in order that their existence may be brought to the test of renewed experiment.

References to original memoirs have been given where important theoretical points are involved; or where doubt exists; and an attempt has been made to guide the reading of students. As a rule, references to recent papers are given; the older references may be found in one of the chemical dictionaries.

WILLIAM RAMSAY.

UNIVERSITY COLLEGE, LONDON,
January, 1891.

CONTENTS.

Part I.

Part II.

THE ELEMENTS.

Part III.

THE HALIDES.

PART IV.

THE OXIDES, SULPHIDES, SELENIDES, AND TELLURIDES.

PART V.

THE BORIDES. THE CARBIDES AND SILICIDES.

PART VI.

THE NITRIDES, PHOSPHIDES, ARSENIDES, AND ANTIMONIDES.

Part VII.

Part VIII.

Part IX.

CONTENTS.

PART IX.

PART I.

CHAPTER I.

INTRODUCTORY AND HISTORICAL.

THE **first** object of the Science of Chemistry is to ascertain the composition of the various things which we see around us. Thus, among familiar objects are air, water, rocks and stones, earth, the bark, wood, and leaves of plants, the flesh, fat, and bones of animals, and so on. Of what do these things consist?

The **second** is to ask, Can such things be made artificially, and, if so, by what methods? Attempts to answer these questions have led to the discovery of many different kinds of matter, some of which have as yet resisted all efforts to split them up into still simpler forms. Such ultimate kinds of matter are termed *elements*. But other kinds of matter can often be produced when two or more of the simpler forms or elements are brought together; the elements are then said to *combine*, and the new substances resulting from their combination are called *compounds*.

The **third** object of the Science of Chemistry is the correct classification of the elements and of their compounds; those substances which are produced in a similar manner, or which act in a similar manner when treated similarly, being placed in the same class.

The **fourth** inquiry relates to the changes which different forms of matter undergo when they unite with each other, or when they split into simpler forms.

Fifthly, the conditions of change are themselves compared with each other and classified; and thus general laws are being deduced, applicable to all such changes.

Lastly, the Science of Chemistry and the sister Science of Physics join in speculations regarding the nature and structure of matter, in the hope that it may ultimately be possible to account for its various forms, the changes which they undergo, and the relations existing between them.

To answer questions such as these, it is obvious that experiments must be made. Each form of matter must be separately exposed to different conditions; heated, for example; or placed under the influence of an electric current; or brought together with other kinds of matter; before it is possible to know what it will do. Now, the ancient philosophers did not perceive this necessity; nor indeed were they much concerned in making the inquiry. Those nations which have left behind them a record of their thoughts, the ancient inhabitants of India, Egypt, Greece, and Rome, devoted their attention, if they aspired to be learned, to oratory, to history, or to poetry. Their only scientific pursuits were politics, ethics, and mathematics. Distinction was to be gained in the forum, in the temple, or in the battlefield; not in wresting secrets from Nature. The practice of such of the arts as were then known was in the hands of slaves and the lower classes of the people, who were content to transmit their methods from father to son, and whose achievements were unchronicled. The citizens of the State, the wealthy and the leisured, despised these low-class arts; and, indeed, it was taught by Socrates and his followers that it was foolish to abandon the study of those things which more nearly concern man, for that of things external to him. It was generally believed that by the exercise of pure thought, without careful observation and experiment, a man could know best the true nature of the objects external to him. Thus Plato says in the 7th Book of the "Republic," "We shall pursue Astronomy with the help of problems, just as we pursue Geometry; but, if it is our design to become really acquainted with Astronomy, we shall let the heavenly bodies alone." Elsewhere he states that, even if we were to ascertain these things, we could neither alter the course of the stars, nor apply our knowledge so as to benefit mankind. And in "Timæus," Plato remarks, "God only has the knowledge and the power which are able to combine many things into one, and to dissolve the one into the many. But no man either is or ever will be able to accomplish either the one or the other operation."

It was impossible, with such a mental attitude towards science, for any accurate knowledge to exist, or for any probable theories to be devised. Yet, as it is interesting to know something of the old ideas concerning matter and its nature, a short sketch will be given here.

The origin of the world was for the ancient philosophers of Egypt, India, and Greece, as it is for ourselves, a subject of the greatest interest; and in attempting to frame some theory to explain the Creation it was necessary to speculate on the nature of matter. The various aspects of matter which we see around us were supposed by **Empedocles** (492 B.C.), and later by **Aristotle** (384 B.C.), to be modifications of one fundamental original material, occurring in various forms, the difference between which was caused by the assumption of certain "elements," or as we should now name them "properties." This original material was imagined by Empedocles to consist of small particles, which he termed *atoms*, or "indivisibles," because they were in his view the ultimate particles into which matter could be divided. Plato imagined such atoms to have the form of triangles of different sizes, equilateral, isosceles, or scalene; and ascribed the "perfection" or "imperfection" of matter to be due to the form of its ultimate particles. But such particles were modified by the "elements" *earth*, *water*, *air*, and *fire;* that is, they assumed a solid, liquid, aeriform, or flaming nature, according to the element which predominated in them. Along with this view, a certain confusion of thought arose which led to the conception that earth, water, air, and fire were actually present in, and constituents of, matter, and that all the elements originated in one, supposed by **Thales** (600 B.C.) to be water, and by **Anaximenes** (about 550 B.C.) to be air or fire. The well-known poem of **Lucretius**, *De rerum naturâ*, is a transcript of these views of the atomic constitution of the universe. But such speculations were wholly without a basis of fact, and led to no new knowledge. These ideas, in all probability, were originally derived from India, where the four elements already mentioned were associated with a fifth and sixth, *ether* and *consciousness*, as appears from the teaching of Bŭddha. The notion that matter was one in kind, modified by certain attributes, developed the belief that by changing the attributes, the matter itself would be transmuted. Thus Timæus is made to say by Plato:—"In the first place, that which we are now calling water, when congealed, becomes stone and earth, as our sight seems to show us [here he refers probably to rock-crystal, a transparent, hard material, which was supposed to be petrified ice]; and this same element, when melted and dispersed, passes into vapour and fire. Air again, when burnt up, becomes fire, and again fire, when condensed and extinguished, passes once more into the form of air; and once more air, when collected and condensed, produces cloud and vapour; and from these, when still more compressed, comes flowing water; and from water come earth and stones once more; and thus generation seems to be transmitted from one to the other in a circle." Here the elements are evidently conceived in their concrete sense; but he goes on to say that certain matter in a state of change partakes of the nature of fire to some extent, and to some extent of the nature of the other elements.

Guided by such considerations, Aristotle gave precision to these speculations by his system of "contraries." The properties shared by all matter in varying proportions were "hotness," "coldness," "moistness," and "dryness." Thus air was hot and moist; fire, hot and dry; water, cold and moist; and earth, cold and dry. By imparting heat to water it becomes steam, that is, air, hot and moist; by taking away its moisture it becomes earth, that is, ice, cold and dry.

The "Timæus" of Plato, which has been quoted several times here, was held in high esteem by the great school of learning which existed at Alexandria during the first centuries of our era. It was here, in all likelihood, that the second great era of chemical theory began. Based on such ideas regarding the constitution of matter, attempts were made to change one substance into another, and above all to transmute the baser metals into gold. The attempt was called *Alchemy* (the Arabic prefix *al* signifying "the"), and from that word, which probably means the dark or secret art, is derived our modern name Chemistry.

In order to realise the attitute of mind which led to the belief in the possibility of the *transmutation of metals*, as the change of one metal into another is called, we must note that it was supposed that the apparent change of one form of matter into another involved the destruction of the first form, and the creation of the second; the properties of the matter were changed, and hence the matter itself was supposed to be changed; no attempt was made, so far as we know, to compare the weights (or masses) of the matter before and after the change had taken place. Pure substances, moreover, were almost unknown, and the separation of an impurity from a compound in many cases entirely altered its properties. Now, the Arabs, who conquered Egypt in the 7th century, and transmitted their knowledge to posterity, possessed a theory of which we learn in the writings of **Geber,** an Arabian alchemist of the 8th century, and in which we can trace a germ of the modern views concerning matter, inasmuch as we find here the first dawn of a conception of a chemical compound, in the modern sense of the word.

Geber, and probably his predecessors the Alexandrians, regarded the metals as alloys of mercury and sulphur in varying proportions. Now-a-days, mercury is the name of a metal which possesses definite unalterable properties; nor does sulphur vary, but is always a distinct substance capable of certain changes, though radically the same throughout these changes. But Geber held that the mercury and the sulphur each varied in kind and in properties, and were not what we should now term definite chemical individuals. His views may best be learned from his own words:—"*It is folly to attempt to extract one substance from another which does not contain it.* But, as all metals consist of mercury and sulphur, it is possible to add to one what is wanting, or to take from another what is in excess." Yet he did not discard the older elements, earth, water, air, and fire, but appears to have regarded them as more remote constituents of matter, while mercury and sulphur were the proximate constituents. The mercury was supposed to impart to metals their brilliancy, their malleability, and their fusibility; while the sulphur which they contained rendered them alterable by fire, which changes many metals to earthy powders. In his writings also we find the first allusion to a connection between the curing of disease and the transmutation of metals, in his illustration, "Bring me the six lepers, and I will heal them," referring to the conversion of six of the metals then known into gold, the seventh.

It is not wonderful that the alchemists should have been led into error by attempts to transmute the metals into gold; for their properties are radically changed by the presence of mere traces of foreign bodies.

Thus the presence of a minute trace of lead or arsenic, for example, renders gold exceedingly brittle, and alters its colour; the presence of a very small quantity of carbon in iron renders it elastic, or if more be present, hard and brittle; a small amount of arsenic in copper colours it white and lowers enormously its power of conducting electricity. These changes, which are still unexplained, received much more attention in the early days of chemistry than of recent years; but it is to be hoped that they will again be exhaustively studied.

For many centuries after Geber's time, although numerous compounds were discovered in the search for gold, no new development of theory can be noticed. The attitude of the Roman Church was hostile to the progress of any knowlege of nature. All learning was in the hands of the priests, and the study of the ancient writers was discouraged or forbidden, as not only useless in itself, but as tending to distract the mind from the higher studies of Divine things. When permitted on sufferance it was with the avowed object of combatting disbelief with its own weapons. **Roger Bacon** (1214—1294) even, who published several works on alchemy, wrote that that science which is not prosecuted with a view of defending the Christian faith leads "to the darkness of hell." Yet after the conquest of Spain by the Arabs, in the beginning of the 8th century, the study of medicine, of mathematics, and of optics slowly grew in the west of Europe. The works of many of the Greek philosophers were known only through Arabic translations. It was not in the nature of the Arabs to originate new theories; they merely preserved those which they had.

In the 15th century, **Basil Valentine**, a Benedictine friar, added one to the two supposed constituents of metals. This was a principle of fixity; something which resisted the action of heat without volatilising into gas. Valentine termed this principle "salt." Again, it is not to be understood that any particular salt is referred to; yet, in the minds of many of Basil Valentine's disciples, the earthy residue obtained by calcining metals such as lead in the air was regarded as the same in essence, from whatever source it had been derived. And this theory was extended to include all matter; it is well described in the words of **Paracelsus** (born 1493, died 1541). "In all things four elements are mingled with each other; among those four, only one is *fixed* and perfect; in that element lies the true 'quintessence.' The other elements are imperfect; yet any one of them is able to tinge and qualify the others, according to its nature. Thus in some the element water predominates; in some fire; in some earth; in others air. In order to separate the predominating element as salt, sulphur, and mercury, each must be broken and destroyed by solution, and calcination, or by such means." "There are various minerals in which the elements are not so firmly locked up as in the metals, and which can be split into their three principles: salt, the fixed element; sulphur, fiery and oily; and mercury, aeriform and watery." (*Wünsch Hüttlein*, Erfurt, 1738, p. 27.)

The language of the mediæval alchemists is most obscure. Not only did they confuse substances now known to be perfectly distinct; not only was their nomenclature ambiguous; but a spirit of mysticism pervaded their writings which led them to believe that it would have been impious to reveal to the common people the processes with which they were acquainted. Chemical elements and compounds form in the pages of their books processions of kings and queens, bridegrooms and brides, lions and dragons, eagles and swans; gold is the Sun; silver the Moon; this king and queen, Apollo and Diana, are devoured, on their bridal eve, by Saturn, (lead), a dragon and serpent which has for ages slept in his rocky cavern. Pluto enters with exceeding heat, expelling the dragon as an eagle with scorched wings, and leaving the royal pair reposing on a bed, white as the mountain snows. Such is Basil Valentine's account of the refining of gold in the second of his "Twelve keys to unlock the door leading to the ancient stone"; among innumerable descriptions of the kind it is one of the few in which the actual processes can be followed under their mystical disguise. We meet with fanciful analogies between the Divine Trinity; the human body, soul, and spirit; and the trio of salt, sulphur, and mercury, in which religion, medicine, and chemistry are mingled in inextricable entanglement. Yet such analogies served a good purpose; they led to the combatting of disease, not as before with charms and incantations, and remedies of a disgusting and fantastic nature, but by the administration of chemical substances as drugs. In spite of all their false theories, connecting certain of the organisms of the bodies with the stars, and these again with the metals, the compounds of which were supposed to act on those organisms with which they were so fancifully related, true progress was made by the only way in which progress is possible—by experiment and deduction; and the virtues of antimony, of mercury, and of other remedies were gradually discovered. This change in the ultimate goal of chemical research was begun by Basil Valentine; its chief advocate, however, was Paracelsus, who boldly announced that "The true scope of chemistry is not to make gold, but to prepare medicines." Yet in his writings there are numerous receipts for the preparation of the *alcahest*, or universal solvent; and of that magic *elixir*, capable not only of converting baser metals into gold, but of conferring on its fortunate possessor long life and eternal youth.

Although no advance in chemical theory was made by the school of alchemists represented by Valentine and Paracelsus, yet the indefatigable labours of these men and of their disciples enriched chemistry by the discovery of many new compounds, and laid a foundation of facts for chemists of a later age.

The era of modern chemistry opens with **Robert Boyle** (1626—1691). In his works, which are very voluminous, we meet with no traces of the spirit of mysticism which prevailed up to his time, but he manifests that aspect of rational inquiry which is typical of modern science. His most important work on chemistry is named "The Sceptical Chymist, or considerations upon the experiments usually produced in favour of the four elements, and of the three chymical principles of the mixed bodies." In it he defines the word element: "The words element and principle are here used as equivalent terms: and signify those primitive and simple bodies of which the mixed ones are said to bo composed, and into which they are ultimately resolved. 'Tis said that a piece of green wood, by burning, discovers the four elements of which mixed bodies are composed, the fire appearing in the flame by its own light; the smoke ascending and readily turning into air, as a river mixes with the sea; the water, in its own form, boiling out at the end of the stick, and the ashes remaining for the element of earth. . . . . . But there are many bodies from whence it seems impossible to extract four elements by fire, and which of these can be obtained from gold by any degree of fire whatsoever?" He then proceeds to consider in detail the supposed evidence for the existence of the Aristotelian elements, and of the principles, salt, sulphur, and mercury; and he finally shows that they cannot be considered as *elements*, using the word in the modern sense of the *constituents* of bodies; and he incidentally points out that compounds, as a rule, do *not* resemble the elements of which they are composed.

Boyle thus successfully combatted the ancient doctrines of the alchemists; and although a belief in alchemy lingered on into the last century, and has even had a few disciples in our own day, yet the formation of the learned societies of Florence (1657), London (1662), Paris (1666), and Vienna, and the open interchange of ideas among men who submitted every doubtful point to the test of experiment, did much to destroy the veil of mysticism which had surrounded the labours of the ancient alchemists, and has ultimately proved the correctness of Boyle's views.

The phenomena of *burning* and *combustion* played a great part in the theories of the mediæval alchemists, and were held to substantiate their views. For when a candle burns, it disappears; the solid is changed into air and flame; a transmutation has taken place, proving the identity of the elements. Many metals, when heated in air, are converted into earthy powders, differing entirely from their originals in properties. Paracelsus seems to have imagined that in certain similar cases, for example when iron pyrites lies exposed to air, "the old demogorgon," as he calls this compound of iron and sulphur, "absorbs the universal salt, whereby it is converted into a greyish crystalline powder." But no consistent theory had been advanced to account for the phenomena of combustion. **John Mayow** (1645—1679), indeed, a contemporary of Boyle's, a medical graduate of Oxford, who practised at Bath, had he lived, would in all probability have advanced the knowledge of chemical theory to the stage which it reached more than a century after his death. During his short life, he anticipated most of the deductions of Lavoisier, who, as we shall shortly see, effected a revolution in the science. Guided by Boyle's researches, and with a rare faculty of devising experiments admirably adapted to decide the points at issue, Mayow pointed out that atmospheric air consists of two kinds, one capable of supporting combustion and life, which he named "*spiritus igno-aerius*," and another, devoid of these properties. He concluded from his experiments that this "*spiritus igno-aerius*," or to give it Lavoisier's name, *oxygen*, was a constituent of nitre or saltpetre, and was also contained in nitric acid; that when oxygen combines with other bodies, such as metals, it increases their weight; that it is the common constituent of acids, sulphuric acid being its compound with sulphur, and nitric acid its compound with the inactive constituent of air, now known as *nitrogen;* and he also devised a method of estimating oxygen by mixing with it one of the compounds of nitrogen and oxygen, *nitric oxide*, a process which was afterwards largely employed, and which has been recently revived. Lastly, he showed the function of oxygen in acid fermentation, and, in his "*Tractatus quinque medico-physici*," in which his investigations and conclusions are recorded, he showed very clearly the part played by oxygen in restoring venous blood to the arterial state, and in maintaining animal heat.

But Mayow was alone in his work; his early death cut short his researches; and his contemporaries and successors did not recognise their merit. **Stephen Hales**, for example, though he prepared in an impure state carbonic acid, nitrogen, hydrogen, and oxygen gases, and also marsh-gas, regarded them all as modifications of air, not as distinct gaseous substances. He ascribed to atmospheric air "a chaotic nature," inasmuch as it was found to be endowed with so many different properties.

But in spite of Mayow's correct surmises regarding the nature of combustion, the opinion which chemists generally held was that when a body was burnt something escaped from it, viz., fire or heat. For although it was well known that combustible substances do not continue to burn in a confined space, this was attributed not to the exclusion of air, but to the prevention of the escape of flame. And in spite of its having been noticed by Boyle and others that metals gain in weight by being calcined, yet no special attention was paid to the fact. So long indeed as what we now know to be different kinds of gases were assumed to be only common air containing impurities, it was impossible to account for the apparent loss of weight which many combustible substances suffer when burnt.

A consistent though erroneous theory of combustion, which served to unite in one group such apparently different processes as the burning of a candle and the conversion of a metal into a "calx" or earthy powder when it was heated in air, was first propounded by **Stahl** (1660—1734). Stahl taught that when a substance burns, it loses something; this he called "phlogiston" (from φλογιστός, inflammable), which signified the common constituent of all combustible bodies. This theory, however, dates from before Stahl's time; phlogiston is identical with the "terra-pinguis" of **Becher** (1635—1682), and the idea that combustible bodies lost a fiery matter, a "sulphur," is even older than Becher. The more readily a substance burns, according to Stahl, the more phlogiston it contains. A substance containing much phlogiston was carbon, or charcoal. And when a metal had lost its phlogiston and had become a "calx," it was possible to restore the lost phlogiston by heating it with charcoal, which would yield up to the calx its phlogiston, again converting it into the metallic state.

But in the meantime the progress of chemistry was furthered by the discovery of many new gases, and the conviction spread not only that gases were not impure atmospheric air, but that matter was capable of existence in three forms, solid, liquid, and gaseous. **Black** (1728—1799) was the first clearly to show (probably about 1755) that carbonic acid gas (carbon dioxide) or "fixed air" was radically distinct from ordinary air, inasmuch as it could combine with or be "fixed" by lime, magnesia, and the caustic alkalies, potash and soda. As acids have this property, Keir suggested that it belonged to the class of acids, and **Bergmann** (1735—1784), following Priestley's suggestion that it was a constituent of air, named it "aerial acid." It is the first substance which was named "gas" (from *geist*, equivalent to *gust*); the name is due to **Van Helmont** (1577—1644), who had noticed that it could be obtained by heating limestone.

The merit of Black's work consists in his having shown that, whereas limestone lost a definite weight by being calcined, its weight is exactly restored if the lime resulting from its calcination is reunited with carbonic acid gas. This was the first successful chemical experiment dealing with *quantities*. A complete investigation of "inflammable air," or, as it is now named, *hydrogen gas*, is due to **Henry Cavendish** (1731—1810). It is the gaseous substance produced when metals such as iron, tin, or zinc are treated with acids. Cavendish, in 1766, proved the identity of the substance from whichever source it was prepared, and examined its properties. He found it to be exceedingly light, and to burn very readily; and it was supposed by some to be the long sought "phlogiston" of Stahl. Cavendish also discovered that its product of combustion was *water*.

But the chemistry of gases, or as it was then termed "pneumatic chemistry," was most advanced by the researches of **Joseph Priestley** (1733—1804). He was the first to devise a convenient method of collecting gases over water or mercury, and his plan is still used in our own day. To him is due the discovery of most of the gaseous substances now known, especially of oxygen gas, on August 1st, 1774, which he named "vital air," owing to its property of supporting life, or "dephlogisticated air," because it was the most ardent supporter of combustion, though not itself combustible. The discovery of oxygen was made independently and almost simultaneously by **Scheele**, a Swede (1742—1786), to whom we also owe the discovery of chlorine.

These discoveries prepared the way for the grand generalisation of Lavoisier. Black, Cavendish, Priestley, and Scheele were all adherents of Stahl's phlogistic theory. But **Lavoisier** (1743—1794), having been shown the method of preparing oxygen by Priestley, who paid him a visit in the autumn of 1774, saw the grand importance of the discovery, and made the great generalisation that, when bodies burn, they *combine* with this constituent of air, to which he gave the name *oxygen*. This discovery laid the foundation of the present science of chemistry; the time was now ripe; and in a very complete series of researches Lavoisier showed first:—that water cannot be converted into earth by boiling, but that it merely dissolves some of the constituents of the glass vessel in which it is boiled, leaving the dissolved matter as a residue after it has evaporated; second, that when tin is heated with air in a closed vessel, although it is changed into a whitish-grey calx, yet the combined weight of the vessel and the tin remains unchanged; thus showing that nothing has escaped from the tin or been lost from the vessel; and that on opening the vessel air entered, so that the whole apparatus increased in weight; and that this increase in weight was practically equal to the increase in weight of the tin due to its conversion into "calx." From this experiment he drew the correct conclusion that the gain in weight of the tin was due to its absorption of one of the constituents of air. Thirdly, he repeated this experiment, substituting the metal mercury for tin; the red powder produced, when heated strongly, yielded up the absorbed gas, identical with the "vital" or "dephlogisticated" air of Priestley, to which Lavoisier gave the name oxygen. Fourthly, he showed that organic matters yield, when burnt, carbonic acid and water; and that carbonic acid, identical with Black's "fixed" air, is produced by the combustion of carbon or charcoal. His views are stated by himself as follows:—

1. Bodies burn only in pure air.
2. This air is used up during combustion, and the gain in weight of the body burned is equal to the loss of weight of the air.
3. The combustible body is generally converted into an acid by its union with pure air; but the metals are converted into *calces* or earthy matters.

To this last statement is due the name "oxygen," or "producer of acids." Up to that date, *acid* (from *acetum*, vinegar) was the name applied to substances with a sour taste, which acted on *calces*, producing crystalline substances, termed salts. Many attempts have since been made to give precision to the conception of the word *acid;* but, however convenient the colloquial use of the word, it has ceased to have a definite chemical signification. It was soon after shown that bodies may possess the defined properties of an acid and yet contain no oxygen.

The discoveries of Lavoisier were owing in great degree to two fundamental conceptions, with regard to which he held the firmest convictions: first, that *heat* was not a substance capable of entering and escaping from bodies like a chemical element, but a condition of matter; and that its gain or loss implied no gain or loss of weight; and second, that matter was indestructible and uncreatable; and that the true measure of its quantity was its mass, or weight; hence the weight of a compound body must equal the sum of the weights of its constituents.

It was many years before Lavoisier's views gained complete acceptance amongst chemists; but the discovery of Cavendish in 1784—85, that the *only* product of the combustion of hydrogen was water, showed the true relations of that important substance to oxygen, and explained many difficulties.

To Lavoisier, too, belongs the merit of having invented a systematic nomenclature, which is still retained in its main features; its convenience and general applicability did much to promote the acceptance of the theory on which it was based.

We have traced the gradual evolution of the science of chemistry from the earliest speculations of the Greek philosophers to the end of last century. With this century opens a new era, which will form the subject of the next chapter.

Note.—The chief works on the history of chemistry are Kopp's *Geschichte der Chemie*, 1843–47; *Entwickelung der Chemie in der neueren Zeit*, 1873; Thomson's *History of Chemistry*, 1830; Meyer's *Geschichte der Chemie*, Leipzig, 1889: the last is specially to be recommended. For short sketches of the subject, see also Muir's *Heroes of Chemistry*, and Picton's *The Story of Chemistry*.

CHAPTER II.

HISTORICAL (CONTINUED).

As most of the common substances which we see around us contain oxygen, their composition could not be determined before it had been shown by Lavoisier that the phlogistic theory was untenable, and before the phenomena of oxidation had received their true explanation. Lavoisier himself showed the true nature of sulphuric acid,* viz., that it was a *compound* of sulphur and oxygen, and not a constituent of sulphur, deprived of phlogiston; and also of carbonic acid,† that it was an *oxide* of carbon, and not carbon deprived of phlogiston. These and similar discoveries of Lavoisier's pointed the way to others, and numerous attempts were made to discover the composition of substances, or to *analyse* them (ἀνάλυσις). And from the time of Lavoisier's enunciation of the true nature of combustion, to the beginning of the 19th century, many analyses were made, and confirmed in many cases also by *synthesis*, that is placing together (σύνθεσις) the constituents of the compounds, so as to reproduce the compound which had been analysed.

At that time very few accurate methods of analysis were known. The *qualitative* composition of compounds was as a rule not difficult to ascertain; but the proportions in which the constituents were contained in the compounds analysed, or their *quantitative* composition, were not accurately determined, and the results of the same experimenter often varied among themselves. It is therefore not to be wondered at that two views were held regarding the composition of compounds: one, of which **Berthollet** (1748—1822) was the author, and which is set forth in his *Essai de Statique Chimique* (1803); he regarded every compound as variable in composition, or, if in some cases its composition was found to be constant, attributed such constancy to the fact that it had been submitted to precisely similar conditions during its

* The name "sulphuric acid" used to be, but is not now, applied to the compound of sulphur and oxygen referred to. According to present nomenclature, the acid contains in addition the elements of water.

† See former note.

preparation at successive times. Berthollet held that the proportion in which elements existed in a compound depended on the relative amounts of the elements present during the change which led to their combination, and on other conditions such as temperature. The other and contrary view, that the same substance had always the same composition, was defended by **Proust** (1755—1826), and the dispute, which was eagerly watched by all chemists, lasted from 1799 to 1808.

But the question had already been decided by **Richter** (1762—1807). The law of "constant proportions," as it is termed, was announced by Richter in the involved language of the phlogistic theory in papers which appeared between 1792 and 1794. Stated in ordinary language, his discovery is as follows:—If two acids, A_1 and A_2, combine with two bases, B_1 and B_2, to form compounds, A_1B_1, A_2B_1, A_1B_2, and A_2B_2, the proportion by weight between A_1 and A_2 in the first two compounds is the same as that between A_1 and A_2 in the second pair if the weight of B_1 and also of B_2 is the same in both cases. Or, to take a particular case:—If 80 grams of sulphuric acid* combine with 62 grams of soda,* or with 94 grams of potash;* and if 108 grams of nitric acid* likewise combine with 62 grams of soda; then 108 grams of nitric acid will combine with 94 grams of potash. Therefore 94 grams of potash are said to be *equivalent* (or of equal value) to 62 grams of soda in their power of combining with acid; and 80 grams of sulphuric acid are equivalent to 108 grams of nitric acid in their power of combining with base. Richter determined and tabulated a number of such "equivalent weights." And **Proust** went still further. In 1799—1801, he showed that tin forms two compounds with oxygen, in which the proportion of oxygen varies not gradually but suddenly; and that iron forms two similar compounds with sulphur; but here he stopped. The discovery of the reason of definite proportions is due to Dalton; it gave a new impetus to the study of chemistry, and has been, in its results, perhaps the most fruitful speculation of any known to science.

John Dalton was born in 1766, at Eaglesfield, in Cumberland. In his younger days he was a schoolmaster at Kendal; he went to Manchester in 1793 as Lecturer on Mathematics and Natural Philosophy in the New College, and afterwards acted as a private mathematical and chemical tutor in Manchester, giving occasional

* These names are used in their old sense of the combinations of the elements sulphur, nitrogen, sodium, and potassium with oxygen. See previous note.

lectures in the larger towns of England and Scotland. He investigated the relations between the temperature and pressure of liquids, the expansion of gases by heat, the solubility of gases in liquids, and other similar subjects; but his discoveries in chemical theory were those which conferred on him a world-wide fame, and have exercised a lasting influence on the science.

It was the habit of the analysts of that time, as it is now, to state their results in parts per 100. Thus Proust gives the following analyses of the compounds of copper and tin with oxygen:—

	"Suboxide of Copper."	"Protoxide of Copper."	"Suboxide of Tin."	"Protoxide of Tin."
Metal........	86·2	80	87	78·4
Oxygen	13·8	20	13	21·6
	100·0	100	100	100·0

It is obvious that, from inspection of the above numbers, no simple relation between the amounts of oxygen in the lower and higher oxides of copper, and in the lower and higher oxides of tin, is evident; yet, if Proust had calculated the ratios, he might have guessed that the proportion of oxygen to copper in the second oxide is nearly double that in the first, viz., 13·8 : 21·5; and similarly with tin, 13 : 24. But still the analyses are not accurate enough to render this proportion self-evident, even if thus stated.

It was during an investigation of two compounds of carbon with hydrogen, viz., marsh gas and olefiant gas, or, as they are now named, methane and ethylene, and two compounds of carbon with oxygen, carbonic oxide and carbonic acid, or, as the latter gas is now called, carbonic anhydride, that Dalton was led to investigate the subject. He found that, if he reckoned the carbon in each the same, then marsh gas contains just twice as much hydrogen as olefiant gas; and carbonic acid just twice as much oxygen as carbonic oxide. He then considered the proportions of hydrogen and oxygen in water, and of hydrogen and nitrogen in ammonia, and having found, first, that *when two elements combine with each other, they do so in constant proportions by weight*, and second, that *when two elements, A and B, form more than one compound with each other, they combine in simple multiple proportions*, he deduced the following laws to account for these facts:—

1. *Each element consists of precisely similar atoms of constant weight.*

2. *Chemical compounds consist of complex "atoms,"* which are*

* As the expression "complex atom" is a contradictory one, it was afterwards replaced by the word "molecule," or "little mass" of atoms.

*produced by the union of the atoms of the constituent elements in simple numerical ratios.**

An example will render these statements clear. Olefiant gas consists of six parts of carbon by weight united with one part of hydrogen; marsh gas of six parts of carbon united with two parts of hydrogen. Similarly, carbonic oxide contains six parts of carbon and eight parts of oxygen; and "carbonic acid," six parts of carbon and 16 of oxygen. The following table shows the relations :—

	Olefiant Gas.	Ratio.	Marsh Gas.	Ratio.
Carbon	85·71 per cent.	6	75·0 per cent.	6
Hydrogen...	14·28 „	1	25·0 „	2

	Carbonic Oxide.	Ratio.	"Carbonic Acid."	Ratio.
Carbon	42·86 per cent.	6	27·27 per cent.	6
Oxygen.....	57·14 „	8	72·72 „	16

It is again evident here that no obvious relation exists between the amounts of hydrogen in marsh gas and olefiant gas, unless they are compared with a uniform weight of carbon. From these results Dalton concluded that olefiant gas consists of one atom of carbon united to one atom of hydrogen, and marsh gas of one atom of carbon united to two atoms of hydrogen; and, similarly, that carbonic oxide is composed of one atom of carbon and one of oxygen, and carbonic acid of one atom of carbon and two of oxygen.

It necessarily follows from this conception that the atom of carbon is six times as heavy as the atom of hydrogen, and that the relative weights of the atoms of carbon and oxygen are as 6 to 8.

Extending these observations to water, the only compound of hydrogen and oxygen then known, the following relation was determined:—

	Water.	Ratio.
Hydrogen..........	11·11 per cent.	1
Oxygen.............	88·88 „	8

Hence Dalton concluded that water is a compound of one atom of hydrogen with one atom of oxygen, and that the atom of oxygen is eight times as heavy as the atom of hydrogen, thus bearing out the conclusions of his former analyses.

Dalton then proceeded to determine the relative weights of the atoms of other elements by similar methods. His numbers are far

* Dalton's *New System of Chemical Philosophy*, 1808; Thomson's *Chemistry* 1807; also edition 1810, Vol. III, p. 441.

from accurate, and indeed, in the above tables, the actual numbers found by him have not been stated, in order to avoid confusion. He next arranged a number of compounds of the elements in classes, according to the number of atoms contained in each class. Thus if only one compound of two elements was known, Dalton assumed it to contain one atom of each element, and named it a *binary compound*, "unless some cause appear to the contrary." If two compounds were known, they were represented as A + B, and as A + 2B; the latter was named a *ternary* compound, because it contained three atoms; and so on with quaternary, &c. Thus he regarded water as a binary compound, in which one atom of hydrogen weighing 1, and one atom of oxygen weighing 8 relatively to the hydrogen were united. Ammonia, a compound of nitrogen and hydrogen, was regarded as also composed of one atom of hydrogen weighing 1 and one atom of nitrogen weighing $4\frac{2}{3}$. Thus he constructed a table of atomic weights; and to render his theory more tangible, he assigned to each element a symbol; thus oxygen was ○, hydrogen ⊙, nitrogen ⦶, sulphur ⊕, and so on; and the symbols of the metals consisted of circles circumscribed round the initial letter of the name of the metal; thus Ⓘ stood for iron, Ⓩ for zinc, and so on. These symbols also stood for the relative weights of the atoms; hence ○⊙ denoted water, ⊙⦶ ammonia, ●⊙ olefiant gas, ⊙●⊙ marsh gas, and so with others.

Now it is evident that Dalton here made a great assumption, inasmuch as he had no sure basis to guide him in assigning such atomic weights. Let us consider his results from another point of view, and we shall see that another set of atomic weights might with equal justice have been adopted.

Turning back to the table on p. 17, it is seen that Dalton assumed that the four substances, marsh gas, olefiant gas, carbonic oxide, and "carbonic acid" each contained one atom of carbon. But it is equally justifiable to assume that each of the first pair contains one atom of hydrogen, and each of the second pair one atom of oxygen. We should then have the ratio:—

	Olefiant Gas.	Ratio.	Marsh Gas.	Ratio.
Carbon	85·71 per cent.	6	75·0 per cent.	3
Hydrogen...	14·28 „	1	25·0 „	1

	Carbonic Oxide.	Ratio.	"Carbonic Acid."	Ratio.
Carbon	42·86 per cent.	6	27·27 per cent.	3
Oxygen	57·14 „	8	72·72 „	8

The smallest amount of carbon in combination is now found

to weigh three times as much as the hydrogen; *i.e.* the atomic weight of carbon is 3. And the first body would then consist of 2 atoms of carbon and 1 of hydrogen; while the second, marsh gas, would contain 1 atom of carbon and 1 of hydrogen. Similarly, carbonic oxide might be composed of 2 atoms of carbon and 1 of oxygen, while "carbonic acid" might consist of 1 atom of carbon and 1 of oxygen.

Dalton himself was quite aware of this difficulty, as is seen by his remarks in the appendix to his second volume, published in 1827. He therefore contented himself by assuming those numbers to be the correct atomic weights which give the simplest proportions between the numbers of atoms contained in all the known compounds of the elements. But Dalton did not possess the analytical skill necessary to determine the composition of the compounds from which such deductions were to be made. In 1808, **Wollaston** published an account of accurate experiments on the carbonates and oxalates of sodium and potassium, in which he showed that the ratio of carbonic acid or oxalic acid in one (the "subcarbonate" or "suboxalate") to the sodium or potassium was half that which it bore in the other (the "supercarbonate" or "superoxalate"). The work of determining the composition of compounds was, however, chiefly undertaken by **Berzelius**, professor of chemistry, medicine, and pharmacy in Stockholm (1779—1848). The aim of this great chemist was to forward the work which had been suggested by Dalton, and, by preparing numerous compounds and analysing them, to determine the ratios of the weights of their atoms. His industry was untiring, and the number of new compounds prepared and analysed by him almost incredible. But it is obvious that for the reasons stated it is impossible, even by comparing all the compounds which one element forms with others, to determine which compound contains only 1 atom of that element. What Dalton and Berzelius really determined was the **equivalents** of the elements, that is, the proportion by weight in which they are capable of combining with or replacing 1 part by weight of hydrogen; they had no data sufficient to enable them to determine what multiple of the equivalent is the true atomic weight. In subsequent chapters the various reasons in favour of the atomic weights at present assigned to the elements will be discussed. We must leave the historical part of the subject at this point, and proceed to discuss the facts of the science, and to arrange the various compounds in an orderly manner.

Assuming, then, that, for reasons to be given hereafter, the relative weights of the atoms are represented by the numbers used in this book, the question arises, what element should be made the standard of comparison? Dalton having found that, of all the elements investigated by him, a smaller weight of hydrogen entered into combination than of any other element, assigned the weight 1 to the atom of that element, and arranged the other atomic weights accordingly. Thus, according to him, the weight of an atom of oxygen was 8 times that of an atom of hydrogen, because water, which he supposed to consist of 1 atom of each, was found on analysis to contain 1 part by weight of hydrogen combined with 8 parts by weight of oxygen. And so with the other elements. There are reasons which will follow in their place (p. 202) for believing that a number between 15·87 and 16·00 (or double the number assigned by Dalton) represents the relative weight of an atom of oxygen referred to hydrogen as unity. But it happens that the equivalents of most of the elements have been determined by synthesising or analysing their compounds with oxygen, or with oxygen and some other element. Hence it appears advisable to accept the atomic weight of oxygen as 16, and to refer the weights of the other elements to that scale. Until the ratio between the atomic weights of hydrogen and oxygen is satisfactorily determined, this appears the best course to pursue; for then the accepted atomic weights of the majority of the elements need not be altered to suit any proposed alteration in the ratio of the accepted atomic weights of hydrogen and oxygen. Moreover this plan has the great advantage that many of the atomic weights are whole numbers, and are therefore more easily remembered. It should here be noticed that if the ratio between the atomic weights of hydrogen and oxygen is really 1 to 15·96, then by placing the atomic weight of oxygen equal to 16, that of hydrogen is no longer 1, but 1·0025, for 15·96 : 16 :: 1 : 1·0025.

A very remarkable relation between the atomic weights of the elements and their chemical and physical properties was pointed out by Mr. **J. A. R. Newlands** in 1864,* and this relation has been further studied by Professors **Mendeléeff**† and **Lothar Meyer**.‡ It is briefly this. If the elements be arranged in the order of their atomic weights in seven double columns, those elements which resemble each other fall in the same column. It is on this principle that the elements and their compounds are classified in this text-book. Such an arrangement is termed a periodic arrangement,

* *Chem. News*, July 30th, 1864; August, 1865; March, 1866; also *On the Discovery of the Periodic Law*, Spon, 1884.

† *Annalen*, Suppl., **8**, 133 (1869).

‡ *Annalen*, Suppl., **7**, 354.

and the following table is named the **periodic table.** The letters, such as H, Li, &c., are abbreviations for the names of the elements; they are termed symbols; and they also represent the numbers which precede or follow them. Thus O represents not merely oxygen, but 16 parts by weight of oxygen; CaO represents not merely a compound of calcium and oxygen, but of 40·08 parts by weight of calcium, and 16 parts by weight of oxygen; $CaCl_2$ represents a compound of 40 parts by weight of calcium with 2 × 35·46 parts by weight of chlorine. Such a representation of compounds by the symbols of the elements which they contain is termed a **formula.**

While most of the elements are represented by the initial letters of their English names, some of the symbols require explanation. The following is a list:—

Na, *Natrium* (connected with the word *nitre*)	Sodium.
K, *Kalium* (from *alkali*, an Arabic name)..........	Potassium.
Cu, *Cuprum* (Latin)	Copper.
Ag, *Argentum* (Latin)	Silver.
Au, *Aurum* (Latin)............................	Gold.
Hg, *Hydrargyrum* (Greek = water-silver)	Mercury.
Sn, *Stannum* (Latin)	Tin.
Pb, *Plumbum* (Latin)	Lead.
Sb, *Stibium* (Latin)...........................	Antimony.
W, *Wolfram*, a mineral containing Tungsten........	Tungsten.
Fe, *Ferrum* (Latin)	Iron.

Note.—For this portion of chemical history, Wurtz's *History of the Atomic Theory*, London, 1880, may be consulted; also Cook's *The New Chemistry*; and the works previously referred to.

The Elements, arranged in the Periodic System.

I.	II.	III.	IV.	V.	VI.	VII.	VIII.
(a) 1 (b) H	(a) — (b)	(a) — (b)	(a) — (b)	(a) — (b)	(a) — (b)	(a) — (b)	
Li 7	Be 9	B 11	C 12	N 14	O 16	F 19	
23 Na	24·5 Mg	27 Al	28·5 Si	31 P	32 S	35·5 Cl	
K 39	Ca 40	Sc 44	Ti 48	V 51·5	Cr 52·5	Mn 55	Fe 56, Co 58·5, Ni 58·5.
(63·5 Cu)	65·5 Zn	70 Ga	72·5 Ge	75 As	79 Se	80 Br	
Rb 85·5	Sr 87·5	Y* 89	Zr 90	Nb 94	Mo 95·5	? 100	Ru 101·5, Rh 103, Pd 106·5.
(108 Ag)	112 Cd	114 In	119 Sn	120·5 Sb	125 Te	127 I	
Cs 133	Ba 137	La* 142·5	Ce 140·5	?‡ 141	?§ 143	?‖ 150	? 152, ? 153, ? 154.
156 ?	158 ?	159 ?	162† ?	166 ?¶	167 ?	169 ?	
? 170	? 172	Yb* 173	? 177	Ta 182·5	W 184	? 190	Os 191·5, Ir, 193, Pt 194·5.
(197 Au)	200 Hg	204 Tl	207 Pb	208 Bi	214 ?	219 ?	
? 221	? 225	? 230	Th 232·5	? 237	U 240	? 244	

* Position doubtful. † Terbium? ‡ Neodymium? § Praseodymium? ‖ Samarium? ¶ Erbium?

NOTE.—The atomic weights are in this table given only to the nearest half unit.

Table of Atomic Weights of Elements (O = 16.).

Aluminium	Al	27·01	Nickel	Ni	58·6
Antimony	Sb	120·30	Niobium	Nb	94
Arsenic	As	75·09	Nitrogen	*N*	14·03
Barium	Ba	137·00	Osmium	Os	191·3
Beryllium	Be	9·1	Oxygen	*O*	16·00
Bismuth	Bi	208·10	Palladium	Pd	106·35
Boron	B	11·0	Phosphorus	P	31·03
Bromine	Br	79·95	Platinum	Pt	194·3
Cadmium	Cd	112·1	Potassium	K	39·14
Cæsium	Cs	132·9	Praseodimium	Prd	143·6
Calcium	Ca	40·08	Rhodium	Rh	103·0
Carbon	C	12·00	Rubidium	Rb	85·5
Cerium	Ce	140·3	Ruthenium	Ru	101·65
Chlorine	*Cl*	35·46	Scandium	Sc	44·1
Chromium	Cr	52·3	Selenium	Se	79·0
Cobalt	Co	58·7	Silicon	Si	28·33
Copper	Cu	63·40	Silver	Ag	107·930
Erbium	Er	166	Sodium	Na	23·043
Fluorine	*F*	19·0	Strontium	Sr	87·5
Gallium	Ga	69·9	Sulphur	S	32·06
Germanium	Ge	72·3	Tantalum	Ta	182·5
Gold	Au	197·22	Tellurium	Te	125 ?
Hydrogen	*H*	1 to 1·0082	Terbium	Tb	162 ?
Indium	In	113·7	Thallium	Tl	204·2
Iodine	I	126·85	Thorium	Th	232·4
Iridium	Ir	193·0	Tin	Sn	119·1
Iron	Fe	56·02	Titanium	Ti	48·12
Lanthanum	La	142·3	Tungsten	W	184·0
Lead	Pb	206·93	Uranium	U	240·0
Lithium	Li	7·02	Vanadium	V	51·4
Magnesium	Mg	24·30	Ytterbium	Yb	173
Manganese	Mn	55·0	Yttrium	Y	89
Mercury	Hg	200·2	Zinc	Zn	65·3
Molybdenum	Mo	95·7	Zirconium	Zr	90
Neodymium	Ndi	140·8			

Solid, Liquid, *Gas*.

Note.—In this table recent determinations have been incorporated with the mean results given by Clarke ("Constants of Nature," Part V, 1882). It is to be understood that the last digit of the figures given may vary within one or two units. Thus zirconium = 90 means that the atomic weight is not certain, and may be 89·5 or 90·5; thallium = 204·2 leaves it uncertain whether the true weight is 204·1 or 204·3; and so on. Where a query (?) is appended, it is to be understood that the weight given may be one or more units wrong. The standard works on the subject are by Clarke, mentioned above; by Lothar Meyer and Seubert, *Die Atomgewichte der Elemente;* and, as a model of research, by Stas, *Recherches sur les Rapports réciproques des Poids atomiques*, Brussels, 1860.

Table of Metric Weights and Measures

Measures of Length.

1 metre = 10 decimetres = 100 centimetres = 1000 millimetres.

1 metre = 1·09363 yard = 3·28090 feet = 39·37079 inches.

Log *n* metres + 0·0388704 = log yards; + 0·5159930 = log feet; + 1·5951743 = log inches.

Log *n* yards + 0·9611296 = log metres; log *n* feet + 0·4840071 = log decimetre; log *n* inches + 0·4048257 = log centimetres.

Measures of Capacity.

1 cubic metre = 1000 litres = 1,000,000 cubic centimetres = 1,000,000,000 cubic millimetres.

1 litre = 61·02705 cubic inches = 0·035317 cubic foot = 1·76077 pints = 0·22097 gallon.

Log *n* litres + 1·7855223 = log cubic inches; + $\bar{2}$·5479838 = log cubic feet; + 0·2457026 = log pints; + $\bar{1}$·3443333 = log gallons.

Log *n* cubic inches + 1·2144774 = log cubic centimetres.

Log *n* cubic feet + 1·4520162 = log litres.

Log *n* gallons + 0·6556667 = log litres.

Measures of Weight.

1 gram = weight of 1 cubic centimetre of water at 4°.

1 kilogram = 1000 grams = 100,000 centigrams = 1,000,000 milligrams.

1 kilogram = 2·2046213 lbs.; = 35·273941 oz. = 15432·35 grains.

Log *n* kilos. + 0·3433340 = log lbs.; log *n* grams + 1·5474540 = log oz.; + 1·1884323 = log grains.

Log *n* lbs. + $\bar{1}$·6566660 = log kilograms; log *n* grains + 2·8115677 log grams.

PART II.—THE ELEMENTS.

CHAPTER III.

HYDROGEN; LITHIUM, SODIUM, POTASSIUM, RUBIDIUM, CÆSIUM; BERYLLIUM, CALCIUM, STRONTIUM, BARIUM; MAGNESIUM, ZINC, CADMIUM; BORON, SCANDIUM, YTTRIUM, LANTHANUM, YTTERBIUM; ALUMINIUM, GALLIUM, INDIUM, THALLIUM.

The elements, it has been seen, when arranged in the order of their atomic weights, fall into certain groups. The various members of these groups resemble each other in their physical and chemical properties, and it is therefore advisable to consider the members of each group in connection with each other. They possess certain properties in common, while exhibiting individual peculiarities. In the following chapters, an account will be given of the **sources** of the elements, whether they occur "free," or "native," that is, as elements, or whether combined with other elements in the form of compounds; of their **properties**; and of the methods of their **preparation**; but fuller details will in some cases be given under the heading of the compounds from which they are prepared.

In the main, the order of the periodic table will be followed; but, as it is still under investigation, and the position of all the elements cannot be regarded as finally settled, certain elements will be grouped together which do not occur near each other in the table.

Group I.—Hydrogen, Lithium, Sodium, Potassium, Rubidium, Cæsium.

Sources.—Hydrogen occurs *free* in the neighbourhood of volcanoes, owing probably to the decomposition of its compound with sulphur, hydrogen sulphide, by the hot lava through which it issues. It has also been proved by the evidence of the spectroscope (see chap. XXXV), to exist as element in the atmosphere of the sun, in certain fixed stars; in nebulæ, and in comets. It has been found associated with iron and nickel in many meteorites. In combination with oxygen it occurs in *water* (hence its name from ὕδωρ, *water*, γεννάω, *I produce*) in the sea, lakes, rivers, in the atmosphere, in many minerals; in all organised matter, animal and vegetable. It is thus one of the most widely distributed and abundant of elements.

Preparation.—1. By heating its compounds with boron, carbon, silicon, nitrogen, phosphorus, arsenic, antimony, sulphur, selenium, tellurium, iodine, or palladium to a red heat; or with oxygen, chlorine, or bromine to a white heat (see these compounds).

2. By the decomposition of its compounds dissolved in water by an electric current (see p. 62).

3. By displacing it from these compounds by means of certain metals. The most usual methods of preparation are by the action (*a*) of *sodium* on *water* (oxide of hydrogen, see p. 192); (*b*) of *iron* on *gaseous water* at a red heat (see p. 255); or (*c*) of *zinc* on *dilute sulphuric* or *hydrochloric acid* (see pp. 415, 112).

Method (*a*). A jar is filled with water, covered with a glass plate, and inverted in a trough of water as shown in figure 1. A piece of the metal sodium

FIG. 1.

not larger than a pea is placed in a spoon made of wire gauze, which is passed quickly under the water beneath the jar, when the hydrogen evolved passes in bubbles into the jar. The sodium melts, moves about, and displaces hydrogen from the water. Other small fragments are successively introduced into the spoon until the jar is full. (Note.—Large pieces must not be used, else an explosion may ensue.)

Method (*b*). A piece of iron gas-pipe, of ½-inch bore, is filled loosely with iron turnings, and closed by stoppers made of asbestos cardboard moistened with water, and moulded round glass tubes, placed as shown in figure 2. The iron tube is then heated in a gas furnace, and the water in the flask is boiled. The iron combines with the oxygen of the steam, setting free the hydrogen, which may be collected in a jar as shown in the figure.

FIG. 2.

Method (*c*). A flask or bottle, as shown in figure 3, is provided with a cork and delivery tube. Some granulated zinc, prepared by pouring melted zinc into water, is placed in the flask A; and a mixture of one volume of hydrochloric acid and four volumes of water, or of one volume of oil of vitriol (sulphuric acid),* and eight volumes of water, is poured through the funnel B. Bubbles begin to appear on the surface of the zinc, and the liquid effervesces. A few minutes must be allowed, so that the hydrogen may displace the air from the bottle. It can then be collected in jars. The zinc displaces the hydrogen from its compound with chlorine in hydrochloric acid, or from its compound with sulphur and oxygen in sulphuric acid. The substances produced are named zinc chloride, or zinc sulphate, according as one or other acid has been used.

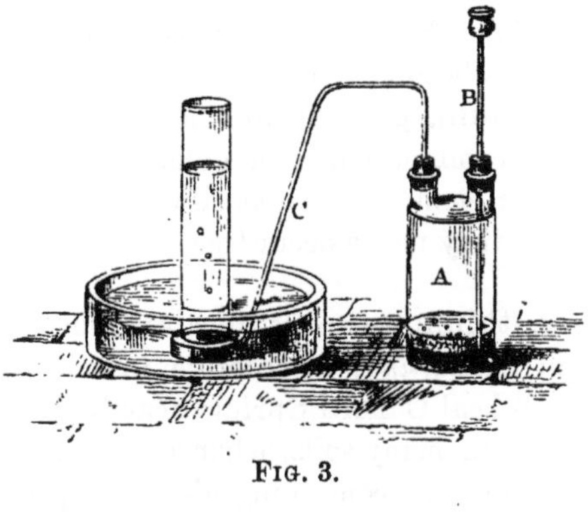

FIG. 3.

* If sulphuric acid be used, sulphur dioxide and hydrogen sulphide are produced if the proportion of water be not a large one.—*Chem. Soc.*, **53**, 54.

Properties.—A colourless, odourless gas; the lightest of all known bodies. As it is nearly fourteen and a half times as light as air, it may be poured *upwards* from one jar into another; or if a light jar or beaker be suspended mouth downwards from the arm of a balance, and counterpoised, and hydrogen be poured into it *from below*, that arm of the balance rises, the heavier air being replaced by the lighter hydrogen. Balloons used to be filled with it; but coal gas is now employed. It burns in air, combining with oxygen to form water, and when mixed with air (about $2\frac{1}{2}$ times its volume) the resulting mixture is explosive (see p. 192). It is sparingly soluble in water; 100 volumes of water absorb 1·93 volumes of hydrogen gas. It is not poisonous, but cannot be respired for any long time, as the oxygen of the air, which is necessary for the support of life, is thereby excluded. Owing to the rate at which it conveys sound, speaking with hydrogen gives a curious shrill tone to the voice. It has never been condensed to the liquid or solid states. Cailletet, and also Pictet, who claim to have condensed it by cooling it to a very low temperature,* and at the same time strongly compressing it, had in their hands impure gas. Its critical temperature, above which it cannot appear as liquid, is probably not above —230°.

It unites directly with the halogens; with oxygen and with sulphur; also with carbon at a very high temperature; and with potassium and sodium. It is absorbed by certain metals, notably by palladium, which can be made to take up 900 times its own volume (see p. 576). From this its density and its specific heat in the solid state have been calculated.†

Lithium, sodium, potassium, rubidium, and **cæsium** are always found in combination with chlorine, or with oxygen and the oxides of other elements such as silicon, carbon, boron, sulphur, phosphorus, &c.; they never occur free. They are named "metals of the alkalies."

Sources.—Lithium occurs as silicate in *lepidolite* and *petallite;* as phosphate in *triphylline;* as chloride in many mineral waters, especially in the Wheal-Clifford Spring, near Redruth, in Cornwall; in sea-water; and in many soils, whence it is absorbed by plants, tobacco-ash, for example, containing about 0·4 per cent. Its compounds are usually prepared from lepidolite.

Sodium forms, in combination with chlorine, *common salt*, or sodium chloride; it occurs in deposits in Chili and Peru as nitrate and iodate, in which its oxide is combined with the oxides of nitrogen

* *Comptes rend.*, **98**, 304.

† *Ibid.*, **78**, 968; also *Phil. Mag.* (4), **47**, 324. See *Palladium*.

and iodine respectively; as sulphate in mineral springs (*Glauber's salts*); as silicate in *soda-felspar* or *albite*; and as fluoride along with aluminium fluoride in *cryolite*; as borate in certain American lakes. It is obtained as carbonate by incinerating sea-plants.

Potassium is found as chloride in mineral deposits at Stassfurth, in N. Germany; the mineral is termed *sylvin*; as nitrate (*saltpetre*, *nitre*), forming an incrustation on the soil in countries where rain seldom falls; and as silicate in many rocks, chiefly in *potash felspar* and *mica*. It is abundant and very widely distributed, being a constituent of every soil. It remains as carbonate on burning to ash all kinds of wood, hence its name, from "pot-ash."

Rubidium and Cæsium are widely distributed, but occur in small amount. They are contained in *lepidolite*, along with lithium and potassium, as silicates; also in *castor* and *pollux*, two rare minerals, found in the Isle of Elba. They also occur in some mineral waters, particularly in a spring at Dürkheim, in the Bavarian Palatinate, from which they were first extracted by Bunsen, their discoverer, in 1860. They are widely distributed in the soil, and are absorbed by some plants to a considerable extent. Thus the ash of beetroot contains 1·75 per mille of rubidium.

Preparation.—These metals are prepared: 1. By passing a current of electricity through their fused hydroxides, chlorides, or cyanides. It was in this way that Davy,* in 1807, obtained potassium and sodium from their hydroxides, which up to that date had not been decomposed; the electrolysis of lithium chloride is still the only method of preparing lithium; and Setterberg,† in 1881, prepared considerable quantities of rubidium and cæsium by electrolysing a fused mixture of their cyanides with cyanide of barium, using as poles strips of aluminium.

To prepare lithium, which may serve as a type of this kind of operation, about 30 grams of lithium chloride are melted over a Bunsen flame in a nickel crucible; when the chloride is quite fused, a piece of gas carbon (the sticks of a Jablochkoff candle answer well), is connected with the positive pole of four or six Bunsen or Grove cells; and a knitting-needle, passing through the hole in the stem of a tobacco pipe, made into a shallow cup at its broken end, is connected with the negative pole; these are dipped in the fused chloride; and when a bead of lithium as large as a small pea has collected on the negative electrode, the fused chloride is allowed to cool, and the bead plunged into rock oil. The bead of lithium is then scraped off with a knife, and the process repeated, until a sufficient quantity has been collected.

---

* *Phil. Trans.*, 1808, 1; 1809, 39; 1810, 16.

† *Annalen*, **211**, 100.

2. Sodium, potassium, and rubidium may be prepared by distilling the hydroxides with carbon.* The carbon unites with the oxygen of the hydroxide, while the hydrogen is liberated and comes off as gas. (A hydroxide, it should be here explained, is a compound of oxygen with hydrogen and with a metal.) The industrial preparation of sodium is thus carried out (see Chapter XXXVIII, p. 651).

Properties.—These elements are all white metals, so soft at the ordinary temperature that they can be cut with a knife, but brittle at low temperatures; they are malleable, and may be squeezed into wire by forcing them through a small hole by means of a screw-press; they may be welded by pressing clean surfaces together; they melt at moderate temperatures, and are all comparatively volatile; hence, lithium excepted, they may be distilled at a bright red heat from a malleable iron tube or retort. They are all lighter than water; lithium, indeed, is the lightest solid known. Each imparts its special colour to a Bunsen or spirit flame; thus compounds of lithium give a splendid crimson light; of sodium a yellow light; potassium compounds colour the flame violet; rubidium red, hence the name of the metal (from *rubidus*); and cæsium blue (*cæsius*). (See **Spectrum Analysis**, Chapter XXXV.) Potassium vapour is green, and sodium vapour, violet. These elements crystallise, when melted and cooled, in the dimetric system.

They all combine readily with the elements chlorine, bromine, iodine (these elements are termed the "halogens"), oxygen, sulphur, phosphorus, &c., with evolution of light and heat; and they all decompose water at the ordinary temperature, liberating hydrogen (see p. 26).

Physical Properties.

	Mass of 1 cub. cent.				
	Solid.	Liquid.	Gas.	Density, H = 1.	Melting-point.
Hydrogen ..	0·62—0·63†	0·025‡	0·0000896	1	Below −230°
Lithium	0·59	?	—	?	180°
Sodium......	0·985	?	—	12·75	95·6°
Potassium ..	0·865	?	—	18·85	62·5°
Rubidium ..	1·50	?		?	38·5
Cæsium	1·88	?		?	26—27°

* Castner, *Chem. News*, **54**, 218.

† Deduced from the mass of 1 c.c. of its alloy with palladium.

‡ At 0°, under a pressure of 275 atmospheres; deduced from the density of a mixture of 1 volume of hydrogen with 8 vols. of carbonic anhydride.

	Boiling-point.	Specific Heat.	Atomic Weight.	Molecular Weight.
Hydrogen......	Below −230°	(Gas) 2·411 (Solid) 5·88	1·0025 ?	2·0
Lithium........	?	0·941	7·02	—
Sodium	742°	0·293	23·04	23·04
Potassium......	667°	0·166	39·14	39·14
Rubidium......	?	?	85·5	—
Cæsium........	?	?	132·9	—

Group II.—**Beryllium** or **Glucinum, Calcium, Strontium, Barium.**

These metals, like those of the previous group, always occur in nature in combination, never in the metallic state. They are found combined with silicon and oxygen, as silicates; with carbon and oxygen, as carbonates; with sulphur and oxygen, as sulphates; and with phosphorus and oxygen, as phosphates. Calcium is also associated with fluorine and with chlorine. They are named "metals of the alkaline earths."

Sources.—**Beryllium** is a somewhat rare element. Its most common sources are: *beryl*, a silicate of beryllium and aluminium, a pale greenish-white mineral, which, when transparent, and of a pale sea-green colour, is named *aquamarine;* and when bright green, *emerald* (the green colour is due to the element chromium); *phenacite*, also a silicate of beryllium; and *chrysoberyl*, a compound of the oxides of beryllium and aluminium.

Calcium is one of the most abundant elements. Its carbonate when pure and crystalline is named *Iceland-spar* or *calc-spar;* earthy and less pure varieties are *limestone*, *chalk*, and *marble*. When associated with magnesium carbonate, the mineral is named *dolomite*. Calcium sulphate is named *gypsum*, *selenite*, and *anhydrite*, according to its state of aggregation. Its phosphate, in which it is combined with phosphorus and oxygen, is named *phosphorite* or *apatite*. The fluoride is named *fluor-* or *Derbyshire-spar;* and its chloride is a constituent of sea-water and many mineral waters. Most natural water contains hydrogen calcium carbonate (bicarbonate) in solution.

Strontium, like calcium, occurs as carbonate, in *strontianite*, and as sulphate in *celestine*. Its name recalls the source in which it was first found—Strontian, a village of Argyllshire, in Scotland.

Barium also occurs as carbonate, *witherite;* and as sulphate, *barytes* or *heavy-spar*, so named from its high specific gravity. Hence the name of the metal, from βαρύς, heavy.

Preparation.—Beryllium, the chloride of which volatilises at a red heat, may for that reason be prepared from that compound by passing its vapour over fused sodium contained in an iron boat.* The sodium combines with the chlorine, which leaves the metal as such. Sodium reacts with cold water, while beryllium does not; hence the sodium may be removed by treatment with water.

Barium, strontium, and **calcium**† are best prepared by passing a current of electricity through their respective chlorides, fused in a porcelain crucible over a blowpipe, using a carbon rod (see lithium) as one electrode, and an iron wire as the other. Solutions of the metals in mercury are easily made by electrolysing strong solutions of the chlorides of the metals, using mercury as the negative electrode. Barium amalgam crystallises out of the mercury; it may be collected, and after washing it with cold water and drying it, the mercury can be distilled off in a vacuum, leaving the barium as a yellowish-white metallic powder, still, however, containing mercury. Another method of preparing an amalgam of mercury and barium (alloys of mercury are termed "amalgams") is to shake up sodium amalgam with a strong solution of barium chloride. The sodium combines with the chlorine, leaving the barium in the mercury. Amalgams of strontium and calcium cannot be made in this manner.

Properties.—Beryllium and calcium are white metals; the other two have a yellow tinge. They melt at a bright red heat, oxidising in presence of air. Calcium and beryllium are brittle; strontium and barium malleable. They are all heavier than water. The compounds of the last three impart characteristic colours to a Bunsen flame, and have well-marked spectra (see Chapter XXXV). The chloride of calcium tinges the flame brick-red; of strontium, bright crimson-red like lithium; and of barium, pale-green. The metals have not been volatilised. They unite readily with the halogens, with oxygen and sulphur, and with phosphorus. Beryllium does not decompose water unless boiled with it; the others act on it at the ordinary temperature, with evolution of hydrogen.

Physical Properties.

	Mass of 1 cub. cent. solid.	Melting-point.	Specific Heat.	Atomic Weight.
Beryllium....	1·85 at 20°	Red heat (above 1230°).	Variable (See Appendix).	9·1
Calcium......	1·58	Bright red	0·167	40.08
Strontium....	2·54	Bright red	?	87·5
Barium	4·0	White	?	137·00

* *Chem. News*, **42**, 297.

† *Annalen*, **183**, 367.

Appendix.—The specific heat of beryllium varies greatly with the temperature. The following results were found by Humpidge.*

Temperature ..	0°	100°	200°	300°	400°	500°
Specific Heat ..	0·3756	0·4702	0·5420	0·5910	0·6172	0·6206

Group III.—Magnesium, Zinc, Cadmium.

Sources.—These three metals are never found native. All three occur as carbonate and as silicate; and the two latter as sulphide. Magnesium sulphide is decomposed by water; hence its non-occurrence in nature. Magnesium occurs also in considerable quantity as sulphate (*Epsom salts*), in sea-water, and in many mineral springs. Its native compounds are named as follows:—Magnesium carbonate, *magnesite;* double carbonate of magnesium and calcium, *dolomite;* it occurs in great rock masses in the range of hills in the Italian Tyrol named the Dolomites. There are many silicates of magnesium and other metals. Among the more important are *talc*, *steatite* or *soap-stone* (French chalk), *serpentine*, and *meerschaum*. *Augite*, *hornblende*, *asbestos*, *olivine*, and *biotite* (a variety of *mica*) are also rich in magnesium (see Silicates, p. 313). *Carnallite*, a chloride of magnesium and potassium, is found at Stassfurth. The commercial sources of metals are named "ores." The ores of zinc are:—*Calamine*, zinc carbonate; *silicious calamine*, the silicate; and *blende*, or "Black Jack," the sulphide. Cadmium always accompanies zinc; the only pure mineral containing it is *greenockite*, cadmium sulphide. The name magnesium is derived from the town of Magnesia, in Asia Minor. Its oxide is sometimes called *magnesia alba*, from its white colour. The word zinc is perhaps connected with the German equivalent for tin, *Zinn*. "Cadmium" is adopted from the name given by Pliny to the sublimate found in brass-founders' furnaces (*cadmia fornacum*).

Preparation.—**Magnesium** is prepared like beryllium; dried *carnallite*, a double chloride of magnesium and potassium combined with water, is heated with sodium. The sodium unites with the chlorine, removing it from the magnesium, which is set free.

The mixture is heated in large iron crucibles to a high temperature. When the reaction is over, the crucible is allowed to cool, and the contents chiselled out. Small globules of magnesium are disseminated throughout the fused mass, and at the bottom of the crucible is a mass of magnesium embedded in

* *Proc. Roy. Soc.*, 39, 1.

flux, as the fused chlorides are termed. The salt with the globules of magnesium is transferred to a crucible, A, the bottom of which is perforated, as shown in the figure, and a tube, B, passes through the bottom, reaching up to near the top of the crucible. The lid is then *luted* on (*i.e.*, fastened on by clay), the

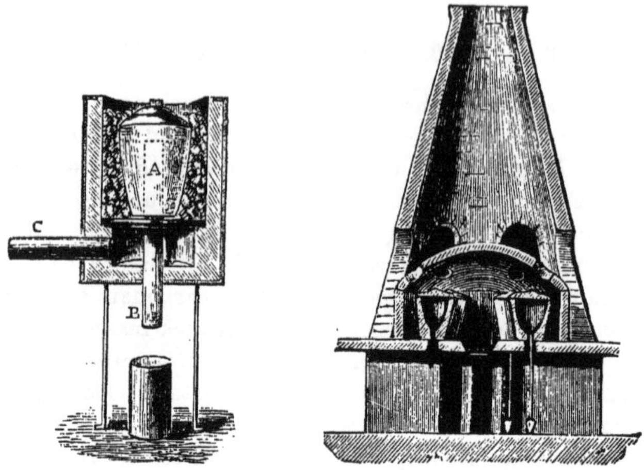

FIG. 4.

top of the tube having been closed by a wooden plug. When the temperature rises to bright redness, the magnesium rises in vapour, and distils down the centre tube, condensing on the lower portion, whence it drops into heavy oil. Hence the old term for this process—"*distillatio per descensum.*"

Zinc is produced by distilling its oxide with coke (carbon) in clay cylinders. The carbon unites with the oxygen, setting free the zinc, which distils over.

The old English method of extracting zinc from its oxide used to be carried out in apparatus like that employed in making magnesium. The roasted zinc ore, consisting of oxide of zinc, was mixed with coke or anthracite coal (carbon), and placed in clay crucibles, similar in construction to the iron one shown in Fig. 4. On raising the temperature to bright redness, the zinc distils over, and drops through the tube which passes through the bottom of the furnace.

The Belgian process, which is now all but universally adopted, consists in distilling the zinc ore with coke from clay cylinders, arranged in tiers. The zinc condenses in conical tubes of cast iron or iron plate, which fit the mouths of the cylinders, and are made tight at the joint by a luting of clay. When the operation is over, these tubes are removed, and the zinc, which forms a crust adhering to their interiors, is chiselled off.

Cadmium accompanies zinc, and as it boils at a lower temperature, the first portions which distil over contain it.

Properties.—These three metals are all white. Zinc, however, has a bluish tinge, and cadmium a yellow tinge. Of the three, magnesium is the hardest, and cadmium the softest; it may be cut with a knife, but with difficulty. Magnesium and zinc are malleable and ductile at a moderately high temperature (zinc at 120°), but are brittle at the ordinary temperature. Zinc is also brittle at 200°, and may be easily powdered in a hot iron mortar. These metals may all be distilled, cadmium most easily, and magnesium at the highest temperature. They are all heavier than water. They combine directly with the halogens; they burn when heated in air, combining with its oxygen. Magnesium gives out a brilliant white light, and it is prepared in the form of ribbon, wire, or dust for signalling, pyrotechnic, and photographical purposes. Zinc burns with a light blue-green flame, and cadmium with a dull flame; they tarnish very slowly in air. They also unite directly with sulphur, phosphorus, &c. When boiled with water, magnesium and zinc slowly decompose it, hydrogen being evolved. Cadmium is without action on water except at a red heat.

Physical Properties.

	Mass of 1 c.c. Solid.	Density, H = 1.	Melting-point.
Magnesium	1·743	?	700—800°
Zinc	7·15	34·5	412°
Cadmium	8·6	52·15	315°

	Boiling-point at 760 mm.	Specific Heat.	Atomic Weight.	Molecular Weight.
Magnesium	About 1000°	0·250	24·30	24·30
Zinc	930° to 942°	0·095	65·43	65·43
Cadmium	About 770°	0·056	112·1	112·1

GROUP IV.—**Boron, Scandium, Yttrium,* Lanthanum,* Ytterbium.***

These elements are never found in the free state. They all exist in nature in combination with oxygen, and other oxides.

Sources.—Boron issues from the earth as hydroxide, or boracic acid, along with steam in the neighbourhood of volcanoes. The hydroxide also occurs as *sassolite;* its other sources are *tincal* or *native borax,* in which its oxide is combined with oxide of sodium and with water (the beds of certain dried up American lakes contain enormous quantities of borax); *boracite,* boron oxide with magnesium oxide and chloride; *boronatrocalcite,*

* It is doubtful if these metals belong to this group.

boron oxide, calcium oxide, and sodium oxide; and *datolite*, boron and silicon oxides with calcium oxide.

The remaining elements of this group are usually associated with cerium, didymium, erbium, terbium, samarium, &c., as oxides, in combination with oxides of silicon, niobium, tantalum, titanium, and other elements. The minerals containing them are named *euxenite*, *orthite*, *columbite*, *gadolinite*, *yttrotantalite*, *samarskite*, and *cerite*. They have been found chiefly at Arendal and Hitterö, in Norway, and in Connecticut, U.S.

Preparation.—**Boron** is obtained by heating with metallic sodium the compound which its fluoride forms with potassium fluoride; or by heating its oxide with potassium, or better, with magnesium dust. The fluorine or oxygen combines with the potassium or magnesium, leaving the boron in the free state. It was by the latter method that it was first prepared in 1808 by Gay-Lussac and Thenard, and later by Deville and Wöhler.*

Metallic **scandium** has not been prepared.

Yttrium was prepared in an impure state, mixed with erbium, by Berzelius, by the action of potassium on the impure chloride. It was a greyish-black lustrous powder.

Lanthanum has been prepared by passing a current of electricity through its fused chloride (see Lithium).

Metallic **ytterbium** has not been obtained.

Properties.—Boron is a brown amorphous (*i.e.*, non-crystalline) powder, which has not been melted even at a white heat. It is insoluble in all solvents which do not act on it chemically. It was for long supposed possible to crystallise it from molten aluminium; the resulting black crystals, however, are not pure boron, but a compound of boron and aluminium. Yellow crystals, obtained by Wöhler and Deville, and also supposed by them to be pure boron, consist of a compound of boron, carbon, and aluminium. The mass of 1 c.c. of pure boron has not been determined. Boron combines with the oxygen and nitrogen of air, burning to oxide and nitride. It is one of the few elements which combine directly with nitrogen. It is also attacked by chlorine and by bromine. Lanthanum is the only one of these elements which has been prepared in a compact state. It resembles iron in colour; is hard, malleable, and ductile. It melts at a lower temperature than silver (below 1000°), and burns with great brilliancy when heated in air; its specific gravity is 6·05 at the ordinary temperature.

The specific heat of boron† undergoes a remarkable change as the temperature is raised. The following results were obtained by Weber:—

* *Annales* (3), **52**, 63.

† *Phil. Mag.* (4), **49**, 161, 276.

Temperature ..	-40°	+27°	77°	126°	177°	233°
Specific Heat ..	0·1915	0·2382	0·2737	0·3069	0·3378	0·3663

In this it resembles beryllium, carbon, and silicon.

GROUP V.—Aluminium, Gallium, Indium, Thallium.

These elements are found only in combination. The sources of aluminium are its oxide, *corundum;* when coloured blue, probably by cobalt, it forms the precious stone the *sapphire*, and when red, coloured by chromium, the *ruby*. Associated with iron oxide, it is named *emery*. Silicate of aluminium is a constituent of many rocks; it exists in *felspar, hornblende, mica*, and numerous other minerals. *China clay* or *kaolin* is a slightly impure silicate of aluminium (see Silicates). The mineral *cryolite*, found in Greenland, is a fluoride of aluminium and sodium. The sulphide of aluminium is decomposed by water; hence its non-occurrence in nature.

The other three elements of this group occur as sulphides. Gallium and indium are found in extremely minute amount in some zinc ores; thallium is contained in some specimens of *iron pyrites* (disulphide of iron) and *copper pyrites.* Zinc sulphide, or *blende*, from the Pyrenees, contains about 0·002 per cent. of gallium; the zinc ores from Freiberg, in Saxony, about 0·05 to 0·1 per cent. of indium.

Preparation. Aluminium is prepared:—

1. By passing the vapour of its chloride over heated sodium;* the sodium unites with the chlorine, while the aluminium remains in the metallic state.

2. By heating its oxide mixed with carbon to an enormously high temperature in the electric arc.† The oxide is thus decomposed, and the carbon unites with the oxygen, while the metal is left. This process is better adapted for preparing the alloys of aluminium than the metal itself.

3. By heating with metallic sodium cryolite, the double fluoride of aluminium and sodium, previously fused with salt.‡

Gallium§ is prepared by passing a current of electricity through a solution of its oxide in caustic potash.

Indium‖ may be obtained by passing a stream of hydrogen

* Wöhler, *Annalen*, **37**, 66; Deville, *Annales* (3), **43**, 5, and **46**, 415. The literature on this subject is now very large.

† *Chem. News*, 1889, 211, 225, 241.

‡ *Brit. Assen.*, 1889.

§ *Comptes rend.*, **82**, 1098; **83**, 636.

‖ *J. prakt. Chem.*, 1863, **89**, 441; **92**, 480; **94**, 1; **95**, 414; **102**, 273.

gas over its oxide heated to a high temperature: the hydrogen combines with the oxygen, producing water, and the indium is left; or by heating its oxide with sodium; or by removing chlorine from indium chloride by placing metallic zinc in a solution of that substance.

Thallium* is most easily obtained by heating its chloride to a red heat with potassium cyanide, a compound of carbon, nitrogen, and potassium. The potassium removes the chlorine, forming potassium chloride; cyanogen, a compound of carbon and nitrogen, escapes as gas; and thallium remains behind as fused metal.

Properties.—Aluminium, gallium, and indium are tin-white metals, while thallium has a duller lustre, resembling that of lead. These metals are moderately malleable and ductile. Indium and thallium are soft, and may be cut with a knife; aluminium and gallium are hard. Thallium and its salts impart a magnificent green colour to the flame of a Bunsen's burner; indium burns with a violet light; aluminium and gallium do not volatilise sufficiently easily to colour the flame.

All these elements unite readily with oxygen at a red heat; aluminium and thallium become tarnished in air at the ordinary temperature. They also combine directly with the halogens and with sulphur. They are not acted on by water at the ordinary temperature, but decompose it at higher temperatures, combining with its oxygen.

Aluminium is contained in *alum*, hence its name; gallium was discovered in 1875 by the French chemist, Lecoq de Boisbaudran, and patriotically named after Gaul; indium derives its name from the blue line in its spectrum (from "indigo"); and thallium was named by its discoverer Crookes, from θαλλός, a green twig, in allusion to the green colour it imparts to the flame.

Of these elements aluminium is the only one which has found a commercial use; the barrels of opera glasses, telescopes, and optical instruments are made of it; and, alloyed with copper, it is employed for cheap jewellery, under the name of "aluminium bronze." Of recent years its manufacture has been greatly increased, and in the near future it will rank as one of the commoner metals.

The metals beryllium, magnesium, zinc, cadmium, lanthanum, didymium, cerium, and aluminium used to be classified together as "metals of the earths;" the so-called earths being their oxides, which are insoluble in water, and hence have not an alkaline reaction like those of calcium, strontium, and barium.

---

* *Chem. News*, **3**, 193, 303; *Proc. Roy. Soc.*, **12**, 150.

Physical Properties.

	Mass of 1 c.c. Solid.	Melting-point.	Specific Heat.	Atomic Weight.	Molecular Weight.
Aluminium..	2·583 at 4°	About 700°	0·2253 from 0° to 100°	27·01	27·01
Gallium	5·94 at 23°	29·5°	Solid 0·079 from 12° to 23° Liquid 0·080 from 106° to 119°	69·9	69·9
Indium	7·42 at 16·8°	176°	0·0565 to 0·0574	113·7	—
Thallium....	11·9........	290°	0·0336.........	204·2	204·2 to 408·4

APPENDIX.

The equations expressing the preparation of the foregoing elements are as follows:

Hydrogen.—(1) $2H_2O = 2H_2 + O_2$.
(2) $2H_2O + 2Na = 2NaOH + H_2$
(3) $4H_2O + 3Fe = Fe_3O_4 + 4H_2$.
(4) $H_2SO_4 + Zn = ZnSO_4 + H_2$.

Lithium, &c.—$2LiCl = 2Li + Cl_2$.

Sodium and Potassium.—$2NaOH + 2C = 2Na + 2CO + H_2$.
$2KOH + 2C = 2K + 2CO + H_2$.

Beryllium.—$BeCl_2 + 2Na = Be + 2NaCl$.

Calcium, Strontium, and Barium.—$BaCl_2 = Ba + Cl_2$.

Magnesium.—$MgCl_2.KCl + 2Na = Mg + 2NaCl + KCl$.

Zinc, Cadmium.—$ZnO + C = Zn + CO$.
$CdO + C = Cd + CO$.

Boron.— (1) $BCl_3 + 3Na = B + 3NaCl$.
(2) $KF.BF_3 + 3Na = B + KF + 3NaF$;
(3) $B_2O_3 + 3Mg = 2B + 3MgO$.

Aluminium.—(1) $AlCl_3 + 3Na = Al + 3NaCl$.
(2) $Al_2O_3 + 3C = 2Al + 3CO$.
(3) $AlF_3.3NaF + 3Na = Al + 6NaF$.

Gallium.—$2Ga_2O_3 = 2Ga + 3O_2$.

Indium.—(1) $In_2O_3 + 3H_2 = 2In + 3H_2O$.
(2) $2InCl_3 + 3\,Zn = 2In + 3ZnCl_2$.

Thallium.—$2TlCl_3 + 6KCN = 2Tl + 6KCl + 3(CN)_2$.

CHAPTER IV.

THE ELEMENTS (CONTINUED).

GROUP VI, THE CHROMIUM GROUP; GROUP VII, THE CARBON GROUP; GROUP VIII, THE SILICON GROUP.

GROUP VI.—**Chromium, Iron, Manganese, Cobalt, Nickel.**

The elements of this group are not, generally speaking, associated in the periodic table, yet they closely resemble each other; and it is convenient to consider them together.

Sources.—They invariably occur in combination with oxygen, when of terrestrial origin. Certain meteorites, however, consist largely of metallic iron and nickel with a little cobalt and a trace of hydrogen. Common proportions are 90 per cent. of iron, 9 per cent. of nickel, and 1 per cent. or less of cobalt.

The chief ore of **chromium** is *chrome iron ore*, or *chromite*; it is a compound of oxygen with chromium and iron (see Chromium, oxides, p. 254). It is found in Silesia, Asia Minor, Hungary, Norway, and N. America. The green colour of the emerald and serpentine is due to traces of chromium.

Compounds of **iron** are very numerous in nature. Its oxides, when found native, are named:—*Hæmatite*, of which varieties are termed *specular iron ore*, *kidney ore*, and *titaniferous ore* (these occur largely in Cumberland, also in the south of Spain); combined with water, *göthite*, *brown iron ore*, *bog iron ore*, and *ake ore*, the latter of which are named from their sources: they are found in Northamptonshire, the Forest of Dean, and Glamorganshire; *magnetic iron ore*, *magnetite* or *loadstone*, an oxide of a different composition (see p. 255): it does not occur largely in England, but is worked in Sweden; the largest deposit of iron ore in the world consists of magnetite: it occurs in Southern Lapland, but is as yet inaccessible. *Spathic ore*, or carbonate of iron, is a white crystalline substance when pure, but is usually interstratified and mixed with clay or shale, when it is termed "*clay-band*" or "*black-band.*" Spathic ores occur in Durham, Cornwall, Devon, and Somerset; *clay iron-stone* in the coal-measures in Staffordshire, Shropshire, Yorkshire, Derbyshire, Denbigh, and South Wales; while *black-band* is mined largely in the Clyde basin, in Scotland.

Iron occurs in combination with sulphur as *pyrites;* it is very widely distributed; perhaps the largest sources are in the south of Spain. At Rio Tinto this ore is worked, not for the iron which it contains, but for its copper (about 3 per cent.) and its sulphur. Iron is also a constituent of most rocks and soils: it is one of the most abundant as well as one of the most widely distributed of elements.

Manganese is nearly always found associated with iron, in combination with oxygen. Its most important source is *pyrolusite* or *black oxide.* Other manganese minerals are *braunite* and *hausmannite,* also oxides; *manganite, psilomelane,* and *wad,* compounds of oxides and water; *manganese-spar,* the carbonate; it also occurs in combination with silicon and oxygen as silicate, and with sulphur as sulphide.

Cobalt and **nickel** are almost invariably associated. As already mentioned, they accompany iron in some meteorites in the state of metals. **Cobalt** occurs as *smaltite* or *tin-white-cobalt,* in combination with arsenic; and as *glance-cobalt,* in combination with arsenic and sulphur.

The chief ore of **nickel** is the oxide and the double silicate of nickel and magnesium, large quantities of which are now imported from New Caledonia, a French convict settlement north-east of Australia. It is found on the continent of Europe chiefly as the *arsenide,* a compound of nickel and arsenic named *Kupfernickel* or *copper-nickel,* from its red colour resembling copper; it is also called *niccolite.* The sulphide, or *capillary pyrites,* also occurs native.

Preparation.—These metals in an impure state may all be prepared by *reducing* (*i.e.,* removing oxygen from) their oxides by means of carbon. Iron and nickel are prepared for commercial purposes; alloys of iron and manganese, and iron and chromium are also produced; and nickel is often deposited by means of an electric current on the surface of other metals, which are then said to be *nickel-plated.*

Chromium, in the pure state, has been prepared by removing chlorine from its chloride, by means of metallic zinc or magnesium.* The chloride is mixed with potassium and sodium chlorides, and

* *Annalen,* 111, 117.

heated with metallic zinc to the boiling-point of the latter metal (about 940°). An alloy of zinc and chromium remains, from which the zinc may be removed by treatment with nitric acid; the chromium remains as a pale-grey crystalline powder. It has also been prepared by decomposing by electricity its chloride in concentrated solution. It then deposits in brittle scales with the lustre of metallic iron.

Iron, in a state of purity, is hardly known. It has been prepared by reducing its oxide by means of hydrogen at a red heat, and heating the resulting greyish-black powder, which consists of pure iron in a state of fine division, to whiteness in a porcelain crucible under a layer of fused calcium fluoride in the oxyhydrogen flame.* It does not fuse, but agglomerates to a sintered mass. It may also be deposited electrically from solution. Ordinary iron contains small quantities of several elements, notably carbon and silicon, which completely alter its properties, and it must, therefore, be considered as a compound. A description of the metallurgy of iron is therefore deferred to Chapter XXXVI.

Manganese, like iron, is almost unknown in a pure state; when produced by the aid of carbon, it combines with that element and acquires peculiar properties. Its metallurgy will be considered along with that of iron. It has recently, however, been prepared in a pure coherent state by heating to redness with magnesium dust a mixture of manganese dichloride with potassium chloride.†

Nickel is prepared in a manner exactly similar to that by which iron is made. Impure nickel can be prepared by heating its oxide with charcoal; the pure metal is obtained by electrolysis. The same remarks apply to **cobalt**.

Properties.—These elements are all greyish-white, with metallic lustre, like iron. Manganese and cobalt have a reddish-tinge; nickel is whiter than iron, but not so white as silver. They all melt at a very high temperature, so high, indeed, that it is reached only by means of the oxyhydrogen blowpipe. The addition of a small amount of carbon, as has been remarked, profoundly modifies their properties; and, indeed, the pure elements are almost unknown in a compact state, owing to the difficulty of melting them into a compact mass in any vessel capable of withstanding the requisite temperature, and not attacked by the metal. The figures in the following table refer, for the most part, to such impure specimens.

They all combine with oxygen, on exposure to moist air, but are

* Troost, *Bull. Soc. Chim.* (2), **9**, 250.

† Glatzel, *Ber. Deutsch. Chem. Ges.*, **22**, 2857.

permanent in dry air; they unite directly with the halogens; with sulphur, selenium, and tellurium; with phosphorus, arsenic, and antimony; with carbon, silicon, and titanium; and they form alloys with each other and with many other metals. Iron and nickel also absorb hydrogen gas to a small extent.

Physical Properties.

	Mass of 1 c.c. Solid.	Specific Heat.	Atomic Weight.	Molecular Weight.
Chromium ..	7·3; 6·81 (at 25°)	Not determined	52·3	?
Iron........	8·00 (at 10°) pure .... 8·14 (at 15·5°) electrolytic)	0·112 (impure)	56·02	?
Manganese..	7·39 at 22°...........	0·122	55·0	55·0
Nickel......	About 9·0	0·109	58·6	?
Cobalt......	About 9·0	0·107	58·7	?

Group VII.—**Carbon, Titanium, Zirconium, Cerium,* Thorium.**

Of these elements, carbon is the only one found in the free state. The others are always found combined with oxygen, and usually with silicon and oxygen as silicates.

The native forms of **carbon** are the *diamond*, *carbonado*, and *graphite*, *black-lead*, or *plumbago*. Diamonds are found *in situ* in pegmatite, or graphic granite, near Bellary, in the Nizam, India, and also in an aqueous magnesian breccia in S. Africa. It is probable that they have been formed simultaneously with these rocks; the conditions of their formation are unknown. *Diamond-fields*, or districts which yield diamonds, occur in Brazil, India, the Cape, California, Borneo, and the Ural Mountains.

Carbonado, a variety of carbon found in the Soap Mountains of Bahia, is a reddish-grey, porous substance; it is evidently closely allied with diamond.

Graphite occurs in nests of trap in the clay slate at Borrowdale, Cumberland, and is also found in certain coal-measures, *e.g.*, at New Brunswick.

Such different forms of an element are said to be *allotropic*, a word which signifies "different forms."

Carbon also occurs in combination with oxygen (the atmosphere contains about 0·04 per cent. by weight of carbon dioxide); and its dioxide, with the oxides of various metals; the most

* It is doubtful if cerium belongs to this group of elements.

important of the carbonates are those of calcium, of magnesium, and of iron (*q.v.*).

Along with hydrogen, oxygen, and nitrogen, it is a constituent of all organised matter; coal, which consists of ancient vegetable matter, agglomerated by pressure and decomposed by heat, contains a large percentage of carbon, anthracite, for instance, containing over 90 per cent.

Titanium occurs only in combination with oxygen, as *rutile*, *anatase*, and *brookite*; and associated with oxides of iron, as *titaniferous iron*; with oxide of calcium in *perowskite*; and with the oxides of silicon and calcium in *sphene*.

Zirconium is found as oxide, in combination with oxide of silicon in *zircon*, and in other rare minerals.

The chief source of **cerium** is *cerite*, a compound of oxide of cerium with oxide of silicon and with water; and it occurs associated with oxides of niobium, tantalum, lanthanum, didymium, &c., in *orthite*, *euxenite*, and *gadolinite*, and other very rare minerals.

Thorium occurs in *thorite* as oxide, in combination with oxide of silicon and with water; it also occurs in *euxenite*, &c., along with cerium.

Preparation.—Carbon is produced by the decomposition by heat of its compounds with hydrogen, sulphur, and nitrogen; at a very high temperature its oxide is also decomposed. It may also be produced by withdrawing chlorine from any of its chlorides by means of metallic sodium, or oxygen from its oxides by metallic potassium. Chlorine also removes hydrogen at a red heat from its compounds with that element, setting free carbon in the form of soot. It is best prepared in a pure state by the first of these processes. Sugar and starch consist of carbon in union with hydrogen and oxygen. On heating these bodies out of contact with air, a large portion of the carbon which they contain remains in the state of element. It is advisable, in order to remove hydrogen completely, to heat to redness in a current of chlorine. The deposit on the upper surface of the interior of retorts during the manufacture of coal-gas by the distillation of coal is named gas-carbon, and is nearly pure. It is thus produced by the decomposition of hydrocarbons (compounds of hydrogen with carbon) by heat. Various impure forms of carbon are prepared for industrial purposes. *Wood charcoal* is obtained by heating wood to redness in absence of air. This used to be the work of "charcoal burners," and the manufacture still survives in Epping Forest. Faggots of wood are piled into a tightly packed heap, covered over with turf, and set on fire, a limited quantity of air being admitted to support combustion.

Most of the wood is thus charred; and when smoke ceases to be emitted more turf is heaped on, so as to extinguish the fire. The mass of charcoal is then allowed to cool, and when cold, the covering of turf is removed, and the billets of charcoal unpiled. The wood yields about 34 per cent. of its weight of charcoal. In the present day, oak or beech wood is distilled from iron retorts, for the production of acetic acid, or vinegar; the retorts are heated with coal, and the charcoal remains in the same form as the logs which are put into the retort. The charcoal made in this way is used chiefly by iron-founders to mix with sand in making moulds for castings. Charcoal for gunpowder is made from willow, dog-wood, or alder.

Coke, the residue on distilling coal, is also impure carbon. The coke forms from 40 to 75 per cent. of the weight of the coal. Coke is largely used as a fuel, especially in iron smelting.

Bone or *animal charcoal*, or *bone-black*, produced by distilling bones, contains about 10 per cent. of carbon, the remainder chiefly consisting of the mineral constituents of bones, calcium phosphate and carbonate. It is used for decolorising solutions of impure sugar, which are filtered through the bone-black, ground to a coarse powder. Its decolorising properties are much increased by dissolving out the calcium compounds by washing it with hydrochloric acid.

Lamp-black, chiefly used for printers' ink, is prepared by burning certain compounds of carbon and hydrogen, especially one constituent of coal-tar oil, named naphthalene. The hydrogen and a portion of the carbon burn, while the greater portion of the carbon is carried away as smoke, and condensed in long flues.

Titanium,* like carbon, may be produced by passing the vapour of its chloride over heated sodium, when the sodium removes the chlorine as sodium chloride, leaving the element, with which sodium does not appear to form a stable compound. It may also be produced by projecting into a red hot crucible potassium, cut into small pieces, along with potassium titanifluoride (a compound of titanium, potassium, and fluorine); the fluorine is removed by the potassium as potassium fluoride, which is soluble in water, and may be separated from the titanium by treatment with water, in which titanium is insoluble, and with which it does not react in the cold.

Zirconium,† like titanium, may be produced by heating potas-

* Wöhler, *Annales* (3), 29, 181.

† Berzelius, *Pogg. Ann.*, 4, 124, and 8, 186. Troost, *Comptes rend.*, 61, 109.

sium zirconifluoride with potassium, or magnesium. The metal aluminium also withdraws fluorine from this compound and the zirconium dissolves in the metal, crystallising from it when it cools. The aluminium is removed by treatment with dilute hydrochloric acid, in which zirconium is insoluble.

Cerium* has been prepared by electrolysing cerium chloride, covered with a layer of ammonium chloride, contained in a porous earthenware cell, placed inside a non-porous crucible, filled with a mixture of sodium and potassium chlorides. The whole arrangement is heated to redness, so as to melt the compounds. On passing an electric current, the cerium deposits on the negative electrode, which is made of iron, inserted through the stem of a clay pipe to prevent its oxidation by the hot air.

Thorium† has also, like carbon and titanium, been prepared by heating with sodium in an iron crucible potassium thorifluoride, covered with a layer of common salt.

Properties.—Carbon, as already mentioned, exists in several different forms, each of which has distinct properties. It is therefore said to display allotropy.

The *diamond* is transparent, crystalline, the hardest of all known substances; it is nearly pure carbon. It is usually colourless, but is occasionally coloured green, brown, or black by mineral matter. It was found to be combustible by the Florentine Academicians, in the 17th century, who succeeded in burning it by concentrating the sun's rays on it by means of a large lens. Lavoisier discovered its identity with carbon. It can be converted into a coke-like substance when exposed to the intense heat of the electric arc. It is used as a gem, also for rock boring, and for cutting glass. Its dust is employed in cutting and polishing precious stones, and in cutting other diamonds.

The weight of diamonds is measured in *carats*‡ (1 carat = 0·205 gram, or 3·165 grains). The value, however, is not proportional to the weight, but to approximately the square of the weight. Among the most remarkable diamonds one of the largest belongs to the Nizam of Hyderabad, and weighs 277 carats; the Crown of Russia possesses another, of a somewhat yellow colour, weighing 194 carats; the Koh-i-Noor, or "mountain of light," belonging to the British Crown, weighs 106 carats. It was

* Hillebrand and Norton, *Pogg. Ann.*, **155**, 633; **156**, 466.

† Nilson, *Berichte*, 1882, 2519 and 2537; 1883, 153.

‡ *Kérat* (Arab.), supposed to be derived from *rati*, the Indian name for the seeds of *Abrus precatorius*.

originally much larger, but was reduced in weight by cutting. The cutting of diamonds is intended to display their great power of refracting light. The two forms in which diamonds are cut are that of the brilliant, which fig. 5 represents, and the table or rosette form, shown in fig. 6. The former is the most valuable.

FIG. 5.

FIG. 6.

Carbonado is a very hard substance, which is also used for rock boring. It has been noticed as a constituent of some meteorites.

Graphite is a blackish-grey, lustrous, soft substance, chiefly used for making very refractory crucibles, when mixed with clay; also for fine iron-castings, and for lead pencils.

No attempt to produce diamonds artificially has succeeded, except perhaps one in which carbon was kept in contact with a large quantity of melted silver; the carbon appears to be slightly soluble in the fused metal, and microscopic crystals which separated out on cooling were said to possess the properties of the diamond.

Graphite may be made artificially by dissolving carbon in molten iron, which dissolves 1 or 2 per cent. When the metal is slowly cooled, part of the carbon separates in this form. It has also been prepared by heating amorphous carbon to an extremely high temperature by passing an electric current of high potential through a rod of carbon, and thus heating it to brilliant incandescence. The temperature at which the change is produced is unknown, but is enormously high.

Carbon is infusible at the ordinary pressure. It is volatile in the electric arc, which when formed, as is usually the case, by passing an electric current between two rods of gas-carbon, always possesses the temperature at which carbon volatilises. What that temperature is has not yet been ascertained.

The diamond is a non-conductor of electricity, like indeed all transparent bodies; but the other forms of carbon conduct, though not so well as metals.

Carbon in one form or another, especially as coal, is the source of all the heat and energy practically utilised by mankind. To utilise this energy stored in carbon it is burned in air, uniting with its oxygen. Charcoal unites very slowly with oxygen at the ordinary temperature, but rapidly at a red heat. The other forms of carbon also burn, but slowly, when heated to redness in oxygen. At a red heat carbon deprives most other oxides of their oxygen, and is therefore used in extracting metals from their oxides. It unites directly with sulphur when heated to redness in sulphur vapour; and with hydrogen at the temperature of the electric arc, to form *acetylene*. It does not appear to combine directly with the other elements, although many compounds have been prepared by indirect methods.

Animal charcoal and, to a less degree, wood charcoal, owing probably to their cellular nature and to the great amount of surface which they possess, have the power of condensing and absorbing gases. The amount of absorption is in the same order to the condensibilities of the gases, those gases which are condensed to liquids by the smallest lowering of temperature being absorbed in greatest amount. Thus 1 volume of boxwood charcoal absorbs 90 volumes of ammonia-gas, which condenses to a liquid at —36°, whereas it absorbs only 1·75 vols. of hydrogen, which is probably not liquid at a temperature of —230°.

Titanium is a dark grey powder, like iron which has been reduced from its oxide by hydrogen. It has not been fused. It unites directly with oxygen and with chlorine, burning when heated in these gases. It has also the rare property of uniting directly with nitrogen when heated in that gas. It decomposes water at 100°, combining with its oxygen, and liberating hydrogen.

Zirconium forms brittle lustrous scales. It fuses at a very high temperature, and does not combine with oxygen at a red heat; but at a white heat it burns to oxide. It unites directly with chlorine.

Thorium is an iron-grey powder, which burns brilliantly when heated in air, forming oxide. Like titanium and zirconium, it is attacked by chlorine, burning brilliantly in the gas; also by bromine and iodine. It unites directly with sulphur.

Physical Properties.

	Mass of 1 c.c. Solid.	Specific Heat.	Atomic Weight.	Molecular Weight.
Carbon (Diamond) ..	3·514 at 18° ..	(see below) ..	12·00	?
(Graphite) ..	2·25 at ? ..	(see below) ..	—	?
(Charcoal) ..	About 1·8 at ?	—	—	—
Titanium	Undetermined	Undetermined	48·13	?
Zirconium..........	4·15 at ? ..	0·0660	90·0	?
Thorium	11·1 at 17° ..	0·0279	232·4	?

Note.—The specific heat of carbon increases very rapidly with rise of temperature (*cf.* beryllium and boron).*

Temperature ..	−50°	−10°	+10°	+33°	+58°	+86°
Sp. heat—						
Diamond ..	0·0635	0·0955	0·1128	0·1318	0·1532	0·1765
Graphite ..	0·1138	0·1437	0·1604	—	0·1990	—
Temperature ..	+140°	+206°	+247°	+600°	+800°	+1000°
Sp. heat—						
Diamond ..	0 2218	0·2733	0·3026	0·4408	0·4489	0·4589
Graphite ..	0·2542	0·2966	—	0·4431	0·4529	0·4670

It is noticeable that, although at low temperatures the specific heat of the diamond differs considerably from that of graphite, yet at high temperatures they nearly coincide.

Group VIII.—**Silicon, Germanium, Tin, Terbium** (?), **Lead.**

The first of these elements, silicon, closely resembles titanium; it is a blackish, lustrous substance. Germanium and tin are white metals, with bright lustre; lead is of a greyer hue. Terbium, as element, has not been prepared in a pure state.

Sources.—The element **silicon**, next to oxygen is the most widely distributed and abundant of the elements on the surface of the globe, forming about 25 per cent. of its total weight. It never occurs in the free state, being attacked by oxygen: hence it exists only as oxide (silica), alone; or in combination with other oxides, as silicates. As such, it is contained in a vast number of minerals. Some of the most typical of these are described under the heading *silica* (p. 300). The more commonly occurring forms of silica are *quartz*, *sandstone*, and *flint;* pure crystallised silica is named *rock-crystal*, *bog diamond*, or *Irish diamond; agate*, *chalcedony*, *opal*, &c., are other forms. *Granite*, *trap*, *basalt*, *porphyry*, *schist*, and *clay* are rocks entirely composed of silicates. The name is derived from *silex*, flint. **Germanium**,† an element patriotically named, has been recently discovered by Winkler in a mineral termed *argyrodite* found in the Himmelsfürst mine, near Freiberg. It contains 6 or 7 per cent. (?) of sulphide of germanium. Euxenite is also said to contain a trace of germanium, —about 0·1 per cent.

Tin is a moderately abundant element, although not widely distributed. The chief mines are in Cornwall; it also occurs in the Erzgebirgé, in Saxony and Bohemia, in the Malay Peninsula, and

* Weber, *Pogg. Ann.*, **154**, 367.

† Winkler, *Berichte*, **19**, 210; *J. prakt. Chem.* (2), **34**, 177.

in Peru. Large deposits of tin-ore have recently been discovered in Australia and in Borneo. It occurs as oxide, in *tin-stone* or *cassiterite*, and as sulphide, a comparatively rare mineral. It is never found as a metal, owing to its tendency to oxidise.

Terbium, the connection of which with this group of elements is open to question, is associated with yttrium (*q.v.*), and is contained in the same minerals as yttrium.

Lead, like tin, is always found in combination, chiefly with sulphur in *galena*, a very widely distributed ore, found in the Isle of Man, in Cornwall, in Derbyshire, in the south of Lanarkshire, and in many foreign countries. Other ores of smaller importance are the carbonate or *cerussite*, the sulphate, phosphate, and arsenate.

Preparation.—Silicon,* like titanium, is produced by withdrawing chlorine from its chloride by passing the vapour of the latter over red hot sodium; or by removing fluorine from its double compound with fluorine and sodium, sodium silicifluoride, by means of metallic sodium; the metal zinc may also be used along with sodium to withdraw the fluorine, when the silicon crystallises from the zinc, which may be removed by dissolving it in weak hydrochloric acid. **Germanium, tin,** and **lead** are reduced from their oxides by heating in hydrogen or with carbon. **Terbium** has not been prepared. For the preparation of tin and lead on the large scale, the chapter on the oxides and sulphides of these metals must be consulted (p. 296, and also Chap. XXXVIII); for their metallurgy involves somewhat intricate operations.

Properties.—Silicon lacks metallic lustre, and is therefore usually classed with the non-metals. It is a blackish brown powder, which, when crystallised from zinc or aluminium, separates either in black lustrous tablets resembling graphite, or in brilliant hard iron-grey prisms. It fuses at a high temperature, and may be cast into rods. It is contained in cast iron, probably however in combination with the iron. The crystalline variety conducts electricity. When heated in oxygen, chlorine, bromine, or sulphur gas, silicon combines with these elements. The crystalline variety can be dissolved only by fusion with caustic potash (see *Silicates*, p. 310).

Germanium is a white metal, somewhat resembling antimony. It is very brittle and can be readily powdered. It may be melted under a layer of borax, which prevents oxidation; it is, however, not very easily oxidised. It melts at a bright red heat. It combines

* Deville and Caron, *Annales* (3), **67**, 435.

directly with oxygen, sulphur, and the halogens when heated in the vapour of these elements.

Tin is a lustrous white metal resembling silver. It is very soft and malleable, and may be hammered into foil (tin foil), but its wire has little tenacity. Up to 100° its malleability increases; but, like zinc, it becomes brittle at higher temperatures, and may be powdered at 200°. Its fracture is crystalline. It melts at a low temperature.

Although not oxidised at the ordinary temperature, it burns in air with a white flame when strongly heated; it also unites directly with the halogens and with sulphur. It forms alloys with many other metals which find commercial use. It is also largely used in tinning iron (see Alloys, p. 583). An allotropic form of tin is produced when tin is cooled to a low temperature, or when it is kept for a long time; it is greyish-red, and exceedingly brittle. When heated to 50° for some hours it is reconverted into ordinary tin.*

Lead has a greyer shade than tin. It is soft, and may be cut with a knife. It may be hammered into foil, and drawn into wire, which however has little tenacity. It is easily fused, and volatilises at a white heat.

Lead combines directly with oxygen at a high temperature, forming "dross"; although not affected by dry oxygen, moist atmospheric air soon tarnishes it. When heated with the halogens or with sulphur, it combines directly with them. It is used largely for pipes, for covering roofs, for bullets, shot, &c.; and its various alloys find a very wide application.

Physical Properties.

	Mass of 1 c.c. Solid.	Melting-point.	Specific Heat.	Atomic Weight.	Molecular Weight.
Silicon—					
Graphitoidal	2·2 at ?	—	(see below)	28·33	?
Adamantine	2·48 at ? ..	About 1100°	,,	—	—
Germanium ..	5·47 at 20·4°	About 900°	0·0758 (100° to 440°)	72·3	?
Tin (solid)....	7·29 at 13°..	226°	0·0562	119·1	119·1
,, ,, ...	7·18 at 226°	—	—	—	—
,, (liquid) ..	6·99 at 226°	—	0·0637	—	—
,, (allotropic)	5·8 to 6·0..	—	0·0545	—	—
Lead (solid) ..	11·35 at 14°..	325°	0·0314	206·93	206·93
,, ,, ..	11·0 at 325°..	—	—	—	—
,, (liquid)..	10·65 ,, ..	—	—	—	—

The specific heat of silicon, like that of beryllium, boron, and carbon, varies

* Fritsche, *Phil. Mag.* (4), **38**, 207.

greatly with the temperature, and attains approximate constancy only at high temperatures. The following determinations were made by Weber.*

Temp.	−40°	+22°	+57°	+86°	+129°	+184°	+232°
Sp. heat ...	0·1360	0·1697	0·1833	0·1901	0·1964	0·2011	0·2029

APPENDIX.

Equations expressing the preparation of elements of Groups VI, VII, and VIII.

Chromium.—$2CrCl_3 + 3Zn = 2Cr + 3ZnCl_2$.
Iron.—$FeO + H_2 = Fe + H_2O$.
Manganese.—$MnO + C = Mn + CO$.
Carbon.—$CCl_4 + 4Na = C + 4NaCl$.
Titanium.—$2KF.TiF_4 + 4K = Ti + 6KF$.
Cerium.—$2CeCl_3 = 2Ce + 3Cl_2$.
Silicon.—$2NaF.SiF_4 + 4Na = Si + 6NaF$.
Germanium.—$GeO_2 + 2H_2 = Ge + 2H_2O$.
Tin.—$SnO_2 + 2C = Sn + 2CO$.
Lead.—$PbO + C = Pb + CO$.

* *Pogg. Ann.*, **154**, 367.

CHAPTER V.

THE ELEMENTS (CONTINUED).

GROUP IX, THE NITROGEN GROUP; GROUP X, THE PHOSPHORUS GROUP; GROUP XI, THE MOLYBDENUM GROUP; GROUP XII, THE OXYGEN AND SULPHUR GROUP.

GROUP IX.—Nitrogen, Vanadium, Niobium (or Columbium), Didymium* (?), Tantalum.

THE first element of this group, like the first of the seventh group, does not outwardly resemble the remaining ones. It is a colourless gas, whereas the others are solids with metallic lustre. It exists free, like carbon, while the others occur only as oxides, because they readily combine with oxygen.

Sources.—Nitrogen forms nearly four-fifths of the volume as well as of the weight of air. It occurs also in small amount in air as ammonia, in which it is combined with hydrogen. Ammonia also exists in the soil, being carried down by the rain, and yields its nitrogen to plants, which use it as food, assimilating it by means of their roots. Nitrogen is essential to the life of plants and animals, and is a constituent of the albuminous matters of which they largely consist. Coal, the relic of a former vegetation, also contains nitrogen in combination with carbon, hydrogen, and oxygen. Lastly, it occurs in combination with oxygen and sodium, and with oxygen and potassium, in sodium and potassium nitrates, which encrust the surface of the soil of dry countries. They are exported from India, and from S. America. Nitrogen has no great tendency to combine with other elements; hence it chiefly occurs in the free state. The spectroscope has also revealed its presence in some nebulæ.

Vanadium is a comparatively rare element. It is found in

* It is doubtful whether didymium belongs to this group of elements. It appears to be a mixture, not a simple substance. See p. 602.

combination with oxygen, along with lead, copper, and zinc oxides, as vanadates of these metals. A crystalline incrustation on the Keuper Sandstone, at Alderley Edge, in Cheshire, in which vanadium is associated with phosphorus and copper, named *mottramite*, is one of its chief sources.

Niobium, tantalum, and **didymium** are associated with rare metals, such as yttrium, cerium, lanthanum, &c., in *euxenite* and similar minerals. The two former are also found in combination with iron and manganese in *niobite* and *tantalite*, minerals found in the United States and in Greenland.

Preparation.—Nitrogen is usually prepared by removing oxygen from air, which consists mainly of these two gases. By heating ammonia, its compound with hydrogen, to a red heat, it is decomposed into its constituents; but the hydrogen is not easily separated from the nitrogen; hence the plan usually adopted is to decompose ammonia by the action of chlorine, or by oxygen at a red heat, both of which unite with the hydrogen, liberating nitrogen. Perhaps the best method is one in which the oxygen of the air is made to combine with the hydrogen of the ammonia; the nitrogen of both air and ammonia is thus collected. The apparatus is shown in the accompanying figure.

A gas-holder, A, is connected with a U-tube, B, filled with weak sulphuric acid, which in its turn communicates by means of indiarubber tubing with a tube of hard glass, C, containing bright copper turnings. The other end of the hard-glass tube is joined

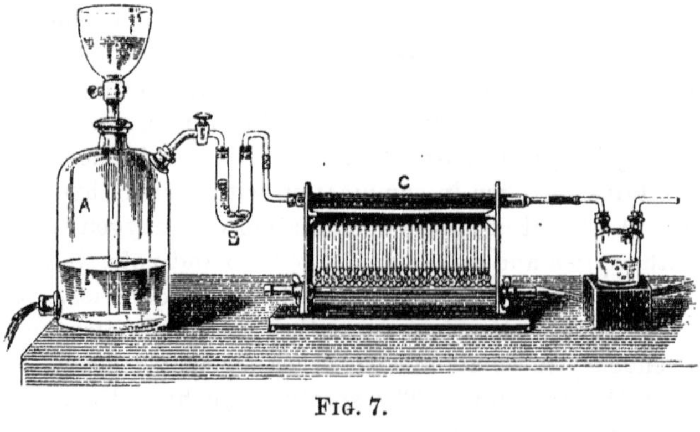

FIG. 7.

to a wash-bottle, half full of strong ammonia solution. The copper is heated to bright redness, and the water in the gas-holder is allowed to escape, a current of air being thus drawn through the ammonia-solution. The gaseous ammonia is carried along with the air over the red-hot copper. The oxygen of the air unites in presence of the red-hot copper with the hydrogen of the ammonia, forming water, and the nitrogen of the air along with the nitrogen from the ammonia both pass on. The sulphuric acid in the U-tube serves to retain excess of ammonia, and pure nitrogen is the product.

Nitrogen may also be prepared by leaving air in contact with any absorbent for oxygen; for example, phosphorus; or a solution of cuprous chloride in ammonia; or potassium pyrogallate.

Vanadium* has been prepared by withdrawing chlorine from one of its compounds with that element by the action of hydrogen a red heat. The utmost precautions must be taken to exclude oxygen and moisture, as vanadium is at once oxidised at a red heat by these substances. As it attacks porcelain, it must be heated in a platinum boat placed in a porcelain tube, during the passage of the hydrogen. The method of preparation of **niobium**† is similar to that of vanadium.

Tantalum is said to have been prepared by a method similar to that which yields silicon, *viz.*, by heating with metallic sodium its compound with potassium and fluorine.

Didymium has been made in the same manner as cerium (*q.v.*). The substance thus called is certainly a mixture of at least two metals (see p. 605).

Properties.—Nitrogen is a colourless, odourless, tasteless gas, somewhat lighter than air. It condenses to a colourless liquid at the very low temperature −193·1,‡ and solidifies to white flakes at −214°, when cooled by boiling oxygen. The liquid is lighter than water.

The only elements with which it combines easily and directly at a red heat are magnesium, boron, titanium, and vanadium. At a higher temperature, that of the electric spark, it combines with hydrogen and with oxygen; indeed combination between oxygen and nitrogen may be caused by burning magnesium in air, when the constituents of air unite to form peroxide of nitrogen; and this gas is suddenly cooled by escaping away from the source of heat, and therefore remains undecomposed.

Vanadium is a white substance with metallic lustre. It does not combine with oxygen at the ordinary temperature, nor even

* Roscoe, *Proc. Roy. Soc.*, **18**, 37 and 316.

† Roscoe, *Chem. News*, **37**, 25.

‡ *Comptes rend.*, **100**, 350.

at 100°, but it takes fire spontaneously and burns in chlorine. It is unaltered by water, except at high temperatures.

Niobium forms an irridescent steel-grey powder.

Didymium is a white metal with a faint yellow tinge.

Tantalum is said to be a black powder, but it is doubtful whether it has been isolated.

Physical Properties.

	Mass of 1 c.c. Solid.	Liquid.	Gas.	Density. H = 1.
Nitrogen	—	0·89 at −194·4°	0·001258	14·08
Vanadium	5·87 at 15°	—	—	—
Niobium	7·06 at 15·5°	—	—	
Didymium	6·54	—		
Tantalum	10·5 ?	—	—	

	Melting-point.	Boiling-point.	Specific Heat.	Atomic Weight.	Molecular Weight.
Nitrogen	−214°	−194·4°	0·2438	14·03	28·06
Vanadium	Not at bright red heat	—	?	51·4	?
Niobium	Very high	—	?	94·0	
Didymium	,,	—	0·0456	Ndi* 140·8; Prdi 143·6	?
Tantalum	,,		?	182·5	?

The critical temperature of nitrogen is −146°, and the critical pressure 35 atmospheres.† Its vapour-pressures are as follows:—

Pressure in atmospheres	35	31	17	1	Very low
Temperature	−146°	−148·2°	−160·5°	−194·4°	−213°

Under a pressure of about 4000 atmospheres, nitrogen has the density 0·8293, at ordinary temperature, compared with water.

Group X.—Phosphorus, Arsenic, Antimony, Erbium,‡ Bismuth.

Owing to the great tendency of phosphorus to unite with oxygen, t is always found in combination with it. Arsenic, too, is seldom found native; it also is easily oxidised. Antimony and bismuth are both found native. Erbium is always asso-

* Neodymium and praseodymium, two bodies into which didymium has been resolved.

† *Comptes rend.*, **99**, 133, 184; **100**, 350.

‡ It is doubtful whether erbium belongs to this group.

ciated with cerium, lanthanum, yttrium, &c. Phosphorus, arsenic, and antimony display allotropy.

Sources.—Phosphorus occurs chiefly in combination with oxygen and calcium, as calcium phosphate, in minerals named *apatite*, in which it is associated with fluorine; *phosphorite*, an earthy variety; and in *coprolites*. It is also found as phosphate of aluminium, or *wavellite* in large deposits; lead and copper phosphates also occur native. It is a constituent of all soils, though in minute amount. From them it is absorbed by plants, and is hence a constituent of all vegetable matter, especially seeds. Through plants it is assimilated by animals, and forms a constituent of the bones (about 58 per cent.) and the nerves. Ignited bones consist mainly of calcium phosphate.

Arsenic occurs most abundantly in combination with iron as arsenical iron, and with nickel and cobalt as *kupfer-nickel* and *smaltite;* also with iron and sulphur in *arsenical pyrites* and *mispickel*. With sulphur it forms *realgar* and *orpiment;* and it is also found combined with oxygen and metals as arsenates. It is sometimes found native.

Antimony is rarely found native; its most abundant ore is *stibnite*, or antimony sulphide; it also occurs as *antimony ochre* or oxide; and it is associated in various minerals with sulphur and lead, mercury, copper, silver, &c.

Bismuth is a comparatively rare metal, and nearly always occurs native. It is sometimes associated with tellurium.

Erbium accompanies yttrium, cerium, &c. (*q.v.*). It is extremely rare.

Preparation. — Phosphorus was originally obtained by Brandt by distilling dried and charred urine at a white heat. The carbon resulting from the decomposition of the animal matter deprived the sodium phosphate of its oxygen, and phosphorus distilled over. It is still made by a somewhat similar process. Metaphosphoric acid, a compound of phosphorus with oxygen and hydrogen, is distilled with powdered coke or charcoal from clay retorts. The carbon deprives this substance of its oxygen, and phosphorus, hydrogen, and oxide of carbon pass over in the gaseous state. The hydrogen and carbonic oxide gases escape, while the phosphorus is condensed and falls into warm water. For a detailed description of the process see Chap. XXXVIII.

Arsenic is produced by distilling *mispickel*, when the arsenic, which is very volatile, distils over, leaving the sulphur and iron behind as ferrous sulphide.

Antimony is prepared by heating its sulphide (*stibnite*) with scrap iron. The iron withdraws the sulphur, and the antimony separates in the metallic state. It is not sufficiently volatile to be conveniently distilled, but it flows down, forming a layer below the sulphide of iron. These operations must all be conducted in absence of air, for phosphorus, arsenic, and antimony readily combine with oxygen.

Bismuth is freed from earthy impurities by melting it in a crucible, when it sinks to the bottom. Arsenic, antimony, and bismuth may also be obtained by heating their oxides in a current of hydrogen. Erbium has not been prepared.

Properties.—Phosphorus exists in two allotropic forms. The variety longest known, called *yellow* or *ordinary* phosphorus, is a yellowish-white, waxy substance, possessing a strong disagreeable smell. It has a great tendency to combine with oxygen even at ordinary temperatures, and shines in the dark owing to slow oxidation; hence the name phosphorus (from φῶς, light, and φέρειν, to bear). It must, therefore, be kept under water. It is easily fusible, but when melted in air it takes fire and burns. It also catches fire when rubbed on a rough surface, owing to the heat produced by friction. Hence its use for lucifer matches. It is soluble in carbon disulphide, a liquid compound of carbon and sulphur, and may be obtained in crystals by the slow evaporation of the disulphide; this solution is named "Greek fire." When the solvent evaporates, the phosphorus is left in a finely divided state, and is spontaneously inflammable. It is also soluble in alcohol, ether, olive oil, turpentine, benzene, and in certain of its own compounds, such as chloride and oxychloride of phosphorus. It is easily distilled at a moderate temperature (290°). Its vapour is yellow.

When heated to 240° for some time in a closed vessel in absence of oxygen, it changes to a red variety, named *red*, or *amorphous*, phosphorus. The change may be brought about more quickly by a higher temperature, or by addition of a trace of iodine to the phosphorus. It is also produced under water on exposure of the yellow variety to light. But red phosphorus, when heated, also changes back to yellow phosphorus. Such a change, which can take place in two directions, is termed a *limited reaction.* Red phosphorus is insoluble in all ordinary solvents; hence it may be purified from yellow phosphorus by digestion with carbon disulphide. It does not combine with oxygen at the ordinary temperature, nor perhaps at any temperature, for it burns in air only when made so hot that the change into the yellow variety begins. The colour of allotropic phosphorus varies, according to the temperature at which it is formed. If prepared at 260° it is deep red, and has a glassy fracture; at 440° it is orange, and has a granular fracture; at 550° it is violet-grey; it fuses at 580°, and solidifies to red crystals, which have a ruby-red fracture.* It is necessary to exclude air and to heat the phosphorus under pressure to produce these changes. It dissolves in melted lead, and separates on cooling in nearly black crystals.

Yellow phosphorus combines directly and very readily with oxygen and the halogens. It also unites with sulphur, selenium, and tellurium, and with most metals. Red phosphorus combines directly with the halogens. Neither variety unites directly with nitrogen or with hydrogen. Yellow phosphorus is poisonous, doses of 1 grain and upwards producing fatal effects, but in small doses it is a powerful remedy for nervous disorders. Yellow phosphorus is a non-conductor of electricity, but red phosphorus conducts.

Arsenic is a very brittle steel-grey substance with metallic lustre on freshly broken surfaces. When heated, it sublimes without melting, and condenses partly in crystals, partly in a black amorphous (*i.e.*, non-crystalline) state. It may, however, be melted under great pressure. The amorphous variety is rendered crystalline by heating it to 360°.† It readily combines with oxygen, and hence loses its lustre on exposure to moist air. It burns when heated in air, spreading a garlic-like smell. It unites directly with oxygen, with the halogens, and with most other elements.

Antimony, like phosphorus and arsenic, also exists in two forms. Ordinary antimony is a bluish-white metal, very brittle and crystalline. It is not oxidised by air at ordinary temperatures, and does not tarnish on exposure. Allotropic antimony‡ is obtained by electrolysing a strong solution of antimony chloride. A greyish deposit is formed on the negative pole, which has the remarkable property of exploding when struck. Its specific gravity is considerably less than that of the ordinary variety. It is said, however, to contain hydrogen. Antimony unites directly with all elements, except nitrogen and carbon.

Bismuth is a greyish-white metal with a red tinge, also very brittle and crystalline. The conductivity for electricity of the three elements arsenic, antimony, and bismuth rises in the order given. Bismuth is the most diamagnetic of the elements.

Erbium has not been isolated.

* *Comptes rend.*, **78**, 748.

† *Ibid.*, **96**, 497 and 1314.

‡ Gore, *Chem. Soc. J.*, **16**, 365; Böttger, *J. prakt. Chem.*, **73**, 484; **107**, 43.

Physical Properties.

	Mass of 1 c.c. Solid.	Mass of 1 c.c. Liquid.	Density, H = 1.	Melting-point.
Phosphorus, yellow ..	1·83 at 10°	1·75 at 40° 1·49 at 278°	65·0 at 1040°	44·4°
" red	2·15 to 2·3 (at 0°)	—	45·4 at 1700°	580°
Arsenic, amorphous ..	4·7 at 14°	—	147·2 at 860°	—
" crystalline ..	5·73 at 14°	—	77·5 at 1736°	—
Antimony, ordinary ..	6·67 at 155°	6·5 ?	155·1 at 1572°	425°
" explosive ..	5·7 to 5·8	—	141·2 at 1640°	—
Bismuth	9·8 at 13·5°	10·0	246·2 at 1700°	268·3°

	Boiling-point.	Specific Heat.	Atomic Weight.	Molecular Weight.
Phosphorus, yellow ..	278·3°	0·01740 0·01895	31·03	62·06 to 124·12
" red	—	0·0170	—	—
Arsenic, amorphous ..	—	0·0758	75·09	150·18 to 300·36
" crystalline ..	—	0·0830	—	—
Antimony, ordinary ..	1300°	0·0508	120·30	120·3 to ?
" explosive ..	—	0·0541	—	—
Bismuth	1640°	0·0308	208·1	208·1 to ?

Group XI.—(**Oxygen, Chromium**).—**Molybdenum, Tungsten, Uranium.**

The resemblance between oxygen and the other four members of this group is a slight one. It is advisable to consider oxygen along with the three elements sulphur, selenium, and tellurium, with which it displays much greater analogy.

The elements molybdenum, tungsten, and uranium present some analogy with chromium, both in their properties as well as in the compounds which they form. But chromium is best considered along with aluminium, iron, and manganese.

Sources.—The chief source of **molybdenum** is the sulphide, *molybdenum glance*, or *molybdenite*, and *wulfenite*, a compound of molybdenum, oxygen, and lead. These are rare minerals; an alloy of lead and molybdenum has also been found native in the State of Utah.

Tungsten occurs in *wolfram*, combined with oxygen, iron, and manganese; and in *scheelite*, with oxygen and calcium.

Uranium is chiefly found as *pitchblende*, in combination with oxygen.

Preparation.—Molybdenum* is prepared by heating its chloride to bright redness in a tube through which a stream of hydrogen gas is passed. The hydrogen unites with the chlorine, forming gaseous hydrogen chloride, leaving the non-volatile molybdenum. It may also be obtained by heating its oxide with charcoal.

Tungsten† can be prepared in a similar manner, or from its oxide by the action of hydrogen ; the hydrogen removes the oxygen as water, which passes off as gas, while the metal remains.

Uranium‡ is best got from its chloride by heating it with metallic sodium in an iron crucible. The sodium and chlorine unite, forming common salt, while the uranium, which does not unite with sodium, sinks to the bottom of the crucible, being heavier than the fused salt.

Properties. — These elements all possess metallic lustre. **Molybdenum** is a brittle silver-white substance, exceedingly hard. It fuses at a high temperature. **Tungsten** is a steel-grey crystalline powder, which fuses at a white heat. **Uranium** is a black powder which is fusible to a grey metallic button of great hardness.

These metals do not combine with oxygen at the ordinary temperature, but are converted into chlorides when thrown into chlorine gas in the state of powder. At a high temperature they burn in air, forming oxides. They also unite with sulphur at a red heat. They are unchanged by water at the ordinary temperature.

Physical Properties.

	Mass of 1 c.c. Solid.	Melting-point.	Specific Heat.	Atomic Weight.	Molecular Weight.
Molybdenum..	8·6	White heat..	0·0722	95·7	Unknown.
Tungsten.....	19·2 (at 12°).	White heat..	0·0334	184	„
Uranium.....	18·7 (at 4°).	Red heat ...	0·0277	240	

Group XII.—**Oxygen, Sulphur, Selenium, Tellurium.**

These elements all occur native, as well as in combination. The first is a gas; the other three are solids at the ordinary temperature, and are often associated with each other.

* Debray, *Comptes rend.*, **46**, 1098.

† Roscoe, *Chem. Soc. J.*, **10**, 286.

‡ Peligot, *Annales* (3), **5**, 53 ; **12**, 549.

Sources.—**Oxygen** is the most abundant and widely distributed of the elements, forming, as has been estimated, 50 per cent. of the earth's crust. About one-fifth of the weight as well as of the volume of air consists of oxygen, the remaining four-fifths being nitrogen, with which the oxygen is *mixed.* It constitutes eight-ninths of the weight of water, and is found in union with every element in nature, except fluorine, chlorine, bromine, platinum and its analogues, and gold, silver, and mercury. Many compounds into which it enters have been already mentioned as sources of the elements.

Sulphur occurs native in the neighbourhood of volcanoes, and coats the surface of the soil in districts of volcanic activity. It is chiefly mined in Italy and Sicily. It also occurs in combination with iron as *iron pyrites*, and with iron and copper as *copper pyrites;* with lead as *galena*, with zinc as *blende*, with mercury as *cinnabar.* It also occurs in union with oxygen and a metal, *e.g.*, in the sulphates of calcium, magnesium, sodium, &c. Its principal sources are *native sulphur;* and *copper pyrites*, of which large mines exist in the South of Spain. It exists also in certain volatile oils, such as oil of mustard, oil of garlic, &c.

Selenium in small quantities almost invariably accompanies sulphur; both native and in its compounds. It is also, but rarely, found in combination with lead and copper; and with nickel, silver, molybdenum, &c.

Tellurium is found in the free state, and also in combination with bismuth, silver, lead, and gold. It is a very rare element.

Preparation.—There is no convenient method of separating nitrogen from air; hence pure **oxygen**, unlike pure nitrogen, cannot be directly prepared from that source. Owing to its tendency to unite with almost all elements, it cannot well be prepared by displacing it from any one of its compounds. The only elements capable of displacing it appear to be fluorine and chlorine, for almost all other elements combine directly with it. It must therefore be prepared by heating certain of its compounds with other elements—certain oxides and double oxides or salts; or by the electrolysis of certain of its compounds, *e.g.*, water. The methods of preparing it may be grouped under three heads:—

1. *The electrolysis of water, or of fused oxides or hydroxides, i.e.*, oxides of hydrogen and another element. Water, however, is a non-conductor of electricity when pure, and it is necessary, in order to make it conduct, to dissolve in it some substances with which it reacts. In practice, the operation is conducted as follows:— A U-tube, *ho*, is filled with dilute sulphuric acid. Through the

lower end of each of these tubes is sealed a piece of platinum wire, connected each with a slip of platinum foil, and the pieces of wire projecting outside are connected by copper wires to the poles of a battery of four Bunsen's or Groves' or bichrome elements (two are sufficient, but the operation is more rapid with four cells). The oxygen is evolved from the electrode connected with the car-

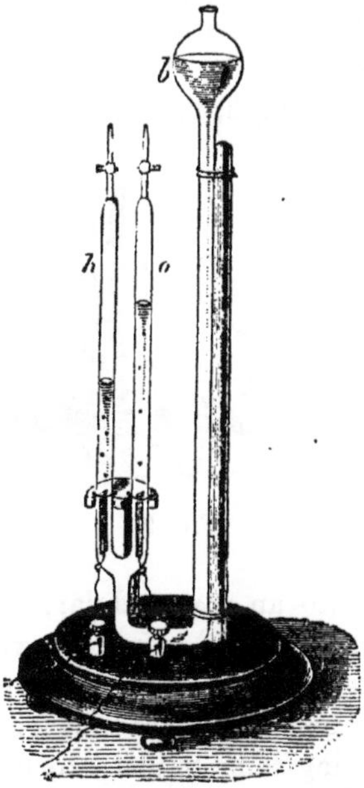

FIG. 8.

bon or platinum plate; the gas issuing from the other electrode is hydrogen. After the current has passed for some time, the tube *o* is partly filled with oxygen gas, and the hydrogen in the tube *h* occupies about twice the volume of the oxygen. On opening the stopcock *o* carefully, the characteristic property possessed by oxygen of rekindling a glowing piece of wood may be shown by allowing the escaping gas to play on it; and the hydrogen may be set on fire as it escapes from the tube *h*.

2. *The heating of certain oxides.*—All compounds with oxygen of the metals of the platinum group; of gold, silver, or mercury; of the chlorine group of elements; of the higher oxides of nitrogen; the higher oxides of the chromium group of elements (*e.g.*, chromium trioxide, chromates, potassium ferrate, manganate or permanganate, manganese dioxide, nickel and cobalt sesquioxides); of the calcium group of elements, and of lead; all these part with oxygen

at a bright red heat, and in many cases at a lower temperature. The action of sulphuric acid on the higher oxides also yields oxygen (see *Manganese Dioxide*, and *Chromate of Potassium*).

Three typical examples are chosen :—

(*a*). *The action of heat on mercuric oxide.*—The apparatus is shown in fig. 9. A is a tube of combustion glass, which is more difficult of fusion than ordinary glass, sealed at one end, and closed at the other end with a perforated indiarubber cork through which a bent glass tube is inserted. This tube dips below the surface of the water in a glass trough, E, and its open end bends upwards,

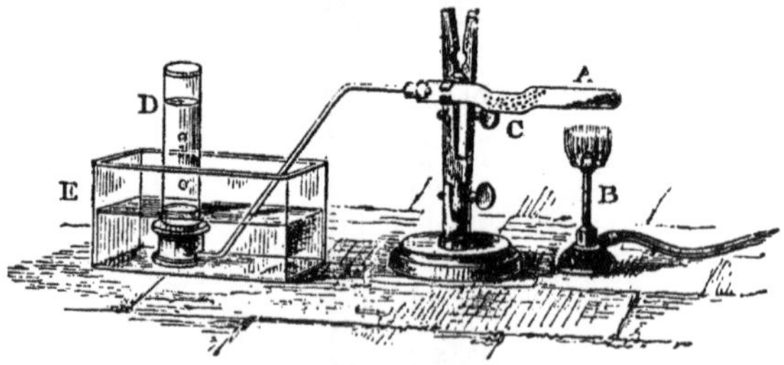

FIG. 9.

so as to deliver gas into an inverted jar, D, full of water. The hard glass tube contains some mercuric oxide. Heat is applied with a Bunsen's burner, B, care being taken to wave about the flame at first, so as to heat the glass tube gradually; else it is apt to crack. After allowing some bubbles to escape, so as to ensure the expulsion of air from the tube, the glass jar is placed above the exit tube, and the gas is collected. The mercury collects in the depression C. It was by this means that Priestley, one of the discoverers of oxygen, first prepared it in 1774. He named it dephlogisticated air (see p. 11).

(*b*). *The action of heat on potassium chlorate.*—This body is a compound of potassium, chlorine, and oxygen. The oxygen is wholly expelled, potassium chloride, a compound of chlorine and potassium, remaining behind. The chlorate is heated in a hard glass flask, by aid of a Bunsen burner (see *Potassium Chlorate*, p. 466). The salt melts and begins to froth, owing to the evolution of oxygen. If some manganese dioxide be mixed with the chlorate, the gas is evolved at a lower temperature, but is not so pure (see *Perchlorates ;* also *Oxides of Manganese*).*

(*c*). *The action of heat on barium dioxide.*—Barium forms two oxides, one, the monoxide, containing less oxygen than the second,

* *Chem. Soc. J.*, 51, 274.

the dioxide. When the monoxide, a grey porous solid, is heated to dull redness in contact with dry air, it absorbs oxygen from the air, producing the dioxide; the absorption is increased by pressure. On decreasing the pressure, the dioxide formed is decomposed; the oxygen may be pumped off by means of an air-pump and

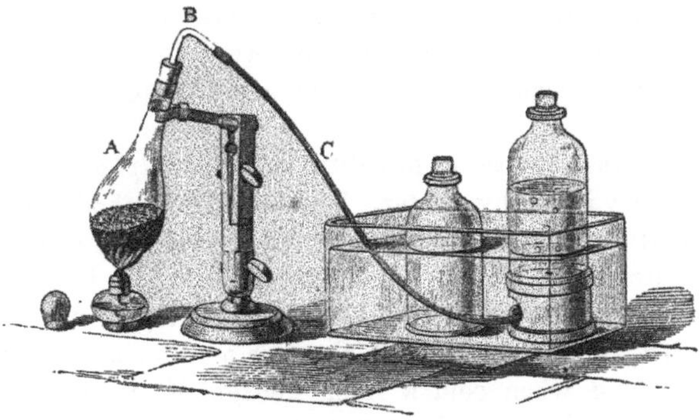

FIG. 10.

forced into iron or steel bottles. This process is now carried out on a large scale, and indeed is the only method by which oxygen is made commercially. The barium dioxide is contained in vertical iron tubes, which are heated with gas from a Siemens's "producer," the temperature being carefully regulated.

3. *By displacement.*—The gaseous element fluorine, which has only recently been prepared, at once acts on water, combining with its hydrogen, and liberating its oxygen (see *Ozone*, p. 387). Chlorine and steam at a red heat react in a similar manner, hydrogen chloride and oxygen being produced. Chlorine gas also slowly acts on water exposed to sunlight, liberating oxygen. The halogens expel oxygen from certain oxides at a red heat; *e.g.*, from the oxides of lead, bismuth, zinc, &c. None of these are practical methods of preparing the gas (see *Oxides of Manganese, Chlorine, Silver, and Lead;* also *Hypochlorites*).

Sulphur.—Sulphur may be prepared (1) by *heating certain sulphides, e.g.*, those of gold and platinum, which part with their sulphur, leaving the metal; or by heating hydrogen sulphide, which splits into sulphur and hydrogen; and (2) by *heating certain persulphides* (compounds of metals with sulphur which form more than one sulphide), the most important of which is *iron pyrites*. As sulphur combines directly with most other elements, there are few methods of preparing it by displacing it from its compounds; yet chlorine, bromine, or iodine dissolved in water combines with the hydrogen of hydrogen sulphide in preference to the sulphur, so that the element is liberated (see also *Polysulphides of Sodium*).

The elements **selenium** and **tellurium** are most readily prepared by displacement. The compounds which they form with oxygen are decomposed by sulphur dioxide, which absorbs their oxygen, itself changing to sulphur trioxide, and liberating the selenium or tellurium (see *Selenium and Tellurium Dioxides*). Their compounds with hydrogen, dissolved in water, are decomposed by atmospheric oxygen, the element falling to the bottom of the solution.

An important source of **sulphur** is *native sulphur*, of which the chief impurity is earthy matter. The modern method of extraction is to melt it under water in a boiler by forcing in steam until the pressure rises to 25 lbs. on the square inch. The temperature of the water is thus raised to over 115°, the melting point of sulphur. The melted sulphur is run off through a stop-cock in the side of the boiler, and when cold a fresh charge of impure sulphur is introduced, and the operation repeated. Sulphur is usually brought into commerce in the form of sticks cast in wooden moulds, and is in this form named *roll sulphur*.

Sulphur is a by-product in the manufacture of alkali, being obtained from *iron* or *copper pyrites* (see Chapter XXXIX).

Selenium is best obtained from certain residues in the manufacture of sulphuric acid, by treating them with nitric acid, and then precipitating the selenium with sulphur dioxide.

Tellurium may be purified by distilling native tellurium at a red heat in a current of hydrogen gas. It is precipitated from its solutions by metallic zinc.

Properties.—Oxygen is a colourless, odourless, tasteless gas, somewhat heavier than atmospheric air. It is very sparingly soluble in water; 100 volumes of water at 4° dissolve 3·7 volumes of oxygen. It has been condensed to a colourless transparent liquid by application of great pressure at a very low temperature, and when still further cooled, it freezes to a white crystalline solid. It was discovered independently by Priestley and by Scheele (see p. 11) in 1774 and 1775; it had previously, however, been recognised as a distinct "air" or gas by Mayow, about 1675 (see p. 9). Its true nature was made public by Lavoisier, as has already been noticed, although Mayow anticipated him in most of his conclusions. Its name, "acid-producer" (ὀξύς γεννάω), was invented by Lavoisier.

Oxygen combines directly with all elements except the halogens, gold, and some metals of the platinum group. Silver and mercury, although they do not readily combine directly with oxygen, can be made to unite under pressure. Many elements, such as carbon, sulphur, and certain metals, do not unite with oxygen except when heated; others, such as sodium, phosphorus, &c., combine at the ordinary temperature. The word "combustion" usually signifies union with oxygen, with evolution of light. All substances which burn in air burn with increased brilliancy in oxygen gas. It is respirable; in its usual dilute state in air, it is when breathed absorbed by the blood, and serves to oxidise the carbon and hydrogen in the body, thereby generating animal heat; if breathed in a pure state, however, oxidation takes place with too great rapidity, and acute febrile symptoms are produced after a short time, followed by death unless the animal is allowed to respire air. The respiration of fishes is sustained by the small amount of oxygen dissolved in the water in which they exist.

When an electric discharge is passed through oxygen, or when the element is liberated by the action of fluorine on water, a portion of it is changed to an allotropic form, which from its strong smell has been named ozone (ὄζειν, to smell). This substance will be considered as an oxide of oxygen, and is described on p. 387.

The remaining three elements of this group, **sulphur, selenium,** and **tellurium,** form a well-marked series. They show progression in their atomic weights: thus S = 32, Se = 79, Te = 125. The atomic weight of selenium is nearly the mean of those of sulphur and tellurium. They show a gradation of colour: sulphur is yellow, selenium red, and tellurium metallic. Sulphur is practically a non-conductor of electricity, selenium conducts when exposed to light, and tellurium is a conductor. No allotropic form of tellurium is known. Selenium is known to exist in three forms: amorphous, which changes to crystalline when fused and kept for some time at 210°; this crystalline variety is insoluble in carbon disulphide; the amorphous variety, produced by precipitating selenium with sulphur dioxide, is a bright-red powder, soluble in carbon disulphide, from which it deposits on evaporation in dark red crystals. Sulphur crystallises in two distinct forms: rhombic crystals (fig. 11), which are found native, and which may be arti-

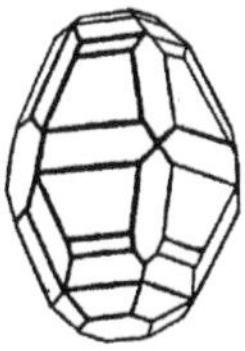

FIG. 11.

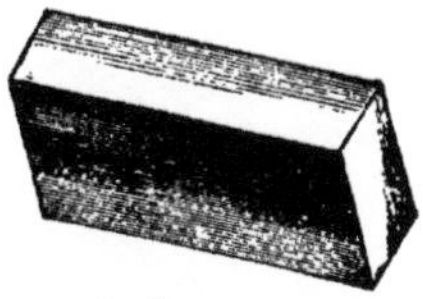

FIG. 12.

ficially produced by crystallising sulphur from carbon disulphide; and monoclinic needles (fig. 12), which may be prepared by melting sulphur, allowing it to cool till the surface has solidified, breaking the solid surface layer, and pouring out the liquid. The interior of the mass is filled with crystals. The monoclinic form also deposits from a solution of sulphur in ether or in benzene. The monoclinic form is less stable than the rhombic; and the crystals, which are clear, transparent, and brownish-yellow, soon become opaque on standing, changing spontaneously into a mass of minute rhombic octohedra. This change is accompanied with evolution of heat. Other crystalline forms have recently been obtained.

Selenium or sulphur, when distilled into a large chamber, condenses in part as a fine powder, named *flowers of sulphur*, or of *selenium*. This is really a mixture of two varieties, one of which is insoluble in carbon disulphide, the other soluble.

Again, in the state of liquid, sulphur exhibits allotropy. It melts at 115° to a clear, pale yellow, mobile liquid. At 200° it turns brown and viscous. When the first variety is poured into water, it at once solidifies to ordinary brittle crystalline sulphur, soluble in carbon disulphide. The viscous variety, however, if poured into water, changes to a curious elastic, indiarubber-like substance, insoluble in carbon disulphide, which only slowly regains its former condition. Between 400° and 446°, its boiling point, sulphur again becomes mobile, still remaining brown. A variety of sulphur soluble in water has recently been discovered.* Sulphur produced by precipitation has a white colour, and its mixture with water is known as *milk of sulphur*.

In the gaseous state also, sulphur displays allotropy. Its density at low temperatures implies a high molecular weight, but at high temperatures the molecule is simpler and weighs less (see p. 614).

These elements unite directly with oxygen, burning in the air when heated; they also combine with each other, with the halogens (the compounds with bromine and with iodine are ill-defined), and with all other elements except nitrogen, when heated in contact with them. They are without action on water at the ordinary temperature.

* *Chem. Soc. J.*, 53, 283.

Physical Properties.

	Mass of 1 c.c. Solid.	Liquid.	Gas.	Density, H = 1.
Oxygen	?	1·24 at −200°	0·001429 (at 0° and 760 mm.)	15·96
Sulphur (rhombic) ..	2·07 at 0°	1·8	—	32·27 (at 1040°)
„ (monoclinic)	1·98			—
„ (plastic)	1·95 at 0°	—	—	—
Selenium, crystallised from fusion	4·4......	4·3	—	82·2 (at 1040°)
Selenium, crystallised from solution	4·8 at 15°	—		—
Selenium, amorphous	4·3......	—		—
Tellurium	6·23 at 0°	—	—	131·4 (at 1440°)

	Melting-point.	Boiling-point.	Specific Heat.	Atomic Weight.	Molecular Weight.
Oxygen	Below −212°	−186°	0·2175	16·0	32
Sulphur (rhombic) ..	115°	446°	0·1776	32·06	64·02
„ (monoclinic)	120°	—	—	—	to
„ (plastic)....	—	—	—	—	252·16 ?
Selenium, crystallised from fusion	217°	665°	0·0746	79·0	158·0 to ?
Selenium, crystallised from solution	—	—	—	—	—
Selenium, amorphous	About 100°	—	—	—	—
Tellurium	Below redness	Bright red heat	0·0483 (crystalline)	125·0	250 to
„	—	—	0·0518 (amorphous)	—	?

*Vapour Pressures of Oxygen at different Temperatures.**

Temperature	−118·8° (crit.)	−121·6°	−125·6°	−129·0°
Pressure, atmos...	50·8 (crit.)	46·7	40·4	34·4
Temperature	−146·8°	−155·6°	−166·1°	−175·4
Pressure, atmos. ..	13·7	8·23	4·25	2·16

Temperature—						
−181·5°	−190°	−192·71°	−196·2°	−198·7°	−200·4°	−211·5°
Pressure, mm.—						
740	160	71	50	20	20	9

These low temperatures were produced by the evaporation of liquid ethylene under reduced pressure. The mass of 1 c.c. of oxygen at its boiling point, −181·4°, under a pressure of 742·1 mm. was found to be 1·124 grams.

* *Comptes rend.*, **98**, 982; **100**, 350, 979; **102**, 1010.

Vapour Densities of Sulphur, Selenium, and Tellurium.—These are discussed on p. 614.

Appendix.—Air.—Air is not a chemical compound, but a mere mixture of nitrogen and oxygen gases. That this is the case is shown by the following considerations:—(1.) There is no heat change on mixing oxygen and nitrogen gases; when a compound is formed, heat is usually evolved. (2.) The density of air is the mean of the densities of the constituent gases; its refractive index for light is also the mean of those of oxygen and nitrogen taken in the proportions in which they occur in air; and so with other physical properties. Such properties, possessed by a compound, are not the mean of those of its constituents. (3.) Oxygen is more soluble in water than nitrogen. On saturating water with air, oxygen dissolves in greater amount than nitrogen; and the gas evolved from the water when it is heated contains a larger proportion of oxygen than does air. (4.) There is no simple relation between the atomic proportions of the nitrogen and oxygen in the air. Such a relation would be characteristic of a compound. The actual composition by weight is, approximately, nitrogen = 77 per cent.; oxygen = 23 per cent. Dividing these numbers by the atomic weights of nitrogen and oxygen respectively, 14 and 16, we obtain the quotients 5·55 and 1·44, representing the relative number of atoms of nitrogen and oxygen in air. The simplest proportion between these numbers is 3·85 to 1; although the ratio approximates to the ratio 4 : 1, yet it is far from being a simple one, as it would be, were air a compound. A substance of the formula N_4O would contain 77·8 per cent. of nitrogen and 22·2 per cent. of oxygen.

Air contains, in addition to nitrogen and oxygen, water-vapour (about 0·84 per cent. by weight, or 1·4 per cent. by volume, on the average), carbon dioxide, from 0·049 to 0·033 per cent. by volume, and a few parts of ammonia per million. Subtracting these from air, the ratio of oxygen to nitrogen by volume approximates to 79·04 volumes of nitrogen, and 20·96 volumes of oxygen. But its composition varies slightly in different places and at different times, although the action of air currents and winds tends to keep it nearly constant.

Air has been liquefied by cooling to −192°; but, as oxygen and nitrogen have not the same boiling points, the less volatile oxygen doubtless liquefies first.

Air is **analysed** (1) by mixing a known volume with a known volume of hydrogen, and exploding the mixture. The oxygen is withdrawn as water, and the residual nitrogen is measured.

2. By passing air deprived of carbon dioxide and moisture over ignited copper. The oxygen unites with the copper, forming oxide; and its amount is ascertained by the gain in weight of the copper; the nitrogen passes on into an empty globe, previously weighed. The gain in weight of the globe gives the weight of nitrogen.

APPENDIX.

Equations expressing the preparation of elements of Groups IX, X, XI, and XII.

Nitrogen.—$2NH_3 = N_2 + 3H_2$.

$2NH_3 + 3Cl_2 = N_2 + 3HCl$.

$4NH_3 + 3O_2 = 2N_2 + 6H_2O$.

Vanadium.—$2VCl_3 + 3H_2 = 2V + 6HCl$.

Phosphorus.—$4HPO_3 + 12C = P_4 + 2H_2 + 12CO$.

Arsenic.—$FeSAs = As + FeS$.

Antimony.—$Sb_2S_3 + 3Fe = 2Sb + 3FeS$.

Molybdenum.—$MoCl_4 + 2H_2 = Mo + 4HCl$.

Tungsten.—$WO_3 + 3H_2 = W + 3H_2O$.

Uranium.—$UCl_4 + 4Na = U + 4NaCl$.

Oxygen.—$2H_2O = O_2 + 2H_2$.

$2HgO = O_2 + 2Hg$.

$2KClO_3 = 3O_2 + 2KCl$.

$2BaO_2 = O_2 + 2BaO$.

$H_2O + Cl_2 = O + 2HCl$.

Sulphur.—$2Au_2S_3 = 3S_2 + 4Au$.

$2FeS_2 = S_2 + 2FeS$.

$2H_2S + O_2 = S_2 + 2H_2O$.

Selenium.—$SeO_2 + 2SO_2 + 2H_2O = Se + 2H_2SO_4$.

CHAPTER VI.

THE ELEMENTS (CONTINUED).

GROUP XIII, THE HALOGEN GROUP; GROUPS XIV AND XV, THE PALLADIUM AND PLATINUM GROUPS; GROUP XVI, THE COPPER GROUP.

GROUP XIII.—**Fluorine, Chlorine, Bromine, Iodine.**

The elements fluorine and manganese present little, if any, analogy. Hence, just as oxygen is best classified along with sulphur, selenium, and tellurium, presenting little, if any, analogy with chromium, so with the elements of this group, manganese and fluorine having little or nothing in common. Both chromium and manganese, it will be remembered, are most conveniently classed with iron, nickel, and cobalt.

The halogens, as these elements are called, are strikingly like each other. They have all low boiling and melting points; and they all combine readily with other elements, oxygen and nitrogen excepted. They all are found in combination; free iodine has been found in the water from Woodhall Spa, near Lincoln.*

Sources.—Fluorine occurs in nature, combined with calcium, in *fluor spar* or *Derbyshire spar;* in *cryolite*, combined with sodium and aluminium; and sometimes in *apatite*, a compound of phosphorus, oxygen, and calcium—calcium phosphate, combined with calcium fluoride. It occurs in small quantity also in the enamel of the teeth and in the bones: it is very widely distributed in nature, although not very abundant.

Chlorine, bromine, and **iodine** are all contained in sea-water, in combination with sodium, potassium, and magnesium. Chlorine also occurs in *rock salt*, of which large mines exist in Cheshire, and in the neighbourhood of the Tyne, in Northumberland. Certain rare ores of silver and mercury contain these metals in combination with chlorine, bromine, and especially with iodine. The chief source of bromine and iodine is sea-weed, which absorbs the compounds of these elements from sea-water.† Iodine is also largely obtained from *Chili saltpetre*, or sodium nitrate,

* *Chem News*, **54**, 300.

† *Dingl. polyt. J.*, **126**, 85.

which contains it in small amount in combination with oxygen and sodium as sodium iodate.*

Preparation.—1. *By electrolysis.*—This is the only method of preparing fluorine; liquid compounds, or solutions of compounds in water, of the other halogens also yield the elements by this process. The preparation of lithium by the electrolysis of its fused chloride (see p. 29) affords an example of the application of electrolysis to a fused compound of chlorine. The gas is evolved at the positive or carbon pole, the metal being deposited on the negative or zinc pole. Concentrated solutions in water of chlorides, bromides, or iodides yield these elements on electrolysis, for such solutions conduct electricity, and the halogens, with exception of fluorine, are not readily acted on by water. Thus hydrogen chloride dissolved in water (hydrochloric acid) splits into chlorine and hydrogen gases when electrolysed. The poles should consist of gas carbon, or platinum, all other substances being attacked, more or

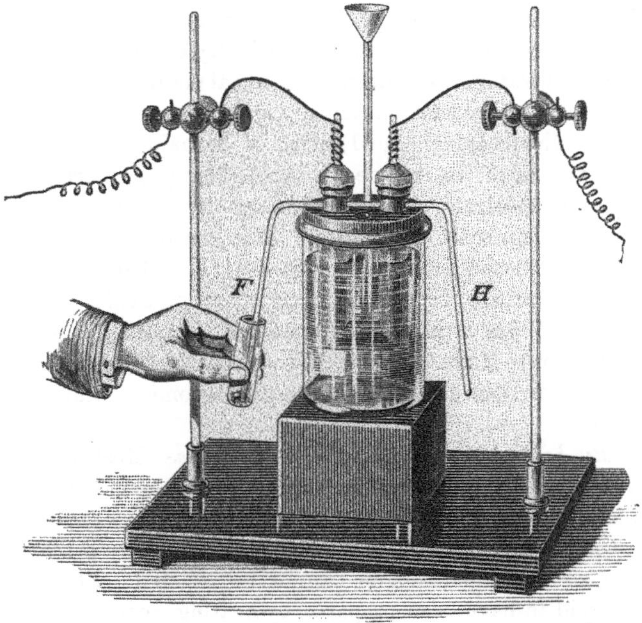

FIG. 13.

less, by the chlorine produced. Fluorine, however, cannot be liberated in presence of water, for it immediately acts on it, liberating oxygen as ozone. Hence, it is prepared by electrolysing in a

* *Dingl. polyt. J.*, **253**, **48**.

U-tube made of an alloy of platinum and iridium, which is but slightly attacked, a solution of potassium fluoride in dry hydrofluoric acid cooled to a low temperature. It is necessary to use such a solution, inasmuch as pure hydrogen fluoride does not conduct electricity, and unless the liquid conduct, electricity cannot pass, and electrolysis cannot take place. The apparatus used by M. Moissan,* who has recently isolated this element, is shown in fig. 13.

2. *By displacement.*—No element appears capable of displacing fluorine from its compounds without combining with it. But chlorine, bromine, and iodine are usually prepared by displacing them from their compounds with potassium, sodium, magnesium, &c., by means of oxygen. The oxygen, however, is not employed in the gaseous state. At a red heat, indeed, such displacement is possible. The action of oxygen, for instance, on red-hot magnesium chloride yields chlorine, while a double compound of chlorine, oxygen, and magnesium (oxychloride) remains behind: again, Deacon's process for producing chlorine, which depends on the interaction of the oxygen of the air and hydrogen chloride at 330° in presence of bricks coated with dry copper chloride, yields chlorine and water as products. The usual method, however, of preparing halogens consists in acting on hydrogen chloride dissolved in water (hydrochloric acid) with a peroxide. The peroxide yields a portion of its oxygen to the hydrogen chloride, forming water and chlorine. The remaining hydrogen chloride subsequently reacts with the oxide. The peroxide generally used is manganese dioxide; but peroxides of lead, barium, &c., potassium permanganate, bichromate, and other peroxides may also be employed. When a mixture of chloride, bromide, and iodide of potassium or sodium is treated so as to liberate the halogens, the iodine is liberated first, then the bromine, and lastly the chlorine.

3. *By heating compounds of the elements.*—Most of the compounds of the halogens are remarkably stable, and although some, such as hydrogen chloride, may be decomposed by exposure to an exceedingly high temperature, yet re-combination takes place on cooling, so that the halogen cannot be isolated. Fluorine, however, is said to have been prepared by heating cerium or lead tetrafluoride.† The chloride, bromide, and iodide of nitrogen are extremely explosive bodies, at once decomposing into their elements when warmed or when exposed to shock; the higher chlorides and bromides of phosphorus and arsenic, when heated, yield lower

* *Comptes rendus*, **102**, 1543; **103**, 202 and 256.

† *Berichte*, **14**, 1944.

compounds and the halogens; compounds of halogens with oxygen are also unstable, and are resolved with explosion into their elements when heated; compounds of the halogens with each other are also easily decomposed by heat. The halogen compounds of gold, platinum, &c., are decomposed into their elements by heat. This type of reaction, however, does not afford a practical method of preparation.

Preparation of chlorine.—About 30 grams of powdered manganese dioxide are placed in a flask closed with a double bored cork; through one hole passes a tube communicating with a wash-bottle full of water; through the other a thistle funnel passes. Strong hydrochloric acid (solution of hydrogen chloride in water)

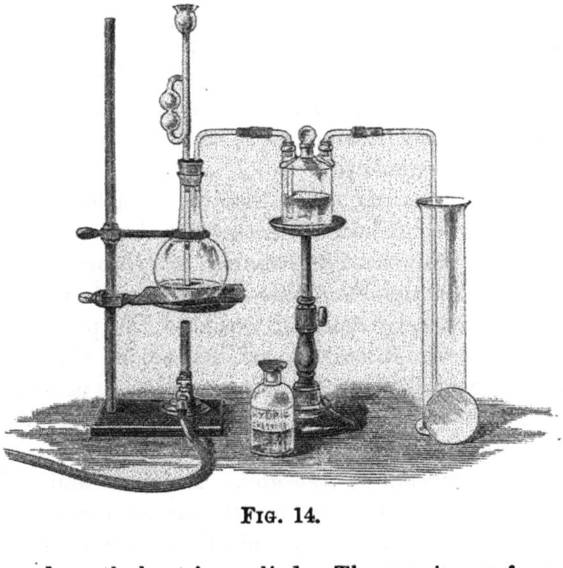

FIG. 14.

is added, and gentle heat is applied. The gas issues from the exit tube of the wash-bottle, and may be collected over *warm* water, in which it is less soluble than in cold; or, better, by downward displacement, for it is heavier than air. The latter arrangement is shown in the figure. To show the tendency of chlorine to combine with other elements, powdered antimony may be thrown into a jar containing it; the metal will burn. A candle will be found to burn in chlorine with a sooty flame; the hydrogen combines, but the carbon is liberated as soot. A solution of chlorine in water acts as a bleaching agent: a coloured rag dipped in such a solution is soon bleached; the chlorine combines with the hydrogen of the water, liberating oxygen, which oxidises the coloured substances to colourless ones. Lastly, some chlorine-water, as a solution of chlorine in water is termed, added to a solution of a bromide or iodide, *e.g.*, potassium bromide or iodide, liberates these elements. Similarly, bromine-water, added to an iodide, liberates iodine.

Properties.—In the gaseous state these elements have all a strong disagreeable smell; that of fluorine, however, is the smell of ozone, for it acts on the moisture in the nose, liberating ozone. **Fluorine** appears to be colourless, **chlorine** is greenish-yellow, **bromine** deep red, and **iodine** violet. The names χλωρός, yellowish-green, βρῶμος, a stench, and ἰοειδής, violet, refer to these properties. As it is impossible to confine fluorine in any vessel which it does not attack, no attempt to liquefy it has been made. Chlorine may be condensed to a greenish-yellow liquid, boiling at a very low temperature; it solidifies to a solid of the same colour.* Bromine is at ordinary temperatures a deep brownish-red liquid, freezing to a blackish-red solid; and iodine at ordinary temperatures is a bluish-black lustrous solid, melting to a brownish-black liquid. It sublimes readily.

Chlorine, bromine, and iodine dissolve in carbon disulphide and tetrachloride, in alcohol and ether, and also in water. One volume of water at 0° absorbs 2·58 volumes of chlorine; and at 15°, 2·36 volumes. Bromine is soluble in 30 times its weight of water at 10°; iodine is very sparingly soluble in pure water. The presence of chlorides, bromides, and iodides in the water greatly increases the solubility of the halogens: it is possible that the solubility of chlorine and bromine in water depends on their interaction with the water. Chlorine and bromine combine with water to form crystalline *hydrates*. Bromine and iodine form compounds with starch; the former has an orange colour, the latter is deep blue. The compound of iodine with starch is used as a delicate test both for iodine and for starch.

These elements combine directly with each other, and at a high temperature, or when moist, with all others except carbon, nitrogen, and oxygen.† The only solid elements which withstand their action even partially are carbon, and iridium, or better, its alloy with platinum. Fluorine attacks glass and porcelain, but the other halogens are without action on these substances, and may be exposed in glass or porcelain vessels to a high temperature.

All these elements tend to combine with hydrogen, whether free or in combination, hence their irritating action on the

* *Monatsh. Chem.*, **5**, 127.

† *Chem. Soc. J.*, **43**, 153.

organism, which chiefly consists of compounds of carbon, hydrogen, and oxygen. They produce catarrh of the mucous membranes when breathed.

No allotropic modifications of these elements are known.

Physical Properties.

	Mass of 1 c.c. Solid.	Mass of 1 c.c. Liquid.	Density, H = 1.	Melting-point.
Fluorine	?	?	18·3 at 15°..	?
Chlorine	?	1·33 at 15·5°	35·4 at 200°	Below −102°
Bromine	?	3·18 at 0° ..	80·0 at 228°	−7·05°
Iodine	4·95	4·00 at m. p.	128·85 at 445°	114·15°

	Boiling-point.	Specific Heat.	Atomic Weight.	Molecular Weight.
Fluorine	?	?	19·0	38·0
Chlorine	−102°	?	35·46	35·46— 70·92
Bromine	58·75°	0·0843 solid	79·95	79·95—159·9
Iodine	184·35°	0·0541	126·85	126·85—253·7

Vapour-densities of Chlorine, Bromine, and Iodine.—These are considered on p. 616.

GROUPS XIV AND XV.—**Ruthenium, Rhodium, Palladium; Osmium, Iridium, Platinum.**

These metals are always associated. They fall into two groups of three, members of the first of which have atomic weights of about 105, and of the second, about 193. They are always found native, or in combination with each other. They are very difficult of attack by any process: even chlorine or oxygen at a red heat produces little effect; hence their occurrence in the free state.

Sources.—(*a*). Metallic particles, consisting chiefly of **platinum** and **palladium**, but containing small quantities of the other metals, occur as flattened grains in the sand of certain rivers in Brazil, Mexico, California, and on the west side of the Ural Mountains.

(*b*). Metallic particles, chiefly consisting of **osmium** and **iridium**, and named *osmiridium*, occur along with the grains of platinum. The complete separation of these metals is a matter of great difficulty (see p. 475).* 3,000 kilos. of platinum ore were exported from the Ural Mountains in 1881.

* Consult *Annales* (3), **56**, 1 and 385; also *Chem. News*, **39**, 175.

Properties.—These elements are all white, with a greyish tinge, and possess strong metallic lustre. They melt only at a very high temperature; in practice they are fused by means of a blowpipe flame of oxygen and hydrogen in crucibles of lime, on which they are without action (see fig. 31, p. 194). Owing to their ability to resist oxidation, an alloy of 90 per cent. of platinum and 10 per cent. of iridium is used for crucibles, retorts for evaporating oil of vitriol, &c., and for standards of length (*e.g.*, the French standard metre). The alloy of osmium and iridium, owing to its extreme hardness, is employed in tipping gold pens, and as bearings for very delicate wheel work. These alloys are very costly, which somewhat limits their use. The metals can be welded.

Platinum and palladium form compounds with hydrogen, in which the last element appears to play the part of a metal in an alloy (see **Alloys**, p. 576).

The name *platinum*, signifying "little silver," was given to the metal by the Spaniards. The name *rhodium* refers to the red colour of its salts. The other names are fanciful, except *osmium*, so called from ὀσμή, a smell, in allusion to the strong odour of its volatile oxide.

Allotropic forms of iridium, ruthenium, and rhodium have been prepared by fusing the metals with zinc or lead, and subsequently dissolving out the zinc or lead with an acid.* The iridium, ruthenium, or rhodium is left as a black powder which explodes on gently warming, being converted into the ordinary form of the metal. Osmium, iridium, and platinum are the heaviest substances known, being more than 21 times as heavy as water.

Physical Properties.

	Mass of 1 c.c. Solid.	Melting-point.	Specific Heat.	Atomic Weight.	Molecular Weight.
Ruthenium	12·26 at 0°..	—	0·0611	101·65	?
Rhodium	12·1 at ? ..	—	0·0580	103·0	?
Palladium........	11·4 at 225° 10·8 (liquid)	—	0·0593	106·35	?
Osmium..........	22·48 at ? ..	—	0·0311	191·3	?
Iridium..........	22·42 at 17·5°	–	0·0326	193·0	?
Platinum	21·50 at 17·6° 18·91 (liquid)	1700 ?	0·0324	194·3	?

* Debray, *Comptes rend.*, **90**, 1195.

GROUP XVI.—**Copper, Silver, Gold, Mercury.**

Of these elements copper, silver, and gold probably belong to the same group: owing to considerable resemblance which mercury bears to them in its compounds, it is convenient to include it in the group.

Sources.—All these metals are found native, for all can resist the action of oxygen at the ordinary temperature. All occur, besides, in combination with sulphur and with arsenic. The chief ore of **copper** is *copper pyrites*, in which it is combined with iron sulphide and sulphur; and other important ores are the oxide, *cuprite*, or *red copper ore*, and the sulphide, *copper-glance;* besides these, it is found in two forms in combination with carbon and oxygen as carbonate, viz., *malachite* and *azurite.*

Silver is mostly found native. But *silver-glance* or sulphide, *pyrargyrite*, *proustite*, and *silver-copper-glance*, in which it is associated with sulphur, antimony, arsenic, and copper, are also important, and it also occurs in combination with the halogens. The chloride is named *horn-silver.*

Gold chiefly occurs native, forming veins and nuggets in quartz-rock; but it also accompanies copper and silver as arsenide and sulphide; and is sometimes associated with tellurium and bismuth. The chief mines are in California, Australia, and the Cape; but it is now mined in Wales, and it is found in upper Lanarkshire, in the Leadhills.

Mercury sometimes occurs free, but its most important ore is *cinnabar*, the sulphide, of which large mines are worked in Austria, Spain, China, and California.

Preparation.—The preparation of **copper** from ores in which it is not associated with sulphur is simple. The ore is powdered and heated with some form of carbon. The carbon unites with the oxygen, forming gaseous carbon monoxide, and the copper fuses, and owing to its greater specific gravity settles at the bottom of the furnace. Copper oxide does not decompose by heat alone; but when heated in an atmosphere of hydrogen it is "reduced," the hydrogen uniting with the oxygen to form water.

If in union with sulphur, one of two methods may be adopted: (1.) The sulphide of copper is roasted in air, whereby it absorbs oxygen, and is converted into sulphate of copper. This is sometimes brought about by leaving the copper ore lying exposed to air for years. The sulphate of copper is treated with water, which dissolves it; and on addition of scrap-iron, the iron replaces the copper in its compound with sulphur and oxygen, forming sulphate of iron, and the copper is precipitated as a metallic sponge. It is then collected, dried, and smelted. This is called the "*wet*" *process* of extraction. The latter part of this process may be shown on a small scale by dipping into a solution of copper sulphate a piece of bright iron, such as the blade of a knife; it will soon become covered with a deposit of metallic copper. (2.) The *dry process* consists in roasting the ore: the iron contained in it combines with oxygen, the copper remaining as sulphide. The oxide of iron is made to unite with sand, or silica, forming a "slag," and by repeating this process several times the copper is finally obtained as sulphide. The sulphide is roasted; both copper and sulphur are oxidised, and a reaction occurs whereby copper separates in the metallic state; the oxygen of the copper oxide unites with the sulphur of the copper sulphide, forming sulphur dioxide, a gas, which escapes, while metallic copper remains. It is melted and brought to market (see Chapter XXXVIII).

Copper chloride loses its chlorine when heated in hydrogen; hydrogen chloride is formed, and the metal remains.

Mercury is easily separated from the sulphur with which it is combined in cinnabar, by roasting in specially constructed furnaces; the oxygen of the air unites with the sulphur, forming gaseous sulphur dioxide, and the mercury passes in the form of gas through a series of cold chambers or earthenware pots, in which it condenses to the metallic state, while the sulphur dioxide escapes. This process may be illustrated by heating in a hard glass tube some powdered cinnabar and aspirating over it a

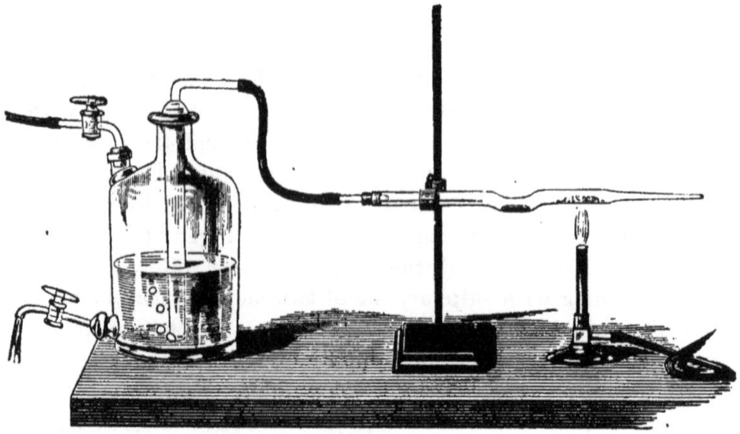

FIG. 15.

current of air. The metallic mercury will condense in the cold part of the tube in small globules, while the sulphur dioxide gas will be carried on into the aspirator (see fig. 15).

Mercury can also be prepared by heating its oxide (see p. 491) although its compounds with the halogens also split into mercury and halogen when heated, yet they recombine on cooling; hence the metal cannot be prepared by this method unless hydrogen, or some other metal, *e.g.*, iron, is present to combine with the halogen.

Mercury may be purified from iron, zinc, lead, and other metals accompanying it by distillation, preferably at a low pressure; and by drawing a slow stream of air for several days through an inclined tube containing the impure metal.

Silver is extracted from its ores, in which it exists chiefly as sulphide, by roasting the ore with common salt, which is a compound of sodium and chlorine. The change is represented as follows:—

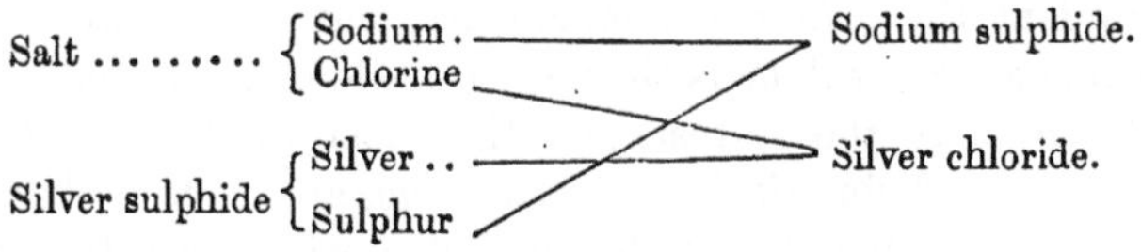

The silver and chlorine combine, and the sulphur and sodium. Such a reaction is termed a "double decomposition." The next stage in the process is to mix the mass with water, and to add scrap-iron, rotating the mixture in wooden barrels. The chlorine and iron combine, the silver separating as a spongy mass. Mercury is added to dissolve the silver; and after renewed rotation of the barrels, the mercury is drawn off, dried, and distilled. The volatile mercury distils away, leaving the much less volatile silver behind.

Silver is also largely extracted from lead ores and from copper ores (see Chapter XXXVIII).*

The process of extracting gold from gold quartz is a mechanical one, for the most part. The ore is stamped to fine powder in mills for the purpose, and washed with water. The fine powder is made to run over a runnel of copper, coated with mercury; the sand is carried on, but the grains of gold unite with the mercury, and are retained. The mercury is then squeezed through chamois-leather: the alloy (or amalgam) of gold and mercury is very sparingly soluble in mercury; hence it remains almost entirely behind. The mercury is then distilled off, and, along with that

* For the preparation of pure silver, see Stas, *Annalen*, Suppl. 4, 168.

portion which had passed through the chamois-leather, used for re-coating the copper plates.

When the gold exists mixed with sulphides, these are roasted to remove sulphur and arsenic, which unite with the oxygen of the air and volatilise away. The residue containing the gold is heated under pressure with chlorine-water, when the gold unites with the chlorine, going into solution as chloride of gold. The gold is then precipitated on metallic copper.

The preparation of mercury, silver, and gold from the chlorides may be shown, (*a*) by placing a piece of bright copper in a solution of mercuric chloride; (*b*) by laying on the top of a bead of fused silver chloride a piece of zinc and adding a little hydrochloric acid; (*c*) by placing a slip of clean copper foil in a solution of chloride of gold.

Properties.—Copper is a red metal; **silver,** brilliant white; **gold,** yellow; and **mercury,** white with a faint grey tinge. Mercury is liquid at the ordinary temperature, but freezes at $-40°$ to a malleable solid. Silver is the most ductile of the remaining three metals; gold is the softest, and the most malleable. Gold and silver leaf, used for *gilding and silvering*, are made by beating the metals into leaves with wooden mallets: when thin they are protected from the direct blow of the mallet by layers of gold-beaters' skin. Copper may also be beaten or rolled into foil and leaf. Gold leaf transmits green light; silver leaf, blue light; and copper leaf, bluish or pink light. All of those metals conduct electricity well. Placing silver equal to 100 at 0°, copper has a conductivity of 77·43 at 18°·8, and gold of 55·19 at 22·7°; mercury follows with a conductivity of 1·63 at 22·8°.

Silver can be distilled at a white heat. Its vapour is bluish-purple. It has the peculiarity of dissolving oxygen (about 22 times its volume) when liquid and discharging it during solidification ("spitting").

Mercury distils about 358°. Its vapour is colourless.

Copper is rendered brittle by slow cooling, and is softened by rapid cooling. The properties of all these elements are very singularly modified by the presence of traces of others. Thus the smallest trace of arsenic renders gold exceedingly brittle; a trace of phosphorus in copper greatly increases its tensile strength; a minute trace of zinc in mercury completely modifies its surface tension.

For the composition of coins, jewellers' metal, &c., see Alloys (p. 587).

Of these elements, none is oxidised on exposure to air, but

copper at a red heat, mercury at the temperature of ebullition, and silver heated in air under a pressure of several atmospheres unite with oxygen. Gold does not combine directly with oxygen. The oxides of the last three are easily decomposed by heat. These metals all unite directly with sulphur, selenium, and tellurium, and with arsenic; with chlorine, bromine, and iodine, and with most metals. They do not unite directly with nitrogen.

Physical Properties.

	Mass of 1 c.c.		Density.	Melting-
	Solid.	Liquid.	H = 1.	point.
Copper	8·90 (at 0°)	8·2	?	1330°
Silver............	10·57	9·5	?	1040°
Gold	19·29	17·1	?	1240°
Mercury..........	14·19 (−40°)	13·596 (at 0°)	100·1	−40°

	Boiling-point.	Specific Heat.	Atomic Weight.	Molecular Weight.
Copper	?	0·0935	63·40	?
Silver..........	White heat	0·0570	107·93	107·93
Gold	?	0·0324	197·22	197·22
Mercury	358·2°	0·0319 (solid)	200·2	200·2

General Remarks on the Elements.

(1.) **Classification.**—It has been customary to divide the elements into two classes: the metals, those which are opaque and which exhibit metallic lustre; such elements are more or less good conductors of electricity and heat; and the non-metals, comprising the remaining elements. Such a division tends to obscure the relations between them; it is, so far as we know, an arbitrary division, and is sanctioned only by long custom. Other objections which might be taken to this division are that a number of elements, such as titanium, arsenic, and tellurium, are difficult to classify, being sometimes considered as metals, sometimes as non-metals: and a still more important objection is that certain elements can exist in both forms. Thus silicon, phosphorus, and carbon exist in compact crystalline forms, with dull metallic lustre, and are then conductors of electricity; while they also exist in forms incapable of conducting, and without metallic lustre. Such reasons have led to the abandonment of this classification here. Still the name metal has generally been applied in this book to those elements which are usually ranked as such; though it is to be understood in a loose, colloquial sense.

It will, however, be convenient to give a list of non-metals, so that the old classification may be understood.

List of non-metals.—Hydrogen (?), boron, carbon, silicon, titanium (?), zirconium (?), nitrogen, phosphorus, vanadium (?), arsenic (?), antimony (?), oxygen, sulphur, selenium, tellurium (?), fluorine, chlorine, bromine, iodine.

The sign (?) denotes that these elements are sometimes included in, sometimes excluded from, the class of non-metals. The remaining elements are classed as metals.

(2.) **Sources.**—As a general rule, those elements are found native which are unaffected by oxygen and moisture in air at the ordinary temperature. Thus carbon, nitrogen, sulphur, selenium, tellurium, the platinum group of metals, and copper, silver, mercury, and gold are among these. It is curious that hydrogen is not found native to any great extent, for it fulfils these conditions. There appears no reason why air should not contain small traces of hydrogen, unless, indeed, its molecular motion may carry it out of the sphere of the earth's attraction.*

Those compounds of elements with the halogens which are not decomposed by water as a rule exist native. Among these are chlorides, bromides, and iodides of sodium, and potassium; of silver, lead, and mercury. From the abundance of oxygen, and the tendency which most elements have to combine with it, the oxides and double oxides are the most widely spread compounds: for example, the silicates, carbonates, phosphates, nitrates, &c. The sulphides rank next in order of distribution; only those stable in presence of air and water, however, occur abundantly. It is indeed probable that the mass of the earth consists largely of sulphides; for the specific gravity of our globe has been found by astronomical measurements to be $5\frac{1}{2}$ times that of water, while the average specific gravity of the crust cannot well exceed 3. It appears not unlikely that the greater density is caused by the presence of the denser sulphides in the interior; and the prevalence of sulphur in volcanic districts, where the interior of the earth is in a state of disturbance, would support this supposition. Some few elements occur in combination with arsenic alone, or with arsenic and sulphur.

3. **Preparation.**—It will have been noticed that there are

* A similar theory would account for the absence of an atmosphere on the moon.

three general methods of preparing elements from their compounds. These are—

(*a.*) **Electrolysis of a liquid compound of the element or of a solution of a solid compound in water.**—It is questionable whether solids or gases can be electrolysed; at all events, the constituents cannot be conveniently collected; hence the limitation to the liquid state. It appears probable that no perfectly pure compound is capable of conducting electricity; those at least, such as pure water, hydrogen chloride, &c., which can be obtained nearly pure do not appear to do so. A liquid mixture, however, is almost always an electrolyte, *i.e.*, capable of yielding its elements under the influence of a current of electricity. In many cases no easily fusible compound of the element required is known, or it is difficult of preparation, or it does not conduct; in other cases the liberated element acts upon water, forming an oxide and liberating hydrogen; hence the method is somewhat limited.

(*b.*) **Heating a compound of the element required.**—It is almost certain that all compounds, if heated to a sufficiently high temperature, would decompose into their elements. But, unless one of the elements possesses a much lower boiling-point than the others with which it is combined, separation cannot be effected, as a rule, for in most instances recombination occurs on cooling. It is owing to the great difference in volatility of mercury and oxygen that the latter can be prepared by heating mercuric oxide; on similar grounds, chlorine can be prepared from gold chloride; or sulphur, by heating platinum sulphide or hydrogen sulphide. In many cases only a portion of one of the combined elements is evolved as gas, as, for instance, oxygen from manganese, barium, or lead dioxides, or from chromium trioxide.

(*c.*) **By displacing one element from a compound by the action of another.**—This method is very largely used. The agents of displacement, however, are limited in number. It is obviously essential to the success of the process that the element used as a displacer shall not combine with the one to be displaced; or, if it do so combine, that it shall be easily expelled from its combination by heat; or that it shall combine much more readily with one of the elements in the compound acted on than with the other.

Thus no metal will displace phosphorus or oxyen from their compound, phosphorus pentoxide, because all metals combine with phosphorus and oxygen. Again, aluminium may be prepared by removing chlorine from its chloride by the action of sodium; for the compound or alloy of aluminium and sodium which is doubtless produced is easily decomposed by heat into sodium, which volatilises away, and aluminium, which remains non-volatile at the temperature employed. And lastly, sulphur may be produced by the action of an insufficient quantity of oxygen on its compound with hydrogen; for hydrogen combines so much more readily with oxygen than sulphur does that water is formed, little of the sulphur combining with the oxygen; and, as another instance, carbon is liberated from its compounds with hydrogen when they burn in chlorine gas, because at the temperature of reaction the chlorides of carbon are decomposed.

In practice the following methods are used:—

1. The action of **carbon** (coal, charcoal) **on the oxide** of the element, or on its compound with oxygen and hydrogen (hydroxide), at a red heat. The most important elements thus prepared are:—

Hydrogen, potassium, rubidium; zinc, cadmium; impure chromium, iron, manganese, nickel and cobalt (these elements combine with a small quantity of the carbon employed in their liberation); germanium, tin, lead; phosphorus, arsenic, antimony, bismuth; molybdenum, tungsten; copper. In many cases this is in reality the action of carbon monoxide on the oxide of the element: the carbon monoxide unites with the oxygen combined with the element, forming carbon dioxide, and the element is liberated.

2. The action of **hydrogen on the oxide** of the element required at a red heat. Elements which may be thus prepared are:—Indium, thallium, tin, lead; nitrogen, arsenic, antimony, bismuth, tungsten; iron, nickel, cobalt, and copper.

3. The action of **hydrogen on the chloride** of the element at a red heat. Examples:—Vanadium, niobium, arsenic, antimony, bismuth, and others.

4. The action of **sodium or potassium,** or of **zinc,** on the fused **chloride,** double chloride, or double fluoride of the element required. Examples:—Magnesium, boron, aluminium, yttrium, carbon, titanium, zirconium, thorium, tantalum, chromium, uranium.

5. The action of **another element** on the **solution of a compound** of the element required. Examples:—Iodine may be prepared by the action of chlorine or bromine on iodide of potassium; bromine, by the action of chlorine on potassium bromide; sulphur, selenium, or tellurium, by the action of atmospheric oxygen on a solution of their compounds with hydrogen; copper, by the action of iron on a solution of copper chloride or sulphate; mercury or silver, by the action of copper on a solution of mercuric or

silver nitrates; gallium, by the action of zinc on a solution of gallium chloride, and many others.

Properties.—The elements, like other forms of matter, exist in the three states of gas, liquid, and solid. Those gaseous at the ordinary temperature are hydrogen, nitrogen, oxygen, fluorine, and chlorine. Two are liquid, viz., bromine and mercury; the remainder are solid.

The mass of one cubic centimetre varies from 0·0000896 gram in the case of hydrogen gas to 22·48 grams in the case of osmium. The variation of this constant with atomic weight will be considered in Chapter XXXVI.

The atomic weights of the elements vary from 1 (hydrogen) to 240, (uranium); and their specific heats from 5·4 (hydrogen alloyed with palladium) to 0·0277 (uranium). It will subsequently be shown that the product of the two is usually a constant number.

It cannot be doubted that many elements remain to be discovered. On referring to the periodic table on p. 23, it will be seen that many atomic weights are accompanied by queries (?). Within the last few years several such gaps have been filled; notably thallium (Crookes), gallium (Lecoq de Boisbaudran), scandium (Cléve), and germanium (Winckler). But this subject will be fully considered in a later chapter.

APPENDIX.

Equations expressing the preparation of elements of Groups XIII and XVI.

Fluorine.—$2HF = H_2 + F_2$.

Chlorine.—$MnO_2 + 4HCl = Cl_2 + MnCl_2 + 2H_2O$.

Bromine.—$2KBr + Cl_2 = Br_2 + 2KCl$.

Iodine.—$2KI + Br_2 = I_2 + 2KBr$.

Copper.—$CuO + C = Cu + CO$.

$$\begin{cases} CuS + 2O_2 = CuSO_4. \\ CuSO_4 + Fe = Cu + FeSO_4. \end{cases}$$

$CuS + 2CuO = 3Cu + SO_2$.

Mercury.—$HgS + O_2 = Hg + SO_2$.

Silver.—$\begin{cases} Ag_2S + 2NaCl = 2AgCl + Na_2S. \\ 2AgCl + Fe = 2Ag + FeCl_2. \end{cases}$

PART III.—THE HALIDES.

CHAPTER VII.

COMPOUNDS :—NOMENCLATURE AND CLASSIFICATION ;—THE STATES OF MATTER.—RELATION OF THE VOLUME OF GASES TO PRESSURE AND TO TEMPERATURE.—METHODS OF DETERMINING DENSITY.

Compounds and Mixtures.

Elements are said *to combine* when on bringing them together a new substance is produced, differing from its constituents and possessing properties which, as a rule, are not the mean of their properties. Such combination is always attended with a rise or fall of temperature, or "heat change;" and, as heat is a form of energy, or power of doing work, elements either gain or lose energy by combination with each other. It appears that direct combination is always attended with loss of energy, heat being evolved. This is illustrated by the combustion of carbon in oxygen, of antimony in chlorine, and by many other instances; and the evolution of heat in many such cases is so great as to raise the substance to the temperature of incandescence, so that it emits light.

Two or more elements may, however, be mingled without sensible evolution of heat. They are then said to constitute a *mixture*. Atmospheric air is an instance in point. On mixing its constituent gases, oxygen and nitrogen, no heat change takes place. But if electric sparks be passed through the air, its constituents are raised to a high temperature and combine; the product is an oxide of nitrogen, possessing a brown-red colour and a strong smell. Certain mixtures are thus definitely distinguished from compounds. But in many cases it is difficult to affirm positively that an element is or is not combined. Some metals mix freely with others, as, for example, tin and lead; but there is no way of absolutely testing whether or not they are combined. Another instance is that of a solution of chlorine in bromine. In such cases, however, the properties of the mixture are apparently the mean of those of its constituents.

The best criterion of a compound is its *definite composition*. With this are associated *definite physical properties*, such as constancy of melting point, of boiling point, and of crystalline form. An *amorphous* condition, *i.e.*, lack of crystalline form, almost always accompanies indefinite composition; but, on the other hand, a substance may possess a definite crystalline form (as, for example, many silicates), and yet have an indefinite composition. Such bodies are, however, usually mixtures of compounds with each other.

Nomenclature.—Chemical nomenclature in its present form was mainly devised by Lavoisier, and, although extended, its principle has not been materially modified since his time. But even he was constrained to adopt certain expressions which had been in use from a very early date, such as "base," "acid," and "salt." These terms are incapable of accurate definition, and must therefore be used loosely. It may be said generally that the word *base* is applied to the oxides of certain elements, either alone or in combination with hydrogen oxide (water); the word *acid*, the oxide of hydrogen and certain other elements, not usually those of which the oxides are called bases; and the word *salt*, a body produced by the interaction of a *basic oxide* with an *acid oxide*. The words *salt* and *acid*, however, are frequently applied to substances containing no oxygen, such as sodium chloride, or hydrochloric acid. In fact, no rule can be given, and the words must be employed in a vague sense, custom alone determining their use.

A compound formed by the union of two elements retains the names of both, one of them, however, acquiring the termination "ide." It is a matter of indifference which receives that ending; but, as most compounds which have been investigated contain one of ten or twelve elements, the names of these are commonly modified. Thus we speak of oxides, sulphides, selenides, tellurides, fluorides, chlorides, bromides, iodides, nitrides, phosphides, arsenides, borides, carbides, and silicides; also of hydrides. The Greek numeral prefixes *mono-*, *di-*, *tri-*, *tetra-*, *penta-*, and the Latin one *sesqui-*, signifying respectively, one, two, three, four, five, and one-and-a-half, are employed, when required, to denote the relative numbers of atoms in the compound.

Many compounds of fluorides, chlorides, bromides, and iodides with each other, and of oxides and sulphides with chlorides, &c., are known. These have generally been named double chlorides, oxychlorides, or basic chlorides, sulphochlorides, &c. Another nomenclature is sometimes used. It is as follows; an example will render it plain. Platinum forms two compounds with chlorine, one containing twice as much chlorine as the other, proportionately to the metal. The one containing least chlorine is named platin*ous* chloride; that containing most, platin*ic* chloride. Each of these forms a compound with potassium chloride; the first is named potassium platin*o*chloride, platin*o*- being contracted from platin*ous*; the second potassium platin*i*chloride, the word platin*i*- standing for platin*ic*. So with ferr*ous* and ferr*ic*, phosphor*ous* and phosphor*ic*, &c.

The double oxides have names which do not show that they contain oxygen. Thus compounds of oxides of chlorine and of a metal are named *hypo*chlor*ites* (hypo = below), chlor*ites*, chlor*ates* or *per*chlor*ates* (*per* = over, from *hyper*), according to the amount of oxygen in combination with the chlorine; so also with compounds of nitrogen, phosphorus, sulphur, gold, &c., &c.

In the case of a few common substances, such as water (hydrogen monoxide), ammonia (hydrogen nitride), vitriol (sulphuric acid or hydrogen sulphate), old and familiar names have been retained. These are fortunately in many cases becoming obsolete.

The Elements

Will be considered in the following order:—

1. **Compounds of the halogens—fluorine, chlorine, bromine, and iodine—with the elements, arranged in groups according to the periodic table, including double compounds.**
2. **Compounds of oxygen, sulphur, selenium, and tellurium with the elements; including oxychlorides, sulphochlorides, &c., and double oxides and sulphides, usually called hydroxides, hydrosulphides, acids, and salts.**
3. **Borides, carbides, silicides.**
4. **Nitrides, phosphides, arsenides, and antimonides. Double compounds.**
5. **Alloys and amalgams.**

Before proceeding with the consideration of the halogen compounds it is necessary, in order to understand the relations between these substances, to study the methods of expressing chemical change, and some of the reasons for assigning definite atomic weights to the elements. This involves a knowledge of the nature of gases, and their behaviour as regards temperature and pressure.

The States of Matter.

Matter is known in three states: the solid, the liquid, and the gaseous.

Solids.—Solids are peculiar in possessing *form;* they have rigidity, enabling them to keep their shape. It is believed that minute particles of which all matter consists, which are named molecules, are so closely packed together in solids as to attract each other powerfully, and to possess very little freedom of motion. Such particles possess symmetrical arrangement in *crystals;* but are heaped together at random in *amorphous* solids. Solids generally expand when their temperature is raised, but only to a small degree. At a sufficiently high temperature, they either melt, volatilise without melting, or decompose. They are very slightly compressible.

Liquids.—Liquids differ from solids in not possessing form, and from gases by possessing a surface. The condition of the liquid matter at the surface differs from that in the interior, and the surface is under a lateral strain, named *surface-tension.* A drop, for example, behaves as if it were covered with a stretched skin or film. The molecules of which liquids consist possess greater freedom of motion than do those of solids; so that they move about, continually gliding past each other, and hence a liquid has no fixity of form. On raising the temperature of a liquid, this motion increases. The motion of the molecules of a liquid is termed *diffusion* or *osmosis.* When liquids are cooled they generally contract, and at a sufficiently low temperature they freeze, or turn to solids; on raising their temperature they expand, and at a sufficiently high temperature they volatilise, changing into gas. Vapour is continually being evolved from the surface of a liquid, and if the liquid be in a closed vessel the pressure which its vapour exerts can be measured. This pressure is termed its *vapour-pressure.* The vapour-pressure increases with rise of temperature; and when it exceeds the pressure of the atmosphere the liquid *boils* and changes wholly into gas, if heat be supplied in sufficient amount.

Gases.—Gases or vapours have neither form nor surface. A solid or a liquid in changing into vapour acquires a greatly increased volume; thus the gas of water occupies about 1700 times the space occupied by its own weight of liquid water at the same temperature and at the same pressure, viz., 100° and 760 mm. pressure. While solids and liquids are but slightly altered in volume by alteration of pressure and temperature, the volumes of gases are greatly changed. The molecules of gases are evidently much more distant from one another than those of solids or liquids, and therefore possess much greater freedom for motion, or *free path.* They occupy but a small portion of the space which they inhabit. And while the molecules of solids and of liquids are so near each other as to exercise great attraction on one another, those of gases are so far apart that the attraction is barely sensible. Hence gases exhibit simple relations to temperature and pressure.

Relation between the volume of a gas and the pressure to which it is exposed.—Boyle's law. The temperature of a gas being kept constant, its volume varies inversely as the pressure to which it is exposed. This law was discovered by Robert Boyle in 1660.

The barometer.—It has been remarked in Chapter II that gases have weight; the weight of a given quantity of matter is not changed by change of state: thus a pound of water weighs a pound, whether it be ice, water, or steam. Air, which is a mixture of nitrogen and oxygen gases, therefore possesses weight; and, the longer or higher a column of air, the greater its weight. A column of air reaching to the upper confines of the atmosphere and resting on the earth at the level of the sea, of 1 square centimetre in section, weighs on the average 1033 grams; or, if 1 square inch in section, about 16 lbs.; but 1033 grams is the weight of a column of mercury at 0° of 1 square centimetre in sectional area and 760 millimetres in length; and 16 lbs. is the approximate weight of a mercury column 1 square inch in sectional area and approximately 30 inches long; or of a column of water about 33 feet in length, also 1 square inch in sectional area. Hence, if it were possible to support the end of such a column of air on one pan of a balance, and to place on the other pan a column of mercury 760 millimetres in length, removing the pressure of the air from its upper surface (else the weight of both air and mercury would press on the other pan), the two columns would balance. Such an operation is actually performed in constructing a barometer. The air is removed from the upper portion of a glass tube, the lower end of which is open and dips in mercury; and the mercury rises in the tube until it balances a column of air of equal sectional area to the tube, rising, in order that it may do so, to a height of 760 millimetres. If the weight of the atmosphere increases, owing to its cooling, or to its compression, the column of mercury rises proportionately, so as to balance it; and, conversely, when the weight of the atmosphere decreases, the balancing column is shorter. The pressure of the atmosphere might be expressed in units of weight for a given sectional area, say, 1 square centimetre; it might be, and indeed sometimes is, measured in fractions or multiples of 1033 grams, just as it is the custom for engineers to express the steam pressure in a boiler, which is closely analogous, in pounds on the square inch of boiler surface; but it is commonly expressed as equal to the pressure of 760 millimetres of mercury, or of a column of greater or less length, according as the weight of the atmosphere varies.

All gases contained in vessels communicating with the atmosphere are therefore under this pressure; hence it must be allowed for in ascertaining the relation between the volume of a gas and the pressure. That Boyle's law is approximately true can be proved by the following experiment:—A U-tube, as shown in fig. 16, about 50 centimetres in length, contains air in its closed

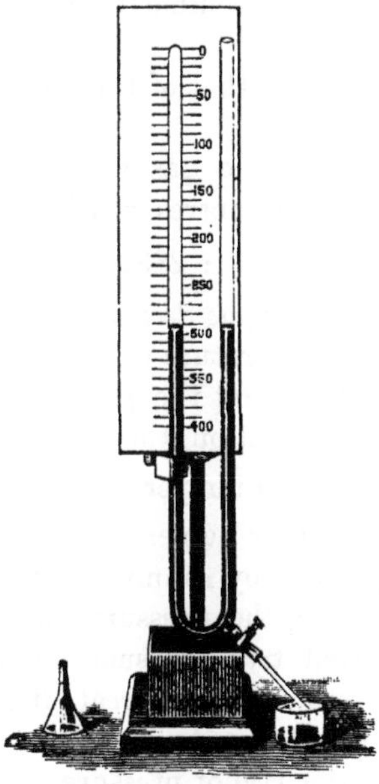

FIG. 16.

limb. The amount of air is adjusted so that when the mercury is level in both limbs it occupies a volume represented by 273 millimetres + a number of millimetres equal to the temperature of the day. Thus, for example, if the temperature of the surrounding atmosphere is 15° C., the length of the column of air enclosed should be 288 millimetres. The reason of this adjustment will appear later. Now mercury is poured into the open limb of the U-tube so as nearly to fill it; the difference in level of the mercury in the open and in the closed limb is read off. The air in the closed limb will be compressed by the weight of the mercury in the open limb, the column being equal in length to the difference in level of the mercury in the open and in the closed limb. For example:—

Distance from top of tube to surface of mercury, that in both limbs being at the same level..	288 mm.
Level of mercury in open limb, after filling it..	0 "
Level of mercury in closed limb, after filling open limb	223 "
Difference in level between mercury in closed and open limbs	223 "

The initial volume of the gas was 288 × x cubic centimetres.

After compression the volume decreased to 223 × x cubic centimetres.

The initial pressure was that of the atmosphere, say, 760 millimetres.*

The final pressure on the gas was that of the atmosphere, 760 millimetres + 223 millimetres = 983 millimetres of mercury.

$$\text{But } 983 : 760 :: 288x : 223x, \text{ nearly.}$$

Hence the *volume of the gas decreases proportionately to the increase of pressure* at a temperature of 15°.

If p_1, p_2, v_1, and v_2 represent the pressures and volumes respectively before and after alteration, then

$$p_1v_1 = p_2v_2, \text{ provided temperature be kept constant.}$$

Similar experiments may be performed, decreasing the amount of mercury in the U-tube, by running out mercury through the stopcock, and so reducing the pressure on the gas; and Boyle's law may thus be proved true for such small alterations of pressure. When the pressure is very great, it ceases to hold: gases become more compressible up to a certain point, and then less compressible with greater rise of pressure.

Gay-Lussac's law.—The volume of a gas increases one two hundred and seventy-third of its volume at 0° (0·00367) for each rise of 1° C., provided pressure remain constant. Thus 1 cubic centimetre of air or other gas measured at 0° becomes, when heated to 1°, $1\frac{1}{273}$ or 1·00367 c.c.; at 2°, $1\frac{2}{273}$, or 1 + (0·00367

* The barometer should be read at the time, and its height substituted here. In the above instance it is supposed to be at its normal height.

$\times$ 2); at 100°, $1\frac{100}{273}$, or 1 + (0·00367 $\times$ 100). This can be illustrated by means of the apparatus used for demonstrating Boyle's law, with a slight addition to allow of an alteration in the temperature of the gas. The closed limb of the U-tube is surrounded by a jacket or mantle of glass, the lower part of which is closed by an indiarubber cork, perforated to allow the limb of the U-tube to pass through. The liquid in the bulb is water. It is boiled by a flame, and the steam jackets the

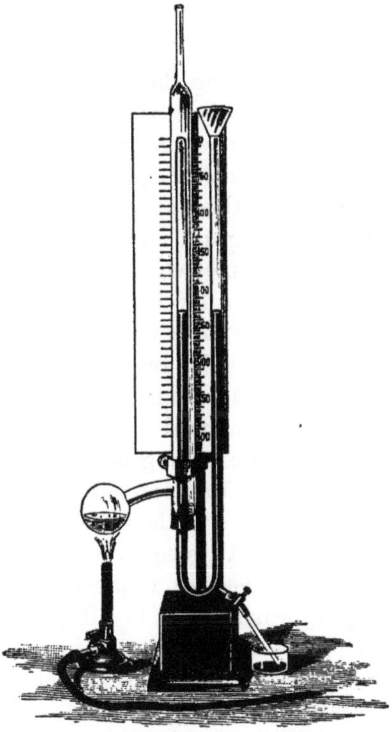

FIG. 17.

U-tube, raising its temperature to 100°, provided the atmospheric pressure is 760 millimetres. If the pressure does not differ much from the normal one, the difference in temperature may be neglected. At 15°, supposed to be the atmospheric temperature of the day, the air in the closed limb of the U-tube is adjusted so as to occupy 288 millimetres of the tube's length, measured from the top downwards, the mercury in both limbs being level. On boiling the water so as to raise the temperature of the gas to 100°, the gas will expand, pushing down the mercury in its own limb, and raising it in the other. When the level is stationary, mercury is run out of the U-tube, so as to restore equal level in both limbs. The gas will then occupy 373 millimetres of the length of the U-tube. Thus:

Initial volume of gas at 15°, at atmospheric pressure, $288x$ c.c.
Final volume of gas at 100°, and at „ „ $373x$ „
Expansion, $373x - 288x = 85x$ c.c.
Rise of temperature, $100° - 15° = 85°$.

The expansion is thus seen to be proportional to the rise of temperature.

It is obvious that by cooling the gas to 0°, by surrounding the tube with melting ice, the volume would contract from $288x$ c.c. to $273x$ c.c. This may also be proved experimentally. It may also be shown that Boyle's law holds equally well at the temperature 100° as at 15° by means of this apparatus.

Such an instrument might be, and indeed with altered construction is, used as a thermometer. It would be convenient to place the number 273° at the level of the mercury when ice surrounds the tube; then the expansion of the gas and the temperature will march *pari passu*. The zero of such a scale will manifestly be at the top of the tube; the degrees are ordinary Centigrade degrees, the interval of temperature between the melting-point of ice and the boiling-point of water under normal pressure being 100°; but on this scale the former is marked 273° and the latter 373°. Such a scale is termed the *absolute scale*; and the temperature −273° C. is equal to 0° *absolute*.

As this behaviour with respect to pressure and temperature is common, speaking approximately, to all gases, it may be conjectured that they possess some property in common, as the cause of their similar changes. This property was discovered in 1811 by Avogadro, and is known as—

Avogadro's law.—Equal volumes of gases, under the same pressure, and at the same temperature, contain equal numbers of molecules. It must be noted that this statement postulates nothing as regards the actual size of the gaseous molecules; it merely asserts that, temperature and pressure being constant, a definite number of molecules of one gas, say, hydrogen, inhabit the same space as the same number of molecules of any other gas, say, oxygen or chlorine. The actual number of molecules is, of course, unknown; and, although attempts to estimate it have been made, they do not concern us here. From the known laws of expansion of gases, and their relation towards pressure, it is possible to compare the weights of equal volumes of different gases, and so to compare the relative weights of the molecules of which they are composed; for it is obvious that if the weight of n molecules of, say, oxygen is 16 times that of n molecules of hydrogen at some temperature and pressure the same for both, the weight of 1 molecule of oxygen is 16 times that of 1 molecule of hydrogen.

It is, therefore, exceedingly important to be able to compare the relative weights of gases, inasmuch as it affords a simple means of comparing the relative weights of their molecules. The term *density* is applied to the weight of a gas relative to hydrogen, the density of which is arbitrarily placed = 1. Sometimes air is chosen as the unit of comparison. The absurdity of this is evident; for it has been repeatedly shown that the composition, and hence the density, of air, which is a chance mixture of the gases oxygen and nitrogen, is not uniform, but varies within small limits. The variation, however, is so small as to be within the usual errors of experiment in determining the density of gases; hence, for practical purposes, as air is about 14·47 times as heavy as hydrogen, densities compared with air may be converted to the hydrogen standard by multiplying the number expressing them by 14·47. The density of a gas which exists as a liquid at ordinary atmospheric temperature is termed a *vapour-density;* there is no real distinction between the words *gas* and *vapour.*

Methods of determining the Density of Gases.

1. **When the substance is a gas at the temperature of the atmosphere.**—Two globes of nearly equal capacity (half a litre to five litres, and which should have as nearly as possible the same weight), provided with tight-fitting stop-cocks, are pumped empty, first by means of a water-pump, and finally with a Sprengel's or other mercury-pump; the stop-cocks are then closed, and they are suspended one from each arm of a balance, as shown in fig. 18, and if not quite equal in weight, counterpoised by addition of weights to one or other pan. The gas to be weighed, is then admitted from a gas-holder into one of the globes, care being taken to dry it, by passing it slowly through U-tubes filled with strong sulphuric acid or phosphorus pentoxide, which has a great tendency to combine with water, and so removes it from the gas. If the gas be soluble in water, it may be passed straight from the

generating flask through the drying tubes into the empty globe, the stop-cock of the latter being opened *slowly* so as to ensure the gas being thoroughly dried. It is again suspended from the hook of the balance pan, and after some hours, the amount of gas which has entered is weighed. The volume of the globe is then ascertained

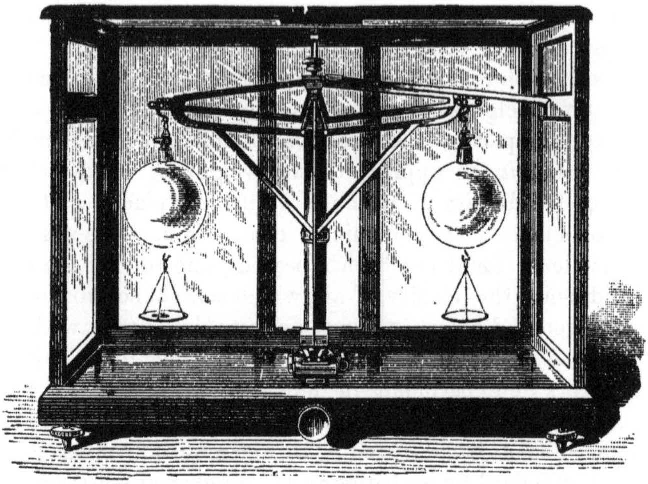

FIG. 18.

by filling it completely with water and weighing it. The difference between the weight of the vacuous globe and the globe full of water gives the weight of water filling the globe. It is sufficient for the present purpose to consider that 1 gram of water occupies 1 cubic centimetre, though for accurate determinations the true volume of the water at 4° must be calculated. Here, also, the expansion of the globe between 0° and atmospheric temperature, and also its diminution of volume when empty of air, due to the presence of the atmosphere, have been neglected.

We have accordingly the data.

Weight of globe full of gas at T° temp., and P mm. pressure	W_2 grams.
Weight of empty globe	W_1 ,,
Weight of V cub. centimetres of gas	W grams.

From this the volume of 1 litre of the gas at 0° and 760 millimetres pressure can be calculated thus:

(*a.*) To ascertain the volume of the gas at 760 millimetres pressure.

Law.—*The volume is inversely as the pressure.* Hence,

$$V_{760} = V_P \times P/760.$$

(*b.*) To ascertain the volume, corrected for pressure, at 0°C.

Law.—*The volume of a gas increases by* 0·00367 *of its volume at* 0° *for each rise of* 1°. Hence,

$$V_{0 \text{ and } 760} = \frac{V_P \times P/760}{1 + (0{\cdot}00367 \times T)}.$$

(*c.*) This volume of gas weighed W grams. To find the weight of 1 litre: $W_{1000 \text{ c.c.}} = 1000\ W/V_{0^\circ \text{ and } 760 \text{ mm.}}$

(*d.*) From Regnault's very accurate experiments we learn that 1000 cubic centimetres of hydrogen weigh 0·0896 gram. Hence, the density of the gas = $W_{1000 \text{ c.c.}}/0{\cdot}0896$.

The relative weight of a molecule of hydrogen is taken as 2, for reasons which will afterwards be considered (p. 109). Hence, the relative weight of a molecule, or the molecular weight of the gas = $2W_{1000 \text{ c.c.}}/0{\cdot}0896$, or is equal to twice its density.

2. **When the substance becomes gaseous at a temperature higher than that of the atmosphere.**—One of the following methods may be employed.

(*a.*) **Dumas' Method.**—This method differs from the method already described only in one particular, viz., in the manner of filling the globe. The globe usually has a capacity of 250 to 500 cubic centimetres. About 10 cubic centimetres of the liquid or solid, of which the density in the gaseous state is required, is introduced into the globe by warming it gently so as to expel air, and dipping the thin neck of the globe into the liquid; or by introducing the solid into the globe before its neck is drawn out. It is then placed in a bath of some liquid or vapour, depending on the temperature required. If the boiling-point is below 100°, water may be used; if between 100° and 250°, olive or castor-oil; and vapour-baths, such as that of boiling mercury (358°), or sulphur (444°), or phosphorus pentasulphide, or stannous chloride, or even the vapours of boiling cadmium or zinc, may be used for higher temperatures, but with the last two the globe must be a porcelain one, for glass softens at about 700°. The liquid or solid begins to evaporate, and its vapour displaces the air from the globe. As soon as vapour ceases to escape, the drawn-out end of the neck of the globe is sealed by means of a hand-blowpipe, or of an oxyhydrogen blowpipe, if a porcelain globe is used (see fig. 19). The globe is then removed, allowed to cool, cleaned, and weighed, balancing it, as before, by a similar globe hung from the other pan of the balance. The calculations are performed exactly as before, but the expansion of the globe must here be allowed for; if of glass, it may be calculated as V + (0·000025t); it is assumed that the gas will remain a gas when cooled to 0°. It would be more

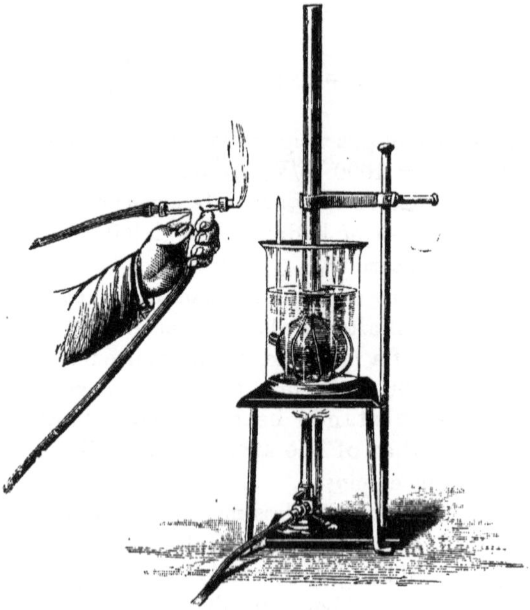

FIG. 19.

rational to compare the weight of the gas with that of an equal volume of hydrogen at the same temperature and pressure as those of the vapour at the time of sealing the globe, but the end result is the same whichever method of calculation be used.

(*b.*) **Hofmann's Method, modified.**—The principle of this method is to ascertain the volume of a known weight of the gas. The apparatus consists of a graduated tube of the form shown in fig. 20. The tube is filled with mercury and inverted into a glass basin containing mercury, and after the jacketing tube has been put on, the apparatus is clamped in a vertical position. The graduated tube passes through a wide hole in an indiarubber cork fitting the jacket; but as this cork is apt to be attacked by the boiling liquid, a little mercury is poured in, so as to cover and protect it. The substance is weighed out in a small bulb, and pushed under the open end of the tube, so that it floats up to the surface of the mercury in the closed end. The temperature is then raised by boiling the liquid, which must be pure, in the bulb of the jacket.

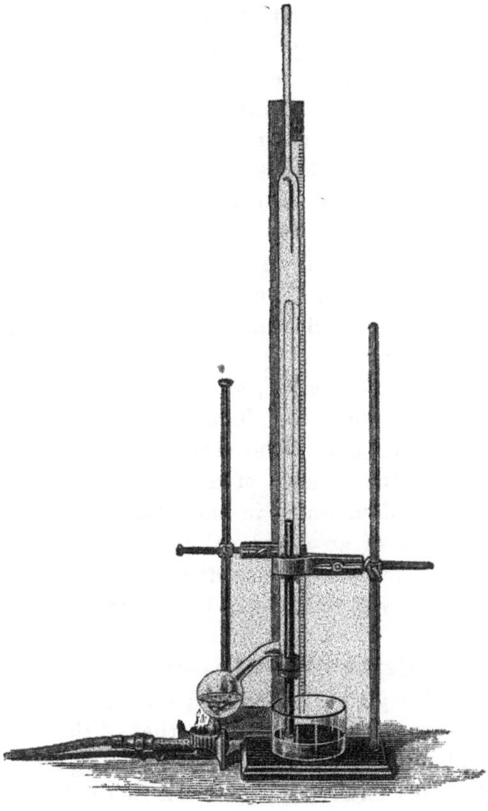

FIG. 20.

The following is a list of convenient substances, with their respective boiling points under a pressure of 760 millimetres:—

	T.	Δ.		T.	Δ
Carbon disulphide.	46·2°	25	Aniline.........	184·5°	20
Alcohol..........	78·3°	30	Chinoline.......	237·5°	17
Chlorobenzene....	132·1°	25	Bromonaphthalene	280·4°	16
Bromobenzene....	156·1°	20			

The column, Δ, represents the average difference in pressure in millimetres per degree at about the pressure 760 millimetres. Thus, if the height of the barometer is 740 millimetres, *i.e.*, 20 millimetres less than 760, the temperature of the carbon disulphide vapour will be not 46·2°, but 46·2° — $\frac{20}{25}$ths of 1° = 45·4°. The mercury in the tube will be displaced by the vapour, and will enter the glass basin in which the tube stands. The volume of the vapour is then read off, if the tube is a graduated one; if not, the level of the mercury in the tube is read on the graduated scale, and also the level of the top of the tube. The volume may be afterwards determined, by inverting the tube, and filling it to the required height with water from a burette. The pressure is that of the atmosphere, diminished by the length of the column of mercury in the tube. But mercury itself, when heated, expands, and a correction must be introduced, because at 0° the length of the mercury column would be less. Again, the gas in the tube consists partly of mercury vapour; its pressure must be calculated and subtracted.* But neglecting these corrections, the plan of calculation is as follows:—

A certain volume of gas, V, has been produced from a known weight of liquid or solid, W. This gas is at the temperature of the jacketing vapour, and under atmospheric pressure diminished by the length of the column of mercury, equal to the distance between the level of mercury in the glass basin and that in the tube. The weight of an equal volume of hydrogen at the same temperature and pressure is calculated, and the weight of the vapour is divided by the weight of the hydrogen. The quotient is the density.

(*c.*) **Victor Meyer's Method.**—In this case not mercury but air (or some other gas) is displaced; and the volume of a known weight of the vapour is deduced from that of the displaced gas, or air. A cylindrical bulb, *c* (fig. 21), with a long stem, *b*, closed by a cork at its upper extremity, as shown in the figure, is heated to some constant temperature by an oil- or vapour-bath, as already described. The air expands while the temperature is rising, and issues through the side tube, *d*, escaping in bubbles through the water in the trough. When bubbles cease to rise the temperature is assumed to be constant. The tube is quickly uncorked, a small tube, full of the liquid or solid whose vapour-density is sought, is dropped in, falling on sand, placed at the bottom of the cylinder, so as to avoid breaking it. The cork is then rapidly replaced. The substance turns to gas, and expels air from the cylindrical bulb. This air is cooled in passing up the stem and through

* The following data are available for this calculation:—

Temperature........	46°	78°	132°	156°	184°	237°	280°
Expansion of 1 c.c. of mercury between 0° and *t*°...........	1·0083	1·0141	1·0240	1·0285	1·0338	1·0438	1·0521
Vapour-pressure of mercury, in mm...	—	0·1	1·0	3·0	10·0	52·5	157·0

the water; it is collected in a graduated tube. Its volume is equal to that of the vapour, supposing the latter to have been cooled to the atmospheric temperature, and to have withstood the process without condensing. We have then a given volume of air at atmospheric temperature and pressure corresponding to that of the vapour; and also the weight of substance which has produced the vapour by which the air has been expelled. From these data it will be seen the density of the vapour may be calculated.

Such are the available means of ascertaining the weights of one litre of various gases and their densities. The processes have been described in some detail, because such determinations have the utmost chemical importance. The deductions to be drawn from them will appear in the next chapter.

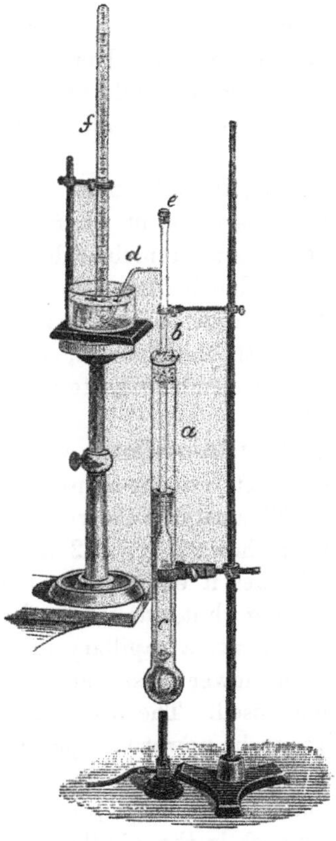

FIG. 21.

CHAPTER VIII.

COMPOUNDS OF THE HALOGENS WITH HYDROGEN, LITHIUM, SODIUM, POTASSIUM, RUBIDIUM, CÆSIUM, AND AMMONIUM. ATOMS AND MOLECULES: FORMULÆ AND EQUATIONS.

Hydrogen Fluoride, Chloride, Bromide, and Iodide.

Only one compound of each of these elements with hydrogen is known.

Sources.—Hydrogen chloride is present in the atmosphere in the neighbourhood of volcanoes; it has been doubtless formed by the action of steam on certain chlorides, easily decomposed by water into oxide of the element and hydrogen chloride. The others do not exist free.

Preparation.—1. **By direct union.** (*a.*) **Hydrogen Fluoride.**—During the preparation of fluorine by Moissan, by the electrolysis of hydrogen potassium fluoride, hydrogen was liberated from the negative, and fluorine from the positive pole (see p. 73). When a bubble of hydrogen escaped round the bend of the U-tube, and mixed with the fluorine, an explosion was heard, showing that these two elements unite at the ordinary temperature, and in the dark.

(*b.*) **Hydrogen Chloride.**—Equal volumes of hydrogen and chlorine gas unite directly on exposure to violet light, or on application of heat. This may be shown as follows:—

A tube of the form shown in fig. 22 is employed. The stop-cock in the middle divides it equally into two halves. The stop-cock in the middle being shut, one side is filled with dry chlorine by downward displacement, a capillary tube serving to conduct the chlorine gas to the lower closed end, as shown in the figure. The stop-cock is then closed. The other half of the tube is then filled with dry hydrogen by upward displacement, for hydrogen is lighter, though chlorine is heavier, than air. The tube is then placed in a dark place, for example, a close fitting drawer, for some hours, the stop-cock in the middle being opened. The two gases will mix, but will not combine. It is then placed for an instant in direct sunlight, or, if that is not available, illumined by burning a piece of magnesium ribbon within a few inches of it. A flash will be seen inside the tube, showing that combination has

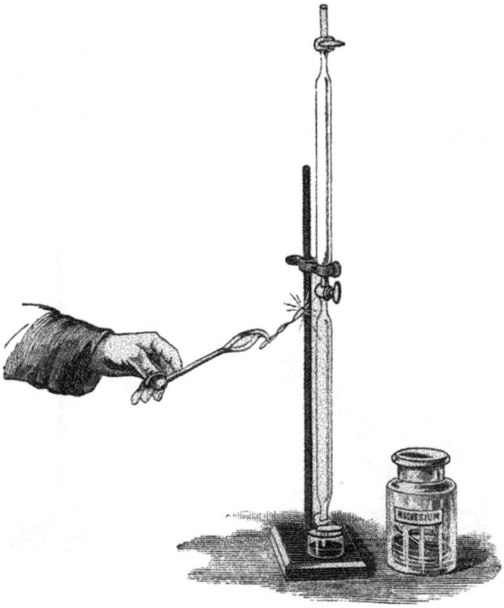

FIG. 22.

taken place, and the green colour of the chlorine will disappear. It is safer, however, to expose the tube for some hours to diffuse daylight. One end of the tube is now dipped in mercury, and the lower stop-cock is opened. The mercury does *not* enter the tube, showing that the hydrogen chloride retains the same volume as its constituents; it does not act on mercury. The stop-cock is again closed, and the lower end of the tube is now dipped in water, and the stop-cock again opened. The water rushes in, and completely fills the tube, provided both compartments were exactly equal, and that all air was displaced on filling it with chlorine and hydrogen. Chlorine is sparingly soluble in water, hydrogen nearly insoluble. Hence a gas has been produced by the combination of equal volumes of hydrogen and chlorine, which occupies the same volume as its two constituents, but which differs from them in properties.

A jet of hydrogen gas may be burned in a jar of chlorine. The hydrogen is lit, and, while burning in the air, a jar of chlorine is brought under it, and raised so that the jet dips into the chlorine.

The hydrogen continues to burn, but with a greenish-white flame. Fumes are produced.

(*c.*) **Hydrogen Bromide.**—Hydrogen and bromine do not combine so readily as hydrogen and fluorine or as hydrogen and chlorine. Their direct combination may be shown as follows:—A bulb tube is connected with an apparatus for generating hydrogen. A few cubic centimetres of bromine are placed in the bulb; the hydrogen passes over the bromine, and carries some with it as gas.

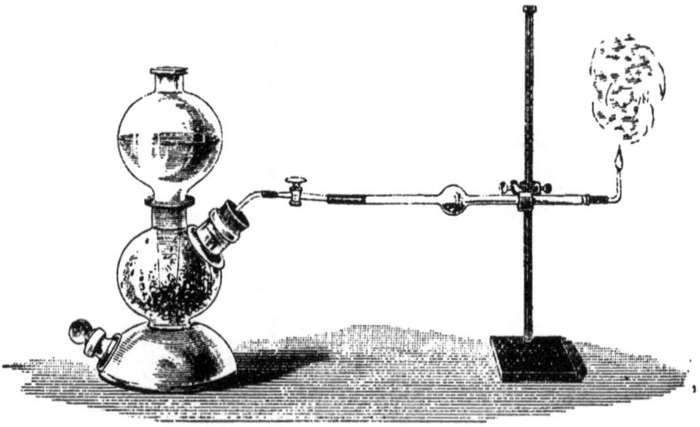

FIG. 23.

The hydrogen is lit, and burns, combining partly with the oxygen of the air, partly with the bromine. The hydrogen bromide formed unites with the water-vapour forming a white cloud of small liquid particles. It is owing to the formation of a similar compound with water that fumes are produced when hydrogen burns in chlorine.

A practical plan of preparing hydrogen bromide is to pass the mixture of hydrogen and bromine, prepared as described, through a glass tube containing a spiral coil of platinum wire, heated to redness by an electric current. The uncombined bromine is absorbed by passing the resulting gas through a tube filled with powdered antimony.

(*d.*) **Hydrogen and Iodine** may be made to combine directly by heating them together in a sealed tube to 440° for many days. Complete combination does not take place, however long the mixture is heated, and about one quarter of the hydrogen and one quarter of the iodine remain uncombined.

2. **By the Action of the Halogen on most Compounds of Hydrogen. Instances.**—(*a.*) **On water.**—A solution of chlorine

gas in water exposed to sunlight yields oxygen and hydrogen chloride; if chlorine and water-gas be led through a red-hot tube, some of the water-gas reacts with the chlorine, yielding hydrogen chloride and oxygen. (*b.*) **On hydrogen sulphide,** dissolved in water. The products are sulphur and the hydrogen compound of the halogen. This is a convenient method of preparing hydrogen iodide. Sulphuretted hydrogen gas (see p. 196) is passed through water in which iodine is suspended. The liquid becomes milky, owing to separation of sulphur, and the colour gradually disappears, owing to the union of the iodine with the hydrogen of the hydrogen sulphide. When the reaction is over, the sulphur is separated by filtration, and the liquid distilled. It is, however, impossible to separate hydrogen iodide from its solution in water by distillation. The aqueous solution is termed *hydriodic acid.* (*c.*) Chlorine, bromine, and iodine act on **ammonia,** yielding nitrogen and the compound of the halogen with hydrogen. Nitrogen combines with the halogen, if the latter is in excess, yielding very explosive bodies (see p. 158). (*d.*) Chlorine and bromine act on **hydrocarbons** (carbides of hydrogen) giving compounds of carbon with both chlorine (or bromine) and hydrogen, and the haloid acid. Generally it may be stated that almost all compounds of hydrogen are decomposed by the halogens, yielding a haloid compound of the element, and hydrogen chloride, bromide, or iodide.

3. **By the Action of Water, or of Double Oxides of Hydrogen and some other Element on Compounds of the Halogens.** *a.* **Action of Water.**—The halogen compounds of boron, silicon, titanium, phosphorus, sulphur, selenium, and tellurium, are at once decomposed by cold water. Hence the halogen added to water in which one of these elements is suspended, combines with part of the hydrogen of the water, the remaining hydrogen and oxygen combining with the element (see these haloid compounds, p. 188). **Instances:**—(*a.*) This is a practical method of preparing hydrogen bromide. The bromine is added very gradually to phosphorus, lying in water in a retort. Phosphorus bromide is produced, and decomposed by the water, forming phosphorous and phosphoric acids, and hydrogen bromide. After all the phosphorus has disappeared, the liquid is distilled. The solution of hydrogen bromide in water is named *hydrobromic acid.*

There is little doubt that all soluble chlorides, bromides, and iodides are decomposed by excess of water, forming the hydroxide of the metal and hydrogen chloride, bromide, or iodide. But in most cases there is no available method of separating the hydroxide from the hydrogen halide, for, on evaporation, the reverse reaction takes place, and water alone escapes. Yet, at a high temperature, magnesium chloride and some other chlorides react with water-gas, giving an oxy-chloride and hydrogen chloride.

(*b.*) This is a recently patented method of manufacturing hydrogen chloride, and promises to be successful. Steam is led over magnesium chloride, heated in tubes; hydrogen chloride is evolved, and a compound of magnesium oxide and chloride remains.*

b. **Action of Hydroxides.**—The hydroxides which react in this manner are termed *acids*. Generally stated, the hydrogen halides can be prepared by the action of any hydroxide which does not react with them. Phosphoric, sulphuric, and selenic acids are such.

c. This is the common method of preparing hydrogen fluoride. The fluoride generally employed is calcium fluoride, or *fluor-spar*, which occurs native; it is treated with sulphuric acid in leaden vessels, and the gas evolved is condensed in a worm of lead and stored in leaden or gutta-percha bottles. It acts on silica, which is a large constituent of glass and porcelain; hence the use of lead, which is but slightly attacked. On a small scale, platinum vessels and potassium fluoride answer better.

d. This is also the best method of preparing hydrogen chloride. On a small scale, about 50 grams of sodium chloride (common salt) are placed in a retort, and covered with a mixture of equal volumes of sulphuric acid and water. On applying a gentle heat the hydrogen chloride comes over in the gaseous state. It may be led into water; the solution is called *hydrochloric acid.*

On a large scale, the operation is conducted in circular furnaces with a revolving bed. The salt and sulphuric acid are introduced from above, and fall on to the middle of a revolving plate of iron covered with fire-clay, which forms the bed of the furnace. The product, sodium sulphate, or "salt-cake," is raked by mechanical means towards the circumference of the plate, and drops through traps for the purpose. The hydrogen chloride is led up brick towers filled with small lumps of coke, kept moist with water from above. The water dissolves the hydrogen chloride, which is sent to market in carboys.

e. As both hydrogen bromide and iodide react with and decompose sulphuric acid (see p. 111), bromine or iodine being liberated, phosphoric acid must be used for their preparation. The method of operation is similar to that of preparing hydrogen chloride.

* *Soc. Chem. Industry*, 1887, 775.

4. **Heating Compounds of the Hydrogen Halide with the Haloid Compounds of other Elements.**—Such compounds always decompose when heated. In practice, this method is employed for the preparation of pure hydrogen fluoride. Its compound with potassium fluoride, after being dried, is heated to redness in a platinum retort, and the hydrogen fluoride which distils over is condensed by passing through a platinum tube surrounded with a freezing mixture, and collected in a platinum bottle. The preparation of pure hydrogen fluoride is exceedingly dangerous, owing to its great corrosive action.

Before considering the properties of these bodies, the nature of the changes which have been described, and the method of representing these changes, must be discussed.

Atoms and Molecules.

It was stated in last chapter that *equal volumes of gases contain equal numbers of molecules.* Now, it has been shown that equal volumes of hydrogen and chlorine unite to form hydrogen chloride. It might be concluded that such a compound consists of 1 molecule of chlorine in union with 1 molecule of hydrogen; but the following considerations will show that such a supposition is inconsistent with Avogadro's law. The actual facts are that 1·0025 gram of hydrogen, occupying at standard temperature and pressure 11·16 litres, combines with 35·46 grams of chlorine, also occupying 11·16 litres, and that the volume of the product is 11·16 × 2, or 22·32 litres. We do not know the actual number of molecules of hydrogen, or of chlorine, in 11·16 litres of these gases; let us call it *n*. Then *n* molecules of hydrogen, on this supposition, unite with *n* molecules of chlorine, and as chemical combination has occurred, *n* molecules of hydrogen chloride are formed. But the volume of the hydrogen chloride is 22·32 litres; hence *n* molecules of hydrogen chloride would thus occupy (11·16 × 2) litres, instead of 11·16; or the requirements of Avogadro's law would not be complied with, inasmuch as there would be only half as many molecules in a given volume of hydrogen chloride as in the same volume of hydrogen or of chlorine. But there is no reason to suppose that hydrogen chloride does not fulfil Avogadro's law; its expansion by rise of temperature and behaviour as regards pressure are practically the same as those of hydrogen and chlorine, hence the conclusion is evidently false. The accepted explanation is as follows:—

A molecule of hydrogen, or a molecule of chlorine, is not a simple thing; it consists of two portions in combination with each other; these portions are named *atoms*. When chlorine and hydrogen combine to form hydrogen chloride, their double atoms or molecules split, each atom of hydrogen leaving its neighbour atom, and uniting to an atom of chlorine, which has also parted with its neighbour atom. The original arrangement may be represented thus:—

(H)——(H) (Cl)——(Cl);

and the final arrangement, thus:—

(H)——(Cl) (H)——(Cl).

In 11·16 litres of hydrogen chloride there is, therefore, the same number of molecules as in an equal volume of hydrogen or of chlorine; but whereas the hydrogen chloride molecules contain an atom of each element, those of hydrogen contain two atoms of hydrogen, and those of chlorine contain two atoms of chlorine.

Symbols are employed to express such changes. The expression of the change is termed an equation; and the above change is written thus:—

$$H_2 + Cl_2 = 2HCl.$$

Where the small numeral follows the letter, it signifies the number of atoms in the molecule, as H_2, Cl_2; where a large numeral precedes the formula, it signifies the number of molecules; thus, $2HCl$. $2H$ would mean two uncombined atoms of hydrogen; H_2 signifies two atoms combined into a molecule. Atoms of hydrogen have not been obtained uncombined with each other; atoms of chlorine, however, exist uncombined, or in the free state, at a sufficiently high temperature.

Such an equation expresses the following facts:—

1. That 22·32 litres of hydrogen react with 22·32 litres of chlorine, producing 44·64 litres of hydrogen chloride; and

2. That 2·005 grams of hydrogen react with 70·92 grams of chlorine, forming 72·925 grams of hydrogen chloride.

It is obvious that 22·32 litres of hydrogen chloride weigh 72·925/2 grams; and as 22·32 litres of hydrogen weigh 2·005 grams, hydrogen chloride is 18·231 times as heavy as hydrogen. This has been found to be the case by direct experiment. Hence the molecular weight of hydrogen chloride = 36·4625 is twice its density compared with hydrogen.

Such formulæ as HCl, H_2, Cl_2, apply only to gases. In this book the symbols for gaseous elements and compounds are printed in italics; those for liquids in ordinary type; and those for solids in bold type. It is still doubtful whether liquids and solids possess such simple formulæ; it is the author's opinion that in many cases they do; but there are certainly many cases in which they possess more complex formulæ. There is, however, as yet no method of determining with certainty the degree of complexity; hence, the simplest formulæ are employed. Liquid hydrogen chloride may have the formula HCl; or it may have the formula (HCl)*n*; but what the value of *n* is, there is no means of determining.

The reactions, whereby the halides of hydrogen are prepared, are represented thus:—

1*a*. $H_2 + F_2 = 2HF$ at high temperatures (see p. 115).

b. $H_2 + Cl_2 = 2HCl$.

c. $H_2 + I_2 = 2HI$.

2*a*. $2H_2O + 2Cl_2 = 4HCl + O_2$, or $2H_2O + 2Cl_2 = 4HCl + O_2$.
Water.

b. $H_2S + \mathbf{I_2} + Aq = 2HI.Aq + \mathbf{S}$. ($Aq$ = *aqua*, water).
Hydrogen sulphide. Hydriodic acid.

c. $2H_3N.Aq + 3Cl_2 = 6HCl.Aq + N_2$.
Ammonia.

d. $CH_4 + Cl_2 = CH_3Cl + HCl$.
Methane. Chloro-methane.

3*a*. $2\mathbf{P} + 5Br_2.Aq + 8H_2O = 2H_3PO_4.Aq + 10HBr.Aq$.
Phosphoric acid.

b. $2\mathbf{MgCl_2} + H_2O = \mathbf{MgCl_2.MgO} + 2HCl$.
Magnesium chloride. Magnesium oxychloride.

c. $\mathbf{CaF_2} + H_2SO_4 + \mathbf{CaSO_4} + 2HF$.
Calcium fluoride. Sulphuric acid. Calcium sulphate.

d. $\mathbf{NaCl} + H_2SO_4 = \mathbf{NaHSO_4} + HCl$.
Sodium chloride. Sulphuric acid. Sodium hydrogen sulphate.

$\mathbf{NaCl} + \mathbf{NaHSO_4} = \mathbf{Na_2SO_4} + HCl$.
Sodium chloride. Sodium hydrogen sulphate. Sodium sulphate.

e. $\mathbf{NaBr} + \mathbf{H_3PO_4} = \mathbf{NaH_2PO_4} + HBr$.
Sodium bromide. Phosphoric acid. Dihydrogen sodium phosphate.

The action of hydrogen bromide or iodide on hot sulphuric acid is represented thus:—

$$H_2SO_4 + 2HBr \text{ (or } 2HI) = SO_2 + 2H_2O + Br_2 \text{ (or } I_2).$$

4. **KF.HF**	=	**KF**	+	*HF.*
Hydrogen potassium fluoride.		Potassium fluoride.		

Properties.—Hydrogen fluoride is a colourless very volatile liquid, boiling at about 19° under atmospheric pressure; hydrogen chloride, bromide, and iodide are all colourless gases. Hydrogen fluoride is fearfully corrosive; a drop on the skin produces a painful sore, and several deaths have occurred through inhaling its vapour. The other three gases are suffocating, but do not produce permanent injury when breathed diluted with air. They condense to liquids at low temperatures. They are exceedingly soluble in water, in all probability forming compounds which mix with excess of water or of the halide. One volume of water at 0° dissolves about 500 times its volume of hydrogen chloride; the solution is about 1·21 times heavier than water, and contains 42 per cent. of its weight of the gas. On cooling a strong solution of hydrogen chloride in water to −18°, and passing into the cold liquid more hydrogen chloride, crystals of the formula $HCl.2H_2O$* separate out. It is probable that, at the ordinary temperature, this compound exists in an aqueous solution of hydrogen chloride, and is decomposed into its constituents to an increasing extent with rise of temperature. Hydrogen fluoride, bromide, and iodide are also exceedingly soluble in water, and their solutions probably contain similar hydrates. The corresponding compound of hydrogen bromide, $HBr.2H_2O$, has been prepared; it melts at −11°. The solutions of these compounds are termed hydrofluoric, hydrochloric, hydrobromic, and hydriodic acids. When saturated, they are colourless fuming liquids; they possess an exceedingly sour taste, and are very corrosive; they change the blue colour of litmus (a substance prepared from a lichen named *lecanora tinctoria*, and itself the calcium salt of a very weak acid) to red, owing to the liberation of the red-coloured acid. This is the usual test for an acid.

The great solubility of hydrogen chloride may be illustrated by help of the apparatus shown in the figure. (Fig. 24.)

The lower flask is filled with water coloured blue with litmus; the upper flask is filled with hydrogen chloride by downward displacement, and inverted over the lower flask. The stopcock is then opened, establishing communication between the two flasks. By blowing through the tube, a little water is forced up into the hydrogen chloride. It immediately dissolves, producing a partial vacuum in the upper flask; and the pressure of the atmosphere causes a fountain of water to enter it. The blue colour of the litmus is at the same time changed to red.

---

* *Comptes rendus*, **86**, 279.

All elements are attacked and dissolved by these acids, hydrogen being liberated, while the halogen combines with the element, with the exception of:—Silver, gold, mercury; boron (attacked by hydrofluoric acid), carbon; silicon, zirconium (both attacked by hydrofluoric acid), lead; nitrogen, vanadium, phos-

FIG. 24.

phorus, arsenic, antimony, bismuth; molybdenum; oxygen, sulphur, selenium, tellurium, and the elements of the platinum group. Mercury and lead are attacked by strong hydriodic acid; moist hydrogen chloride, bromide, and iodide are decomposed by light in presence of oxygen.* The first two are not decomposed when dry; dry hydriodic acid, however, yields water and iodine.

Uses.—Hydrofluoric acid is employed for etching on glass. The glass is protected by a coating of beeswax, and a pattern is drawn on the wax. The article is then dipped in the strong acid, and the pattern remains after removing the wax, the glass appearing frosted where the acid has attacked it. Hydrochloric acid is used for many purposes, one of the chief of which is the manufacture of chlorine and the chlorides of metals.

* *Chem. Soc.*, **51**, 800.

Proofs of the Volume-Composition of the Halides of Hydrogen.

It has already been shown that hydrogen chloride consists of equal volumes of hydrogen and chlorine united without contraction; it may be shown to contain its own volume of hydrogen by the following experiment:—A U-tube, as shown in fig. 25, is filled with mercury, which is then displaced in the

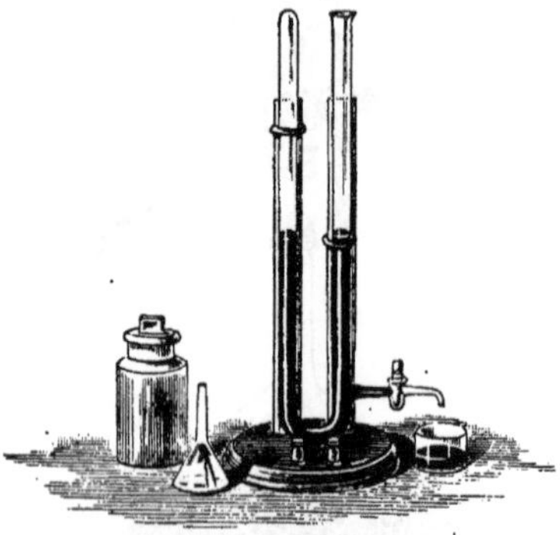

Fig. 25.

closed limb by gaseous hydrogen chloride. The level of the mercury is made equal in the two limbs, and the position marked. The open limb is then filled with liquid sodium amalgam (an alloy of mercury and sodium containing about 2 per cent. of sodium) and closed with the thumb. The tube is then inverted, so as to bring the gas into contact with the sodium amalgam. The sodium reacts with the hydrogen chloride, liberating hydrogen, thus:—

$$2HCl + 2\text{Na} = 2\mathbf{NaCl} + H_2.$$

The hydrogen is then again transferred into the closed limb by inclining the tube, and the levels again equalised; it will be seen to occupy half the volume originally occupied by the hydrogen chloride.

That hydrogen bromide and iodide yield half their volume of hydrogen when similarly treated has also been proved. Hydrogen fluoride has been synthesised by heating silver fluoride with hydrogen gas. The product occupied at 100° twice the volume of the hydrogen employed for its formation. The equation is $2\mathbf{AgF} + H_2 = 2HF + 2\mathbf{Ag}$, silver being set free as metal.

It is argued that hydrogen fluoride, bromide, and iodide possess respectively the formulæ *HF*, *HBr*, and *HI*, from these experiments, from their densities, and from their similarity to hydrogen chloride. Recent experiments have, however, shown that at low temperatures gaseous hydrogen fluoride has a greater molecular weight than that expressed by the formula HF; but the actual degree of complexity is not yet certain (see below).

Physical Properties.

	Mass of 1 c.c. Solid.	Liquid.	Gas.	Density. H = 1.
Hydrogen fluoride ..	?	0·988 at 12·7°	See below	See below
Hydrogen chloride ..	?	?	0·001633*	18·23*
Hydrogen bromide ..	?	?	0·003626*	40·47*
Hydrogen iodide	?	?	0·005727*	63·92*

	Melting-point.	Boiling-point.	Specific Heat.	Molecular Weight.
Hydrogen fluoride..	−92·3°	19·4°	?	20 (see below)
Hydrogen chloride .	−112·5°	−102°	0·1304 (gas)	36·46
Hydrogen bromide . .	−73°	−87°	?	80·95
Hydrogen iodide...	−55°	?	?	127·85

Heat of formation.—$H_2 + Cl_2 = 2HCl + 440K.$
$H_2 + Br_2 = 2HBr + 242K.$
$H_2 + I_2 = 2HI + 0K$ at about 184°.

Note.—Molecular weight of hydrogen fluoride.† The vapour-density of hydrogen fluoride increases with fall of temperature, implying the association of molecules of HF to form $(HF)n$. (The value of n appears to be 4.) The highest density was found at atmospheric pressure, and at 26·4°, to be 25·59, implying the molecular weight of 51·18. This corresponds to a mixture of 81·24 per cent. molecules of H_4F_4 and 18·76 per cent. of molecules of HF. At 100° and above, the density is normal, and corresponds to the formula HF.

Compounds of the Halogens with Lithium, Sodium, Potassium, Rubidium, and Cæsium (Ammonium).

Sources.—Sodium fluoride occurs native in Greenland in combination with aluminium fluoride, as *cryolite*, $3NaF.AlF_3$. Lithium, sodium, and potassium chlorides, bromides, and iodides occur in sea-water; sodium chloride in by far the greatest amount—about 3·5 per cent.; and also in many mineral springs. That at Dürkheim, in the Bavarian Palatinate, is comparatively rich in cæsium and rubidium chlorides, and was the source from which Bunsen and

* These numbers are calculated.
† Thorpe, *Chem. Soc.*, **53**, 765; **55**, 163.

Kirchhoff extracted these elements for the first time.* The Wheal Clifford spring, in Cornwall, is specially rich in lithium chloride. Sodium chloride also occurs as rock-salt in mines, in various parts of the world; the largest in Britain are in Cheshire, but recently other deposits have been discovered near the Tyne. Very large deposits of potassium chloride occur at Stassfurth, near Magdeburg, in N. Germany. It also occurs in *kelp*, the ash of *fucus palmatus*, species of seaweed. The ash of the beetroot contains about 0·17 per cent. of rubidium chloride.

Preparation.—1. By direct union of the elements.—This takes place with great loss of energy (*i.e.*, evolution of heat); the elements take fire and burn in chlorine gas. Perfectly dry chlorine, bromine, or iodine, however, does not act on sodium in the cold.† A subchloride of a purple colour is said to be produced by the action of chlorine on metallic potassium.

2. **By double decomposition.** (*a.*) **Action of the halogen acids on the oxides, hydroxides,** or **carbonates** of the metals; in the first two cases, the hydrogen of the halogen acid unites with the oxygen of the oxide, or the hydroxyl (a name applied to the group OH) of the hydroxides; in the third case, carbon dioxide and water are liberated. Examples of this action are given in the following equations:—

$$\mathbf{KOH} + \mathit{HF} = \mathbf{KF} + H{\cdot}OH.$$
$$\mathbf{Na_2O} + 2\mathit{HCl} = 2\mathbf{NaCl} + H_2O.$$
$$\mathbf{Li_2CO_3} + 2\mathit{HI} = 2\mathbf{LiI} + H_2O + \mathit{CO_2}.$$

These reactions also occur in solution.

On adding to a solution of a hydroxide, containing an unknown quantity of the hydroxide, a solution of a hydrogen halide, the completion of the reaction, or the "point of neutralisation," may be ascertained by the addition of litmus, or of phenol-phthaleïn, to the hydroxide; the former gives a blue, the latter a cherry-red colour with these hydroxides; when the colour is on the point of changing to brick-red, with litmus, or being discharged entirely, with phenol-phthaleïn, the reaction is complete, and there is no excess either of acid, or of alkali, as such hydroxides are named.

If carbonates be used, the solution must be boiled during the addition of acid, so as to expel carbon dioxide gas, else it will produce a colour change.

(*b.*) **By certain other "double decompositions;"** thus sodium chloride is obtained as a by-product in the manufacture

* *Pogg. Ann.*, **110**, 161; **113**, 337; **119**, 1; *Annales* (3), **64**, 290.

† *Berichte*, **6**, 1518; *Chem. Soc.*, **43**, 155.

of potassium nitrate from sodium nitrate and potassium chloride :—

$$KCl.Aq + NaNO_3.Aq = KNO_3.Aq + \mathbf{NaCl}.$$

The sodium chloride, being much less soluble in water than potassium nitrate, separates in crystals on evaporation. The sulphides and hydrosulphides of these metals also yield halides on treatment with halogen acids.

3. **By heating compounds of these metals with oxygen and with the halogens,** *e.g.*, chlorates, iodates, &c. (see p. 466).

4. **Compounds of ammonium with the halogens** are prepared by addition of the halogen acid to a solution of ammonia in water. Direct combination ensues, thus :—

$$NH_3.Aq + HCl.Aq = NH_4Cl.Aq.$$

Ammonium, NH_4.

The group of elements to which the name ammonium has been given exhibits the greatest similarity to metals of the sodium group, and is usually classed along with them. It has never been isolated (see, however, pp. 577, 578). But ammonia, consisting of one atom of nitrogen and three atoms of hydrogen, NH_3 (see p. 512), has the power of union with acids (as well as with oxides and double oxides); compounds of ammonium with the halogens differ from those of sodium and the other metals by splitting up when heated into ammonia and the hydrogen halide.

The union of ammonia with a halide of hydrogen may be illustrated by placing a jar filled with ammonia gas (see p. 512) mouth to mouth over a jar of hydrogen chloride, both being covered with glass plates; when the plates are withdrawn, dense white fumes of ammonium chloride are seen; they gradually settle in the lower jar as a white powder.

The decomposition of this compound by heat may be shown by applying heat to a fragment in a platinum basin; it will volatilize completely, being decomposed into its constituents—ammonia, NH_3, and hydrogen chloride, HCl; they unite when cooled by the air, forming dense white fumes.

Special methods of extraction and preparation.—Owing to their importance, the following compounds require consideration:—Common salt, or sodium chloride, is produced by the evaporation of sea-water in "salt pans," shallow ponds exposed to the air. To promote evaporation, the salt water is sometimes allowed to trickle over ledges, running into gutters which lead it to the ponds. When a portion of the water has been thus removed, it is boiled down in shallow iron pans. Rapid evaporation produces fine-grained salt, such as is used for the table; slow evaporation causes the salt to separate in larger crystals; it is used for curing fish, &c.

In Cheshire, water is run into the mines, and the brine is pumped up and evaporated. In cold climates, the salt is sometimes extracted from sea-water by freezing; the ice which separates is nearly pure, while the salt remains dissolved in the last portions of water.

Potassium bromide and iodide are prepared (*a*) by the action of bromine or iodine on a solution of potassium carbonate; the water is removed by evaporation, and the residue is heated to redness (see p. 467); or (*b*) by treating iron filings with bromine or iodine, producing ferrous bromide or iodide, to which a solution of potassium carbonate is then added. The resulting ferrous carbonate is insoluble in water; it is removed by filtration, and the filtrate is evaporated to dryness. The equations are:—

$$\mathbf{Fe} + Br_2 + Aq = FeBr_2.Aq;$$
$$FeBr_2.Aq + K_2CO_3.Aq = 2KBr.Aq + \mathbf{FeCO_3}.$$

The equation for the preparation of potassium iodide is similar.

Properties.—These substances are all white solids, crystallising in the cubical system, with the exception of cæsium chloride, which crystallises in rhombohedra. They are all soluble in water; lithium chloride, sodium bromide, and sodium and potassium iodides are also soluble in alcohol.

100 grams of water dissolve at the ordinary temperature (about 15°)—

	Fluoride.	Chloride.	Bromide.	Iodide.
Lithium	trace	—	—	—
Sodium	4	36	88	173
Potassium	—	33	65	143
Ammonium	—	37	72	—

grams of these salts.

They all form double compounds with water, *e.g.*, $\mathbf{NaCl.2H_2O}$, crystallising at a low temperature. They melt at a red heat and volatilise at a bright red heat; ammonium chloride dissociates at 339°, under ordinary atmospheric pressure, into hydrogen chloride and ammonia; the other compounds of ammonium behave similarly.

	Melting-points.				Mass of 1 c.c.			
	F.	Cl.	Br.	I.	F.	Cl.	Br.	I.
Lithium	801°	598°	547°	446°	2·29	2·00	3·10	3·48
Sodium	902°	772°	708°	628°	2·56	2·16	3·08	3·65
Potassium	789°	734°	699°	634°	2·10	1·98	2·60	3·01
Rubidium	753°	710°	683°	642°	3·20	2·80	3·36	3·57
Cæsium	—	—	—	—	?	4·00	4·46	4·54
Ammonium	—	—	—	—	?	1·52	2·46	2·44

The vapour densities of the following compounds have recently been determined at about 1200° by V. Meyer's method:—

Found........	KI = 184·1; RbCl = 139·4; RbI = 221·6
Calculated....	KI = 166·0; RbCl = 121·0; RbI = 212·3

Found	CsCl = 179·2; CsI = 267;
Calculated......	CsCl = 168·4; CsI = 259·8*

These numbers represent molecular weights, *i.e.*, vapour-densities multiplied by two.

It may be concluded from analogy that the other halides, in the gaseous state, have also simple formulæ, such as NaCl; at present we know nothing about the molecular weights of these bodies in the liquid or solid state.

Double compounds.—1. With halogens.—Iodine unites directly with potassium iodide in aqueous or alcoholic solution, and forms dark lustrous prisms, possessing the formula $\mathbf{KI_3}$.† The mass of 1 c.c. is 3·50 grams at 15°. It melts at 45°. Chlorine and bromine are more soluble in solutions of chlorides and bromides than in pure water, owing probably to the formation of similar compounds, which are partially dissociated at the ordinary temperature. Ammonium tri-iodide and tribromide, $\mathbf{(NH_4)I_3}$ and $\mathbf{(NH_4)Br_3}$, have been prepared by a similar method, and are closely analogous.

2. **With hydrogen halides.**—Potassium fluoride unites with hydrogen fluoride in three proportions, forming (*a*) **KF.HF**, (*b*) **KF.2HF**, and (*c*) **KF.3HF**.‡ They are all stable in dry air, but decompose when heated into potassium and hydrogen fluorides. No doubt, similar compounds of the other halogen salts would prove stable at low temperatures.

For compounds of the formula $4NH_3.HCl$, and $7NH_3.HCl$, see p. 525.

Heats of formation—

Li + *Cl* = **LiCl** + 938K + Aq = +84K.§
Na + *Cl* = **NaCl** + 976K + Aq = −11·8K.
Na + Br = **NaBr** + 858K + Aq = −1·9K.
Na + **I** = **NaI** + 691K + Aq = −12K.
K + *Cl* = **KCl** + 1056K + Aq = −44·1K.
K + Br = **KBr** + 951K + Aq = −50·8K.
K + **I** = **KI** + 801K + Aq = −51·1K.

---

* Scott, *Brit. Assn.*, 1887, 668; *Proc. Roy. Soc. Edin.*, **14**.
† *Chem. Soc.*, **31**, 249; **33**, 397; *Berichte*, **14**, 2398.
‡ *Comptes rendus*, **106**, 547.
§ For an explanation of "K," see p. 127.

CHAPTER IX.

COMPOUNDS OF THE HALOGENS WITH BERYLLIUM, CALCIUM, STRONTIUM, AND BARIUM; WITH MAGNESIUM, ZINC, AND CADMIUM. DOUBLE HALIDES. SPECIFIC AND ATOMIC HEATS. REASONS FOR MOLECULAR FORMULÆ. VALENCY.

Beryllium, Calcium, Strontium, and Barium Halides.

Sources.—Calcium fluoride, or *fluor-spar*, CaF_2.—This beautiful mineral, crystallising in cubes, sometimes showing octahedral modifications, occurs in granite and porphyry rocks, especially where the veins border other strata. It forms the gangue of the lead-veins which intersect the coal-formations of Northumberland, Cumberland, Durham, and Yorkshire; it is abundant in Derbyshire and also in Cornwall, where the veins intersect much older rocks. A large vein occurs in Jefferson Co., New York State, in granular limestone. It often possesses a pink, amethyst, or green colour, from the presence of certain metallic fluorides.

Calcium chloride is a constituent of all natural waters, and exists in small amount in sea-water. Traces of the chlorides of strontium and barium are also found in some mineral waters.

Preparation.—The methods of preparation are similar to those of the halides of the alkali-metals.

1. **By direct union of the elements.**—The metals of this group are so difficult to prepare that the method is impracticable.

2. **By double decomposition.**—(*a.*) The action of the haloid acid on the oxides, hydroxides, sulphides, or hydrosulphides, or on double oxides, such as carbonates, silicates, &c. This method serves for the production of the chlorides, bromides, and iodides; not well for the fluorides, for the fluorides of calcium, strontium, and barium, are insoluble in water, and the hydroxide or carbonate becomes coated over with the insoluble fluoride, and action ceases. The reactions may be typified by the following equations:—

$$\mathbf{BeO} + 2HCl = \mathbf{BeCl_2} + H_2O.$$
$$\mathbf{Ca(OH)_2} + 2HCl = \mathbf{CaCl_2} + 2H_2O.$$
$$\mathbf{SrCO_3} + 2HCl = \mathbf{SrCl_2} + H_2O + CO_2.$$
$$\mathbf{BaS} + 2HCl = \mathbf{BaCl_2} + H_2S.$$

These reactions occur both in solution and with the dry materials.

This process is practically made use of in preparing strontium and barium chlorides, from their carbonates and sulphides.

(*b.*) The fluorides of calcium, strontium, and barium, being insoluble in water, may be precipitated by adding a soluble fluoride, such as potassium fluoride, to a soluble salt of one of these metals, such as calcium chloride, barium iodide, &c. The reaction is, for example:—

$$CaCl_2.Aq + 2KF.Aq = \mathbf{CaF_2} + 2KCl.Aq.$$

Potassium chloride is soluble in water, and may be separated from the insoluble calcium fluoride by filtration.

Doubtless similar reactions occur on mixing soluble compounds of the other halogens with soluble compounds of these metals; thus it may be supposed that

$$2KI.Aq + BaCl_2.Aq = 2KCl.Aq + BaI_2.Aq.$$

But as all the compounds concerned in the change are soluble in water, they cannot be separated. It is probable that such changes are only partial; *i.e.*, not all the potassium iodide is converted into chloride, nor all the barium chloride converted into iodide, but that after mixture the solution contains all four compounds.

This method of "double decomposition," *i.e.*, reciprocal exchange, is also practically applied in the preparation of strontium and barium chlorides on a large scale. The chief sources of these metals are the sulphates of strontium and barium (see p. 422). These substances are heated to redness with calcium chloride, when the calcium transfers its chlorine to the strontium or barium, itself being converted into sulphate, thus:—

$$\mathbf{BaSO_4} + \mathbf{CaCl_2} = \mathbf{BaCl_2} + \mathbf{CaSO_4}.$$

On treatment with water the insoluble calcium sulphate remains, while the soluble strontium or barium chloride dissolves, and may be purified by crystallisation from water.

Properties.—Beryllium fluoride has not been prepared free from water; on attempting to dry the gummy mass obtained by its evaporation it reacts with the water (see below).

The fluorides of calcium, strontium, and barium are white crystalline powders, insoluble in water.

The remaining halides of this group are all white solids, soluble in water. They unite with water, forming crystalline compounds. Among these are $\mathbf{BeCl_2.2H_2O}$; $\mathbf{CaCl_2.6H_2O}$; $\mathbf{SrCl_2.3H_2O}$; $\mathbf{BaBr_2.2H_2O}$; and $\mathbf{BaI_2.7H_2O}$. The only one of the halides which has been volatilised is beryllium chloride, which becomes vapour somewhat below 520° under ordinary pressure. At higher temperatures (812°) it has the vapour-density 40·42, implying the molecular weight 80·02.* The compounds of beryllium have a sweet, disagreeable taste; the soluble compounds of the other elements are saline and burning.

Uses.—Calcium fluoride is employed as a *flux*, or material to be added to metals to make them flow (*fluo*) when they are being fused. It probably acts by dissolving a film of oxide encrusting the globules, and thereby causes the metallic surfaces to come in contact and unite. It is also a source of hydrogen fluoride (see p. 106). Calcium chloride is employed on a small scale for drying gases, and liquid compounds of carbon; it has a great tendency to unite with water, hence it deliquesces on exposure to moist air, attracting so much moisture as to dissolve.

Some of these substances react with water; hence beryllium halides, calcium chloride, bromide, and iodide, and strontium and barium bromides and iodides, cannot be prepared pure in an anhydrous state by evaporating their solutions. The reaction is a partial one. With calcium bromide, for instance, it is:—

$$\mathbf{CaBr_2} + H_2O = \mathbf{CaO} + 2HBr.$$

But the calcium bromide and oxide unite, forming various oxybromides, which remain, while a portion of the hydrogen bromide escapes.

Physical Properties.

	Melting-points.				Mass of 1 c.c.			
	F.	Cl.	Br.	I.	F.	Cl.	Br.	I.
Beryllium	?	600°	600°	?	?	?	?	?
Calcium	902°	719°	676°	631°	3·14	2·20	3·32	?
Strontium	902°	825°	630°	507°	4·21	3·05	3·98	4·41
Barium	908°	860°	812°	?	4·83	3·82	4·23	4·92

* Nilson and Petterssen, *Comptes rendus*, **98**, 988.

Heats of formation:—

Ca	+	Cl_2	=	$\mathbf{CaCl_2}$	+	1698K	+	Aq	=	+174K.
Ca	+	Br_2	=	$\mathbf{CaBr_2}$	+	1409K	+	Aq	=	+256K.
Ca	+	I_2	=	$\mathbf{CaI_2}$	+	1073K	+	Aq	=	+277K.
Sr	+	Cl_2	=	$\mathbf{SrCl_2}$	+	1846K	+	Aq	=	+111K.
Sr	+	Br_2	=	$\mathbf{SrBr_2}$	+	1577K	+	Aq	=	+161K.
Ba	+	Cl_2	=	$\mathbf{BaCl_2}$	+	1947K	+	Aq	=	+ 21K.
Ba	+	Br_2	=	**BaBr**	+	1700K	+	Aq	=	+ 50K.

Double compounds.—The seare all prepared by direct addition. Among them may be mentioned:—$\mathbf{BeF_2.2KF}$, $\mathbf{BeCl_2.2KCl}$, and similar compounds with sodium and ammonium chlorides, and $\mathbf{BaF_2.BaCl_2}$. The solubility of barium and strontium fluorides in hydrofluoric acid is probably due to the formation of double compounds with hydrogen fluoride.

Magnesium, Zinc, and Cadmium Halides.

Sources.—Magnesium chloride, bromide, and iodide are contained in sea-water, and in many mineral springs. *Carnallite*, $MgCl_2.KCl.6H_2O$, occurs in large quantities at Stassfurth, and is a valuable source of magnesium and potassium compounds.

Preparation.—1. **By direct union.**—The halogens unite with these metals directly, even in the cold, to produce halides. In presence of water, solutions are obtained.

2. **By the action of the halogen acid on the metal** hydrogen is evolved, and the halide of the metal is formed.

3. **By double decomposition.**—(*a*.) By the action of the halogen acid on the oxides, hydroxides, sulphides, and on some double oxides, such as carbonates, borates, &c. This process yields solutions of the halides (except in the case of magnesium fluoride, which is insoluble in water). But the water cannot be removed completely by heat, for it reacts with the chlorides, forming oxychlorides. The double chlorides with ammonium chloride, however, are unacted on when evaporated with water, hence anhydrous magnesium chloride may be produced by heating the compound, $MgCl_2.2NH_4Cl$, to redness; the ammonium chloride sublimes (see p. 117), leaving the anhydrous magnesium chloride. It can also be prepared by heating the aqueous chloride in a current of hydrogen chloride. Similar methods would probably succeed with the bromides and iodides.

(*b*.) Other methods of double decomposition may be sometimes employed; *e.g.*, $MgSO_4.Aq + BaCl_2.Aq = MgCl_2.Aq + \mathbf{BaSO_4}$. Barium sulphate is insoluble, and may be removed by filtration. Another method, which succeeds on a large scale, is to heat, under pressure, magnesium carbonate with a solution of calcium chloride; the equation—

$$\mathbf{MgCO_3} + CaCl_2.Aq = MgCl_2.Aq + \mathbf{CaCO_3}$$

represents the reaction, the insoluble calcium carbonate being removed by filtration.

Typical Equations—

1. $\mathbf{Zn} + \mathit{Cl_2} = \mathbf{ZnCl_2}$.
2. $\mathbf{Cd} + 2HI.Aq = CdI_2.Aq + \mathit{H_2}$.
3. $\mathbf{MgO} + 2HBr.Aq = MgBr_2.Aq + H_2O$.
   $\mathbf{ZnS} + 2HCl.Aq = ZnCl_2.Aq + \mathit{H_2S}$.
   $\mathbf{CdCO_3} + 2HF.Aq = CdF_2.Aq + H_2O + \mathit{CO_2}$.

Properties.—With the exception of magnesium fluoride, the halides of these metals are soluble in water. They are white and crystalline. The fluorides excepted, they are all volatile and are decomposed at a red heat by atmospheric oxygen, yielding the halogens and oxyhalides. This has been proposed as an effective method of manufacturing chlorine. They also react with water at a red heat; the products are oxyhalide and hydrogen halide. This method is in operation for the preparation of hydrogen chloride; the equation has been given on p. 111. They all unite with water, forming crystalline compounds; for example, $\mathbf{MgCl_2.6H_2O}$; $\mathbf{MgBr_2.3H_2O}$; $\mathbf{ZnF_2.4H_2O}$; $\mathbf{ZnCl_2.H_2O}$; $\mathbf{CdCl_2.2H_2O}$; $\mathbf{CdBr_2.H_2O}$; $\mathbf{CdI_2}$ crystallises as such from water. Zinc chloride has such a strong tendency to combine with water as to be able to withdraw the elements, hydrogen and oxygen, from compounds in which they do not exist as water; thus it chars wood and destroys the skin; it is therefore used in surgery as a caustic. They all, except magnesium fluoride, attract moisture from moist air, and deliquesce.

Uses.—Magnesium chloride is employed as a disinfectant, and is also used fraudulently for "weighting" flannel and cotton goods. Zinc chloride is also employed as a disinfectant under the name of "Burnett's Disinfecting Fluid." Cadmium bromide and iodide are used in photography.

Physical Properties.

	Mass of 1 c.c. solid.				Melting-point.				Boiling-point.			
	F.	Cl.	Br.	I.	F.	Cl.	Br.	I.	F.	Cl.	Br.	I.
Magnesium	2·86	2·18	?	?	?	708°	695°	?	?	?	?	?
Zinc	4·60	2·75	3·64	4·7	734°	262°	394°	446°	?	680°	{695°, 699°}	624°
Cadmium	6·00	3·62	4·8	5·7	520°	541°	570°	404°	?	{861°, 954°}	{806°, 812°}	{708°, 719°}

Another variety of cadmium iodide is known, with the specific gravity 4·6 or 4·7; it has a brownish colour, whereas the usual variety is white. It is converted at 50° into the usual modification.*

Heats of formation.—
$$Mg + Cl_2 = MgCl_2 + 1510K + Aq = +359K.$$
$$Zn + Cl_2 = ZnCl_2 + 972K + Aq = +156K.$$
$$Zn + Br_2 = ZnBr + 760K + Aq = +150K.$$
$$Zn + I_2 = ZnI_2 + 492K + Aq = +113K.$$
$$Cd + Cl_2 = CdCl_2 + 932K + Aq = + 30K.$$
$$Cd + Br_2 = CdBr_2 + 952K + Aq = + 4·4K.$$
$$Cd + I_2 = CdI_2 + 488K + Aq = - 9·6K.$$

Molecular weights.—The vapour-densities of zinc chloride and of cadmium chloride, bromide, and iodide nearly correspond to the formula $ZnCl_2$ and $CdCl_2$;† there is slight dissociation at the temperatures employed (898° and 1200°); cadmium iodide undergoes considerable dissociation at the higher temperature.

Double compounds.—1. With hydrogen halides.—

$$2ZnCl_2.HCl.2H_2O \text{ and } ZnCl_2.HCl.2H_2O$$

are produced in crystals by saturating a concentrated aqueous solution of zinc chloride with hydrogen chloride. They decompose on rise of temperature.‡

2. **With halides of the alkali metals.—**

(*a.*) *Fluorides.*—$MgF_2.NaF$; $ZnF_2.2KF$.

(*b.*) *Chlorides.*—$MgCl_2.NaCl.H_2O$; $MgCl_2.KCl.6H_2O$. $ZnCl_2.NH_4Cl$; $ZnCl_2$ $2NH_4Cl$; $ZnCl_2.3NH_4Cl$; $ZnCl_2.2KCl$; $ZnCl_2.2NaCl.3H_2O$; $2CdCl_2.2KCl.H_2O$; $CdCl_2.2NaCl.3H_2O$; $CdCl_2.2NH_4Cl.H_2O$; $CdCl_2.4NH_4Cl$; $CdCl_2.4KCl$.

(*c.*) *Bromides.*—$CdBr_2.KBr.H_2O$; $2CdBr_2.2NaBr.5H_2O$; $2CdBr_2.2NH_4Br.H_2O$; $CdBr_2.4KBr$; $CdBr_2.4NH_4Br$.

(*d.*) *Iodides.*—$ZnI_2.KI$; $ZnI_2.2NH_4I$. $CdI_2.KI.H_2O$; $CdI_2.2NaI.6H_2O$; $CdI_2.2KI.2H_2O$; $CdI_2.2NH_4I.2H_2O$.

3. **With calcium, strontium, and barium halides—**

$2CdCl_2.CaCl_2.7H_2O$; $CdCl_2.SrCl_2.7H_2O$;
$CdCl_2.BaCl_2.4H_2O$; $CdCl_2.2CaCl_2.2H_2O$.
$2ZnBr_2.BaBr_2$; $CdBr_2.BaBr_2.2H_2O$
$2ZnI_2.BaI_2$; $2CdI_2.ZnI_2.8H_2O$; $2CdI_2.BaI_2$.

4. **With each other**—$MgCl_2.ZnCl_2.6H_2O$; $MgCl_2.2CdCl_2.12H_2O$.

These are some of the numerous compounds which have been prepared. The ratios between the numbers of atoms of chlorine in the constituents appear to be:—2 : 1; 2 : 2; 2 : 3; 2 : 4; and 4 : 1.

---

* *Amer. Chem. Jour.*, 5, 235.

† *Brit. Assn.*, 1887, 668; *Berichte*, 12, 1195.

‡ *Compt. rend.*, 102, 1068.

As examples we may select:—2 : 1; $MgCl_2.NaCl$; $2CdCl_2.CaCl_2$;
2 : 2—$CdCl_2.2NaCl$; $CdBr_2.BaBr_2$; 2 : 3—$ZnCl_2.3NH_4Cl$;
2 : 4—$CdCl_2.4KCl$; $CdCl_2.2CaCl_2$; 4 : 1—$2ZnCl_2.HCl$.

These bodies are all prepared by direct addition, concentrated aqueous solutions of their constituents being added to one another.

Concluding remarks on these groups.—Molecular formulæ.—It has been seen that whereas the metals of the alkalies combine with the halogens in the ratio 1 : 1, as a rule, *e.g.*, **NaCl**, those of the beryllium and magnesium groups display the ratio 2 : 1, as for example, $CaCl_2$, $BeCl_2$. The inquiry may here be made: How is this known to be the case? To take a specific instance:—We know, from the densities of gaseous *HCl*, *HBr*, *KBr*, *RbI*, &c., that these compounds contain an atom of each element; the vapour-density of zinc chloride has been found to correspond to the molecular weight 136·37; now subtracting 35·46 × 2, corresponding to the weight of two atoms of chlorine, the remainder, 65·45, is the relative weight of an atom of zinc, *provided the compound contains only one atom of zinc.* But how is this known? Might not its formula be Zn_2Cl_2? In which case 65·45 would represent the relative weight of two atoms of zinc, and 32·72 that of one. And if such a question may be asked in the case of zinc, where we know the molecular weight of one of its compounds in the gaseous state, the uncertainty in the case of barium would appear to be much greater, for in this instance no compound has ever been gasified.

The answer to this question is to be found (1) in a study of the specific heats of these elements, and (2) in their position in the periodic table. These will now be considered in their order.

1. Specific Heats of Elements.

The data for these have been given in the tables of physical properties appended to the description of the groups of elements.

The **specific heat** of a body is defined as **the amount of heat required to raise the temperature through 1°, compared with the amount of heat required to raise the temperature of an equal weight of water through 1°.** Or, as water is chosen as unit of weight as well as of specific heat, specific heat may be defined as **the amount of heat required to raise the temperature of 1 gram of a body through 1°.** But the specific heat of water is not constant; more heat is required to raise a gram of water from 99° to 100°, than from 0° to 1°. Hence the unit is now generally accepted to be the hundredth part of the heat required to raise the temperature of 1 gram of water from 0° to 100°. This happens nearly to coincide with the value of 1 heat unit at the temperature 18°. Such a heat unit is termed a *calory*, and its abbreviated symbol is c. Where large amounts of heat are in question a unit of 100 calories is often used, and is represented by the letter K. This unit is convenient in expressing heat changes which take place during chemical action.

In 1819, a simple relation was discovered by Dulong and Petit to exist between the amount of heat required to raise the temperature of 1 gram of each of the following thirteen elements through 1°:—copper, gold, iron, lead, nickel, platinum, sulphur, tin, zinc, bismuth, cobalt, silver, and tellurium.

Dulong and Petit's law.—The specific heats of the elements are inversely proportional to their atomic weights, approximately, or

$$(\text{Sp. Ht.})_A \times (\text{At. Wt.})_A = (\text{Sp. Ht.})_B \times (\text{At. Wt.})_B.$$

Now the product of the specific heat of an element, or heat required to raise the temperature of 1 gram of the element through 1° into its atomic weight, is termed its atomic heat. For instance, the atomic weight of sodium is 23, and its specific heat 0·293; and the atomic weight of lithium is 7, and its specific heat 0·941. The product of the first pair, 23 × 0·293 = 6·74 calories, represents the amount of heat necessary to raise the temperature of 23 grams of sodium through 1°; and the product of the second pair, 7 × 0·941 = 6·59 calories, is similarly the amount of heat required to raise the temperature of 7 grams of lithium through 1°. But 23 and 7 are the relative weights of the atoms of sodium and lithium; and to raise these relative weights expressed in grams through 1° requires 6·74 and 6·59 calories respectively; these numbers are approximately equal. Hence the conclusion from this and similar instances, that **the atomic heats of the elements are approximately equal.**

This law is not without apparent exceptions, as, for example, in the cases of beryllium, boron, carbon, and silicon, but it holds closely enough to be a valuable guide in selecting the true atomic weights. It appears also to apply only to solids. As regards the real meaning of this law, we have at present no knowledge. We can form no probable conception of the change in the motion or position of the atoms in a molecule due to their rise of temperature; but it is a valuable empirical adjunct for the purpose mentioned.

The product of atomic weight and specific heat, in the instances given, is approximately 6·5; in other cases it falls as low as 5·5.

It may be stated then, that this product is approximately a constant, not differing much from the number 6. Hence 6/specific heat of any element should approximately equal its atomic weight; and conversely 6/atomic weight, should give an approximation to its specific heat.

As the atomic weight of hydrogen is 1, its atomic heat should be 6, and should be identical with its specific heat. Solid hydrogen, however, has never been prepared. But it forms a solid alloy with palladium; and as the specific heat of an alloy is the mean of those of its constituents, that of solid hydrogen has been indirectly determined. It has been found equal to 5·88, a sufficiently close approximation to 6.

To return now to the atomic weights of members of the beryllium and magnesium groups; the following table gives their atomic heats:—

Name.	Atomic Weight		Specific Heat.		Atomic Heat.
Beryllium....	9·1	×	0·6206 (at 500°)*	=	5·65
Calcium......	40·08	×	0·167	=	6·69
Strontium....	87·5	×	?	=	?
Barium......	137·00	×	?	=	?
Magnesium ...	24·30	×	0·250	=	6·07
Zinc.........	65·43	×	0·095	=	6·22
Cadmium	112·1	×	0·056	=	6·28

At 100° the specific heat of beryllium is 0·4702; its atomic heat is therefore 4·28. It was for long doubtful whether beryllium had not the atomic weight 13·65, *i.e.*, $3 \times \frac{9{\cdot}1}{2}$; the formula of its chloride would then have been $BeCl_3$, and its atomic heat $0{\cdot}4702 \times 13{\cdot}65 = 6{\cdot}42$, agreeing with those of many other elements; but its vapour-density decided the question. A substance of the formula $BeCl_3$ should have had the vapour-density $\{13{\cdot}65 + (3 \times 35{\cdot}46)\}2 = 60{\cdot}01$. Actual experiment gave 40·42 (see p. 122), hence its molecular weight is 80·84 $(9{\cdot}1 + (2 \times 35{\cdot}46) = 80{\cdot}02)$.†

The atomic weights of calcium, magnesium, zinc, and cadmium given in the table, correspond, it will be seen, with the usual atomic heat.

2. The similarity of the metals calcium, strontium, and barium, and of their compounds, lead to the inference that they belong to the same group of elements, hence they find their position in the

* See p. 33. † *Comptes rend.*, **98**, 988.

periodic table. The atomic weights are deduced from this similarity, and from their position in the table (see p. 22).

For these reasons it is concluded that the general formula of the halides of this group of elements is MX_2, where M stands for metal and X for halogen; that of the members of the lithium group is MX. Lithium and its congeners are termed *monad* or *monovalent* elements in these compounds; beryllium, magnesium, and elements of their groups, are termed *dyad* or *divalent* in their compounds. But it has been amply shown that *valency*, as the property of acting as a monad, a dyad, a triad element is termed, is not a constant quality of any element; nor in such compounds as KI_3, or in the double halides mentioned, can we tell how the atoms are held together, whether the metal attracts halogen, or halogen attracts halogen, or both attracts both. We are at present without any satisfactory theory to account for such compounds, and must, in the meantime, simply accept the fact of their existence.

The specific heats of some elements may be simply determined with fair approximation by the "method of mixture," and Dulong and Petit's law may be easily illustrated. A cylindrical can of thin sheet brass serves as a calorimeter (fig. 26). It should have a capacity of about 300 cubic centimetres. Having placed in it 200 cubic centimetres of water, the temperature of the water is accurately ascertained by a delicate thermometer, graduated in tenths of a degree. Three small hemispheres of zinc, tin, and lead, each weighing 100 grams, are suspended in a bath of boiling water by thin wires. The zinc is

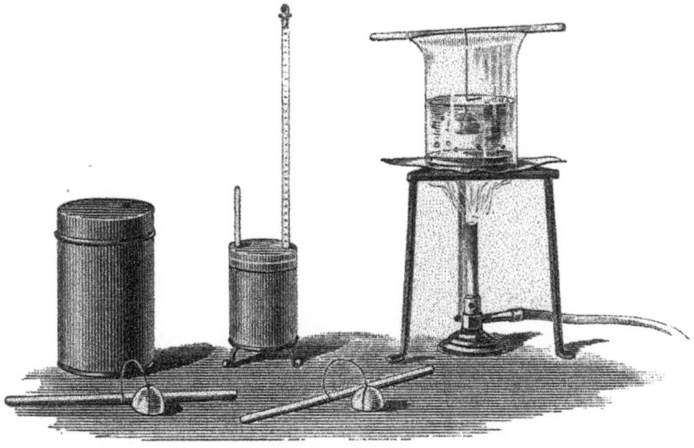

FIG. 26.

quickly lifted out and dropped into the calorimeter; the water is stirred with the thermometer or with a special stirrer, as shown in the figure, and its temperature ascertained. Similarly, the amount of heat given up to fresh supplies of cold water by the other two metals, tin and lead, is found. Their specific heats may be calculated as follows:—

Rise of temperature of the water × 200 = heat given up to the water by 100 grams of metal in cooling from 100° to the final temperature of the water. Hence, if $t - t'$ rise of temperature, then $(t - t')\ 200 = (100 - t)x$, where x = capacity for heat of the metal; and $x/100$ = specific heat of the metal.

This experimental illustration, rough as it is, yields fairly good results, probably because the errors neutralise each other in part. The sources of error are—(1) Hot water is carried over by the metal into the calorimeter; (2) heat is lost by the metal during its transit; (3) no allowance is made for the capacity for heat of the metal of the calorimeter; and (4) no correction is made for the loss of heat of the calorimeter by radiation.

CHAPTER X.

COMPOUNDS OF THE HALOGENS WITH BORON, SCANDIUM, YTTRIUM, LANTHANUM, AND YTTERBIUM; WITH ALUMINIUM, GALLIUM, INDIUM, AND THALLIUM; WITH CHROMIUM, IRON, MANGANESE, COBALT, AND NICKEL.—DOUBLE HALIDES OF ELEMENTS OF THESE GROUPS.

Boron, Scandium, Yttrium, Lanthanum, and Ytterbium Halides.

Of these elements boron is the only one the halides of which are well known.

Sources.—None of the haloid compounds of these elements exist in nature.

Preparation.—1. By direct union.—Boron burns when heated in chlorine gas, producing the chloride BCl_3; the bromide may also be prepared by passing bromine vapour through a tube in which amorphous boron is heated to redness. The iodide is unknown.

2. **By the simultaneous action of chlorine or bromine and carbon (charcoal) on the oxide at a bright red heat.**—The carbon withdraws the oxygen, producing carbon monoxide, while the halogen unites with the boron; thus:—

$$\mathbf{B_2O_3} + \mathbf{3C} + 3Cl_2 = 2BCl_3 + 3CO.$$

An intimate mixture of sugar-charcoal, oil, and boron oxide is made into balls, and ignited to carbonise the oil out of contact with the air. They are then heated to bright redness in an atmosphere of halogen.

Carbon monoxide is a gas, very difficult to condense; boron chloride and bromide are liquids at the ordinary temperature; hence by leading the products through a freezing-mixture, the halide condenses. The halides of the other elements may be similarly prepared; but as they are solids, difficult to volatilise, they remain mixed with the surplus carbon.

3. **By double decomposition.**—(*a.*) The action of the halogen acid on the oxides or hydroxides.—This is the usual method of

preparing boron fluoride. The hydrogen fluoride is prepared from calcium fluoride and sulphuric acid (see p. 108), and while being formed acts on boron oxide contained in the mixture.

The first action is $3\mathbf{CaF_2} + 3H_2SO_4 = 3\mathbf{CaSO_4} + 6\mathit{HF}$; and the second $\mathbf{B_2O_3} + 6\mathit{HF} = 2\mathit{BF_3} + 3H_2O$. The water produced would decompose the boron fluoride, were it not that it combines with the sulphuric acid (see p. 415), and it is thus withdrawn from the action. The other hydrogen halides have no action on boron trioxide. With other oxides of the group, and with the hydroxides, aqueous solutions of the halogen acids yield halides.

(*b.*) Boron chloride may be produced by heating together phosphorus pentachloride, PCl_5, and boron trioxide in sealed tubes to 150°. The equation $6\mathbf{PCl_5} + 5\mathbf{B_2O_3} = 3\mathbf{P_2O_5} + 10\mathit{BCl_3}$ expresses the change.

Properties.—Boron fluoride is a colourless gas very soluble in water (1059 volumes at 0°). Boron chloride and bromide are volatile colourless liquids, the former boiling at 18·23°, the latter at 90·5°; they react at once with water, forming the hydroxide and hydrogen halide, thus:—$\mathit{BCl_3} + 3H_2O = B(OH)_3 + 3\mathit{HCl}$. Boron fluoride has such a tendency to combine with water that it withdraws hydrogen and oxygen from carbon compounds containing them, liberating carbon, and in this respect resembling zinc chloride. It also reacts with water; the first stage of the reaction is $2\mathit{BF_3} + 3H_2O = B_2O_3.6HF$.* On heating the solution, $\mathit{BF_3}$ and $\mathit{H_2O}$ are evolved, and the compound $HBO_2.3HF$ named *fluoboric acid* remains (see p. 236). On dilution with water, boron hydroxide deposits and hydroborofluoric acid is formed, thus:—

$$4(HBO_2.3HF) = B(OH)_3 + 3HF.BF_3 + 5H_2O.\dagger$$

The halides of the other elements of this group are white crystalline substances soluble in water, and decomposed on evaporation with water. They are not easily volatile, hence they may be produced anhydrous by evaporation with ammonium chloride, as anhydrous magnesium chloride is prepared (see p. 123). Yttrium iodide is unstable in moist air.

Heat of formation.—$\mathbf{B} + \mathit{Cl_3} = \mathit{BCl_3} + 1040K$.

Double halides.—The double halides of boron fluoride only have been studied. It was mentioned above that on heating a solution of boron fluoride, some fluoride escapes, but some reacts with the water, giving $HF.BF_3$, named *hydroborofluoric acid.* It is also produced by dissolving boron oxide, $\mathbf{B_2O_3}$, in hydrofluoric acid. It is known only in aqueous solution, for on concentration

* Basarois, *Comptes rend.*, **78**, 1698.

† Considerable doubt exists regarding these changes (see p. 236).

hydrogen fluoride is evolved, while boron hydroxide, $B(OH)_3$, remains in solution, thus:—$HF.BF_3 + 3H_2O = B(OH)_3 + 4HF$. Compounds with other fluorides can also be produced by direct union of boron fluoride with the fluorides of these elements; but such compounds are also formed by the action of hydroborofluoric acid on the oxides, hydroxides, or carbonates of the metals. They are almost all soluble in water and crystalline. The potassium compound has the formula **$KF.BF_3$**; the barium compound **$BaF_2.2BF_3.H_2O$**. The zinc compound may be prepared by the action of the hydrogen compound on metallic zinc, when hydrogen is evolved, thus:—$2HF.BF_3.Aq + Zn = ZnF_2.2BF_3 + H_2$. These bodies are commonly termed salts of *hydroborofluoric acid* or *borofluorides*.

Aluminium, Gallium, Indium, and Thallium Halides.

Sources.—The only important compound found native is aluninium fluoride, which, in combination with sodium fluoride, forms the white crystalline mineral *cryolite*, **$3NaF.AlF_3$**.

Formation.—These elements combine with the halogens in several proportions, as seen in the following table:—

	Fluorine.	Chlorine.	Bromine.	Iodine.
Aluminium.	AlF_2*; AlF_3	— $AlCl_3$	— $AlBr_3$	— AlI_3.
Gallium ...	? GaF_3	$GaCl_2$; $GaCl_3$†	? $GaBr_3$	? GaI_3.
Indium	? InF_3	$InCl$; $InCl_2$; $InCl_3$	? $InBr_3$	? InI_3.
Thallium ..	TlF; TlF_3	$TlCl$; $TlCl_2$; $TlCl_3$	$TlBr$; $TlBr_2$; $TlBr_3$	TlI; TlI_3.

Preparation.—1. **By direct union.**—The compounds of the general formula MX_3 are formed in this way.

2. **By replacement.**—AlX_3, GaX_3, and InX_3, are produced by dissolving the respective metals in the haloid acid; hydrogen is evolved; thallium dissolves very slowly, being protected by a layer of sparingly soluble halide, forming a thallous salt, **TlX**. By heating indium in dry hydrogen chloride, however, **$InCl_2$** is produced.

3. The lower chlorides, **$GaCl_2$**, **$InCl_2$**, and **$InCl$**, have been produced by heating the higher chlorides with the respective metals.

4. **By double decomposition.**—(*a.*) *Solution of the respective oxides, hydroxides, or sulphides in the haloid acid.*

* Only known in the compound $2NaF.AlF_2$.

† *Comptes rend.*, **93**, 294 and 329.

Thus:—$\mathbf{Al_2O_3} + 6HCl.Aq = 2AlCl_3.Aq + 3H_2O$;
$\mathbf{Tl_2O} + 2HCl.Aq = 2TlCl.Aq + H_2O$;
$\mathbf{Tl_2O_3} + 6HCl.Aq = 2TlCl_3.Aq + 3H_2O$;
$\mathbf{In_2S_3} + 6HCl.Aq = 2InCl_3.Aq + 3H_2S$.

Thallous carbonate dissolves in haloid acids, giving thallous salts.

(*b.*) *By precipitation.*—The chloride, bromide, and iodide of thallium being nearly insoluble in water, may be prepared by treating a soluble compound, *e.g.*, the nitrate, $TlNO_3$, with a soluble halide; thus—

$$TlNO_3.Aq + KI.Aq \rightleftharpoons \mathbf{TlI} + KNO_3.Aq.$$

(*c.*) Aluminium chloride and bromide, like the corresponding halides of boron, may be produced by passing chlorine over a mixture of the oxide and charcoal heated to redness; or by passing the vapour of carbon tetrachloride, CCl_4, over red-hot alumina. The equations are:—

$$\mathbf{Al_2O_3} + 3\mathbf{C} + 3Cl_2 = 2AlCl_3 + 3CO; \text{ and}$$
$$2\mathbf{Al_2O_3} + 3CCl_4 = 4AlCl_3 + 3CO_2.$$

Properties.—MX_3.—These compounds, with the exception of $\mathbf{InI_3}$, which is yellow, $\mathbf{TlF_3}$, green (?), $\mathbf{TlBr_3}$, yellow, and $\mathbf{TlI_3}$, red, are colourless crystals; they are all soluble in water. They melt and sublime at comparatively low temperatures. They crystallise from water with water of crystallisation. Their solutions, when evaporated, decompose, halogen acid being liberated, and an oxyhalide being left. The anhydrous halides all attract atmospheric moisture.

MX_2.—Gallium and indium dichlorides are white; that of thallium pale yellow, as also its dibromide. They are attacked by water, indium and gallium dichlorides apparently decomposing into mono- and trichlorides, thus:—

$$2\mathbf{InCl_2} + Aq = \mathbf{InCl} + InCl_3.Aq.$$

The monochloride in contact with water deposits the metal, trichloride remaining in solution, thus:—

$$3\mathbf{InCl} + Aq = InCl_3.Aq + 2\mathbf{In}.$$

MX.—**InCl*** is reddish-yellow, and is decomposed by water (see above). **TlF, TlCl,** and **TlBr,** are white crystalline bodies; **TlI** is yellow. The fluoride is the most soluble, the iodide almost insoluble in cold water. They all crystallise from solution in hot water, and do not react with it on evaporation.

* *Chem. Soc.*, **53**, 820.

Molecular weights.*—The chloride of aluminium at temperatures between 218° and 440°, and at pressures varying from 300 to 760 mms. has the formula Al_2Cl_6, and similarly the bromide and iodide possess the respective formulæ Al_2Br_6 and Al_2I_6, as shown by their respective vapour-densities. As the temperature rises above 440° these molecules dissociate: thus $Al_2Cl_6 = 2AlCl_3$. The vapour-density therefore falls with rise of temperature, an ever-increasing number of simpler molecules being produced by the splitting up of the more complex ones; till at 800–900° the density reveals the fact that the gas consists wholly of molecules of the formula $AlCl_3$. At still higher temperatures chlorine gas is liberated, possibly owing to the formation of a lower chloride, possibly owing to the separation of aluminium. Gallium trichloride† at temperatures below 270°, and at atmospheric pressure, appears also to possess the formula Ga_2Cl_6; its density likewise decreases with rise of temperature and with fall of pressure, and at 440° and higher temperatures its density corresponds to the formula $GaCl_3$. On the other hand, indium trichloride does not gasify till it has nearly reached its temperature of complete dissociation; at 850° and upwards its formula is $InCl_3$. It is not known whether the other chlorides possess the formulæ M_2Cl_4, M_2Cl_2, or not; for at the temperature at which they gasify, they are already resolved into the simpler molecules, MCl_2 and MCl. It appears then that just as these halides form double compounds with the halides of other metals, so they form double compounds with themselves, acquiring thereby a double molecular formula. Thallous chloride has a vapour-density corresponding to the formula $TlCl$.

The atomic heats of these elements is normal. The results are:—

Aluminium.	Gallium.	Indium.	Thallium.
6·08	5·52	6·42	6·86

Hence the molecular formula of these compounds.

That of boron, it will be seen on reference to p. 37, increases rapidly with rise of temperature; at 233° it is 4·03; but at still higher temperatures it would doubtless become normal.

* *Annales* (3), **58**, 257.

† *Zeitschr. Phys. Chem.*, **1**, 460; and **2**, 659; *Comptes. rend.*, **106**, 1764, and **107**, 306; *Chem. Soc.*, **53**, 814.

Physical Properties.

	Mass of 1 c.c. solid.				Melting-point.				Boiling-point.			
	F_3.	Cl_3.	Br_3.	I_3.	F.	Cl.	Br.	I.	F.	Cl.	Br.	I.
Boron (liquid compounds)	?	1·35	2·69	—	?	?	?	—	?	18·2°	90·5°	—
Aluminium..	3·1	?	2·54	2·63	?	?*	93°	125°	?	*	264°	350°
Gallium	?	2·36 at 80°	?	?	?	75·5°	?	?	?	220°	?	?
Indium	?	?	?	?	?	*	?	?	?	*	?	?
Thallium,TlX	?	7·0	7·54	7·8	?	427'	458°	439°	719°	?	?	800°

$GaCl_2$: m.-p., 164°; b.-p., c. 535°.

Heats of formation:—

1. **Al** + 3*Cl* = $\mathbf{AlCl_3}$ + 1610K + Aq = 768K.
   Al + 3Br = $\mathbf{AlBr_3}$ + 1197K + Aq = 853K.
   Al + 3I = $\mathbf{AlI_3}$ + 704K + Aq = 890K.
2. **Tl** + Br = **TlBr** + 413K.
   Tl + **I** = **TlI** + 302K.

Double halides.—Of these, only the compounds of aluminium and thallium seem to have been prepared. They are all obtained by direct addition, sometimes, however, being prepared in presence of water, sometimes by fusion.

1. **Derivatives of** $\mathbf{MX_3}$.

$\mathbf{AlF_3.3NaF}$.	$\mathbf{AlF_3.2KF}$.	$\mathbf{AlCl_3.NaCl}$	Similar iodides are said to exist.
$\mathbf{AlF_3.3KF}$.	$\mathbf{AlF_3.2NaF}$.	$\mathbf{AlBr_3.KBr}$	
$\mathbf{TlCl_3.3NH_4Cl}$.	$\mathbf{TlCl_3.2KCl}$.	$\mathbf{TlBr_3.NH_4Br}$.	
$\mathbf{TlCl_3.3TlCl}$.	—	$\mathbf{TlCl_3.TlCl}$.	
$\mathbf{TlBr_3.3TlBr}$.	—	$\mathbf{TlBr_3.TlBr}$.	
		$\mathbf{TlI_3.KI}$.	

Besides these are known:—$\mathbf{TlI_3.5TlI}$; $\mathbf{2TlBr_3.3KBr}$; and its analogue, $\mathbf{2TlI_3.3KI}$; also $\mathbf{4AlF_3.MgF_2.NaF}$, a mineral named *ralstonite*. The most important of these is the mineral *cryolite*, $\mathbf{AlF_3.3NaF}$, which is mined at Evigtok, in West Greenland, where it forms a deposit 80 × 300 feet in depth and length. It is used as a source of fluorine, of pure alumina, and of caustic soda.

2. **Derivatives of** $\mathbf{MX_2}$.—The compound $AlF_2.2NaF$, belonging to this group, is an interesting one, inasmuch as it is the only one in which aluminium is combined with two atoms of a halogen, or, more comprehensively, the only one in which aluminium functions as a dyad (see p. 129). It has recently been prepared by heating cryolite with metallic aluminium to redness, in an iron

* Sublimes without fusing.

crucible in a current of hydrogen. It is a white insoluble substance, evolving hydrogen on treatment with hydrochloric acid.*

3. **Derivatives of MX.**—The only representative known is **TlF.HF**, which is produced by direct addition. It resembles its potassium analogue, **KF.HF** (see p. 119), in being decomposed by heat.

It has been shown that the compound TlI_3KI may equally well be produced from **TlI** and KI_3.† We cannot therefore regard it as necessarily composed of thall*ic* iodide and potassium iodide; it may equally well be viewed as a compound of potassium triiodide, KI_3, and thall*ous* iodide, **TlI**. In fact we have to confess our complete ignorance of the manner of combination of the atoms in the molecule. It might therefore be better to write the formula $KTlI_4$, thus committing ourselves to neither view; but simplicity of arrangement is certainly aided by the method adopted.

Chromium, Iron, Manganese, Cobalt, and Nickel Halides.

Sources.—None of these compounds is found native except ferric chloride, Fe_2Cl_6, which sometimes occurs in the waters of volcanic districts.

These elements, generally speaking, combine with the halogens in two proportions, as shown in the following table:—

	Fluorine.	Chlorine.	Bromine.	Iodine.
Chromium ...	— CrF_3.	$CrCl_2$; $CrCl_3$.	$CrBr_2$; $CrBr_3$.	— CrI_3.
Iron	FeF_2; FeF_3.	$FeCl_2$; $FeCl_3$.	$FeBr_2$; $FeBr_3$.	FeI_2; FeI_3.
Manganese...	MnF_2; MnF_3. MnF_4.	$MnCl_2$; $MnCl_3$‡	$MnBr_2$; —	MnI_2; —
Cobalt......	CoF_2; —	$CoCl_2$; $CoCl_3$‡	$CoBr_2$; —	CoI_2; —
Nickel	NiF_2; —	$NiCl_2$; —	$NiBr_2$; —	NiI_2; —

Manganese forms a tetrachloride, stable in ethereal solution; chromium a hexafluoride, CrF_6.

Preparation.—1. **By direct union.**—Chromium and iron form dihalides, if the halogen be not in excess; and trihalides with excess of halogen; manganese, nickel, and cobalt, form only dihalides.

2. **By the action of the halogen acid on the metals with or without presence of water.**—In all cases the dihalide is formed, thus:—

$$Fe + 2HCl = FeCl_2 + H_2.$$

3. **By double decomposition.—The action of the halogen**

* *Chem. News*, **59**, 75.
† Johnson, *Chem. Soc.*, **33**, 183.
‡ Known only in solution.

acid on the oxide, hydroxide, sulphide, carbonate, sulphite, &c.—With oxides, sulphides, &c., in which the metal acts as a *dyad*, the dihalides are formed, thus:—

$$\mathbf{FeO} + 2HCl.Aq = FeCl_2.Aq + H_2O.$$
$$\mathbf{Mn(OH)_2} + 2HCl.Aq = MnCl_2.Aq + H_2O.$$
$$\mathbf{NiS} + 2HCl.Aq = NiCl_2.Aq + H_2S.$$
$$\mathbf{CoCO_3} + 2HCl.Aq = CoCl_2.Aq + \mathit{CO_2} + H_2O.$$

If the sesquioxide, dry or hydrated (hydroxide), be employed, the trihalides are produced when capable of existence; if not, the halogen is evolved, thus:—

$$\mathbf{Fe_2O_3} + 6HCl.Aq = 2FeCl_3.Aq + 3H_2O.$$
$$\mathbf{Cr(OH)_3}.Aq + 3HCl.Aq = CrCl_3.Aq + 3H_2O.$$
$$\mathbf{Ni_2O_3} + 6HCl.Aq = 2NiCl_2.Aq + 3H_2O + \mathit{Cl_2}.$$
$$\mathbf{Mn_2O_3} + 6HBr.Aq = 2MnBr_2.Aq + 3H_2O + Br_2.$$

With a higher oxide of the metal, or a double oxide containing such a higher oxide, the highest halide capable of existence at the temperature of action is produced, and the halogen is liberated: thus, if the solution be cold,

$$\mathbf{2MnO_2} + 4HCl.Aq = 2MnCl_3.Aq + 4H_2O + \mathit{Cl_2};\text{ but if hot,}$$
$$\mathbf{MnO_2} + 4HCl.Aq = MnCl_2.Aq + 2H_2O + \mathit{Cl_2}.$$

Similarly, $2CrO_3.Aq + 12HI.Aq = 2CrI_3.Aq + 6H_2O + 3\mathbf{I_2}$; and

$$K_2Cr_2O_7.Aq (= K_2O.2CrO_3) + 14HCl.Aq = 2KCl.Aq + 2CrCl_3.Aq + 7H_2O + 3\mathit{Cl_2}.$$

Also, $2KMnO_4.Aq (= K_2O.Mn_2O_7) + 16HCl.Aq = 2KCl.Aq + 2MnCl_2.Aq + 8H_2O + 5\mathit{Cl_2}$.

These last methods, involving the use of higher oxides, are the practical methods of preparing the elements—chlorine, bromine, and iodine (see p. 75). Fluorine cannot be thus liberated. Hydrogen fluoride either is without action, or it liberates oxygen as ozone, or (in the case of manganese dioxide or of chromium trioxide), higher fluorides are produced (see p. 142).

4. With chromium alone, the action of **hydrogen at a low red heat** on the trihalide produces the dihalide, thus:—

$$\mathbf{2CrCl_3} + \mathit{H_2} = \mathbf{2CrCl_2} + 2\mathit{HCl}.$$

This is best carried out practically by heating a mixture of chromic chloride and ammonium chloride to bright redness in a porcelain retort.

On treatment with hydrogen at a red heat, the other chlorides are reduced to metal; as is that of chromium at a high temperature.

5. **By the action of the halogen on a red-hot mixture of the oxide and carbon.**—This method is specially used for preparing the trihalides of chromium, for the metal is difficult to prepare. The halide volatilises, and is thus separated from the excess of carbon.

Properties.—Dihalides.—These compounds, if anhydrous, crystallise in lustrous scales. Their colours are:—

	Chromium.	Iron.	Manganese.	Nickel.	Cobalt.
Fluoride..	?	White	?	?	?
Chloride..	White	White	Rose	Yellow	Blue.
Bromide .	White	Yellowish	Pale-red	Yellow	Green.
Iodide ...	?	Grey	White?	Dark, metallic	Black, lustrous.

They are all deliquescent, and dissolve in water, heat being evolved by the union. They also dissolve in alcohol. They crystallise from such solutions, with more or less water of crystallisation. They cannot be dried, for they react with water, giving oxyhalides. The colours of these compounds with water are:—

	Chromium.	Iron.	Manganese.	Nickel.	Cobalt.
Fluoride........	?	Colourless	Amethyst	Green	Rose.
Chloride........	Blue	Blue-green	Rose	Green	Pink.
Bromide..	Blue	Green	Red	Green	Red.
Iodide	?	Green	White	Green	Green.

Manganous fluoride is insoluble in water, but dissolves in aqueous hydrofluoric acid, doubtless forming a double fluoride. Almost all these compounds are soluble in alcohol; manganous chloride dissolves with a green colour. The halides of nickel and of cobalt undergo a curious change on concentration, or on addition of halogen acid; those of nickel turn yellow; those of cobalt blue, or green. This is probably due to the formation of the anhydrous chloride. The solutions are used as "sympathetic inks."

When the paper on which they are traced as ink is warmed, a change of colour takes place. A very curious effect may be produced by combination of ordinary water-colours with such sympathetic inks; a landscape, cleverly painted, may be made to show a transition from a winter to a summer scene when held before the fire.

The chromous and ferrous halides, on exposure to air, combine with its oxygen, forming chromic or ferric oxyhalides (see p. 257). Their solutions, especially those of the chromium halides, rapidly absorb oxygen; the oxidation being accompanied by a change of colour—to green, in the case of chromium, and to brown-yellow, in the case of iron. Such substances are said to have power of "reduction," meaning that they tend to absorb oxygen from bodies capable of parting with it, they themselves being "oxidised." In presence of halogen acid, such a reaction as this occurs:—$2FeCl_2 + 2HCl + O = 2FeCl_3 + H_2O$; the oxygen being derived from the air, or from any substance capable of yielding it. Hence, chromous and ferrous halides are converted into chromic or ferric halides, by the action of the halogen in presence of water.

Physical Properties.

	Mass of 1 c.c.				Melting-points.	Boiling-points.
	F.	Cl.	Br.	I.		
Chromium..	?	2·75	?	?	Unknown.	Unknown.
Iron	?	2·53	?	?		
Manganese..	?	2·48	?	?		
Cobalt	?	2·94	?	?		
Nickel	2·86	2·56	?	?		

Hydrated:—$\mathbf{NiCl_2.4H_2O}$, 2·01; $\mathbf{FeCl_2.4H_2O}$, 1·93; $\mathbf{CoCl_2.6H_2O}$, 1·84.

Heats of formation:—

$$\mathbf{Cr} + Cl_2 = \mathbf{CrCl_2} + \ ? \ + Aq = ?$$
$$\mathbf{Fe} + Cl_2 = \mathbf{FeCl_2} + 821K + Aq = 179K.$$
$$\mathbf{Mn} + Cl_2 = \mathbf{MnCl_2} + 1120K + Aq = 160K.$$
$$\mathbf{Ni} + Cl_2 = \mathbf{NiCl_2} + 745K + Aq = 192K.$$
$$\mathbf{Co} + Cl_2 = \mathbf{CoCl_2} + 765K + Aq = 183K.$$

Double compounds of the dihalides.—One hydrochloride is known, viz., $\mathbf{2HCl.3CrCl_2.13H_2O}$; and crystals, too unstable to be collected, have also been obtained by passing hydrogen chloride into a cold solution of cobaltous chloride. The other double salts may be divided into two groups, of which instances are $\mathbf{FeF_2.2KF}$, $\mathbf{FeCl_2.2KCl.2H_2O}$, $\mathbf{MnCl_2.2NH_4Cl}$; also $\mathbf{NiCl_2.NH_4Cl}$, and $\mathbf{MnCl_2.NH_4Cl}$.

Not many such compounds have been prepared.

Trihalides.—The anhydrous trihalides also form lustrous scales. Their colours are—

	Chromium.	Iron.
Fluoride..........	Dark green	Pale yellow.
Chloride..........	Pale violet	Black.
Bromide	Dark olive green	Black. ?
Iodide	?	Black.

Chromic chloride, after sublimation, is insoluble in cold water, but dissolves after long boiling. If prepared by drying the hydrated chloride in a current of hydrogen chloride, it is soluble; as soon as it has been sublimed, it is insoluble. The presence of a trace of chromous chloride causes the insoluble variety to dissolve at once. The other halides are deliquescent, and readily soluble in water. They also, like the dihalides, react with water, forming oxyhalides (see p. 257).

The trihalides of manganese and cobalt are unknown in the anhydrous state.

The aqueous solutions have different colours, owing, no doubt, to the presence in solution of a compound with water. They are—

	Chromium.	Iron.	Manganese.	Cobalt.
Fluoride	Green	Colourless	Ruby	?
Chloride	Green	Yellow	Brown-yellow	Brown
Bromide	Green	Brown-red	?	?
Iodide	Green	Brown	?	?

Chromic chloride exists in two modifications, green and violet. The green solution has possibly a more complex molecule than the violet one. The violet modification is produced from the violet sulphate (see p. 426) by double decomposition with barium chloride, thus, $Cr_2 3SO_4.Aq + 3BaCl_2.Aq = 2CrCl_3.Aq + 3\mathbf{BaSO_4}$; or by dissolving the grey modification of the hydroxide (see p. 252) in hydrochloric acid. These chlorides probably all react with water, giving oxychlorides. That of manganese, indeed, if much water be added, gives a precipitate of sesquioxide, thus:—

$$2MnCl_3.Aq + 3H_2O = \mathbf{Mn_2O_3.Aq} + 6HCl.Aq.$$

Manganic fluoride, when heated with water, gives off oxygen, and hydrogen fluoride, thus: $2MnF_3.Aq + H_2O = 2MnF_2.Aq + 2\mathit{HF} + \mathit{O}_2$. Manganese and cobalt trichlorides are very unstable, evolving chlorine at the ordinary temperature, thus: $2MnCl_3.Aq = 2MnCl_2.Aq + \mathit{Cl}_2$. Ferric chloride is more stable, but it may be reduced or deprived of chlorine by means of *nascent* hydrogen, *i.e.*, hydrogen in process of formation. Hydrogen gas may be passed through a solution of ferric chloride without action; but if the hydrogen be prepared in a solution of ferric chloride by the action of zinc and hydrochloric acid for example (see p. 27), the ferric chloride is changed to ferrous chloride, thus:— $FeCl_3.Aq + H = FeCl_2.Aq + HCl.Aq$. It is supposed, with great probability, that the hydrogen is liberated in the atomic condition. In presence of ferric chloride it unites with chlorine; but if no reducible substance is present, it combines with itself to form molecular hydrogen, H_2, which is then without action. Chromic chloride cannot be easily reduced in aqueous solution.

Physical Properties.

	Mass of 1 c.c. solid.				Melting-point.				Boiling-point.			
	F.	Cl.	Br.	I.	F.	Cl.	Br.	I.	F.	Cl.	Br.	I.
Chromium	?	2·76	?	?	Unknown.				Unknown.			
Iron..........	?	2·80	?	?								

Heat of formation :—

$$\mathbf{Fe} + Cl_3 = \mathbf{FeCl_3} + 961K; + Aq = FeCl_3.Aq + 633K.$$

Double compounds of the trihalides.—These are made by direct addition and belong to the following four types :—

1. $\mathbf{CrCl_3.KCl}$; $\mathbf{CrBr_3.KBr}$; $\mathbf{CrI_3.KI}$; $\mathbf{FeF_3.KF}$.

These are stable in presence of excess of the hydrogen-halide, but decompose with water.

2. $\mathbf{CrF_3.2KF}$; $\mathbf{FeF_3.2KF}$; $\mathbf{FeCl_3.2KCl}$; $\mathbf{FeCl_3.2NH_4Cl}$; $\mathbf{FeCl_3.MgCl_2}$; $\mathbf{MnF_3.2KF}$; $\mathbf{MnF_3.2NH_4F}$; $\mathbf{MnF_3.2NaF}$; $\mathbf{MnF_3.2AgF}$.
3. $\mathbf{CrF_3.3KF}$.
4. $\mathbf{2FeI_3.FeI_2}$; $\mathbf{2MnF_3.MnF_2}$.

The green modifications of chromic halides do not form double compounds. They are possibly combinations of molecules of the chromium halides with each other.

Higher halides.—Manganese tetrafluoride, MnF_4, is produced by treating manganese dioxide with aqueous hydrogen fluoride, thus :—

$$\mathbf{MnO_2} + 4HF.Aq = MnF_4.Aq + 2H_2O.$$

It is soluble in alcohol and in ether. Its aqueous solution, when warmed, decomposes, depositing the dioxide, $\mathbf{MnO_2.Aq}$. On addition of a solution of potassium fluoride it forms the double compound, $\mathbf{2KF.MnF_4}$, as a rose-coloured precipitate.

Manganese dioxide, suspended in ether, and saturated with hydrogen chloride, gives a green solution of $MnCl_4$.

Chromium hexafluoride, CrF_6, is produced by the action of hydrogen fluoride on chromium trioxide, CrO_3, in presence of anhydro-sulphuric acid to absorb the resulting water, thus :—

$$CrO_3 + 6HF + 3H_2S_2O_7 = CrF_6 + 6H_2SO_4.$$

It is a fuming volatile liquid, of a blood-red colour, which attacks silicon oxide, and hence cannot be kept in glass vessels.*

General remarks.—The elements of this group combine with

* This substance is also said to be an oxyfluoride of the formula CrO_2F_2 (*Gazzetta chimica italiana*, **16**, 218).

halogens in four different proportions, thus : MX_2, MX_3, MX_4, and MX_6. The higher members are most stable with chromium, and the lower ones most stable with nickel. The molecular formulæ of these bodies have given rise to much dispute. Chromium dichloride appears to exist partly as $CrCl_2$, partly as Cr_2Cl_4, in the gaseous state at 1600°; at 1400–1500°, ferrous chloride possesses the simpler formulæ, $FeCl_2$. Chromic chloride, above its volatilising-point, about 1060°, has the formula, $CrCl_3$; ferric chloride, at temperatures below 620°, is Fe_2Cl_6;* but as temperature rises, these complex molecules dissociate, and at 750° and upwards, its density shows it to have the formula, $FeCl_3$.† The molecular weights of the double compounds of these halides are unknown, but it appears probable that they possess the simpler formulæ given them.

The formulæ of these compounds are deduced—

1. From the simplicity of the ratios of metal and halogen :—viz., 1 : 2; 1 : 3; 1 : 4; and 1 : 6.
2. From the vapour-densities.
3. From the atomic heat of the metals. These are :—

Cr.	Fe.	Mn.	Ni.	Co.
?	6·27	6·69	6·43	6·31

* *Comptes rend.*, **107**, 301.

† *Zeitschr. Phys. Chem.*, **2**, 659; *Chem. Soc.*, **53**, 814.

CHAPTER XI.

COMPOUNDS OF THE HALOGENS WITH CARBON, TITANIUM, ZIRCONIUM, CERIUM, AND THORIUM; WITH SILICON, GERMANIUM (TERBIUM), TIN, AND LEAD.—DOUBLE HALIDES OF ELEMENTS OF THESE GROUPS. —PROOF OF THEIR MOLECULAR FORMULÆ.

Carbon, Titanium, Zirconium, Cerium, and Thorium Halides.

The halides of carbon differ from those of the remaining elements of this group, in being more numerous, and in being insoluble in water. It appears advisable, in the present state of our knowledge, to include cerium in this group, although its halides do not closely resemble those of the other elements of the group.

Sources.—None of these halides occur native, except *fluocerite*, to which Berzelius gave the formula $\mathbf{CeF_3}$, and *tysonite*, $4\mathbf{CeF_3}$, $3\mathbf{LaF_3}$.

These elements form the following compounds with the halogens:—

	Fluorine.	Chlorine.	Bromine.	Iodine.
Carbon....	CF_4	CCl_4.$\mathbf{C_2Cl_6}$; C_2Cl_4, &c.	CBr_4; C_2Br_6; C_2Br_4.	CI_4.
Titanium..	$\mathbf{TiF_3}$; TiF_4	$\mathbf{TiCl_2}$; $\mathbf{Ti_2Cl_6}$; $TiCl_4$	$TiBr_4$	$\mathbf{TiI_4}$.
Zirconium.	$\mathbf{ZrF_4}$	$\mathbf{ZrCl_4}$	$ZrBr_4$*	?
Cerium....	CeF_3;* CeF_4*	$\mathbf{CeCl_3}$	$CeBr_3$*	CeI_3*
Thorium ..	$\mathbf{ThF_4}$	$\mathbf{ThCl_4}$	$ThBr_4$*	ThI_4*.

Preparation.—1. By direct union.—Carbon does not combine directly with halogens, except with fluorine. The other elements are converted into those compounds which contain the largest amount of halogen.

2. **By the action of the halogen on a red-hot mixture of the oxide with charcoal.**—By this means, $TiCl_4$, $TiBr_4$, $\mathbf{ZrCl_4}$, and $\mathbf{ThCl_4}$ have been prepared. The preparation of chloride of titanium may serve as a type of the rest:—

$$\mathbf{TiO_2} + 2\mathbf{C} + 2Cl_2 = TiCl_4 + 2CO.\dagger$$

* These have been obtained only in combination with water.

† *Chem. Soc.*, **47**, 119; *Comptes rend.*, **104**, 111; **106**, 1074. Carbon tetrachloride may be substituted for free carbon and free chlorine.

$CeCl_3$ has also been prepared by passing a mixture of carbon monoxide, CO, and chlorine over the ignited oxide; and $TiCl_4$, by the action of CCl_4 on ignited **TiO_2**.

3. **By the action of the halogen on the hydride or sulphide of the element.**—This is the method by which carbon tetrachloride, CCl_4, is commercially prepared. The disulphide (see p. 282), mixed with chlorine, is passed through a tube filled with pumice-stone and heated to redness. The chlorine combines with both carbon and sulphur, thus:—

$$CS_2 + 3Cl_2 = CCl_4 + S_2Cl_2.$$

The chloride of sulphur is afterwards decomposed by the action of lime-water (see p. 167), and the carbon tetrachloride purified by distillation.

Methane or marsh gas (hydrogen carbide), CH_4 (see p. 560), is also converted by the prolonged action of chlorine into the tetrachloride, thus:—

$$CH_4 + 4Cl_2 = CCl_4 + 4HCl.$$

There are, however, three intermediate stages, CH_3Cl, CH_2Cl_2, and $CHCl_3$.

Similarly, C_2H_6 can be converted into **C_2Cl_6**, through the following stages:—

$$C_2H_5Cl;\ C_2H_4Cl_2;\ C_2H_3Cl_3;\ C_2H_2Cl_4;\ C_2HCl_5, \text{ and } \mathbf{C_2Cl_6}.$$

4. **By the action of the hydrogen halide on the element.**—By this method $TiCl_3$, ZrF_4, CeF_3, $CeCl_3$, $CeBr_3$, CeI_3, and $ThCl_4$, have been produced in solution. Hydrogen is evolved.

5. **By the action of heat on CCl_4** other chlorides are produced, thus:—$2CCl_4 = C_2Cl_6 + Cl_2$; $2CCl_4 = C_2Cl_4 + 2Cl_2$; $6CCl_4 = C_6Cl_6 + 12Cl_2$. Special names are given to these bodies, viz., CCl_4, tetrachloromethane; **C_2Cl_6**, hexachlorethane; C_2Cl_4, tetrachlorethylene; **C_6Cl_6**, hexachlorobenzene.

6. **By the action of hydrogen at a red heat** on titanium tetrachloride or tetrafluoride they yield the trifluoride or trichloride. The dichloride is produced by the further action of hydrogen on the trichloride.

7. **Double decomposition.** — (*a.*) **The action of the hydrogen halide on the oxide or hydroxide of the element.**—All the fluorides, except that of carbon, have been thus prepared in solution; also solutions of $ZrCl_4$, $ZrBr_4$, $CeCl_3$, $CeBr_3$, CeI_3, $ThCl_4$, $ThBr_4$, and ThI_4. These substances, in solution, react with water on evaporation. Cerium chloride has been dried in the same manner as magnesium chloride, viz., by preparing the double salt

with ammonium chloride, and, after drying it, igniting it to remove ammonium chloride; also by passing a mixture of chlorine and carbon monoxide over the sesquioxide at a red heat. It is probable that the others could be obtained anhydrous in a similar manner.

(*b.*) This process is applied to the preparation of carbon bromide and iodide from the tetrachloride. A mixture of aluminium bromide or iodide and carbon tetrachloride, all diluted with carbon disulphide, yields carbon tetrabromide or iodide on heating; carbon tetrafluoride, CF_4, is produced by heating silver fluoride, AgF, in a sealed tube with carbon tetrachloride. Cerous fluoride, $\mathbf{CeF_3}$, which is an insoluble white substance, is also prepared by this general method by the interaction between solutions of sodium fluoride and cerium chloride, thus: $CeCl_3.Aq + 3NaF.Aq = \mathbf{2CeF_3.H_2O} + 3NaCl.Aq$.

Properties.—The tetrahalides are all volatile at comparatively low temperatures. Carbon tetrafluoride is a gas; carbon tetrachloride, bromide, and iodide, titanium tetrachloride, and tetrachlorethylene are colourless liquids; hexachlorethane, zirconium chloride, cerium trichloride, and thorium chloride are colourless solids, which can be sublimed. Titanium dichloride is a black powder,* which rapidly decomposes water, with evolution of hydrogen, combining with the oxygen to form an oxychloride. Titanium trifluoride and trichloride consist of violet scales, soluble in water with a violet colour. Titanium tetrabromide is a red liquid; and the tetriodide forms brown needle-shaped crystals. Ceric fluoride is not known in the anhydrous state. Combined with water as $\mathbf{CeF_4.H_2O}$, it is a brown insoluble powder, produced by treating the hydrated dioxide with aqueous hydrofluoric acid. It is doubtful whether the substance described as thorium fluoride is not in reality an oxyfluoride, $\mathbf{ThOF_2}$.

Carbon tetriodide decomposes when heated, or when exposed to air. With the exception of the carbon compounds, cerium tetrafluoride, and possibly thorium fluoride, these substances are deliquescent, and soluble in water, probably reacting with it to form oxyhalides; this change certainly takes place on evaporation, in some cases an oxyhalide, in others the oxide, being produced. Carbon tetrabromide occurs as an impurity in commercial bromine.

* Friedel and Guerin, *Annales* (5), 7; 24.

Physical Properties of Bodies of the Formula MX_4.

	Mass of 1 c.c. solid or liquid.				Melting-points.				Boiling-points.			
	F.	Cl.	Br.	I.	F.	Cl.	Br.	I.	F.	Cl.	Br.	I.
Carbon...	—	1·632 at 0°	3·42 at 14°	4·34 at 20°	?	?	91°	100°*	?	76·7°	189·5°	—
Titanium.	?	1·761 at 0°	2·6	?	?	?	39°	150°	?	136·4°	230°	360°*
Zirconium.	?	?	?	?	?	?	?	—	white heat	?	?	?
Cerium..	?	—	—	—	?	?	?	?	?	?	?	?
Thorium.	?	?	?	?	?	?	?	?	?	?	?	?

Of the other halides:—

	Mass of 1 c.c.	Melting-point.	Boiling-point.
C_2Cl_6	1·62	187°	187°
C_2Br_6	?	170°	—
C_2Cl_4	1·65 at 0°	−18°	121°
C_2Br_4	?	50°	decomposed
TiF_3	?	?	—
Ti_2Cl_6	?	?	—
$CeCl_3$	?	not at bright redness	—

Heats of formation.—The following only have been determined:—

$$C + 2Cl_2 = CCl_4 + 210K.$$
$$2C + 2Cl_2 = C_2Cl_4 - 12K.$$

The vapour-densities of many of these compounds have been determined, and it may be safely concluded that, in the gaseous state, most of them possess the molecular formulæ given above.

Double halides.—These are for the most part produced by mixing solutions of the two halides and crystallisation. Those of carbon are produced by substitution of chlorine for bromine, or by addition of bromine to a chloride (*e.g.*, $C_2Cl_4 + Br_2 = C_2Cl_4Br_2$), or of chlorine to a bromide.

Carbon compounds. CCl_3Br; a liquid boiling at 104·3°. CCl_2Br_2 boils at a higher temperature. $C_2Cl_4Br_2$ exists in two forms, isomeric with each other, one produced by direct addition of bromine to C_2Cl_4; the other by the action of bromine on C_2HCl_5. There are also known:—$C_2Br_4Cl_2$; C_2Br_3Cl; and $C_2Br_2Cl_2$. These bodies have vapour-densities corresponding with the formulæ given.

The other halides combine in varying amount with halides of other elements. As instances, the following compounds may be given:—

8 : 1.—$2ThCl_4.KCl.18H_2O$.

6 : 1.—$3TiCl_4.2PH_4Cl$.

4 : 1.—$ZrF_4.KF$; $ThF_4.KF$.

4 : 2.—$TiF_4.2HF$; $TiF_4.2KF$; $TiF_4.2NH_4F$; $TiF_4.CaF_2$; $TiF_4.CaF_2$; $TiF_4.NiF_2$; $ZrF_4.2KF$; $ZrF_4.MgF_2$; $ZrF_4.MnF_2$; $ZrCl_4.2NaCl$; $ThF_4.2KF$.

8 : 3.—$2CeF_4.3KF$.

* Melts with decomposition.

4 : 3.—$TiCl_4.3NH_4Cl$; $ZrF_4.3KF$; $2ZrF_4.3CuF_2$.
4 : 4.—$ZrF_4.2ZnF_2$; $ZrF_4.2CdF_2$; $ZrF_4.2MnF_2$; $ZrF_4.2NiF_2$; $2ZrF_4.2KF.NiF_2$.
4 : 6.—$TiF_4.2FeF_3$; $TiCl_4.6NH_4Cl$.
4 : 8.—$ThCl_4.8NH_4Cl$.
3 : 3.—$TiF_3.3NH_4F$.

These halides are able to combine with others in many proportions. The products are crystalline substances often combined with water, sometimes anhydrous. As regards their molecular weights, nothing is known; hence the simplest possible formulæ have been assigned to them.

Halides of Silicon, Germanium, Tin, Terbium, and Lead.

It has been already remarked as doubtful whether terbium belongs to this group of elements. These bodies, like those of the last group, show a decrease of volatility with increase of the atomic weight of the metallic element.

Sources.—The only native halide is lead chloride, $PbCl_2$, which was found in the crater of Vesuvius, after the eruption of 1822. A chloride and carbonate of lead also occurs native, though rarely, as *corneous lead;* its formula is $PbCO_3.PbCl_2$.

The following compounds are known:—

	Fluorine.	Chlorine.	Bromine.	Iodine.
Silicon	Si_2F_6; SiF_4.	Si_2Cl_4; Si_2Cl_6; $SiCl_4$.	Si_2Br_6; $SiBr_4$.	SiI_2; Si_2I_6; SiI_4.
Germanium	? GeF_4.	$GeCl_2$? $GeCl_4$.	?	GeI_4.
Tin	SnF_2; SnF_4.*	$SnCl_2$; $SnCl_4$.	$SnBr_2$; $SnBr_4$.	SnI_2; SnI_4.
Terbium	—	$TbCl_3$?*	—	—
Lead	PbF_2.	$PbCl_2$; $PbCl_4$?*	$PbBr_2$.	PbI_2.

Preparation.—1. By direct union.—These elements readily combine with the halogens, when they are heated together, forming the compounds containing the greatest amount of halogen.

Silicon takes fire in fluorine gas, burning to silicon fluoride.

This is the only method of preparing silicon tetriodide, SiI_4.

2. **By the action of the halogen on a red-hot mixture of the oxide with charcoal** (see p. 131).—This is the most convenient method of preparing silicon tetrachloride and tetrabromide. It is necessary to take the utmost precaution to exclude moisture by scrupulously drying the halogen; for the chloride and bromide are instantly decomposed by water. The silicon chloride or

* Not known in the anhydrous state.

bromide is condensed in a U-tube, cooled by a freezing mixture. The equation is:—$\mathbf{SiO_2} + 2\mathbf{C} + 2Cl_2 = SiCl_4 + 2CO$.

3. **By the action of the hydrogen halide on the element.**—By this means germanium fluoride and tin dichloride, bromide, and iodide may be conveniently prepared. Silicon fluoride may also be formed thus. Hydrogen gas is in every case evolved. It is believed that hydrogen chloride, at a red heat, converts germanium into the dichloride, $\mathbf{GeCl_2}$.

The usual method of preparing stannic chloride, which bears a close analogy to the action of a haloid acid on the element, is by distilling a mixture of granulated tin with mercuric chloride. The stannic chloride distils over, leaving the mercury in combination with the excess of tin, thus:—

$$\mathbf{2HgCl_2} + \mathbf{Sn} = 2Hg + SnCl_4.$$

4. **By double decomposition.**—(*a*.) This is the usual and easiest method of preparing the halides of lead, a solution of the nitrate or the acetate of lead being treated with a solution of any soluble halide, for example, with the nitrate, $Pb(NO_3)_2.Aq + 2KF.Aq = \mathbf{PbF_2} + 2KNO_3.Aq$; and with the acetate, $Pb(C_2H_3O_2)_2.Aq + 2HCl.Aq = \mathbf{PbCl_2} + 2C_2H_4O_2.Aq$.

(*b*.) **The action of the hydrogen halide on the oxide or hydroxide of the element.**—Silicon tetrafluoride, the halides of tin, and terbium chloride have been thus produced. The oxides of lead are attacked superficially by the halogen acids; but, the halides of lead being sparingly soluble, a coating of halide is formed, which renders the action slow. By alternately boiling lead oxide with the halogen acid, and with water, in order to dissolve this coating, complete conversion into halide may be accomplished.

Lead dioxide, thus treated with solutions of hydrogen chloride, bromide, or iodide, undergoes the following reactions, half the halogen being liberated:—

$$\mathbf{PbO_2} + 4HCl.Aq = \mathbf{PbCl_2} + 2H_2O + Cl_2 + Aq.$$

Hydrogen fluoride is without action on lead dioxide.

5. **By the action of the element at a red heat on the tetrahalide** the disilicon hexahalide has been prepared, thus:—

$$6SiCl_4 + 2Si = 4Si_2Cl_6.$$

As examples of these methods of preparation, the following instances may be chosen:—

1. Tin, melted in a deflagrating spoon, and plunged into a jar of chlorine gas, burns to the **tetrachloride**.

2. A mixture of silica and carbon, made into a paste with starch, and moulded into balls, and then strongly ignited, is heated in a porcelain tube by means of a Fletcher's tube-furnace, provided with a blast, in a current of chlorine, perfectly dried by passing through tubes filled with phosphorus pentoxide.

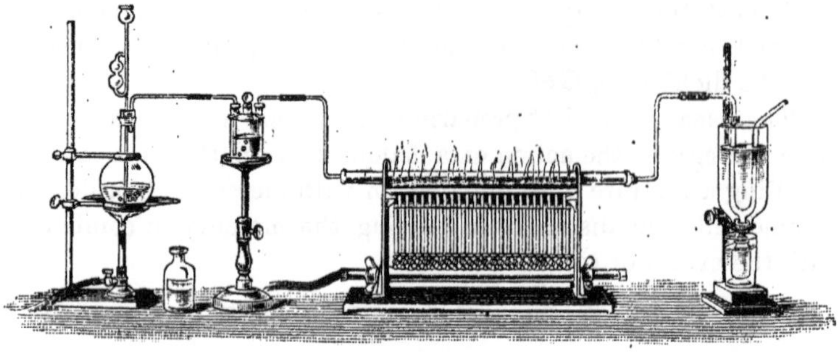

FIG. 27.

The **silicon chloride** produced must be condensed in a U-tube dipping in a freezing-mixture. The preparation is not easy, and is not well adapted for a lecture experiment.

3. Tin, granulated by pouring the melted metal into water, is boiled in a flask with strong hydrochloric acid, a few pieces of platinum-foil being added to form a galvanic couple and assist solution. It slowly dissolves, forming **stannous chloride.**

4. **Silicon tetrafluoride** may be prepared by heating in a glass flask a mixture of equal parts of fine sand and powdered fluorspar with excess of sulphuric acid. The hydrogen fluoride liberated attacks the sand, forming water, which unites with the sulphuric acid, and hence does not exercise a decomposing action on the silicon fluoride. The latter escapes as a colourless gas. It may be made to react with water, by causing the exit-tube to dip into a little mercury in a beaker, the beaker being filled up with water. The mercury is required, else the exit-tube would be soon blocked by deposition of silicon hydroxide (or silicic acid), resulting from the decomposition of the fluoride (see, p. 153).

The action of lead dioxide on the halides of hydrogen may be easily shown by warming in a test-tube a few grams with some hydriodic acid. Violet fumes of iodine escape, and the dioxide is converted into yellow iodide.

5. The formation of the **halides of lead** may be shown, as in 4*a*.

Properties.—Tetrahalides.—These compounds boil at comparatively low temperatures. Silicon tetrafluoride is a colourless gas at ordinary temperatures, the chloride and bromide are volatile liquids; and the iodide a white solid. Germanium chloride* is a colourless volatile liquid; and tin tetrachloride is also mobile and colourless, boiling at a somewhat higher temperature. Germanium

* *J. prakt. Chem.* (2), **34**, 177.

bromide and fluoride do not appear to have been prepared; the iodide is a yellow solid, giving a yellow vapour. It dissociates somewhat below 658°. Tin tetrafluoride has not been obtained in the anhydrous condition; the bromide forms volatile white crystals, and the iodide is yellowish-red, and also volatile. All these substances react with water, forming oxides, or oxyhalides; hence, being volatile, they all fume in the air. The vapour-densities of most of them have been determined, and correspond to the simple formulæ MX_4.

Bodies of the formula M_2X_6.—These are only known to exist as compounds of silicon. The fluoride, Si_2F_6 (?), is a white powder (probably an oxyfluoride). The iodide, Si_2I_6, produced by the action of finely-divided silver on the tetriodide, is separated from the excess of silver by solution in carbon disulphide, from which it deposits in colourless prisms. By warming it with mercuric chloride it is converted into the corresponding chloride, Si_2Cl_6, which is a colourless mobile liquid. The corresponding bromide is produced by shaking a solution of the iodide with bromine dissolved in carbon disulphide, and removing the iodine by agitation with mercury. It forms white crystals. A determination of the vapour-density of the chloride, Si_2Cl_6, showed it to possess the molecular weight corresponding to that formula.*

Dihalides.—Silicon dichloride is a liquid, which has not yet been obtained pure; the di-iodide remains as an orange-coloured residue on distillation of the compound Si_2I_6, which splits into the tetriodide and di-iodide, thus:—$Si_2I_6 = SiI_4 + SiI_2$. It is insoluble in all known solvents, and is decomposed by water.

Germanium dichloride is a colourless liquid. Its formula is as yet uncertain, and it may possibly be $GeHCl_3$, for it has not been analysed.

Tin difluoride has not been obtained anhydrous. It crystallises from water in small opaque prisms. The dichloride crystallised from water is known as "tin-salt." On evaporation of its solution, a portion reacts with water, forming oxychloride and hydrogen chloride. The excess of water evaporates along with the hydrogen chloride. On raising the temperature the undecomposed stannous chloride distils over, leaving the oxychloride. It forms a white lustrous crystalline mass. With a large quantity of water it gives a precipitate of oxychloride, $SnCl_2.SnO.2H_2O$. Its solution is a powerful reducing agent, for it tends to take chlorine from hydrogen chloride or oxygen from water, liberating hydrogen,

* *Annales* (4), 9, 5; 19, 334; 23, 430; 27, 416; (5), 19, 390.

when there is any substance present with which the hydrogen can combine. The dibromide is similar to the dichloride. The di-iodide is a dark-red mass; its iodine is replaced by oxygen when it is heated in air.

Lead difluoride, dichloride, and dibromide are white solids, sparingly soluble in boiling water and crystallising therefrom in long needles. The iodide is yellow and crystallises in golden-yellow spangles.

From the vapour-density of stannous chloride it would appear that these bodies in the state of gas have, at temperatures not far removed above their boiling-points, the double formula, *e.g.*, Sn_2Cl_4*; but that, as the temperature rises, the complex molecule dissociates into two simpler ones, viz., $SnCl_2$ (see N_2O_4, p. 333). Lead chloride appears to dissociate before its volatilises, for its density corresponds to the simple formula $PbCl_2$.†

Physical Properties.

Tetra-halides.	Mass of 1 c.c. liquid. F.	Cl.	Br.	I.	Melting-point. F.	Cl.	Br.	I.	Boiling-point. F.	Cl.	Br.	I.
Silicon	?	1·524 at 0°	2·823 at 0°	?	−102°‡	?	−12°	?	?	57·6°	153°	?
Germanium.	?	1·887 at 18°	?	?	?	?	?	144°	?	86°	?	350–400°
Tin	?	2·379 at 0°	?	4·696 at 11°	?	?	30°	146°	?	114°	201°	295°

Hexahalides:—Si_2Cl_6, sp. gr. 1·58 at 0°; m.-p. −1°; b.-p. 146—148°. Si_2Br_6, b.-p. about 240°. Si_2I_6, m.-p. about 250°, with decomposition.

Dihalides:—	$SnCl_2$	$SnBr_2$	SnI_2
Sp. gr.:	?.	5·117 at 17°.	?.
M.-p.:	249·3°.	215·5°	316°.
B.-p.:	601°.	620°	?

Sp. gr.: PbF_2, 8·24 at 2°; $PbCl_2$, 5·80 at 15°; $PbBr_2$, 6·60 at 7·5°; PbI_2, 6·06 at ?.

M.-p.: $PbCl_2$, 498°; $PbBr_2$, 499°; PbI_2, 383°.

B.-p.: $PbCl_2$, 900°; $PbBr_2$, above 861°; PbI_2, 861—954°.

---

* *Zeitschr. phys. Chem.*, **2**, 184. The author differs entirely from the concluding words of this memoir regarding the non-existence of Sn_2Cl_4 in the state of gas.

† *Brit. Assn.*, 1887, 668.

‡ Volatilises without melting. This behaviour is explained as follows:—The boiling-point of a liquid is dependent on the pressure. By lowering the pressure, the boiling-point is lowered, whereas the melting-point is almost unaffected by small alteration of pressure. It is evident that by a sufficient reduction of pressure the boiling-point may be lowered till it occurs at a temperature below the melting-point. Such bodies as silicon fluoride, hexachlorethane, C_2Cl_6, and many others are in this condition under ordinary atmospheric pressure. By increasing the pressure, so as to raise their boiling-points, they can be melted.

Heats of formation:—

$$\mathbf{Sn} + Cl_2 = \mathbf{SnCl_2} + 808K; \ + Aq = 811K.$$
$$\mathbf{Sn} + 2Cl_2 = \mathbf{SnCl_4} + 1273K; + Aq = 299K.$$

The last number implies decomposition when solution takes place:—

$$\mathbf{Pb} + Cl_2 = \mathbf{PbCl_2} + 828K; \ + Aq = -68K.$$
$$\mathbf{Pb} + Br_2 = \mathbf{PbBr_2} + 645K; \ + Aq = -100K\,(?).$$
$$\mathbf{Pb} + \mathbf{I_2} = \mathbf{PbI_2} + 398K; \ + Aq = -160K\,(?).$$

Double halides.—Silicon, like carbon, forms double halides, of which the molecular weights have been determined in many cases. For example, by the action of bromine on the compound $SiHCl_3$, named silicon chloroform (see p. 501), three chloro-bromides have been obtained: one has the formula $SiCl_3Br$, the second, $SiCl_2Br_2$, and the third, $SiClBr_3$.* They are all liquids: the first boiling at 80°, the second at about 100°, and the third at 140—141°. There appear to be similar chlorobromides of tin, which, however, are not stable in the gaseous state.

The **tetrahalides** form numerous double salts. Those of silicon tetrafluoride have been most carefully studied; they are named silicifluorides. Germanifluorides and stannifluorides have also been prepared. The following is a list the more important ones:—

$SiF_4.2HF.Aq$; $\mathbf{SiF_4.2KF}$; $\mathbf{SiF_4.BaF_2}$; &c.
$\mathbf{GeF_4.2KF}$.
$\mathbf{SnF_4.2KF}$; $\mathbf{SnF_4.BaF_2}$.
$\mathbf{SnCl_4.2HCl}$; $\mathbf{SnCl_4.2KCl}$; $\mathbf{SnCl_4.2CaCl_2}$; $\mathbf{SnCl_4.BaCl_2}$. $\mathbf{SnCl_4.2NH_4Cl}$.
$SnBr_4.2HBr$; $\mathbf{SnBr_4.2NaBr}$; $\mathbf{SnBr_4.MgBr_2}$, and others.
$PbCl_4.2HCl.Aq$ (?); $\mathbf{PbCl_4.9NaCl}$; $\mathbf{PbCl_4.16CaCl_2}$.

The compound $\mathbf{SnCl_4.2NH_4Cl}$ is known as "pink salt," being used as a means of fixing pink dyes.

These compounds are mostly prepared by direct addition; but those of silicon may also be produced by the action of $SiF_4.2HF.Aq$ on the oxides, hydroxides, or carbonates of the metals. When silicon fluoride is passed into water the following reaction takes place (see Borofluorides, p. 132)—$3SiF_4 + 3H_2O + Aq = \mathbf{H_2SiO_3} + 2H_2SiF_6.Aq$. The gelatinous precipitate formed when silicon tetrafluoride is passed through water consists of silicic acid, $\mathbf{H_2SiO_3}$; the aqueous solution contains the body H_2SiF_6, **hydrosilicifluoric acid**; its formula is deduced from that of its salts, as it decomposes on evaporation into hydrofluoric acid and silicon fluoride, a portion of which reacts with the water to form more silicic acid.

The more important compounds of this acid are **potassium silicifluoride**, $\mathbf{K_2SiF_6}$, which is one of the few sparingly soluble salts of potassium; it is used as a source of silicon (see p. 50); and the **barium salt**, which is insoluble in water, the corresponding salts of strontium and calcium being soluble. This is utilised as a method of separating barium from these metals.

* *Chem. Soc.*, 51, 590.

Cæsium stannichloride, $SnCl_4 2CsCl$, being nearly insoluble in water, may be separated as such from the corresponding compounds of sodium, potassium, and rubidium.

All these double salts crystallise in the same form, and are therefore termed *isomorphous.*

Many double halides are known of the **dihalides** of tin and lead. None have been gasified; hence their molecular weights are unknown; the simpler formulæ are therefore given as a rule.

Compounds containing two different halogens:—

$SnClI$; $PbFCl$; $PbClBr$; $PbBrI$.
$2PbCl_2.PbI_2$; $3PbBr_2.PbI_2$; $6PbBr_2.PbI_2$.

Compounds with the halides of other elements:—

2 : 1.—$SnCl_2.HCl$; $SnCl_2KCl$; $SnBr_2.KBr$; $SnBr_2.NH_4Br$; $SnI_2.KI$; $SnI_2.NH_4I$; $PbI_2.KI$.
2 : 2.—$SnCl_2.2KCl$; $SnCl_2.BaCl_2$; $SnBr_2.2NH_4Br$; $SnI_2.2KI$; $PbI_2.2HI$; $PbI_2.2KI$; $PbI_2.2NH_4Cl$.

Many more complex ratios have also been noticed among the lead halides, *e.g.* :—2 : 3, $PbI_2.3NH_4Cl$; 2 : 4, $PbI_2.4KI$; 2 : 6, $PbBr_2.6NH_4Br$; 2 : 7, $PbBr_2.7NH_4Br$; 2 : 9, $PbCl_2.9NH_4Cl$; 2 : 10, $PbCl_2.10NH_4Cl$, and others still more complex. These last bodies possess the qualifications usually attributed to definite chemical compounds, viz., definite crystalline form, coupled with constant composition.

The **formulæ** of these halides of the carbon and silicon groups have been determined :—

1. **From the vapour-densities of many of the compounds,** and from the analogy of those of which the vapour-densities have not been determined with those in which that constant is known.

2. **By the method of replacement.**—It is argued, for instance, that the formula of the compound CCl_3Br implies the existence of four atoms of chlorine in the compound CCl_4, inasmuch as one-fourth of the total amount of chlorine it contains has been replaced by bromine. In this case, and in that of the similar silicon compounds, $SiCl_3Br$, $SiCl_2Br_2$, and $SiClBr_3$, this view is confirmed by the vapour-densities of the bodies. But there is no means of ascertaining whether such a body as $SnClI$ possesses that formula or the formula $SnCl_2.SnI_2$, for it has never been gasified. Indeed, judging from the vapour-density of Sn_2Cl_4, the latter formula would appear the more probable; and no simpler formula than $2PbCl_2PbI_2$ is possible in the case of the tetrachlorodiiodide of lead.

3. **The atomic heats** of carbon and silicon present special anomalies. It has been shown by Weber,* however, that, like

* *Pogg. Ann.*, **154**, 367.

those of beryllium and boron, they approach constancy at high temperatures, and become approximately normal. They are as follows:—

		T.	−50°.	−10°.	+10°.	33°.	58°.	86°.
C.	Diamond.	Sp. ht.	0·0635	0·0955	0·1128	0·1318	0·1532	0·1765
	Graphite.	,,	0·1138	0·1437	0·1604	—	0·1990	—

		T.	140°.	206°.	247°.	600°.	800°.	1000°.
C.	Diamond.	Sp. ht.	0·2218	0·2733	0·3026	0·4408	0·4489	0·4589
	Graphite.	,,	0·2542	0·2966	—	0·4431	0·4529	0·4670.

	T.	−40°.	+22°.	57°.	86°.	129°.	184°.	232°.
Si.	Sp. ht.	0·1360	0·1697	0·1833	0·1901	0·1964	0·2011	0·2029

The atomic weights of carbon and silicon have been deduced from their atomic heats at 1000° and 232° respectively, which are, for carbon, 5·608, and for silicon, 5·671.

A few words must be added as to the views which are held regarding the **nature of the atomic combination** in the compounds C_2Cl_6, Si_2Cl_6, C_2Cl_4, Sn_2Cl_4, and analogous bodies. These views are based on the behaviour of the compound of carbon and hydrogen named ethane, C_2H_6, which is analogous to $\mathbf{C_2Cl_6}$, and which, indeed, can be converted into the latter by the continuous action of chlorine, whereby all the hydrogen atoms are successively replaced by an equal number of atoms of the halogen. The compound CH_3I, named iodomethane, when acted on by sodium, loses its iodine, sodium iodide being produced. But the vapour-density of the resulting gas shows it to possess not the formula CH_3, but the double formula $CH_3—CH_3$, or C_2H_6. This is also borne out by the fact that the hydrogen in ethane, C_2H_6, may be replaced by chlorine in sixths, giving C_2H_5Cl, monochlorethane, $C_2H_4Cl_2$, dichlorethane, &c.

It is argued that the group CH_3 may be regarded as replacing an atom of chlorine in CH_3Cl, or of iodine in CH_3I, and that the compounds CH_3Cl and $CH_3—CH_3$ are in that sense analogous. Hence the formula of $\mathbf{C_2Cl_6}$ may be written $\mathbf{CCl_3—CCl_3}$; and of Si_2Cl_6, $SiCl_3—SiCl_3$. And by similar reasoning it is argued that the compound C_2Cl_4 may be regarded as composed of two separate portions, viz., $CCl_2{=}CCl_2$, the two horizontal lines expressing the hypothesis that the group CCl_2 replaces *two* atoms of chlorine in the compound CCl_4. And the vapour-density of these compounds C_2Cl_6 and C_2Cl_4, and of their hydrogen analogues, C_2H_6 and C_2H_4, even at the highest temperatures to which they can be submitted without decomposition, shows that they still possess the formulæ given. On the other hand, there can be no doubt that stannous chloride, Sn_2Cl_4, at a sufficiently high temperature, has a vapour-density corresponding to the simpler formula $SnCl_2$. The conclusion appears, therefore, to follow, that, if it were possible to subject C_2Cl_4 to a sufficiently high temperature without inducing decomposition, it, too, would possess the formula CCl_2. Si_2Cl_6, when heated, splits into $SiCl_4$ and $SiCl_2$; it is, therefore, extremely improbable that any member of this group, at any temperature, will be found to have the formula MX_3, for more stable forms of union exist. But, in the chromium group, chlorides of both the general formulæ $\mathbf{MCl_2}$ and $\mathbf{MCl_3}$ are known; and these appear capable of existence in the two molecular states, MCl_2 and MCl_3, and M_2Cl_6 and M_2Cl_4, respectively; it will be remembered that, in the chromium group, chlorides of the general formula MCl_4 are exceedingly unstable, the only representative definitely known being MnF_4, and that only in aqueous solution. Hence the stability of compounds with the simpler molecular form MF_3.

CHAPTER XII.

HALIDES OF NITROGEN, VANADIUM, NIOBIUM, AND TANTALUM; OF PHOSPHORUS, ARSENIC, ANTIMONY, AND BISMUTH; OF MOLYBDENUM, TUNGSTEN, AND URANIUM; OF SULPHUR, SELENIUM, AND TELLURIUM.—DOUBLE HALIDES OF THESE ELEMENTS.—PROOFS OF THEIR MOLECULAR FORMULÆ.

Halides of Nitrogen, Vanadium, Niobium, Tantalum (Neodymium, see p. 605).

Again it is to be noticed that the compounds of nitrogen, the first element of this group, differ considerably from those of the other members. While the halogen compounds of nitrogen are exceedingly explosive, those of the other elements are stable, though decomposed by water. For these reasons none of them are found in nature. The following table shows the compounds known:—

	Fluorine.	Chlorine.	Bromine.	Iodine.
Nitrogen ...	—	NCl_3.	NBr_3?	NI_3.
Vanadium ..	VF_4?*	VCl_2; VCl_3; VCl_4.	VBr_3.	VI_4?*
Niobium....	NbF_5.*	$NbCl_3$; $NbCl_5$.	$NbBr_5$.	—
Tantalum ..	TaF_5.*	$TaCl_5$.	$TaBr_5$.	—

The iodides of niobium and tantalum, though probably capable of existence, have not been prepared.

Preparation.—1. By direct union.—Nitrogen will not combine directly with the halogens. Vanadium tetrachloride and tribromide are prepared by passing the vapour of the halogen over the heated element; and tantalum pentachloride has also been thus obtained.

2. **By the action of the halogen on a red-hot mixture of the oxide with charcoal.**—This is the method of preparation of niobium and of tantalum pentachloride and pentabromide. Vanadium oxytrichloride, $VOCl_3$ (see p. 332), when passed over red-hot charcoal along with chlorine, also yields the trichloride.

* Known only in solution.

3. **By heating a higher halide.**—Vanadium tetrachloride by distillation alone splits up into chlorine and the trichloride; along with hydrogen, the dichloride is formed; and niobium pentachloride, passed through a red-hot tube, yields the trichloride and chlorine.

4. **By the action of the halogen on a compound of the element.**—Ammonia (hydrogen nitride, NH_3) on treatment with *excess* of chlorine, bromine, or iodine, yields exceedingly explosive bodies. The method of preparation is as follows:—

A flat leaden dish, in which a smaller thick dish is placed, is filled with a strong solution of ammonium chloride. A small jar, of about 200 cubic centimetres capacity, provided with a neck, is placed in the solution, standing on the smaller leaden dish. The neck is closed with a cork, through which a tube passes, which is connected with an apparatus for generating chlorine by means of a short piece of india-rubber tubing, on which a clip is placed. The solution of ammonium chloride is drawn up into the jar by suction, and when the jar is full the clip is closed. The chlorine apparatus is then connected, and by opening the clip the jar is quickly filled with chlorine. The chlorine is absorbed by the solution, while oily drops collect on the surface, and sink, collecting in the leaden dish. Air is then admitted by disconnecting the chlorine apparatus and opening the clip, and the jar is removed. These drops, when touched with an oiled feather tied to the end of a long stick, explode with the greatest violence, shooting a column of water into the air and flattening the leaden vessel.

Recent analysis* has shown that the hydrogen of the ammonia is replaced by stages, exactly as in the case of the hydrogen of methane, CH_4 (see p. 145). By passing chlorine for half an hour into water in which these drops are suspended, the trichloride is finally formed. The equations are these:—

$$NH_4Cl.Aq + Cl_2 = 2HCl.Aq + NH_2Cl.$$
$$NH_4Cl.Aq + 2Cl_2 = 3HCl.Aq + NHCl_2.$$
$$NHCl_2 + Aq + Cl_2 = HCl.Aq + NCl_3.$$

The corresponding bromine compounds have been little investigated, but are made by treating ammonia with excess of bromine. Aqueous ammonia reacts with iodine dissolved in alcohol giving NI_3; but with a weaker solution of ammonia NHI_2 is produced. The action of chlorine or bromine on vanadium nitride, VN, at a red heat gives the trichloride or dibromide, and nitrogen.

The oxygen in niobium oxytrichloride, $NbOCl_3$, is replaced by chlorine when its vapour mixed with chlorine is passed through a red-hot tube.

5. **By double decomposition. Action of the hydrogen halide on the oxide of the element.**—Vanadium tetroxide dis-

* *Berichte*, 21, 751.

solves in hydrofluoric acid, yielding a blue solution, which on evaporation deposits green crystals. This oxide is also soluble in the other haloid acids, giving similar solutions. The pentoxide, when boiled with hydrochloric acid, yields chlorine. Tantalum pentoxide, if hydrated, likewise dissolves in hydrofluoric acid, and the solution on evaporation is said to evolve the fluoride, leaving a residue of oxyfluoride. Niobium pentoxide is also soluble in hydrofluoric acid.

Properties.—The halides of nitrogen are exceedingly explosive, and the preparation of more than a drop or two of the chloride and bromide is attended by great danger. They are oily yellow liquids, insoluble in water, which slowly decompose when left in contact with water or solution of ammonia. The iodide is a brownish black powder, of which it is also advisable to prepare only a few decigrams at a time. They explode on contact with an oiled feather, or indeed by the slightest impact, and often without any apparent cause. The pure chloride has been heated to 90° without decomposition, but at 95° a violent explosion occurred.

Vanadium dichloride forms apple-green crystalline plates. The element may be obtained from it by heating it to redness in a current of very carefully dried hydrogen. Vanadium trichloride closely resembles chromium trichloride in appearance. When heated in air, its chlorine is replaced by oxygen, and the pentoxide is formed by further absorption of oxygen. The tribromide is a greyish-black amorphous mass; it is very unstable. The tetrachloride is a reddish-brown volatile liquid, soluble in water with a blue colour.

Niobium and tantalum pentafluorides form colourless solutions. Niobium trichloride closely resembles iodine in appearance; it is unaffected by water. Niobium and tantalum pentachlorides form yellow volatile crystals; the bromides are similar in appearance, but of a darker colour.

Physical Properties.

		Mass of 1 c.c.	Melting-point.	Boiling-point.
Dichloride.	Vanadium ..	3·23 c.c. at 18°	?	?
Trichlorides.	Nitrogen ...	1·65 „ „	?	Above 90°.
	Vanadium ..	3·00 „ „	Decomposes.	Decomposes.
	Niobium....	?	?	?
Tetrachloride.	Vanadium ..	1·858 at 0°.	Below −18°.	154°.
Pentachlorides.	Niobium ...	?	194°	240·5°.
	Tantalum ..	?	211°	242°.

The properties of the other halides have not been determined.

Double halides.—Although double halides of the oxyfluorides of vanadium, niobium, and tantalum have been studied (see p. 336), the tantalifluorides are the only compounds of any of the halides of this group with the halides of other elements. They have all the general formula $TaF_5.2MF$. They are produced by direct union of the respective fluorides in aqueous solution, and crystallise well. They are soluble in water. The following have been prepared:—

$TaF_5.2NH_4F$; $TaF_5.2KF$; $TaF_5.2NaF$; $TaF_5.CuF_2$; and $TaF_5.ZnF_2$.

Halides of Phosphorus, Arsenic, Antimony, (Erbium), and Bismuth.

Sources.—None of these compounds are found in nature.

The halogen compounds known are given in the following table:—

	Fluorine.	Chlorine.	Bromine.	Iodine.
Phosphorus ..	PF_3; PF_5.	PCl_3; **PCl_5**.	PBr_3; **PBr_5**.	**P_2I_4; PI_3**.
Arsenic......	AsF_3.	$AsCl_3$.	**$AsBr_3$**.	**As_2I_4; AsI_3**.
Antimony....	**SbF_3; SbF_5**.	**$SbCl_3$**; $SbCl_5$.	**$SbBr_3$**.	**SbI_3**.
(Erbium)	ErF_3.*	$ErCl_3$.*	$ErBr_3$.*	—
Bismuth	**BiF_3**.	**Bi_2Cl_4; $BiCl_3$**.	**Bi_2Br_2; $BiBr_3$**.	**BiI_3**.

Preparation.—1. By direct union.—All these bodies are best prepared thus, except the fluorine compounds; phosphorus, arsenic, and antimony take fire spontaneously in fluorine and chlorine gases, and all combine with the halogens with great evolution of heat. With excess of halogen, the higher halide is formed where they exist; with excess of the other element, the lower halides.

As examples of this method of formation, the following types may be chosen:—

(*a*.) A retort is half filled with dry sand, and on it are placed a few pieces of phosphorus. Dry chlorine is led into the retort, so as to impinge on the phos-

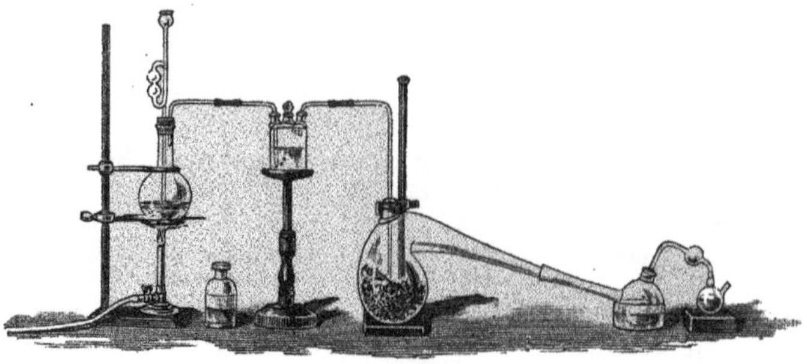

FIG. 28.

* Known only in aqueous solution.

phorus from the generating flask. The phosphorus burns with a greenish flame, and the liquid chloride distils over, and may be condensed in the receiver. It is purified by distillation.

(*b.*) A little powdered antimony is thrown into a jar of dry chlorine. It burns with scintillation, and on standing, the fumes condense to crystals of the pentachloride.

(*c.*) A mixture is made of 1 volume of bromine and 3 volumes of carbon disulphide and placed in a flask. Powdered antimony is added in small quantities at a time, and the flask is warmed gently with continuous shaking over a water-bath, taking care to have no flame near, for fear the disulphide vapour should inflame. It is then allowed to cool, when the tribromide separates in crystals.

2. **By heating a higher halide.**—Phosphorus and antimony pentachlorides yield the trichloride when heated; phosphorus pentabromide behaves similarly. Phosphorus pentafluoride, on the other hand, is stable, showing no decomposition, even at the high temperature of the electric spark. The decomposition of phosphorus pentachloride may be well seen by heating it in a flask; its vapour has a greenish-yellow colour, due to the presence of free chlorine. Phosphorus tri-iodide gives off iodine when heated.

3. **By heating a lower halide.**—Bismuth dichloride at 330°, and the dibromide at a temperature not much above that of the atmosphere decompose into bismuth and the trihalide.

4. **By the action of the halogen on a compound of the element.**—This method is not generally employed; yet hydrogen phosphide, arsenide, and antimonide are at once acted on by chlorine or bromine, yielding hydrogen halide, and the halide of the element. It is certain that all other compounds, except perhaps the oxides, would behave similarly.

5. **By double decomposition.**—(*a.*) **The action of the hydrogen halide on the oxide or sulphide of the element.**—The oxides of phosphorus are not attacked. But the oxides and sulphides of the other elements yield the respective halides, either when heated in a current of the hydrogen halide, or when treated with a halogen acid. In presence of a great excess of water, the halides are decomposed; hence the acids must not be too dilute.

Arsenious fluoride is prepared by heating together arsenious oxide, $\mathbf{As_4O_6}$, fluorspar, $\mathbf{CaF_2}$, and sulphuric acid, H_2SO_4.* The hydrogen fluoride liberated by the action of the sulphuric acid on the calcium fluoride (see p. 108) attacks the arsenious oxide, producing arsenious fluoride and water, which combines with the excess of sulphuric acid, thus:—

$$\mathbf{As_2O_3} + 6HF + 3H_2SO_4 = 3(H_2SO_4.H_2O) + 2AsF_3.$$

* *Comptes rend.*, 99, 874.

The chloride may be similarly prepared from sulphuric acid, sodium chloride, and arsenious oxide. It may also be obtained by distilling arsenic with mercuric chloride, thus:—

$$2As + 3HgCl_2 = 2\mathit{AsCl_3} + 3Hg.$$

(*b*.) Phosphorus trichloride or pentachloride reacts with arsenic trifluoride, yielding the trifluoride* or pentafluoride† of phosphorus, thus:—

$$PCl_3 + AsF_3 = \mathit{PF_3} + AsCl_3;$$
$$\text{and } 3PCl_5 + 5AsF_3 = 3\mathit{PF_5} + 5AsCl_3.$$

Properties.—The pentafluorides of phosphorus and arsenic are gases at the ordinary temperature; the trichloride and tribromide of phosphorus, the trifluoride and trichloride of arsenic, and the pentachloride of antimony are colourless liquids, fuming in air, owing to their reacting with the water-vapour which it contains; the remaining trifluorides, chlorides, and bromides are colourless crystalline solids. Phosphorus di-iodide forms orange-coloured, and tri-iodide, red, crystals. Arsenic di-iodide, produced by melting arsenic with iodine in theoretical proportion, forms a dark cherry-red mass, which crystallises from carbon disulphide in prisms, and which decomposes into arsenic and the tri-iodide on addition of water. The tri-iodide forms red tablets. Antimony tri-iodide exists in three forms: when crystallised from carbon disulphide, it forms red hexagonal crystals; when sublimed below 114°, yellow trimetric crystals; and from its solution in carbon disulphide exposed to sunlight, in monoclinic crystals. The last variety is converted into the hexagonal modification at 125°. Bismuth tri-iodide forms a greyish mass with metallic lustre.

Phosphorus pentachloride and pentabromide are yellowish crystalline solids; antimony pentachloride is a colourless fuming liquid. These three substances dissociate when heated into the trihalides and two atoms of the halide; hence, their vapour-densities do not correspond to their formulæ. In excess of trichloride, however, the decomposition of phosphorus pentachloride is prevented, and it volatilises as $\mathit{PCl_5}$.‡ Bismuth dichloride, on the other hand, decomposes into bismuth and the trichloride when heated, thus:—

$$3Bi_2Cl_4 = 4BiCl_3 + 2Bi.$$

These substances are all, with the exception of bismuth trifluoride, deliquescent, attracting water, and reacting with it to

* *Comptes rend.*, **99**, 655; **100**, 272.

† *Proc. Roy. Soc.*, **25**, 122; *Comptes rend.*, **101**, 1496.

‡ *Comptes rend.*, **76**, 601.

form oxyhalides or hydroxides (acids). They are all soluble in carbon disulphide, benzene, &c. The erbium halides form colourless solutions.

Physical Properties. Trihalides.

	Mass of 1 c.c. solid or liquid.				Melting-point.			
	F.	Cl.	Br.	I.	F.	Cl.	Br.	I.
Phosphorus.	?	1·613 at 0°	2·923 at 0°	?	?	?	?	55°
Arsenic....	2·66 at 0°	2·205 at 0°	3·66 at 15°	4·39 at 13°	?	−18°	20–25°	146°
Antimony..	?	3·064 at 26°	4·148 at 23°	4·85* at 24°	292°	73·2°	90°	167°
Bismuth ...	5·32 at 20°	4·56 at 11°	5·4 at 20°	5·64 at 20°	?	227°	200°	†

	Boiling-point.			
	F.	Cl.	Br.	I.
Phosphorus ..	—	76·0°	172·9°	?
Arsenic	60·4°	130·2°	220°	394–414°
Antimony	?	223·5°	275·4°	401°
Bismuth	?	427–439°	454–498°	†

$SbCl_5$: mass of 1 c.c. 2·346 at 20°; m.-p. −6°. $\mathbf{PCl_5}$: m.-p. 148°, under increased pressure; volatilises at 148°.

$\mathbf{Bi_2Cl_4}$: m.-p. 176°. $\mathbf{Bi_2Br_4}$: m.-p. 202° (uncorr.).

Heat of formation :—

P + 3*Cl* = PCl_3 + 755K.	**P** + 3Br = PBr_3 + 448K.
P + 5*Cl* = $\mathbf{PCl_5}$ + 1050K.	**P** + 5Br = $\mathbf{PBr_5}$ + 591K.
P + 2I = $\mathbf{PI_2}$ + 99K.	**P** + 3I = $\mathbf{PI_3}$ + 109K.
As + 3*Cl* = $AsCl_3$ + 715K.	**As** + 3Br = $AsBr_3$ + 449K.
As + 3I = $\mathbf{AsI_3}$ + 127K.	
Sb + 3*Cl* = $\mathbf{SbCl_3}$ + 914K.	**Sb** + 5*Cl* = $\mathbf{SbCl_5}$ + 1049K.
Bi + 3*Cl* = $\mathbf{BiCl_3}$ + 906K.	

The vapour-densities of phosphorus tri- and penta-fluorides, tri- and penta-chlorides, tribromide, and tri-iodide have been determined; also those of the trihalides of arsenic and of antimony and bismuth trichlorides; their molecular weights are represented by the formulæ given. Diphosphorus tetriodide has a vapour-density corresponding to the formula P_2I_4; moreover, the analogous compounds of bismuth easily decompose into the trihalide and metal; hence, the more complex formulæ have been chosen, although the choice is not justified by any absolute proof in the case of bismuth.

* Hexagonal; monoclinic: 4·77 at 22°.

† Decomposed.

Double halides.—1. Compounds containing two halogens.—These are known only in the case of phosphorus. They are, for the most part, made by adding the halogen to the halide dissolved in carbon disulphide, and crystallising from that solvent. Their molecular weights are unknown.

They are as follows :—*

PF_3Br_2.	PCl_5ICl.	PCl_2Br_7.	PCl_3Br_8.
PCl_3Br_2.	PCl_3Br_4.		
PCl_4Br.	PCl_2Br_5.		

The following compounds with the halides of other elements are known :—

Phosphorus pentahalides:
$PCl_5.FeCl_3$. $PCl_5.SnCl_4$. $PCl_5.SbCl_5$. $PCl_5.3HgCl_2$.
$PCl_5.AlCl_3$. $PCl_5.SeCl_4$. $PCl_5.UCl_5$.
$PCl_5.CrCl_3$. $PCl_5.WCl_4$.
$PCl_5.AsCl_3$. $PCl_5.MoCl_4$.

Arsenic pentahalides .... $AsF_5.KF$. $AsF_5.2KF$.

Antimony pentahalides:
$SbF_5.NaF$. $SbF_5.2KF$. $SbCl_5.SCl_4$. $SbCl_5.5HCl.10H_2O$.
$SbF_5.KF$. $SbF_5.2NH_4F$. $SbCl_5.SeCl_4$.
$SbF_5.NH_4F$.

Trihalides. Phosphorus. $PCl_3.AuCl$.

Antimony. $SbF_3.KF$. $SbF_3.2KF$. $SbF_3.3NaF$. $2SbI_3.3KI$.
$2SbCl_3.HCl.2H_2O$. $SbI_3.KI$. $SbCl_3.2NH_4Cl$. $SbCl_3.3KCl$.
$SbI_3.NH_4I$. $SbCl_3.BaCl_2$. $SbCl_3.3KBr$.†
$SbI_3.BaI_2$. $SbBr_3.3KCl$.†

Bismuth.... $2BiCl_3.NH_4Cl$. $BiI_3.HI$. $BiCl_3.2NaCl$. $BiCl_3.3NH_4Cl$.
$BiCl_3.4NH_4Cl$.
$2BiI_3.3NaI$. $BiF_3.HF$. $BiI_3.BaI_2$. $BiF_3.3HF$.
$2BiCl_3.HCl.3H_2O$. $BiI_3.KI$. $BiCl_3.2KBr$.

These are some of the compounds known. It will be noticed that the ratios of the number of atoms of halogen in the two components vary between 6 : 1 and 3 : 4. All these substances react with water, producing oxyhalides (see p. 385).

Halides of Molybdenum, Tungsten, and Uranium.

These bodies present some analogy with the halides of chromium, which, indeed, in the periodic table, falls in this group.

Sources.—None of these halides occurs native.

The following is a list of the known compounds :—

	Fluorine.			Chlorine.				
Molybdenum	MoF_3;‡	MoF_4;‡	MoF_6.	$MoCl_2$;	$MoCl_3$;	$MoCl_4$;	$MoCl_5$	—
Tungsten ..	WF_2‡	—	WF_6.	WCl_2;	—	WCl_4	WCl_5;	WCl_6.
Uranium	—	UF_4	—	—	UCl_3;	UCl_4;	UCl_5	—

* *Chem. Soc.*, **49**, 815.

† These bodies are identical, although prepared by direct addition.

‡ Known only in solution.

	Bromine.				Iodine.
Molybdenum...	$MoBr_2$;	$MoBr_3$;	$MoBr_4$	—	MoI_2;* MoI_4.*
Tungsten......	WBr_2	—	WBr_4;	WBr_5.	WI_2.
Uranium......	—	—	UBr_4	—	UI_2.

Preparation.—1. **By direct union.**—It is important to avoid the presence of air and water-vapour, else oxyhalides are obtained. This process yields molybdenum and uranium penta-chlorides, tungsten hexachloride, molybdenum tetrabromide, and tungsten pentabromide, all of which are volatile.

2. **By the action of the halogen on a mixture of the oxide and charcoal.**—By this means, molybdenum tribromide and uranium pentachloride have been prepared; it is doubtless adapted for the production of any of the higher halides.

3. **By heating the higher halides.**—Molybdenum tri- and tetra-bromides, when distilled, undergo decomposition into bromine and the dibromide, $MoBr_2$. Tungsten hexachloride, between 360° and 440°, dissociates into pentachloride and free chlorine. In other cases, the distillation of a halide yields a mixture of two halides; for example, molybdenum trichloride, sublimed in dry carbon dioxide, splits into the di- and tetra-chlorides, thus:— $2MoCl_3 = MoCl_2 + MoCl_4$. And the tetrachloride is also unstable when distilled, giving tri- and penta-chlorides, $2MoCl_4 = MoCl_5 + MoCl_3$.

4. **By the action of hydrogen on the heated halide.**—Molybdenum pentachloride yields hydrogen chloride and the tri-choride at 250°. Tungsten hexachloride and uranium penta-chloride also yield a mixture of lower chlorides when thus treated.

5. **By double decomposition.—The action of a halide on the oxide.**—(*a.*) The fluorides are all thus prepared from the corresponding oxides by the action of aqueous hydrofluoric acid. Solutions of many of the other halides may also be prepared thus.

(*b.*) Tungsten hexachloride is produced by heating in a sealed tube to 200° a mixture of tungsten trioxide and phosphoric chloride.

Properties.—1. **Dihalides.**—Molybdenum dihalides when prepared in the dry way are insoluble in water; but when obtained from the oxides they form brown or purple solutions. The dichloride is a sulphur-yellow powder. Tungsten dichloride is a loose grey powder; the fluoride forms a yellow solution.

2. **Trihalides.**—Molybdenum trichloride is a red powder like

* Known only in solution.

amorphous phosphorus; the tribromide forms dark needles; both are insoluble in water. Uranium trichloride is dark brown.

3. **Tetrahalides.**—Molybdenum tetrafluoride forms a red solution; and uranium tetrafluoride an insoluble green powder. Molybdenum tetrachloride is a volatile brown substance; that of tungsten a greyish-brown crystalline powder; while uranium tetrachloride forms magnificent dark green octohedra, and yields a red vapour. The tetrabromides form brown or black crystals. These compounds are deliquescent, and soluble in water.

4. **Pentahalides.**—Molybdenum pentachloride is a black substance yielding a brown-red vapour; those of tungsten and uranium consist of black needles.

5. **Hexahalide.**—Tungsten hexachloride volatilises in bluish-black needles, resembling iodine.

Many of these compounds require further investigation. As has been seen, they are very numerous, and their reactions have by no means been exhaustively studied.

Physical Properties.

The mass of 1 cubic centimetre has not been determined for any one of these halides. The following melting- and boiling-points and vapour-densities have been determined:—

$MoCl_5$. M.-p., 194°; b.-p., 268°; vap.-dens. at 350°, 136·0 to 137·9. Calc. 136·5.
WCl_5. M.-p., 248°; b.-p., 275·6°; vap.-dens. at 360°, 182·8. Calc. 180 65.
WCl_6. M.-p., 275° b.p., ?; vap.-dens. at 360°, 190·9. Calc. 208·65.
UCl_5 dissociates when its vapour, mixed with carbon dioxide, is heated. Dissociation begins at 120° and is complete at 235°.

The heats of combination are undetermined.

Double Halides.—These again have been very little studied. Some compounds of molybdenum containing two halogens are known, *e.g.*, $2MoCl_2.MoBr_2$, $2MoBr_2.MoI_2$, &c., and one compound of the formula $2MoCl_2.MoBr_2.KBr$; a compound of the tetrachloride has also been prepared, viz., $3MoCl_4.2KCl$. No similar compounds of tungsten have been prepared, and only one of uranium, viz., $UF_4.KF$. These bodies much require investigation.

The atomic weights of these elements have been determined from the equivalents, and by the vapour-densities given above.

Halides of Oxygen, Sulphur, Selenium, and Tellurium.

The halides of oxygen are best considered as oxides of the halogens, *q. v.* (p. 459). Those of the three other elements form a well-marked group. None of them occurs in nature. They are as follows:—

	Fluorine.	Chlorine.	Bromine.	Iodine.
Sulphur....	?	S_2Cl_2; SCl_2; SCl_4.	?	S_2I_2?
Selenium...	?	Se_2Cl_2; $SeCl_4$.	Se_2Br_2; $SeBr_4$.	?
Tellurium..	?	$TeCl_2$; $TeCl_4$.	$TeBr_2$; $TeBr_4$.	TeI_2; TeI_4?

Preparation.—1. **By direct union.**—This is the general method of preparing these bodies. Sulphur is said to burn in fluorine. When chlorine is led over sulphur contained in a retort it grows warm, and disulphur dichloride is formed; by keeping sulphur in excess it is the only product. Diselenium dichloride is similarly produced, and may be obtained fairly pure by distillation in presence of selenium. But it dissociates to some extent during distillation, with formation of tetrachloride and free selenium. Sulphur dichloride, produced by saturating S_2Cl_2 with chlorine, is stable up to nearly 20°, but above that temperature dissociation proceeds rapidly, so that at 120° it has nearly all decomposed into the compound S_2Cl_2 and chlorine. The tetrachloride is still more unstable; at −22° it can exist, but at +6° it has wholly split up into dichloride and chlorine. It will be remembered that it forms a double chloride with antimony trichloride, which crystallises and has a definite composition, $2SbCl_3.3SCl_4$.

Selenium tetrachloride is freed from accompanying dichloride by washing it with carbon disulphide, in which it is sparingly soluble. It may be volatilised without decomposition.

Sulphur and bromine mix in all proportions with evolution of heat, but no definite compound has been isolated. It is not improbable that the resulting liquid is a mixture of the compounds S_2Br_2 and SBr_4 with excess of uncombined sulphur and bromine.

Sulphur and iodine, and selenium and iodine, mix in all proportions when melted together, but no products of definite composition have been isolated. Tellurium di-iodide is similarly prepared; the excess of iodine is volatilised away by gentle heat.

2. **By double decomposition.**—Sulphur and selenium fluorides are said to have been prepared by distilling a mixture of dry lead fluoride and sulphur or selenium. They have not been analysed. Tellurium dioxide dissolves in hydrofluoric acid, but no definite compound has been isolated. With hydriodic acid tellurium yields the tetriodide as a soft black powder.

Properties.—The chlorides of sulphur are yellow-brown oily liquids decomposed by water with separation of sulphur, thus:—

$$2S_2Cl_2 + 2H_2O + Aq = H_2SO_3.Aq + 4HCl.Aq + 3\mathbf{S}.$$
$$2SCl_2 + 2H_2O + Aq = H_2SO_3.Aq + 4HCl.Aq + \mathbf{S}.$$
$$SCl_4 + 2H_2O + Aq = H_2SO_3.Aq + 4HCl.Aq.$$

Diselenium dichloride and dibromide are dark-brown liquids, which, when vaporised, dissociate partially into free selenium and tetrahalide. The tetrachloride and tetrabromide form yellow crystals. Tellurium dichloride is a black amorphous solid, melting to a black liquid, and giving a yellow vapour. The dibromide forms black needles, and the diiodide black flocks. The tetrachloride is a yellow crystalline mass, melting to a yellow liquid; it is volatile without decomposition. The tetrabromide sublimes in pale-yellow needles, which melt to a red liquid. The tetraiodide is a black powder. All these bodies are decomposed by water.

Physical Properties.

The following determinations have been made of the mass of 1 c.c. of these compounds:—

S_2Cl_2: 1·709 grams at 0°. Se_2Cl_2: 2·906 grams at 17·5°
S_2Br_2: 2·628 „ at 4°. Se_2Br_2; 3·604 „ at 15°.

The other known constants are as follows:—

Melting and boiling points:—S_2Cl_2: b.-p. 138°; $TeCl_2$: m.-p. 175°, b.-p. 324°; TeI_2: m.-p. 160°; $TeCl_4$: m.-p. 209°, b.-p. 380°. The vapour-densities of S_2Cl_2, $SeCl_4$, $SeCl_3Br$, $TeCl_2$, and $TeCl_4$ have been determined, and are normal, corresponding to the formulæ given.

Heats of combination:—

$$2S + 2Cl = S_2Cl_2 + 143K.$$
$$2Se + 2Cl = Se_2Cl_2 + 222K; \quad Se + 4Cl = SeCl_4 + 351K.$$
$$Te + 4Cl = TeCl_4 + 774K.$$

It is thus seen that the more stable compounds are formed with greatest evolution of heat.

Double halides.—1. $SeClBr_3$, $SeCl_2Br_2$, and $SeCl_3Br$, have been prepared by **addition**. They are yellowish powders.*

2. **By acting with chlorine on sulphides**, the following bodies have been obtained:—

$SCl_4.2AlCl_3$; $3SCl_4.2SbCl_3$; $2SCl_4.SnCl_4$; and $2SCl_4.TiCl_4$.

3. **By mixing aqueous solutions of the constituent halides**, tellurium halides combine thus:—

$TeF_4.KF$; $2TeF_4.BaF_2$. $TeCl_4.2KCl$; $TeCl_4.2AlCl_3$. $TeBr_4.2KBr$; $TeI_4.2KI$.

These compounds form reddish crystals. Few attempts have been made to prepare double halides.

---

* *Chem. Soc.*, **45**, 70.

CHAPTER XIII.

COMPOUNDS OF THE HALOGENS WITH EACH OTHER; WITH RHODIUM, RUTHENIUM, AND PALLADIUM; WITH OSMIUM, IRIDIUM, AND PLATINUM; AND WITH COPPER, SILVER, GOLD, AND MERCURY.

Compounds of the Halogen Elements with each other.

These compounds have no great stability. Fluorides of chlorine and bromine are unknown. Iodine is said by Moissan to unite with fluorine when exposed to it, and to be a colourless fuming liquid. Chlorine and bromine mix, but yield no definite compound; similarly, iodine dissolves in bromine, but separates on distillation. No attempts are recorded of cooling mixtures of these elements, but it is highly probable that evidence of combination would be obtained if the experiment were made. The only compounds investigated are the chlorides of iodine. They do not occur in nature. They are two in number, **ICl**, of which two modifications exist, and **ICl_3**.

Preparation.—1. **By direct union.**—Iodine heated in chlorine yields the monochloride with iodine in excess; with excess of chlorine, the trichloride.

2. **By displacement and subsequent combination.**—This is accomplished by heating a mixture of iodine and potassium chlorate, **$KClO_3$**. This body decomposes thus: $\mathbf{K_2O.Cl_2O_5 + I_2 = K_2O} + 5O + 2\mathit{ICl}$. Subsidiary reactions take place, thus:—$\mathbf{K_2O.Cl_2O_5} = \mathbf{2KCl} + 6O$; $\mathbf{KCl} + 4O = \mathbf{KClO_4}$; $\mathbf{KI} + 3O = \mathbf{KIO_3}$, perchlorate and iodate of potassium being simultaneously formed. The reaction is a violent one, and the iodine monochloride distils over very rapidly; hence the arrangements for condensing it must be complete.

3. **By double decomposition.**—The trichloride is thus formed by treating iodine pentoxide with dry hydrogen chloride, thus:—$\mathbf{I_2O_5} + 10\mathit{HCl} = 5H_2O + 2\mathit{Cl_2} + 2\mathit{ICl_3}$. The higher chloride, $2I_2Cl_5$, presumably formed for an instant, is unstable, and decomposes, liberating chlorine.

Properties.—**Monochloride**, ICl. The liquid product, if cooled to −25°, solidifies in long dark-red needles, melting at 27·2°. This is the α-modification. The β-modification is sometimes obtained as dark-red plates, melting at 13·9°, on crystallising the liquid between +5° and −10°. On cooling it below −12° it changes into the α form.*

The **trichloride** forms yellow needles, melting under pressure at 101°. The monochloride is only slightly decomposed at 80°, boiling with partial dissociation between 102° and 106°; whereas the trichloride dissociates when gasified.

Heats of combination.—

$$I + Cl = ICl + 58K.$$
$$I + Cl_3 = ICl_3 + 215K.$$

Both of these bodies react with water, forming iodic acid, HIO_3, hydrogen chloride, and free iodine. Among the products a yellow body of the formula **ICl.HCl**† is said to exist, soluble in ether.

Halides of Ruthenium, Rhodium, and Palladium.

Sources.—These substances do not occur native.

The following compounds are known:—

	Fluorine.	Chlorine.	Bromine.	Iodine.
Ruthenium..	—	$RuCl_2$; $RuCl_3$ ($RuCl_4$)‡	—	—
Rhodium ...	—	— $RhCl_3$ —	—	—
Palladium...	PdF_2.	$PdCl_2$ — $PdCl_4$.	$PdBr_4$.	PdI_2.

Preparation.—1. **By direct union.**—The respective metals, heated in chlorine, yield $RuCl_2$, $RuCl_3$, and $RhCl_3$.

2. **By the action of hydrogen chloride on the metal.**—The presence of nitric acid, HNO_3, is necessary to furnish oxygen, with which the hydrogen of the hydrogen chloride may combine. By this means, $PdCl_2$ is formed; with excess of nitric acid the product is $PdCl_4$.

3. **By heating a higher halide.**—$PdCl_4$ loses chlorine, yielding $PdCl_2$.

4. **By removing chlorine from a higher chloride.**—A solution of $RhCl_3$, on treatment with hydrogen sulphide, yields $RhCl_2$.

5. **By double decomposition.**—(*a*.) The action of the hydro-

* *Rec. trav. chim.*, **7**, 152.
† *Compt. Rend.*, **84**, 389.
‡ Known only in combination with KCl.

gen halide on the hydrated oxide, in presence of water. This is the method of preparing PdF_2 (from PdO), $RuCl_3$ (from $Ru_2O_3.nH_2O$), and $RuCl_4.2KCl$ (from $RuO_2.nH_2O$), in presence of KCl; also $RhCl_3$ (from Rh_2O_3,nH_2O); and $PdBr_4$.

(*b*.) **By double decomposition.**—On adding a solution of an iodide to that of a soluble compound of palladium, *e.g.*, the nitrate, $Pd(NO_3)_2$, palladous iodide, $\mathbf{PdI_2}$, is precipitated in a gelatinous form.

Properties.—Ruthenium dichloride remains as a black crystalline powder when chlorine is passed over ruthenium, while the trichloride volatilises. The trichloride, prepared in the wet way, is a yellow-brown crystalline substance. On passing hydrogen sulphide through its solution it is converted into ruthenious chloride, thus:—$2RuCl_3.Aq + H_2S = 2RuCl_2.Aq + 2HCl + \mathbf{S}$. The dichloride forms a blue solution.

Rhodium trichloride, prepared in the dry way, is a reddish-brown insoluble body. Prepared from the hydrated oxide, it forms a red solution.

Palladium difluoride, obtained by evaporating palladous nitrate, $Pd(NO_3)_2$, with hydrogen fluoride, forms colourless soluble crystals. The dichloride fuses to a black mass. The tetrachloride and tetrabromide are said to form dark-brown solutions. It is probable that they are really compounds with hydrogen chloride and bromide, $PdCl_4.2HCl$ and $PdBr_4.2HBr$. The di-iodide is a black gelatinous precipitate, drying to a black powder, and decomposing into its elements at 300—360°. The only one of these compounds which finds practical application is palladium di-iodide, which is insoluble, the corresponding chlorides and bromide being soluble. It is therefore used as a means of separating iodine from the other halogens. The physical constants and molecular weights of these bodies are unknown.

Double halides.—Palladium fluoride is said to form double compounds with fluorides of potassium, sodium, and ammonium. The following compounds of the other halides have been prepared:—

$\mathbf{PdCl_2.2KCl}$.	$\mathbf{RuCl_3.2KCl}$.	$\mathbf{RuCl_4.2KCl}$.
	$\mathbf{RhCl_3.2KCl}$.	$\mathbf{PdCl_4.2KCl}$.
	$\mathbf{RhCl_3.3KCl}$.	$\mathbf{PdBr_4.2KBr}$.

These bodies are generally prepared by addition. The ruthenic chloride, $\mathbf{RuCl_4.2KCl}$, is obtained by dissolving the hydrated dioxide in hydrogen chloride in presence of potassium chloride, and evaporating the solution. Compounds with some other chlorides have also been prepared. The corresponding palladium salt is very unstable, decomposing even when its solution is warmed, with evolution of chlorine.

Halides of Osmium, Iridium, and Platinum.

None of these compounds is found in nature.

The following is a list of the known compounds:—

	Fluorine.	Chlorine.	Bromine.	Iodine.
Osmium...	—	$OsCl_2$; $OsCl_3$; $OsCl_4$.	—	—
Iridium...	—	$IrCl_2$?; $IrCl_3$; $IrCl_4$.	$IrBr_3$; $IrBr_4$.	IrI_2; IrI_3; IrI_4.
Platinum..	PtF_4	$PtCl_2$; — $PtCl_4$.	$PtBr_2$; $PtBr_4$.	PtI_2; — PtI_4.

Preparation.—1. **By direct action of the halogen** on the metal at a red heat, osmium dichloride, trichloride, and tetrachloride, iridium trichloride, and platinum tetrafluoride and tetrachloride have been prepared. The double chlorides are in many cases produced by the action of chlorine, at a red heat, on a mixture of chloride of potassium, &c., with the metal. Platinum tetrabromide is formed by the action of bromine and hydrobromic acid on spongy platinum at 180° in a sealed tube.

2. **By the action of nitro-hydrochloric acid on the metal.**—The action between nitric and hydrochloric acids generates free chlorine, thus:—

$$HNO_3 + 3HCl.Aq = 2H_2O.Aq + \mathit{NOCl} + \mathit{Cl_2}.$$

Metallic iridium and platinum dissolve in *aqua regia*, as the mixture is called, with formation of the double compounds of hydrogen chloride with tetrachlorides. Platinum tetrachloride, $PtCl_4.4H_2O$, is produced by dissolving the calculated amount of platinic oxide in this solution. Similarly, a mixture of nitric and hydrobromic acids yields the tetrabromides in solution.

3. **By heating higher halides.**—Iridium trichloride and tribromide have been obtained from the tetrachloride and tetrabromide by heat. Platinic chloride (*i.e.*, the tetrachloride) yields the dichloride at 440°. There is little doubt that in every case the application of heat to a tetrahalide would be followed by the formation of a lower halide; but in many cases it appears to be difficult to avoid complete loss of halogen and reduction to metal.

4. **Double decomposition.**—(*a.*) **The action of a halogen acid on the corresponding oxide or hydroxide of the metal.**—Osmium dichloride in solution has been thus prepared from osmium monoxide, OsO; similarly, iridium trichloride is produced from Ir_2O_3.

(*b.*) Iridium tetriodide is produced by mixing solutions of the tetrachloride and potassium iodide. On mixing it with ammonium iodide, the tetraiodide is probably formed at first, but it loses iodine, yielding the tri-iodide. Platinum tetrafluoride is produced by adding silver fluoride to platinum tetrachloride, filtering from the precipitated silver chloride, and evaporating the solution. Platinum di- and tetra-iodides are formed on addition of potassium iodide to the di- and tetra-chlorides. Iridium tetrabromide may be similarly produced by the action of potassium bromide on the tetrachloride.

5. **By reduction of a higher halide.**—Various reducing agents may be used to prepare a lower from a higher halide. The one commonly used is sulphurous acid, which absorbs oxygen from water, liberating hydrogen, which combines with a portion of the halogen. By this means osmium di- and tri-chlorides and iridium di-iodide are produced. The last reaction is as follows:—

$$IrI_4.Aq + H_2O + H_2SO_3.Aq = \mathbf{IrI_2.Aq} + H_2SO_4.Aq + 2HI.Aq.$$

Properties.—Most of these bodies are non-crystalline powders. Iridium trichloride, tetrachloride, tri-iodide, and tetra-iodide are black powders. Osmium dichloride is blue-black. It is very unstable, but its compound with chloride of potassium is more permanent. The trichloride is known only in solution. The tetrachloride is a red mass. Iridium dichloride is an olive-green, and the di-iodide a brown, powder. The tribromide forms olive-green crystals. Platinum tetrafluoride is a buff-yellow crystalline deliquescent mass. The tetrachloride forms orange-brown crystals containing water. The tetrabromide is a non-deliquescent black mass, soluble with brown colour. The dichloride and dibromide are greenish-brown masses. These substances are all easily decomposed by heat. The following are soluble in water:—$OsCl_2$, dark-violet; $OsCl_3$, green; $OsCl_4$, red. $IrCl_3$ and $IrCl_4$ are deliquescent; PtF_4, yellow; $PtCl_2$, orange; $PtCl_4$, orange-brown; $IrBr_3$, olive-green; $IrBr_4$, red. Osmium tetrachloride decomposes on addition of much water.

Double halides.—These bodies are, as a rule, crystalline in this group, and are more stable than the simple halides. The following is a list:—

Fluorides.		Chlorides.		
$Pt.F_4nKF$.	$OsCl_2.nKF$.	$IrCl_2.nKCl$.	$PtCl_2.KCl$.	$PtCl_4.2KCl$.
	$OsCl_3.3KCl$.	$IrCl_2.3KCl$.	$PtCl_2.2HCl$.	$PtCl_4.2NH_4Cl$.
	$OsCl_4.2KCl$.	$IrCl_3.3AgCl$.	$PtCl_2.2KCl$.	$PtCl_4.BaCl_2$.
	$OsCl_4.2NaCl$.	$IrCl_4.2KCl$.	$3PtCl_2.2AlCl_3$.	$PtCl_4.AlCl_3$.
	$OsCl_4.2AgCl$.		$2PtCl_2.SnCl_4$.	$PtCl_4.FeCl_3$.
				$PtCl_4.SnCl_4$.
				$PtCl_4.SeCl_4$.

Bromides and Iodides.

$PtBr_2.2KBr$.	$IrBr_3.3HBr$.	$IrBr_4.2KBr$.	$IrCl_4.NH_4I$.
$PtBr_2.CuBr_2$.	$IrBr_3.3KBr$.	$PtBr_4.2HBr$.	$IrI_4.2NH_4I$.
$IrI_2.2NH_4I$.	$IrI_3.3KI$.	$PtBr_4.2KBr$.	$PtI_4.2KI$.
	$IrI_3.3AgI$.	$PtBr_4.BaBr_2$.	

Also, $PtCl_4.PtI_4$, or $PtCl_2I_2$ is known.

A compound of platinum dichloride with phosphorus trichloride is formed by heating to 250° spongy platinum with phosphorus pentachloride; its formula is $PtCl_2.PCl_3$. The resulting crystals melt at 170°, and are soluble in carbon tetrachloride and in chloroform. It combines with chlorine to form the double compound $PtCl_3.PCl_4$.

The most important of these compounds are $PtCl_4.2HCl$, produced by direct addition, and the corresponding potassium and ammonium compounds, produced by double decomposition, thus:—

$$PtCl_4.2HCl.Aq + 2KCl.Aq = \mathbf{PtCl_4.2KCl} + 2HCl.Aq.$$

These compounds are yellow crystalline powders, sparingly soluble in water, and nearly insoluble in a mixture of alcohol and ether. As the similar sodium platinichloride dissolves in these solvents, potassium and ammonium are usually separated from sodium by precipitation as platinichlorides, and weighed as such. The ammonium salt at a red heat yields *spongy platinum* as a porous grey metallic mass. All these compounds, indeed, lose halogen when heated, leaving a mixture of the metal of the platinum group with the halide of the conjoined metal.

The mass of 1 c.c. of platinum dichloride is 0·87 gram at 11°. The mass of 1 c.c. of many of the platinichlorides has also been determined, but with these exceptions the physical constants are unknown.

Halides of Copper, Silver, Gold, and Mercury.

These elements resemble each other in their monohalides. The monochlorides, bromides, and iodides are all insoluble in water. They have a certain analogy with the compounds of the palladium and platinum groups, and in their formulæ correspond with those of the elements of the potassium group, in which the first three members are classed. Mercury, in the periodic table, is the last element in the magnesium group, which it resembles in the formulæ of its dihalide compounds.

Sources.—Silver chloride, **AgCl**, occurs native as *horn silver*, or *kerargyrite*, in waxy translucent masses. *Bromargyrite* is the name of native silver bromide, a lustrous yellow or greenish mineral. Chlorobromides of silver of the formulæ **3AgCl.AgBr, 3AgCl.2AgBr, 3AgBr.AgCl, 5AgCl.4AgBr,** and **3AgCl.AgBr** also occur native. Native iodide or *iodargyrite* is also found in yellow-green masses. **AgCl.AgBr.AgI** has also been found native.

Mercurous chloride, **HgCl**, or *horn quicksilver*, accompanies *cinnabar*, **HgS**, occurring in dirty-white crystals.

The following halides are known:—

	Fluorine.	Chlorine.	Bromine.	Iodine.
Copper..	Cu_2F_2; CuF_2.	Cu_2Cl_2; $CuCl_2$	Cu_2Br_2; $CuBr_2$.	Cu_2I_2; CuI_2.
Silver ..	AgF.	AgCl.	AgBr.	AgI.
Mercury.	HgF; HgF_2.	HgCl; $HgCl_2$.	HgBr; $HgBr_2$.	HgI; HgI_2.
Gold ...	—	AuCl; $AuCl_2$; $AuCl_3$.	AuBr — $AuBr_3$.	AuI — AuI_3.

Preparation.—1. **By direct union.**—Fluorine, chlorine, bromine, and iodine attack these elements when finely divided in the cold, but the action is promoted by heat. In this way cuprous and cupric chlorides and bromides, and cuprous iodide have been prepared, the monohalide being formed in presence of a small amount of halogen; but the dihalide with excess of halogen. Silver chloride, bromide, and iodide, mercurous and mercuric chloride and iodide and mercurous bromide, and gold dichloride, $AuCl_2$ or Au_2Cl_4, and the corresponding bromide, $AuBr_2$ or Au_2Br_4, have also been thus obtained.

The higher halides are often prepared by the action of a mixture of nitric and hydrochloric, or nitric and hydrobromic, acids on the elements (see p. 172). The free halogen attacks the metal, forming the halide. Thus mercuric chloride, $HgCl_2$, cupric chloride, $CuCl_2$, and auric chloride and bromide, $AuCl_3$ and $AuBr_3$, are produced in solution by this means: $3Hg + 6HCl.Aq + 2HNO_3.Aq = 3HgCl_2.Aq + 4H_2O + 2NO$; $Au + 3HBr.Aq + HNO_3.Aq = AuBr_3.Aq + 2H_2O + NO$.

2. **By the action of the halide of hydrogen on the metal.**—A solution of hydrogen iodide dissolves silver, forming the double halide, AgI.HI. Hydrochloric acid dissolves copper in presence of air: $Cu + O + 2HCl.Aq = CuCl_2.Aq + H_2O$.

3. **By heating a higher halide.**—Cupric chloride and bromide, $CuCl_2$ and $CuBr_2$, when heated, yield cuprous halide, Cu_2Cl_2 and Cu_2Br_2; and cupric iodide decomposes spontaneously into cuprous iodide, Cu_2I_2, and iodine. Aurous chloride is produced at 185° from auric chloride, and auric bromide yields

aurous bromide at 115°. Auric iodide decomposes spontaneously into aurous iodide and iodine.

4. **By the action of the metal on the higher halide.**—A solution of cupric chloride in hydrochloric acid, when shaken with scraps of metallic copper, is converted into the dichloride, thus:— **Cu** + $CuCl_2.nHCl.Aq = Cu_2Cl_2.nHCl.Aq$. Mercuric chloride or bromide triturated with mercury yields mercurous chloride or bromide.

5. **By double decomposition.**—(*a.*) **By the action of the halogen acid on the oxide or carbonate of the metal.**—All these compounds may be thus prepared. It is, however, not convenient for the preparation of insoluble compounds, inasmuch as the oxides, being insoluble, become coated over with a film of the insoluble halide and protected from the further action of the halogen acid. The following compounds have been prepared thus:—Cu_2F_2, CuF_2, Cu_2Cl_2, $CuCl_2$, $CuBr_2$, AgF, HgF (by the action of HF on Hg_2O); HgF_2, $HgCl_2$, $HgBr_2$, AuI (from Au_2O_3 and HI, thus:—$\mathbf{Au_2O_3} + 6HI = \mathbf{2AuI} + 3H_2O + \mathbf{I_2}$; the auric iodide, AuI_3, decomposing at the moment of its formation).

(*b.*) Other cases of preparation by double decomposition:—

$\mathbf{Cu_2Cl_2}$. This is the best method of preparation. A strong solution of copper sulphate, $CuSO_4$, and sodium chloride, NaCl, in equivalent proportions, is saturated with sulphur dioxide. The sulphur dioxide liberates hydrogen from water, itself forming sulphuric acid; and the nascent hydrogen removes chlorine from cupric chloride, produced by the interaction of copper sulphate and sodium chloride, precipitating cuprous chloride, thus:—

$$CuSO_4.Aq + 2NaCl.Aq = CuCl_2.Aq + Na_2SO_4.Aq; \text{ and}$$
$$2CuCl_2.Aq + 2H_2O + SO_2.Aq = \mathbf{Cu_2Cl_2} + H_2SO_4.Aq + 2HCl.Aq$$

$\mathbf{Hg_2Cl_2}$ may be similarly prepared from mercuric chloride, $HgCl_2$, and sulphur dioxide.

$\mathbf{Cu_2I_2}$. Copper sulphate, or any other soluble salt of copper, reacts with potassium iodide, giving in very dilute solution a blue solution of cupric iodide; in strong solution the cupric iodide decomposes into cuprous iodide and free iodine. The reactions are as follows:—

$$CuSO_4.Aq + 2KI.Aq = CuI_2.Aq + K_2SO_4.Aq; \text{ and}$$
$$2CuI_2.Aq = \mathbf{Cu_2I_2} + \mathbf{I_2} + Aq.$$

AgCl, **AgBr**, and **AgI**. These are prepared by adding a soluble salt of silver, *e.g.*, the nitrate, to the required halide of hydrogen, or to any other soluble halide, thus :—

$$AgNO_3.Aq + KI.Aq = \mathbf{AgI} + KNO_3.Aq.$$

AuF_3? An attempt to prepare auric fluoride by adding silver fluoride to auric chloride resulted in the precipitation of auric oxide, Au_2O_3, through the action of water on the fluoride, thus :—

$$2AuF_3 + 3H_2O = \mathbf{Au_2O_3} + 6HF.$$

AuI_3. Auric iodide is formed by the addition of auric chloride to potassium iodide, thus :—

$$AuCl_3.Aq + 4KI.Aq = AuI_3.KI.Aq + 3KCl.Aq.$$

The double iodide is decomposed on addition of more auric chloride, with precipitation of auric iodide :—

$$3KI.AuI_3.Aq + AuCl_3.Aq = 4AuI_3 + 3KCl.Aq.$$

HgF. Mercurous chloride, digested with silver fluoride, yields mercurous fluoride and silver chloride, thus :—

$$AgF.Aq + \mathbf{HgCl} = \mathbf{AgCl} + HgF.Aq.$$

HgCl, **HgBr**, and **HgI**. **By precipitation.**—Mercurous nitrate, $Hg(NO_3)$, and a soluble halide yield mercurous halide and a soluble nitrate, *e.g.*, $HgNO_3.Aq + NaCl.Aq = \mathbf{HgCl} + NaNO_3.Aq$. Another method of preparing **HgCl** is to sublime mercurous sulphate, **Hg_2SO_4**, with salt, **NaCl** :—

$$\mathbf{Hg_2SO_4} + \mathbf{2NaCl} = \mathbf{2HgCl} + \mathbf{Na_2SO_4}.$$

$HgCl_2$. Mercuric sulphate, **$HgSO_4$**, and salt yield mercuric chloride on sublimation; hence its name *corrosive sublimate*.

HgI_2. Mercuric iodide, being insoluble, is precipitated by addition of mercuric chloride to potassium iodide. The sesquiiodide, **$HgI_2.HgI$**, is similarly precipitated from a mixture of mercurous and mercuric nitrates by potassium iodide.

Properties.—These substances are all solid. The cuprous and mercurous, and the silver and aurous compounds are all insoluble in water, but dissolve in concentrated halogen acids; mercurous and aurous halides are decomposed when boiled with acids. Cuprous fluoride is a red powder, fusing to a black mass; when prepared by precipitation it is white. The chloride is also white, but is affected by light, which turns it dirty violet; it appears to lose chlorine. The bromide is greenish-brown, and the iodide brownish-white. Silver fluoride is a white soluble mass; the chloride is white, but turns purple on exposure to light. This is said to be owing to the formation of a subchloride, Ag_2Cl, inasmuch as the purple substance is not dissolved by nitric acid, in which silver itself is soluble. The bromide is pale-yellow, and the iodide darker yellow. These substances are used to detect and estimate the halogens, for they are almost absolutely insoluble in water. They melt to horny masses. Mercurous fluoride is a light-yellow crystalline powder, partly decomposed on boiling with water, and decomposed by heat. The chloride, the common name for which is *calomel*, is dirty white in colour, and also partially decomposes when volatilised, but its constituents recombine on cooling; hence it can be sublimed. It condenses as a fibrous, translucent, very heavy solid. It is quite insoluble in water. The bromide is also a fibrous yellow mass. The iodide is a greenish-yellow powder, sparingly soluble in water.

Aurous chloride, $\mathbf{AuCl}$, is white, insoluble in water, but decomposed on boiling with water into gold, and auric chloride, $\mathbf{AuCl_3}$. The bromide is also insoluble in water, and yellowish-grey in colour. It is decomposed by hydrobromic acid, thus:—

$$3\mathbf{AuBr} + HBr.Aq = AuBr_3.HBr.Aq + 2\mathbf{Au}.$$

Aurous iodide is an insoluble yellow powder, soluble in hydriodic acid.

The higher halides are all soluble in water. Those of mercury and cupric chloride are also soluble in alcohol and in ether.

Cupric fluoride forms sparingly soluble blue crystals; mercuric fluoride is a white crystalline mass.

Cupric chloride is a brownish-yellow deliquescent powder; it dissolves in water with a blue colour, and deposits blue crystals of $\mathbf{CuCl_2.2H_2O}$. The bromide consists of iron-black crystals, soluble in water with a brown colour.

Gold dichloride,* $\mathbf{Au_2Cl_4}$, is regarded as a compound of $AuCl_3$ with AuCl. Its molecular weight, however, is unknown. It is a hard dark-red substance, decomposed by water into $AuCl_3$ and AuCl. The trichloride, $\mathbf{AuCl_3}$, forms dark-red crystals, and is soluble in water, alcohol, and ether. The dibromide is a black substance, which reacts with water like the corresponding chloride, yielding monobromide and tribromide. The latter is

* *J. prakt. Chem.* (2), **37**, 105.

dark-brown and dissolves in water, alcohol, and ether. Auric iodide, **AuI_3**, is a dark-green precipitate, decomposing spontaneously into aurous iodide and iodine.

Mercuric chloride, or *corrosive sublimate*, is a white crystalline substance; 100 parts of water dissolve 7·4 parts at 20°; 100 parts of alcohol dissolve 40 parts at the ordinary temperature. The bromide crystallises in soft white laminæ. The iodide is a scarlet powder, sparingly soluble in water, more soluble in alcohol and ether. It crystallises from aqueous potassium iodide in red octahedra. When sublimed, it condenses in yellow prisms, which, when rubbed, suddenly change into red octahedra.

Physical Properties.

	Mass of 1 c.c.				Melting-point.				Boiling-point.			
	F.	Cl.	Br.	I.	F.	Cl.	Br.	I.	F.	Cl.	Br.	I.
Copper.	—	? (ous)	4·72	5·70	908°	434°	504°	601°	?	954†	861°	759°
Silver..	—	5·505 at 0°	6·215 at 17°	5·67	?	451°	427°	527°	?	?	?	White heat.
Gold...	—	?	?	?	?	250°*	115°*	*	*	*	*	*
Mercury	—	6·56 (ous)	7·31	7·64	?	‡	405°	290°	?	400°	‡?	310°
		5·45 (ic)	5·73	6·30	130°*	288°	244°	238°	?	303°	319°	339°

Double halides.—Cupric fluoride is said to combine with the fluorides of the alkaline metals to form black compounds. The following compounds of the other halides have been prepared:—

$Cu_2Cl_2.4HCl$.	**$HgCl_2.KCl$.**	**$CuCl_2.2HCl$.**	**$AuCl_3.KCl$.**
$Cu_2I_2.HgI_2$.	**$2HgCl_2.CaCl_2$.**	**$CuCl_2.2KCl$.**	**$AuCl_3.NaCl$.**
$AgF.HF$.	**$HgBr_2.KBr$.**	**$CuCl_2.2NH_4Cl$.**	**$2AuCl_3.CaCl_2$.**
$AgCl.NH_4Cl$.	**$2HgBr_2.SrBr_2$.**	**$HgCl_2.2NH_4Cl$.**	**$2AuCl_3.ZnCl_2$.**
$AgCl.KCl$.	**$HgI_2.KI$.**	**$HgI_2.2NH_4I$.**	**$AuBr_3.HBr$.**
$AgI.HI$.	**$HgCl_2.NH_4Cl$.**	**$HgCl_2.2KCl$.**	**$AuBr_3.KBr$.**
$AgI.KI$.	**$HgI_2.HgI$.**	**$HgI_2.2KI$.**	**$AuI_3.KI$.**
$AgI.2KI$.	**$2HgI_2.BaI_2$.**	**$HgI_2.HgCl_2$.**	
$2HgCl.SrCl_2$.	**$2HgCl_2.HgI_2$.**		
$2HgCl.SCl_2$.			

Besides these, $2HgCl_2.KCl$, $3HgCl_2.MgCl_2$, and $5HgCl_2.CaCl_2$ are known, in which the mercuric chloride bears a larger ratio to the other chloride than in the tabulated examples. The name aurichlorides (sometimes, but incorrectly, "chloraurates") has been applied to the compounds of auric chloride. The compound $Cu_2I_2.HgI_2$ is a red body, and has the curious property of turning black when heated. It has been used as a means of indicating whether the axles of engines become superheated. The compound $HgCl_2.2NH_4Cl$ has been

* Decomposes.
† Between 954° and 1032°; $CuCl_2$, 498°; Cu_2Br_2, 861–954°.
‡ Sublimes between 400° and 500° without melting.

known since the times of the alchemist, and was termed by them *sal alembroth.* All these bodies are prepared by direct addition. Those of silver are decomposed on dilution, giving precipitates of halides. The compound $HgCl_2.SnCl_2$ is produced by subliming an alloy of tin and mercury with mercurous chloride.

The molecular weights of some of these compounds have been determined. The density of cuprous chloride, Cu_2Cl_2, was found to be 102·0, while the calculated number for that formula is 106·86.* Silver chloride gave a density corresponding to the molecular weight 160·8, instead of the theoretical one, 143·39, for the formula AgCl.† As regards mercurous chloride, it is most probable that the molecular weight is that equivalent to the formula HgCl. It is not difficult to vaporise mercurous chloride; the difficulty has been to ascertain whether it decomposes, in the state of gas, into mercuric chloride, $HgCl_2$, and mercury, or is stable. In each case the density found corresponded to the formula HgCl, not to the formula Hg_2Cl_2. The actual number was 231·8; the calculated molecular weight, 235·4. The density was determined in presence of an atmosphere of mercuric chloride, and under these circumstances little or no dissociation takes place.‡

The molecular weights of the remaining halides are unknown, but the formulæ have been made to accord with those of which the value has been ascertained.

* *Berichte,* 2, 1116.
† *Proc. Roy. Soc. Edin.*, vol. 14.
‡ *Gazzetta,* 1881, 341; *Chem. Soc. Abs.,* 42, 466.

CHAPTER XIV.

REVIEW OF THE HALIDES; THEIR SOURCES, PREPARATION AND PROPERTIES, PHYSICAL AND CHEMICAL.—THEIR COMBINATIONS.—THEIR REACTIONS WITH WATER AND HYDROXIDES.—CONSIDERATION OF THEIR MOLECULAR FORMULÆ.

Having concluded the description of the compounds of the halogens with other elements, and with each other, it may be here advisable to give a summary of their leading features. This will be done in the same order as that observed in the special description of each class of compounds, viz., their sources, their preparation, and their properties.

1. **Sources.**—If a compound occur free in nature, it must either be unacted on by substances around it at the temperature at which it exists, or must have only an ephemeral existence. The two most important and widely spread agents are the oxygen of the air and water. It must, therefore, be able to resist the combined action of both of these substances.

As an instance of a compound produced under certain unusual circumstances, hydrogen chloride may be named. It is found in the air and water in the neighbourhood of volcanoes; but, although not altered by air or water, it soon is dissolved by the rain, and reacts with the constituents of the soil, forming chlorides of calcium, sodium, potassium, &c., which ultimately find their way into the sea, being carried down by rivers. It is, therefore, only found in the locality where it is formed before it has been exposed to those influences. Ferric chloride, $\mathbf{Fe_2Cl_6}$, occurs under similar conditions.

The chlorides, bromides, and iodides of lithium, sodium, potassium, calcium, and magnesium are all soluble in water. It is not improbable that they are partially decomposed by solution; thus, for example, $NaCl + H_2O = NaHO + HCl$. But when such a solution is evaporated, the reaction, if there is one, occurs in the inverse sense, and the water evaporates, leaving the chloride. By the evaporation of inland lakes, such as the Dead Sea, these salts are deposited. Such has doubtless been the case where mines of rock salt exist; and at Stassfurth, in N. Germany, the layers of salt are found in the order of their solubility, the least soluble forming the lowest layers.

Insoluble salts, such as fluorspar (calcium fluoride), cryolite (aluminium sodium fluoride, $\mathbf{AlF_3.3NaF}$), silver chloride, bromide, and iodide, lead chloride, &c., which are not attacked by water or oxygen, are also found in nature.

Preparation.—The general methods of preparation may be summed up as follows:—

1. **Direct union.**—The halides may, as a rule, be thus prepared. Fluorine appears to act on all elements, oxygen and nitrogen excepted, at the ordinary temperature. The metals iridium and platinum are, perhaps, the least affected of any in the cold; hence the use of an alloy of these metals in forming the vessel in which fluorine was isolated by electrolysis. Chlorine, when dry and cold, appears not to attack some metals, such as sodium and zinc, which are readily acted on when hot; but, as a rule, the elements combine with chlorine, bromine, and iodine when heated in contact with them. Those which do not combine, even at a red heat, are carbon, nitrogen, and oxygen.

2. **Replacement.**—Action of a compound of the halogen on the element; or action of the halogen on a compound of the element. The most common instance of the first method is the action of the halide of hydrogen on a metal. A list of the elements not thus attacked is given on p. 112. But there are many other processes involving similar reactions, where the method is not used as a means of preparing a halide, but of liberating the element with which the halogen was in combination. The elements magnesium, boron, aluminium, silicon, and others are prepared by the action of sodium or potassium on their halides, which, of course, results in the formation of sodium or potassium halides. The action of the halogen on a compound of the element, of which the halide is required, is also a method not frequently employed; for, owing to the fact that there are few elements which do not combine with the halogen, a mixture of two halides is thus obtained, which are often not easily separated. An instance of its application, however, is found in the preparation of hydrogen iodide, by the action of iodine and water on hydrogen sulphide; and of carbon tetrachloride, by the action of chlorine on carbon disulphide. The preparation of nitrogen, too, by the action of chlorine on ammonia would also come under this head, yielding hydrogen chloride.

3. **Double decomposition.—Mutual action of two compounds on each other, one containing halogen.**—This is,

perhaps, the most usual method of preparing compounds of the halogens. As a rule, the resulting halide must be gaseous or solid, or water or hydrogen sulphide must be the product of the action. Instances of such action are very numerous. Among them may be mentioned the action of sulphuric or phosphoric acid on halides of the metals, whereby the hydrogen halide is formed; the action of the halides of boron, silicon, phosphorus, &c., on water; the action of a halide of hydrogen on oxides, hydroxides, sulphides, or carbonates of the metals; the action of calcium chloride on barium sulphate at a red heat; the precipitation of calcium fluoride; the preparation of magnesium chloride; of boron fluoride; boron chloride; and many other cases. The method is almost universally applicable; but it does not yield halides of nitrogen or of oxygen.

A special method, applicable to the preparation of aluminium chloride, is the action of the vapour of carbon tetrachloride on the red-hot oxide. The simultaneous action of carbon and chlorine on the oxides of silicon, boron, &c., at a red heat can hardly be considered double decomposition, inasmuch as the chlorine and carbon are not combined, but it is difficult to classify such actions elsewhere, unless they be regarded as cases of direct union.

To distinguish the halogens when all three may be present, the mixture is distilled with strong sulphuric acid and potassium dichromate. If chlorine be present, the volatile chromyl chloride, $CrOCl_2$, is produced, and distils over. If the distillate contains chlorine, chromium will be found therein. To detect bromine and iodine in presence of each other, chlorine-water is gradually added to the solution of their sodium or potassium salts, and the liquid is shaken with carbon disulphide or chloroform, which do not mix with water. If iodine be present, a violet solution is obtained; if bromine be also present, further addition of chlorine-water will destroy the violet colour of the chloroform or carbon disulphide, and it will be replaced by an orange-red colour.

4. If two or more halides exist, the compound containing most halogen may almost always be prepared by **heating the one containing less with the required halogen.** Thus, iron dichloride yields the trichloride when heated in chlorine; mercurous is converted into mercuric chloride; stannous into stannic, &c.

5. **By heating the higher halide,** in certain cases, the halogen is evolved, and the lower halide is left. Thus, thallic chloride, $TlCl_3$, yields thallous chloride, $TlCl$, when heated; and auric chloride similarly gives aurous chloride, two atoms of chlorine being lost.

Sometimes, but rarely, the lower halide decomposes into the element and the higher halide. This is the case with bismuth dichloride, **$BiCl_2$**. It is sometimes necessary to heat in contact with some element capable of combining with the halogen. For example, aluminous sodium fluoride, **$AlF_2.2NaF$**, is prepared by heating cryolite with metallic aluminium; the compounds **$GaCl_2$**, **$InCl_2$**, and **InCl**, by heating **$GaCl_3$** and **$InCl_3$**, with gallium and indium respectively; disilicon hexachloride is similarly prepared from the tetrachloride; and chromous chloride, **$CrCl_2$**, results from the action of hydrogen at a red heat on **$CrCl_3$**; the lower chlorides of titanium, molybdenum, and tungsten are also prepared thus.

Sometimes the removal of halogen from the higher halide may be accomplished in solution. Thus, the familiar operation of "reducing" ferric chloride in solution by means of the hydrogen generated from zinc and hydrochloric acid, or by sulphur dioxide, or by stannous chloride, falls under this head; also the formation of mercurous from mercuric chloride, and that of osmium di- and trichlorides, and iridium di-iodide. Hydrogen sulphide is also used as a reducing agent for ferric halides, for rhodium trichloride, &c.

Properties.—(*a.*) **Physical properties**:—*Colour.*—The colour of objects is due to their absorbing light rays of certain wave-lengths in the visible part of the spectrum. It is to be noticed that the iodides of those metals which form white fluorides, chlorides, and bromides often are yellow or red; as examples, the cases of thallium, silver, mercury, &c., may be noticed. In general, those halides with higher molecular weights towards the end of the periodic table display colour. But substances which appear colourless to our eyes have the power of absorbing vibrations of wave-lengths which do not affect our sight, and to eyes sensitive to other scales of vibration than ours such bodies would appear coloured. It may also be generally stated that halides containing a large proportion of halogen display colour when those containing less are colourless.

Form.—The halides are almost without exception **crystalline**, but up to the present their crystalline form has not yet been connected with their chemical nature (see Isomorphism, Chap. XXXV).

State of aggregation.—Compared with the oxides and sulphides, the halides may generally be said to be easily fusible and volatile. This is probably due to their simplicity of structure and low molecular weight. The fluorides, however, have, as a rule, greater complexity than the chlorides, bromides, and iodides. For example, hydrogen fluoride is known to have a more complex molecule than hydrogen chloride, even in the gaseous state (see p. 115); and the non-volatility of many fluorides, compared with the volatility of the corresponding chlorides, would lead to the inference that their molecules are complex. Some fluorides, however, such as those of boron and silicon, have undoubtedly simple formulæ; and it is to be remarked that these bodies are very stable. The comparative insolubility of many fluorides, *e.g.*, those of calcium, strontium, barium, magnesium, tin, &c., may also point to complex molecular structure; and further evidence may be derived from the fact that the fluorides form double compounds more easily than the other halides.

The solubility of a compound, however, may perhaps partly depend on its chemical action on the solvent, though probably not invariably. It certainly appears to be connected with simplicity of molecular structure, implying low molecular weight.

The mass of one cubic centimetre of the halides also shows regularity. The iodides are, as a rule, specifically heavier than the bromides; the bromides than the chlorides; the chlorides, however, are not always heavier than the fluorides; but, again, this may depend on molecular complexity, contraction always occurring when chemical union occurs, even between molecules of the same kind. It is also to be noticed that, in each group of elements, the halides of those which possess the highest atomic weights are specifically heavier than the earlier members of each series.

(*b.*) **Chemical properties.**—Some halides, when heated, decompose into their elements, or into lower halides and halogen. It is probable, indeed, that at a sufficiently high temperature all chemical compounds would decompose thus. In certain cases, for example, the halides of nitrogen, oxygen, and carbon, when the elements are once apart, they do not again combine. The halides of oxygen and nitrogen are formed, not, as usual, with evolution of heat, but with absorption, and such compounds are always readily decomposed. Those of nitrogen and of oxygen are exceedingly explosive, and cannot be produced by direct union. Other halides, such as those of gold, platinum, &c., decompose into their elements when heated, but if kept in contact the elements would again recombine. But, as the metallic element is volatile only at a very high temperature, the halogen, which is easily volatile, distils away, leaving the metal. Other halides, such as the higher ones of selenium, phosphorus, and antimony, are also decomposed, but the lower halide is not so different in volatility from the halogen itself; hence, the two are difficult to separate. When a compound decomposes into constituents which reunite on cooling, it is said to *dissociate.* The term *decomposition* includes dissociation, but may be employed in the stricter sense of splitting up without recombination. There is a temperature of decomposition peculiar to each compound, at which, if recombination does not occur, after sufficient time, all the compound would be decomposed; whereas, if recombination is possible, a state of balance is maintained, the relative proportions of the constituents depending on the temperature, on the pressure, and on the relative amounts of the constituents. Excess of one constituent prevents decomposition. Thus, phosphorus pentachloride is stable in the gaseous form in presence of excess of chlorine or of phosphorus trichloride, and mercurous chloride can exist as gas in presence of mercuric chloride. These statements probably also apply to bodies in solution.

The halogens are capable of replacing each other. Here, again, the relative amounts have a great influence on the result. Bromine replaces iodine from its compounds with elements of the potassium, calcium, and magnesium groups dissolved in water; and chlorine replaces bromine and iodine. But a current of bromine vapour led over hot potassium chloride results in the formation of potassium bromide. Again, on digesting precipitated silver chloride with bromine-water, silver bromide is formed; and iodine, under similar circumstances, replaces both chlorine and bromine. Yet, on heating silver iodide in a current of chlorine or bromine, the iodine is expelled, and replaced by chlorine or bromine. In these cases, the mass of the halogen acting on the halide has the effect of reversing the process which takes place in presence of water.

Combinations.—The halides of the elements in most cases combine with water to form crystalline compounds containing water of crystallisation. It is sometimes, but not always, possible to expel such water by heat; in many cases, the water reacts with the halide, forming hydroxide, oxide, or oxyhalide. The crystalline form is altered by the presence of the water, and when several hydrates exist, they have usually different crystalline forms. The lower the temperature, the greater the amount of water with which the substance will combine. A halide crystallising without water at the ordinary temperature sometimes forms a hydrate at low temperatures, as is the case with sodium chloride. The remarkable change of colour of some halides, *e.g.*, those of nickel, cobalt, iron, &c., when hydrated appears to point to some profound modification in molecular structure by hydration; and the persistence of this colour in dilute solution leads to the inference that the hydrate exists dissolved in the water. It has been pointed out that compounds of halides with hydrogen halides invariably contain two molecules of water of crystallisation for every molecule of hydrogen halide present.

Double halides.—The halides in almost all cases, as has been seen, combine with each other, forming double compounds. These are usually prepared by mixing solutions of the two halides of which it is desired to form a compound, and evaporating the mixture, best at the ordinary temperature, for a low temperature is favourable to combination. The compounds with halides of hydrogen are generally, but not always, called acids. In many cases they are exceedingly unstable, and mere removal from the presence of a strong solution of the halogen acid is sufficient to decompose them, the hydrogen halide escaping as gas. They usually crystallise with water, if, indeed, they can be obtained crystalline; the anhydrous compounds are rare. Of the four halogens fluorine is most prone to form double compounds. This is probably connected with the tendency of its compounds to *polymerise*, *i.e.*, the tendency for several molecules to enter into combination with each other. It is probable, indeed, that there is no difference in kind between compounds of two molecules of the same halide, such as $\mathbf{Fe_2Cl_6}$ (which may be regarded as a compound with each other of two molecules of $FeCl_3$), and compounds produced by the union of the halides of two different elements, such as $\mathbf{PtCl_4.2KCl}$, $\mathbf{SbCl_5.SCl_4}$, &c.; such bodies, however, exhibit very different degrees of stability, certain of them withstanding a fairly high temperature without decomposition, so far as can be ascertained, while others exist only at a low temperature. If one of the halide constituents of a double halide is easily decomposed by heat, it is usually rendered more stable by combination; although on heating such a double halide, the more easily decomposable halide is decomposed, while the more stable one resists decomposition. An instance is given above; $\mathbf{SbCl_5.SCl_4}$ is stable at the ordinary temperature, while SCl_4 can exist only below $-22°$; but on heating the double halide chlorine is evolved, while the stable chloride of sulphur, S_2Cl_2, is formed, the antimony pentachloride remaining unaffected. Similarly the other double halide mentioned above, $\mathbf{PtCl_4.2KCl}$, when heated, decomposes, a mixture of metallic platinum and potassium chloride being left, while chlorine is evolved. Here, again, the comparatively unstable platinum tetrachloride is decomposed, the stable potassium chloride resisting decomposition. It is said that solution in water decomposes such double halides into their constituent halides. But it appears more likely that the degree of decomposition depends on the relative proportion of water and double halide, and on the temperature of the solutions; and that such a solution really contains in many cases both the double halide and the two simple halides. With increase of solvent, or with rise of temperature, it is probable that the relative amount of the double halide decreases, while that of the single halides increases. These are matters, however, still involved in considerably obscurity.

Action of water.—The action of water on many of the halides is to decompose them, hydrogen halide and the oxyhalide or hydroxide of the element being produced. The following halides are *known* to be thus decomposed by water:—(*a.*) At the ordinary temperature:—Halides of boron, silicon, zirconium, germanium; tetrahalides of tin; halides of phosphorus, arsenic, antimony, bismuth, vanadium, niobium, tantalum, molybdenum, tungsten, uranium, sulphur, selenium, and tellurium. In certain cases the halide is not decomposed in presence of great excess of hydrogen halide, even although water be present, possibly owing to the formation of a double halide of the element and hydrogen. This is known to be the case with the fluorides of boron and of silicon, which form the compounds $BF_3.HF$, and $SiF_4.2HF$, which are stable even in presence of a large amount of water. Arsenic, antimony, and bismuth trihalides dissolve in excess of halogen acid, probably forming similar stable compounds. (*b.*) At a red heat, most of the halides react with water-gas to form the oxides, those of lithium, sodium, potassium, rubidium, and caesium excepted.

It is, however, not improbable that, as has been already stated, solutions of all halides in water are partially decomposed by the water, sodium chloride, for example, reacting to form sodium hydroxide and hydrogen chloride, thus:—$NaCl + H_2O = NaOH + HCl$; and so with other chlorides. The degree of this decomposition depends, no doubt, largely on the relative amounts of water and halide, and on the temperature, and varies for each salt. The presence of a second halide appears in many cases to retard or diminish such decomposition, and to render salts stable in solution which would decompose or react with water in their absence.

Action of hydroxides.—Halides which are not decomposed by water, so that their constituents can be separated, and which are not re-formed on alteration of temperature, dilution, &c., can in most cases be decomposed by a soluble hydroxide. Thus sodium or potassium hydroxides react with almost all halides producing hydroxides, that is, oxides in combination with water. Ammonia dissolved in water has in most cases a similar action, the solution acting as if it were hydroxide of ammonium, NH_4OH. In many instances, particularly if the element belongs to the class generally termed "non-metals," the hydroxide produced combines with the reacting hydroxide, forming a double oxide, or salt, and water. Oxides such as these are termed "acid-forming oxides," or "chlorous" oxides; those which have less tendency to such combination being named "basic" or "basylous" oxides. The following instances will exemplify what has been stated:—

The action of potassium hydroxide on cupric chloride is to form potassium chloride and cupric hydroxide, thus:—

$$CuCl_2.Aq + 2KOH.Aq = \mathbf{Cu(OH)_2} + 2KCl.Aq.$$

Cupric hydroxide may be viewed as a distinct individual, or as a compound of cupric oxide, CuO, with water. This point will be discussed later. A great excess of caustic potash, KHO, develops the slight power of combination of copper oxide, which dissolves with a blue colour, forming, no doubt, some compound such as $CuO.K_2O$, or $Cu(OK)_2$. Such a compound is certainly formed by the action of zinc chloride, $ZnCl_2$, on caustic potash, KOH, the body $Zn(OK)_2$ being produced. But this kind of change is the usual and normal one of the chlorides of those elements whose halides are decomposed by water; thus phosphorous chloride at once gives with water phosphorous acid, H_3PO_3, or $P(OH)_3$ (?), and with caustic potash, KOH, potassium phosphite, the caustic potash reacting thus with the phosphorous acid:—

$$2KOH.Aq + H_3PO_3.Aq = HK_2PO_3.Aq + 2H_2O.$$

As the hydroxides when heated are as a rule transformed into oxides with loss of water, this forms one of the most convenient methods of preparing hydroxides and oxides, as will soon appear.

The formulæ of the halides are, as a rule, undoubtedly simple. It has already been remarked that we do not know with certainty the formulæ of liquids and of solids, inasmuch as their molecular complexity is unknown. But it is probable that mere change of physical state from gas to liquid, or from liquid to solid, is not necessarily accompanied by chemical aggregation. Thus, if the formula of hydrogen chloride as gas is *HCl*, and if no sign of aggregation is seen on its approaching its temperature of liquefaction; that is, if its contraction on cooling runs *pari passu* with that of hydrogen, there would appear to be no good reason to suppose that merely because it has liquefied its formula is thereby rendered more complex; but where, as in the case of hydrogen fluoride, distinct signs of molecular aggregation are to be noticed as the temperature falls, no doubt can be entertained as regards the fact that the molecular structure is complex in liquid hydrogen fluoride; but that it begins to occur before the liquid state is reached would appear to negative the supposition that it is directly connected with change of state. In the present state of our knowledge, therefore, it may be concluded that the formula possessed by a halide in the gaseous state also represents its molecular weight in the liquid condition, although there may well be examples of aggregation beginning in the liquid or solid states with fall of temperature, which are not to be detected by determination of the density of the gas. A full discussion of this point is better reserved until the oxides and sulphides have been studied; for there is strong ground for the belief that their molecular structure is complex.

In every case, however, where the molecular complexity of a compound is unknown, the simplest formulæ have been adopted.

These formulæ are deducible :—

1. **From the results of analysis, which yields the equivalents of the elements, but gives no information as regards their atomic weights.**
2. **By the law of simplicity, as applied by Dalton and Berzelius.**
3. **By use of Avogadro's law, that equal volumes of gases contain equal numbers of molecules: the chief method of investigation being the method depending on the vapour-densities of compounds.**
4. **From the atomic heats of the elements (Dulong and Petit's law).**

Other methods will be considered in a subsequent chapter.

Detection and Estimation of the Halogens.—Fluorine is detected by heating the suspected fluoride with strong sulphuric acid, and trying if the gas evolved will etch glass, *i.e.*, will produce silicon fluoride. Chlorine, bromine, and iodine, when in combination, are detected by adding to a solution of the suspected compound in nitric acid a solution of silver nitrate. A chloride gives a white precipitate; a bromide, a yellowish precipitate; an iodide, a yellow precipitate. These may be further distinguished by addition of excess of aqueous ammonia. Silver chloride easily dissolves; the bromide is sparingly soluble; and the iodide insoluble.

PART IV.—THE OXIDES, SULPHIDES, SELENIDES, AND TELLURIDES.

CHAPTER XV.

OXIDES, SULPHIDES, SELENIDE, AND TELLURIDE OF HYDROGEN.—VOLUME-COMPOSITION. — PHYSICAL PROPERTIES. — ATTEMPTS TO ASCERTAIN THE QUANTITATIVE COMPOSITION OF WATER.—DOUBLE COMPOUNDS.

The elements oxygen, sulphur, selenium, and tellurium, like the elements fluorine, chlorine, bromine, and iodine, combine readily with other elements, and many of their compounds have been carefully studied. Like the halogens, these four elements bear a marked resemblance to each other, oxygen being the analogue of fluorine, while the other three elements correspond more or less closely to chlorine, bromine, and iodine. The previous arrangement of matter will be adhered to; but additional paragraphs must be added, describing the double compounds of the elements of this group with those of the halogens and with each other.

Compounds of Oxygen, Sulphur, Selenium, and Tellurium with Hydrogen.

Hydrogen oxides, sulphides, selenide, and telluride; H_2O; H_2O_2; H_2S; H_2S_3; H_2Se; H_2Te.

Sources.—Water, H_2O, is the most widely distributed of compounds, and occurs in larger proportion in nature than any other. It forms the sea, lakes, and rivers; as ice it caps the tops of high mountains, and covers the land in the neighbourhood of the North and South Poles; in the form of small liquid particles it forms clouds, fogs, and mist; its vapour is always present in the atmosphere in greater or less amount, and is known as "humidity." It is a constituent of many minerals, and of all organised beings, vegetable and animal, forming from 70 to 95 per cent. of their weight. It is conjectured, from the appearance of the planets

Mars and Venus, that their atmospheres contain water-vapour, and that their land is intersected by seas. It has not been proved to exist in the Moon, and it probably does not exist as such in the Sun.

Hydrogen dioxide, H_2O_2, is present in minute amount in rain and snow, and in all natural waters,* and, being a body prone to give up oxygen, probably plays an important part in oxidising dead vegetable and animal matter. It appears to be produced by the evaporation or exposure to light of water in which oxygen gas is dissolved.

Hydrogen sulphide, H_2S, escapes from fissures in the earth in volcanic districts, and is a constituent of many mineral springs; such waters are termed "hepatic," and are used as a cure for diseases of the skin. The wells at Harrogate are much frequented on this account. It is not widely spread, being slowly oxidised on exposure to air. Hydrogen selenide and telluride do not occur in nature.

Preparation.—1. By direct union.—(*a.*) **Water.**—A mixture of hydrogen and oxygen gases in the proportion of two volumes of the former to one volume of the latter explodes violently when heated to its igniting point at the ordinary pressure, forming water. The fact that by the union of hydrogen with oxygen water is the *sole* product was first proved by Cavendish, though its true nature was first determined by Lavoisier.

The combination may be easily shown by filling a strong soda-water bottle two-thirds full of hydrogen and one-third with oxygen, and after wrapping it in a cloth, for fear of the glass being shattered to fragments by the explosion, applying a lighted taper to the mouth. A violent explosion will occur, owing to the sudden expansion of the water-gas caused by the heat evolved by the union of its constituents.

The quantitative relations between the volumes of the gases and their product, water-gas, may be shown in a more instructive manner as follows:—

A is a strong U-tube, of about 15 mm. in internal diameter, with platinum wires sealed through its upper end, surrounded by a jacketing tube, B, in the bulb of which water is boiled. A is filled with dry mercury, and placed in position in a mercury trough. A mixture, obtained by electrolysing water (see below), of oxygen and hydrogen in the approximate proportions of two volumes of hydrogen to one volume of oxygen is introduced into the tube A, so as to fill it about one-third. The water is then boiled so as to jacket the inner tube, A, with steam. The mixed gases expand, and when the temperature has become constant the mercury is run out by opening the stop-cock C until it is level in both limbs of the U-tube. The level of the gases in then marked by an india-rubber

* Schöne, *Berichte*, 7, 1693.

ring, and mercury is again allowed to flow out so as to reduce the pressure on the gas. A spark from an induction coil is then caused to pass between the platinum wires sealed through the glass. The gases are heated to their temperature of ignition; the portions thus heated unite, and the heat evolved by the union raises the neighbouring portions to their ignition-point. An explosion takes place, but owing to the increased volume of the gas, it is not so violent as it would be at atmospheric pressure and ordinary temperature.

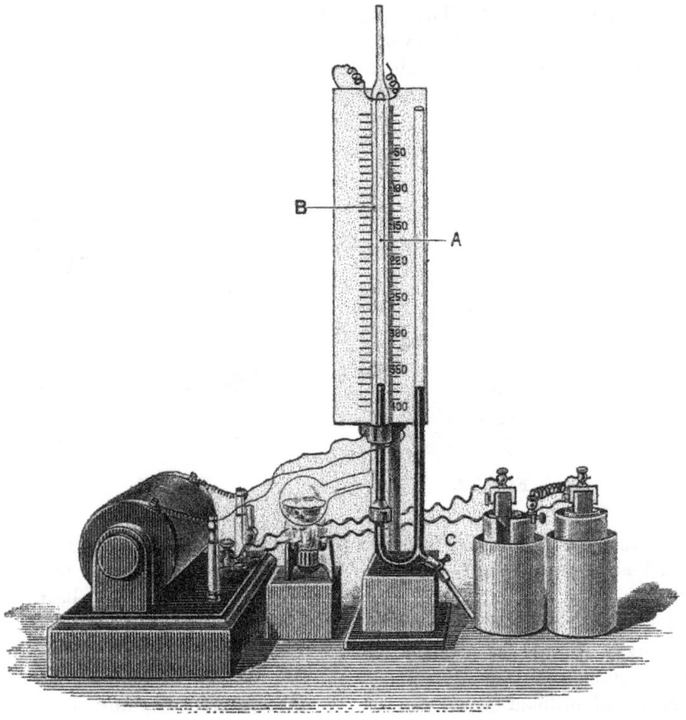

FIG. 29.

The gases after combination contract, and, to bring them back to atmospheric pressure, mercury is poured into the open limb of the U-tube until it stands at equal height in both limbs. The volume of the water-gas is seen to be about two-thirds of that of the mixed gases before combination; three volumes have become two. This experiment is adapted only as an illustration; it is inaccurate owing to the non-introduction of various corrections; for example, a mixture of oxygen and hydrogen, prepared by electrolysis, contains ozone (see p. 387), and hence occupies too small a volume; and some water-vapour condenses on the glass, and hence possesses a smaller volume than it ought to occupy.

Oxyhydrogen blowpipe.—By forcing mixed hydrogen and oxygen gases through a narrow tube and setting them on fire, a pointed flame is produced of a very high temperature. But the rate of explosion of a mixture of these gases is very rapid, and there is great danger of the explosion travelling back through the narrow tube and inflaming the mixture. Hence a special form of blowpipe must be employed. The temperature of ignition of the mixed gases is a high one; probably 600° to 700° at the ordinary pressure. By cooling the gases below this temperature they will not ignite. The cooling is effected by passing the mixed gases through a tube filled with copper gauze, or packed with fragments of copper wire. The explosion cannot travel back through such a tube, for the flame is extinguished owing to its giving up its heat to the copper, which is a good conductor of heat. The danger of explosion can be thus avoided. An almost equally hot flame, however, may be produced without danger by urging oxygen under pressure through a flame of hydrogen gas by a blowpipe of the form shown in fig. 30. The temperature of such a flame is estimated at

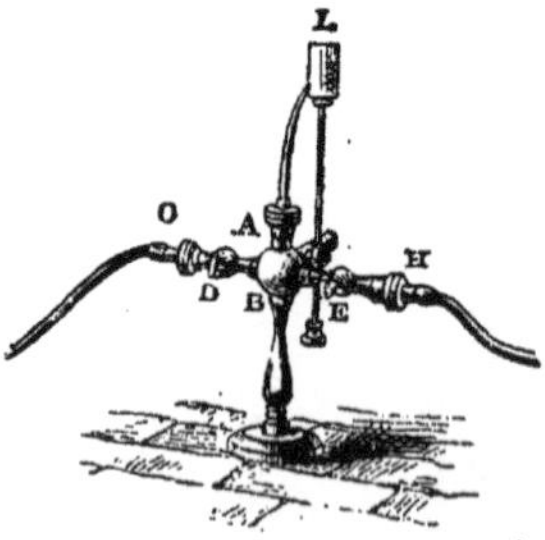

FIG. 30.

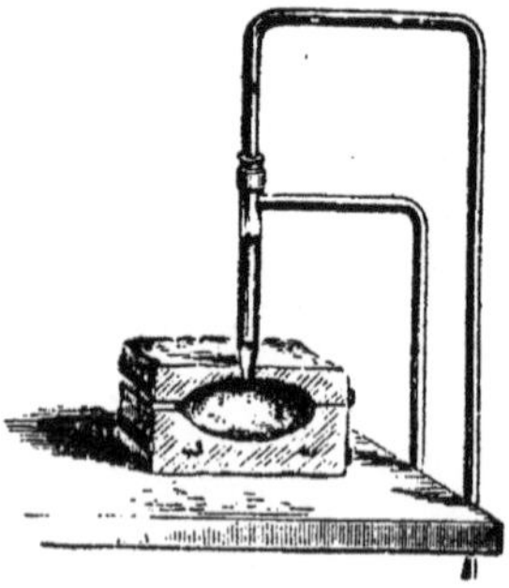

FIG. 31.

2200° to 2400°. The very infusible metal, platinum, can be melted, and even boiled when thus heated; silica can be melted and drawn into threads like glass; and the stem of a pipe, which is composed of aluminium silicate, can be softened and bent. With such a flame the hardest glass (combustion glass) can be worked as easily as ordinary glass; and when directed on a piece of lime or of zirconium oxide, a dazzling light is emitted, the solid being raised to the temperature of brilliant incandescence. Coal-gas, which contains about 50 per cent. of hydrogen, is usually substituted for hydrogen in such experiments: the temperature, though not quite so high, is still high enough for practical purposes. The applications of such a blowpipe are the fusion of platinum and iridium, and the production of the lime-light, or, as it is named from its discoverer, Captain Drummond, the "Drummond" light (Fig. 30). The crucible shown in Fig. 31 is made of lime, which is almost the only material capable of withstanding such a high temperature without softening. In it such metals as platinum, iridium, &c., can be melted.

Hydrogen dioxide, H_2O_2, is also formed in small amount when water is evaporated; it exists in very minute quantity in all natural waters, and is apparently produced by the action of heat and light on water containing oxygen in solution.

Hydrogen sulphide.—Hydrogen burns in sulphur vapour,

with formation of monosulphide, H_2S. This may be shown by boiling sulphur in a flask, and introducing a jet of burning hydrogen into the vapour; the hydrogen continues to burn feebly in the sulphur gas. **Selenium** and **tellurium** also unite directly with hydrogen at about 500°.

2. **By replacement.**—(*a.*) **Action of hydrogen on an oxide.**—This process has already been alluded to as a means of obtaining the elements indium, iron, germanium, tin, and lead, nitrogen, arsenic, antimony, and bismuth, tungsten, the metals of the platinum group, and copper, silver, mercury, and gold. The method consists in heating the solid oxide to redness in a tube through which a current of hydrogen is passing, when the hydrogen unites with the oxygen of the oxide, forming water, and the reduced element is left. In some cases higher oxides, such as manganese dioxide, chromium trioxide, &c., are reduced not to the state of element, but only to lower oxides. Similar experiments on sulphides, selenides, and tellurides have not been thus carried out, but would doubtless prove efficient in many cases.

(*b.*) **Action of oxygen, sulphur, &c., on a compound of hydrogen.**—All compounds of hydrogen, excepting hydrogen fluoride, are thus decomposed by oxygen. This is the principle of Deacon's chlorine process (p. 74), and of the manufacture of lampblack (p. 45); while a useful method of preparing hydrogen sulphide consists in heating a mixture of paraffin wax (a mixture of compounds of carbon and hydrogen) with sulphur. The sulphur replaces some of the hydrogen, which combines with excess of sulphur to form hydrogen sulphide. Similarly, by heating selenium with colophene, hydrogen selenide is continuously evolved.

3. **By double decomposition.**—**Water** is produced by innumerable interactions of this kind. For example, when many oxides, hydroxides, carbonates, silicates, &c., are treated with halogen acids, halides are formed together with water. This is also the usual and only available method of manufacturing **hydrogen dioxide.*** For this purpose barium dioxide is dissolved in dilute hydrochloric acid until the acid is nearly neutralised. Dilute baryta-water is then added to the filtered and cooled solution in order to precipitate foreign oxides and silica, which are often present as impurities in commercial barium dioxide. The solution, again filtered, is again treated with a strong solution of barium hydroxide, which throws down a precipitate of hydrated

* Thénard, *Annales* (2), **9**, **441**; **10**, 114, and 335; **11**, 208. *Berichte*, **7**, 73; *Annalen*, **192**, 257.

barium peroxide. This precipitate is filtered and washed until free from hydrogen chloride. It is then added to dilute sulphuric acid (1 part H_2SO_4 to 5 parts H_2O) with constant stirring, until the acid is nearly neutralised. The precipitated barium sulphate, which is practically insoluble in water, is then removed by filtration, and the small trace of sulphuric acid remaining is precipitated by careful addition of dilute baryta-water. The slight precipitate is allowed to settle, and the clear liquid decanted and evaporated in a vacuum over strong sulphuric acid. The equations are as follows:—

$$\mathbf{BaO_2} + 2HCl.Aq = BaCl_2.Aq + H_2O_2.Aq;$$
$$Ba(OH)_2.Aq + H_2O_2.Aq = \mathbf{BaO_2.8H_2O} + Aq;$$
$$\mathbf{BaO_2.8H_2O} + H_2SO_4.Aq = \mathbf{BaSO_4} + H_2O_2.Aq.$$

Hydrogen sulphide, selenide, and **telluride** are also usually prepared by double decomposition. Sulphide of iron, **FeS**, is treated with dilute sulphuric acid; or sulphide of antimony, $\mathbf{Sb_2S_3}$, or selenide of zinc or telluride of magnesium, **ZnSe** or **MgTe**, are treated with hydrochloric acid, thus:—

$$\mathbf{FeS} + H_2SO_4.Aq = FeSO_4.Aq + \mathit{H_2S};$$
$$\mathbf{Sb_2S_3} + 6HCl.Aq = 2SbCl_3.Aq + 3\mathit{H_2S};$$
$$\mathbf{ZnSe} + H_2SO_4.Aq = ZnSO_4.Aq + \mathit{H_2Se}.$$

Hydrogen sulphide, prepared from crude ferrous sulphide containing metallic iron, obtained by heating together iron and sulphur, always contains hydrogen. The pure gas may be produced from antimony sulphide. Many other sulphides are similarly attacked; among those which resist the action of acids (dilute sulphuric or hydrochloric) are the sulphides of tin, lead, arsenic, bismuth, platinum, &c., copper, silver, mercury, and gold. Certain sulphides and hydrosulphides are decomposed by water alone; among these are sulphides of magnesium, aluminium, boron, silicon, phosphorus, chlorine, &c. The heating of a solution of magnesium hydrosulphide to 100° causes such a reaction:—$Mg(SH)_2.Aq = \mathbf{Mg(OH)_2} + \mathit{H_2S} + Aq$. This method yields pure hydrogen sulphide. The selenide and telluride could doubtless be similarly prepared. The gases are best collected by downward displacement.

Hydrogen trisulphide is prepared in an impure state by pouring into cold hydrochloric acid a solution of sodium polysulphide. The resulting yellow oil does not correspond to the formula H_2S_3, for it contains sulphur in solution. An orange-coloured compound with the alkaloid strychnine is, however, known, which on treatment with strong sulphuric acid yields colourless drops of the trisulphide, H_2S_3. No persulphides of selenium or tellurium are known.

Properties.—**Water** is a liquid at ordinary temperatures, colourless in thin layers, but blue when a white light is passed through a stratum 6 feet long contained in a blackened tube. Ice, when seen in thick masses, has also a bluish-green colour. The vapour of water also appears to be blue. **Hydrogen sulphide, selenide,** and **telluride** are colourless gases; the first has been condensed to a clear liquid, and frozen to a colourless solid. Water, when pure, possesses no smell or taste; hydrogen sulphide has the smell of rotten eggs, being produced by the decomposition of the albumen of eggs, which contains sulphur; the odour of hydrogen selenide and telluride is not so offensive as that of the sulphide, but they produce exceedingly disagreeable nervous effects. The sulphide, selenide, and telluride are exceedingly poisonous; when breathed undiluted with air, instant insensibility is produced. **Hydrogen dioxide** is a colourless viscid liquid, miscible in all proportions with water. It has a faint pungent smell, and a sharp metallic taste. Hydrogen trisulphide has a pungent smell, and is insoluble in water.

These compounds are of very different degrees of stability. While water decomposes only at a very high temperature—that of melted platinum, for example—into its elements, hydrogen sulphide is resolved into hydrogen and sulphur at a low red heat, and hydrogen selenide and telluride slowly decompose at the ordinary temperature.

The dissociation of water may be shown by passing steam through a tube containing a spiral of platinum wire heated to whiteness by an electric current. The hydrogen and oxygen produced by the dissociation mix with the steam, and are cooled below the temperature of ignition; and a test-tube full of explosive gas may thus easily be collected. The dissociation of sulphuretted hydrogen may be shown by passing the gas through a red-hot glass tube, when sulphur deposits on the cool part of the tube.

Hydrogen dioxide and hydrogen trisulphide are very unstable bodies. The former, even at 18° or 20°, begins to decompose into water and oxygen. It thus dilutes itself, and in dilute solution it is more stable. On warming even a very dilute solution, however, it decomposes, bubbles of oxygen being evolved. Many substances of a porous consistency cause this decomposition to take place at the ordinary temperature; and it reacts with certain oxides and peroxides, depriving them of oxygen, while it also loses oxygen. Silver oxide, manganese dioxide, and potassium permanganate have an action of this nature. With silver oxide, for example, the action is shown by the equation $\mathbf{Ag_2O} + H_2O_2.Aq = 2\mathbf{Ag} + H_2O.Aq + O_2$. The tendency of the oxygen of the silver oxide to combine with one atom of the oxygen of the dioxide so as to form a molecule of oxygen, O_2, causes the change to take place. Hydrogen dioxide cannot be vaporised appreciably without decomposition, but the fact of its possessing a smell points to its being able to exist for some time as gas.*

Hydrogen trisulphide† when heated, at once splits up into sulphur and hydrogen sulphide. This decomposition occurs spontaneously when hydrogen trisulphide is kept in a sealed tube, and pressure rises until the resulting hydrogen sulphide is liquefied, solid sulphur separating out.

Many instances have already been given of the decomposition of water by elements. Some, such as sodium and calcium, decompose it at the ordinary temperature; others, such as magnesium, iron, copper, carbon, phosphorus, &c., act on it at a high temperature. In all such cases hydrogen is evolved, while the element combines with the oxygen; the resulting oxide often itself combines with the excess of water, forming a hydroxide or an acid. Sulphuretted, seleniuretted, and telluretted hydrogen are similarly decomposed, yielding sulphides, selenides, and tellurides of the elements, with evolution of hydrogen. But when fluorine or chlorine acts on water, oxygen is evolved.

Hydrogen sulphide, selenide, and telluride are soluble in water, but their solutions soon decompose on exposure to air. A solution of the first is largely employed as a reagent in qualitative and quantitative analysis.

The presence of water can be detected and estimated by heating the substance containing it in a current of dry air, and leading the current through a weighed tube containing dry calcium chloride, phosphorus pentoxide, or strong sulphuric acid, all of which bodies are hygroscopic. The amount of water present is determined by weighing the absorbing tube a second time. Hydrogen dioxide may be detected‡ by adding to the liquid containing it a little ether, and one drop of a solution of potassium bichromate; on shaking, the ether is tinged blue, if dioxide be present, by a compound of chromium of the formula $CrO_3.H_2O_2$, produced by the union of the hydrogen dioxide with the chromium trioxide,

* *Comptes rend.*, **100**, 57.

† *Comptes rendus*, **66**, 1095; *Chem. Soc.*, **27**, 857.

‡ *Annales* (3), **20**, 364.

CrO_3, of the bichromate. Another very delicate test is freshly prepared titanium hydroxide, with which the peroxide gives a yellow colour. Hydrogen sulphide is recognised by its smell and its blackening a piece of paper soaked in a solution of lead acetate; black sulphide of lead is formed.

Physical properties of water.—As water, owing to its abundance, and the ease with which it can be purified, serves as the standard substance for many physical constants, a somewhat detailed description of its physical properties is necessary.

(*a.*) **Mass of 1 cubic centimetre.**—The mass of 1 cubic centimetre of water at 4° is accepted as the unit of weight, 1 gram. Ice is specifically lighter than water. 1 cubic centimetre of ice at 0° weighs 0·917 gram; hence ice floats in water with about 9/10ths of its bulk submerged.

(*b.*) **Expansion.**—Water, unlike other liquids, has a point of maximum density at 4°; when cooled below that temperature, or warmed above it, it expands. It is possible to cool water a few degrees below 0° without its freezing; it continues to expand on fall of temperature, instead of contracting as all other known substances do.

(*c.*) **Vapour-pressures.**—At 100° Centigrade, 80° Réaumur, or 212° Fahrenheit, water-vapour exerts a pressure equal to that of 760 millimetres of mercury; it is then at its boiling-point under normal atmospheric pressure. With decrease of temperature its vapour-pressure decreases, and at 0° its vapour-pressure is equal to that of 4·6 millimetres of mercury. When pressure is reduced by pumping out air, its temperature falls, that portion of water which evaporates withdrawing heat from the remainder, until at a pressure of 4·6 millimetres its temperature is 0°. On still further reducing pressure, its temperature falls still lower, but it is difficult to prevent freezing. It is, however, possible to lower temperature to −5° or −7° without freezing. Ice has also a vapour-pressure. At 0° it is equal to that of water at the same temperature, viz., 4·6 millimetres; on reducing the pressure still further, the temperature of the ice falls by evaporation, exactly as with water, owing to its cooling itself by evolving vapour; if heat be communicated to the ice, it does not raise the temperature of the ice, provided the pressure does not rise, but is entirely expended in evaporating the ice, which passes directly from the state of solid to that of vapour. The vapour-pressures of water are as follows:—

T.	0°.	10°.	20°.	30°.	40°.	50°.	60°.	70°.
P. mm. ..	4·60	9·16	17·40	31·55	54·91	91·98	148·79	233·09

T.	80°.	90°.	100°.	110°.	120°.	130°.	140°.	150°.
P. mm. ..	354·64	525·45	760·0	1075·4	1484	2019	2694	3568

T.	160°.	170°.	180°.	190°.	200°.	210°.	220°.
P. mm.	4652	5937	7478	9403	11625	14240	17365

T.	230°.	240°.	250°.	260°.	270°.
P. mm.	20936	25019	29734	35059	41101

(*d.*) **Specific heat.**—The amount of heat required to raise the temperature of 1 gram of water through 1° is termed a *calory*. But the specific heat of water, like that of other substances, is not a constant; hence the hundredth part of the heat required to raise the temperature of a gram of water from 0° to 100° is generally accepted as the value of a calory. This amount is practically coincident with the amount required to raise the temperature of 1 gram from 18° to 19°. A unit of 100 calories is employed in this book under the symbol *K*. It is better adapted to express large amounts of heat, such as are evolved or absorbed during chemical reactions. The specific heat of ice between −78° and 0° is 0·474 calory per degree; that of water-gas at constant volume is 0·4805 calory.

(*e.*) **Heat of fusion of ice.**—To melt 1 gram of ice, 80 calories are absorbed; hence to melt 18 grams (or 1 gram-molecule) of ice requires 14·4 K at atmospheric pressure.

(*f.*) **Heat of evaporation of water.**—To evaporate 1 gram of water at 100° into steam of that temperature requires an absorption of 537 calories; hence to evaporate 18 grams, or 1 gram-molecule requires $(537 \times 18)/100 = 96{\cdot}66$ K. To convert 1 gram of water at 0° into steam at t° requires an absorption of heat of $(606{\cdot}5 + 0{\cdot}305t)$ calories.

(*g.*) **Volumes of saturated steam.**—From direct measurements the following numbers have been obtained:—

Temperature	140°.	150°.	160°.	170°.	180°.	190°.
Vol. of 1 gram; c.c...	506·0	392·4	307·9	246·4	197·1	160·9

Temperature	200°.	210°.	220°.	230°.	240°.	250°.
Vol. of 1 gram; c.c...	129·8	108·7	89·2	73·8	62·1	52·1

Physical properties of water, hydrogen sulphide, hydrogen selenide, and hydrogen telluride.

	Mass of 1 c.c.				Melting-point.			
	H_2O.	H_2S.	H_2Se.	H_2Te.	H_2O.	H_2S.	H_2Se.	H_2Te.
Solid....	0·917 at 0°	?	?	?	0° at 760 mm.	−85°	?	?
Liquid ..	1·00 at 4°	1·19 at ?	?	?	—	—	—	—

	Boiling-point.			
	H_2O.	H_2S.	H_2Se.	H_2Te.
Liquid	100°	?	?	?

Heats of combination:—

$$2H + O = H_2O + 684K; \; + O + Aq = H_2O_2.Aq - 231K.$$
$$2H + S = H_2S + 47K; \; H_2S + Aq = H_2S.Aq + 46K.$$
$$2H + Se = H_2Se - 111K.$$

Proofs of the composition of the oxide, sulphide, selenide, and telluride of hydrogen.—We have seen that two volumes of hydrogen and one volume of oxygen unite to form two volumes of water-gas. An experiment has also been described on p. 62, whereby it is shown that when water is electrolysed, it decomposes into two volumes of hydrogen and one volume of oxygen approximately. From Avogadro's law it may therefore be concluded that the reaction occurs between 2 molecules of hydrogen and 1 molecule of oxygen, 2 molecules of water-gas being formed, thus:—

$$2H_2 + O_2 = 2H_2O,$$

or in gram-molecules, 4 grams of hydrogen, occupying 11·16 × 4 = 44·64 litres, unite with 32 grams of oxygen, occupying 11·16 × 2 = 22·32 litres, to form 44·64 litres of water-gas weighing 36 grams. Hence, as the weight of 11·16 litres of hydrogen is 1 gram, water-gas under similar conditions of pressure and temperature weighs 36/4 = 9 times as much as hydrogen. Its molecular weight is therefore 18; that is, a molecule of water-gas weighs 18 times as much as an atom of hydrogen.

Similarly the weight of 22·32 litres of hydrogen sulphide is 34 grams, and its specific gravity 17; and the specific gravities of hydrogen selenide and telluride have been found equal to 40·5 and 64·3 respectively, giving molecular weights of 81 and 128·6.

The fact that hydrogen sulphide contains approximately its own volume of hydrogen may be shown by heating in a tube, by means of a spiral of platinum wire traversed by a current, a known volume of hydrogen sulphide. The gas is decomposed into hydrogen and sulphur, and on opening the tube under water no contraction takes place.

The **exact quantitative composition of water** has been the subject of numerous researches, and is even now by no means certain. The processes for ascertaining the composition may be grouped in two divisions: (1) **Determination of the relative weights of oxygen and hydrogen gases, and of the exact proportions**

by volume in which they combine; and (2), **Synthesis of water by passing a known weight of hydrogen over a weighed quantity of red-hot copper oxide, CuO, and estimating its loss of weight, the weight of the water produced being also determined.**

1. By the second method, Erdmann and Marchand,* in 1842, established the ratio between the weights of hydrogen and oxygen in water as 2 : 16.

2. In the same year, Dumas also obtained the ratio 2 : 16, and therefore the ratio between the atomic weights of hydrogen and oxygen of 1 : 16.†

3. Stas, in 1867, determined the ratio between the atomic weight of silver, and the molecular weights of ammonium chloride and bromide, by precipitating the chlorine and bromine contained in weighed quantities of these compounds by silver nitrate produced from pure silver. As he had previously determined the ratios of the atomic weights of silver, chlorine, bromine, and nitrogen to oxygen (these numbers are given on p. 23), the ratio of hydrogen to oxygen could be calculated. He found H : O : : 1 : 15·885.‡

4. Regnault, in 1847, found the relative densities of hydrogen and oxygen 1 : 15·964.§ Applying a correction overlooked by him, but necessary on account of the decrease of the volume of the vacuous globe, owing to the external pressure of the atmosphere, the ratio is reduced to 15·939.

5. Scott, in 1887–8,|| redetermined the ratios between the volumes of hydrogen and oxygen combining with one another, and found it to be O = 1, H = 1·994; applying this correction to Regnault's results, the ratio 1 : 16·01 is obtained.

6. Van der Plaats, in 1886, found the ratio 1 : 15·95 by oxidising a known volume of hydrogen.

7. Lord Rayleigh, in 1888 and 1889,¶ found the ratio 1 : 15·89, from the relative weights of the gases.

8. Cooke and Richards, in 1888,** by weighing the water

* *J. pr. Chem.*, **26**, 461.

† *Annales* (3), **8**, 189.

‡ *Récherches sur les Rapports reciproques des Poids atomiques*, Brussels, 1860.

§ *Relations des Expériences*, Paris, 1847, 151.

|| *Proc. Roy. Soc.*, **42**, 396; *Brit. Assn. Rep.*, 1888, 631. Scott has since found the ratio to exceed 1 : 2.

¶ *Proc. Roy. Soc.*, **43**, 356.

** *Amer. Chem. Jour.*, **10**, 81.

produced by the combustion of known weights of hydrogen, obtained the number 15·869. Lastly,

9. Keiser, in 1888,* weighed hydrogen in combination with palladium, and after combining it with oxygen, found the ratio 1 : 15·949.

These numbers vary between 15·869 and 16·01; their difference amounts to nearly 1 per cent., and the question cannot be regarded as settled. Hence, as remarked on p. 20, seeing that most atomic weights have been determined by the analysis of oxides, it is advisable to assume as the basis of atomic weight, O = 16, leaving the exact ratio between hydrogen and oxygen to the test of further experiment.

Compounds of water with halides.—The compounds of water with halides are very numerous. The water thus combined is generally termed "water of crystallisation," and compounds containing water are said to be "hydrated." To give a complete list of such compounds would occupy too much space. In some instances, the amount of water has been stated in the formulæ given. The same salt may crystallise with several different amounts of water; thus, ferric chloride, $\mathbf{Fe_2Cl_6}$, forms the hydrates, $\mathbf{Fe_2Cl_6.10H_2O}$ and $\mathbf{Fe_2Cl_6.5H_2O}$; calcium chloride combines with water in the proportions $\mathbf{CaCl_2.6H_2O}$, and $\mathbf{2H_2O}$; and so with other halides. It may generally be stated that the lower the temperature, the larger the amount of water of crystallisation with which the halide will combine. The halides of hydrogen also form compounds with water (see p. 112), which are partially decomposed at the ordinary temperature; but when distilled, an acid of a definite strength always comes over; the relative amounts of halide and water depend, however, on the pressure.

Some double halides are unstable, and are not known in a solid state unless combined with water. This is particularly the case with the double halides of hydrogen with those of other elements. The compounds $SiF_4.2HF$, $PtCl_4.2HCl$, and many others, are unknown except in combination with water. Their formulæ are deduced from those of their salts, *i.e.*, from compounds such as $\mathbf{SiF_4.2KF}$, $\mathbf{PtCl_4.2KCl}$, &c., which can be dried. Such hydrochlorides appear to be unstable unless for every molecule of hydrogen chloride two molecules of water are present.

This water tends to leave the substance with which it is combined, evaporating into the air. Its vapour, therefore, exerts a definite pressure. If the pressure of the water-vapour in the air

* *Berichte*, **20**, 2323.

be equal to or but little greater than that of the water of crystallisation, evaporation is balanced by assimilation of water, and no change occurs. If, however, it be greater, the compound turns wet, and is said to "deliquesce;" such substances are termed "hygroscopic;" if less, the compound loses water, turns opaque and lustreless, and is said to "effloresce." Water of crystallisation is usually expelled by heating to 100°, but a much higher temperature is often required.

Compounds of hydrogen sulphide, selenide, and telluride with the halides are unknown.

Compound of hydrogen sulphide with water.—Crystals of the compound $\mathbf{H_2S.7H_2O}$ are deposited when a saturated solution of hydrogen sulphide in water, under a pressure slightly higher than that of the atmosphere, is cooled to 0°.

CHAPTER XVI.

THE OXIDES.—CLASSIFICATION.—THE DUALISTIC THEORY.—HYDROXYL; THE THEORY OF SUBSTITUTION.—CONSTITUTIONAL FORMULÆ.—MOLECULAR AND ATOMIC COMPOUNDS.—OXIDES, SULPHIDES, SELENIDES, AND TELLURIDES OF LITHIUM, SODIUM, POTASSIUM, RUBIDIUM, CÆSIUM, AND AMMONIUM.—HYDROXIDES AND HYDROSULPHIDES.—PREPARATION OF SODIUM.

The Oxides, Sulphides, Selenides, and Tellurides.

Like the halogens, oxygen, sulphur, selenium, and tellurium form many double compounds. But (and this is especially true of the double oxides) such compounds have been usually placed in a different class, and viewed in a different manner from the double halides. Many of the double halides are decomposed into their constituent single halides on treatment with water; but there is no obvious sign of decomposition with most of the double oxides. Water, also, is an oxide, and enters into combination with other oxides, as, indeed, it does with halides; but it is often expelled only at a high temperature, and, in one or two cases, cannot apparently be expelled at any temperature short of that of the electric arc, in which the constituent oxide is itself decomposed into oxygen and element. But, besides such firmly bound water, some oxides crystallise with water, and such "water of crystallisation" is expelled with more or less readiness at a moderate temperature, as it is from the double halides also united with water of crystallisation. Some double sulphides, selenides, and tellurides are also known, but they, unlike the double oxides, are often unstable in presence of water, tending, indeed, to react with the water in which they are dissolved, forming hydrogen sulphide and an oxide. The sulphides, moreover, do not, as a rule, form stable compounds with hydrogen sulphide, and the few compounds which exist have been little investigated.

Classification of oxides.—The oxides of the commoner elements have long been divided into two classes; those of the one class chiefly consist of the oxides of elements of low atomic weight, with some marked exceptions, and have been termed *acids* or *acid-forming oxides;* elements forming such oxides are generally termed *non-metals;* while those of the other class which yield compounds with acid oxides have been termed *bases*, or *basic oxides*. Examples of the first class are: $\mathbf{B_2O_3}$, boron oxide; $\mathbf{SiO_2}$, silica, or silicon dioxide; CO_2, carbon dioxide; SO_2 and $\mathbf{SO_3}$, sulphur di- and tri-oxides; and of the second, $\mathbf{Na_2O}$, sodium oxide; $\mathbf{CaO}$, calcium oxide; $\mathbf{Al_2O_3}$, aluminium oxide; $\mathbf{Fe_2O_3}$, iron sesquioxide, &c. In certain cases, an oxide may belong to both of these classes, as, for example, $\mathbf{Al_2O_3}$, which combines with basic oxides, on the one hand, to form compounds such as $\mathbf{Al_2O_3.K_2O}$, or $\mathbf{KAlO_2}$; and, on the other, with acid oxides, such as $\mathbf{SO_3}$, to form such compounds as $\mathbf{Al_2O_3.3SO_3}$, or $\mathbf{Al_23SO_4}$. And with some elements, which combine with oxygen in several proportions, basic properties are displayed by those oxides containing least oxygen, as, for example, $\mathbf{Cr_2O_3}$; while the higher oxides show acid properties, for instance, $\mathbf{CrO_3}$.

The Dualistic Theory.—Such properties led Lavoisier to assign the nomenclature to bodies which he did, and suggested to Davy* the theory of "dualism," as it was subsequently termed by Berzelius, its great expositor.† Inasmuch as an oxide, decomposed by the electric current, yields up its oxygen at the positive pole, and the other constituent element at the negative pole of the battery, Berzelius supposed that the atoms of oxygen were negatively, and the atoms of the element with which it is in combination, positively electrified. When combinations of such oxides are electrolysed, it was supposed by Berzelius that they also decompose in like manner, the electro-negative constituent of the double oxide being attracted to the positive pole, and the electro-positive constituent to the negative pole of the battery. Thus, as examples, the oxides

$$\overset{+}{\mathrm{Fe}}\overset{-}{\mathrm{O}},\quad \overset{+}{\mathrm{Ba}}\overset{-}{\mathrm{O}},\quad \overset{+}{\mathrm{S}}\overset{-}{\mathrm{O}}_3,\quad \overset{+}{\mathrm{C}}\overset{-}{\mathrm{O}}_2,$$

were supposed to be constituted of electro-positive and electro-negative atoms respectively, while the compounds

$$\overset{+}{\mathrm{BaO}}.\overset{-}{\mathrm{SO}_3},\ \text{and}\ \overset{+}{\mathrm{FeO}}.\overset{-}{\mathrm{CO}_2},$$

were likewise imagined to consist of groups of atoms, which, taken as a whole, themselves displayed positive or negative electrification. On these grounds, he explained the dualistic theory, namely, that every chemical compound is composed of two con-

* *Phil. Trans.*, 1807, 1.

† *Schwaigger's Jour.*, 6, 119.

stituents, one electro-negative and one electro-positive, in combination with each other.

But among the reasons which led to the abandonment of this view, two are of special importance. First, many compounds exist, especially of the element carbon, which cannot be represented on the dualistic system.* Such compounds are, for example, CCl_3Br, C_2H_5Cl, and numerous others, the molecular weights of which are established by their vapour-densities; hence they have not such formulæ as $3CCl_4.CBr_4$, $5C_2H_6.C_2Cl_6$, &c. Second, on electrolysis of solutions of compounds, such as Na_2SO_4, or $Na_2O.SO_3$, the basic oxide does not accumulate at the negative, and the acid oxide at the positive pole, but the compound splits into the element sodium and the group SO_4, neither of which are stable in the presence of water, but react with it, sodium combining with its oxygen and half its hydrogen, liberating the other half; while the group, SO_4, parts with a fourth of its oxygen, remaining as SO_3. It cannot, therefore, be supposed that compounds such as sodium sulphate, $Na_2O.SO_3$, really consist of two distinct portions Na_2O and SO_3; but its molecule exists as a complete individual, Na_2SO_4. In further support of the second argument, it has also been adduced that a similar compound, $PbSO_4$, lead sulphate, may be produced by the following methods: union of PbO and SO_3; union of PbO_2 and SO_2; and union of PbS with $4O$.

The first argument is termed the **argument from substitution**; it was suggested by the French chemist, Dumas, and by the Swiss chemists, Laurent and Gerhardt, and its development has led to the classification of the compounds of carbon, and to the discovery of an enormous number of new bodies.†

This view of the constitution of chemical compounds has also been extended to include compounds other than those of carbon, and compounds of which the molecular weight is absolutely unknown. Thus, sodium monoxide has a composition most simply expressible by the formula $\mathbf{Na_2O}$. This oxide unites with water with great readiness, producing the compound $\mathbf{Na_2O.H_2O}$. But the same compound may be produced by the action of the metal sodium on water; the equation is—

$$2Na + 2H_2O = 2\mathbf{NaHO} + \mathit{H}_2.$$

An atom of sodium expels and *replaces* an atom of hydrogen from water. The secondary action of the union of two atoms of

* Dumas, *Annales* (2), **56**, 113 and 140.

† References are not introduced, as they refer almost exclusively to the compounds of carbon. See E. v. Meyer's *Geschichte der Chemie*, Leipsig, 1889.

hydrogen to form a molecule at once occurs, and ordinary hydrogen is evolved. The formula NaHO is identical with the formula $Na_2O.H_2O$, so far as concerns the expression of the composition of the body, for $Na_2O.H_2O = 2NaHO$; but, as the further action of sodium on fused NaHO is to yield Na_2O and hydrogen, thus:—

$$2NaHO + 2\mathbf{Na} = \mathbf{Na_2O} + \mathit{H_2},$$

the reactions are adduced as a proof that water contains two atoms of hydrogen, inasmuch as the hydrogen can be replaced by sodium in two stages, the series of compounds being

$$H_2O,\ \mathbf{NaHO},\ \mathbf{Na_2O}.$$

Again, many chlorides on treatment with water exchange their chlorine for oxygen. Thus, $\mathbf{PCl_5}$ with a small quantity of water forms PCl_3O, thus:—

$$\mathbf{PCl_5} + H_2O = PCl_3O + 2\mathit{HCl},$$

one atom of oxygen taking the place of two atoms of chlorine; and that PCl_3O is really the formula of the compound is proved by its vapour-density; it is not $3PCl_5.P_2O_5$, which would express the same percentage composition. And so with many other instances.

Constitutional or rational formulæ.—The analogy between the halides and the *hydroxides*, as bodies such as **NaOH** are termed (the word being a contraction for "hydrogen-oxides"), is also a close one. Thus we have **NaCl**, **NaOH**; $\mathbf{CaCl_2}$, $\mathbf{Ca(OH)_2}$; $\mathbf{SiCl_4}$, $\mathbf{Si(OH)_4}$, and so on; and although no hydroxide is volatile enough at high temperatures, or indeed, as a rule, stable enough to make it possible to determine its molecular weight by means of its vapour-density, the analogy is an instructive one. The molecule of chlorine, moreover, $\mathit{Cl_2}$, finds its analogue in hydrogen peroxide, or dihydroxyl, $(OH)_2$.

The action of halides of hydrogen on the hydroxides can also be well represented on the scheme of replacement. Thus we have $\mathbf{NaOH} + \mathit{HCl} = \mathbf{NaCl} + H.OH$; sodium hydroxide being converted into sodium chloride, while hydrogen chloride is changed to hydrogen hydroxide or water; and so with $\mathbf{Ca(OH)_2} + 2\mathit{HCl} = \mathbf{CaCl_2} + 2H.OH$.

An example of the reverse action, viz., replacement of chlorine by hydroxyl, is given in the action of water on phosphorus trichloride, PCl_3, thus:—

$$P\begin{cases}Cl\\Cl\\Cl\end{cases} + \begin{matrix}H.OH\\H.OH\\H.OH\end{matrix} = P\begin{cases}OH\\OH\\OH\end{cases} + \begin{matrix}\mathit{H.Cl}\\\mathit{H.Cl}\\\mathit{H.Cl}\end{matrix}.$$

(See, however, p. 375). Certain oxychlorides of known molecular weight undergo similar changes, for instance:—

$$SO_2\begin{cases}Cl\\Cl\end{cases} + \begin{matrix}H.OH\\H.OH\end{matrix} = SO_2\begin{cases}OH\\OH\end{cases} + \begin{matrix}HCl\\HCl\end{matrix};$$

and so with many other examples. Such formulæ as those given above are termed *constitutional* or *rational* formulæ, in contradistinction to *empirical* formulæ, such as H_3PO_3, H_2SO_4, by which the percentage composition of the body only is expressed, and not the possible functions which it may exhibit.

The action of such compounds on hydroxides may also be similarly represented. Thus, the formation of sodium sulphate by the action of sulphuric acid on sodium hydroxide is represented empirically:—

$$H_2SO_4 + 2NaHO = Na_2SO_4 + 2H_2O.$$

Its rational representation is:—

$$SO_2<\begin{matrix}OH\\OH\end{matrix} + \begin{matrix}NaOH\\NaOH\end{matrix} = SO_2<\begin{matrix}ONa\\ONa\end{matrix} + \begin{matrix}H.OH\\H.OH\end{matrix}.$$

In both instances, however, the exchange of hydrogen for sodium and of sodium for hydrogen is obvious. The name "sodoxyl" may be given to the group (ONa), and it may be supposed to exist in combination with itself in sodium peroxide $(ONa)_2$, or Na_2O_2.

Intermediate compounds are also known, such as $SO_2<\begin{matrix}Cl\\OH\end{matrix}$, chlorosulphonic acid, half chloride, half hydroxide; and $SO_2<\begin{matrix}ONa\\OH\end{matrix}$, sodium hydrogen sulphate, only half of the hydrogen being expelled by sodium (see p. 421).

This method of representation has evidently great advantages; it permits an insight, if only a limited one, into the constitution of such double oxides and chlorides; and it has been almost universally adopted, save among certain French chemists. It has been founded largely on the behaviour of compounds of carbon, the constitution of which is elucidated in a similar manner and in a much more extended degree.

Molecular Compounds.—The universal acceptance of this system, however, has not been wholly good. There are many compounds which cannot be thus classified, and which have consequently been relegated to the position of so-called "molecular" compounds. Such is the case with the double halides described in previous chapters. The name "molecular" has been applied to all double compounds the formation of which cannot be represented by the device of replacement, and it has been attempted to draw a distinction between "atomic" compounds, such as the simple halides, and compounds such as those represented above, and "molecular" compounds. Thus, NaCl, $CaCl_2$, $FeCl_3$, CCl_4, PF_5, are regarded as atomic compounds, the halogen being in direct combination with its neighbour element; and such elements are termed monad, dyad, triad, tetrad, or pentad, according as they combine with one, two, three, four, or five atoms of halogen. And compounds such as SO_2Cl_2, $SO_2(OH)_2$, $SO_2(ONa)_2$, $POCl_3$, $PO(OH)_3$, &c., are also regarded as atomic compounds, inasmuch as they fulfil the required condition of replacement. But compounds like $BF_3.HF$, $AlF_3.3NaF$, $FeCl_3.2KCl$, and of double oxides with each other, such as $MgSO_4.K_2SO_4$ (although the latter compounds may often be represented as formed by replacement) have been regarded as molecular or addition compounds. The water which often accompanies crystalline salts, commonly called water of crystallisation, has also been regarded as molecularly combined.

Now it is questionable whether it is permissible to arbitrarily divide compounds into two classes without sufficient reason. And there is justice in the view that a uniform system of representation should be adopted. Yet, as we know nothing of the true internal arrangement of atoms in a molecule, any systems which contribute towards classification of like compounds, and representation of like changes which they undergo, may be made use of in arranging compound bodies. The method of representing compounds constitutionally often serves a useful purpose, and likewise the method of representation of compounds as addition-products. There is advantage to be gained by representing sodium sulphate as $SO_2(OH)_2$, inasmuch as its analogy with SO_2Cl_2 and $SO_2<^{Cl}_{OH}$ is thereby brought out: and there is also advantage in representing it as $SO_3.H_2O$, inasmuch as reactions occur in which the group SO_3 remains unaltered, while the group H_2O is affected. For example, on distillation with phosphorus pentoxide, the compound SO_3 is liberated as such, while the water combines with the phosphoric oxide. Both systems of representation will therefore be employed as occasion offers.

With these preliminary remarks, which apply *mutatis mutandis* to the sulphides, selenides, and tellurides, we proceed to the consideration of the compounds of elements of the sodium group.

Compounds of Oxygen, Sulphur, Selenium, and Tellurium, with Lithium, Sodium, Potassium, Rubidium, Cæsium, and Ammonium.

The following table gives a list of these compounds:—

	Oxygen.	Sulphur.	Selenium.	Tellurium.
Lithium....	Li_2O; Li_2O_2?	Li_2S?	?	?
Sodium....	Na_2O; Na_2O_2.*	Na_2S; Na_2S_2; Na_2S_3. Na_2S_4; Na_2S_5.†	Na_2Se.	?
Potassium..	K_2O; K_2O_2. K_2O_3; K_2O_4.*	K_2S; K_2S_2; K_2S_3. K_2S_4; K_2S_5.†	K_2Se.	K_2Te?
Rubidium..	Rb_2O?	Rb_2S?	Rb_2Se?	Rb_2Te?
Cæsium ...	Cs_2O?	Cs_2S?	Cs_2Se?	Cs_2Te?
Ammonium.	—	$(NH_4)_2S$; S_2; S_3; S_4; S_5; and S_7.‡	?	?

It will be seen that the compounds of potassium, sodium, and ammonium alone have been investigated with any degree of completeness.

Sources.—None of these compounds occurs free in nature; the monoxides of the type M_2O occur in combination with other oxides, especially with CO_2, SiO_2, N_2O_5, and SO_3, as carbonates, silicates, nitrates, and sulphates.

Preparation.—1. **By direct union.**—The monoxides are produced when thin slices of the metals are exposed to dry oxygen. At higher temperatures higher oxides are formed when the metals are heated in oxygen or nitrous oxide, N_2O; this process yields K_2O_2, Na_2O_2, and higher oxides. The formula of lithium monoxide is conjectural; the monoxides of potassium and sodium have been analysed.

A mixture of sulphides is produced on heating the metals with sulphur, unless excess of sulphur is used, when the pentasulphides are formed.

Ammonium monosulphide is produced by the union of ammonia and hydrogen sulphide at a temperature not higher than $-18°$, thus:—

$$2NH_3 + H_2S = (NH_4)_2S.$$

2. **By expelling or withdrawing an element from a compound.**—Sodium and potassium monoxides have been produced

* *Chem. Soc.*, **14**, 267; **30**, 565.
† *Pogg. Ann.*, **131**, 380,
‡ *J. prakt. Chem.*, **24**, 460.

by heating the hydroxides **NaOH** and **KOH** with the metal, thus:—

$$2NaOH + 2Na = 2\mathbf{Na_2O} + H_2.$$

The higher oxides of potassium are formed on exposing the dioxide to moist air; a portion of the potassium is converted into hydroxide, and the remainder stays in combination with oxygen as trioxide and tetroxide, thus:—

$$3\mathbf{K_2O_2} + 2H_2O = 2\mathbf{KOH} + \mathit{H_2} + 2\mathbf{K_2O_3};$$
$$2\mathbf{K_2O_2} + 2H_2O = 2\mathbf{KOH} + \mathit{H_2} + \mathbf{K_2O_4}.$$

The hydrosulphides on exposure to air yield the polysulphides, the hydrogen uniting with atmospheric oxygen, thus:—

$$2\mathbf{KSH} + \mathit{O} = \mathbf{K_2S_2} + H_2O.$$

The sulphates, selenates, and tellurates, when heated, do not lose oxygen as the chlorates, bromates, and iodates do, leaving sulphide, selenide, or telluride as the halogen-compounds leave halide; but if hydrogen or carbon is present, oxygen is lost at a red heat, thus:—

$$\mathbf{Na_2SO_4} + 4\mathit{H_2} = 4H_2O + \mathbf{Na_2S};$$
$$\text{or } \mathbf{Na_2SO_4} + 4\mathbf{C} = 4\mathit{CO} + \mathbf{Na_2S}.$$

The action of heat on ammonium pentasulphide, $\mathbf{(NH_4)_2S_5}$, yields ammonium mono- and heptasulphides, thus:—$3\mathbf{(NH_4)_2S_5} = 2(NH_4)_2S_7 + \mathit{(NH_4)_2S}$. The sulphide being unstable at temperatures above $-18°$, decomposes into hydrosulphide and ammonia, thus:—

$$\mathit{(NH_4)_2\,S} = \mathit{NH_3} + \mathit{NH_4SH}.$$

3. **By double decomposition.**—Hydrogen sulphide passed over fused sodium chloride produces monosulphide, $\mathit{H_2S} + 2NaCl = \mathbf{Na_2S} + 2\mathit{HCl}$. The sulphides of potassium have also been produced by double decomposition; the trisulphide by exposing red hot potassium carbonate to the vapour of carbon disulphide, thus:—

$$2\mathbf{K_2CO_3} + 3\mathit{CS_2} = 2\mathbf{K_2S_3} + 4\mathit{CO} + \mathit{CO_2}.$$

And the tetrasulphide by similar treatment of the sulphate:—

$$\mathbf{K_2SO_4} + 2\mathit{CS_2} = \mathbf{K_2S_4} + 2\mathit{CO} + \mathit{SO_2}\,(?).$$

The existence of this compound is doubtful.

By distillation of ammonium chloride with a sulphide of potassium, the corresponding ammonium sulphide is produced, *e.g.*,

$K_2S_2 + 2NH_4Cl = 2KCl + (NH_4)_2S_2$. In this manner $(NH_4)_2S_2$, $(NH_4)_2S_3$, $(NH_4)_2S_4$, and $(NH_4)_2S_5$ have been prepared.

"Liver of sulphur" or "*hepar sulphuris*," a substance which has been long known, is produced by fusing 4 gram molecules of potassium carbonate with 10 gram atoms of sulphur, thus :—

$$4K_2CO_3 + 10S = K_2SO_4 + 3K_2S_3 + 4CO_2.$$

It is a mixture of sulphate and trisulphide.

Properties.—The monoxides, so far as they have been prepared, are white or grey solids. Lithium monoxide is said to be non-volatile at a white heat; the others melt with difficultly and volatilise at a very high temperature. Ammonium monoxide is incapable of existence, decomposing at once into ammonia and water.

Sodium dioxide is a white, and potassium dioxide a brownish-yellow solid. Potassium trioxide is lemon-yellow, and the tetroxide sulphur-yellow ; both fuse to orange-red liquids, turning black with rise of temperature, but returning to yellow on solidification.

The sulphides of potassium, sodium, and ammonium are all yellow or brownish-yellow solids which have a peculiar "hepatic" smell. Ammonium heptasulphide is a deep-red substance, volatilising without dissociation at 300°. With acids, the polysulphides give off hydrogen sulphide, while sulphur separates as a white emulsion (milk of sulphur).

Potassium selenide is a greyish or brownish mass; the telluride is a brittle substance with metallic lustre. Both are soluble in water and deposit selenium or tellurium on exposure to air.

All these substances are soluble in water, in all probability combining with it. The union of the monoxides with water takes place with great evolution of heat, and the water cannot be expelled on ignition (see Hydroxides, below). But water may be expelled from solutions of sodium dioxide, and of sodium and potassium monoselenides and disulphides, the anhydrous salts being left on evaporation. Hydrated sulphides are known of the formulæ $K_2S.2H_2O$, $K_2S.5H_2O$, $2Na_2S.9H_2O$, $Na_2S.5H_2O$, and sulphide, selenide, and telluride of sodium with $9H_2O$.

Little is known of the **physical properties** of these substances. The following data, however, are approximate :—

Mass of 1 *c.c.*—Li_2O, 2·102 at 15°; Na_2O, 2·805; K_2O, 2·656; Na_2S, 2·471; K_2S, 2·130.

Volatility.—Li_2O has not been volatilised; K_2O volatilises at a red heat; Na_2O melts at a red heat and volatilises with difficultly. The sulphides appear to be difficultly volatile; potassium pentasulphide melts at a red heat.

Heats of formation :—

$2Na + O = Na_2O + 804K$; $+H_2O = 2NaOH + 352K$; $+ Aq = 198K$.
$2Na + S = Na_2S + 870K$; $+ Aq = Na_2S.Aq + 150K$.

K_2O has not been investigated.

$$2K + S = K_2S + 1012K; \quad + Aq = K_2S.Aq + 100K.$$

None of these substances has been gasified; their molecular weights are therefore unknown.

Double Compounds.—Double oxides of potassium are known of the formula $K_2O_2.K_2O$; $K_2O_2.2K_2O$; and $K_2O_2.3K_2O$. These are bluish solids produced by heating potassium in oxygen or nitrous oxide. They melt to deep red liquids.

Hydroxides, Hydrosulphides, Hydroselenides, and Hydrotellurides.

These names are given to compounds of the oxides with water, or of the sulphides, &c., with hydrogen sulphide, selenide, or telluride. None of these compounds occurs free in nature. The double selenides and tellurides have not been investigated.

Monoxides and Monosulphides; Monohydrates and Monosulphydrates.

H_2O.	$Li_2O.H_2O$;	$Na_2O.H_2O$;	$K_2O.H_2O$;	$Rb_2O.H_2O$;	$Cs_2O.H_2O$.	—
H_2S.	—	$Na_2S.H_2S$.	$K_2S.H_2S$.	—	—	$(NH_4)_2S.H_2S$.

Polyhydrates and Polysulphydrates.

$Li_2O.3H_2O$;	$Na_2O.5H_2O$.	$K_2O.5H_2O$.
—	$Na_2O.8H_2O$.	—
	$Na_2S.5H_2O$.	$K_2S.2H_2O$.
	$Na_2S.9H_2O$.	$K_2S.5H_2O$.
	$Na_2S.H_2S.12H_2O$.	$K_2S.H_2S.H_2O$.

Hydrated Polyoxides and Polysulphides.

$Na_2O_2.2H_2O$.	$Na_2S_2.5H_2O$.	$Na_2S_3.3H_2O$.	$Na_2S_4.8H_2O$.	$Na_2S_5 8H_2O$.
$Na_2O_2.8H_2O$.	—	—	$K_2S_4.2H_2O$.	—

Preparation.—1. **By direct addition.**—All of these substances may be thus prepared. As has been remarked, it is still an open question whether the formula of sodium hydroxide is **NaOH**, one atom of sodium replacing one atom of hydrogen in water; or $Na_2O.H_2O$, which may be viewed as an additive product. If the second view be chosen, the analogy with the halides is concealed, and the substances should be named *hydrates:* if the first, the compounds with more molecules of water are difficult to classify;

and there appears no good reason for preferring one method of representation to another. These remarks apply also to the sulphides. Similar compounds of the selenides and tellurides have not been investigated.

The compound $\mathbf{Na_2O.5H_2O}$ is prepared by crystallising a solution of $\mathbf{Na_2O.H_2O}$ from alcohol containing 2 per cent. of water; the similar compound of potassium separates from water; this compound, when treated with metallic sodium, gives a liquid alloy of potassium and sodium.

The hydrate, $\mathbf{Na_2O.8H_2O}$, crystallises from water.

The compound, $\mathbf{Na_2O_2.8H_2O}$, crystallises from water, and when dried over sulphuric acid it loses water, and has then the formula $\mathbf{Na_2O_2\ 2H_2O}$.

The hydrates of the mono- and polysulphides are all obtained by crystallising them from water. In most cases the water may be evaporated by heat, leaving the anhydrous sulphides.

Ammonium hydrosulphide, $\mathbf{NH_4HS}$, is produced by direct addition of ammonia to hydrogen sulphide above $-18°$.

2. **By double decomposition.**—The hydrates are prepared by (*a*) the action of barium hydroxide on the sulphate, thus:—

$$Li_2SO_4.Aq + Ba(OH)_2.Aq = 2LiOH.Aq + \mathbf{BaSO_4};$$

the barium sulphate being insoluble, it may be separated by filtration; (*b*), the action of calcium hydroxide on the carbonate—

$$Na_2CO_3.Aq + Ca(OH)_2.Aq = 2NaOH.Aq + \mathbf{CaCO_3};$$

or (*c*) by the action of silver hydroxide on the chloride, bromide, or iodide—

$$KCl.Aq + \mathbf{AgOH} = KOH.Aq + \mathbf{AgCl}.$$

The second method (*b*) has been long made use of in *causticising* soda or potash, *i.e.*, in converting the carbonate into the hydroxide, named *caustic soda* or *caustic potash;* a solution of the carbonate is boiled with milk of lime (*i.e.*, calcium hydroxide stirred up with water) in an iron, nickel, or silver vessel, for vessels of other metals or of glass or china are attacked by the soluble hydroxide.

Potassium hydrosulphide has been prepared by passing a stream of hydrogen sulphide over red-hot potassium hydroxide or carbonate, thus:—

$$\mathbf{KOH} + H_2S = \mathbf{KSH} + H_2O;$$
$$\mathbf{K_2O.CO_2} + 2H_2S = \mathbf{2K_2S.H_2S} + CO_2 + H_2O.$$

Various other methods of preparing caustic soda and caustic potash (NaOH and KOH) have been employed on a manufacturing scale. The most important of these, which yields a mixture of hydroxide and carbonate, is the Leblanc process. The principle of this process is the simultaneous action of calcium oxide and carbon on sodium sulphate. The reaction may be conceived to take place in two stages, which, however, are not separated in practice :—

$$\mathbf{Na_2SO_4} + 2C = \mathbf{Na_2S} + 2CO_2; \text{ and}$$
$$\mathbf{Na_2S} + \mathbf{CaO} = \mathbf{Na_2O} + \mathbf{CaS}.$$

The product is termed "black-ash." On treatment with lukewarm water in tanks, the hydroxide dissolves and the calcium sulphide remains insoluble.

If the mixture were boiled, the hydroxide of sodium would react with the calcium sulphide, reversing the second of these equations, thus:—$2NaOH.Aq + \mathbf{CaS} = 2NaSH.Aq + Ca(OH)_2.Aq$. But the solution is separated from the solid as quickly as possible and concentrated by evaporation. During evaporation chloride and sulphate of sodium contained as impurities separate out; they are "fished" out with perforated ladles, and hence are termed "fished salts;" while the solution is concentrated, freed from carbonate by addition of lime, and finally evaporated in hemispherical iron pots till fused caustic soda, NaOH, remains. It is then run into iron drums and brought to market. The principle of the manufacture of caustic potash is similar.

Properties.—The hydroxides of the metals lithium, sodium, potassium, rubidium, and cæsium of the general formula **MOH** have been termed "caustic" lithia, soda, &c., owing to their corrosive and solvent properties (*καίω*, I burn). They are all white soluble solids melting at a red heat and volatilising at a white heat. When dissolved in water, great heat is evolved owing to combination. When fused, they attack glass and porcelain, dissolving the silica of the glass and the silica and alumina of the porcelain; they act on metals, converting them into oxides, with exception of nickel, iron, silver, and gold. Cæsium hydroxide is most, and lithium hydroxide least volatile.

Sodium and potassium hydroxides usually contain as impurities sulphates, carbonates, and chlorides. A partial purification may be effected by treatment with absolute alcohol in which the hydroxides dissolve, while the salts are insoluble. The clear solution is decanted from the undissolved salts, the alcohol is removed by distillation, and the residue fused.

If absolutely pure hydroxides are required, however, they are best prepared from the metals by throwing small pieces into water, and subsequently evaporating the solution of hydroxide in a silver basin.

The hydrosulphides are white crystalline bodies, which fuse to black liquids, but turn white again on solidification. They may be obtained in solution by saturating solutions of the hydroxide with hydrogen sulphide, thus :—

$$\mathrm{NaHO.Aq} + H_2S = \mathrm{NaHS.Aq} + \mathrm{H_2O}.$$

The hydroxides volatilise as such when heated ; but hydrosulphides lose hydrogen sulphide, and leave the sulphides.

Ammonium hydrosulphide dissociates into ammonia and hydrogen sulphide at 50°, and above, and on cooling, the constituents re-unite.* It forms colourless crystals.

Appendix.—Manufacture of sodium and potassium. An indication of the method of preparing these metals was given on p. 30. As they are now prepared from the hydroxides, by a process devised by Mr. Castner,† a short sketch of the manufacture is here appended.

The following reaction takes place at a red heat between carbon and the hydroxide :—$6\mathrm{NaOH} + 2\mathrm{C} = 2\mathrm{Na_2CO_3} + 3H_2 + 2Na$. But if carbon is heated with caustic soda, the hydroxide melts, and the carbon, which is lighter than the soda, floats to the surface, and is for the most part unacted on. Hence it is necessary to weight the carbon so as to cause it to sink, or else to add some substance to prevent the caustic alkali fusing completely, so that the carbon may remain mixed with it. The old plan consisted in adding lime ; but the temperature at which the metal distilled off was rendered so high that the yield was small, and the destruction of the wrought-iron tubes used as stills was enormous. The new method is to heat a mixture of pitch and finely-divided iron (spongy iron) to redness. Compounds of hydrogen and carbon distil off, and an intimate mixture of iron and carbon is left in a porous state. This mixture is introduced along with caustic soda into cast-iron crucibles provided with tight lids, from each of which a tube conveys the metallic vapour to the condensers, which themselves are tubes about 5 inches in diameter and 3 feet long, and which are placed in a sloping position so that the melted metal runs down into a small pot through a hole about 20 inches from the nozzle. The crucibles are heated by means of gas to about 1000° ; and when the distillation is over, in about an hour and a quarter, the crucible is lowered in the furnace, so as to separate it from the lid which is stationary; it is then withdrawn, emptied, recharged while still hot, and replaced. It is next lifted by hydraulic power till it again meets its lid, and the operation again commences. The mixture of sodium carbonate and spongy iron emptied from the crucible after each distillation is treated with water, the iron is recharged with carbon, and the sodium carbonate is converted by means of lime into caustic soda to be used in a subsequent operation.

The metals potassium and rubidium can be similarly prepared ; but lithium and cæsium must be obtained by electrolysis.

* Engel and Moitessièr, *Comptes rend.*, **88**, 1353.

† *Chem. News*, **54**, 218.

CHAPTER XVII.

OXIDES, SULPHIDES, SELENIDES, AND TELLURIDES OF THE BERYLLIUM GROUP.—HYDROXIDES AND HYDROSULPHIDES.—DOUBLE COMPOUNDS WITH HALIDES.—OXIDES, SULPHIDES, SELENIDES, AND TELLURIDES OF THE MAGNESIUM GROUP.—HYDROXIDES AND HYDROSULPHIDES.—DOUBLE COMPOUNDS.

Oxides, Sulphides, Selenides, and Tellurides of Beryllium, Calcium, Strontium, and Barium.

The compounds of beryllium differ from those of the other metals; those of calcium, strontium, and barium strongly resemble each other.

Sources.—These compounds are never found free; but the oxides occur in combination with carbon dioxide, silica, and sulphur trioxide, as carbonates, silicates, and sulphates.

List.	Oxygen.	Sulphur.	Selenium.	Tellurium.
Beryllium..	**BeO.**	**BeS.** (?)	**BeSe.**	?
Calcium....	**CaO; CaO_2.**	**CaS; CaS_2; CaS_5.***	**CaSe.**	?
Strontium..	**SrO; SrO_2.**	**SrS; SrS_4.**	**SrSe.**	?
Barium....	**BaO; BaO_2.**	**BaS; BaS_3; BaS_5.†**	**BaSe.**	?

Preparation.—1. By direct union.—All of these metals readily oxidise when exposed to air, and burn when heated in air or oxygen, producing monoxides. They would also in all probability combine with sulphur, selenium, and tellurium.

Barium dioxide is produced when the monoxide is heated to 450° in a current of pure dry air; the polysulphides of these metals are also formed when the hydrosulphides are boiled with sulphur, thus:—$Ca(SH)_2.Aq + \mathbf{S} = CaS_2.Aq + H_2S$; and similarly with others; also by heating the monosulphides with sulphur.

2. **By heating hydroxides, nitrates, or carbonates.**—These compounds may be viewed as compounds of the oxides with oxides of hydrogen, nitrogen, carbon, or iodine, thus:

* *Chem. Soc.*, **47**, 478.

† *Ibid.*, **49**, 369.

$CaO.H_2O$; $CaO.N_2O_5$; $CaO.CO_2$. At a red or white heat, the water, nitrogen pentoxide (which splits into lower oxides of nitrogen and oxygen), or carbon dioxide, are evolved as gas, while the non-volatile oxide of the metal remains. The loss of water takes place readily with beryllium hydroxide; slowly, beginning at 100°, or even lower, with calcium hydroxide, and at a very high temperature with strontium and barium hydroxides. The loss of N_2O_5 takes place at a red heat in all cases. This method is adopted as the only practical one in preparing barium oxide, which is now made on a large scale. To ensure thorough expulsion of oxides of nitrogen, the partially decomposed oxide is heated in a vacuum.* Beryllium carbonate is decomposed at low redness; calcium carbonate begins to decompose below 400°; and provided the carbon dioxide be removed by a current of air or steam, so that recombination cannot take place, the decomposition, if sufficient time be given, is complete at that temperature.

The decomposition of calcium carbonate (limestone) by heat, termed "lime-burning" is carried out in "lime-kilns," towers open above, with a door below, into which alternate layers of lime and coal are introduced from above. The coal is set on fire, and the "burnt" or "quick" lime is withdrawn below, after all carbon dioxide has been expelled, and when cold. Strontium and barium oxides may also be produced from their carbonates, but at a higher temperature; it is well to mix them with a little coal, which reduces the carbon dioxide to monoxide, so that no recombination takes place.

Calcium sulphide is similarly formed by heating calcium hydrosulphide, **$Ca(SH)_2 = CaS.H_2S$**, in a current of hydrogen sulphide. Strontium and barium sulphides could no doubt be obtained in an analogous manner.

The monoxides of calcium, strontium, and barium are also obtainable by heating the dioxides to 450° under reduced pressure, or to a higher temperature. This process is made use of in producing oxygen on a large scale (see p. 65).

Calcium dioxide is also said to be produced in small amount when the carbonate is heated to low redness. The hydrated dioxides may be dried by moderate heat.

3. **By double decomposition. Monoxides.**—Barium monoxide is prepared by heating together barium sulphide and copper or zinc oxide. On treatment with water, barium hydroxide goes into solution.

* Boussingault, *Annales* (5), **19**, 464.

Sulphides.—The hydroxides, when heated in a current of hydrogen sulphide yield the monosulphides, thus:—

$$\mathbf{Ca(OH)_2} + H_2S = \mathbf{CaS} + 2H_2O.$$

4. **By removing oxygen from the sulphates, selenates, or selenites** by heating to redness with carbon or carbon monoxides; the sulphides or selenides are left. The sulphides of calcium, strontium, and barium are thus prepared. The selenides are similarly prepared by heating selenates or selenites to dull redness in a current of hydrogen. It is in this way that barium compounds are produced from the insoluble sulphate, which is mixed with bituminous coal and heated to redness. The sulphide thus produced is converted into the chloride by treatment with hydrochloric acid, or into the oxide, by heating with copper or zinc oxide. The soluble hydroxide is produced on treatment with water.

Properties.—**Monoxides.**—These are white powders, or hard, white, or greyish-white masses. They all unite with water with evolution of much heat. Beryllium oxide forms the least stable, and barium oxide the most stable compound. Beryllium oxide is said to volatilise at a high temperature; calcium oxide melts only in the electric arc, while strontium and barium oxides melt at a white heat. The oxides are crystalline when prepared by heating the nitrates in covered porcelain crucibles. Beryllium oxide crystallises from its solution in fused sulphate of beryllium and potassium, or in fused boron oxide.

The **dioxides** are white substances, which evolve oxygen when heated, calcium dioxide most readily, barium dioxide at a bright-red heat; barium dioxide is said to fuse before evolving oxygen (?). They dissolve in water with moderate ease, forming compounds.

The **monosulphides** are white amorphous powders, very sparingly soluble in water, but reacting with it (see below).

The **monoselenides** are also white, sparingly soluble powders, which turn red on exposure to air, owing to the expulsion of selenium by oxygen; the monosulphides turn yellow, owing to the formation of polysulphides. The tellurides have not been examined.

The **polysulphides** are yellow solids, soluble in water. Barium monosulphide, when heated in a current of steam, decomposes it, hydrogen being evolved, and barium sulphate remaining. The impure monosulphides, produced by heating the powdered carbonates with sulphur, or the sulphate with carbon, possess the curious property of remaining luminous in the dark, after having been exposed to light. Such substances used to be called *phos-*

phori. The calcium compound used to be known as "Canton's phosphorus," and the barium compound as "Bolognian phosphorus." The modern "luminous paint" owes its property to this peculiarity.

All these oxides are converted into chlorides when heated in a current of chlorine.

Uses.—Calcium oxide (lime) when heated to whiteness in the oxy-hydrogen flame evolves a brilliant light (Drummond's light); barium oxide and dioxide are employed in the commercial manufacture of oxygen.

Physical Properties.—The melting and boiling-points of these bodies are unknown.

Mass of one cubic centimetre—

BeO.	CaO.	SrO.	BaO.	BaO_2.
3·18 at 14°	3·25	4·75	5·72	4·96

Heats of formation:—

Ca + O = **CaO** + 1310K; + H_2O = 155K; + Aq = 30K.
Sr + O = **SrO** + 1284K; + H_2O = 177K; + Aq = 116K.
Ba + O = **BaO** + 1242 (?)K; + H_2O = 223K; + Aq = 122K.
BaO + O = BaO_2 + 172K; + H_2O_2 = 102K.
Ca + **S** = **CaS** + 869K.
Sr + **S** = **SrS** + 974K.
Ba + **S** = **BaS** + 983 (?)K.

Double Compounds.

(*a.*) **With water, &c.** The following bodies are known:—

	Oxides with water.		Dioxides with water.	Dioxide with hydrogen dioxide.
Beryllium..	*$3BeO.10H_2O$.	*$2BeO.3H_2O$.	—	—
	*$BeO.4H_2O$.	$BeO.H_2O$.		
Calcium ...	$CaO.H_2O$. = $Ca(HO)_2$.		$CaO_2.8H_2O$.	—
Strontium..	$SrO.H_2O$ = $Sr(OH)_2$.		$SrO_2.8H_2O$.	—
	$SrO.9H_2O$ = $Sr(OH)_2.8H_2O$.			
Barium....	$BaO.H_2O$ = $Ba(OH)_2$.		$BaO_2.8H_2O$.	$BaO_2.H_2O_2$.
	$BaO.9H_2O$ = $Ba(OH)_2.8H_2O$.			

	Sulphides with water.	Sulphides with hydrogen sulphide.
Beryllium .	BeS (?)H_2O.	?
Calcium ...	$CaS.H_2O$ = $Ca(SH)(OH)$.	$CaS.H_2S$ = $Ca(SH)_2$.†
	$CaS.4H_2O$ = $Ca(SH)(OH).3H_2O$.	
Strontium..	$SrS.H_2O$ = $Sr(SH)(OH)$. ?	$SrS.H_2S$ = $Sr(SH)_2$?
Barium....	$BaS.H_2O$ = $Ba(SH)(OH)$?	$BaS.H_2S$ = $Ba(SH)_2$?

* The existence of these compounds is doubtful.

† *Chem. Soc.*, **45**, 271 and 696.

Sulphide with water and hydrogen sulphide—
Calcium .... $\mathbf{CaS.H_2S.6H_2O = Ca(SH)_2.6H_2O}$.

Hydrated polysulphides.—$\mathbf{Ca_2S_2.3H_2O}$; $\mathbf{SrS_4.6H_2O}$, and others.

Preparation.—Hydrated oxides, and hydroxides. 1. By direct addition.—All of these oxides unite with water directly; beryllium oxide shows least tendency; calcium oxide unites with great evolution of heat; the water is at first absorbed, and then the lumps of lime grow so hot as to evolve clouds of steam, and break up into a bulky white powder. This is the familiar operation of "slaking lime." The product is termed "slaked lime." Barium oxide unites with water with so great an evolution of heat as to turn red hot when thrown into water. Calcium hydroxide is sparingly soluble in water, and the solubility diminishes with rise of temperature. At 15°, 1 gram of calcium *oxide* dissolves in 779 grams of water; at 20°, in 791 grams; and at 95°, in 1650 grams. It would thus appear that calcium hydroxide loses water when heated even in contact with water, and hence shows no tendency towards further hydration. Strontium and barium hydroxides, on the other hand, dissolve to some extent in hot water, and on cooling, crystals of $\mathbf{Sr(OH)_2.8H_2O}$, or $\mathbf{Ba(OH)_2.8H_2O}$ separate. At 15°, 1 gram of barium hydroxide dissolves in about 20 grams of water; and at 100°, in 2 grams. Strontium hydroxide is less soluble. Calcium hydroxide, $\mathbf{Ca(OH)_2}$, separates in crystals when its solution is evaporated *in vacuo*. The hydrated peroxides are also formed by dissolving the peroxides in water and crystallising. The compound $\mathbf{BaO_2.H_2O_2}$, separates from a solution of BaO_2 in H_2O_2 containing water. A possible, though improbable, view of the constitution of the compound $\mathbf{Ca(SH)(OH).3H_2O}$ is that it consists of $\mathbf{CaO.H_2S.4H_2O}$. It is produced by passing sulphuretted hydrogen into a paste of calcium hydroxide and water.

2. **By double decomposition.**—(*a.*) By addition of a soluble hydroxide (*e.g.*, of lithium, sodium, potassium, &c., or ammonia and water) to a soluble compound of beryllium, calcium, strontium, or barium, thus:—

$$CaCl_2.Aq + 2KOH.Aq = \mathbf{Ca(OH)_2} + 2KCl.Aq.$$

No doubt this change always takes place to a greater or less extent. But as strontium and barium hydroxides are fairly soluble in water, they separate only when the solution is a concentrated one. With beryllium, the hydroxide produced by heating any oluble salt, such as the chloride, sulphate, or nitrate, with potassium hydroxide, thus:—$BeCl_2.Aq + 2KOH.Aq = \mathbf{Be(OH)_2.2H_2O} + 2KCl.Aq$, redissolves in excess of the potassium hydroxide, doubtless producing a soluble double oxide of beryllium and potassium; but the solution of this substance, when boiled, decomposes into beryllium hydroxide, $\mathbf{Be(OH)_2}$, which precipitates, and potassium hydroxide, which remains in solution.

Solutions of strontium and barium hydroxides give precipitates with soluble salts of beryllium and calcium, owing to the greater insolubility of the hydroxides of the latter metals.

The hydrated peroxides may be similarly produced by addition of some dioxide, such as hydrogen or sodium dioxide, to a solution of the hydroxide of the metal, thus:—

$$Ca(OH)_2.Aq + H_2O_2.Aq = \mathbf{CaO_2.8H_2O} + Aq.$$

As they are sparingly soluble they are precipitated.

(*b.*) **By the action of hydrogen sulphide on the hydroxides**, the hydrated sulphides are formed, and in presence of excess of hydrogen sulphide the sulphydrated sulphides. With calcium, for example, the action is as follows:—

$$Ca(OH)_2.Aq + \mathit{H_2S} = Ca(SH)(OH).Aq + H_2O; \text{ or}$$
$$CaO.H_2O.Aq + \mathit{H_2S} = CaS.Aq + H_2O;$$

and further,

$$Ca(SH)(OH).Aq + \mathit{H_2S} = Ca(SH)_2.Aq + H_2O.$$

If the solutions are strong and cold, the substances—

$$\mathbf{Ca(SH)(OH)3H_2O} \ (= \mathbf{CaS.4H_2O}) \text{ and } \mathbf{Ca(SH)_2.6H_2O} \ (= \mathbf{CaS.H_2S.6H_2O})$$

separate in crystals.

The calcium compounds are the only ones which have been carefully investigated as regards their behaviour with hydrogen sulphide; similar compounds no doubt exist with beryllium, strontium and barium, and also with hydrogen selenide and telluride.

The hydrosulphide, $\mathbf{Ca(SH)_2}$, when heated with water (as it cannot be obtained free from the six molecules of water with which it crystallises, this water reacts), gives off hydrogen sulphide, and the hydroxy-hydrosulphide remains, thus:—

$$Ca(SH)_2.Aq. + H_2O = Ca(SH)(OH).Aq + \mathit{H_2S}.$$

The hydrosulphide, when treated with sulphur, evolves hydrogen sulphide with formation of a polysulphide. Such polysulphides are known only in solution.

Properties.—The hydroxides are white powders; that of beryllium is insoluble in water, but dissolves in a solution of ammonium carbonate, and is reprecipitated on boiling. This reaction serves to separate it from aluminium hydroxide, which is insoluble in aqueous ammonium carbonate. The hydroxide of calcium is sparingly soluble in water (see p. 222), that of strontium more soluble, and barium hydroxide easily soluble. The hydrates of strontium and barium, $\mathbf{Sr(OH)_2.8H_2O}$, and $\mathbf{Ba(OH)_2.8H_2O}$, are white crystalline bodies, rapidly turning opaque on exposure to air, owing to absorption of carbon dioxide. When heated to 75°, 7 molecules of water are lost, and the eighth only at a red heat. From this it would appear that the compound $\mathbf{BaO.2H_2O}$ is not much inferior in stability to $\mathbf{BaO.H_2O}$, and that the formula $\mathbf{Ba(OH)_2}$ for the latter does not express any exceptionally stable form of combination between water and oxide.

The hydrated dioxides are crystalline powders, which may be dried *in vacuo* to the dioxides. That of barium, indeed, may be heated to over 300°, without loss of oxygen.

The hydrosulphides are very unstable bodies, capable of existence only when cooled by ice in presence of hydrogen sulphide. When placed in water at the ordinary temperature, hydrogen sulphide is evolved, and the hydroxy-hydrosulphide,

$$\mathbf{Ca(SH)(OH).3H_2O},$$

is left.

There appear to be various compounds of oxides and sulphides of these metals (the existence of which, however, requires further proof), *e.g.*, $\mathbf{2CaO.CaS_2}$, $\mathbf{3CaO.CaS_2}$, $\mathbf{3CaO.CaS_3}$, &c., in combination with water.

On boiling solutions of the hydroxides, calcium, strontium, or barium, with sulphur, polysulphides are formed, together with thiosulphates, thus:—

$$3Ca(OH)_2.Aq + 2\mathbf{S} + n\mathbf{S} = CaS_2O_3.Aq + 2CaS_{n/2}.$$

The polysulphide formed depends on the amount of sulphur present. A deep yellow solution is obtained from which the thiosulphate separates in crystals.

(The slaking of lime, the precipitation of calcium hydroxide with sodium hydroxide, the crystallisation of barium hydroxide from a hot solution; the preparation of calcium sulphide by the action of hydrogen sulphide or calcium hydroxide; the formation of polysulphides of calcium on boiling "milk of lime" with sulphur; and the precipitation of "milk of sulphur" on addition of sulphuric acid to the orange solution form suitable lecture experiments.)

(*c.*) **Double compounds with halides.**—These are few in number. $BeCl_2.BeO$, is said to be obtained on evaporating an aqueous solution of beryllium chloride. $CaCl_2.3CaO.15H_2O$ is prepared by boiling calcium hydroxide in a solution of calcium chloride, and filtering while hot; $BaCl_2.BaO.5H_2O$, $BaBr_2BaO.5H_2O$, and $BaI_2.BaO.5H_2O$ are similarly prepared.

There appear also to be indications of similar calcium and strontium compounds, $SrCl_2.SrO.9H_2O$ having been prepared.

It is possible to regard these compounds as hydroxychlorides, thus:—$Ba\!<\!{}^{Cl}_{OH}.2H_2O$, &c. Although somewhat similar formulæ could be constructed for more complex compounds, as, for example, $Cl{-}Ca{-}O{-}Ca{-}O{-}Ca{-}O{-}Ca{-}Cl.15H_2O$; yet, inasmuch as similar double halides exist in number, which cannot in reason be similarly represented, it appears advisable, in the present state of our knowledge, to adhere to the simpler and older methods of representation.

Oxides, Sulphides, Selenides, and Tellurides of Magnesium, Zinc, and Cadmium.

As many of these compounds are unaffected by air and carbon dioxide, and do not react or combine with water, they occur native.

Sources.—Magnesium oxide occurs as *periclase;* also, in combination with water, $Mg(OH)_2$, or magnesium hydroxide, as *brucite*, in white rhombohedra. It also occurs in combination with carbon dioxide, silicon dioxide, &c. Zinc oxide, ZnO, is named *zincite* or *red zinc ore;* it is red owing to its containing ferric oxide in small quantity; it is also found in combination with oxides of iron and manganese as *franklinite.* Zinc sulphide occurs as *blende*, associated with many other sulphides, both in crystalline and in sedimentary rocks. It is the chief ore of zinc. It has usually a black colour, but is white when pure. Cadmium sulphide, CdS, occurs as the rare mineral *greenockite.* Zinc oxide also occurs in combination with carbon dioxide and with silica.

List.	Oxygen.	Sulphur.	Selenium.	Tellurium.
Magnesium ..	MgO.	MgS.	$MgSe$?	$MgTe$?
Zinc	ZnO; ZnO_2?	ZnS; ZnS_5?	$ZnSe$.	$ZnTe$?
Cadmium....	CdO; CdO_2?	CdS.	$CdSe$.	$CdTe$.

Preparation.—1. **By direct union.**—These elements all burn in oxygen, or when heated to a high temperature in air. Magnesium burns with a brilliant white flame, but if the supply of air is limited, the nitride, $\mathbf{Mg_3N_2}$, is simultaneously produced. The metal is sold in the form of thin ribbon for purposes of signalling, photographing dark chambers, &c.; and in fine dust, for signalling. A little powder, when thrown into a flame, gives a brilliant flash of light. Zinc burns with a green flame, giving off filmy clouds of oxide. Cadmium also burns to a brown oxide.

The sulphides are also produced by throwing sulphur on to the red-hot metals. Zinc and cadmium do not readily combine with selenium; if the metal be fused with selenium, the latter distils off, leaving the metal coated with a crust of selenide. But with tellurium, tellurides are produced, the boiling-point of that element being higher.

2. **By heating a compound.**—The hydroxides, carbonates, nitrates, or sulphates of these metals, when heated, leave the oxide. The hydroxides and carbonates are decomposed at a low red heat; the nitrates and sulphates require a higher temperature.

3. **By double decomposition.**—Sulphides of these metals are produced by heating the oxides in a current of hydrogen sulphide or carbon disulphide, thus :—

$$\mathbf{MgO} + H_2S = \mathbf{MgS} + H_2O\text{; and}$$
$$2\mathbf{MgO} + CS_2 = 2\mathbf{MgS} + CO_2.$$

Zinc and cadmium selenides have been similarly prepared.

Inasmuch as the sulphides, selenides, and tellurides of zinc and cadmium are insoluble in water, they may be produced by precipitation, viz., by passing a current of hydrogen sulphide through a solution of a soluble salt of the metals; thus :—

$$\mathrm{ZnSO_4.Aq} + H_2S = \mathbf{ZnS} + \mathrm{H_2SO_4.Aq}.$$

There appear good grounds for believing that this reaction gives not a sulphide such as **ZnS**, but a hydrosulphide, $\mathbf{ZnS}.n\mathbf{H_2S}$. The body produced contains more sulphur than corresponds to the formula **ZnS**, and gives off hydrogen sulphide on heating. The precipitate produced as above is soluble in many acids; hence, to ensure thorough precipitation, the acid must be neutralised by an alkali, *e.g.*, by soda or ammonia. Acetic acid, however, has no solvent action; hence precipitation is complete from a solution of zinc acetate. Cadmium sulphide, prepared in a similar manner, is also probably a hydrosulphide. It is, unlike zinc sulphide, insoluble in dilute acids; but dissolves in moderately strong hydrochloric acid.

Magnesium sulphide cannot be thus prepared; if the hydroxide is employed the hydrosulphide is produced.

The selenides and tellurides of zinc and cadmium may be similarly obtained.

Zinc and cadmium peroxides, and probably also magnesium peroxide, are formed by addition of hydrogen dioxide to the hydroxides. They appear to be compounds of dioxide with monoxide in proportions as yet unascertained. The pentasulphide of zinc is produced when a zinc salt is treated with a solution of potassium pentasulphide.

Properties.—Magnesium and zinc oxides and sulphides are white; cadmium oxide brown, and its sulphide yellow. When prepared by the union of the metal with oxygen, magnesium oxide is dense, and has the specific gravity 3·6. *Magnesia usta*, or calcined magnesia, is a very loose white powder produced by gently glowing the hydroxycarbonate, known as *magnesia alba*. When produced from the native carbonate, *magnesite*, it is dense and hard, and is made use of as a lining for the interior of Bessemer converters. It is known as "basic lining." It is very sparingly soluble in water, 50,000 parts of water dissolving only one part of oxide; it probably dissolves as hydroxide. It unites slowly with water, when it has not been strongly ignited; and also attracts carbon dioxide from the air, if moist. It is soluble in all acids.

Zinc oxide is also a soft white powder. When produced by burning zinc, it is sometimes named "*lana philosophica*," on account of its woolly texture. When heated it turns yellow, but its white colour returns on cooling. It is insoluble in and does not combine directly with water, nor does it unite with carbon dioxide.

Cadmium oxide is a soft brown powder.

None of these bodies are easily volatilised, nor do they melt easily.

Magnesium sulphide reacts with water, giving hydroxyhydrosulphide (?) or hydroxide and hydrosulphide. It is an amorphous pinkish body, infusible, and burning when heated in air to oxide and sulphur dioxide.

Zinc sulphide, as *blende*, forms compact masses of various colours due to impurities; it is usually black, and is known to miners as "black-jack." It is translucent and crystalline. When "roasted" or heated in air, it changes to oxide and sulphur dioxide. Prepared artificially, by precipitation and subsequent heating, it forms a white infusible powder. It is employed as a pigment under the name of "zinc-white." Its "covering power" is not so great as that of white lead (see Carbonates, p. 289), but it has the advantage of not turning black on exposure to hydrogen sulphide as white lead does, zinc sulphide being white. Cadmium sulphide, as *greenockite*, occurs in yellow transparent crystals; prepared by precipitation, it is a yellow powder, and is used as an artist's colour, under the name of "cadmium yellow," or "*jaune brillant.*" It is not permanent, being easily oxidised by moist air. When heated to redness it turns first brownish, then carmine-red. It fuses at a white heat, and crystallises in scales on cooling.

The oxygen of these oxides is displaced at a red heat by chlorine.

The peroxides of zinc, magnesium, and cadmium are white powders. They do not contain enough oxygen to correspond to the formulæ MgO_2, &c., and are either mixtures or compounds of higher oxides with the monoxides.

Zinc pentasulphide is a flesh-coloured precipitate, which, on treatment with hydrochloric acid, dissolves with effervescence of hydrogen sulphide, sulphur being deposited.

Zinc selenide, **ZnSe**, is a yellow amorphous powder, which changes into yellow crystals when heated in a current of hydrogen. Cadmium selenide forms deep reddish-black crystals. The amorphous telluride has metallic lustre, but forms a red powder. When heated in hydrogen it forms ruby-red crystals; cadmium telluride is also a metallic-looking substance giving black crystals. These bodies are probably decomposed by hydrogen into the elements, which recombine in the cooler part of the tube. It is improbable that they are volatile as compounds.

Physical Properties.

Mass of one cubic centimetre :—

	Oxygen.	Sulphur.	Selenium.	Tellurium.
Magnesium....	3·636* (crystallised)	?	?	?
Zinc	5·78 at 15°	4·05	5·4 at 15°	6·34 at 15°
Cadmium	8·11 (crystalline)	4·5 (precipitated)	5·8 at 15°	6·2 at 15°

Heats of formation :—

$$\mathbf{Mg} + O = \mathbf{MgO} + 1440K.; + H_2O = 50K.$$
$$\mathbf{Zn} + O = \mathbf{ZnO} + 853K; + H_2O = -26K.$$
$$\mathbf{Cd} + O = \mathbf{CdO} + 755K; + H_2O = -98K.$$
$$\mathbf{Mg} + \mathbf{S} = \mathbf{MgS} + 776K.$$
$$\mathbf{Zn} + \mathbf{S} = \mathbf{ZnS} + 396K.$$
$$\mathbf{Cd} + \mathbf{S} = \mathbf{CdS} + 324K.$$

* The density increases on calcination; magnesia produced by igniting the carbonate has the density 3·19 at 0°.

Double compounds.—(*a.*) **With water:** hydrates or hydroxides.—The mineral brucite, $\mathbf{MgO.H_2O}$, or $\mathbf{Mg(OH)_2}$, occurs native, usually in masses of serpentine. It crystallises in rhombohedra. Magnesium oxide, when prepared from the nitrate or carbonate at a low red heat, unites with water, forming a solid translucent substance harder than marble. After being heated to whiteness, it loses the property of combination with water. Zinc and cadmium oxides do not combine with water directly.

Soluble salts of magnesium, zinc, and cadmium, on treatment with hydroxides of sodium, potassium, or barium give gelatinous precipitates of the hydrates. Ammonia in water (equivalent to ammonium hydroxide) also produces precipitates, but redissolves them if added in excess. Magnesium hydroxide does not react with excess of sodium or potassium hydroxides, whereas zinc and cadmium hydroxides are soluble in excess of the precipitant, forming double compounds (see *infra*).

Crystals of $\mathbf{ZnO.H_2O}$ and of $\mathbf{CdO.H_2O}$ are produced after some time by placing a stick of zinc or cadmium in aqueous ammonia, in contact with iron, lead, or copper. The zinc compound forms rhombic prisms, of 2·68 specific gravity. And octahedral crystals of $\mathbf{ZnO.2H_2O}$ have been formed by allowing a solution of ZnO_2K_2 to stand for some months. The following bodies are thus known:—

$$\mathbf{MgO.H_2O = Mg(OH)_2;\ ZnO.H_2O = Zn(OH)_2;}$$
$$\mathbf{CdO.H_2O = Cd(OH)_2;\ ZnO.2H_2O.}$$

(*b.*) **With hydrogen sulphide.**—Zinc and cadmium sulphides do not appear to combine with hydrogen sulphide. But if a stream of that gas is led through water in which magnesium oxide or carbonate is suspended, a soluble compound is formed, which has not been obtained solid, but which is supposed to have the formula $MgS.H_2S = Mg(SH)_2$, and to be magnesium hydrosulphide. When gently warmed, this solution evolves hydrogen sulphide, thus:—$Mg(SH)_2.Aq + 2H_2O = \mathbf{Mg(OH)_2.Aq} + 2H_2S$. This solution dissolves sulphur with a yellow colour, and may then contain polysulphides of magnesium.

The selenides and tellurides have not been investigated.

(*c.*) **Compounds of oxides with oxides.**—White crystals of $\mathbf{ZnO.K_2O}$ and $\mathbf{ZnO.Na_2O}$ [= $\mathbf{Zn(OK)_2}$, and $\mathbf{Zn(ONa)_2}$] separate from solutions of zinc hydroxide in caustic alkali. Metallic zinc dissolves in boiling caustic potash or soda, with evolution of hydrogen, thus:—$\mathbf{Zn} + 2NaOH.Aq = Zn(ONa)_2.Aq + H_2$. A similar cadmium compound is formed by dissolving cadmium oxide in fused potassium hydroxide. On treating a solution of zinc hydroxide in caustic soda with alcohol, the compound **$ZnO.Na_2O.8H_2O$** is thrown down in crystals. These bodies correspond to the hydroxides, the hydrogen being wholly or partially replaced by sodium or potassium.

(*d.*) **Compounds of sulphides with sulphides.**—Zinc sulphide is said to be wholly dissolved when added to a solution of sodium sulphide containing a weight of sulphur equal to that contained in the zinc sulphide. The inference is that the compound **$ZnS.Na_2S$** is produced. Cadmium sulphide is also sparingly soluble in excess of alkaline sulphides.

Cadmium sulphide is supposed to polymerise when boiled with acids or with sodium sulphide; and the sulphide produced by treating with hydrogen sulphide cadmium hydroxide which has been boiled with water is vermilion-coloured. Cadmium sulphide may also be obtained dissolved in water by washing the precipitated sulphide thoroughly, and treatment with solution of hydrogen sulphide. A yellow solution is produced, which coagulates on treatment with weak solutions of salts, especially those of cadmium.

(*e.*) **Compounds of sulphides with oxides.**—Magnesium oxide heated in a mixture of carbon dioxide and disulphide is converted into **$MgO.MgS$**. The corresponding zinc compound has been prepared by heating zinc sulphate, **$ZnSO_4$**, in hydrogen; and the cadmium compound, **$CdO.CdS.H_2O$**, is thrown down as a red precipitate when hydrogen sulphide is passed through a boiling solution of a cadmium salt. The compound **$4ZnO.ZnS$** has been found in zinc furnaces.

(*f.*) **Compounds of oxides with halides.**—The following "basic" halides have been prepared by the reaction of water at a high temperature on the halides:—

$2MgCl_2.MgO$; $MgCl_2.MgO$; $MgCl_2.5MgO$; $MgCl_2.9MgO$; $MgCl_2.10MgO$.
$ZnCl_2.3ZnO$; $ZnCl_2.6ZnO$; $ZnCl_2.9ZnO$.
$CdCl_2.CdO$; $CdBr_2.CdO$.

These bodies crystallise with varying amounts of water; thus crystals of $MgCl_2.5MgO$ have been obtained with 17, 14, 8, and $6H_2O$. Zinc oxychlorides possess the property of dissolving silk, but not wool or cotton, and their solutions are employed as a means of separating the constituents of mixed fabrics. The zinc oxychlorides are used by dentists as a stopping for teeth.

Physical Properties.

Mass of 1 cubic centimetre :—

	Be.	Ca.	Sr.	Ba.
O......	3·02	3·16—3·32*	4·5—4·75*	5·32—5·72*
$(OH)_2$..	—	2·8	3·63	4·49
S......	—	—	—	—
Se.....	—	—	—	
Te.....	—	—	—	—

	Mg.	Zn.	Cd.
O......	3·20—3·75*	5·47—5·78*	6·95—8·11*
$(OH)_2$..	2·36*	2·68—3·05	4·79
S......	—	3·92—4·07*	4·50—4·91*
Se......	—	5·40	5·8—8·9
Te.....	—	6·34	6·20

The asterisked higher numbers usually refer to the crystallised varieties, but are sometimes the results of different experimenters.

Heats of formation :—

$$\mathbf{Ca} + O = \mathbf{CaO} + 1310K; + H_2O = \mathbf{Ca(OH)_2} + 155K.$$
$$\mathbf{Sr} + O = \mathbf{SrO} + 1284K; + H_2O = \mathbf{Sr(OH)_2} + 177K.$$
$$\mathbf{Ba} + O = \mathbf{BaO} + 1242K; + H_2O = \mathbf{Ba(OH)_2} + 223K.$$
$$\mathbf{Mg} + O = \mathbf{MgO} + 1440K; + H_2O = \mathbf{Mg(OH)_2} + 50K.$$
$$\mathbf{Zn} + O = \mathbf{ZnO} + 853K; + H_2O = \mathbf{Zn(OH)_2} - 26K.$$
$$\mathbf{Cd} + O \quad .. \quad .. \quad + H_2O = \mathbf{Cd(OH)_2} + 657K.$$

$$\mathbf{Ca} + \mathbf{S} = \mathbf{CaS} + 896K; \mathbf{Mg} + \mathbf{S} = \mathbf{MgS} + 776K.$$
$$\mathbf{Sr} + \mathbf{S} = \mathbf{SrS} + 974K. \mathbf{Zn} + \mathbf{S} = \mathbf{ZnS} + 396K.$$
$$\mathbf{Ba} + \mathbf{S} = \mathbf{BaS} + 983K; \mathbf{Cd} + \mathbf{S} = \mathbf{CdS} + 324K.$$

CHAPTER XVIII.

OXIDES AND SULPHIDES OF ELEMENTS OF THE BORON GROUP.—DOUBLE COMPOUNDS WITH WATER AND OXIDES; BORACIC ACID AND BORATES; HYDROXIDES OF SCANDIUM, YTTRIUM, LANTHANUM, AND YTTERBIUM.—OXYHALIDES; FLUOBORATES.—OXIDES AND SULPHIDES OF ELEMENTS OF THE ALUMINIUM GROUP.—HYDROXIDES, HYDROSULPHIDES; DOUBLE OXIDES AND SULPHIDES.—OXYHALIDES.

Oxides and Sulphides of Boron, Scandium, Yttrium, Lanthanum, and Ytterbium.

Of these, boron oxide and sulphide, and the oxides of the remaining elements of the group have alone been investigated. The selenides and tellurides are unknown.

Sources.—These compounds do not occur native. Boron oxide is found in combination with water, as $B_2O_3.3H_2O$, as *sassolite;* with sodium oxide as *borax*, $2B_2O_3.Na_2O.10H_2O$; with magnesium oxide and chloride as *boracite*, $8B_2O_3.6MgO.MgCl_2$; and with silicon and calcium oxides as *datolite*,

$$3SiO_2.B_2O_3.2CaO.H_2O.$$

Scandium, yttrium, and ytterbium oxides are found in combination with silica in *gadolinite*, and with niobium and tantalum oxides in *yttrotantalite*, *samarskite*, and *euxenite;* while lanthanum oxide accompanies cerium and didymium oxide in *cerite*, in combination with silica.

List.	Boron.	Scandium.	Yttrium.	Lanthanum.	Ytterbium.
Oxygen..	B_2O_3.	Sc_2O_3.	Y_2O_3; Y_4O_9.	La_2O_3; La_4O_9.	Yb_2O_3
Sulphur .	B_2S_3.	—	—	—	—

Preparation.—1. **By direct combination.**—Boron burns in oxygen or nitric oxide, *NO.* Yttrium is also oxidised when heated in air, and lanthanum becomes covered with a steel-blue film. When strongly heated it takes fire and burns. The other elements of this group have not been prepared. Boron unites with sulphur at a white heat.

2. **By heating the hydroxides, &c.**—This is the usual method of preparation. These substances part with water at a red heat, leaving the oxides. The oxalates, carbonates, and nitrates of scandium, yttrium, lanthanum, and ytterbium also yield the oxides when heated to redness.

3. **By double decomposition.**—Boron oxide mixed with carbon, and heated to redness in a stream of carbon disulphide gas, yields the sulphide.

Properties.—**Boron trioxide,** $\mathbf{B_2O_3}$, is a non-volatile glass, melting to a viscid liquid at a red heat. It reacts with and dissolves in alcohol and in water. When fused with the oxides of metals they are dissolved, forming borates, *i.e.*, double oxides of boron and the metal. The **sulphide,** $\mathbf{B_2S_3}$, is a whitish-yellow substance, volatile when heated in a stream of hydrogen sulphide, and melting at a red heat. It is decomposed by water, yielding boracic acid and hydrogen sulphide.

The oxides of scandium, yttrium, lanthanum, and ytterbium are white powders, insoluble in water, and soluble with difficulty in acids. They do not react with alkaline hydroxides, nor do they fuse in the oxyhydrogen flame. The peroxides of yttrium and lanthanum are also white powders, which part with the excess of oxygen when heated.

Mass of 1 cubic centimetre:—$\mathbf{B_2O_3}$, 1·85 grams at 14·4°; $\mathbf{Sc_2O_3}$, 3·8 grams; $\mathbf{Y_2O_3}$, 5·03 grams at 22°; $\mathbf{La_2O_3}$, 6·5 grams at 17°; $\mathbf{Yb_2O_3}$, 9·2 grams.
Heat of formation:—$\mathbf{B_2} + 3O = \mathbf{B_2O_3} + 3172K$; $+ Aq = 180K$.

Double compounds.—(*a.*) **With water. Preparation.**—Boron trioxide dissolves in water with evolution of heat, combining with it to form the compound $\mathbf{B_2O_3.3H_2O}$, or $\mathbf{H_3BO_3}$, commonly called **boracic acid.** The same compound can also be prepared by addition of sulphuric acid to a solution of borax or some other borate in water, when the sodium of the borax is replaced by hydrogen, thus:—

$$Na_2B_4O_7.Aq + H_2SO_4.Aq + 5H_2O = \mathbf{4H_3BO_3} + Na_2SO_4.Aq.$$

The boracic acid separates in pearly-white scales, which have a bitterish cooling taste. Boracic acid is also obtainable by the action of moist air on boron; also by boiling boron with nitro-hydrochloric acid, when it unites simultaneously with oxygen and water.

The hydrated oxides of scandium, ytterbium, lanthanum, and didymium, are produced, like those of magnesium, by adding sodium hydroxide or any soluble hydroxide to solutions of the chlorides, or any other soluble compounds of the metals. They are insoluble in and do not combine with these hydroxides to form compounds undecomposed by water.

Boracic acid is a natural product, obtained in volcanic districts, especially in Tuscany, and in the Lipari Islands. The native form is named *sassolite*. Steam containing vapour of boracic acid issues from jets in the ground called *soffioni*. The steam from these jets is made to blow into artificial basins or *lagoni*, where the boracic acid condenses along with the steam. The solution is concentrated by causing it to flow over long sheets of lead, heated by the waste steam of the *soffioni*. It finally runs into crystallising tanks, where the boracic acid separates out on cooling. The crude product contains about 76 per cent. of boracic acid; it is purified by recrystallisation. Other compounds of boron trioxide with water are produced by heating H_3BO_3; these are $B_2O_3.H_2O$ and $2B_2O_3.H_2O$. The first remains on heating to 100°; the second is left at 160°; while at 270° the compound $8B_2O_3.H_2O$ is said to remain.

Properties.—Boracic acid, H_3BO_3 ($B_2O_3.3H_2O$) crystallises in nacreous laminæ; the other compounds are glassy substances. The hydrates of scandium, &c., are white gelatinous precipitates. Their exact composition has not been ascertained. Boracic acid is volatile with steam; and it reacts also with ethyl and especially with methyl alcohol, forming volatile compounds. It is estimated by distilling with sulphuric acid and methyl alcohol; the distillate is evaporated to dryness with a known weight of lime. It is used as an antiseptic, and is employed as a preservative of milk, fish, &c. A flame held in the steam evolved from a boiling solution is tinged green; if alcohol be present, it burns with a green flame. This constitutes the usual qualitative test for boron.

(*b.*) **With hydrogen sulphide.**—None of these possible compounds has been investigated.

(*c.*) **Compounds of oxides with oxides.**—No compounds of scandia, &c., are known with the oxides of elements preceding them in the periodic table. They combine with sulphur trioxide, forming sulphates, colourless crystalline bodies; with nitric pentoxide, forming nitrates, &c. These compounds are considered later.

Boron trioxide combines with other oxides when they are heated together. The resulting compounds are termed **borates.** The most important of these is borax, sodium borate. The following is a list of the more typical of these compounds; in this classification the combined water has not been included, as there is no evidence that it replaces either oxide of boron or oxide of the combined metal. The ratios are very numerous and complex. The metal, in the following table, has been considered analogous to calcium oxide, CaO, and has been termed MO in the heading. It would correspond to $\frac{1}{3}M_2O_3$, or to M_2O. The amount of water in the salts which have been prepared has been placed in brackets; if another classification is adopted (see *Silicates*, p. 308), it often becomes an integral portion of the formula. The question of these formulæ will be treated of further on, under silicates, phosphates, &c. The ratio given is that of the oxygen in the boron trioxide to the oxygen in the metallic oxide, the water, as before stated, being neglected.

Ratio 2 : 5 ($2B_2O_3.15MO$).	$2B_2O_3.6MgO.3Fe_2O_3$.
„ 2 : 4 ($2B_2O_3.12MO$).	$2B_2O_3.4Al_2O_3$ ($6H_2O$; also anhydrous).
„ 2 : 3 ($2B_2O_3.9MO$).	$2B_2O_3.3Al_2O_3$.($7H_2O$).
„ 1 : 1 ($2B_2O_3.6MO$).	$B_2O_3.3Na_2O$; $B_2O_3.3CaO.CaCl_2$; $B_2O_3.3CdO(3H_2O)$.
„ 6 : 5 ($2B_2O_3.5MO$).	$2B_2O_3.5BaO$.
„ 3 : 2 ($2B_2O_3.4MO$).	$B_2O_3.2BaO$; $B_2O_3.2MgO$.
„ 2 : 1 ($2B_2O_3.3MO$).	$2B_2O_3.3CaO$; $2B_2O_3.3SrO$; $2B_2O_3.3CoO(4H_2O)$.
„ 3 : 1 ($2B_2O_3.2MO$).	$B_2O_3.Na_2O$($3H_2O$, also $4H_2O$); K_2O; CaO($2H_2O$, also anhydrous); SrO; BaO ($10H_2O$, also H_2O) MgO($4H_2O$, also $8H_2O$); CdO; $3B_2O_3.Fe_2O_3$.($3H_2O$); $B_2O_3.NiO(2H_2O)$; $PbO(H_2O)$; $Ag_2O(H_2O)$; also $B_2O_3.PbO.PbCl_2(H_2O)$.
„ 4 : 1 ($4B_2O_3.3MO$).	$4B_2O_3.3Ag_2O$.
„ 5 : 1 ($5B_2O_3.3MO$).	$5B_2O_3.3SrO(7H_2O)$.
„ 6 : 1 ($2B_2O_3.MO$).	$2B_2O_3.Li_2O.5H_2O$; $2B_2O_3.Na_2O$($10H_2O$, *borax;* $6H_2O$; $5H_2O$; also $3H_2O$). $K_2O(5H_2O)$; $(NH_4)_2O$($3H_2O$, also $4H_2O$); $BaO.(H_2O)$; SrO($4H_2O$, also anhydrous); BaO($5H_2O$, also anhydrous); $PbO(4H_2O)$.
„ 9 : 1 ($3B_2O_3.MO$).	$3B_2O_3.Li_2O.6H_2O$; $3B_2O_3.K_2O(8H_2O)$; $BaO(14H_2O)$; $MgO(8H_2O)$.
„ 12 : 1 ($4B_2O_3.MO$).	$4B_2O_3.Li_2O.10H_2O$; $4B_2O_3.Na_2O(10H_2O)$; $(NH_4)_2O(6H_2O)$; $CaO(9H_2O)$; $SrO(6H_2O)$; $MgO(11H_2O)$.
„ 15 : 1 ($5B_2O_3.MO$).	$5B_2O_3.Na_2O(10H_2O)$; $(NH_4)_2O(6H_2O)$.
„ 18 : 1 ($6B_2O_3.MO$).	$6B_2O_3.K_2O(9H_2O)$; $(NH_4)_2O(9H_2O)$; $MgO.(18H_2O)$.

This list comprises nearly all the known borates. They are prepared by one of three methods:—(1.) By mixing a solution of boracic acid with the hydroxide or carbonate of the metal, evaporating, and crystallising. This method applies only to the borates of the elements of the sodium group; their hydroxides and carbonates, as also their borates, are soluble. (2.) By heating the oxide or carbonate, or even the nitrate or sulphate, of the metal with boron trioxide to a high temperature. The mass often crystallises on cooling. The borates of many oxides such as those of copper, nickel, &c., are coloured. Few of them have been analysed. (3.) By adding a soluble borate such as sodium borate to a soluble salt of the metal. A precipitate is formed with all elements except those of the sodium group. These precipitates, when washed with water, are decomposed, the boracic acid being washed out, and the hydroxide of the metal remaining behind. They are thus unstable compounds, largely or wholly decomposed by water.

The compounds containing water are almost always crystalline; those produced by fusion are also often crystalline, but are sometimes, like glass, amorphous; those produced by precipitation are of doubtful existence, inasmuch as a mixture of hydroxide and borate might on analysis give numbers which would lead to a definite formula.

The most important of these bodies is *borax.* It occurs as an incrustration on the soil of districts in Central Asia, and is known as *tincal;* it is found most abundantly, however, in lakes in California, 450 miles S.E. of San Francisco, the most important of which is 12 miles in length and 8 miles broad; the greater part of "Borax Lake" is dry, and the surface is charged with borax, common salt, sodium and magnesium sulphates, and salts of ammonium. These salts are collected and purified by recrystallisation. A solution of borax dissolves many substances insoluble in water, such as stearic acid, resins, arsenious oxide, &c. It is chiefly employed for glazing porcelain and for soldering metals; the film of oxide coating the heated metal dissolves in melted borax, and clean surfaces of the metal can thus be brought in contact. It has also considerable antiseptic and detergent properties.

(*d.*) **Double compounds of sulphides, selenides, and tellurides** are unknown, also (*e.*) **compounds of sulphides and oxides.**

(*f.*) **Compounds of oxides with halides.**—The only compounds which have been prepared are the double fluorides and oxides of boron and metals, and an oxychloride. Boron trioxide dissolves in hydrofluoric acid, and the solution, when concentrated by standing over sulphuric acid, is a syrup, which contains B_2O_3 and HF in the ratio $B_2O_3.6HF.H_2O$; it has been named fluoboric acid. The same liquid is obtained by saturating water with boron fluoride, BF_3, and distilling. The existence of this body as a definite substance appears to be questionable. It is decomposed by water into boracic and hydrofluoric acids.*

The oxychloride, BOCl, is produced by heating to 150° a mixture of $\mathbf{B_2O_3}$ and $2BCl_3$. It is a fuming liquid. With water it yields boracic and hydrochloric acids.

* Bassarow, *Comptes rend.*, **78**, 1698.

Oxides, Sulphides, Selenides, and Tellurides of Aluminium, Gallium, Indium, and Thallium.

These are as follows:—

	Oxygen.	Sulphur.	Selenium.	Tellurium.
Aluminium......	Al_2O_3.	Al_2S_3.	?	?
Gallium	GaO(?); Ga_2O_3.	Ga_2S_3(?),	?	?
Indium.........	In_4O_3?; In_2O_3.	In_2S_3.	?	?
Thallium	Tl_2O; Tl_2O_3; (TlO_2)*.	Tl_2S; Tl_2S_3.	Tl_2Se.	?

Sources.—Aluminium oxide, Al_2O_3, occurs native in a pure state as *corundum;* contaminated with ferric oxide as *emery;* coloured blue by cobalt oxide as *sapphire;* coloured red by chromium oxide as *ruby;* coloured purple by manganese sesquioxide, as *amethyst;* and yellow by ferric oxide, as *topaz.* It also occurs in combination with water, with silica, and with other oxides (see below; Silicates, p. 303; and Spinels, p. 241). Gallium and indium sulphides accompany zinc in some blendes; and thallium is found in the "flue-dust" of pyrites burners, being contained in certain samples of *iron pyrites*, FeS_2.

Preparation.—1. **By direct union.**—The metals all oxidise when heated in air, but not very readily. Fused aluminium becomes coated with a film of its oxide, Al_2O_3; gallium, too, oxidises only on its surface, even when strongly heated; indium forms a film of pale-yellow In_2O_3, and thallium becomes covered with a layer of a mixture of Tl_2O and Tl_2O_3. The sulphides and selenides may also be prepared by direct union; Tl_2S_3 can be prepared only thus.

2. **By heating compounds.**—(1.) The hydrates, when heated, yield the oxides. Aluminium hydrate loses all its water at 360°; indium hydrate at 655°; and thallium hydrate at 230°. (2.) The compound of indium sulphide, In_2S_3, with hydrogen sulphide loses hydrogen sulphide when heated. (3.) Aluminium oxide has been prepared by heating potash alum, $K_2SO_4.Al_2(SO_4)_3$, to whiteness; a mixture of potassium sulphate and alumina remains, sulphuric anhydride escaping; the potassium sulphate is dissolved out with water, leaving the alumina. (4.) Gallium oxide has been prepared by heating the nitrate.

3. **By double decomposition.**—Gallium sulphide, Ga_2S_3, is produced by addition of a soluble sulphide (ammonium sulphide has been used) to a soluble salt of gallium; indium sulphide, In_2S_3, is precipitated by hydrogen sulphide. Solutions of thallous

* In combination.

salts, such as $TlNO_3$, or Tl_2SO_4, give with hydrogen sulphide a precipitate of $\mathbf{Tl_2S}$. If a thallic salt be used, it is first reduced to a thallous salt by the hydrogen sulphide, with separation of sulphur, and thallous sulphide is then precipitated, thus :—

$$TlCl_3.Aq + H_2S = \mathbf{TlCl} + 2HCl.Aq + \mathbf{S};$$
$$2TlCl.Aq + H_2S = \mathbf{Tl_2S} + 2HCl.Aq.$$

When carbon disulphide gas is passed over red-hot alumina, some of the oxide is converted into sulphide. A similar action takes place with hydrogen sulphide. Indium sulphide, $\mathbf{In_2S_3}$, may be produced in scales like mosaic gold, by fusion of indium trioxide, $\mathbf{In_2O_3}$, with sodium carbonate and sulphur. No doubt sodium sulphate is formed at the expense of the oxygen of the indium oxide, and the indium combines with the excess of sulphur.

4. **By the action of heat,** in a current of hydrogen, gallium trioxide gives a bluish-grey sublimate, supposed to be monoxide; and indium trioxide, $\mathbf{In_2O_3}$, similarly treated, gives a mixture of oxides, one of which is said to have the formula $\mathbf{In_4O_3}$. It is probably a mixture or a compound of $\mathbf{In_2O}$ with $\mathbf{In_2O_3}$. When thallic oxide, $\mathbf{Tl_2O_3}$, is heated to 360° it begins to lose oxygen, giving the compound $\mathbf{3Tl_2O_3.Tl_2O}$, which is perfectly stable up to 565°; at higher temperatures, up to 815°, Tl_2O volatilises away; and the residue $\mathbf{Tl_2O_3}$ is stable in presence of air above that temperature. The monoxide, $\mathbf{Tl_2O}$, when heated in air is partially oxidised to $\mathbf{Tl_2O_3}$.

By removing oxygen from thallous sulphate, $\mathbf{Tl_2SO_4}$, thallous sulphide is left. This action is analogous to the loss of oxygen which sodium, and barium sulphates, &c., suffer when heated in hydrogen or with carbon. In the case of thallium, the sulphate is heated with potassium cyanide, KCN, which is doubtless converted into cyanate, KCNO.

5. **Special methods.**—Crystalline alumina has been produced by fusing the amorphous variety in the oxyhydrogen flame; by heating the oxide along with aqueous hydrochloric acid to 350° in a sealed tube; and by melting together at a white heat aluminium oxide with lead monoxide (litharge), or with barium fluoride. The last two processes yield artificial corundum; and if a trace of cobalt oxide or chromium oxide be added, artificial sapphires or rubies are produced.*

Properties.—Trioxides.—Aluminium and gallium trioxides are white powders, or friable masses; indium trioxide has a tinge

* *Compt. rend.*, **104**, 737.

of yellow, especially when hot; and thallium trioxide is a brown powder. Crystalline aluminium trioxide is exceedingly hard, and is insoluble in acids. The amorphous trioxide, when ignited, appears also to alter its structure, probably polymerising (*i.e.*, several molecules unite to one), for it is then almost unattacked by acids. It can still be dissolved, however, by boiling sulphuric or strong hydrochloric acid, forming the sulphate or chloride; the crystalline variety is totally insoluble. All the trioxides are without action on water.

Trisulphides, &c.—Aluminium trisulphide forms yellow crystals, which turn dark when heated; the selenide and telluride are black non-volatile powders; gallium trisulphide has not been described; indium trisulphide is a brown powder, or gold-coloured scales; and thallium trisulphide, a black, ropy substance, brittle below 12°. Aluminium sulphide is decomposed by water, giving the hydrate and hydrogen sulphide, thus:—

$$\mathbf{Al_2S_3} + 3H_2O.Aq = \mathbf{Al_2O_3.}n\mathbf{H_2O} + 3H_2S.$$

The other three are unchanged by water, but decompose when boiled with acids.

Other oxides and sulphides.—There are no lower oxides or sulphides of aluminium; the lower oxide of gallium, produced by heating the trioxide in hydrogen, is a bluish-grey substance. The lower oxides of indium are black powders.

Thallium monoxide is a reddish-black substance, melting about 300°, and is volatile between 585° and 800°. When heated with sulphur, the oxygen is replaced by sulphur. It combines directly with water, forming the hydrate, $\mathbf{Tl_2O.H_2O}$, and absorbs carbon dioxide from moist air. It has thus some resemblance to the hydroxides of the metals of the sodium group. Thallous sulphide, when precipitated, forms greyish or brownish flocks; from a hot, slightly acid solution it comes down in blue-black crystals. It may be fused to a black lustrous mass like plumbago. The selenide is a black crystalline body.

Physical Properties.

1. Mass of 1 cubic centimetre:—$\mathbf{Al_2O_3}$: 3·98 grams at 14°; $\mathbf{In_2O_3}$: 7·18; $\mathbf{Tl_2S}$: 8·0.
2. Melting-point:—$\mathbf{Tl_2O_3}$: 759°.
3. Heats of formation:—$\mathbf{2Al + 3S = Al_2S_3}$ + 1224K.
   $\mathbf{2Tl} + O = \mathbf{Tl_2O}$ + 423K; + H_2O = 33K.
   $\mathbf{2Tl + S = Tl_2S}$ + 197K.

Double compounds.—(*a.*) **With water: hydrates or hydroxides.**—The hydrated trioxides are produced by addition of a soluble hydroxide, such as that of sodium, potassium, or barium; or even of thallium (TlOH), to solutions of soluble salts of the metals. A solution of ammonia in water acts in a similar manner, as if it contained ammonium hydroxide, $NH_4.OH$. The reaction is as follows, *e.g.*, with aluminium:—

$$Al_2Cl_6.Aq + 6KOH.Aq = \mathbf{Al_2O_3nH_2O} + 6KCl.Aq.$$

Excess of precipitant (except ammonia) dissolves the hydrates of aluminium and gallium; gallium hydroxide is soluble even in solution of ammonia. Solution takes place owing to the formation of soluble double compounds (see below).

Aluminium hydroxide may also be produced by passing a current of carbon dioxide into a solution of potassium aluminate ($Al_2O_3 K_2O$). Potassium carbonate is formed, and the hydrate of aluminium precipitated. Aluminium sulphide, $\mathbf{Al_2S_3}$, reacts with water, giving the hydrate and hydrogen sulphide. Hence, when solution of ammonium sulphide is added to a soluble aluminium compound, the hydrate is precipitated, whilst sulphuretted hydrogen is evolved.

The sulphides are not known to form compounds with water.

Thallium monoxide, $\mathbf{Tl_2O}$, dissolves in water, and on cooling, or on evaporation, the solution deposits yellow needles of $\mathbf{Tl_2O.H_2O = 2TlOH}$. Its solution absorbs carbon dioxide from the air. Aluminium hydrate, prepared by precipitation, forms gelatinous flocks, and when dried at ordinary temperature in air, has approximately the formula $\mathbf{Al_2O_3.5H_2O}$. This is a hard, horny mass; when heated it gives up its water. Up to 65° the loss is rapid, and at that temperature the hydrate has approximately the formula $\mathbf{Al_2O_3,3H_2O}$. The rate of loss of water diminishes as the temperature rises to 150°, and increases up to 160°, diminishing again up to 200°. The composition of the hydrate between 160° and 200° is nearest the formula $\mathbf{Al_2O_3.2H_2O}$. From 200° to 250° the rate of loss of water is rapid, but is much slower between 250° and 290°, and here the formula approximates to $\mathbf{Al_2O_3.H_2O}$. Complete dehydration does not occur till 850° is reached. It is probable that there are many hydrates of alumina, but that no one is stable over any considerable range of temperature.

The action of excess of water, however, on aluminium amalgam yields a crystalline hydrate of the formula $\mathbf{Al(OH)_3 = Al_2O_3.3H_2O}$.

Three natural hydrates are known, *gibbsite*, $\mathbf{Al_2O_3.3H_2O}$, *bauxite*, $\mathbf{Al_2O_3.2H_2O}$, and *diaspore*, $\mathbf{Al_2O_3.H_2O}$. Artificial crystals of gibbsite have been produced by the slow action of the carbonic acid of the air on a solution of aluminate of potassium; and by boiling aluminium acetate or hydroxide for a long time with water, the dihydrate is said to be precipitated.

Indium hydrate is a gelatinous white precipitate, which when air-dried has approximately the formula $\mathbf{In_2O_3.6H_2O}$. When heated, it loses water gradually up to 150°. The rate of loss then increases to 160°, again to slacken. The composition between 150° and 160° nearly corresponds to the formula $\mathbf{In_2O_3.3H_2O}$. It is not dehydrated till 655°; and there are no signs of other hydrates.

Air-dried hydrate of thallium has the formula $\mathbf{Tl_2O_3.H_2O}$. At higher temperatures it is dehydrated.

(*b.*) **With hydrogen sulphide.**—Indium sulphide, $\mathbf{In_2S_3}$, when precipitated from soluble compounds of indium with hydrogen sulphide, has a deep yellow colour. It can be dried in air, but when heated it evolves hydrogen sulphide, leaving the sulphide. It is probably a compound of the nature of a hydrate: $\mathbf{In_2S_3.}n\mathbf{H_2S}$. The white precipitate produced by ammonium sulphide with salts of indium is also probably of this nature. It is soluble in solution of ammonium sulphide.

(*c.*) **Compounds of oxides with oxides.**—On adding a solution of potassium hydroxide to aluminium hydrate, complete solution occurs when the ratio of the alumina to the potash is as $Al_2O_3 : K_2O$; the same compound is precipitated when a solution in excess of hydroxide is mixed with alcohol, in which caustic potash is soluble, but not the aluminate. It has the formula $\mathbf{Al_2O_3.K_2O = 2KAlO_2}$. A similar sodium compound has been prepared. The compounds $Al_2O_3.2Na_2O$ and $Al_2O_3.3Na_2O$ are also said to have been prepared. By dissolving hydrate of alumina in solution of barium hydroxide and evaporating, crystals of $\mathbf{Al_2O_3.BaO.6H_2O}$, $\mathbf{Al_2O_3.2BaO.5H_2O}$, and $\mathbf{Al_2O_3.3BaO.11H_2O}$ are successively deposited.* These bodies may be compared with the borates.

The mineral named *spinel* is a compound of alumina with magnesium oxide, $\mathbf{Al_2O_3.MgO}$. It crystallises in octahedra, and has been prepared artificially. *Gahnite* is a similar compound with zinc oxide of the formula $\mathbf{Al_2O_3.ZnO}$, and *chrysoberyl* with beryllium oxide $\mathbf{Al_2O_3.BeO}$.

Two compounds with barium oxide and chloride are also known, viz., $\mathbf{Al_2O_3.BaO.BaCl_2}$ and $\mathbf{Al_2O_3.BaO.3BaCl_2}$.

Gallium oxide would, no doubt, enter into similar combinations, but these have not been investigated.

* *Berichte*, **14**, 2151; *J. prakt. Chem.*, **26**, 385, **474**; *Chem. News*, **42**, 29.

A higher oxide of thallium in combination with barium oxide is produced by passing a rapid current of chlorine through potassium hydroxide, in which thallic hydrate is suspended. The solution turns violet, and with barium nitrate gives a violet precipitate which contains the oxide TlO_2.*

(*d.*) **Compounds of sulphides with sulphides.**—Indium sulphide forms with potassium and sodium sulphides red crystalline compounds of the formulæ $\mathbf{In_2S_3.K_2S}$, and $\mathbf{In_2S_3.Na_2S}$. A silver compound of similar formula is produced on addition of silver nitrate to their solutions. Thallic sulphide, $\mathbf{Tl_2S_3}$, also unites with thallous sulphide, $\mathbf{Tl_2S}$, giving black crystalline bodies.

(*e.*) No compounds of **oxides with sulphides** are known.

(*f.*) **Compounds with halides.**—On evaporating an aqueous solution of aluminium chloride, it is probable that oxychlorides are produced, inasmuch as hydrogen chloride is evolved. On repeated evaporation, all aluminium remains as hydroxide. Similar compounds, but somewhat indefinite, have been produced by the action of aluminium chloride on aluminium in presence of air. Gallium chloride, on addition of water, gives a white precipitate of oxychloride, the formula of which is unknown.

Uses.—The chief use of alumina is as a mordant. When a salt of aluminium in solution is boiled in contact with animal or vegetable fibre, it splits into acid, and hydrate of alumina, the latter depositing on the fibre. The fibre has the power of absorbing and "fixing" colouring matters, when boiled with their solutions. If the colouring matter be dissolved in water along with a salt of aluminium, and the solution be boiled, the hydrated alumina often comes down in combination with the colour, giving a "lake." Such lakes are made use of as paints.

Physical Properties.

Mass of 1 cubic centimetre :—

	B.	Sc.	Y.	La.	Yb.	Al.	Ga.	In.	Tl.
O........	1·85	3·8	5·0	6·5	9·2	3·90—4·0	—	7·18	—
OH	1·49	—	—	—	—	2·39†	—	—	—
S	—	—	—	—	—	—	—	—	8·00‡

Heats of formation.

$2B + 3O = \mathbf{B_2O_3} + 2 \times 1586K;\ + 3H_2O = \mathbf{2B(OH)_3} + 2 \times 60K.$

$2Al + 3O + 3H_2O = \mathbf{2Al(OH)_3} + 2 \times 1945K.$

$2Tl + O = \mathbf{Tl_2O} + 423K;\ + H_2O = \mathbf{2TlOH} + 33K;\ + Aq = -32K.$

$2Tl + 3O + 3H_2O = \mathbf{2Tl_2(OH)_3} + 2 \times 432K.$

$2Al + 3S = \mathbf{Al_2S_3} + 2 \times 612K.$

$2Tl + S = \mathbf{Tl_2S} + 2 \times 98{\cdot}5K.$

* *Gazzetta*, 17, 450.

† *Gibbsite*, $Al(OH)_3$. *Diaspore*, $AlO(OH) = 3{\cdot}4$.

‡ Tl_2S.

CHAPTER XIX.

OXIDES, SULPHIDES, SELENIDES, AND TELLURIDES OF ELEMENTS OF THE CHROMIUM GROUP.—HYDROXIDES.—DOUBLE OXIDES AND SULPHIDES.—THE SPINELS.—OXYHALIDES.—CHROMATES, FERRATES, AND MANGANATES.—PERMANGANATES.—CHROMYL AND MANGANYL CHLORIDES; CHLORO-CHROMATES.

Oxides, Sulphides, Selenides, and Tellurides of Chromium, Iron, Manganese, Cobalt, and Nickel.

These compounds may be divided into five well-defined groups: (1) the **monoxides, monosulphides,** &c., such as **FeO, FeS,** &c.; (2) the **sesquioxides, sequisulphides,** &c., for example, Fe_2O_3, Fe_2S_3, &c.; (3) the **dioxides,** such as MnO_2; (4) the **trioxides,** of which CrO_3 is an instance; and (5) the **heptoxides,** of which compounds are known in the case of manganese, Mn_2O_7. The double compounds will be considered in connection with each group. As these bodies or their compounds are very numerous, it is advisable to consider them in the order of the above groups. It may be noticed generally that in formula, preparation, and properties, the monoxides, &c., show a certain resemblance to those of magnesium, zinc, and cadmium; while the sesquioxides, &c., are comparable with those of aluminium. The trioxides find their closest analogues in the sulphur group; and the compounds of manganese heptoxide have the same crystalline form as the perchlorates.

I. **Monoxides, monosulphides, monoselenides,** and **monotellurides**:—

List.	Oxygen.	Sulphur.	Selenium.	Tellurium.
Chromium ...	—	**CrS.**	**CrSe.**	?
Iron.........	**FeO.**	**(Fe_2S)FeS.**	**FeSe.**	**FeTe**?
Manganese ...	**MnO.**	**MnS.**	?	?
Cobalt	**CoO.**	**CoS.**	**CoSe.**	?
Nickel	**NiO.**	**(Ni_2S)NiS.**	**NiSe.**	?

Sources.—CrO is said to exist in combination with Cr_2O_3 in some chrome ores. **FeO** exists in combination with CO_2 as carbonate in *spathic iron ore*, and with Fe_2O_3 in *magnetite*. **MnO** has been found native. It forms crystals which reflect green, and transmit red light. **NiO** is also found native. **FeS** is sometimes found in meteoric iron, in combination with dinickel sulphide, Ni_2S, as $2FeS.Ni_2S$. Manganous sulphide, **MnS**, occurs as *manganese blende*, or *alabandine*, in iron-black lustrous cubes or octahedra. Native cobalt sulphide is known as *syepoorite*. It forms steel-grey to yellow crystals, and is used by Indian gold workers to give a rose-colour in burnishing gold. Nickel sulphide, **NiS**, occurs in nature as long brass-yellow needles, and is named *capillary pyrites*, or *millerite*. Nickel oxide, **NiO**, along with magnesium oxide, occurs as a silicate in the ore from New Caledonia. The ore contains about 18 per cent. of nickel oxide.

Preparation.—1. **By direct union.**—Higher oxides are produced by the union of chromium, iron, manganese, and cobalt with oxygen; but nickel burns to **NiO**. Iron, manganese, cobalt, and nickel unite directly with sulphur, selenium, and probably tellurium, forming monosulphides, &c.

The union with sulphur may be illustrated by heating an intimate mixture of iron filings with sulphur in a test tube; the mixture glows throughout, and is converted into ferrous sulphide.

2. **By heating double compounds.**—Iron, manganese, cobalt, and nickel oxides may be obtained by heating the oxalates, thus:—

$$\mathbf{FeC_2O_4} = \mathbf{FeO} + CO + CO_2.$$

Manganese, cobalt, and nickel monoxides are produced when their carbonates or hydroxides are heated, thus: $\mathbf{MnCO_3} = \mathbf{MnO} + CO_2$; $\mathbf{Ni(OH)_2} = \mathbf{NiO} + H_2O$. Air must be excluded, except in the case of nickel. Nickel monoxide alone is produced on igniting the nitrate; with the other metals higher oxides are formed. We here see a proof of the comparative stability of the higher oxides; those of chromium being most, and those of nickel least stable.

3. **By reducing a higher oxide or sulphide.**—Iron sesquioxide, Fe_2O_3, heated in a mixture of carbon monoxide and dioxide, such as is produced by the action of sulphuric acid or oxalic acid, is reduced to the monoxide. It is also produced in a crystalline form by heating iron to redness in a current of carbon dioxide; and by heating the sesquioxide, Fe_2O_3, in hydrogen; between the temperatures 330° and 440° magnetic oxide, Fe_3O_4, is produced; but from 500° to 600° the product is **FeO**. At still higher temperatures metallic iron is formed.* The higher oxides of cobalt

* *Chem. Soc.*, **33**, 1, 506; **37**, 790.

and nickel lose oxygen when heated alone, the former at a white heat, the latter at a red heat.

Chromium monosulphide and monoselenide, **CrS** and **CrSe**, have been produced by heating the sesquisulphide or selenide to redness in hydrogen. Ferric sulphate, $\mathbf{Fe_2(SO_4)_3}$, heated in hydrogen is said to give $\mathbf{Fe_8S}$; and ferrous sulphate, $\mathbf{FeSO_4}$, heated in sulphur vapour, $\mathbf{Fe_2S}$. As both these bodies are strongly magnetic, there appears reason to suspect that they contain metallic iron; they are blackish-grey powders. When heated with carbon, $\mathbf{FeSO_4}$ is said to yield **FeS**; cobalt sulphate behaves similarly.

Ferrous sulphide, **FeS**, is produced by heating to redness the disulphide, iron pyrites, $\mathbf{FeS_2}$, or magnetic pyrites, $\mathbf{Fe_2S_3.3FeS}$; sulphur volatilises; it may also be formed by heating pyrites, $\mathbf{FeS_2}$, with metallic iron. Cobalt sulphate, $\mathbf{CoSO_4}$, heated in hydrogen, gives an oxysulphide, **CoO.CoS** (see below); but nickel sulphate yields dinickel sulphide, $\mathbf{Ni_2S}$.

4. **By double decomposition.**—Manganous oxide is most easily prepared by heating the dichloride, $\mathbf{MnCl_2}$, with sodium carbonate, $\mathbf{Na_2CO_3}$, and a little ammonium chloride. The reaction is as follows, $\mathbf{MnCl_2 + Na_2CO_3 = MnO + 2NaCl} + CO_2$. The oxide is really formed by decomposition of the carbonate produced by double decomposition. The fused mass is deprived of sodium chloride by treatment with water. Higher oxides of iron or manganese, when heated with sulphur to a high temperature, yield the monosulphide; the sulphur combining with, as well as replacing oxygen. Thus $\mathbf{Fe_2O_3}$ and $\mathbf{Fe_3O_4}$, $\mathbf{Mn_2O_3}$ and $\mathbf{MnO_2}$ yield monosulphides, and sulphur dioxide; both reduction and double decomposition proceed simultaneously. Manganese dioxide is also converted into sulphide when it is heated in vapour of carbon disulphide, the carbon removing oxygen while manganese and sulphur unite. Cobalt and nickel sulphides have also been produced by heating the oxides in a current of hydrogen sulphide or sulphur gas. All monosulphides (and probably also monoselenides and tellurides) are precipitated on adding to a soluble chromous, ferrous, manganous, cobalt, or nickel compound a soluble sulphide (selenide or telluride). Ammonium sulphide is commonly employed. Manganese, cobalt, and nickel sulphides are also precipitated from solutions of their acetates by hydrogen sulphide. The typical equations are:—

$$\mathrm{FeSO_4.Aq} + \mathrm{(NH_4)_2S.Aq} = \mathbf{FeS} + \mathrm{(NH_4)_2SO_4.Aq};$$
$$\mathrm{Mn(C_2H_3O_2)_2.Aq} + H_2S = \mathbf{MnS} + \mathrm{2C_2H_4O_2.Aq}.$$

Properties.—Ferrous oxide is a black amorphous powder,

pyrophoric, *i.e.*, igniting and glowing like tinder on exposure to air it decomposes water, slowly at the ordinary temperature, quickly on boiling, liberating hydrogen. When prepared by the action of carbon dioxide on metallic iron, it forms small black lustrous crystals. Manganous oxide is a greyish-green powder, melting about 1500° to a green mass. When heated to redness in a current of hydrogen chloride it is converted into transparent emerald-green octahedra. Cobalt monoxide is an olive-green, and nickel monoxide a greyish-green, powder. The latter has been obtained in crystals by fusing a mixture of nickel sulphate and potassium sulphate; sulphur trioxide and its decomposition products, sulphur dioxide and oxygen escape, and crystals of nickel oxide remain disseminated through the potassium sulphate, the latter of which can be removed by solution in water. These bodies are all insoluble in water, and are not easily attacked by acids.

Chromous, ferrous, cobaltous, and nickelous sulphides when prepared by precipitation are black flocculent masses; manganous sulphide, similarly obtained, is pale yellowish-pink. Very finely divided iron sulphide is green when suspended in water. Pink manganous sulphide when heated in a sealed tube with yellow ammonium sulphide (polysulphide) changes to green, owing probably to some molecular change. When prepared in the dry way, chromous and cobalt sulphides are grey, ferrous aad nickel sulphides brass-yellow, and the native form of manganous sulphide iron-black. They all exhibit dull metallic lustre. Manganous sulphide changes to yellow-green hexagonal prisms when heated to redness in a current of hydrogen sulphide. The selenides are white, yellow, or grey bodies, also with dull metallic lustre. All these substances are insoluble in water; they react with acids, giving, for example, with hydrochloric or sulphuric acids, the chloride or sulphate of the metal and hydrogen sulphide. Hydrochloric acid, if dilute, does not attack nickel or cobalt sulphides unless it is heated. The action of dilute hydrochloric or sulphuric acid on ferrous sulphide is the usual method of preparing hydrogen sulphide.

A wet mixture of iron filings, sulphur, and ammonium chloride turns hot, owing to the combination of the iron and sulphur. Such a mixture is employed in cementing iron, for example in the construction of submarine piers for bridges. The sulphides can all be fused at a white heat. Dinickel sulphide, Ni_2S, can even be melted in glass vessels.

Double compounds.—(*a.*) **With water.—Hydrates or hydroxides.** These substances are prepared, as usual, by the action of a soluble hydroxide or of ammonia dissolved in water on some soluble compound of the metal, *e.g.*, the chloride or sulphate. With chromium, iron, and, in a less degree, manganese and cobalt, great care must be taken to exclude oxygen; the water in which the precipitant is dissolved must be boiled *in vacuo*, to remove dissolved oxygen, and the precipitation, filtration, &c., conducted in an atmosphere of hydrogen. Chromous compounds are best prepared from the acetate, which is made by the action of nascent hydrogen from zinc and hydrochloric acid on a solution of chromium trichloride. On adding potassium acetate, chromous acetate is precipitated as a red powder. On treatment with potassium hydroxide, it yields chromous hydrate, $2CrO.H_2O$, as a substance yellow when wet, turning brown when dried.* When boiled with water, hydrogen is evolved, and chromic hydrate is produced. The water which it contains cannot be removed by heat, for the reaction takes place $\mathbf{2CrO.H_2O = Cr_2O_3} + H_2$.

Ferrous hydrate, $\mathbf{FeO.H_2O}$ (?), is a white precipitate, which becomes much denser on standing in a solution of potassium hydroxide. It is sparingly soluble in water (1 in 150,000). It absorbs carbon dioxide from air; and when dry it turns hot and oxidises on exposure to air. The wet hydrate, on atmospheric oxidation, turns first green, then rust coloured.

Manganous hydrate is also white, and turns brown on exposure to air. It is said to contain 24 per cent. of water, and hence must have approximately the formula $\mathbf{3MnO.4H_2O}$. It can be produced by boiling manganous sulphide, $\mathbf{MnS}$, with caustic potash. Cobaltous hydrate is a dingy-red powder, prepared by *boiling* a solution of cobalt dichloride with caustic potash, and collecting and drying the precipitate. In the cold, a blue oxychloride is precipitated. The hydrate of nickel, $\mathbf{NiO.2H_2O}$, occurs native, in small emerald-green prisms; and $\mathbf{NiO.H_2O = Ni(OH)_2}$ is an apple-green precipitate. By leaving a solution of nickel carbonate in excess of ammonia to crystallise, this hydrate separates in green crystals.

It appears probable that the precipitated sulphides of these metals are in reality compounds of the sulphides with water.

(*b.*) No hydrosulphides are known; and (*c.*) No double oxides

(*d.*) **Compounds of Sulphides with Sulphides:—**

$\mathbf{2FeS.K_2S}$; obtained by igniting $\mathbf{Fe_2S_3,K_2S}$ in hydrogen.

$\mathbf{3MnS.K_2S}$; dark red scales, produced by heating together man-

* *Annales* (5), **25**, 416; *Comptes rend.*, **92**, 792; 1051.

ganese sulphate, $MnSO_4$, with potassium sulphide and carbon, and dissolving out the excess of potassium sulphide with water.

$3MnS.Na_2S$. Light red needles, similarly prepared. Also **$MnS.2Na_2S$.**

$NiS.2FeS$. A double sulphide of nickel and iron, named *pentlandite*, which forms bronze-yellow crystals.

$FeS.2ZnS$, known as *christophite*, occurs native; also **$CoS.CuS$**, *carrolite.*

(*e.*) Compound of oxide and sulphide.—**$CoS.CoO$**, a dark grey sintered mass, produced by heating cobalt sulphate, **$CoSO_4$**, in hydrogen.

(*f*). Compounds with halides.—Chromous chloride is said to give a light grey oxychloride; and cobalt chloride heated with water a greenish-blue oxychloride. Similar bodies, green and insoluble, are produced, when nickel chloride or iodide is heated with nickel hydroxide. Their formulæ are unknown.

II. **Sesquioxides, sesquisulphides, sesquiselenides** (the tellurides have not been investigated).

List.	Oxygen.	Sulphur.	Selenium.
Chromium	Cr_2O_3.	Cr_2S_3.	Cr_2Se_3.
Iron..............	Fe_2O_3.	Fe_2S_3.	Fe_2Se_3.
Manganese	Mn_2O_3.	—	—
Cobalt	Co_2O_3.	Co_2S_3.	
Nickel	Ni_2O_3.	—	

Sources.—Chromium sesquioxide exists in combination with ferrous oxide in *chrome iron ore* or *chromite*, the chief source of chromium. It occurs in veins in serpentine rock. As *chrome-ochre* it forms a yellow-green earthy deposit, which is found in Shetland. **Iron sesquioxide** is very widely distributed, and occurs as *red hæmatite* or *specular ore* in large deposits in Cumberland and Lancashire in early formations; in carboniferous strata as *brown hæmatite* or *limonite* in the Forest of Dean, in Glamorganshire, or associated with oolitic rocks as the earthy hæmatite of Northamptonshire and Lincolnshire. More recent deposits of *limonite* occur as *bog-iron-ore* in Ireland and North Germany. *Magnetic ore*, or *magnetite*, **$Fe_2O_3.FeO$**, is also very widely distributed. It is, perhaps, the purest form of iron ore, and occurs as sand in Sweden. From it the celebrated Swedish iron is made. *Magnetic pyrites*, **$Fe_2S_3.FeS$**, and **$2Fe_2S_3.FeS$**, and *copper pyrites*, **$Fe_2S_3.Cu_2S$**, are made use of as sources of sulphur. **Manganese sesquioxide, Mn_2O_3**, occurs as *braunite*, and hydrated, **$Mn_2O_3.H_2O$**, as *grey maganese ore*. *Wad* is a mixture of oxides of manganese, probably consisting largely of **Mn_2O_3**. In combination with MnO, it forms *hausmannite*, **$Mn_2O_3.MnO$** (see Spinels). Cobalt and nickel sesquioxides do not occur native, but $Co_2S_3.CoS$ is known as *linnæite.*

Preparation.—1. By direct union.—Chromium, heated in air, forms Cr_2O_3; but iron, manganese, and cobalt burn to compounds of the sesquioxides and monoxides, depending on the temperature. A steel watch-spring set on fire by being tipped with burning sulphur, burns in oxygen with brilliant scintillations to $Fe_2O_3.FeO$, or magnetic oxide, which fuses and drops from the wire.

FIG. 32.

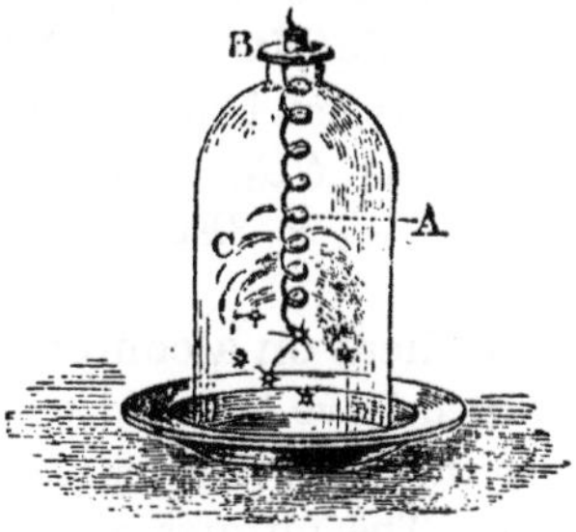

This forms a telling experiment, and illustrates well the direct union of metals of this group with oxygen. The jar in which the combustion takes place should stand in an iron tray, or in a plate full of water, for the fused oxide is certain to crack any glass tray on which it falls.

Iron filings, heated to dull redness in a current of sulphur gas, forms Fe_2S_3; and the corresponding selenide, Fe_2Se_3, has been similarly made.

2. **By reducing a higher compound.**—Chromium trioxide, CrO_3, when strongly ignited, loses oxygen, forming the sesquioxide. Compounds of the trioxide, such as mercurous chromate, Hg_2CrO_4, ammonium dichromate, $(NH_4)_2Cr_2O_7$, and others also yield the sesquioxide on ignition. Chromates, such as bichrome, $K_2Cr_2O_7$, at a white heat give neutral chromate, chromium sesquioxide, and oxygen, thus:—$2K_2Cr_2O_7 = 2K_2CrO_4 + Cr_2O_3 + 3O$. Manganese dioxide, at a dull-red heat, likewise loses oxygen, giving Mn_2O_3.

3. **By oxidising a lower compound.**—Ferrous sulphate, $FeSO_4$, when distilled for the manufacture of anhydrosulphuric acid (see p. 433), leaves a residue of sesquioxide. It may be supposed that the ferrous oxide decomposes water-gas, arising from water still combined with the ferrous sulphate, producing the sesquioxide. Ferrous carbonate heated gently in air yields ferrous oxide, FeO, which unites with oxygen, forming the sesquioxide.

It is also produced by heating ferrous sulphate with a little nitre, $\mathbf{KNO_3}$, to supply oxygen. Ferrous oxalate, $\mathbf{FeC_2O_4}$, yields the monoxide on ignition, and in air the sesquioxide is produced. The lower oxides of manganese, **MnO** and $\mathbf{Mn_3O_4}$, when heated in oxygen give the sesquioxide, when the pressure of oxygen is greater than 0·26 of an atmosphere. As the pressure of the oxygen in ordinary air is approximately one-fifth of an atmosphere, such oxidation does not occur in air, unless it be compressed. The nitrates of these metals, when heated, yield the sesquioxides. This is a case of simultaneous decomposition and oxidation. The nitrate is decomposed into monoxide and nitric pentoxide, thus:—$\mathbf{Fe(NO_3)_2} = \mathbf{FeO} + N_2O_5$; but the pentoxide parts with its oxygen, being itself converted into lower oxides of nitrogen, NO and NO_2, thus: $2\mathbf{FeO} + N_2O_5 = \mathbf{Fe_2O_3} + 2NO_2$; and $6\mathbf{FeO} + N_2O_5 = 3\mathbf{Fe_2O_3} + 2NO$. And similarly with the other metals.

4. **By the action of heat on a compound.**—The hydrates of these metals when heated leave the oxides. Ferric hydrate, when boiled for a long time in water, is ultimately dehydrated, and dry ferric oxide settles out. The nitrates and sulphates, &c., are also decomposed by heat, and also the borates. The excess of boracic acid is removed by weak hydrochloric acid.

5. **By double decomposition.**—Ferric oxide is produced in a crystalline form when ferric chloride and lime are heated to redness, or when ferrous sulphate and sodium chloride are heated together in air. The ferrous oxide is oxidised by the air, and crystallises from the salt. The sulphides are generally prepared by double decomposition. Chromium sesquisulphide is obtained when chromium trioxide is heated to whiteness in a current of carbon disulphide gas; heated in sulphur gas or in hydrogen sulphide, it suffers no change; but the chloride is converted into sulphide or selenide by hydrogen sulphide or selenide at a red heat, and the hydrate, when heated to 440° in sulphur gas, or to a higher temperature in selenium vapour, yields the sulphide or selenide. Cobaltic hydrate gently heated in hydrogen sulphide also gives cobalt sesquisulphide. Nickel sesquisulphide is unknown.

Properties.—Oxides.—Chromium sesquioxide is an amorphous green powder; when crystalline it forms green tablets, or, if produced at a high temperature, brown crystals. The amorphous variety, if it has not been exposed to a high temperature during its formation, becomes incandescent when gently heated, no doubt owing to polymerisation, several molecules uniting to form one. It is then practically insoluble in acids. This behaviour is also seen with aluminium, manganese, and iron sesquioxides. The crystalline varieties of chromium oxide are produced in presence of chlorine, or by some solvent for the oxide. Thus chromium oxychloride, CrO_2Cl_2, when passed through a red-hot tube, deposits crystalline oxide; similarly, potassium dichromate, heated in chlorine, gives a mixture of crystalline oxide and potassium chloride, the excess of oxygen being expelled. **Iron sesquioxide** may be obtained in crystals by fusing the amorphous variety with calcium chloride, or by heating it in a current of hydrogen chloride. It would appear that in such cases the volatile chloride is formed; and that it is decomposed by oxygen, yielding oxide, which is deposited in crystals. Crystalline varieties of the sesquioxides of cobalt and nickel, owing to their easy decomposition, have not been obtained. That of manganese has not been prepared artificially. Amorphous ferric sesquioxide is brown-red or red, according to the method of preparation. If prepared from ignited ferrous sulphate it has a fine colour, and is used as a paint, under the name "Venetian red." It is also used under the name of "rouge" for grinding and polishing glass objects, such as the lenses of telescopes, &c., and as "crocus" to produce shades from purple-red to yellow, according to the amount, on porcelain, in combination with silica. The crystalline variety is black. When native, as *specular ore*, it forms very lustrous rhombohedra; another crystalline variety, *martite*, occurs in octahedra; while *hæmatite* consists of kidney-shaped (botryoidal) masses, with a radiated crystalline structure. **Manganese sesquioxide**, when amorphous, is a black powder; as *braunite* it forms brownish-black lustrous quadratic pyramids. **Cobalt sesquioxide**, prepared by heating the hydrated compound to 600—700°, is a black powder, as is also **nickel sesquioxide.**

These bodies show different degrees of stability. While chromium sesquioxide can be fused at a white heat without change, iron sesquioxide is converted into Fe_3O_4, and at a bright-red heat, manganic sesquioxide gives Mn_3O_4. Cobalt and nickel sesquioxide lose oxygen at a dull-red heat, giving Co_3O_4 and NiO respectively. Cobaltic oxide, as borate, is made use of as a black pigment in enamel painting. **Chromium sesquisulphide** and **sesquiselenide** form brilliant black plates; iron **sulphide** and **selenide** are yellowish-grey with metallic lustre; and cobalt **sesquisulphide** forms a dark iron-grey mass.

Double compounds.—(*a*.) **With water: hydrates or hydroxides.**—These are produced as usual from a soluble salt of the hydroxide. Those of cobalt and nickel are formed by the action of an alkaline solution of sodium or potassium hypochlorite on a salt of the metal. Hydrated monoxide is produced and further oxidised by the hypochlorite, thus :—

$$2\mathbf{CoO}.n\mathbf{H_2O} + NaClO.Aq = \mathbf{Co_2O_3}.n\mathbf{H_2O} + NaCl.Aq.$$

Cobalt is more easily oxidised than nickel, for chlorine water converts the hydrated monoxide into the sesquioxide, thus :—

$$2\mathbf{CoO}.n\mathbf{H_2O} + Cl_2.Aq + H_2O = \mathbf{Co_2O_3}.n\mathbf{H_2O} + 2HCl.Aq.$$

Hydrated chromium sesquioxide is dissolved by excess of cold caustic potash or soda, but is precipitated on warming (see below).

There are two varieties of chromic salts, which are respectively green and violet. Both varieties give with alkalis a grey-green precipitate. By varying the conditions, the following hydrates have been prepared :—

$\mathbf{Cr_2O_3.9H_2O}$. Grey-violet powder.

$\mathbf{Cr_2O_3.7H_2O}$. Greyish-green; soluble in alkali with violet colour.

$\mathbf{Cr_2O_3.6H_2O}$. Green, gelatinous, drying to a hard black mass.

$\mathbf{Cr_2O_3.5H_2O}$. Similar to last.

$\mathbf{Cr_2O_3.4H_2O}$. Green; by boiling chromic chloride and caustic alkali.

$\mathbf{Cr_2O_3.2H_2O}$. *Guignet's or Pannetier's green;* produced by heating bichrome, $\mathbf{K_2Cr_2O_7}$, and borax. Oxygen is lost, and a borate of chromium and alkali is formed. On treatment with water, the borate is decomposed, leaving the hydrate. This body is a fine green pigment.

These hydrates dissolve in cold acids, giving violet salts, the solutions of which turn green when warmed, most probably owing to the formation of a basic salt. Thus chromic sulphate, $Cr_2(SO_4)_3.Aq$ (or $Cr_2O_3.3SO_3.Aq$), when warmed, is supposed to give $Cr_2O.(SO_4)_2.Aq$ (or $Cr_2O_3.2SO_3.Aq$), losing the elements of sulphuric anhydride.

No native form of chromium hydrate is known.

Ferric hydrates are found native. *Brown* or *yellow clay iron ore* is supposed to be the trihydrate, $\mathbf{Fe_2O_3.3H_2O}$ or $\mathbf{Fe(OH)_3}$; *xanthosiderite* is $\mathbf{Fe_2O_3.2H_2O}$, or $\mathbf{Fe_2O(OH)_4}$; and *göthite* or *needle iron ore*, $\mathbf{Fe_2O_3.H_2O}$ or $\mathbf{FeO.(OH)}$. *Limonite* is $\mathbf{2Fe_2O_3.3H_2O}$; and *turgite*, $\mathbf{2Fe_2O_3.H_2O}$. Precipitated hydrate, dried in air, possesses the approximate formula $\mathbf{Fe_2O_3.5H_2O}$; when heated, water is gradually lost, no sign of formation of intermediate hydrates being found. It is probable that there are many hydrates, each of which is stable within a very limited range of temperature; hence, on drying, indefinite mixtures are produced.* By prolonged boiling in water, the hydrate $\mathbf{Fe_2O_3.H_2O}$ is produced, and after a long time the precipitate consists of anhydrous sesquioxide; it appears, therefore, that the hydrate may lose water even in presence of great excess of water at 100°. Hydrate of iron is used as a mordant (see aluminium hydrate, p. 242). It produces stains of "iron-mould" on linen; these can be removed by oxalic acid, and a little metallic tin to reduce the iron from sesquioxide to monoxide, which is more easily soluble.

Hydrated manganese sesquioxide occurs native as *manganite* or *grey manganese ore;* its formula is $\mathbf{Mn_2O_3.H_2O}$. *Wad*, a mixture of oxides of manganese, probably contains some other hydrates. Both ferrous and manganous hydrates, suspended in water, when shaken with oxygen or air, are converted into hydrated sesquioxides. That of iron is rust-brown, and of manganese dark-brown.

Hydrated sesquioxides of cobalt and nickel are black precipitates. That of nickel is said to have the formula $\mathbf{Ni_2O_3.3H_2O}$.

It is probable that the sesquisulphides, produced by precipitation, are also hydrated. A green flocculent precipitate is produced by addition of a polysulphide of ammonium, $(NH_4)_2S_n$ (yellow sulphide), to a solution of ferric chloride, to which a small quantity of chlorine water or solution of bleaching powder has been added. With excess, it is oxidised and dissolved. This green precipitate is soluble in ammonia with a green colour, possibly giving a double sulphide. Its formula is said to be $\mathbf{2Fe_2S_3.3H_2O}$. A cobaltic salt gives, with hydrogen sulphide, a dark-grey precipitate of **cobalt sesquisulphide**, also probably hydrated. No similar nickel compound has been prepared.

(*b.*) No compounds with hydrogen sulphide are known.

(*c.*) **Compounds of oxides with oxides.**—As has been stated, hydrated chromium sesquioxide dissolves in cold solutions of the hydroxides of potassium of sodium, but is reprecipitated on warming. This behaviour is so far analogous to that of aluminium hydrate; the double oxide of aluminium and alkali-metal, however, is more stable than that formed by chromium, for its solution can be boiled without change. The other hydrates of this group are insoluble in alkalis.

The Spinels.—Compounds of these sesquioxides with monoxides of dyad metals form a very important group of minerals,

* *Chem. Soc.*, **53**, 50.

crystallising in octahedra, or in rhombic dodecahedra, named **spinels**, the name spinel being generally applied, but being specially applicable to the oxide of aluminium and magnesium, $Al_2O_3.MgO$. The following is a list:—

$Cr_2O_3.FeO$; *chromite*, or *chrome-iron ore.*	$Al_2O_3.ZnO$; *gahnite.*
$Fe_2O_3.FeO$; *magnetite*, or *magnetic iron ore.*	$Al_2O_3.FeO$; *zeilanite.*
$Fe_2O_3.MgO$; *magnesio-ferrite.*	$Al_2O_3.BeO$; *chrysoberyl.*
$Fe_2O_3.ZnO$; *franklinite.*	$Mn_2O_3.ZnO$; *hetaerolite.*
$Al_2O_3.MgO$; *spinel.*	$Mn_2O_3.MnO$; *hausmannite.*

Besides these, $Cr_2O_3.ZnO$, $Cr_2O_3.CrO$, $Cr_2O_3.MnO$, $Fe_2O_3.CaO$, $Co_2O_3.CoO$, and $Ni_2O_3.NiO$ have been made artificially.*

Chromous hydrate, $Cr(OH)_2$, made by addition of caustic soda to chromium dichloride, when exposed to air, changes to a snuff-coloured powder, of the formula $Cr_2O_3.CrO$. It has not been obtained crystalline. When iron wire and lime are heated to whiteness in presence of air, black crystals of $Fe_2O_3.CaO$ are produced; the same compound is formed by strongly igniting a mixture of hæmatite and chalk. *Franklinite*, $Fe_2O_3.ZnO$, has been produced by strongly igniting a mixture of iron sesquioxide, zinc sulphate, and sodium sulphate. The zinc oxide remaining after decomposition of the sulphate combines with the oxide of iron. The sodium sulphate may act as a solvent. Iron, manganese, and cobalt sesquioxides lose oxygen, the first at a white heat, the second at bright redness, the last at a dull-red heat, giving these complex oxides. That of iron is the important *magnetic iron ore*, occurring largely in Sweden. **Manganoso-manganic oxide** is a reddish-brown powder, which turns black when heated, but recovers its red colour on cooling. **Cobaltoso-cobaltic** and **nickeloso-nickelic oxides** form grey octahedra with metallic lustre. That of cobalt may be produced by heating the nitrate or oxalate to redness, and boiling the residue with hydrochloric acid; and that of nickel by heating nickel dichloride, $NiCl_2$, to 350—400° in a current of moist oxygen. Manganese dichloride, on exposure to moist air, is also changed into the crystalline oxide; and it may also be produced by heating manganous oxide, MnO, to redness in water-gas.

These bodies are also known in a hydrated condition.

The snuff-coloured powder, obtained as described, from chromous oxide in air, is probably hydrated. A dingy green hydrate of ferroso-ferric oxide is produced by oxidation of ferrous hydrate; and black hydrates are precipitated by addition of an alkali to a

* *Comptes rend.*, **104**, 580.

mixture in molecular proportions of a ferrous and ferric salt, thus:—

$$FeCl_2.Aq + 2FeCl_3.Aq + 8KOH.Aq = 8KCl.Aq + \mathbf{Fe_3O_4.}n\mathbf{H_2O}.$$

Like the anhydrous oxide, $\mathbf{Fe_3O_4}$, these hydrates are magnetic. A solution of manganoso-manganic oxide in phosphoric acid gives a brown precipitate with potash, doubtless of hydrate.

A few other double compounds are known, in which the sesquioxide and protoxide are present in different ratios. Thus, by addition of ammonia solution to a solution of a mixture of calcium chloride and chromium trichloride, the body $\mathbf{Cr_2O_3.2CaO}$ is precipitated. A somewhat similar compound, but containing calcium chloride in addition, of the formula $\mathbf{Fe_2O_3.2CaO.CaCl_2}$, crystallises from a solution of iron sesquioxide and lime in fused calcium chloride, in shining black prisms. And lastly, by heating a mixture of hydrated ferric oxide, potassium carbonate, and potassium chloride, till the latter is volatilised, ferric oxide, in combination with a small quantity of water and potassium oxide, remains as transparent red-brown crystals.

"**Smithy scales**" are produced by heating iron to redness in air. Two layers are formed; the outer layer has approximately the composition $\mathbf{Fe_3O_4}$; the inner layer forms a blackish-grey, porous, brittle mass, and has the formula $\mathbf{Fe_2O_3.6FeO}$. Ferroso-ferric oxide is produced also when iron burns in oxygen, when iron is heated in water-gas, or when the monoxide is heated in a current of hydrogen chloride.

It is possible to take two views of the constitution of these oxides; the first is that the sesquioxides are chemical individuals, derived from the corresponding trichlorides; and there appears little doubt that this is the case with chromium and iron sesquioxides, $\mathbf{Cr_2O_3}$ and $\mathbf{Fe_2O_3}$, being easily derived from and convertible into $\mathbf{Cr_2Cl_6}$ and $\mathbf{Fe_2Cl_6}$ respectively. Similarly, their compounds with the protoxides would justly have the formulæ $\mathbf{Cr_2O_3.CrO}$ and $\mathbf{Fe_2O_3.FeO}$. But in the case of manganese there appears to be some evidence of the existence of two bodies of like formula, but of different properties, implying different constitution. There is little doubt that the fact that such bodies become much more dense and insoluble in acids on ignition, sometimes, indeed, themselves evolving heat when gently warmed, is due to polymerisation, *i.e.*, the association of several simple molecules to form a more complex one. But the evidence as regards manganese sesquioxide points to a different cause. That body may be regarded as either a chemical individual, $\mathbf{Mn_2O_3}$, and the derived manganoso-manganic oxide as **$Mn_2O_3.MnO$**; or it may be conceived to be **$MnO_2.MnO$**, a compound of dioxide and monoxide, or a manganite of manganese; and the substance, **Mn_3O_4**, might be **$MnO_2.2MnO$**. Now manganese sesquioxide, when treated with dilute nitric acid, gives a solution of manganous nitrate, $Mn(NO_3)_2$, and a residue of **MnO_2**. With sulphuric acid oxygen is evolved, and manganous sulphate, $MnSO_4$, dissolves. The acetate, phosphate, &c., of Mn_2O_3 can, however, be prepared; and it is very unlikely that such bodies are mixtures of manganous salts and salts of manganese dioxide; salts of the latter being almost unknown. On addition of alkali to such salts a brown precipitate is produced, soluble in acids with formation of salts of the sesquioxide; whereas the hydrated sesquioxide, **$Mn(OH)_3$**, produced by oxidation in air of manganous hydrate is split by nitric acid into manganous nitrate and insoluble hydrated dioxide, **$MnO_2.nH_2O$**; and it is insoluble in dilute sulphuric acid. These facts would lead us to conclude that two bodies of the formula **Mn_2O_3** exist, one of which, however, has the constitution **$MnO_2.MnO$**. The oxides would well repay study in this direction.

(*d.*) **Compounds of oxides with sulphides.**—Iron sesquioxide, heated in sulphur gas, gives the compound **$Fe_2O_3.3Fe_2S_3$**. No other compounds of this nature have been prepared in this group.

(*e.*) **Compounds of sulphides with sulphides.**—The following is a list:—*

$Cr_2S_3.Na_2S$:	brick-red powder.	**$Cr_2S_3.MnS$**:	chocolate-coloured powder.
$Cr_2S_3.CrS$:	grey-black powder.	**$Fe_2S_3.Cu_2S$**.	*Copper-pyrites.*
$Cr_2S_3.ZnS$:	dark brown powder.	**$Co_2S_3.CoS$**.	*Linnæite.*
$Cr_2S_3.FeS$:	*Daubreelite;* black.	**$Ni_2S_3.NiS$**.	*Berychite.*

In these compounds, as in the spinels, one metal may replace another without reference to atomic weight. If any one molecule be considered, it of course possesses a definite formula, such as **$Co_2S_3.CoS$**. But the mineral named *nickel-linnæite* contains some **$Ni_2S_3.NiS$**; or, perhaps, **$Co_2S_3.NiS$**, along with the former. The atomic ratio of metal to sulphur is a constant one, but as these bodies have the same crystalline form, and as their molecules occupy approximately the same volume, they can replace one another in any crystal. The usual way of denoting such replacement in any proportion is to write the formula, for example, of *nickel-linnæite*, thus: **$(Co,Ni)_3S_4$**.

* *Wien. Akad. Ber.* (2), **81**, 531; *Monatsh. f. Chem.*, **2**, 266.

The same peculiarity is noticeable in the spinels, where aluminium, chromium, iron, and manganese may replace each other as sesquioxides; and beryllium, magnesium, zinc, &c., as monoxides. This will be again referred to in treating of the silicates.

The double sulphides which have been prepared artificially have been obtained by passing hydrogen sulphide over a heated mixture of the hydrates of the respective metals; thus, a mixture of chromic hydrate and zinc hydrate thus treated, gives a mass which, when boiled with hydrochloric acid, leaves a dark brown powder of the formula $Cr_2S_3.ZnS$.

More complex sulphides of iron are found native, and are generally termed *magnetic pyrites*. They have the formulæ $Fe_2S_3.3MS$, $Fe_2S_3.4MS$, $Fe_2S_3.5MS$, and $Fe_2S_3.6MS$, **M** representing iron, cobalt, or nickel. They form yellow crystals with metallic lustre, *Copper pyrites*, *barnhardtite*, and *chalcopyrrhotite* are similar bodies, containing copper, and have respectively the formulæ $Fe_2S_3.Cu_2S$, $Fe_2S_3.2Cu_2S$, and $Fe_2S_3.2CuS.FeS$. *Purple copper ore* is a similar compound of uncertain formula. By fusing iron with sulphur and potassium carbonate, purple-brown needles, of the formula $KFeS_2$, are formed. By ignition in hydrogen it yields $2FeS.K_2S$.

(*f*.) **Compounds with halides.**—These bodies, as usual, are formed either by evaporating or heating an aqueous solution of the trichlorides, or by heating a mixture of chloride and hydrate.

The following have been prepared :—

$Cr_2O_3.8CrCl_3.24H_2O$.
$Cr_2O_3.4CrCl_3.8H_2O$, and $3H_2O$.
$Cr_2O_3.2CrCl_3.2H_2O$.
$Cr_2O_3.CrCl_3\ 3H_2O$.

The corresponding compounds of iron are not so definite. Weak solutions of ferric chloride, when heated, give 1, soluble ferric hydrate and hydrogen chloride, which recombine slowly on cooling.

2. From stronger solutions mixtures of oxychlorides separate.

3. At high temperatures the ferric hydrate loses water, and ferric oxide is deposited.

Dark red plates, of the formula $9Fe_2O_3.FeCl_3$, separate from a strong solution of ferric hydrate in ferric chloride on evaporation *in vacuo*.

Oxychlorides are also produced when solutions of ferrous chloride are exposed to air.

Oxychlorides of manganese, cobalt, and nickel are unknown.

III. Dioxides and disulphides.

List.	Chromium.	Iron.	Manganese.	Cobalt.	Nickel.
Oxygen....	CrO_2.	—	MnO_2.	(CoO_2).*	(NiO_2).*
Sulphur...	—	FeS_2.	—	—	NiS_2.

Sources.—Manganese dioxide, or *pyrolusite* (from πῦρ, fire, and λύειν, to loose, referring to its action in removing the green and brown tints of glass coloured by iron, owing to the complementary action of its purple colour), is one of the chief ores of manganese. It is an iron-black or grey mineral, very hard, and somewhat brittle, with fibrous texture. It is largely employed for making chlorine.

Nodules containing manganese dioxide are of common occurrence on the sea-bottom; they have been dredged from the bed of the Pacific and Atlantic Oceans, and are found in the Firth of Clyde.

Iron pyrites or *mundic*, FeS_2, is a golden-yellow mineral crystallising in cubes. It is very hard and brittle, and was formerly used as a means of striking fire, whence its name. *Marcasite* is a whitish mineral with metallic lustre, of the same formula, crystallising in the trimetric system. Both of these minerals occur in slate, coal, shale, &c. They oxidise on exposure to moist air, and furnish the sulphuric acid necessary for alum in alum shale. They are used as a source of sulphur.

Preparation.—1. **By direct union.**—Hydrated chromium sesquioxide, $Cr_2O_3.nH_2O$, heated in air to 200°, is oxidised to the hydrated dioxide, $CrO_2.H_2O$; the hydrated compound is dried at 253°. Iron and sulphur combine below redness to form FeS_2; and lower sulphides of iron unite with sulphur when gently heated in a current of hydrogen sulphide.

2. **By heating a compound.**—The hydrated dioxides can be dried at 200—250°, yielding the dioxides. Chromium nitrate, when heated, yields the dioxide, and manganese dioxide is produced by heating manganous nitrate, or manganous carbonate and potassium chlorate. The oxygen is derived from the nitric anhydride, or from the potassium chlorate.

3. **By double decomposition.**—Oxides of iron, heated in hydrogen sulphide to above 100°, are converted into disulphide; and an alkaline polysulphide reacts with ferrous chloride or sulphate at 180°, yielding disulphide. Nickel disulphide, is

* In combination, as $2CoO_2.CoO$ and $2NiO_2.NiO$. See also Cobalt-amines.

produced by heating a mixture of nickel carbonate, potassium carbonate, and sulphur to dull redness.

Properties.—Chromium dioxide is a black powder, giving off oxgen at 350°. It is insoluble in water, but soluble in acids, and reprecipitated from its solution as hydrate by ammonia. **Manganese dioxide** is a black powder when prepared artificially. It dissolves in strong sulphuric acid, yielding a yellow sulphate, $MnO_2.2SO_3$. On diluting this solution, the hydrated dioxide is precipitated. **Iron disulphide** when prepared artificially is a black powder, or sometimes yellow cubes like the native form, insoluble in acids; and **nickel disulphide** is a steel-grey powder.

Double compounds. (*a.*) **With water.**—Hydrated chromium dioxide is produced, as before mentioned, by the spontaneous oxidation of the hydrated sesquioxide at 200°; and also by reducing chromium trioxide or its compounds. Thus, by passing a current of nitric oxide, *NO*, through a dilute solution of potassium dichromate, $K_2Cr_2O_7$, it is deprived of part of its oxygen, and gives a flocculent brown precipitate of the hydrated dioxide. The reduction may be effected by ammonia, as when a solution of ammonium dichromate, $(NH_4)_2Cr_2O_7.Aq$, is boiled, the oxygen going to oxidise the hydrogen of the ammonia; or by means of a chromic compound, *e.g.*, by heating together a solution of chromium trichloride, $CrCl_3$, with potassium dichromate, $K_2Cr_2O_7$ or $K_2O.2CrO_3$; chromium hydrate may be supposed to be formed by the action of water on chromium trichloride, thus:—$2CrCl_3.Aq + 3H_2O = Cr_2O_3.Aq + 6HCl.Aq$; and the hydrate then acts on the trioxide combined with potassium oxide in the dichromate, thus:—$Cr_2O_3.Aq + CrO_3.Aq = 3CrO_2.nH_2O$. The complete equation is:—$4CrCl_3.Aq + 5H_2O + K_2Cr_2O_7.Aq = 2KCl.Aq + 6CrO_2.nH_2O + 10HCl.Aq$. Heat alone expels oxygen from chromium trioxide, but the resulting substance is said to be $\mathbf{3CrO_2.Cr_2O_3}$. Oxalic acid, $H_2C_2O_4$, or alcohol may also be used to effect the reduction.

It is still a question whether this body is not a chromate of chromium, $CrO_3.Cr_2O_3$. Against this view, it may be stated that while chromates, when distilled with sodium chloride and strong sulphuric acid, give chromyl dichloride, CrO_2Cl_2 (see p. 268), this substance does not do so; and that it dissolves in acids as a whole, and is reprecipitated by alkalis, as it would be, were it a definite individual. Yet, on boiling with alkalies, hydrated chromium sesquioxide is precipitated, and the trioxide combines with the alkali, forming a chromate.

The compounds $\mathbf{MnO_2.2H_2O}$, $\mathbf{MnO_2.H_2O}$, $\mathbf{2MnO_2.H_2O}$, $\mathbf{3MnO_2.H_2O}$ and $\mathbf{4MnO_2.H_2O}$ are known. They are all brownish-black or black powders. The last of these is produced by treating $\mathbf{Mn_3O_4}$ or $\mathbf{Mn_2O_3}$ with strong nitric acid, whence the conclusion that these bodies are compounds of MnO_2 with 2MnO and MnO respectively. The monohydrate, $\mathbf{MnO_2.H_2O}$, is formed by the spontaneous decomposition of a solution of potassium permanganate, $KMnO_4$ or $K_2O.Mn_2O_7$, or by the action of chlorine on manganous carbonate suspended in water. The compound $\mathbf{2MnO_2.H_2O}$ is precipitated by addition of potassium hypochlorite to a manganous salt in presence of excess of ferric chloride; and the compound $\mathbf{3MnO_2.H_2O}$ by evaporating a solution of manganous bromate. The dihydrate, $\mathbf{MnO_2.2H_2O}$, is precipitated on addition of water to the sulphate $MnO_2.2SO_3$; the existence of this sulphate appears to lend support to the theory that the dioxide is a chemical individual, and not a manganate of manganese, $\mathbf{MnO_3.MnO}$. It need hardly be pointed out that the molecular weights of all these bodies are unknown.

(*b*.) **Double oxides.**—Several manganese compounds are known, viz.:—$\mathbf{MnO_2.MnO}$, $\mathbf{MnO_2.CaO}$, $\mathbf{2(MnO_2).K_2O}$, $\mathbf{2(MnO_2).CaO}$. These substances are formed by the action of air on (1) warm hydrated manganese monoxide precipitated from the dichloride $MnCl_2$ by its equivalent of calcium hydrate; (2) by the same process, twice the equivalent of lime being added, thus:—$MnCl_2.Aq + 2Ca(OH)_2 + O = \mathbf{MnO_2.CaO} + CaCl_2.Aq + 2H_2O$, and (3) by the action of manganese dichloride on the former compounds, thus:—

$$2(\mathbf{MnO_2.CaO}) + MnCl_2Aq = 2(\mathbf{MnO_2}).\mathbf{CaO} + \mathbf{Mn(OH)_2} + CaCl_2Aq.$$

These bodies are all hydrated, but the amount of combined water is unknown. Their formation is the principle of "Weldon's manganese-recovery process" whereby manganese dioxide which has been used for the manufacture of chlorine, and converted into dichloride, is restored to the state of dioxide, and thereby again rendered available for preparing chlorine (see p. 75). Such bodies as $\mathbf{MnO_2.CaO}$ are termed manganites. The compound $\mathbf{2(MnO_2).K_2O}$ is a black powder; others containing less oxide, *e.g.*, $\mathbf{12(MnO_2).Na_2O}$, $\mathbf{5(MnO_2).Na_2O}$, &c., are produced by heating manganous chloride with sodium hydrate and sodium chloride.*

Compounds of the formula $\mathbf{5(MnO_2)M''O}$, where M″ stands for calcium, strontium, zinc, or lead, may be produced by heating

* *Bull. Soc. Chim.* (2), **30**, 110; *Dingl. polyt. Jour.*, 129, **51**; *Chem. Soc. J.*, **37**, 22; 591; *Compt. rend.*, **101**, 167; **103**, 261.

chlorides of these metals with potassium permanganate. They form black crystals. At higher temperatures, **$2(MnO_2).M''O$** and, at still higher, **$MnO_2.M''O$** are produced.

Similar cobalt and nickel compounds, **$2(CoO_2).CoO$** and **$7CoO_2.4CoO$** (with water of hydration from $4H_2O$ to H_2O), also **$3NiO_2.5NiO.9H_2O$** are produced by adding sodium hypochlorite, NaClO, to a mixture of the hydrate of cobalt or nickel and excess of soda. A cobalt compound of the formula

$$3(2CoO_2.3CoO).K_2O.3H_2O$$

is produced by heating the monoxide with caustic potash in presence of air. No doubt, double compounds with other metals could be prepared.

(*c.*) **Oxyhalides.**—An oxyfluoride of the formula **$MnO_2.MnF_4$** or **$MnOF_2$** is said to be produced by adding manganese tetrachloride to a boiling solution of potassium fluoride. It is a rose-coloured powder, and combines with potassium fluoride, forming the compound **$MnOF_2.2KF$**. The trifluoride is said to yield similar double salts, *e.g.*, **$Mn_2F_4O.4KF$**. These bodies are produced by treating potassium permanganate, $KMnO_4$, with aqueous hydrogen fluoride.

IV. **Trioxides.**—(*a.*) The only trioxides known in the free state are chromium trioxide, or chromic anhydride, **CrO_3**, and manganese trioxide, **MnO_3**. Iron trioxide exists in combination with potassium monoxide in potassium ferrate, and that of manganese in potassium manganate.

Preparation.—By double decomposition.—Chromyl fluoride (see p. 268), led into a crucible slightly damp and loosely covered with damp paper, reacts with the water, depositing long needles of the trioxide; thus:—

$$CrO_2F_2 + \mathit{H_2O} = \mathbf{CrO_3} + \mathit{2HF}.$$

By the action of sulphuric acid on a chromate, a sulphate and chromic anhydride are formed. On pouring 1 volume of a saturated solution of potassium dichromate, $K_2Cr_2O_7$, into $1\frac{1}{2}$ volumes of strong sulphuric acid, long needles of chromic anhydride hydride deposit on cooling. They are difficult to free from sulphuric acid; and the present method of preparing the trioxide for commercial use is by adding to strontium chromate exactly enough sulphuric acid to precipitate the strontium as sulphate, to decant the solution of trioxide from the insoluble sulphate, and to evaporate to dryness.*

Manganese trioxide is obtained by dropping a solution of

* The author has tried the process, but without success.

potassium permanganate in strong sulphuric acid on to sodium bicarbonate; MnO_3 is liberated, and is carried on in the solid state by the carbon dioxide.

Properties.—Chromium trioxide forms a red crystalline powder, a mass of loose woolly crystals, or scarlet crystals. It melts at 190°, and begins to decompose at 250°, losing oxygen. It is soluble in water, and the solution contains chromic acid, $CrO_3.H_2O$, or H_2CrO_4,* or $H_2Cr_2O_7$. Its compounds with other oxides are called *chromates*. The blue solution obtained by shaking a dilute solution of chromium trioxide with hydrogen dioxide, and extracting with ether, is said to be a compound of the formula $CrO_3.H_2O_2$. On evaporation of the ether, it remains as a blue oil.†

Manganese trioxide is a reddish, amorphous, deliquescent substance, unstable at the ordinary temperature.

(*b.*) **Compounds with other oxides.—Chromates, ferrates, and manganates.** Of these the chromates are the most stable, and have been best investigated. They may be divided into four classes:—

1. **Basic chromates**; those in which the number of atoms of oxygen in the base exceeds one-third of that in the chromic anhydride. These compounds are orange, red, or brown in colour. They are produced by double decomposition, a solution of a soluble chromate, such as potassium chromate, K_2CrO_4, being added to a soluble salt of the metal; in such a case, uncombined chromic anhydride exists in solution; or by digesting a chromate, such as **$PbCrO_4$ = $PbO.CrO_3$** with alkali, or with excess of base. They are as follows:—

Ratio, 3 : 9 :—**$CrO_3.3Bi_2O_3$**. Ratio, 6 : 9 :—**$2CrO_3.3Bi_2O_3$**.
„ 3 : 4 :—**$CrO_3.4ZnO,3H_2O$: $CrO_3.4CuO$**.
„ 3 : 3 :—**$CrO_3Al_2O_3$; $CrO_3.Cr_2O_3$**(?) (this body is **CrO_2**); **$CrO_3.Fe_2O_3$; $CrO_3.3NiO.3H_2O$; $CrO_3.Bi_2O_3$; $CrO_3.3CuO$; $CrO_3.3HgO$**.
„ 6 : 5 :—**$2CrO_3.5NiO.12H_2O$**. Ratio, 21 : 9 :—**$7CrO_3.3Bi_2O_3$**.
„ 3 : 2 :—**$CrO_3.2ZnO.H_2O$; $CrO_3.2CdO.H_2O$; $CrO_3.2MnO.2H_2O$; $CrO_3.2CoO.2H_2O$; $CrO_3.2NiO.6H_2O$; $CrO_3.2PbO$; $CrO_3.2CuO$; $CrO_3.2HgO$; $CrO_3.2Hg_2O$; $2CrO_3.3CuO.K_2O.3H_2O$**.
„ 6 : 3 :—**$2CrO_3.3PbO$; $2CrO_3.Bi_2O_3$; $2CrO_3.CuO.2PbO$**.

These compounds are orange, red, or brown powders, and are insoluble in water, or nearly so; they dissolve in acids, being converted into chromates containing a larger proportion of trioxide of chromium. The most important of them is the chromate **$CrO_3.2PbO$**; it is named "chrome-red" or "Persian-red." It is pro-

* *Comptes rend.*, **98**, 1581.
† *Ibid.*, **97**, 96.

duced by addition of lead oxide to the monoplumbic chromate, **$PbCrO_4$**, or **$CrO_3.PbO$**; or with a purer shade by heating that body with potassium nitrate; the potassium oxide withdraws chromic anhydride, and on washing with water, excess of potassium nitrate and chromate are withdrawn and the basic chromate is left as a red powder. Cloth on which a precipitate of yellow **$PbCrO_4$** has been formed, may be changed to a brown-red by plunging it into a bath of boiling milk of lime ($Ca(OH)_2.Aq$), which withdraws half the chromic anhydride. **$2CrO_3.3PbO$** occurs in scarlet crystals as *melanochroite;* and **$2CrO_3.CuO.2PbO$** as a yellowish-brown mineral named *vauquelinite.*

2. The second class of chromates is often termed "**neutral.**" This name was originially applied to those substances incapable of affecting the colour of litmus. But most of these chromates are insoluble; moreover, the typical "neutral" chromate of potassium, $K_2CrO_4(CrO_3.K_2O)$ has an alkaline reaction and turns red litmus blue. It is better therefore to discard the misleading name. The oxygen of the chromic anhydride bears to that of the base the ratio 3 : 1. The following is a list:—

Ratio 3 : 1.—
$CrO_3.H_2O$ (chromic acid); **$CrO_3.Li_2O.2H_2O$**; **$CrO_3.Na_2O.10H_2O$** (crystallised above 30°, this body is anhydrous); **$CrO_3.K_2O$**; **$CrO_3.K(NH_4)O$** (= **KNH_4CrO_4**); **$CrO_3MgO.7H_2O$**; **$CrO_3.CaO.4H_2O$**; **$CrO_3.SrO$**; **$CrO_3.BaO$**; **$CrO_3.Tl_2O$**; **$CrO_3.PbO$**; **$CrO_3.CuO$**; **$CrO_3.Ag_2O$**; **$CrO_3.HgO$**; **$CrO_3.Hg_2O$**.

Of these, the hydrogen, lithium, sodium, potassium, magnesium, calcium, copper, and mercuric compounds are soluble in water. **Hydrogen chromate, H_2CrO_4,** is produced by dissolving chromium trioxide in water, and cooling with melting ice. It forms small red deliquescent crystals, which readily part with water. **Potassium chromate, K_2CrO_4,** has a light-yellow colour, and a bitter, cooling taste; it is exceedingly poisonous; it is insoluble in alcohol, but soluble in water (100 grams of water at 15° dissolve 48·3 grams of chromate). It melts at a low red heat, and crystallises in double hexagonal pyramids. **Strontium chromate, $SrCrO_4$,** is sparingly soluble. It is the one from which chromic anhydride is now made commercially by addition of sulphuric acid. It is found to be the only available chromate from which the chromium trioxide is completely expelled by its equivalent of sulphur trioxide, by the action of sulphuric acid; hence its use. **Barium chromate, $BaCrO_4$,** is an insoluble yellow powder, used as a pigment under the name, "yellow ultramarine." **Lead chromate, $PbCrO_4$,** is found native, as *red lead-ore* or *crocoïsite.* It crystallises in monoclinic prisms. It is a translucent yellow body, and occurs in decomposed granite or gneiss. Prepared by addition of potassium chromate or dichromate to a soluble salt of lead, it is a yellow powder, and is known as "chrome-yellow" and used as a pigment. It fuses to a brown liquid, and solidifies to a brown-yellow mass. It is made use of in estimating carbon and hydrogen in carbon compounds. It is practically insoluble in acids, but dissolves easily in potassium hydrate, forming chromate and plumbite of potassium. **Silver chromate, Ag_2CrO_4** is a deep-red precipitate, crystalline in structure; the individual crystals transmit green light.

3. **Dichromates.**—These bodies are often called "acid" chromates, and their solutions have an acid reaction with litmus. They are produced by adding some acid, *e.g.*, chromic acid, or more often nitric acid to the monochromates. They are as follows:—

Ratio 6 : 1. $2CrO_3.Li_2O$; $2CrO_3.Na_2O.2H_2O$; $2CrO_3.K_2O$; $2CrO_3.(NH_4)_2O$; $2CrO_3.CaO.3H_2O$; $2CrO_3.BaO$; $2CrO_3.Tl_2O$; $2CrO_3.PbO$; $2CrO_3.Ag_2O$.

The most important of these is **potassium dichromate,** or "bichrome," which is prepared on a manufacturing scale. It is produced by acidifying the monochromate, K_2CrO_4, with sulphuric acid, thus:—$2K_2CrO_4.Aq + H_2SO_4.Aq = K_2SO_4.Aq + K_2Cr_2O_7.Aq + H_2O$. It forms deep orange-red tables or prisms. It is insoluble in alcohol, but soluble in water (100 grams dissolve at 20° 12·4 grams of bichrome). It melts at a dull red heat, and decomposes at a white heat into potassium chromate, chromium sesquioxide, and oxygen. It is affected by light, and has the curious property of rendering gelatine impregnated with it insoluble in water after exposure to light, and it thus finds an application in photography. It is largely used as an oxidising agent, and for making chrome-yellow, &c.

The dichromates are decomposed by much water, excepting those of sodium, potassium, and ammonium.

The name *anhydrochromates* is sometimes applied to these bodies, the view being taken that they are compounds of monochromate and anhydride, thus:—$K_2CrO_4.CrO_3$.

4. **Polychromates; tri-, tetra-, &c.**

Ratio 9 : 2. $3CrO_3.2ZnO$, soluble, crystalline; $3CrO_3.2Tl_2O$.
Ratio 9 : 1. $3CrO_3.K_2O$; $3CrO_3.(NH_4)_2O$; $3CrO_3.Tl_2O$.

These bodies are deep-red crystals, formed on crystallising the dichromates from strong nitric acid.

Ratio 12 : 1.—$4CrO_3.K_2O$, similarly prepared. The polychromates decompose on treatment with much water.

Ferrates.—Of these, only the potassium, sodium, and barium salts are known. Their formulæ are supposed to be $FeO_3.K_2O$; $FeO_3.Na_2O$; and $FeO_3.BaO$; but the potassium and sodium salts are stable only in presence of a large excess of alkali, and the barium salt has not been analysed. The ratio of oxygen to iron in the iron trioxide has been determined; hence the deduction of the formula, FeO_3.

Sodium or **potassium ferrate** is formed by heating iron-filings and sodium or potassium nitrate to dull redness; by igniting iron sesquioxide with sodium or potassium hydrates in an open crucible, better with addition of sodium or potassium nitrate; by passing chlorine through a very strong solution of sodium or potassium hydrates in which ferric hydrate is suspended; the ferrate, being insoluble in the strong alkali, is precipitated as a black powder; and by electrolysing a strong solution of potash or soda with iron poles; the ferrate crystallises on the positive pole. The production of ferrate may be shown as a lecture experiment by adding a few lumps of potassium hydrate to some solution of ferric chloride, and adding bromine and warming.

The potassium ferrate may be dried on a porous plate; it cannot be filtered through paper, as it at once loses oxygen. It forms a fine cherry-red solution, but it soon decomposes with loss of oxygen. **Barium ferrate** is a purple precipitate produced by adding a solution of barium hydroxide to the solution of potassium ferrate. The ferrates at once lose oxygen on addition of an acid.

Manganates.—Of these, only the sodium, potassium, calcium, and barium salts are known. They are prepared by heating manganese dioxide with sodium or potassium hydroxides or carbonates; manganate and a lower oxide of manganese are formed; or nitrate or chlorate of calcium or barium with manganese dioxide. The yield may be increased by adding sodium or potassium nitrate to the hydroxides. On treatment with cold water, they form a deep green solution, and when it is evaporated in a vacuum, crystals are deposited. These crystals have the formula $\mathbf{K_2MnO_4} = \mathbf{MnO_3.K_2O}$; the barium, calcium, and sodium manganates are supposed to have similar formulæ. On leaving a strong solution of potassium manganate exposed to air, crystals of dimanganate have been formed, $\mathbf{2MnO_3.K_2O.H_2O}$, the carbon dioxide of the air having withdrawn half the potash.

Potassium manganate is stable only in presence of excess of alkali, and is decomposed by pure water with formation of permanganate and dioxide, thus:—

$$3(MnO_3.K_2O) + 2H_2O + Aq = Mn_2O_7.K_2O.Aq + 4KOH.Aq + \mathbf{MnO_2.nH_2O}.$$

Owing to this change of colour from green to purple, the old name for potassium manganate was " mineral chameleon."

Manganate of barium is known as " baryta-green." Potassium manganate having been produced by gradually adding manganese dioxide to a fused mixture of two parts of potassium hydrate and one part of potassium nitrate, the cooled mass is treated with water and filtered. On addition of barium nitrate to the filtrate, a violet precipitate of barium manganate is produced, which is heated to redness with solid barium hydrate till it assumes a bright-green colour. It is then treated with water to remove barium hydrate. The green colour is in all cases probably due to basic manganates.

Perchromates and permanganates.—These bodies are compounds of oxides with the heptoxides of chromium or manganese, Cr_2O_7, or Mn_2O_7. Those of chromium are very unstable, if, indeed, they are capable of existence. If hydrogen dioxide, H_2O_2, be added to a solution of chromic acid, or of potassium chromate and sulphuric acid, a dark-brown colour is produced. On shaking the solution with ether, the upper layer of ether has a fine blue colour; on evaporation at $-20°$, a deep indigo-blue oily liquid is left; this is possibly perchromic anhydride, or chromium heptoxide, Cr_2O_7, but is also said to be a compound of the formula $CrO_3.H_2O_2$. Its salts are unknown. This reaction affords a very delicate test both for chromium trioxide and for hydrogen dioxide (see p. 197).

Potassium permanganate, $\mathbf{Mn_2O_7.K_2O}$, or **$\mathbf{KMnO_4}$**, is produced by acidifying potassium manganate. It may be supposed that the manganic acid, $MnO_3.H_2O$, decomposes at the moment of its liberation, yielding manganese dioxide and permanganic acid, thus:—$3MnO_3.H_2O.Aq = Mn_2O_7.H_2O.Aq + \mathbf{MnO_2.nH_2O}$. The same change is produced by boiling a solution of potassium manganate, or by treating sodium manganate with magnesium sulphate, thus:—$3(MnO_3.Na_2O)Aq + 2(SO_3.MgO)Aq + 2H_2O = Mn_2O_7.Na_2O.Aq + 2(SO_3.Na_2O)Aq + 2(\mathbf{MgO.H_2O}) + \mathbf{MnO_2.nH_2O}$; magnesium manganate being unstable. Manganate may also be converted into permanganate without separation of dioxide by means of chlorine, thus:—$2(MnO_3.K_2O)Aq + Cl_2 = 2KCl.Aq + Mn_2O_7.K_2O.Aq$.

The following permanganates are known:—

$Mn_2O_7.H_2O.Aq$, or $HMnO_4.Aq$; $\mathbf{Mn_2O_7.K_2O}$, or $\mathbf{KMnO_4}$; $\mathbf{NH_4MnO_4}$; $\mathbf{Ba(MnO_4)_2}$; $\mathbf{Pb(MnO_4)_2}$; and $\mathbf{AgMnO_4}$.

The **barium salt** is made by the action of carbon dioxide on barium manganate; and from it the **free acid**, $HMnO_4$, may be separated by addition of monohydrated sulphuric acid, $H_2SO_4.H_2O$. It forms a greenish-yellow solution, and deposits slowly a dark, reddish-brown liquid, not solidifying at $-20°$; it is said to be manganese heptoxide, or permanganic anhydride, Mn_2O_7. It is non-volatile.* This liquid dissolves in strong sulphuric acid with a yellow-green colour; it explodes when strongly heated. The yellow-green solution contains $(MnO_3)_2SO_4$. On adding water the colour changes to violet—that of permanganic acid.

The **silver** and **lead salts** are formed by adding soluble salts of silver or lead to potassium permanganate. They are dark-coloured precipitates. The **ammonium salt** is made by mixing the silver salt with ammonium chloride. Potassium permanganate, with excess of potassium hydrate, turns green with formation of manganate, oxygen being evolved, thus:—$2KMnO_4.Aq + 2KOH.Aq = 2K_2MnO_4.Aq + H_2O + O$.

Potassium permanganate forms dark-red, almost black, crystals, with greenish reflection; its solution is sold as a disinfectant under the name of "Condy's fluid," and has a splendid purple colour. The dichromate and permanganate of potassium are used as means of oxidising substances in presence of water. Bichrome does not readily part with its oxygen, even to an easily oxidisable body, unless an acid be present; when it does, chromium dioxide is produced. Thus:—

$$2CrO_3.K_2O.Aq + H_2O = \mathbf{2CrO_2.}n\mathbf{H_2O} + 2KOH.Aq + 2O.$$

Potassium permanganate acts similarly, thus:—

$$Mn_2O_7.K_2O.Aq + H_2O = \mathbf{2MnO_2.}n\mathbf{H_2O} + 2KOH.Aq + 3O.$$

In presence of an acid (usually sulphuric acid) a salt of chromium or manganese is produced, thus:—$2CrO_3 = Cr_2O_3 + 3O$; and $Cr_2O_3 + 3H_2SO_4 = Cr_2(SO_4)_3 + 3H_2O$. Also $Mn_2O_7 = 2MnO + 5O$; and $MnO + H_2SO_4 = MnSO_4 + H_2O$. The complete equations are:—

$$K_2Cr_2O_7.Aq + 4H_2SO_4 = K_2SO_4.Aq + Cr_2(SO_4)_3.Aq + 4H_2O + 3O;$$
$$\text{and } 2KMnO_4.Aq + 3H_2SO_4.Aq = K_2SO_4.Aq + 2MnSO_4.Aq + 3H_2O + 5O.$$

The oxygen, being in the nascent or atomic state, is available for oxidation of compounds of carbon, &c.

* See *Chem. Soc.*, **53**, 175.

Compounds of oxides with halides.—These are as follow:—CrO_2Cl_2, chromyl dichloride; CrO_2F_2, chromyl difluoride, and, possibly, MnO_2Cl_2, manganyl chloride. They are formed by distilling a mixture of sodium chloride or fluoride, potassium dichromate or permanganate, and strong sulphuric acid. The reaction takes place between the liberated chromium trioxide or manganese heptoxide and the hydrogen halide, the sulphuric acid combining with the water produced, which would otherwise decompose the chromyl or manganyl halide, thus:—

$$CrO_3 + 2HCl + H_2SO_4 = CrO_2Cl_2 + H_2SO_4.H_2O.$$

Chromyl dichloride may indeed be obtained by the direct action of dry hydrogen chloride on pure chromium trioxide. Hydrogen bromide and iodide are decomposed with liberation of bromine or iodine. Chromyl chloride is a deep-red liquid, closely resembling bromine in appearance; it boils at 118°, and gives a deep-red vapour. It mixes in all proportions with carbon disulphide and with chloroform. The manganese compound is said to be a purple vapour, condensing at a very low temperature; but it requires re-investigation. Chromyl fluoride may be made by a similar process. Chromyl chloride reacts with water, forming chromium trioxide or chromic acid, thus:—

$$CrO_2{<}^{Cl}_{Cl} + {}^{H.OH}_{H.OH} = CrO_2{<}^{OH}_{OH} + {}^{HCl}_{HCl}.$$

As its vapour-density shows it to have the formula CrO_2Cl_2, it is concluded that chromic acid is analogously constituted, and may be represented by the structural formula $CrO_2(OH)_2$, and chromates as $CrO_2(OM')_2$. It is obvious that an intermediate compound between CrO_2Cl_2 and $CrO_2(OM')_2$ should exist of the formula $CrO_2{<}^{OM'}_{Cl}$. Such a body is known, and is termed a chloro-chromate. The potassium salt is produced by saturating a hot solution of dichromate of potassium with hydrogen chloride and leaving it to crystallise. Flat rectangular prisms of the compound $\mathbf{CrO_2{<}^{OK}_{Cl}}$ are deposited; on treatment with water they decompose. The mercuric salt is also known. Compounds have also been prepared of the formulæ $\mathbf{2CrO_3.KF}$ and $\mathbf{2CrO_3.NH_4F}$; they are produced by adding aqueous hydrofluoric acid to potassium or ammonium dichromates; they may be constituted thus:—

$$\begin{matrix} CrO_2{<}^{OK}_{O} \\ CrO_2{<}_{F} \end{matrix}.$$

On heating chromyl dichloride to 180—190° in a sealed tube chlorine separates, and the compound $Cr_3Cl_2O_6$ remains as a black powder. Its constitution may be thus represented:—

$$Cl—CrO_2—CrO_2—CrO_2—Cl.$$

The corresponding potassium salt is produced by saturating potassium chlorochromate with ammonia, and the ammonium salt by saturating a solution of chromyl dichloride in chloroform with ammonia. Their constitutional formulæ may be:—

$$KO—CrO_2—CrO_2—CrO_2—OK; \text{ and}$$
$$(NH_4)O—CrO_2—CrO_2—CrO_2—O(NH_4).$$

Such constitutional formulæ will be further referred to in treating of silicates, phosphates, and sulphates.

No compounds of bromine or iodine analogous to chromyl chloride are known; bromine or iodine are invariably liberated. The volatility of the chlorine compound serves to identify chlorine in presence of bromine or iodine; on distilling a mixture of halides with bichrome and sulphuric acid, if chromium is found in the distillate, the presence of chlorine in the mixture is proved.

Physical Properties.—1. *Weight of* 1 *cubic centimetre.*

MnO, 5·1; CoO, 5·6; NiO, 5·6; 6·8 (crystallised).
FeS, 4·8; MnS, 4·0; NiS, 5·6.
Cr_2O_3, 6·2 (crys.); Fe_2O_3, 5·3 (native); Mn_2O_3 (*braunite*), 4·75; Co_2O_3, 4·8.
Cr_2S_3, 4·1; Fe_2S_3, 4·3; Co_2S_3, 4·8.
Ni_2O_3, 4·8.
Cr_5O_9, 4·0; Fe_3O_4 5·12 (*magnetite*); Mn_3O_4, 4·85 (native); Co_3O_4, 6·3.
MnO_2, 4·83;
MnS_2, 3·46; FeS_2, 5·04 (*pyrites*); 4·8 (*marcasite*).
CrO_3, 2·8.

Heats of formation.

$Cr_2O_3 + 3O = 2CrO_3 + 143K.$

$Fe + O + H_2O = Fe(OH)_2 + 683K.$

$2Fe + 3O + 3H_2O = Fe(OH)_3 + 1912K.$

$3Fe + 4O = Fe_3O_4 + 2647K.$

$Fe + S + nH_2O = FeS.nH_2O + 238K.$

$2Co(OH)_2 + O + H_2O = Co_2O_3.3H_2O + 223K$

$2Ni(OH)_2 + O + H_2O = Ni_2O_3.3H_2O + 13K.$

$Mn + O + 2H_2O = Mn(OH_2) + 948K.$

$Mn + 2O + H_2O = MnO_2.H_2O + 1164K.$

$Mn + S + nH_2O = MnS.nH_2O + 444K.$

$Co + O + H_2O = Co(OH)_2 + 634K.$

$Co + S + nH_2O = CoS.nH_2O + 197K.$

$Ni + O + H_2O = Ni(OH)_2 + 608K.$

$Ni + S + nH_2O = NiS.nH_2O + 174K.$

CHAPTER XX.

OXIDES, SULPHIDES, SELENIDES, AND TELLURIDES OF ELEMENTS OF THE CARBON GROUP; FORMIC AND OXALIC ACIDS; CARBONATES, TITANATES, ZIRCONATES, AND THORATES; SULPHOCARBONATES AND OXYSULPHOCARBONATES, OXYHALIDES, AND SULPHOHALIDES.

Oxides, Selenides, and Tellurides of Carbon, Titanium, Zirconium, Cerium, and Thorium.

This group gives representatives of **monoxides, sesquioxides, dioxides,** and **peroxides.** The monoxides show little tendency towards combination; the dioxides form compounds with the oxides of other elements, which are named **carbonates, titanates,** and **zirconates.** Some similar compounds of the sulphides have also been prepared.

	Carbon.	Titanium.	Zirconium.
Oxygen....	CO; CO_2.	TiO;* Ti_2O_3; TiO_2; TiO_3.	ZrO_2; Zr_2O_5.
Sulphur ...	CS; CS_2.	TiS; Ti_2S_3; TiS_2.	ZrS_2?

	Cerium.	Thorium.
Oxygen.........	Ce_2O_3; CeO_2; CeO_3.	ThO_2; Th_2O_7.
Sulphur	Ce_2S_3; —	ThS_2.

I. **Monoxides** and **monosulphides** (selenides and tellurides have not been prepared).

Sources.—Carbon monoxide is produced by the decay of organic matter, and by the incomplete combustion of fuel.

Preparation.—By direct union.—Carbon is said to combine with oxygen to form monoxide; it appears more likely that the dioxide is first formed, and by its contact with red-hot carbon is converted into monoxide, thus $CO_2 + C = 2CO$.

2. **By replacement.**—Steam, led over white-hot carbon, yields a mixture of hydrogen and carbon monoxide. This mixture is well-adapted for heating purposes, and is commercially termed "water-gas." It is frequently employed in driving gas-engines. Carbon

* As hydrate, $Ti(OH)_2$.

withdraws oxygen from sodium sulphate, Na_2SO_4, forming monoxide and sodium sulphide. Carbon withdraws oxygen from many oxides, carbon monoxide being formed.

3. **By reduction.**—Zinc or copper withdraws oxygen from carbon dioxide, producing monoxide; heating a mixture of magnesium carbonate and zinc dust is an available method of preparation. Carbon monosulphide is deposited from carbon disulphide, after long exposure to light; and titanium monosulphide is produced by the action of hydrogen on the red-hot disulphide.

4. **By decomposition of a compound.**—The oxide C_2O_3 appears to be incapable of existence, but oxalic acid, $\mathbf{C_2O_4H_2}$, may be viewed as its compound with water. On depriving oxalic acid of water by the action of concentrated sulphuric acid, a mixture of carbon monoxide and dioxide is evolved, thus—

$$\mathbf{C_2O_4H_2} + H_2SO_4 = \mathit{CO} + \mathit{CO_2} + H_2SO_4.H_2O.$$

Similarly, if the elements of water are withdrawn from formic acid, CO_2H_2, by strong sulphuric acid, carbon monoxide is produced. This is by far the most convenient, though not the cheapest, method of preparation, and yields perfectly pure monoxide.

5. **By double decomposition.**—Hydrocyanic acid, HCN (see p. 559), liberated in presence of fairly strong sulphuric acid, takes up water, forming carbon monoxide, and ammonia which combines with the sulphuric acid, thus—

$$HCN + H_2SO_4.H_2O = \mathit{CO} + (NH_4)HSO_4.$$

The hydrocyanic acid is conveniently produced from potassium ferrocyanide.

Properties.—Carbon monoxide is a colourless gas at ordinary temperatures; it condenses to a liquid at −190°, and the white solid produced by its evaporation melts at −199°. Its critical temperature is about −139·5°; and its critical pressure is 35·5 atmospheres. It is soluble in alcohol; 100 volumes dissolve about 20 volumes; but it is very sparingly soluble in water, 100 volumes dissolving only 3 volumes of the gas. It has a faint smell, but no taste. It is poisonous, forming a compound with the hæmoglobin of the blood which gives a spectrum closely resembling that of oxyhæmoglobin; but, while the latter is at once altered by ammonium sulphide, the spectrum due to carbon monoxide lasts after such treatment for several days. It is absorbed by potassium (see below), and by compounds of silver, and gold; also by cuprous chloride. When left long in contact with potassium or sodium hydroxide it combines, forming formate of potassium or sodium.

Carbon monosulphide is a red powder, sparingly soluble in carbon disulphide and in ether; it dissolves in solution of potassium hydrate, and is reprecipitated by acids; it decomposes at 200° into carbon and sulphur. It is probably a polymeride of CS. **Titanium monosulphide** is a black insoluble substance, decomposed only by fusion with alkalis.

Compounds with water.—It is sometimes stated that carbon monoxide is the anhydride of **formic acid,** CO_2H_2, and if their formulæ alone be considered such might be the case. But there can be no doubt that formic acid has the constitution $H—\overset{O}{\overset{\|}{C}}—OH$, and that it is partly a carbide of hydrogen, and is derived from tetrad carbon. The true acid derived from carbon monoxide is unknown; its formula should be HO—C—OH. Hence carbon monoxide reacts slowly with potassium hydroxide, a molecular rearrangement being effected in order to produce potassium formate, $H—\overset{O}{\overset{\|}{C}}—OK$. The explosive grey compound produced by direct combination of carbon monoxide with potassium, which has the formula $K_nC_nO_n$ is probably also partly a carbide of potassium.

Ti(OH)$_2$ is said to be produced by the action of sodium amalgam on the tetrachloride, $TiCl_4$, in presence of water. Titanium dichloride, $TiCl_2$, decomposes water, giving a mixture of trichloride and sesquioxide. No compounds of these monoxides with oxides or sulphides are known.

Compounds with chlorides.—Carbon monoxide combines directly with platinous chloride, to form the body **$PtCl_2$ 2CO,** with platinic chloride to form **$PtCl_4$.3CO**, and with cuprous chloride to form **Cu_2Cl_2.2CO**. These are insoluble crystalline compounds. The last is formed when carbon monoxide is shaken with a solution of cuprous chloride in hydrochloric acid, and is used as a means of separating carbon monoxide from other gases with which it may be mixed.

A compound of the formula **TiO.$TiCl_3$** is also known; it is produced by the action of oxygen on titanium tetrachloride, $TiCl_4$, at a red heat. **2CeO.$CeCl_2$** is formed by the action of steam and nitrogen on a mixture of cerium and sodium chlorides; it forms silvery scales.

II. **Sesquioxides and sesquisulphides.—Carbon sesquioxide** is unknown; its compound with water is **oxalic acid. Carbon sesquisulphide** is said to be produced by the action of sodium amalgam on the disulphide; it is a red-brown powder.

Titanium sesquioxide is formed when the dioxide is heated in hydrogen, or during the preparation of the trichloride (see page 145) due to the action of air. It forms copper-coloured crystals, and has the same crystalline form as *specular iron*, $\mathbf{Fe_2O_3}$. **Titanium sesquisulphide** is produced by the action of a *moist* mixture of hydrogen sulphide and carbon disulphide on the dioxide, $\mathbf{TiO_2}$, at a bright red heat. It is a black powder. **Cerium sesquioxide** is produced by heating the oxalate in a current of hydrogen. It is a grey solid, reacting with acids forming salts. The **sesquisulphide,*** produced by the action of dry hydrogen sulphide on red hot cerium dioxide, or by passing that gas over a fused mixture of cerium trichloride and sodium chloride, is a crystalline vermilion or black compound, according to the temperature. It is slowly decomposed by warm water. Similar compounds of zirconium and thorium have not been prepared.

Compounds with water.—Oxalic acid may be regarded as the hydrate of the unknown carbon sesquioxide. It has the formula $C_2O_4H_2$, and not CO_2H, as can be shown by the following synthesis :—Ethylene is known to possess the formula C_2H_4 from its vapour-density. On bringing ethylene and bromine together, direct addition takes place, and ethylene dibromide, $C_2H_4Br_2$, is formed. This body, on treatment with silver hydroxide, exchanges bromine for hydroxyl, thus :—$C_2H_4Br_2 + 2\mathbf{AgOH} = C_2H_4(OH)_2 + 2\mathbf{AgBr}$. Glycol, as the substance $C_2H_4(OH)_2$ is named, on oxidation yields oxalic acid, thus :—$C_2H_4(OH)_2 + 4O = C_2O_2(OH)_2 + 2H_2O$. It is therefore concluded that oxalic acid contains two atoms of carbon. Its constitutional formula is written

$$\begin{array}{l} O{=}C{-}OH \\ \quad | \\ O{=}C{-}OH \end{array}$$

, and it would thus appear that the atom of carbon is here capable of combining with four monads, and is a tetrad. As carbon tetrachloride possesses the formula CCl_4, and carbon hexachloride is C_2Cl_6 (see p. 155), it is seen that two atoms of carbon possess the property of combining with each other. Now, in contrasting this with the behaviour of members of the previous group, such as iron, it must be remembered that ferric chloride possesses the formula $FeCl_3$, as shown by its vapour-density at high temperatures. At low temperatures, its formula is Fe_2Cl_6, and it has been supposed that iron at low temperatures, like carbon under almost all circumstances, is a tetrad. The hydroxide, $\mathbf{Fe_2O_3.H_2O}$, has probably a high molecular weight, for the sesquioxide, $\mathbf{Fe_2O_3}$, has the power of combining with a

* *Comptes rend.*, **100**, 1461.

large number of molecules of other oxides, and presumably combines with itself to form considerable molecular aggregates. But ignoring this, the formula of this hydroxide may be **O=Fe—OH**, or it may have a constitution analogous to that of oxalic acid, viz., $\genfrac{}{}{0pt}{}{O}{HO}\!>\!Fe{-}Fe\!<\!\genfrac{}{}{0pt}{}{O}{OH}$, in which case the atoms of iron would be tetrad. At present there is no means of deciding the point, though the opinion of chemists favours the triad nature of the atom of iron.

The study of oxalic acid and its compounds belongs to the domain of Organic Chemistry; and these are so numerous that their formation and relationship would occupy too large a space in such a book as this.

Titanium tetrachloride on treatment with metallic copper or silver in a state of fine division yields the hexachloride, Ti_2Cl_6; and on addition of an alkali, to its solution in water, a brown precipitate of the hydroxide, $\mathbf{Ti_2O_3.3H_2O}$, is produced. It is soluble in acids giving violet salts.

Hydrated cerium sesquioxide is formed by addition of an alkali to a solution of the trichloride. It rapidly oxidises on exposure to air.

Compounds with halides.—On treating trichloride of titanium with a little water, the body $Ti_2O_2Cl_2$ is produced. Supposing it to be constituted like oxalic acid, its formula would be $\genfrac{}{}{0pt}{}{O}{Cl}\!>\!Ti{-}Ti\!<\!\genfrac{}{}{0pt}{}{O}{Cl}$. It is also formed by the action of a mixture of hydrogen and titanium tetrachloride on the red hot dioxide, thus:— $TiCl_4 + H_2 + \mathbf{TiO_2} = 2HCl + \mathbf{Ti_2O_2Cl_2}$. It forms reddish-brown laminæ.

A compound of the formula $\mathbf{Ce_2O_3.Ce_2Cl_6}$ is produced by the action of sodium hydroxide, and subsequently of water on the trichloride, or of a mixture of steam and nitrogen on the trichloride. It is an insoluble dark purple powder.

III. **Dioxides.—Sources.**—All of these dioxides, that of cerium excepted, are found native. **Carbon dioxide** occurs in air. Ordinary country air contains somewhat under 4 volumes per 10,000 of air; in cities, owing to its evolution from chimneys and from respiration, it is present in somewhat higher amount, and in fogs may amount to 6 volumes. It issues from the ground in volcanic districts. The "Grotto del Cane," near Naples, is well known in this respect; the gas in the depression in the ground contains from 60 to 70 per cent. of carbon dioxide. It is a frequent constituent of mineral waters, and is present in small quantity in all natural water, including sea-water. It is the source from which plants derive their carbon, and is produced by the decay of all organic matter. Some specimens of quartz contain cavities filled with liquid carbon dioxide. In combination with other oxides, especially with lime, as carbonate, it forms a great portion of the earth's crust. **Titanium dioxide** seems native in dimetric prisms, as *rutile*, in granite, gneiss, or mica slate; also as *anatase*, in acute rhombohedra; and as *brookite* in trimetric crystals.—**Zirconium dioxide** occurs in combination with silica as *zircon*, or *hyacinth*, $\mathbf{ZrO_2.SiO_2}$, and as *malacone*, in some granites. **Thorium dioxide** occurs as *thorite*, $3(\mathbf{ThO_2.SiO_2}).4\mathbf{H_2O}$, and is also combined with niobic and tantalic pentoxides in *euxenite*.

Preparation.—In considering the methods of preparation of these compounds it must be remembered that carbon dioxide is a gas, while the dioxides of the other elements are non-volatile solids.

1. **By direct union.**—The elements all burn in oxygen, forming dioxides, with exception of cerium. In presence of excess of the element, carbon forms monoxide, and titanium forms sesquioxide. Cerium yields, not dioxide, but sesquioxide. Compounds of carbon also burn, giving carbon dioxide. Carbon unites with sulphur at a red heat, forming disulphide; but it does not combine directly with selenium or tellurium; and zirconium and thorium also form disulphides when heated in sulphur gas. The selenides and tellurides of the other elements have not been prepared.

The combustion of carbon in oxygen may be shown by heating a piece of charcoal to redness in a Bunsen's flame, and plunging it into oxygen gas, as shown in fig. 33. The charcoal continues to burn brightly, and the product is carbon

FIG. 33.

dioxide. The combustion of a diamond may also be shown, as in fig. 34, by wrapping up a fragment of diamond in a small spiral of thin platinum wire connected with two stout copper wires which pass through an indiarubber cork closing the end of a wide test-tube. The test-tube is filled with oxygen, and by means of an electric current from four Bunsen's cells, the thin platinum wire

FIG. 34.

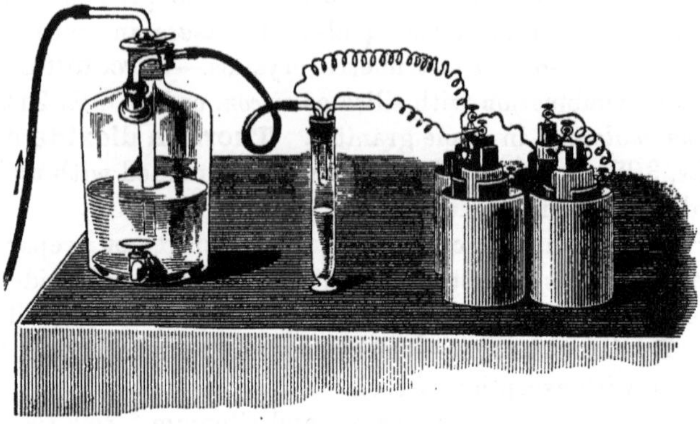

is heated to whiteness. The diamond is thus raised to its point of ignition, and on discontinuing the current it continues to glow until it is finally totally consumed. That carbon dioxide is the product of combustion may be shown by shaking the contents of the tube with a little baryta-water ($Ba(OH)_2.Aq$), when a white precipitate of barium carbonate, $BaCO_3$, is formed. Another instructive

FIG. 35.

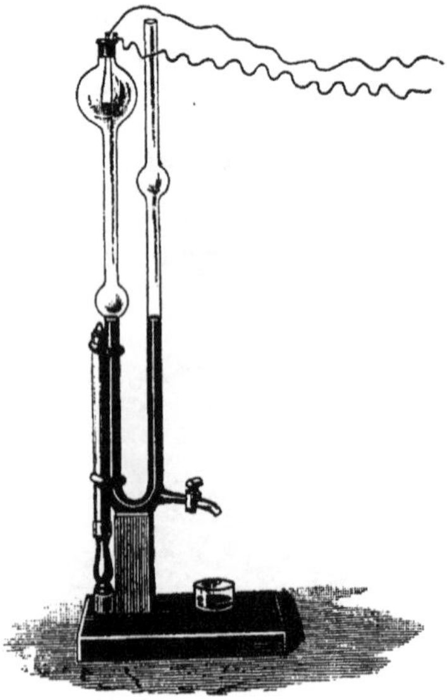

experiment is devised to show that the volume of carbon dioxide produced by the union with carbon of a known volume of oxygen is equal to that of the oxygen. The oxygen is contained in the bulb, and confined over mercury. The carbon is wrapped in a piece of platinum wire, and, as in the case of the diamond. heated to its point of ignition. The gas expands at first, of course, owing to its temperature being raised, but on cooling, the mercury in the two limbs of the U-tube returns to its original level, showing that the volume of gas is the same after it has been converted into carbon dioxide. (See fig. 35.)

Carbon also withdraws oxygen from its compounds with other elements, combining with it, a mixture of monoxide and dioxide being usually formed. Carbon heated to bright redness in steam gives a mixture of monoxide, dioxide, and hydrogen ("water-gas") There is little doubt that the oxides of the other elements of this group could be similarly formed.

Compounds of carbon with hydrogen and oxygen also burn in oxygen forming dioxide. Thus when a candle, consisting chiefly of carbon and hydrogen, burns, both its carbon and hydrogen unite with oxygen. The union takes place more rapidly in oxygen gas than in air, but the total amount of heat evolved is the same whichever be employed. But owing to the greater rapidity of combination, the *temperature* is higher during combustion in oxygen than in air. The oxidation of the blood of animals is also a slow combustion, taking place in the capillary bloodvessels, the oxygen being derived from the inspired air.

2. **By union of a lower oxide or sulphide with oxygen or sulphur.**—Carbon monoxide burns in air or oxygen to form dioxide. A mixture of the two gases explodes on passing a spark, provided they are moist. No explosion takes places when they are dry, although combination occurs in the space through which the spark passes. Carbon monoxide also withdraws oxygen from oxides of many other elements, such as those of iron, copper, &c., to form dioxide. When heated to whiteness with steam, a portion is converted into dioxide. Titanium sesquioxide and sesquisulphide readily unite with oxygen or sulphur, forming dioxide or disulphide.

3. **By the action of heat on a compound.**—All carbonates, those of lithium, sodium, potassium, rubidium, and cæsium excepted, lose carbon dioxide when heated. Barium carbonate requires a white heat; strontium carbonate a bright red heat, and calcium carbonate a red heat. These carbonates decompose more readily if heated in a current of some indifferent gas, such as air or steam. Compounds of the other dioxides have not been thus decomposed, owing to the non-volatility of the dioxides. But the sulphocarbonates, like the carbonates, are decomposed by

heat into sulphides and carbon disulphide. Calcium compounds, for example, decompose thus—

$$\mathbf{CaCO_3 = CaO} + CO_2; \mathbf{CaCS_3 = CaS} + CS_2.$$

The dioxides of titanium, zirconium, cerium, and thorium are produced by heating their hydrates or sulphates, and that of thorium by heating its oxalate.

4. **By displacement.**—This method, as a rule, yields the hydrates; but as carbonic acid (the hydrate of carbon dioxide) is very unstable, it is produced thus: for example, a carbonate, treated with sulphuric acid, yields a sulphate, carbon dioxide, and water:—$Na_2CO_3.Aq + H_2SO_4.Aq = Na_2SO_4.Aq + CO_2 + H_2O$; or the reaction may be thus written:—$CO_2.Na_2O + SO_3.H_2O = SO_3.Na_2O + CO_2 + H_2O$. There is no tendency to form a compound between carbon dioxide and sulphur trioxide.

In actual practice, carbon dioxide is prepared on a large scale by burning carbon in air, or by treating calcium carbonate with sulphuric or hydrochloric acid. When the last acid is used, some spray of hydrogen chloride is apt to be carried over with the carbon dioxide, hence it is advisable to wash it by leading it through a solution of hydrogen sodium carbonate. If sulphuric acid is employed, the calcium carbonate must be in the state of fine powder, else it becomes coated with an insoluble layer of sulphate which hinders further action. It is by this method that carbon dioxide is usually made in the manufacture of "aerated water."

Cerium tetrafluoride, when heated in air, loses fluorine, and yields the dioxide. This is probably due to the moisture in the air, forming hydrogen fluoride, and would come under the next heading

5. **By double decomposition.**—Carbon disulphide has been produced by heating carbon tetrachloride to 200° with phosphorus pentasulphide; substituting selenium for sulphur, a liquid was produced containing about 2 per cent. of diselenide.

Special method.—Carbon dioxide is produced by the decomposition of grape-sugar, $C_6H_{12}O_6$, under the action of the yeast ferment (*Saccharomyces cerevisiæ*), when ethyl alcohol, $C_2H_5.OH$, and carbon dioxide are the chief products.

The starch contained in grain is converted during the process of "malting," or incipient germination, during which the grain is kept warm and moist on the "malting-floors," into grape-sugar, by aid of the ferment *diastase*, contained in the grain. The growth is then stopped by heating the malt; it is crushed, and is known as "grist;" it is transferred to the "mash-tun," a large cask or vat, where it is treated with warm water. The solution of grape-sugar thus obtained is called the "wort;" it is mixed with yeast, and left to ferment, when the change already mentioned takes place. The carbon dioxide fills the vat and escapes into the air. The equation is—$C_6H_{12}O_6 = 2C_2H_5.OH + 2CO_2$.

Properties.—At the ordinary temperature carbon dioxide is a gas. Its boiling point under normal pressure is about −79°. Its melting point is nearly the same as its boiling point; it is given as −78·5°, hence the liquid easily freezes by its own evaporation. It may be condensed to a liquid at a pressure of about 36 atmospheres at 10°. The gas is colourless, has a faint sweetish smell and taste, and is much heavier than air, hence it is best collected by downward displacement. Its great density (22, compared to air=14·47) permits of its being poured from one vessel to another without much loss. Its density is easily shown by pouring it into a light beaker, suspended from the beam of a balance, and counterpoised.

FIG. 36.

Carbon dioxide supports the combustion of the elements potassium, sodium, and magnesium. They deprive it of a portion of its oxygen, forming oxides and carbon monoxide, as well as some free carbon; the oxide then unites with excess of carbon dioxide, forming a carbonate. Carbon may also be said to burn in carbon dioxide, inasmuch as when the dioxide is led over red hot carbon, the monoxide is formed; but because the heat evolved by this reaction is comparatively small, the carbon is not thereby kept at its temperature of incandescence, and action ceases, unless a supply of heat be added from without. When carbon burns in oxygen, therefore, the whole of the oxygen is not converted into carbon dioxide; the action ceases when the dioxide formed bears a certain proportion to the total gas present; the reverse action then tends to begin. Hence a candle, burning in air, goes out when the carbon dioxide formed reaches a certain proportion of the total gas; and for the same reason, an animal dies when breathing a confined atmosphere, long before it has completely deprived it of oxygen. A man can breathe, however, for some time in an atmosphere in which a candle refuses to burn, as was shown by the late Dr. Angus Smith. Carbon dioxide is decomposed by the green colouring matter of plants in sunshine; the exact nature of this decomposition is not known; there are grounds for supposing that it consists in a reaction occurring between carbon dioxide and water, as follows:

$$CO_2 + H_2O = H_2CO + O_2.$$

The substance H_2CO is named formic aldehyde, and it has been recently shown to be easily transformable into a kind of sugar, $C_6H_{12}O_6$, named formose. There may be some connection between this transformation and the formation of sugar in plants. The carbon dioxide is absorbed by the *stomata* or "small mouths" in the under surface of the leaves of plants, and oxygen is evolved. This may be experimentally shown by placing some blades of grass in a jar of water inverted over a trough. The oxygen gas collects in the upper portion of the jar during several days' exposure to sunlight, and may be recognised by the usual tests.

Liquid carbon dioxide is heavier than water, and does not mix with it. It is a non-conductor of electricity. Above the temperature 30·9°, the critical point of carbon dioxide, the gas cannot be made to assume the liquid state by compression. The solid dioxide is a loose white powder, like snow, produced by allowing the liquid to escape into a thin flannel bag; the liquid absorbs heat during its conversion into gas, and a portion solidifies, owing to its being thus cooled. A mixture of solid carbon dioxide with ether gives a temperature of −100°.

It has been recently shown that carbon and carbonic oxide do not unite with perfectly dry oxygen, unless they be kept exposed to a very high temperature. The presence of water or some other compound containing hydrogen being necessary, it is supposed that the carbon or carbonic oxide reacts with the water liberating hydrogen, thus—$CO + H_2O = CO_2 + 2H$, or $\mathbf{C} + H_2O$

$= CO + 2H$, and that the hydrogen then unites with the oxygen, to form water, which is again acted on.*

The test for carbon dioxide is its combination with calcium or barium oxide, when shaken with a solution of the respective hydroxide, to form carbonate, in either case a white powder, which effervesces with acids.

The presence of carbon dioxide in expired air may be demonstrated by the arrangement shown in the figure :—

FIG. 37.

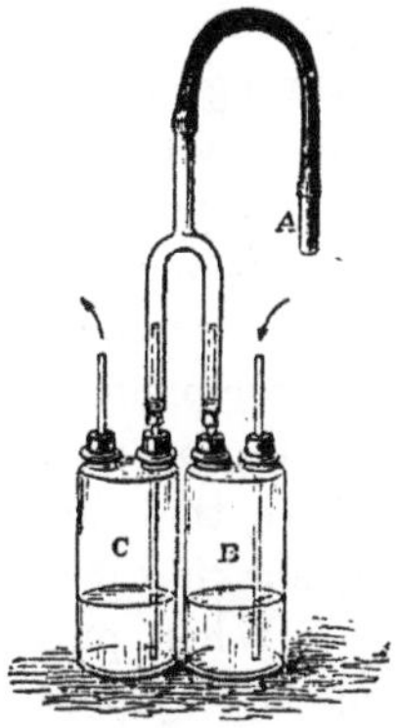

The air entering the lungs passes through lime-water in the bottle on the right hand side; as ordinary air contains only 4 volumes of carbon dioxide in 10,000, a turbidity is not seen for some time. The exhaled air passes through the lime-water in the left hand bottle and soon turns it turbid.

The amount of carbon dioxide in atmospheric air may be estimated comparatively by measuring the amount required to produce incipient turbidity in baryta water.

The little apparatus is shown in fig. 38. The indiarubber ball is squeezed, the air escaping through the opening. The opening is then closed with the finger, and, on allowing the ball to expand, air is drawn through the baryta

FIG. 38.

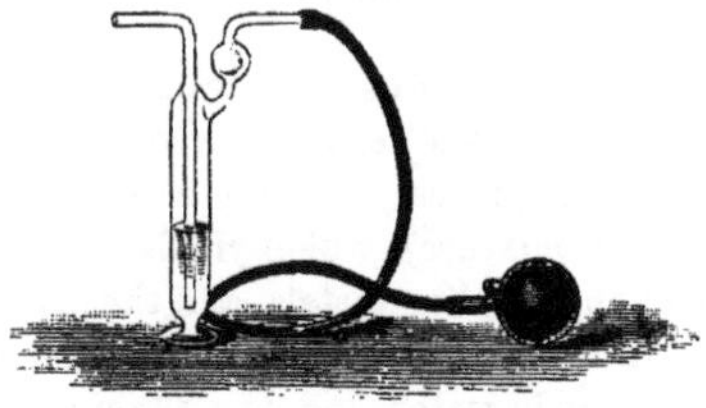

* Dixon, *Chem. Soc.*, **49**, 94.

water. On removing the finger the ball is again squeezed empty, and air is again drawn through the baryta water. Having found the number of charges of the ball which must pass through the baryta water to produce a turbidity with ordinary air, it may be assumed with fair correctness that the normal amount is present, viz., 4 volumes in 10,000. On applying the same test to vitiated air, fewer charges are required, and the amount of carbon dioxide may be calculated by simple proportion.

Carbon dioxide rapidly combines with the hydroxides of sodium and potassium, as well as with those of calcium and barium. The method of absorbing it from gaseous mixtures is to shake them with a strong solution of potassium hydroxide. It may also easily be absorbed by passing it through a solid mixture of hydroxides of calcium and sodium, commonly termed "soda-lime."

Carbon disulphide is a limpid colourless liquid, heavier than water and not mixible with it, melting at $-110°$ and boiling at 46·04°. In the crude state it contains hydrogen sulphide and disagreeably-smelling sulphur compounds. It may be purified from hydrogen sulphide by shaking it with a solution of potassium permanganate, which oxidises that impurity, and from sulphur-compounds and sulphur by shaking it with mercuric chloride and mercury and distilling it. When pure it has a not unpleasant ethereal odour. Its vapour is very poisonous when breathed. Its vapour ignites very easily when mixed with air (at 149°), hence it must be kept away from a flame and distilled by aid of a water-bath. It is decomposed by light, acquiring thereby a disagreeable smell. It is slightly soluble in water. Its vapour explodes when exposed to the shock of decomposing fulminate of mercury, being resolved into the elements carbon and sulphur. It is formed with absorption of heat, hence its instability; heat is evolved when it is exploded. It mixes easily with alcohol, ether, and oils, and is used for extracting oils and fats from acids, animal refuse, wool, &c., and as a solvent for sulphur chloride in vulcanising caoutchouc.

It unites with sulphides, giving sulphocarbonates (see below), and when passed through a hot tube with chlorine it yields sulphur chloride (S_2Cl_2) and carbon tetrachloride (see p. 145).

In preparing the pure **dioxides** of **titanium**, **zirconium**, **cerium**, and **thorium**, the chief difficulty is the separation from the oxides of other elements, especially from silica. The process is, fusion with a mixture of potassium and sodium carbonates (fusion-mixture), which yields in each case silicate, titanate, zirconate, or thorate of the alkaline metals, and the oxides of the other metals, if these are present. In the case of titanium, hydrogen fluoride is added to the solution of the fused mass in water, and the titanium thrown down as double fluoride of titanium and potassium, **$TiF_4.2KF$**. These crystals are afterwards dissolved in water, and on addition of ammonia the titanium is thrown down as hydrate. With zirconium, the fused mass, consisting of silicate and zirconate of sodium and potassium, is mixed with excess of hydrochloric acid, and evaporated to dryness. This gives a mixture of silica and oxychlorides of zirconium. On treatment with hydrochloric acid, the silica, not being thus converted into chloride, does not dissolve, but the zirconium dissolves as chloride, along with iron, &c. The solution is boiled with thiosulphate of sodium, which precipitates the zirconium, leaving the iron in solution. On ignition of the thiosulphate of zirconium, pure zirconia, ZrO_2, is left.

Cerium is similarly separated,* but it is precipitated as oxalate, and on ignition the oxide Ce_2O_3 is left. Thorium is precipitated as oxalate, from its solution in hydrochloric acid, after separation of silica; and from a solution of the oxalate in hydrochloric acid by a strong solution of potassium sulphate, with which it combines, forming a double sulphate of thorium and potassium see p. 428).† It also yields an insoluble thiosulphate.

Titanium dioxide, native as *rutile*, forms reddish-brown crystals; artificially prepared it is a reddish-brown powder. It is insoluble in water and does not react with acids, except with strong sulphuric acid or fused bisulphates.

It melts in the oxyhydrogen flame. It has been artificially crystallised by passing vapours of titanium tetrachloride and steam through a red-hot tube.

Zirconia, or **zirconium dioxide,** is a white powder; it is obtained in small quadratic prisms by crystallisation from fused borax.

Cerium dioxide is a pale-yellow insoluble substance, which also crystallises from fused borax in tesseral crystals. On boiling with hydrochloric acid, chlorine is evolved, and the trichloride is produced, $CeCl_3$. With sulphuric acid it also dissolves, the sulphate $Ce_2(SO_4)_3.Ce(SO_4)_2$ being formed, with evolution of oxygen. It is soluble in nitric acid.

Thorium dioxide, or **thoria,** is a white powder, separating from its solution in borax in transparent quadratic crystals.

Compounds with water and hydrogen sulphide.—Carbon dioxide as gas dissolves to some extent in water; 100 volumes of water at 20° dissolve about 90 volumes, and at 15° about 100 volumes. The solution has a pleasant sharp taste, and is usually called "soda-water." The carbon dioxide is, however, forced in under a pressure of several atmospheres. The gas escapes quickly if the pressure is decreased immediately; but after some days or weeks it appears to have entered into combination to some extent with the water, and does not then escape so

* For details regarding cerium compounds see Brauner, *Chem. Soc.*, **47**, 879; references to other papers are given.

† Clève, *Bull. Soc. Chim.* (2), **21**, 115.

readily. The solution turns litmus solution claret coloured. It acts on zinc, iron, and magnesium, forming carbonates and liberating hydrogen. It probably consists of a weak solution of **carbonic acid**, H_2CO_3, with carbon dioxide uncombined but mixed with the water.

Carbon disulphide does not unite directly with hydrogen sulphide, but **sulphocarbonic acid,** as the compound is named, H_2CS_3, is produced on addition of weak hydrochloric acid to a solution of a sulphocarbonate, *e.g.*, Na_2CS_3 (see below). It is a dark-yellow oil, with a pungent odour, and on rise of temperature it rapidly decomposes into carbon disulphide, CS_2, and hydrogen sulphide, H_2S.

Many **hydrates of titanium dioxide** have been described, but the data regarding them are as a rule contradictory. On heating titanic hydrate thrown down from its chloride by an alkali it loses water gradually, with rise of temperature, and shows no sign of any definite hydrates. It is probable that there are many, and that no one is stable over any large range of temperature.

The **hydrates of zirconium dioxide** appear also to be numerous. The only sudden break in drying the hydrate precipitated from the chloride by ammonia is at 400°. On reaching this temperature the body suddenly turns incandescent, and all water is expelled. It has then become difficult to dissolve in acids, and it is believed that sudden polymerisation has occurred, many molecules of ZrO_2 having united to form one complex molecule.

Cerium hydrate at 600° has the formula $\mathbf{CeO_2.2H_2O}$. At lower temperatures it is brownish-yellow, but at that temperature and above it is bright yellow; as it dries further, its colour changes to a salmon-pink. It is produced by the action of sodium hypochlorite on $\mathbf{Ce_2O_3}$.

Thorium hydrate is a gelatinous mass; it probably resembles titanium hydrate.

Compounds with Oxides and Sulphides—Carbonates, Titanates, Zirconates, Thorates—Carbon Oxysulphide, Oxysulphocarbonates, and Sulphocarbonates.

These compounds may be divided into two classes: (1) **normal compounds**, those in which the ratio of the number of oxygen atoms in the dioxide to that of the oxide of the metal is as 2 : 1; and (2) **basic compounds**, those in which the ratio is less than 2 : 1; no acid compounds, those in which the ratio is greater than 2 : 1, are known. The normal compounds are most numerous.

But before considering these bodies it is advisable to describe **carbon oxysulphide**, of which the formula is *COS*,* as shown by its gaseous density. This body, therefore, cannot be regarded as a compound of carbon dioxide and carbon disulphide, $CO_2.CS_2$, but as carbon dioxide, of which one atom of oxygen is *replaced* by sulphur. It may be produced by leading a mixture of carbon dioxide and carbon disulphide gases through a tube filled with platinum black, *i.e.*, finely-divided platinum, or by the union of carbon monoxide with sulphur. But it is most easily produced by the reaction between sulphocyanide of hydrogen and water. The compound KCNS, on treatment with sulphuric acid, yields the acid HCNS. If the sulphuric acid be moderately strong and warm, it combines with the ammonia produced by the decomposition of the acid, thus:—$HCNS + H_2O = NH_3 + COS$. Carbon oxysulphide is a not infrequent constituent of mineral springs, but, as a rule, it has for the most part reacted with water to form carbon dioxide and hydrogen sulphide, thus:—$COS + H_2O = CO_2 + H_2S$. It is a colourless gas, without odour or taste when pure, somewhat soluble in water, and combustible to dioxides of carbon and sulphur. It is hardly affected by aqueous potash, but is easily absorbed by an alcoholic solution. Its physiological effects resemble those of nitrous oxide.†

There are thus three bodies, all of which form compounds with oxides and sulphides, viz.: CO_2, carbon dioxide; *COS*, carbon oxysulphide; and CS_2, carbon disulphide.

Compounds of Carbon Dioxide with Oxides.

1. **Normal carbonates.**—Ratio of oxygen in carbon dioxide to oxygen in combined oxide, 2 : 1.

The following is a list of the known compounds:—

Simple carbonates:—Li_2CO_3; Na_2CO_3 with 15, 10, 8, 7, 6, 5, 2, and $1H_2O$; K_2CO_3 with $2H_2O$ and H_2O; $Rb_2CO_3.H_2O$; Cs_2CO_3; $(NH_4)_2CO_3.H_2O$.

Complex carbonates:—$HNaCO_3$; $H_2Na_4(CO_3)_3.3H_2O$; $HKCO_3.H_2O$; $HRbCO_3$; $HCsCO_3$; HNH_4CO_3; $H_2(NH_4)_4(CO_3)_3.2H_2O$.

These carbonates are all made by the action of carbon dioxide on a solution of hydroxide of the metal, thus:—$2NaHO.Aq + CO_2 = Na_2CO_3.Aq + H_2O$.

* Than, *Annalen*, Suppl. 1, 236.

† *J. prakt. Chem.* (2), 36, 64.

Of these, lithium carbonate, Li_2CO_3, occurs in mineral waters; it is sparingly soluble in water (about 1·4 grams in 100 of water at 20°), and may be produced by addition of a concentrated solution of sodium carbonate to a soluble salt of lithium. For the preparation of sodium and potassium carbonates, see p. 671.

Sodium carbonate is a constituent of certain "soda-lakes" in Egypt and Hungary; it also occurs in volcanic springs.

The ordinary name for the carbonate $\mathbf{Na_2CO_3}$ is *soda-ash;* for the crystalline salt, $\mathbf{Na_2CO_3.10H_2O}$, *soda crystals* or "washing-soda;" and for hydrogen sodium carbonate $\mathbf{HNaCO_3}$, *bicarbonate*, or "baking-soda." The latter is produced by treating the normal carbonate (crystals) with carbon dioxide, thus:—$\mathbf{Na_2CO_3} + CO_2 + H_2O = \mathbf{2NaHCO_3}$. Hydrogen sodium carbonate is less soluble than sodium carbonate.

Carbonate of sodium melts at about 818°, and of potassium at about 830°. On heating hydrogen sodium carbonate it loses water and carbon dioxide, and yields sodium carbonate, thus:—$\mathbf{2HNaCO_3} = H_2O + CO_2 + \mathbf{Na_2CO_3}$. The simple carbonates, except those of ammonium (see p. 533), volatilise unchanged at a bright red or white heat, and are not decomposed into carbon dioxide and metallic oxide. It is probable that a carbonate of sodium and potassium also exists, of the formula $\mathbf{NaKCO_3}$; a mixture of the two is named "fusion mixture," and is used in the decomposition of silicates, &c. It has a much lower melting point than either of the pure salts. The compound $\mathbf{H_2Na_4(CO_3)_3.3H_2O}$ occurs native, and is known as *trona* or *urao;* in old times it used to be an important source of soda. These bodies have all an alkaline, cooling taste; the ammonium compound smells of ammonia, owing to its decomposing, on exposure, into ammonia, carbon dioxide, and water. Hydrogen ammonium carbonate is found in guano deposits.

Simple carbonates:—$\mathbf{BeCO_3.4H_2O}$; $\mathbf{CaCO_3}$, also $\mathbf{5H_2O}$; $\mathbf{SrCO_3}$; $\mathbf{BaCO_3}$; $\mathbf{MgCO_3}$; also $\mathbf{3H_2O}$ and $\mathbf{5H_2O}$; $\mathbf{ZnCO_3}$; $\mathbf{CdCO_3}$.

Complex carbonates:—$H_2Ca(CO_3)_2.Aq$ (?); $\mathbf{Na_2Ca(CO_3)_2.5H_2O}$; $H_2Mg(CO_3)_2.Aq$; $\mathbf{Na_2Mg(CO_3)_2}$; $\mathbf{HKMg(CO_3)_2.4H_2O}$; $\mathbf{(NH_4)_2Mg(CO_3)_2.4H_2O}$; $\mathbf{H_2K_8Zn_6(CO_3)_{11}.7H_2O}$: $\mathbf{Na_6Zn_8(CO_3)_{11}.8H_2O}$. $\mathbf{Na_6Zn_8(CO_3)_{11}.8H_2O}$.

These carbonates are all white solids. They are decomposed by heat (see the respective oxides), barium carbonate requiring the highest temperature. The following are found native:—Calcium carbonate, $CaCO_3$, as *calcspar* or *Iceland-spar*, in hexagonal rhombohedra; as *arrayonite* in trimetric rhombic prisms; as *marble, limestone, chalk;* a constituent of shells, bones, &c. It may be produced in the form of calcspar by crystallisation from a mixture of fused sodium and potassium chlorides; by precipitation from solution below 30°; and as arragonite by precipitation above 90°.* Between these temperatures mixtures of microscopic crystals of the two are precipitated by addition of sodium or ammonium carbonate to a solution of a soluble salt of calcium. When heated to redness in a closed iron tube, calcium carbonate fuses, and then yields a crystalline mass resembling marble. The carbonates of calcium, strontium, and barium are formed by direct union of oxide with carbon dioxide; the union is attended with great evolution of heat, causing the oxide to become incandescent; the product with lime has the formula $\mathbf{CO_2.2CaO}$. The compound $\mathbf{Na_2Ca(CO_3)_2.5H_2O}$ is named *gaylussite;* strontium carbonate, $\mathbf{SrCO_3}$, is found native as *strontianite;* and barium carbonate, $\mathbf{BaCO_3}$, as *witherite.* $\mathbf{MgCO_3}$ is *magnesite*, and a double carbonate of calcium and magnesium, in which indefinite amounts of both metals are present, is *dolomite;* it forms large mountain ranges, named "The Dolomites," in northern Italy. Zinc carbonate, $\mathbf{ZnCO_3}$, occurs native as *calamine*, and is accompanied by cadmium carbonate, $\mathbf{CdCO_3}$.

The so-called acid carbonates, *e.g.*, $H_2Ca(CO_3)_2$, $H_2Mg(CO_3)_2$, and similar compounds of barium, strontium, &c., have not been isolated. Their existence is assumed because the normal carbonates dissolve freely in a solution of carbonic acid. On warming the solution, they are decomposed with evolution of carbon dioxide and precipitation of the simple carbonates.

Carbonates of boron and scandium are unknown; of this group only $\mathbf{Y_2(CO_3)_3.12H_2O}$, and $\mathbf{2H_2O}$; and $\mathbf{La_2(CO_3)_2}$, found native as *lanthanite*, are known. The existence of a carbonate of aluminium is doubtful; carbonate of gallium is unknown. $\mathbf{In_2(CO_2)_3}$, however, has been prepared. These are insoluble white bodies, which lose carbon dioxide when heated, leaving the oxides.

Thallium forms no thallic carbonate, but thallous carbonate, $\mathbf{Tl_2CO_3}$, is produced by precipitation. There is some evidence of a hydrogen thallium carbonate, $\mathbf{HTl(CO_3)}$.

Chromic, ferric, and manganic carbonates are unknown. On addition of a soluble carbonate to their soluble salts, *e.g.*, chlorides, the hydrates are precipitated, and carbon dioxide escapes, thus:—

$$2CrCl_3.Aq + 3Na_2CO_3.Aq = Cr_2O_3.Aq + 6NaCl.Aq + 3CO_2.$$

* *Comptes rend.*, **92**, 189.

The carbonates derived from the monoxides of these metals are as follows:—

Simple carbonates:—$CrCO_3$; $FeCO_3$; $MnCO_3$; $CoCO_3$; $NiCO_3$.
Complex carbonates:—$HKCo(CO_3)_2.2H_2O$; $H_2Na_2Co(CO_3)_3.4H_2O$; $Na_2Co(CO_3)_2.10H_2O$; $HKNi(CO_3)_2.4H_2O$; $K_2Ni(CO_3)_2.4H_2O$; $Na_2NiCo(CO_3)_3.10H_2O$.

Of these, chromous carbonate is produced by mixing a solution of chromous chloride with sodium carbonate; $FeCO_3$ is found native, and named *spathic iron ore* or *siderite;* in an impure state, mixed with clay or shale, it is termed *clay-band* or *black-band*, and forms one of the most important ores of iron. When pure it is a whitish crystalline rock. It is soluble in water containing carbon dioxide; such a solution may contain hydrogen ferrous carbonate, $H_2Fe(CO_3)_2$, which, however, has not been isolated. It is in this form a constituent of iron springs, and, on exposure to air, it loses carbon dioxide, and the iron oxidises to ferric hydrate, and deposits on the bed of the stream. A hydrated carbonate, $FeCO_3.H_2O$, also occurs native. Manganese carbonate, $MnCO_3$, occurs native as *manganese spar*.

No carbonates of **titanium** or **zirconium** are known.

Cerium hydrate, however, on exposure to air, absorbs carbon dioxide, yielding $Ce_2(CO_3)_3.9H_2O$. **Silicon** and **germanium** do not yield carbonates, but **tin** forms a basic carbonate (see below). **Lead carbonate**, $PbCO_3$, occurs native, and is known as *cerussite*. Lead oxide sometimes replaces calcium oxide in native calcium carbonate, to the extent of 3 or 4 per cent.; the compound is called *plumbocalcite*. *Plumbo-arragonite* has also been found native at the lead hills in Lanarkshire. A chlorocarbonate, of the formula $PbCO_3.PbCl_2$, may be produced by boiling lead carbonate and chloride together in water. It is an insoluble white substance. It also occurs native as *corneous lead*. When heated, it loses carbon dioxide, leaving $PbO.PbCl_2$.

Carbonates of nitrogen, vanadium, niobium, and tantalum are unknown, and also carbonates of phosphorus, arsenic, and antimony; a basic carbonate of bismuth has been prepared (see below).

Carbonates of molybdenum and tungsten do not appear to exist, but several double carbonates of uranyl (UO_2) (see p. 407) have been prepared. These are $Na_4(UO_2)(CO)_3$, $K_4(UO_2)(CO_3)_3$, and $(NH_4)_4(UO_2)(CO_3)_3$. A calcium compound occurs native; its formula is $Ca(UO_2)(CO_3)_2.10H_2O$. It is seen that the group UO_2, or uranyl, acts like a dyad metal.

Normal carbonate of copper is unknown. The only known normal compound has the formula $\mathbf{Na_2Cu(CO_3)_2.6H_2O}$. **Silver carbonate,** $\mathbf{Ag_2CO_3}$, is a yellowish-white powder, produced by precipitation; it loses carbon dioxide at 200°; $\mathbf{KAgCO_3}$ is formed if $\mathbf{HKCO_3}$ be used; it is white. **Mercurous carbonate,** $\mathbf{Hg_2CO_3}$, is a very unstable brown precipitate. Carbonates of gold and of the metals of the palladium and platinum groups are too unstable to exist.

Considering these carbonates as a whole, it may be noticed (1) that with exception of those of the sodium group of metals all are decomposed by heat into oxide and carbon dioxide; (2) those of the sodium, calcium, and magnesium groups, and thallous and cerium carbonates are formed by direct union of the hydroxides and carbon dioxide; (3) that the oxides, except those of the sodium group, do not combine directly with carbon dioxide; calcium oxide, however, begins to combine at 415°; and (4) that the carbonates of calcium, strontium, barium, and silver, are the only normal ones produced by precipitation by addition of a soluble carbonate to a soluble salt of the metals. In all other cases, basic carbonates are precipitated. These will now be considered.

2. **Basic carbonates.**—These bodies contain a greater proportion of the oxide of the metal than is represented by the ratio given before. The oxygen of the metallic oxide bears a larger ratio to that of the carbon dioxide than 1 : 2. Their formulæ are most conveniently stated as addition-formulæ; the relations then appear most clearly. They are unknown in the sodium group of elements.

$\mathbf{CO_2.5BeO.5H_2O}$; $\mathbf{6CO_2.3K_2O.4BeO}$; $\mathbf{CO_2.2CaO.H_2O}$; $\mathbf{CO_2.2CaO}$ (produced by heating $CaCO_3$; by heating CaO in CO_2, the mass turning incandescent during union; or by exposure of $Ca(OH)_2$ to air); $\mathbf{3CO_2.4CaO}$, also produced by direct union; $\mathbf{CO_2.2SrO}$; $\mathbf{CO_2.2BaO}$; $\mathbf{4CO_2.5MgO}$ (precipitated hot); $\mathbf{3CO_2.4MgO}$ (native; *hydromagnesite*); $\mathbf{CO_2.2ZnO.H_2O}$; $\mathbf{CO_2.3ZnO.3H_2O}$ (native; *zinc-bloom*); $\mathbf{CO_2.5ZnO.6H_2O}$; $\mathbf{2CO_2.3ZnO.H_2O}$; $\mathbf{2CO_2.5ZnO.5H_2O}$; $\mathbf{4CO_2.5ZnO.H_2O}$; $\mathbf{4CO_2.9ZnO.6H_2O}$; $\mathbf{CO_2.3CdO}$. These substances (the cadmium, strontium, and barium compounds excepted) are produced by precipitation under various conditions of temperature and dilution.

$\mathbf{2CO_2.4ThO_2.8H_2O}$ appears to exist, but there are no corresponding compounds of tin or lead. $\mathbf{CO_2.2SnO}$, however, is thrown down on addition of sodium carbonate to stannous chloride, $SnCl_2$, as a white precipitate.

The substance known as *white-lead* is probably a mixture of basic carbonates of lead. Seen under the microscope, it consists of small spherical masses, each of which is opaque and reflects white light. Hence its use as a paint. It possesses great "covering power," owing to its not transmitting light. It is produced by the action of acetic acid, carbon dioxide and water on metallic lead; similar basic carbonates, which however, have not the same opaque quality, are produced by precipitation. The following have been analysed:—

$2CO_2.3PbO.H_2O$; $3CO_2.4PbO.H_2O$; $5CO_2.6PbO.H_2O$; $6CO_2.7PbO.2H_2O$; and $5CO_2.8PbO.3H_2O$.

Ferrous salts, on treatment with a soluble carbonate, give a white precipitate of presumably basic carbonate. This precipitate rapidly turns green, absorbing oxygen: it has not been analysed. With manganese, cobalt and nickel the following compounds are known:—$CO_2.3CoO.3H_2O$; $2CO_2.5CoO.4H_2O$; $CO_2.3NiO.6H_2O$ (found native, and named *emerald nickel*); $2CO_2.5NiO.7H_2O$.

Copper and mercury also form basic carbonates. $CO_2.2CuO.H_2O$ occurs native as *malachite*, a beautiful green mineral, and $2CO_2.3CuO.H_2O$, as *azurite*, which has a splendid blue colour. By precipitation $CO_2.6CuO$ and $CO_2.8CuO.5H_2O$ are formed as light blue precipitates. Mercuric salts with soluble carbonates give a reddish precipitate of $CO_2.4HgO$.

Titanates, zirconates, and thorates.—These have been little investigated. The compounds which have been prepared are:—

Na_2TiO_3; K_2TiO_3; $MgTiO_3$; $FeTiO_3$; $CaTiO_3$; and $ZnTiO_3$; also $TiO_2.2ZnO$; $2TiO_2.3ZnO$; $5TiO_2.4ZnO$.

The titanates of sodium and potassium are yellowish, fibrous masses, produced by heating titanic oxide with excess of carbonate of sodium or potassium. On treatment with water they decompose, a sparingly soluble (acid ?) salt being precipitated, while a (basic ?) salt remains in solution. Obviously these compounds have little stability. Magnesium and iron titanates are produced by heating titanium oxide with magnesium chloride, or with a mixture of ferrous fluoride and sodium chloride. The iron titanate forms long thin steel-grey needles. It is formed native as *ilmenite*: it is isomorphous with and crystallises along with iron sesquioxide. The compounds $TiO_2.2MgO$ and $TiO_2.2FeO$ are similarly prepared. Calcium titanate, $CaTiO_3$, occurs native as *perowskite*.

By igniting together zirconia and sodium carbonate, the compound Na_2ZrO_3 is formed. It is decomposed by water into zirconium hydrate and sodium hydrate. A larger amount of carbonate yields $ZrO_2.2Na_2O$; it is also decomposed by water, and deposits hexagonal crystals of a salt of the formula $8ZrO_2.Na_2O.12H_2O$. Magnesium zirconate has also been prepared by fusing zirconium dioxide and magnesium oxide in presence of ammonium chloride. It is a powder consisting of transparent crystals.

Although thorium dioxide dissolves in alkalies, and probably unites with oxides, no thorates have been analysed.

Compounds of sulphides with sulphides.—These bodies have been investigated only in the compounds of carbon. They are named *sulphocarbonates* or *thiocarbonates*, from the Greek word for sulphur, *θεῖον*. They are produced by the action of carbon disulphide on sulphides, which is analogous to that of carbon dioxide on oxides. Those which have been prepared and analysed are:—

$LiCS_3$; $NaCS_3$; K_2CS_3; $(NH_4)_2CS_3$; $MgCS_3$; $CaCS_3$; $SrCS_3$; $BaCS_3$.

Precipitates are produced by potassium sulphocarbonate in solutions of zinc, cadmium, chromium, iron, manganese, cobalt, nickel, tin, lead, bismuth, platinum, silver, gold, and mercury. These require further investigation.

Potassium sulphocarbonate consists of yellow deliquescent crystals; it is formed by digesting an aqueous or alcoholic solution of potassium sulphide, K_2S, with carbon disulphide; the crystals contain water, which is expelled at 80°, leaving a brown-red solid. On heating it, potassium trisulphide, $\mathbf{K_2S_3}$, remains, mixed with carbon. The ammonium salt is produced along with ammonium sulphocyanide, by digesting carbon disulphide with alcoholic ammonia, thus:—$2CS_2 + 4NH_3 = (NH_4)_2CS_3 + NH_4CNS$. It forms yellow crystals, insoluble in alcohol, but soluble in water. The calcium and barium salts are prepared like the potassium salt. Milk of lime and carbon disulphide give an orange-red basic salt, $\mathbf{CaCS_3.2CaO.8H_2O}$; at 30° it melts to a red liquid, from which $\mathbf{CaCS_3.3CaO.10H_2O}$ separates. The action of carbon oxysulphide on sulphides requires investigation.

Compounds of oxides with halides.—It has already been stated that carbon monoxide and chlorine combine directly; the product is **carbonyl chloride**, or carbon oxychloride, $COCl_2$. Its vapour-density shows it to have that formula, and not to be a compound of CO_2 and CCl_4. It is produced on exposing a mixture of the two gases to sunlight, hence its old name, *phosgene* gas, a gas produced by light (φῶς). It is more easily prepared by passing carbon monoxide through hot antimony pentachloride, $SbCl_5$, which loses two atoms of chlorine; or by passing a mixture of the gases through a tube filled with hot animal charcoal. It condenses to a liquid boiling at 8·4°.* When treated with water it produces carbon dioxide and hydrogen chloride. Assuming the carbon dioxide to remain in combination with water, as carbonic acid, the change may be thus represented:—

$$CO{<}^{Cl}_{Cl} + \begin{matrix}H.OH\\H.OH\end{matrix} = CO{<}^{OH}_{OH} + \begin{matrix}HCl\\HCl.\end{matrix}$$

Light is thus thrown on the constitution of carbonic acid. It appears to consist of carbon monoxide in combination with hydroxyl; and the normal carbonates may be similarly represented; for example, sodium carbonate as $CO{<}^{ONa}_{ONa}$; hydrogen sodium carbonate as $CO{<}^{OH}_{ONa}$; calcium carbonate as $CO{<}^{O}_{O}{>}Ca$; basic copper carbonate as $CO{<}^{O—Cu}_{O—Cu}{>}O$, each atom

* For sulphochlorides, see P. Klason, *Berichte*, **20**, 2376.

of copper being half oxide, half carbonate. The more complex basic carbonates may also be similarly represented; *e.g.*, basic lead carbonate may be written $\begin{array}{l}\mathrm{CO}<\begin{array}{l}\mathrm{OPb}\diagdown\mathrm{O}\\ \mathrm{O}\end{array}\\ \mathrm{CO}<\begin{array}{l}\mathrm{O}\\ \mathrm{OPb}\diagup\mathrm{O}\end{array}\end{array}>\mathrm{Pb}$. But such complicated formulæ are not confirmed by any other considerations, and should be sparingly used; moreover, it is impossible to represent the various amounts of water in combination with such compounds in any way but by simple addition.

The **oxychlorides of titanium** have recently been investigated, and their formulæ appear capable of similar modes of expression. Titanium tetrachloride may be supposed to react with water forming the hydroxide $Ti(OH)_4$, which, however, appears to be unstable (see p. 284). The corresponding carbon hydroxide, $C(OH)_4$, is certainly incapable of existence, but if, instead of hydrogen, it contain certain hydrocarbon groups, such as ethyl, C_2H_5, it becomes stable. For example, the body $C(OC_2H_5)_4$ is known, and is named ethyl orthocarbonate, the name orthocarbonic acid being applied to the unknown $C(OH)_4$. If water containing hydrogen chloride in solution (36 per cent. HCl) be mixed with titanium chloride, a violent reaction occurs, and a yellow, spongy, very deliquescent mass is produced, which has the formula $\mathbf{Ti(OH)Cl_3}$. It is tolerably stable in aqueous solution. On adding titanium tetrachloride to very cold water in theoretical amount, the dihydroxydichloride, $\mathbf{Ti(OH)_2Cl_2}$, is produced. It is a yellow deliquescent substance; and may also be mixed with water. On exposing the di- or tri-chloride to moist air for some time, the trihydroxymonochloride is formed, $\mathbf{Ti(OH)_3Cl}$. It has been obtained in a crystalline form. It is insoluble in water, but soluble in weak hydrochloric acid. We have thus the series :—

$$\mathbf{TiCl_4};\ \mathbf{Ti(OH)Cl_3};\ \mathbf{Ti(OH)_2Cl_2};\ \mathbf{Ti(OH)_3Cl};\ \text{and}\ \mathbf{Ti(OH)_4}.$$

All of these compounds, when heated alone, evolve titanium tetrachloride or hydrogen chloride, leaving a residue of dioxide.

Oxychloride of zirconium, $\mathbf{ZrOCl_2}$, separates in tetragonal crystals from a hydrochloric solution of the oxychloride in water; a similar bromide is known.

A higher oxide of titanium is produced on treating titanium hydrate with hydrogen dioxide.* It is a yellow substance, the formula of which approximates to $\mathbf{TiO_3.3H_2O}$. It appears to form compounds with $\mathbf{TiO_2}$ in the ratios $\mathbf{4TiO_2.TiO_3}$; $\mathbf{3TiO_2.TiO_3}$; $\mathbf{2TiO_2.TiO_3}$; and $\mathbf{TiO_2.TiO_3}$.

* *Chem. Soc.*, **49**, 150, 484.

Certain fluorine derivatives of this body in combination are also known. They are as follows :—

$\mathbf{TiO_2F_2.2HF}$; $\mathbf{TiO_2F_2.2KF}$; $\mathbf{TiO_2F_2.BaF_2}$; $\mathbf{2TiO_2F_2.3BaF_2}$; $\mathbf{TiOF_4.BaF_2}$; and $\mathbf{TiO_2F_2.3NH_4F}$.

Attempts to prepare similar zirconium compounds yielded $\mathbf{Zr_2O_5}n\mathbf{H_2O}$, as a white precipitate;* and cerium trioxide has been thus prepared as an orange-red precipitate.† Thorium yields an oxide of the formula $\mathbf{Th_2O_7}$ by similar treatment.

Physical Properties.

Mass of 1 cubic centimetre :—

	C.	Ti.	Zr.	Ce.	Th.
Monoxides.......	?	—	—	—	—
Dioxides	1·2—1·6‡	4·25§	5·85	6·93—7·09	10·22
Hydrated dioxides	—	—	—	—	—
Monosulphides...	1·66	—	—	5·1‖	—
Disulphides......	1·29 (0°)	—	—	—	8·29

Carbonates.	Li.	Na.	K.	Ca.	Sr.	Ba.	Mg.	Zn.	Cd.
	1·79	2·4	2·1	2·9	3·6	4·3	3·0	4·4	4·3
	Tl.	Pb.	Mn.	Fe.	Ag.				
	7·2	6·5	3·6	3·8	6·0				

Heats of formation :—

$\mathbf{C} + O = CO + 290\mathrm{K}$; $CO + O = CO_2 + 680\mathrm{K}$; $\mathbf{C} + 2O = CO_2 + 970\mathrm{K}$.

$CO + Cl_2 = COCl_2 + 261\mathrm{K}$; $\mathbf{C} + O + \mathbf{S} = COS + 370\mathrm{K}$; $\mathbf{C} + 2\mathbf{S} = CS_2 - 260\mathrm{K}$.

$2\mathrm{NaOH.Aq} + CO_2 = \mathrm{Na_2CO_3.Aq} + 261\mathrm{K}$; $2\mathrm{KOH.Aq} + CO_2 = \mathrm{K_2CO_3.Aq} + 261\mathrm{K}$.

$\mathbf{CaO} + CO_2 = \mathbf{CaCO_3} + 308\mathrm{K}$ (?); $\mathbf{SrO} + CO_2 = \mathbf{SrCO_3} + 958\mathrm{K}$; $\mathbf{BaO} + CO_2 = \mathbf{BaCO_3} + 622\mathrm{K}$.

$\mathbf{Ag_2O} + CO_2 = \mathbf{Ag_2CO_3} + 200\mathrm{K}$; $\mathbf{PbO} + CO_2 = \mathbf{PbCO_3} - 744\mathrm{K}$.

---

* *Berichte*, **15**, 2599.

† *Comptes rend.*, **100**, 605.

‡ Solid.

§ Artificial; *Rutile*, 4·42; *Brookite*, 3·89–4·22; *Anatase*, 3·75–4·06.

‖ Ce_2S_3.

CHAPTER XXI.

OXIDES, SULPHIDES, SELENIDES, AND TELLURIDES OF MEMBERS OF THE SILICON GROUP.—SILICATES, STANNATES, AND PLUMBATES.—OXYHALIDES.

Oxides, Sulphides, Selenides, and Tellurides of Silicon, Germanium, Tin, and Lead.

The formulæ of the compounds of this group of elements resemble those of carbon and titanium. There are monoxides, sesquioxides, dioxides, and intermediate combinations, but no peroxides have been prepared. But a noticeable difference is that the monoxides, except that of silicon, form compounds; the compounds of the dioxides are very numerous; and we again meet with resemblances between the first number of the previous group with that of this group; *i.e.*, between compounds of carbon and silicon, as we do between those of beryllium and magnesium, and of boron and aluminium.

I. **Monoxides, monosulphides, selenides, and tellurides.**

	Silicon.	Germanium.	Tin.	Lead.
Oxygen	**SiO** (?).	**GeO.***	**SnO.**	**PbO.**
Sulphur	**SiS.**	**GeS.**	**SnS.**	**PbS.**
Selenium	?	?	**SnSe.**	**PbSe.**
Tellurium	—	—	—	**PbTe.**

Sources.—Lead monoxide occurs as *lead ochre*, a yellow earthy mineral found sparingly among lead ores. The sulphide is the chief ore of lead. It is named *galena*. It occurs in crystals derived from the cubical system, usually rhombic dodecahedra. It has a very distinct cubical cleavage, and forms leaden-coloured masses with brilliant metallic lustre. It is found in the Isle of Man, at the lead hills in Lanarkshire, in Cornwall, in the mountain limestone of Derbyshire, and in the lower silurian strata of Cardiganshire and Montgomeryshire. It is also found in combina-

* *J. prakt. Chem.* (2), **34**, 177; **36**, 177; *Chem. Centralbl.*, 1887, 329.

tion with the sulphides of arsenic, antimony, and copper. Lead selenide occurs as *clausthalite*, and the telluride as *altaite*.

Preparation.—1. Direct union.—The only monoxide obtainable thus is that of lead. It is prepared as *massicot* by heating lead in a reverberatory furnace to dull redness, taking care that the resulting oxide shall not fuse, and raking it away as it is formed. If the oxide fuses, it forms *litharge*. The monosulphides of tin and lead are also produced directly, by melting the metal and adding sulphur. In the case of lead, the mixture becomes incandescent owing to the heat liberated during combination. Lead selenide is similarly prepared.

2. **By heating a compound.**—Germanous, stannous, and lead hydrates, heated in a current of carbon dioxide, lose water, leaving the monoxides. If heated in hydrogen, the temperature must not exceed 80°, else reduction to metal takes place. The dehydration of stannous hydrate takes place on boiling water in which it is suspended, the condition being the absence of ammonia. Lead hydrate suspended in water loses water on exposure to sunlight, forming crystalline monoxide. Tin oxalate and lead oxalate, carbonate, or nitrate, when heated, yield monoxides.

3. **By reducing a higher compound.**—Silicon monoxide is said to have been formed as one of the products of heating silica in the Cowles' electric furnace, which is lined with carbon. No doubt it would be possible to prepare germanium and tin monoxides from the dioxides by careful heating in hydrogen gas; but the reduction is apt to go too far, and to produce metal. Lead dioxide and its compounds, when strongly heated, yield monoxides.

Silicon disulphide, when heated to whiteness, loses sulphur, and yields monosulphide; and germanium disulphide is reduced to monosulphide by heating in hydrogen.

Stannic sulphide, SnS_2, loses sulphur at a red heat, forming monosulphide; also the sesquisulphide, Sn_2S_3.

4. **By double decomposition.**—Tin and lead monoxides are produced by heating their corresponding chlorides, $SnCl_2$ or $PbCl_2$, with sodium carbonate. It may be supposed that the carbonates first formed are decomposed, leaving the monoxides.

The sulphides are produced by heating the oxides in vapour of carbon disulphide, or in the case of germanium, tin, and lead, by treating a solution of a salt of the metal or of the hydroxide in potassium hydroxide with hydrogen sulphide or some other soluble sulphide.

Stannous selenide is best prepared by the action of selenium on hot stannous chloride.

Properties.—**Silicon monoxide** is said to be an amorphous greenish powder; those of **germanium** and of **tin** blackish powders. Tin monoxide may be obtained crystalline by heating a mixture of the hydroxide and acetate to 133°; and of a vermilion colour by evaporating a solution of ammonium chloride in which the hydrate is suspended. **Lead monoxide**, in the form of *massicot*, is lemon-yellow; it may be prepared pure by strongly heating lead carbonate or nitrate; and in the form of *litharge* as a yellowish-red laminated mass of crystals. A red variety is produced by heating the hydroxide to 110°. It, too, can be obtained crystalline by fusion with caustic potash; it separates out in cubes on slow cooling; if it is allowed to deposit from an aqueous solution of potassium hydroxide it separates first in yellow laminæ, and afterwards in red scales.

Of these oxides, lead oxide is the only one soluble in water; it requires 7000 times its weight of water for solution.

The monoxides of silicon, germanium, and tin appear to have very high melting points; lead oxide melts at a red heat.

Silicon monosulphide is a volatile yellow body; that of **germanium**, when obtained by precipitation, forms a reddish-brown amorphous powder; but when prepared in the dry way it consists of thin plates, transparent and transmitting red light; but grey, opaque, and exhibiting metallic lustre by reflected light. Its vapour-density is normal, corresponding to the formula **GeS**. It volatilises easily.

Tin monosulphide is a leaden-grey crystalline substance, exhibiting metallic lustre. It has also been prepared by electrolysis of a solution in alkaline sulphide, and then forms metallic-looking cubes. The precipitated variety is brown and amorphous, and is sparingly soluble in alkaline sulphides. It dissolves in and crystallises from fused stannous chloride, $SnCl_2$. The **selenide** forms steel-grey prisms.

The appearance of **lead sulphide** as galena has been already described. When heated it melts, and volatilises at a high temperature. Prepared by precipitation, it is a black amorphous powder if the solution be cold; and if warm, greyish and crystalline.

Lead sulphide and oxide react together when heated, yielding metallic lead and sulphur dioxide, thus—

$$\mathbf{PbS} + \mathbf{2PbO} = 3\dot{P}b + SO_2.$$

This reaction is made use of in the extraction of lead from its ores. The sulphide when roasted is converted partially into the oxide; and on raising the temperature, metallic lead is produced.

The **selenide** is a grey porous mass when artificially prepared; native as *clausthallite* it forms leaden grey crystals with metallic lustre.

Compounds of the monoxides, &c. (*a.*) **With water.**—Silicon monoxide has not been obtained in combination with water. The hydrate of germanium monoxide has not been analysed; it is a white precipitate produced on boiling germanium dichloride with caustic potash. That of tin monoxide is produced by adding sodium carbonate to a solution of tin dichloride; this precipitate is also said to consist of a basic carbonate of the formula $\mathbf{CO_2.2SnO}$ (see p. 289).

Hydrate of lead monoxide, prepared by precipitation and dried in air, has the formula $\mathbf{2PbO.H_2O}$; and after standing for some weeks over sulphuric acid, so as further to dry it, its formula is $\mathbf{3PbO.H_2O}$. The first of these hydrates forms microscopic crystals, and the second, lustrous octahedra.

A mixture of lead hydrate and basic carbonate is produced on exposing metallic lead to the action of water and air. Water alone has no effect on lead, nor has oxygen; but together they attack it, and as the metal lead is commonly used for water-pipes, the slight solubility of the oxide is apt to cause it to contaminate the water. It is found that the presence of sulphates, carbonates, and chlorides stops this action.

(*b.*) **Compounds with oxides.**—No compounds have been prepared with silicon or germanium monoxides; but hydrated tin and lead monoxides are soluble in sodium, potassium, calcium and barium hydroxides, no doubt forming compounds. The calcium compound is said to form sparingly soluble needles. A yellow precipitate of the formula $\mathbf{2PbO.Ag_2O}$ is produced by adding caustic potash to a mixture of two soluble lead and silver salts; it is probably analogous to the compounds with the former oxides. On boiling a solution of stannous hydrate in caustic potash, metallic tin is deposited, and a stannate (see below) is formed.

(*c.*) **Compounds of sulphides with sulphides.**—Monosulphides of silicon, germanium, tin, and lead are insoluble in solutions of monosulphides of the alkalies, and no compounds are known. Compounds of lead sulphide with those of arsenic and antimony occur in nature, and will be described later on.

(*d.*) **Compounds with halides.**—No compounds of silicon or germanium monoxides with halides are known; but stannous chloride, $SnCl_2$, if dissolved in much water, deposits a white powder of the formula $\mathbf{SnO.SnCl_2.2H_2O}$, according to the equation—

$$2SnCl_2.Aq + 3H_2O = \mathbf{SnO.SnCl_2.2H_2O} + 2HCl.Aq.$$

The same compound is produced by the action of atmospheric oxygen on a solution of stannous chloride—

$$3SnCl_2.Aq + O = SnCl_4.Aq + \mathbf{SnO.SnCl_2.2H_2O}.$$

The decomposition may be prevented by addition of a soluble chloride, such as hydrogen or ammonium chloride, which forms a double salt with stannous chloride not decomposed by air, and not altered by water (see p. 154).

The **oxyhalides of lead** are pretty numerous. A fusible oxyfluoride is produced by heating together fluoride and oxide. Five oxychlorides are known, viz.:—

$\mathbf{PbO.3PbCl_2}$, a laminar pearl-grey substance, obtained by fusion of oxide with chloride, and treatment with water.

$\mathbf{PbO.PbCl_2}$, found native as *matlockite* in yellowish translucent crystals; and produced by fusing together lead chloride and carbonate, or by heating lead chloride in air. It is manufactured as a pigment by adding to a hot solution of lead chloride, lime water in quantity sufficient to remove half the chlorine as calcium chloride. It has a white colour, and good covering power.

$\mathbf{2PbO.PbCl_2}$, a mineral known as *mendipite*, forming white, translucent crystals.

$\mathbf{3PbO.PbCl_2}$, prepared by fusion; or by adding a solution of sodium chloride to lead oxide. It is a yellow substance, and is used as a pigment under the name of *Turner's yellow*.

$\mathbf{5PbO.PbCl_2}$, produced by fusion, is a deep yellow powder, and $\mathbf{7PbO.PbCl_2}$, prepared by heating together litharge and ammonium chloride, forms cubical crystals. It is a fine yellow pigment, and is known as *Cassel yellow*.

Two oxybromides have also been produced, by decomposition of the double bromide of lead and ammonium (see p. 154) with water, viz., $\mathbf{PbO.PbBr_2.H_2O}$, and $\mathbf{2(PbO.PbBr_2)3H_2O}$. The same compounds are produced by the action of atmospheric oxygen on fused lead bromide, $PbBr_2$, but anhydrous. Oxyiodides of the formulæ $\mathbf{PbO.PbI_2}$, $\mathbf{2PbO.PbI_2.H_2O}$, $\mathbf{3PbO.PbI_2.2H_2O}$, and $\mathbf{5PbO.PbI_2}$, are produced by similar reactions. The first of these, when prepared by the action of hydrated lead peroxide on potassium iodide in contact with air, combines with the potassium carbonate produced by the action of the carbon dioxide of the air on the resulting potassium hydroxide, giving compounds of the formulæ $\mathbf{PbO.PbI_2.K_2CO_3 2H_2O}$, $\mathbf{2(PbO.PbI_2)3K_2CO_3.2H_2O}$, and $\mathbf{PbI_2.2KI.K_2CO_3.3H_2O}$; and by mixing together potassium iodide and lead carbonate, the compound $\mathbf{PbO.PbI_2.CO_2}$ is produced.

It appears possible also to obtain mixed halides; one of these produced by the action of lead oxide on zinc chloride has the formula $PbO.ZnCl_2$.

II. **Sesquioxides and sesquisulphides.**—Of these compounds, hydrated sesquioxides of silicon and tin, sesquioxide of lead, and tin sesquisulphide are the only representatives. Their formulæ are $Si_2O_3.H_2O$, $Sn_2O_3nH_2O$, Pb_2O_3, and Sn_2S_3. The first, $Si_2O_3.H_2O$, from its analogy to the corresponding carbon compound, oxalic acid, is sometimes named silico-oxalic acid. The constitution of oxalic acid has been noticed on p. 273, and it is probable that the analogous silicon compound is similarly constituted. It is produced by the action of ice-cold water on silicon hexachloride; and its formation may be represented graphically thus:—

$$Si\begin{matrix}Cl\\Cl\\Cl\end{matrix} + O\begin{matrix}H\\H\end{matrix} \quad \begin{matrix}H.OH\\H.OH\end{matrix} \quad Si\begin{matrix}O\\OH\end{matrix} \quad \begin{matrix}2HCl\\HCl\end{matrix}$$
$$Si\begin{matrix}Cl\\Cl\\Cl\end{matrix} + O\begin{matrix}H\\H\end{matrix} = Si\begin{matrix}OH\\O\end{matrix} + \begin{matrix}HCl\\2HCl\end{matrix}$$

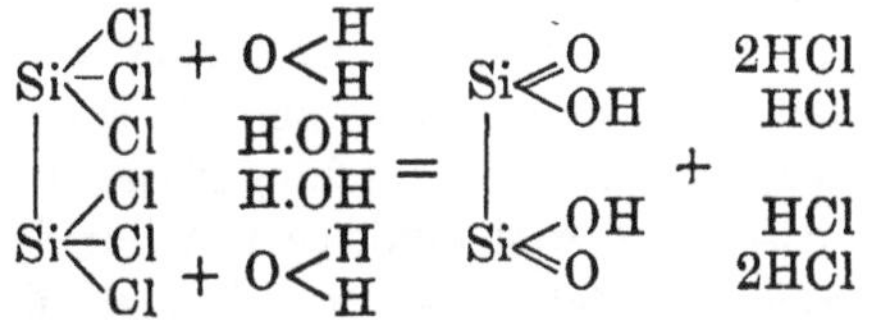

four atoms of chlorine being replaced by two atoms of oxygen, and two by hydroxyl $(OH)'$. It is a white mass; but unlike oxalic acid the remaining hydrogen of the hydroxyl cannot be replaced by metals. It is, therefore, said to be "devoid of acid properties." When treated with any soluble hydroxide, it gives a silicate with evolution of hydrogen. The compound is, however, of considerable interest, inasmuch as it displays the analogy between silicon and carbon.

Hydrated sesquioxide of tin is said to be produced by boiling together hydrated ferric sesquioxide, $Fe_2O_3.nH_2O$, and stannous chloride, $SnCl_2$. It is a slimy grey precipitate.

Lead sesquioxide, Pb_2O_3, is produced by the action of sodium hypochlorite, $NaOCl.Aq$, on salts of lead, or on a solution of lead hydrate in caustic soda; and also by the action of alkalies on a solution of red lead in acetic acid. The last action will be noticed below, in treating of red lead. The sesquioxide is a reddish-yellow insoluble powder; it dissolves for a moment in hydrochloric acid, but almost at once chlorine is evolved, and the dichloride precipitated. No double compounds of sesquioxides are known.

Tin sesquisulphide is produced by heating three parts of the monosulphide with one part of sulphur to dull redness. It has a greyish-yellow metallic lustre, and at high temperatures decomposes into monosulphide and sulphur.

III. Dioxides, disulphides, diselenides, and ditellurides.

List.	Oxygen.	Sulphur.	Selenium.	Tellurium.
Silicon............	SiO_2	SiS_2	$SiSe_2$?	$SiTe_2$
Germanium	GeO_2	GeS_2	?	?
Tin	SnO_2	SnS_2	$SnSe_2$	?
Lead	PbO_2	—	—	

These are the most stable compounds with silicon, germanium, and tin; lead dioxide, however, easily loses oxygen.

Sources.—Silicon dioxide occurs native in hexagonal prisms, capped by hexagonal pyramids, as *rock-crystal*, *bog-diamond*, or *Irish diamond.* When coloured yellow or orange by sesquioxide of iron it is named *cairngorm;* it also also occurs with an amethyst colour due to manganese sesquioxide; and of a rose-red colour (*rose-quartz*). It is very hard, easily scratching glass. It frequently contains small cavities, filled with liquid carbon dioxide, often containing a minute cubical crystal of sodium chloride. *Quartz* is a name applied to less perfectly crystalline silica, and usually occurs in white translucent masses. When perfectly transparent it is used for the lenses of spectacles, being harder and less easily scratched than glass. It is cut into slices by a copper disc, moistened with emery and oil, then ground and polished. *Flint* and *chert* are forms of silica found embedded in chalk, or older limestones, and are due to the siliceous spicules of sponges, now extinct. It has usually a dull grey-brown colour, owing probably to its containing some free carbon, derived doubtless from the animal matter of the shell-fish, the remains of which constitute the chalk, for it turns white on ignition. *Chalcedony* is a variety of quartz, not displaying definite crystalline structure, but showing a fibro-radial structure, and occurring in kidney-shaped, translucent masses. Varieties of chalcedony are named *agate*, *hornstone*, *onyx*, *carnelian*, *catseye*, *chrysoprase*, &c. *Sandstone* consists mainly of water- or air-rolled grains of quartz, bound together by a little lime.

Silica also occurs in combination with many other oxides, as silicates. With water, it occurs as *opal*, an amorphous translucent substance, which has been deposited in thin layers. This produces in some specimens a brilliant play of colours, owing to the refraction and interference of the light which it reflects. Opal is soluble in a hot solution of potassium hydrate; it is thus distinguished from quartz. The other silicates will be considered later.

Germanium disulphide, in combination with silver sulphide, forms the mineral *argyrodite*, found in the Himmelsfürst mine at Freiberg. It is almost the only mineral in which germanium has been found.

Tin dioxide, named *cassiterite*, or *tinstone*, is the only important source of tin. It occurs in veins, traversing the primitive granite and slate of Cornwall; it is also exported from Melbourne. It forms translucent white, grey, or brownish quadratic crystals. Its crystalline form is the same as that of *anatase*, one of the forms of titanium dioxide.

Stannic sulphide, SnS_2, occurs in combination with sulphides of iron and copper, and is named *tin pyrites*.

Preparation.—1. **By direct union.**—Silicon dioxide, or silica, is formed when silicon burns in air or oxygen. Germanium dioxide and stannic oxides are similarly produced. The oxides thus prepared are amorphous. Lead unites with oxygen to form monoxide, PbO, as already mentioned. The highest stage of oxidation produced directly is that of red lead, $Pb_3O_4 = PbO_2.2PbO$. Stannous oxide also unites directly with oxygen to produce the dioxide.

2. **By decomposing a double compound.**—All these oxides remain on heating to redness their various hydrates; germanium dioxide has also been prepared from its sulphate, $\mathbf{Ge(SO_4)_2}$, which loses sulphur trioxide at a red heat.

3. **From lower compounds.**—Lead monoxide heated to dull redness with potassium chlorate is oxidised to the dioxide. The potassium chloride and excess of chlorate are dissolved out by water. It is also formed by fusing lead monoxide with potassium hydroxide. Hydrogen is evolved, and potassium plumbate is produced; on treatment with water the dioxide remains in black hexagonal tables.

Tin disulphide and diselenide are prepared by a somewhat curious method. When tin and sulphur are heated together, the sesquisulphide is the highest sulphide formed. But if ammonium or mercuric chloride be heated in a glass retort with the mixture of tin and sulphur the disulphide is produced. It is supposed that a double chloride of tin and ammonium, or of tin and mercury, is first formed, and that this reacts with sulphur, thus:— $2(\mathbf{SnCl_2.2NH_4Cl}) + 2S = \mathbf{SnS_2} + \mathbf{SnCl_4.2NH_4Cl} + 2\mathbf{NH_4Cl}$. Diselenide of tin is produced by the action of iodine on the monoselenide, in presence of carbon disulphide, thus:—$2\mathbf{SnSe} + 2I_2 = SnI_4 + \mathbf{SnSe_2}$; at the same time some selenium is liberated, according to the equation, $\mathbf{SnSe} + 2I_2 = SnI_4 + \mathbf{Se}$. The tin tetriodide dissolves in the carbon disulphide, leaving the diselenide, which is insoluble.

4. **By double decomposition.**—Tin dioxide is produced in a crystalline form by passing the vapours of stannic chloride, $SnCl_4$, and water through a red-hot tube. The crystals produced are of the same form as *brookite* (TiO_2): quadratic crystals are formed by the action of hydrogen chloride on the red-hot dioxide. The disulphides of silicon and germanium are both produced by double decomposition. To prepare the former, silica, or a mixture of carbon and silica, is exposed at a white heat to the action of carbon disulphide; the monosulphide is simultaneously produced, probably owing to the decomposition of the disulphide. The disulphides of germanium and of tin are precipitated from solutions of the dioxides by hydrogen sulphide. Tin disulphide is also produced by passing a mixture of hydrogen sulphide and gaseous tin tetrachloride through a tube heated to dull redness.

Properties.—The properties of native **silica** have been already described. It fuses at a white heat in the oxyhydrogen flame to a glassy mass, which can be drawn into threads. In this form it furnishes one of the best insulators for electricity, and has been used to suspend the needles of galvanometers. Such threads have great tenacity and are very elastic. Even when moist they do not conduct. Amorphous silica, produced by heating the hydrate, is a loose white powder; it is said to volatilise when heated to whiteness in water-vapour, resembling boron oxide in this respect. While the crystalline form is not attacked by *solutions* of potassium or sodium hydroxide, the amorphous variety dissolves slowly. Crystalline silica is attacked only by hydrofluoric acid.

Germanium dioxide is a dense, gritty, white powder, sparingly soluble in water, and crystallising from it in small rhombohedra. Its solubility is:—1 gram of GeO_2 dissolves in 247·1 grams of water at 20°, and in 95·3 grams at 100°.

Tin dioxide is a white or yellowish powder, insoluble in water. It turns dark yellow when heated, but again becomes white on cooling. Under the name "putty powder" it finds commercial use in polishing stone, glass, steel, &c.

Lead dioxide is a soft brown powder, insoluble in water; when heated to redness it loses oxygen, leaving a residue of monoxide.

Silicon disulphide forms long white volatile needles. It is remarkable that the oxide is so non-volatile, while the sulphide can be sublimed; it leads to the supposition that while the sulphide has the formula assigned to it, SiS_2, the formula of the oxide, as we know it, is really a high multiple of SiO_2. And on comparing the silicon and carbon compounds, this conclusion is strengthened. For while the boiling-points of carbon dioxide, disulphide, and tetrachloride are respectively −78·5°, 46°, and 76·7°, an ascending series, we have with silicon, the dioxide melting at a white heat, the sulphide easily volatile, and the chloride boiling at 58°. The order of volatility is reversed. And as it is found with compounds of carbon and hydrogen, that the more complex the molecule, the higher the boiling-point, we may conclude that the non-volatility of silica is due to its molecular complexity. There is at present, however, no means of ascertaining how many molecules of SiO_2 are contained in the complex molecule of ordinary silica, the formula of which must therefore be written $(SiO_2)_n$.

Germanium disulphide is described as a white precipitate, sparingly soluble in 221·9 parts of water, and also soluble in sulphides. It appears not to decompose on solution; but silicon disulphide reacts with water, forming hydrogen sulphide and a hydrate of silica.

Tin disulphide, prepared by precipitation, is a buff-yellow powder, insoluble in water. When obtained by the dry process it forms golden-yellow scales, and is named "mosaic gold." The diselenide is a red-brown crystalline powder.

Double compounds.—It is convenient to group these as follows:—(*a*) Compounds of the oxides with water and oxides; (*b*) oxyhalides and sulphohalides; (*c*) compounds of sulphides with other sulphides; and (*d*) oxysulphides.

(*a*.) **Compounds with water and oxides.**—The most important of these compounds are the silicic acids and the silicates; allied to them are the stannates and plumbates, and there appears to be indications of the existence of germanates.

General Remarks on the Silicates.—The ratios between the oxygen of the silica and the oxygen of the metallic oxides combined with it are very numerous. The silicates form a very large portion of the crust of the earth, and they have very varied composition. Among the native silicates the ratio of oxygen in silica to that in oxide of metal may vary for monad and dyad metals, such as potassium or calcium, between 2 : 4 and 4 : 1; or to take hypothetical specific instances,—between $SiO_2.4K_2O$, or $SiO_2.4CaO$ and $2SiO_2.K_2O$, or $2SiO_2.CaO$; and for silicates of triad metals, such as aluminium or iron, between 2 : 6, as in $SiO_2.2Al_2O_3$, and 12 to 3, as in $6SiO_2.Al_2O_3$. It must be remembered that the native silicates have almost always been formed in a matrix containing compounds of many elements; hence it is rare to find among the silicates pure compounds such as those of which the formulæ have been given above. For instance, it is not unusual to find a silicate containing both the metals, potassium and calcium, as oxides combined with silica; or the oxides of the metals, iron and aluminium, or of calcium and aluminium, and that not in atomic proportion; for we may have a silicate of aluminium containing only a trace of iron, or a silicate of calcium containing only a trace of magnesium or ferrous iron, the crystalline form of which does not differ from that of the pure silicate. It is not to be conceived that in such instances any given molecule has not, as is usual among compounds, a perfectly definite formula; but it would appear that it is possible for an apparently homogeneous crystal to be made up of molecules of silicate of aluminium and silicate of iron, or of silicate of magnesium and silicate of calcium in juxtaposition; so that, to take a suppositious case, a crystal containing 1000 molecules might consist of 999 molecules of magnesium silicate and one molecule of calcium silicate, or of 998 molecules of magnesium silicate and two molecules of calcium silicate, and so on; oxides of magnesium and calcium being mutually replaceable in any proportion whatever. And similarly with the compounds of silica with the sesquioxides of iron and aluminium. But all oxides are not capable of mutually replacing each other. While beryllium, calcium, magnesium, iron, manganese, nickel, cobalt, sodium, and potassium monoxides mutually replace one another, and while the sesquioxides of aluminium, iron, manganese, chromium, &c., are also mutually replaceable, it is found that the place of a monoxide is not taken by a sesquioxide, and *vice versâ*. But a silicate may contain at once a mixture of sesquioxides and a mixture of monoxides in combination with silica.

To deduce the formula of a natural silicate from its percentage composition is a problem of some difficulty. It is solved by ascertaining the ratio of all the oxygen combined with dyad metals, whatever they may be, to that combined with triad metals, and to that contained in the silica. An example will render this clear.

On analysis, a specimen of the felspar named *orthoclase* (which is essentially a silicate of aluminium and potassium, but which may contain iron sesquioxide replacing alumina, and sodium, magnesium, and calcium oxides replacing potassium oxide) gave the following numbers:—

SiO_2.	Al_2O_3.	CaO.	K_2O.	Na_2O.	
65·69	17·97	1·34	13·99	1·01	= 100·00 per cent.

These constituents contain oxygen in the following proportions:—

$$\frac{32}{60{\cdot}33} \quad \frac{48}{102{\cdot}02} \quad \frac{16}{56{\cdot}08} \quad \frac{16}{94{\cdot}28} \quad \frac{16}{62{\cdot}09},$$

the denominators being the molecular weights of the oxides, and the numerators the oxygen contained in these weights. The ratios are, therefore, as follows:—

SiO_2.	Al_2O_3.	CaO.	K_2O.	Na_2O.
$\frac{65{\cdot}69 \times 32}{60{\cdot}33}$	$\frac{17{\cdot}97 \times 48}{102{\cdot}02}$	$\frac{1{\cdot}34 \times 16}{56{\cdot}08}$	$\frac{13{\cdot}99 \times 16}{94{\cdot}28}$	$\frac{1{\cdot}01 \times 16}{62{\cdot}09}$,

or $34{\cdot}84 + 8{\cdot}45 + 0{\cdot}38 + 2{\cdot}37 + 0{\cdot}26 = 46{\cdot}30$ per cent. of oxygen.

We have, therefore, the ratio:—

	Oxygen in silica.		Oxygen in alumina.		Oxygen in lime, potash, and soda.
	34·84	:	8·45		0·38 + 2·37 + 0·26 = 3·01*
or	12	.	3	:	1, nearly.

Hence the formula is $6SiO_2.Al_2O_3.M_2O$, where M stands for calcium, potassium, or sodium. It is usually written thus:— $6SiO_2.Al_2O_3.(Ca, K_2, Na_2)O$, the comma between those symbols enclosed within brackets signifying that they are mutually replaceable in any proportions. Had iron sesquioxide been present, the oxygen contained in it would have been added to that of the alumina, and the formula would then have been written,

$$6SiO_2(Al,Fe)_2O_3(Ca, K_2, Na_2)O.$$

As with the borates, chromates, and carbonates, there are two methods of representing the formulæ of the silicates. The first method is to consider them as addition compounds cf silica with other oxides, and the formula of orthoclase, as written above, is constructed on that principle. It must, however, clearly be understood that, inasmuch as we know almost nothing regarding the internal constitution of such compounds, we can only guess at their structure from analogy with the hydrocarbons and their derivatives.

The method of writing given above does not imply that the compound contains *as such* the molecular group SiO_2 united with

* The calculated ratio of oxygen in the above compound is—

SiO_2.		Al_2O_3.		M_2O.
34·39	:	8 69	:	2·87.

molecular groups Al_2O_3 and K_2O. It is merely a method of showing the proportions of ingredients which the compound contains in an orderly manner, and is better than if we were to write the formula, $Al_2K_2Si_6O_{16}$.

The second method starts from the fact that in such compounds silicon is a tetrad element; that analogous to its compounds with fluorine or chlorine, SiF_4 or $SiCl_4$, the typical silicic acid has the formula $Si(OH)_4$. This substance is named *orthosilicic acid.* Its salts may be supposed to be formed by replacing the hydrogen of the hydroxyl groups by metals; thus the potassium salt has the formula $Si(OK)_4$, the calcium salt, $SiO_4Ca^{II}_2$, or $Si\begin{cases}O \\ O\end{cases}>Ca \begin{cases}O \\ O\end{cases}>Ca$, and the aluminium salt, $3(SiO_4)^{IV}Al_4^{III}$. These are the same as $SiO_2.2K_2O$, $SiO_2.2CaO$, and $3SiO_2.2Al_2O_3$, and are named orthosilicates.

Silicates of the formula, $SiO_2.K_2O$, $SiO_2.CaO$, &c., are also known, and in them the oxygen of the silica bears the ratio to that of the oxide as 2 : 1. These may be supposed to be derived from the hydroxide $SiO_2.H_2O$, which is named *metasilicic acid*, and which may be regarded as orthosilicic acid deprived of a molecule of water; its constitution may be represented $Si\begin{cases}=O \\ -OH \\ -OH\end{cases}$, and its potassium and calcium salts as $Si\begin{cases}=O \\ -OK \\ -OK\end{cases}$, and $Si\begin{cases}=O \\ -O \\ -O\end{cases}>Ca$.

It will be remembered that an analogy was drawn between chromyl dichloride CrO_2Cl_2, and chromic acid $CrO_2(OH)_2$ (see p. 268), and it was pointed out that the substance $CrO_2<^{Cl}_{OK}$ might be regarded as partaking of the nature both of the dichloride and of potassium chromate, $CrO_2<^{OK}_{OK}$, being, in fact, an intermediate stage. We should expect, therefore, intermediate compounds between silicon tetrachloride, $SiCl_4$, and silicon tetrahydroxide, $Si(OH)_4$. Only one such body is known, viz., $SiCl_3.SH$, in which hydrosulphuryl replaces hydroxyl. But derivatives of the elements of this group are known, which represent similar compounds connected with metasilicic acid, $SiO(OH)_2$. Although the corresponding chloride $SiOCl_2$ is unknown, yet it is represented by $GeOCl_2$; and although $Si\begin{cases}=O \\ -OH \\ -Cl\end{cases}$ is also unknown, it finds a representative in the compound of tin, $Sn{\displaystyle{\nwarrow}\atop\displaystyle{\swarrow}}{O\atop OH}\atop Cl$. This method of repre-

sentation, which may be termed the *method of substitution*, is, there-fore, justified.

But we may go still further   Hitherto we have been dealing with compounds containing only one atom of silicon. It is, how-ever, conceivable that two molecules of orthosilicic acid may form an anhydride, water being lost between them, thus :—

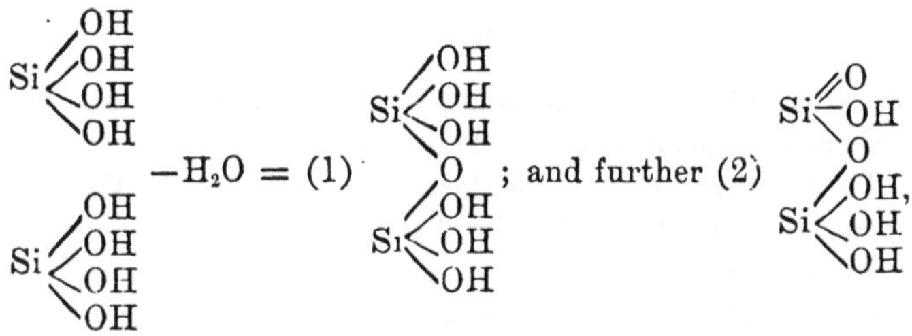

$$
\begin{array}{c}
Si{<}{OH\atop OH\atop OH\atop OH} \\
Si{<}{OH\atop OH\atop OH\atop OH}
\end{array}
\ -H_2O = (1)\
\begin{array}{c}
Si{<}{OH\atop OH\atop OH} \\
O \\
Si{<}{OH\atop OH\atop OH}
\end{array}
\ ;\ \text{and further (2)}\
\begin{array}{c}
Si{<}{O\atop OH\atop O} \\
Si{<}{OH\atop OH\atop OH}
\end{array},
$$

$$
\text{and (3)}\
\begin{array}{c}
Si{<}{O\atop OH} \\
O \\
Si{<}{OH\atop O}
\end{array}.
$$

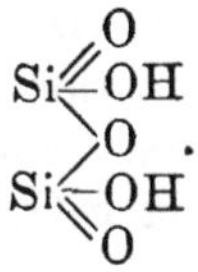

The compound (1) is termed disilicic acid; (2) is the first, and (3) the second anhydride of disilicic acid. A representative of (1) is *serpentine*, $2SiO_2.3MgO$; *wollastonite*, $2SiO_2.2CaO$, may be a representative of (2), although its formula may equally well be the simpler one, $SiO_2.CaO$, or $SiO(O_2)Ca$; and *okenite*, $2SiO_2.CaO$, may represent the calcium salt of (3).

A chlorine-derivative of (1), however, is known, viz., Si_2OCl_6, in which all hydroxyl is replaced by chlorine. That it possesses that simple formula is shown by its vapour-density.

In a similar manner, a trisilicic acid may be derived from three molecules of orthosilicic acid, by loss of two molecules of water; it in its turn will yield three anhydrosilicic acids; and a tetrasilicic acid may be supposed to exist, of which four anhydro-acids are theoretically capable of existence. Of this tetrasilicic acid three chlorine-derivatives have been prepared of the formula, $Si_4O_3Cl_{10}$, $Si_4O_4Cl_8$, and $Si_4O_5Cl_6$, corresponding to the respective acids, $Si_4O_3(OH)_{10}$, $Si_4O_4(OH)_8$, and $Si_4O_5(OH)_6$, as shown by their vapour-densities. The first is tetrasilic acid itself; the second and third its first and second anhydrides respectively. Salts of even more condensed silicic acids may exist.

Many silicates are known, containing more base than that

contained in orthosilicates, in which the ratio is $SiO_2.2M''O$. For example, *collyrite* has the formula $SiO_2.2Al_2O_3$, the ratio of oxygen in the silica to that in the oxide being 2 : 6. Such silicates are termed *basic*. Their formulæ may be written in an analogous manner, on the supposition that the metal exists partly as oxide, partly as silicate. Thus the above compound may be represented thus :—

$$Si\begin{cases} O-Al\equiv O \\ O-Al\equiv O \\ O-Al\equiv O \\ O-Al\equiv O \end{cases};$$

each atom of aluminium being one-third ortho-silicate, and two-thirds oxide. And so with other instances.

These remarks must be held to apply also to the titanates, zirconates, stannates, and plumbates; but similar compounds of tin and lead are not numerous.

One point must still be noticed before proceeding with a description of the silicates, viz., the question as to whether or not water, occurring in combination with a silicate or stannate, should be included in the formula. For example, by including water, a compound of the formula $SiO_2.CaO.H_2O$ may be represented as an orthosilicate, $Si\begin{cases} OH \\ O \\ O \\ OH \end{cases}\!\!\!\!{>}Ca$, or, excluding the water, as a meta-silicate, $Si\begin{cases} {=}O \\ O \\ O \end{cases}\!\!\!\!{>}Ca.H_2O$, the water being regarded as water of crystallisation. There is no rule for guidance in discriminating water of crystallisation from combined water; and indeed we have no reason to regard water of crystallisation as combined in any other fashion than other oxides. At present, however, no satisfactory theory has been devised whereby water of crystallisation can be rendered a part of the formula, like the molecule of water in the first example given above; and in the present state of our knowledge the only course is to exercise discretion as regards its inclusion or exclusion.

Silicates, Stannates, and Plumbates.

$SiO_2.2H_2O$ (?) = $Si(OH)_4$; $SiO_2.H_2O$ = $SiO(OH)_2$.— $GeO_2.nH_2O$.
$SnO_2.4H_2O$ (?) = $Sn(OH)_4$; $SnO_2.H_2O$ (?) = $SnO(OH)_2$; $3SnO_2.H_2O$; $7SnO_2.2H_2O$; $5SnO_2.5H_2O$.
$PbO_2.H_2O$; $3PbO_2.H_2O$.

These compounds are very indefinite. On addition of dilute hydrochloric acid to a dilute solution of sodium or potassium silicate, no precipitate is produced. Placing this solution in a dialyser—a circular frame, like a tambourine, covered with parchment or parchment paper or bladder, and floated on water—the crystalline sodium chloride passes through the diaphragm, while the colloid (glue-like) non-crystalline silicic acid remains behind for the most part. It was suggested by Graham, the discoverer of this method of separation, that the molecules of the colloid body are much more complicated and larger than those of the crystalline substance, and hence pass much more slowly through the very minute pores of the dialyser. To such passage Graham gave the name *osmosis*, and the general phenomenon is termed *diffusion*. Recent researches appear to confirm this view, and to show that the molecules of colloid bodies are very complex. It is supposed that the silicic hydrate thus remaining soluble is orthosilicic acid, $Si(OH)_4$, inasmuch as it is produced from an orthosilicate. To obtain it pure, the water outside the dialyser must be frequently renewed. A solution of silicic acid containing 5 per cent. of SiO_2 may thus be prepared; and by placing it in a dry atmosphere over sulphuric acid, it is slowly concentrated until it reaches a strength of 14 per cent.

It forms a limpid colourless liquid, with a feeble acid reaction. When warmed, it gelatinises; this change is retarded by the presence of a small amount of hydrochloric acid, or of caustic soda or potash; but is furthered by the presence of a carbonate.

Similar results were obtained from a stannate mixed with dilute hydrochloric acid, and also from a titanate. The solutions have similar properties.

It is supposed that the gelatinous substance produced from orthosilicic acid is *metasilicic acid*, $SiO(OH)_2$. When dried for several months over strong sulphuric acid, it corresponds with that formula. This hydrate is also supposed to be produced when a halide of silicon reacts with water. A convenient method of preparation consists in leading silicon fluoride into water (see p. 153). It is said to have been obtained in crystals of the formula $\mathbf{SiO(OH)_2.3H_2O}$, by the action of hydrochloric acid on a siliceous limestone.

On drying precipitated silica for five months over sulphuric acid, it had the approximate formula $\mathbf{3SiO_2.4H_2O}$. When the temperature was raised, it lost water gradually, but no evidence of any definite hydrate was obtained; no point could be found at which a small rise of temperature did not produce a further loss of water. The same remarks apply to stannic hydrate. But about 360°, the substance, which had the composition $\mathbf{3SnO_2.H_2O}$, and a dirty brown colour, displayed a sudden change of colour to pale yellow, and had then the formula $\mathbf{7SnO_2.2H_2O}$.

When metallic tin is treated with strong nitric acid, it is oxidised; copious red fumes are evolved, and a white powder is produced. While ordinary hydrate, prepared by precipitation, is soluble in acids, this white substance is not; and after drying in a vacuum or at 100° it has the formula $\mathbf{5SnO_2.5H_2O}$ (see below).

Hydrated lead peroxide, dried in air, has the formula

$$\mathbf{3PbO_2.H_2O}.$$

On further heating, water and then oxygen are lost.

By passing a current of electricity between two lead plates, dipping in dilute sulphuric acid, hydrated peroxide of lead is formed on the positive, while hydrogen is evolved at the negative plate. This hydrate has the formula $\mathbf{PbO_2.H_2O}$. Such an arrangement gives a very powerful current, lead peroxide being very strongly electro-positive; and it is made use of for "storage batteries."

$\mathbf{SiO_2.2Li_2O}$; $\mathbf{SiO_2.Li_2O}$; $\mathbf{5SiO_2.Li_2O}$.— $\mathbf{SiO_2.2Na_2O}$(?) ; $\mathbf{SiO_2.Na_2O.8H_2O}$; $\mathbf{5SiO_2.2Na_2O}$; $\mathbf{4SiO_2\,Na_2O}$.— $\mathbf{SiO_2.2K_2O}$(?) ; $\mathbf{SiO_2.K_2O}$; $\mathbf{9SiO_2.2K_2O}$, or perhaps $\mathbf{4SiO_2.K_2O}$.

$\mathbf{SnO_2.Na_2O.3}$, 8, 9, and $\mathbf{10H_2O}$: $\mathbf{SnO_2.K_2O.3H_2O}$; $\mathbf{2SnO_2.(NH_4)_2O.}n\mathbf{H_2O}$. $\mathbf{5SnO_2.Na_2O.4H_2O}$; $\mathbf{5SnO_2.K_2O.4H_2O}$; $\mathbf{7SnO_2.K_2O.3H_2O}$.

$\mathbf{PbO_2.K_2O.3H_2O}$, and others.

When silica is fused with a carbonate or hydroxide of lithium, sodium, or potassium, a glass is formed of indefinite composition, depending on the proportions taken. The lithium glass, however, dissolves in fused lithium chloride, and crystallises out on cooling. The lithium chloride withdraws lithia from the silicate, forming oxychloride; and by keeping the mass fused for different times, the three compounds given above are formed.

Soluble glass, or *water-glass*, is a silicate of sodium or potassium. It is prepared as mentioned; or by heating silica (quartz, flint, sand, &c.) with solution of caustic soda or potash under pressure. The proportion of silica and potash usually corresponds with the formula $4SiO_2.K_2O$; on treating the solution with alcohol, a substance of that formula is thrown down; it is suggested that the more probable formula is $9SiO_2.2K_2O$. It is probably a mixture. If the sodium silicate be saturated with silica, $4SiO_2.Na_2O$, is produced.

Soluble glass is a syrupy liquid, obtained by dissolving the product of fusion in water, or by evaporating the solution of silica in alkaline hydroxide. It is used to form artificial stone; for it reacts with calcium hydrate or carbonate, giving insoluble calcium silicate, which may be used to bind together large amounts of sand into a coherent mass. It is also employed in mural painting; and it is added to cheap soaps.

Hydrated silica dissolves to some extent in solution of ammonia, but no solid compound has been obtained.

Decomposition of silicates.—The usual method of decomposing insoluble silicates is by fusing them with a mixture of sodium and potassium carbonates, named "fusion-mixture." Carbonates or oxides of the metals remain, and the silica combines with the sodium and potassium oxides, forming a mixture of silicates. This mixture is now treated with water, when the silicates of the alkalies pass into solution, and may be removed from the insoluble oxides by filtration. But as it is usually required to separate the silica, it is more common to add hydrochloric acid, which, if the solution be strong, precipitates gelatinous silicic acid, and converts the oxides of the metals into chlorides. On evaporation to dryness and heating, the silicic acid is decomposed into water and silica, and after re-evaporation with a little hydrochloric acid, it is insoluble in dilute hydrochloric acid, which dissolves the chlorides of the metals, and they may then be separated by filtration. On ignition, the silica remains as a white loose powder, and if required it may be weighed.

The corresponding **stannates** are prepared by fusing tin dioxide with hydroxide, sulphide, nitrate, or chloride of sodium or potassium; or by heating metallic tin with a mixture of hydroxide and nitrate, from the latter of which it derives its oxygen. On treatment with water the mass dissolves, and on evaporation deposits crystals containing 3, 8, 9, or 10 molecules of water, according to circumstances. The salt with $3H_2O$ may also be precipitated by adding alcohol.

Stannic hydrate is soluble in ammonia, forming a jelly, in which the ratio of SnO_2 to ammonia corresponds with the formula $2SnO_2.(NH_4)_2O$.

Metastannates.—By boiling the product of the action of nitric acid on tin, $5SnO_2.5H_2O$, with sodium or potassium hydroxide, a solution is obtained, from which, if caustic soda be used, granular white crystals deposit on cooling, of the formula

$$5SnO_2.Na_2O.4H_2O.$$

If potash be used, a similar compound, $5SnO_2.K_2O.4H_2O$, is precipitated by addition of excess of potash, in which it is insoluble. It is a gummy non-crystalline substance. Both of these compounds are decomposed by boiling water into alkali and metastannic acid. It is the fact that one-fifth of the water is replaced by sodium or potassium oxide, which leads to the formula $5SnO_2.5H_2O$ for metastannic acid, instead of $SnO_2.H_2O$, which would more simply represent its percentage composition.

On mixing metastannic acid, dissolved in hydrochloric acid, with caustic potash, until the precipitate at first produced redissolves, and then adding alcohol, a precipitate of the formula

$$7SnO_2.K_2O.3H_2O$$

is produced. There appear also to be other analogous substances.

Plumbates.—By fusing lead dioxide with excess of caustic potash, it dissolves; the solution of the product, in a little water, deposits octahedral crystals of the formula $PbO_2.K_2O.3H_2O$ analogous to the stannate. By fusing litharge with potassium hydroxide, the compounds $PbO_2.K_2O$ and $3PbO_2.2K_2O$, are formed with absorption of oxygen. These salts are decomposed, on treatment with water, into potassium hydroxide and hydrated lead dioxide; they are stable only in presence of excess of alkali.

$SiO_2.2BeO$ (*phenacite, beryl, emerald*); $SiO_2.CaO$ (*wollastonite*); $2SiO_2.CaO.2H_2O$ (*okenite*); $3SiO_2.2CaO.H_2O$ (*gyrolite*); $SnO_2.CaO$, also $5H_2O$; $2SnO_2.SrO.10H_2O$; $SnO_2.2BaO.10H_2O$.*

These silicates are found native; they are well crystallised minerals. By adding to solutions of calcium, strontium, or barium chlorides a solution of sodium or potassium silicate, white curdy insoluble precipitates are produced of the respective silicates, the composition of which is analogous to that of the alkaline silicate from which they are produced.

Of the native silicates, *phenacite* is an orthosilicate; *wollastonite* probably a metasilicate; *okenite* a salt of disilicic acid, $Si_2O(OH)_6$; and *gyrolite*, of the second anhydride of trisilicic acid, $Si_3O_4(OH)_4$. And with the stannates, we have barium orthostannate; calcium metastannate (rejecting the water); and the strontium salt of distannic acid.

* *Comptes rend.*, 96, 701.

Two compounds are known, the first occurring native, a titanate and silicate of calcium, named *sphene;* and the second, of similar crystalline form (monoclinic prisms) produced by heating a mixture of silica and tin dioxide with excess of calcium chloride to bright redness for eight hours. These bodies are derived from a compound analogous to the second anhydride of disilicic acid. Their formulæ are probably—

$$\mathrm{Ca}<^{O}_{O}>\mathrm{Si}<^{O}_{O}>\mathrm{Ti}<^{O}_{O}>\mathrm{Ca}, \text{ and } \mathrm{Ca}<^{O}_{O}>\mathrm{Si}<^{O}_{O}>\mathrm{Sn}<^{O}_{O}>\mathrm{Ca},$$

Similar silico-zirconates occur native.

Ordinary Mortar is made by mixing sand with slaked lime. The rapid setting of the mortar is, however, not due to the combination of the calcium and silicon oxides, but to the formation of the compound $CO_2.2CaO$, by absorption of carbon dioxide from the air. But, after the lapse of years, combination of the silica does take place, and very old mortars contain much calcium silicate.

Hydraulic mortars, as those mortars are named which "set" under water, on the other hand, cannot be produced from anhydrous silica. A mixture of precipitate silica or of crushed opal and lime hardens under water; but the best hydraulic mortars are made from hydrated silicates of alumina. The celebrated *pozzolana* of Naples is such a substance. When mixed with lime, there is formed a silicate of aluminium and calcium, which is rapidly produced, and perfectly insoluble in water. Tufa, pumice, and clay-slate form similar insoluble mortars. Marl, a mixture of clay and calcium carbonate, after ignition, sets when moistened. This is probably in the first instance due to hydration, and subsequently to the formation of a silicate of aluminium and calcium.

$SiO_2.2(Mg, Fe)O$ (*chrysolite, olivine*); $SiO_2.(Mg, Fe'', Mn'', Ca)O$ (*augite* and *hornblende;* these differ in crystalline form, but are both monoclinic); $2SiO_2.3(Mg, Fe'')O.2H_2O$ (*serpentine,* sometimes containing Na_2, Mn'', and Ni''); $3SiO_2.2MgO.2H_2O$, or $4H_2O$ (*meerschaum*); $5SiO_2.4MgO$ (*talc;* contains a little water.—$SiO_2.2ZnO.H_2O$ (*siliceous calamine*) : $SiO_2.2ZnO$ (*willemite*).—$2SnO_2.3ZnO.10H_2O$.

These silicates are all found native and, as a rule, crystalline. *Chrysolite* and *willemite* are orthosilicates; *siliceous calamine* possibly a basic metasilicate of the formula $SiO.(OZnOH)_2$; *augite* and *hornblende* are metasilicates, but one is probably a polymeride of the other, possibly a derivative of the disilicic acid, $^{HO}_{HO}>\mathrm{Si}<^{O}_{O}>\mathrm{Si}<^{OH}_{OH}$, like *sphene*, with which, however, neither is isomorphous. *Serpentine* is a derivative of disilicic acid, and *meerschaum* and *talc* of tri- and penta-silicic acids respectively.

The silicates of boron, aluminium, ferric iron, &c., are very numerous, and it is here impossible to do more than give a sketch of their nature.

Datolite has the formula $SiO_2.B_2O_3.CaO$; and *botryolite* contains, in addition, two molecules of water. They are doubtless derived silicates of boron and calcium, and may be constituted thus :—

$$B{<}^{O}_{O} \qquad\qquad B{<}^{OH}_{OH}$$
$$B{<}^{O}_{O}{>}Si{<}^{O}_{O}{>}Ca, \text{ and } \qquad B{<}^{O}_{O}{>}Si{<}^{O-CaOH}_{OH}.$$

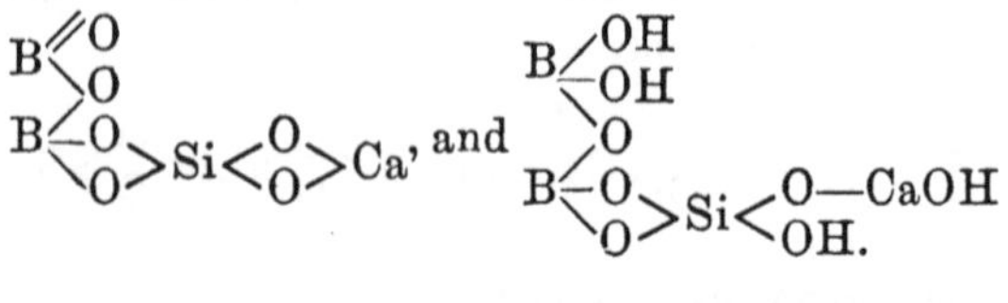

Xenolite is aluminium orthosilicate, $3SiO_2.2Al_2O_3$. A number of minerals, including *fibrolite, topaz, muscovite, paragonite* and *eucryptite* (varieties of mica), *dumortierite, grossularite* (a lime alumina garnet), *prehnite,* and *natrolite* (or *soda mesotype*), may be simply derived from it; the following structural formulæ show their relations (Clarke) :—

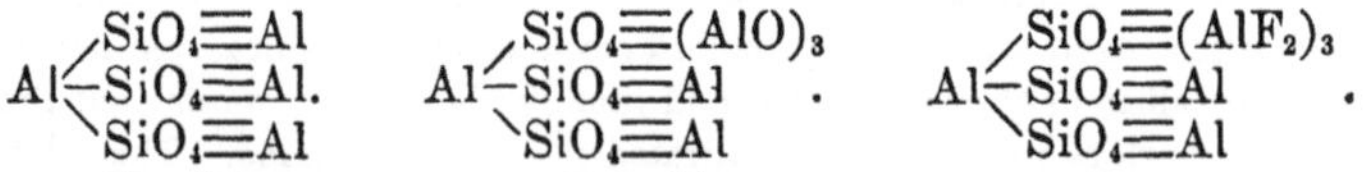

$$Al{<}\begin{matrix}SiO_4{\equiv}Al\\SiO_4{\equiv}Al.\\SiO_4{\equiv}Al\end{matrix}$$
Xenolite.

$$Al{<}\begin{matrix}SiO_4{\equiv}(AlO)_3\\SiO_4{\equiv}Al\\SiO_4{\equiv}Al\end{matrix} .$$
Fibrolite.

$$Al{<}\begin{matrix}SiO_4{\equiv}(AlF_2)_3\\SiO_4{\equiv}Al\\SiO_4{\equiv}Al\end{matrix} .$$
Topaz.

$$Al{<}\begin{matrix}SiO_4{\equiv}(AlO)_3\\SiO_4{\equiv}(AlO)_3.\\SiO_4{\equiv}Al\end{matrix}$$
Dumortierite.

$$Al{<}\begin{matrix}SiO_4{\equiv}KH_2\\SiO_4{\equiv}Al\\SiO_4{\equiv}Al\end{matrix} .$$
Muscovite.

$$Al{<}\begin{matrix}SiO_4{\equiv}NaH_2\\SiO_4{\equiv}Al.\\SiO_4{\equiv}Al\end{matrix}$$
Paragonite.

$$Al{<}\begin{matrix}SiO_4{\equiv}Li_3\\SiO_4{\equiv}Al.\\SiO_4{\equiv}Al\end{matrix}$$
Eucryptite.

$$Al{<}\begin{matrix}SiO_4{\equiv}NaH_2\\SiO_4{\equiv}NaH_2.\\SiO_4{\equiv}Al\end{matrix}$$
Natrolite.

$$Al{<}\begin{matrix}SiO_4{\equiv}\\SiO_4{\equiv}\\SiO_4{\equiv}Al\end{matrix}{\Big\}}Ca_3.$$
Grossularite.

$$Al{<}\begin{matrix}SiO_4{\equiv}CaH\\SiO_4{\equiv}CaH.\\SiO_4{\equiv}Al\end{matrix}$$
Prehnite.

$$Al{<}\begin{matrix}OH\\SiO_4{\equiv}H_3.\\SiO_4{\equiv}Al\end{matrix}$$
Kaolin.

$$Al{<}\begin{matrix}BO_2\\SiO_4{\equiv}R_3.\\SiO_4{\equiv}R_3\end{matrix}$$

Kaolin or china-clay, is, it will be seen, partly hydrate of aluminium. R in the last formula may be calcium, iron (dyad), magnesium, sodium, or potassium, or generally a mixture.

The metasilicates may be similarly represented. Among these are *pyrophyllite*, $4SiO_2.Al_2O_3.H_2O$ and *spodumene, jade,* and *leucite* containing lithium, sodium, and potassium respectively. They may all be represented by the formulæ $Al{<}^{SiO_3}_{SiO_3-R}$, where R stands for H, Li, Na, or K.

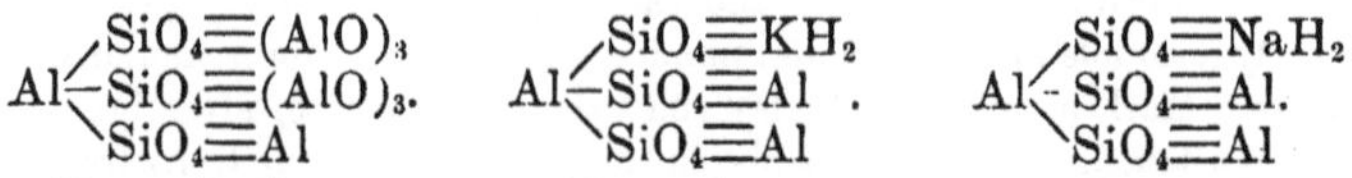

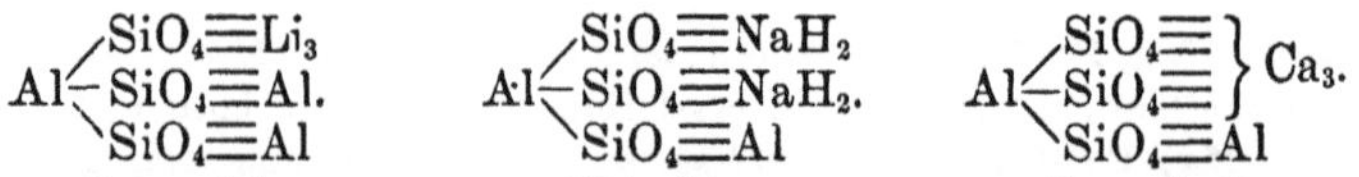

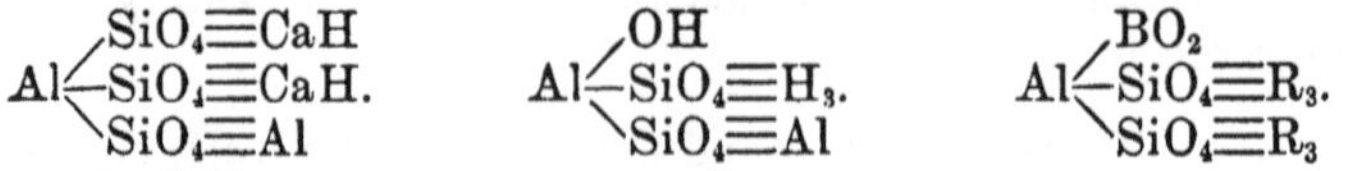

There are at least two other silicic acids, $2SiO_2.H_2O$, or $Si_2O_3(OH)_2$, the second anhydride of disilicic acid, and $3SiO_2.2H_2O$, or $Si_3O_4(OH)_4$, the second anhydride of trisilicic acid, which yield salts. *Petalite*, $(Si_2O_5)_2Al.Li$, is a salt of the first, and the felspars *albite*,

$$Si_3O_4\left\langle\begin{array}{l}ONa\\O\\O\\O\end{array}\right\rangle Al^{\prime}$$ and *orthoclase*, $$Si_3O_4\left\langle\begin{array}{l}OK\\O\\O\\O\end{array}\right\rangle Al^{\prime}$$ of the second.

By trebling these formulæ we obtain groups analogous to those of the orthosilicates; and this shows a striking analogy between these and other minerals, otherwise difficult to classify. Thus analogous to $Al\left\langle\begin{array}{l}SiO_4\equiv\\SiO_4\equiv Al\\SiO_4\equiv Al\end{array}\right.$, we have $Al\left\langle\begin{array}{l}Si_3O_8\equiv\\Si_3O_8\equiv Al\\Si_3O_8\equiv Al\end{array}\right.$. The calcium salt corresponding to the first formula and the sodium salt of the second are respectively the minerals—

$$Al\left\langle\begin{array}{l}SiO_4\equiv\ Ca_3\ \equiv SiO_4\\SiO_4\equiv Al\ Al\equiv SiO_4\\SiO_4\equiv Al\ Al\equiv SiO_4\end{array}\right\rangle Al,\quad Al\left\langle\begin{array}{l}Si_3O_8\equiv Na_3\\Si_3O_8\equiv Al\\Si_3O_8\equiv Al\end{array}\right., \text{ and}$$

Anorthite. *Albite.*

$$Al\left\langle\begin{array}{l}Si_3O_8\equiv(Al(OM)_2)_3\\Si_3O_8\equiv Al\\Si_3O_8\equiv Al\end{array}\right..$$

Labradorite.

If potassium replaces the sodium of albite, the mineral is *orthoclase*, or potash felspar.*

Lastly, an instructive analogy is pointed out by Clarke, which promises to throw light on a curious compound of a brilliant blue colour, found native, and named *lapis lazuli*, which is now manufactured in large quantities as a paint under the name of *ultramarine*, by heating together sodium sulphate, sulphur, felspar, and some carbon compound such as resin. The mineral *sodalite* has the formula

$$4SiO_2.4Al_2O_3.2Na_2O.NaCl.$$

Ultramarine may be represented as the sodo-sulphuryl (SNa) compound of which sodalite is a chloride; and the analogy is strengthened inasmuch as the constitution of another mineral, *nosean*, closely allied to ultramarine, is thereby represented. The formulæ are :—

* F. W. Clarke, *Amer. Jour. of Science*, Nov., 1886; Aug., 1887: *Amer. Chem. J.*, **10**, 120; **38**, 384.

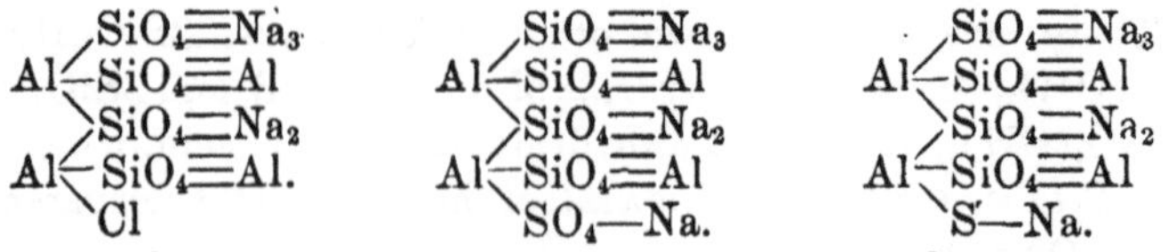

$$Al \Big\langle \begin{matrix} SiO_4 \equiv Na_3 \\ SiO_4 \equiv Al \\ SiO_4 \equiv Na_2 \end{matrix} \\ Al \Big\langle \begin{matrix} SiO_4 \equiv Al. \\ Cl \end{matrix}$$

Sodalite.

$$Al \Big\langle \begin{matrix} SiO_4 \equiv Na_3 \\ SiO_4 \equiv Al \\ SiO_4 \equiv Na_2 \end{matrix} \\ Al \Big\langle \begin{matrix} SiO_4 \equiv Al \\ SO_4 - Na. \end{matrix}$$

Nosean.

$$Al \Big\langle \begin{matrix} SiO_4 \equiv Na_3 \\ SiO_4 \equiv Al \\ SiO_4 \equiv Na_2 \end{matrix} \\ Al \Big\langle \begin{matrix} SiO_4 \equiv Al \\ S - Na. \end{matrix}$$

Ultramarine.

Such are some of the attempts which have been made to classify these complex silicates. Whether they are justified or not, if they serve to connect together bodies resembling each other, and to point the way to new researches, they have their use.

A few other silicates have still to be mentioned.

$SiO_2.2MnO$ (*tephroite*) ; $SiO_2.MnO$ (*rhodonite*) ; $SiO_2.CuO.2H_2O$ (*chrysocolla*) ; $3SiO_2.2Ce_2O_3$ (*cerite*) ; $3SiO_2.2(Y, Ce, Fe, Mn, &c.)_2O_3$ (*gadolinite*) ; $3(SiO_2.ThO_2)4H_2O$ (*thorite*).

After what has been said, these may easily be grouped in their respective classes. Other stannates have also been prepared, for example—

$SnO_2.NiO.5H_2O$; $SnO_2.CuO.6H_2O$; $SnO_2.CuO.4H_2O$; $SnO_2.CuO(NH_4)_2O.2H_2O$; $SnO_2.Ag_2O$.

The germanates have not been investigated. But as dioxide of germanium is soluble in excess of caustic alkali, they are, without doubt, capable of existence. It has lately been announced that germanium exists in small amount in *euxenite* ; and it is present, no doubt, in the form of a germanate.

Double Compounds of the Sulphides and Selenides.

Those of tin alone have been investigated.

Stannous sulphide, **SnS**, when treated with a very strong solution of potassium sulphide, K_2S, dissolves; while tin precipitates in the metallic state. The equation is—

$$2SnS + K_2S.Aq = SnS_2.K_2S.Aq + Sn.$$

By further action, hydrogen is evolved, thus—

$$Sn + 3K_2S.Aq + 4H_2O = SnS_2.K_2S.Aq + 4KOH.Aq + 2H_2.$$

The same compound is also produced by warming stannous sulphide with the polysulphide of an alkali, *e.g.*, $K_2S_5.Aq$, or $(NH_4)_2S_5.Aq$; the monosulphide is then converted into the disulphide which dissolves in the solution of sulphide; or, more simply still, tin disulphide may be dissolved in a solution of potassium sulphide.

The hydrogen salt of sulphostannic acid, **$SnS_2.H_2S$**, or **$SnS(SH)_2$** (it will be noticed that this is a meta-acid), is produced on adding an acid to a sulphostannate, as a yellow precipitate, which becomes dark-coloured on exposure to air. The following salts exist; they are all prepared thus, and are soluble in water:—

$SnS(SNa)_2.3H_2O$, also $2H_2O$; $SnS(SK)_2.10H_2O$, also $3H_2O$; $3SnS_2.(NH_4)_2S.6H_2O$; $SnS(S_2)Ba.14H_2O$; $SnS(S_2)Sr.12H_2O$; and $SnS(S_2)Ba.8H_2O$.

$SnSe(SeK)_2.3H_2O$ has been analysed; and a mixed compound obtained by digesting potassium sulphide with tin and selenium has the formula $SnSe_2.K_2S.3H_2O$.* It would appear that two isomerides might here exist, viz., $SnSe<^{SeK}_{SK}$, and $SnS<^{SeK}_{SeK}$; but they have not been identified.

A native sulphostannate of copper, iron, and zinc is known as *tin-pyrites*. Its formula is $SnS_2.(Cu_2, Fe, Zn)S$. It is also a meta-compound.

(*b.*) **Compounds with halides.**—These are, as a rule, difficult to prepare, for almost all are acted on by water. No compound of the formula $SiOCl_2$ is known. The corresponding germanium compound, $GeOCl_2$, is produced by distilling germanium tetrachloride in contact with air. It is a colourless, fuming, oily liquid boiling above 100°. By passing a mixture of silicon tetrachloride and hydrogen sulphide through a red-hot tube the substance $SiCl_3.SH$ is formed. It boils at 196°. The sulphochloride, $SiSCl_2$, is said to be formed by the same process; probably the former compound dissociates at a high temperature into hydrogen chloride and the latter.

Silicon tetrachloride, $SiCl_4$, led over fragments of felspar contained in a white-hot porcelain tube, deprives the felspar of oxygen, and yields the oxychloride $(SiCl_3)_2O$. It is a liquid boiling at 136—139°; and this compound, passed through a hot glass tube along with oxygen, yields a liquid from which, on fractionation, the following compounds have been isolated:—Si_2OCl_6 (136—139°); $Si_4O_3Cl_{10}$ (152—154°); $Si_4O_4Cl_8$ (198—202°). These substances give vapour-densities which lead to the formulæ ascribed. A fourth is formed which, on analysis, gives the numbers for $Si_4O_5Cl_6$; the molecular weight of such a body should be 405; that deduced from its vapour-density was 614; its formula is therefore doubtful. It boiled at a very high temperature.†

A somewhat analogous body to the compound $SiCl_3(SH)$ is the substituted orthostannic acid produced by the action of water on stannic chloride. Its formula is $SnO<^{OH}_{Cl}$. It may be regarded

* *Comptes rend.*, **95**, 641.

† Troost and Hautefeuille, *Annales* (5), **7**, 453.

as metastannic acid, with one hydroxyl-group replaced by chlorine. On treatment with ammonia it yields the salt $SnO<^{ONH_4}_{Cl}$.

In conclusion, a set of curious compounds of carbon and silicon with oxygen and sulphur may be mentioned, which require further investigation.* The first of these is a greenish-white mass produced by the action of carbon dioxide on white-hot silicon. Its formula is SiCO. Vapours of hydrocarbons passed over silicon, heated in a porcelain tube, yield a bottle-green substance of the formula $SiCO_2$, the oxygen being derived from the tube. By substituting a mixture of carbon dioxide and hydrogen the substance Si_2C_3O is produced; and by the action of silicon and carbon at a white heat on porcelain, a body of the formula $Si_2C_3O_2$ is formed. No clue has been obtained regarding the constitution of these bodies.

Here also may be mentioned a very remarkable compound of carbon monoxide with nickel, produced by passing that gas over hot finely-divided nickel, and condensing by means of a freezing mixture. It has the empirical formula $Ni(CO)_4$, and is a colourless liquid, boiling about 45°. Its vapour density corresponds with the formula given. It deposits metallic nickel when heated to 180°, as a brilliant mirror.†

Physical Properties.

Weights of 1 cubic centimetre :—

	Si.	Ge.	Sn.	Pb.		Si.	Ge.	Sn.	Pb.
O ..	2·89	—	6·0—6·6	8·74‡—9·29§	O_2 ..	2·65‖	—	6·7	8·9
S ..	—	—	5·0	7·5	S_2 .	—	—	4·6	6·3¶
Se..	—	—	6·2	8·1	Se_2..	—	—	5·1	—
Te..	—	—	6·5	8·1	Te_2..	—	—	—	—

Heats of formation :—

$$Si + 2O + Aq = SiO_2.Aq + 1779K\ (?).$$
$$Sn + 2O = SnO_2 \ldots\ldots + 1400K\ (?).$$
$$Sn + O = SnO \ldots\ldots + 700K\ (?).$$
$$Sn + O + H_2O = SnO_2H_2 + 681K.$$
$$Pb + O = PbO \ldots\ldots + 503K.$$
$$Pb + S = PbS \ldots\ldots + 184K.$$

* *Comptes rend.*, **93**, 1508. † *Chem. Soc.*, **57**, 749. ‡ Red. § Yellow.
‖ Rock crystal at 10° :—Tridymite, 2·3; fused to glass, 2·22. ¶ Pb_2S_3.

CHAPTER XXII.

OXIDES, SULPHIDES, &C., OF ELEMENTS OF THE NITROGEN GROUP.—THE PENTOXIDES AND PENTASULPHIDES, AND THEIR COMPOUNDS; NITRATES, VANADATES, SULPHOVANADATES, NIOBATES, AND TANTALATES.—TETROXIDES.—TRIOXIDES; NITRITES AND VANADITES.—NITRIC OXIDE AND SULPHIDE.—NITROUS OXIDE AND HYPONITRITES.

Oxides, Sulphides, Selenides, and Tellurides of Nitrogen, Vanadium, Niobium, and Tantalum.

These are very numerous. The compounds of nitrogen are not formed by direct union, for heat is absorbed during their formation and they therefore are easily decomposed. Those of vanadium niobium, and tantalum, on the other hand, are very stable.

List of Oxides.

	Nitrogen.	Vanadium.	Niobium.	Tantalum.
Monoxides	N_2O	—	—	—
Dioxides..........	NO	$\mathbf{VO}$*	$\mathbf{NbO}$	
Trioxides	N_2O_3	$\mathbf{V_2O_3}$	—	—
Tetroxides	NO_2; N_2O_4	$\mathbf{VO_2}$*	$\mathbf{NbO_2}$	$\mathbf{TaO_2}$
Pentoxides........	$\mathbf{N_2O_5}$	$\mathbf{V_2O_5}$	$\mathbf{Nb_2O_5}$	$\mathbf{Ta_2O_5}$
Hexoxide	$\mathbf{N_2O_6}$	—	—	—

List of sulphides, selenides, and tellurides:—

$\mathbf{NS}$; $\mathbf{NSe}$; $\mathbf{VS_2}$; $\mathbf{V_2S_5}$; $\mathbf{TaS_2}$(?).

Sources.—None of these bodies occurs native. The pentoxides occur in combination with the oxides of metals in the nitrates, vanadates, niobates, and tantalates, which will be described later. Among the most important are nitrates of sodium and of potassium, named respectively *Chili saltpetre* and *saltpetre* or *nitre; vanadinite,* a vanadate and chloride of lead; *pyrochlore,* a niobate of calcium, cerium, &c.; *euxenite,* a niobate and tantalate of cerium, yttrium, &c.; and *tantalite,* a tantalate of iron and manganese.

Preparation.—The starting-point for the preparation of all

* As the molecular weight of these bodies is unknown their simplest formulæ are given.

the oxides of the members of this group is the compounds of the pentoxides with other oxides. For nitrogen oxides, the nitrates of potassium and sodium; for vanadium oxides, the vanadate of lead; for the oxides of niobium and tantalum, the niobates and tantalates of yttrium, lanthanum, iron, manganese, &c. On treatment of these compounds with strong sulphuric acid, hydrates of the pentoxides are set free. This may be regarded as the displacement of an oxide by another oxide, viz., SO_3. As nitric acid, $N_2O_5.H_2O$, or as its vapour-density shows us, HNO_3, is a liquid, volatile without decomposition, it can be distilled away from the solid sulphate of sodium or potassium; the vanadate of lead, on treatment with sulphuric acid, or, better, on fusion with hydrogen potassium sulphate, $HKSO_4$, is decomposed, lead sulphate, which is insoluble in water, being left behind; and on treatment with water vanadate of potassium is dissolved, from which strong nitric acid sets free vanadic acid, $V_2O_5.H_2O$, as a reddish precipitate. The pentoxides of niobium and tantalum are also produced by fusing the ores with hydrogen potassium sulphate, and after cooling, boiling the fused mass with water; the iron, yttrium, &c., all go into solution as sulphates, and the pentoxides remain as insoluble hydrates.

We shall begin—reversing the usual order—with the pentoxides, because they form the sources of the lower oxides.

Pentoxides.—Nitrogen pentoxide is produced by the action on nitric acid of phosphoric anhydride, $\mathbf{P_2O_5}$, a body which has a great tendency to combine with water, and which, therefore, withdraws it from nitric acid. The acid cannot be dehydrated by heat alone, for the pentoxide easily decomposes into the tetroxide, losing oxygen. Phosphorus pentoxide is gradually added to ice-cold, pure nitric acid, and the syrupy liquid is distilled at a low temperature. The liquid distillate consists of two layers, the upper one being the pentoxide, mixed with a little of the compound $2N_2O_5.H_2O$; the lower consisting of the latter compound. The upper layer is separated, and cooled with a freezing mixture, when the pentoxide deposits in crystals. The equation is:—

$$2HNO_3 + \mathbf{P_2O_5} = \mathbf{2HPO_3} + \mathbf{N_2O_5}.$$

This substance may also be prepared by heating silver nitrate, $\mathbf{AgNO_3}$, to 58—68° in a current of perfectly dry chlorine. This reaction should yield a hexoxide, N_2O_6, thus, $\mathbf{AgNO_3} + Cl_2 = \mathbf{2AgCl} + N_2O_6$; but the hexoxide is unstable, and decomposes at the moment of liberation into pentoxide and oxygen. The hexoxide is said, however, to be produced by passing an electric discharge through a mixture of nitrogen and oxygen at $-23°$, and to form a volatile crystalline powder.

Another method, which appears to act well, is to pass a mixture of nitric peroxide, NO_2, and chlorine over dry silver nitrate at 60—70°. The equation is $NO_2 + Cl + \mathbf{AgNO_3} = \mathbf{AgCl} + N_2O_5$. The pentoxide must be condensed in a U-tube, surrounded by a freezing mixture; and the most scrupulous care must be taken to exclude moisture, by drying the apparatus and materials perfectly before use, and by preventing the access of moist air.

Vanadium, niobium, and **tantalum pentoxides** are produced (1) by burning the elements in air, or by the oxidation of the lower oxides when heated in air. (2) By heating their hydroxides (acids), or in the case of vanadium by heating ammonium vanadate, $\mathbf{2NH_4VO_3} = 2NH_3 + H_2O + \mathbf{V_2O_5}$; or by heating a solution of the oxide VO_2 in strong sulphuric acid; the first reaction is the formation of the sulphate, $\mathbf{V_2O_5.2SO_3.H_2O}$, a portion of the sulphuric acid losing oxygen to oxidise the tetroxide to pentoxide, thus $\mathbf{2VO_2} + H_2SO_4 = \mathbf{V_2O_5} + H_2O + SO_2$; the pentoxide then forms the above sulphate; $\mathbf{V_2O_5} + 2H_2SO_4 = \mathbf{V_2O_5.2SO_3} + 2H_2O$. The sulphate is decomposed on further ignition into V_2O_5 and SO_3. (3) By the action of water on the pentahalide or oxyhalides. This yields the hydroxides, from which the oxides are obtainable.

Properties.—Nitric pentoxide forms brilliant, colourless, transparent rhombic prisms; it melts at 30°, and boils about 45°. It is very unstable, forming nitric peroxide with loss of oxygen, but can be preserved for several days at 10° in a dry atmosphere. It hisses when dropped into water, forming hydrated nitric acid.

Vanadium pentoxide is a reddish-yellow solid, sparingly soluble in water, to which it imparts a yellow colour. The solution is tasteless, but has an acid reaction. It melts when heated to redness, and on solidifying it turns incandescent, probably displaying the phenomenon of superfusion.

Niobium pentoxide is a white insoluble solid, turning yellow when heated, but regaining its whiteness on cooling. It has been fused at a white heat. After ignition it is insoluble in acids.

Tantalum pentoxide is also a white insoluble powder, which has not been fused. It is also insoluble in acids.

Vanadium is the only element of which a **pentasulphide** is known. It is produced by adding ammonium sulphide to a solution of the pentoxide, and precipitating with hydrochloric acid. It is a brown precipitate, which turns black on drying. It is soluble in sulphides of sodium and potassium, forming sulphovanadates (see below).

Compounds with water and oxides.—Of these oxides that of nitrogen is the only one which readily dissolves in water, forming a compound. That of vanadium is slightly soluble; but the pentoxides of niobium and tantalum do not combine with water. The name "**acid**" is applied to the hydrates of these oxides, because the hydrogen of the combined water is replaceable by metals, when the compound in solution is treated with hydroxides of the metals, or heated with the carbonates. These acids are as follows:—

$$2N_2O_5.H_2O;\ N_2O_5.H_2O, \text{ or } HNO_3;\ V_2O_5.H_2O, \text{ or } HVO_3;$$

(this body contains another molecule of water, which is easily expelled by heat, and which is therefore not regarded as essential to its composition); $Nb_2O_5.nH_2O$, and $Ta_2O_5.nH_2O$, the value of n being unknown.

There are two classes of nitrates, the ordinary nitrates, and the basic nitrates; and many classes of vanadates, niobates, and tantalates.

Nitric acid and nitrates.—Preparation.—The method of preparation of nitric acid is by distillation of sodium or potassium nitrate with excess of sulphuric acid. The reaction is as follows—

$$\mathbf{KNO_3} + H_2SO_4 = \mathbf{HKSO_4} + HNO_3.$$

It would appear as if economy of sulphuric acid might be attained by using the proportions $2KNO_3 + H_2SO_4 = K_2SO_4 + 2HNO_3$; but at the temperature at which hydrogen potassium sulphate attacks a nitrate, nitric acid is largely decomposed. On the small scale, the distillation is carried out in a glass retort (see Fig. 39); on the large scale in one of iron. The iron is protected by a film

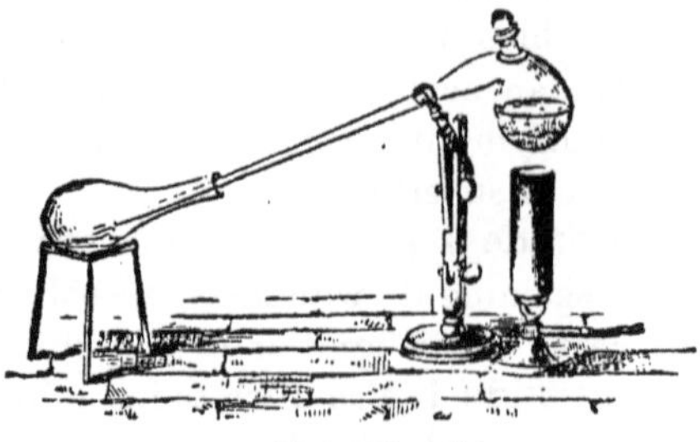

FIG. 39.

of ferroso-ferric oxide, $\mathbf{Fe_3O_4}$, which is at once formed on the surface, and on which nitric acid is without action. The worm of the condenser and the receivers are usually made of stoneware.

Nitric acid is also produced along with nitrous acid by the action of water on nitric peroxide, N_2O_4 or NO_2, thus—$N_2O_4 + H_2O$

$= HNO_3 + HNO_2$; also by heating a solution of nitrous acid, $3HNO_2 = HNO_3 + H_2O + 2NO$.

When prepared by distillation it usually has a yellow colour, owing to its containing peroxide, NO_2, in solution. This substance is easily volatile, and may be removed by passing a current of air through the acid for some hours.

Properties.—Nitric acid, when pure, is a colourless liquid, fuming slightly in the air, being somewhat volatile at the ordinary temperature. It freezes at −55°, and boils at 86°, partially decomposing into tetroxide, N_2O_4, oxygen and water, a weaker acid remaining behind. It is completely decomposed when heated in a sealed tube to 256°. Its density corresponds with the formula HNO_3. It absorbs water from the air, forming, no doubt, a hydrate, which, however, has not been isolated, although it is stated to have the formula $2HNO_3.3H_2O$, or $N_2O_5.5H_2O$.

The hydrate $2N_2O_5.H_2O$ is produced during the preparation of nitric anhydride, N_2O_5, by use of phosphorus pentoxide. It is the lower layer, into which the distillate separates, and it crystallises out when cooled by a freezing mixture; and it can also be prepared by adding nitric anhydride to nitric acid. At the ordinary temperature it is a liquid, but it turns solid at about −5°.

A solution of vanadium pentoxide in water perhaps contains the compound $V_2O_5 3H_2O$, or H_3VO_4; but the hydrate best known is $\mathbf{V_2O_5.H_2O}$, or $\mathbf{HVO_3}$, corresponding to nitric acid. This substance is a brown-red powder, prepared by adding nitric acid to a solution of one of its salts, *e.g.*, $\mathbf{V_2O_5.K_2O}$, or $\mathbf{KVO_3}$. It is also formed by heating a mixture of solutions of copper sulphate with vanadic acid and a large excess of ammonium chloride to 75°. The copper vanadate decomposes, depositing golden-yellow scales of metavanadic acid. It contains a molecule of water in addition, $\mathbf{V_2O_5.2H_2O}$, but as the second molecule is lost when it is dried over strong sulphuric acid, it must be very loosely combined. It is also produced by the action of water or vanadium pentachloride, or oxychloride, $VOCl_3$. It is a reddish-yellow powder or golden-yellow scales; it is very hygroscopic.

Niobic and tantalic acids are precipitated as white powders on adding hydrochloric acid to a solution of sodium or potassium niobate or tantalate; or by the action of water on the pentachloride of niobium or tantalum. When heated they lose water, and leave the pentoxide.

Nitrates, vanadates, niobates, tantalates.—These salts are all produced by the action of nitric, vanadic, niobic or tantalic acid, in presence of water, on hydrates, oxides, or carbonates, or by fusion of the pentoxides of the three last with hydrates or carbonates of lithium, sodium, potassium, &c. The following equations may be taken as typical :—

$$HNO_3.Aq + KOH.Aq = KNO_3.Aq + H_2O;$$
$$2HNO_3.Aq + \mathbf{CuO} = Cu(NO_3)_2.Aq + H_2O.$$
$$2HNO_3.Aq + \mathbf{CaCO_3} = Ca(NO_3)_2.Aq + H_2O + \mathit{CO_2}.$$
$$\mathbf{V_2O_5} + 2KOH.Aq = 2KVO_3.Aq + H_2O;$$
$$\mathbf{V_2O_5 + 3Na_2CO_3 = 2Na_3VO_4} + 3\mathit{CO_2}.$$

These equations are rendered more simple by the older method of representation, thus :—

$$N_2O_5.Aq + K_2O.Aq = N_2O_5.K_2O.Aq;$$
$$N_2O_5.Aq + \mathbf{CuO} = N_2O_5CuO.Aq.$$
$$N_2O_5.Aq + \mathbf{CO_2\,CaO} = N_2O_5.CaO.Aq + \mathit{CO_2}.$$
$$\mathbf{V_2O_5 + K_2O.H_2O = V_2O_5.K_2O} + \mathit{H_2O}:$$
$$\mathbf{V_2O_5} + \mathbf{3CO_2.Na_2O = V_2O_5.3Na_2O} + 3\mathit{CO_2}.$$

All nitrates are soluble, and hence cannot be produced by precipitation, unless the solution be a concentrated one. It is possible to prepare certain nitrates, however, such as those of lead, silver, and barium, by addition of much nitric acid to a soluble salt of such metals; for the nitrates produced are sparingly soluble in nitric acid. Thus :—

$$BaCl_2 + 2HNO_3 = \mathbf{Ba(NO_3)_2} + 2HCl;$$
$$Ag_2SO_4 + 2HNO_3 = \mathbf{2AgNO_3} + H_2SO_4.$$

The nitrate of barium or silver is precipitated as a crystalline powder. Many vanadates, niobates, and tantalates are produced by precipitation, *e.g.*, those of lead and silver.

Nitrates, vanadates, niobates, and tantalates.

$\mathbf{LiNO_3}$; below 10°, $\mathbf{2LiNO_3.3H_2O}$; $\mathbf{NaNO_3}$; $\mathbf{KNO_3}$; $\mathbf{KNO_3.2HNO_3}$ melting at −3°.
$\mathbf{RbNO_3}$: $\mathbf{RbNO_3.5HNO_3}$. $\mathbf{CsNO_3}$. $\mathbf{NH_4NO_3}$; $\mathbf{NH_4NO_3.2HNO_3}$: $\mathbf{NH_4NO_3.HNO_3}$.

These are all white, soluble salts. Those containing excess of nitric acid are made by mixture and cooling. With the exception of ammonium nitrate, the action of heat on which is peculiar, and will be fully treated of later in this chapter, these salts when heated to bright redness fuse and give off oxygen, forming at first the corresponding nitrites $\mathbf{MNO_2}$; at very high temperatures they give off nitrogen and oxygen, and leave oxides and peroxides. They cannot be strongly ignited even in gold, silver, or platinum vessels without attacking them, forming oxides.

Two of them, sodium and potassium nitrates, are commercially important. **Sodium nitrate,** named "Chili saltpetre," does not occur in Chili, but forms immense beds, several feet thick, at Tarapaca, in Northern Peru. Its local name is "caliche." Its crystalline form is nearly cubic, but in reality it forms very obtuse rhombohedra; it is often erroneously named "cubic saltpetre." One gram of the salt dissolves in 1·4 grams of water at 15°; it is also soluble in alcohol. It is largely used as a manure.

Potassium nitrate, also called "nitre" or "saltpetre," is present in most soils, being especially abundant in chalk or marl. It also occurs in the leaves of many plants, especially in those of the tobacco-plant. It is found as an efflorescence on the soil of hot countries, being formed by the action of a ferment or ammonia in presence of bases, the ammonia being derived from decomposing animal or vegetable matter.* The nitrate ferment is a minute organism similar to those which produce fermentation. Nitrification, as the process of transformation of ammonia into nitric acid is called, goes on beneath the surface of the soil, the necessary conditions being apparently presence of air and absence of light. It ceases and does not recommence if the soil be kept for some time at 100°, the organism being destroyed; but after exposure to the atmosphere fresh germs enter, and it again proceeds. The manufacture of nitre by this process has been carried out for ages in Upper India; stable-manure and limestone are exposed to air for several months, and the resulting nitrate of calcium is converted into nitrate of potassium by treatment with potassium carbonate or sulphate; the soluble potassium nitrate is easily separated from the insoluble calcium carbonate or sulphate. The process is also carried out in France and elsewhere.

Potassium nitrate is now largely prepared from the Peruvian sodium nitrate by mixing the latter with potassium chloride. Sodium chloride is formed, and, as it is much less soluble in hot water than potassium nitrate, it separates out on evaporation. The mother-liquor, after removal of most of the salt, deposits crystals of nitre.

Potassium nitrate crystallises in two forms: in trimetric prisms, and in rhombohedra, like calcspar. It has a cooling taste; it is soluble in 3½ parts of water at 18°, but insoluble in alcohol. It melts at 339°.†

* *Chem. Soc.*, **35**, 454.

† For lists of melting points, see Carnelley and Williams, *Chem. Soc.*, **33**, 279.

Ammonium nitrate is prepared by mixing nitric acid and ammonia, and evaporating till the water is expelled. It dissolves in half its weight of water at 18°, and is also soluble in alcohol. It melts at 108°, and can be distilled at 180°, splitting into nitric acid and ammonia, which recombine on cooling (?). At a higher temperature it decomposes into nitrous oxide and water. It is formed in solution by the action of dilute nitric acid on some metals, especially on tin.

Orthovanadates: Na_3VO_4; K_3VO_4; also with $3H_2O$ and $2H_2O$. Pyrovanadate: $V_2O_5.2K_2O.3H_2O$. Metavanadates: $LiVO_3$; $NaVO_3$; also with $2H_2O$; $KVO_3.2H_2O$; $NH_4.VO_3$. **Acid Vanadates:** $2V_2O_5.Li_2O$; also with $9H_2O$; $2V_2O_5.Na_2O$; $2V_2O_5.K_2O$.3, 4, and $7H_2O$; $2V_2O_5.(NH_4)_2O.4H_2O$; $3V_2O_5.2Na_2O.9H_2O$; $3V_2O_5.K_2O.6H_2O$ (insoluble); $3V_2O_5.Na_2O.9H_2O$; $3V_2O_5.(NH_4)_2O.6H_2O$; $4V_2O_5.6Na_2O$.

The **orthovanadates** are produced by fusing vanadium pentoxide with carbonates in theoretical proportions. With sodium carbonate the pyrovanadate, $Na_4V_2O_7$, is formed first. They are soluble in water, but decompose slowly at the ordinary temperature, rapidly on warming, into sodium or potassium hydroxides and pyrovanadates or metavanadates. They are yellowish solids.

The **metavanadates** are white, soluble, earthy bodies which, on acidifying with acetic acid, turn orange, and on evaporation deposit orange-yellow crystals of the acid vanadates. Ammonium metavanadate is produced by addition of ammonia in excess to vanadic acid; the acid vanadate, $2V_2O_5.(NH_4)_2O$, by saturating ammonia with vanadic acid; and on acidifying with acetic acid the body $3V_2O_5.(NH_4)_2O$ is produced.

Niobates.—$3Nb_2O_5.4K_2O.16H_2O$; $7Nb_2O_5.3K_2O.32H_2O$; $2Nb_2O_5.3K_2O.13H_2O$; $3Nb_2O_5.K_2O.5H_2O$; $4Nb_2O_5.2K_2O.11H_2O$; $Nb_2O_5.2K_2O.11H_2O$; $Nb_2O_5.Na_2O.6H_2O$; $3Nb_2O_5.2Na_2O.9H_2O$.

The first of these is obtained by fusion of niobic pentoxide with potassium carbonate solution in water, and crystallisation; the solution also deposits crystals of the second compound; and the third is formed by addition of potassium hydroxide to one of the former, and crystallisation. The fourth is produced by boiling potassium fluoxyniobate, $NbOF_3.2KF.H_2O$, with hydrogen potassium carbonate; it is nearly insoluble in water. These compounds are all white, and crystallise. The sodium salts are easily decomposed by water into hydrated niobic pentoxide and sodium hydroxide.

Tantalates.—$3Ta_2O_5.4Na_2O.25H_2O$; $3Ta_2O_5.4K_2O.16H_2O$; $Ta_2O_5.Na_2O$; $Ta_2O_5.K_2O$.

On fusing tantalum pentoxide with excess of caustic potash or soda, and washing out excess of the alkali with alcohol, the salts of the formula $3Ta_2O_5.4M_2O$ remain as crystalline powders. Their solutions, when warmed, deposit the other salts of the formula $Ta_2O_3.M_2O$ as insoluble precipitates.

$Be(NO_3)_2.3H_2O$; $Be(OH).NO_3.H_2O$; $Ca(NO_3)_2.4H_2O$; $Sr(NO_3)_2.5H_2O$; $Ba(NO_3)_2$.

These are also white soluble salts. The basic **beryllium nitrate** is produced by digesting a solution of the ordinary nitrate with beryllium hydrate. **Calcium nitrate** often occurs as an efflorescence on caverns frequented by bats and birds, and in stables, &c., where animal matter decomposes in presence of calcium carbonate. It is easily soluble in water, and in alcohol, and may be fused without decomposition. **Strontium nitrate** is also an easily soluble salt; it is used to produce red fire in pyrotechny. **Barium nitrate** is one of the important salts of barium. It is formed by dissolving barium sulphide (*q.v.*) or carbonate in dilute nitric acid, or on account of its sparing solubility (1 part in 11·7 of water at 20°) by addition of potassium nitrate to a strong solution of barium chloride. It is insoluble in strong nitric acid and also in alcohol. These nitrates, when heated, yield nitrites, and then oxides at a bright red heat.

$Ca(VO_3)_2$; $Sr(VO_3)_2$; $Ba(VO_3)_2.H_2O$; $2Ca_2V_2O_7.5H_2O$; $Ba_2V_2O_7$; $2V_2O_5.CaO$; $2V_2O_5.BaO$; $2V_2O_5.3BaO.19H_2O$.

The three first are yellowish-white gelatinous precipitates formed by adding ammonium metavanadate to soluble salts of the metals; the three last are orange-coloured, and are produced by acidifying the former with acetic acid. The other vanadates are insoluble and are formed on adding to a soluble salt of the metal potassium orthovanadate. They have not been analysed. The pyrovanadates are produced by precipitation.

$Nb_2O_5.2CaO$; $Nb_2O_5.CaO$.

These are prepared by fusing niobium pentoxide with calcium chloride, or with calcium fluoride and potassium chloride. They are insoluble. Other niobates and tantalates are formed as insoluble precipitates on adding a soluble niobate or tantalate to a soluble salt of calcium, strontium, or barium. They have not been analysed.

$Mg(NO_3)_2.6H_2O$; $Zn(NO_3)_2.6H_2O$; $Cd(NO_3)_2.4H_2O$.—Basic nitrates:— $3N_2O_5.4ZnO.3H_2O$; $N_2O_5.2ZnO.3H_2O$; $N_2O_5.2CdO.3H_2O$; and $N_2O_5.8ZnO$.

Magnesium, zinc, and cadmium nitrates are white deliquescent crystals, soluble in alcohol. The basic nitrates of zinc, produced by digesting the ordinary nitrate with zinc hydrate, are non-crystalline soluble masses. $Nb_2O_5.4MgO$ and $Nb_2O_5.3MgO$ are also known.

$Sc(NO_3)_3$; $Y(NO_3)_3.4H_2O$ and $12H_2O$; $La(NO_3)_3.6H_2O$ and $4H_2O$.

These are colourless soluble deliquescent salts. A crystalline vanadate of boron has been produced by fusion.

$Al(NO_3)_3.9H_2O$; $Ga(NO_3)_3$; $In(NO_3)_3.3H_2O$; $TlNO_3$; $TlNO_3.3HNO_3$

Aluminium nitrate is deliquesent; when digested with hydroxide, or when heated, it forms basic salts, similar to those of zinc. Indium also forms basic nitrates.

Thallous nitrate is insoluble in alcohol; the acid salt crystallises from strong nitric acid. All these salts are colourless.

Tl_3VO_4 ; $Tl_4V_2O_7$; $TlVO_3$; also $4V_2O_5.Tl_2O$; $5V_2O_5.6Tl_2O$; $7V_2O_5.6Tl_2O$.

The first of these, prepared by fusion, is a red substance; the second is precipitated by addition of *ortho*vanadate of sodium, Na_3VO_4 to a thallous salt, as a yellowish powder; and the third is produced by fusion; it forms dark scales. The pyrovanadate is formed with liberation of alkali; if it be warmed with water, more and more alkali goes into solution, and the other acid vanadates are produced.

$Cr(NO_3)_3.9H_2O$; $Fe(NO_3)_3.9H_2O$.—Basic salts, $2N_2O_5.Cr_2O_3$; $3N_2O_5.2Cr_2O_3$; also many basic ferric salts. Two basic salts of chromium should be here included, viz.: $Cr\begin{cases}NO_3\\OH\\Cl\end{cases}$ and $Cr\begin{cases}NO_3\\NO_3\\Cl\end{cases}$, produced by treating the compound $Cr(OH)_2Cl$ with nitric acid.

Chromium nitrate is a violet crystalline substance; and the ferric salt lavender-blue; both are very soluble. The basic salts of chromium are green; those of iron orange-yellow.

$Fe(NO_3)_2.6H_2O$; $Mn(NO_3)_2.6H_2O$, and $3H_2O$; $Co(NO_3)_2.6H_2O$; $Ni(NO_3)_2.6H_2O$.—*Basic salt*:—$N_2O_5.6CoO.5H_2O$.

The solution of ferrous nitrate must be evaporated in the cold; when heated, oxygen from the nitric group oxidises the iron to ferric nitrate, and a basic substance is formed. The ordinary nitrate of iron is green; that of manganese, pink; that of cobalt, red; and of nickel, grass-green. They are all soluble in alcohol.

Basic cobalt nitrate is produced as a blue precipitate by adding a solution of ammonia to the normal nitrate; and nickel nitrate, similarly treated, gives a green basic salt.

Hydrated titanium and zirconium dioxides are soluble in nitric acid; but on warming the solution of the former, the hydrate separates out. Zirconium nitrate can be evaporated to dryness; it leaves a gummy mass. Cerium sesquioxide dissolves in nitric acid, and on evaporation a crystalline mass of $Ce(NO_3)_3.6H_2O$ is left. The dioxide also forms an orange-yellow solution in nitric acid.

Thorium nitrate, $\mathbf{Th(NO_3)_4}$, is a crystalline salt; it also forms a double salt with potassium nitrate $\mathbf{Th(NO_3)_4.KNO_3}$.

Silica, recently precipitated, is sparingly soluble in nitric acid. The nitrate of germanium has not been prepared; that of tin, $\mathbf{Sn(NO_3)_4}$, is obtained by dissolving stannic hydrate, $Sn(OH)_4$, in dilute nitric acid; on rise of temperature it easily decomposes into metastannic acid, $5\mathbf{SnO_2.5H_2O}$, and nitric peroxide, NO_2. If ammonium nitrate be present, the decomposition does not occur, probably because it forms a double salt.

Stannous nitrate, $\mathbf{Sn(NO_3)_2}$, is produced by dissolving tin in dilute cold nitric acid. It also is easily decomposed when heated, giving metastannic acid. Lead nitrate, $\mathbf{Pb(NO_3)_2}$., forms octahedra; when crystallised below 16°, it contains $2H_2O$; it is insoluble in alcohol. By digesting it with lead hydrate, or by adding ammonia to ordinary lead nitrate, the basic salts, $\mathbf{N_2O_5.2PbO.H_2O}$ $(= \mathbf{NO_3 - Pb - OH})$; $\mathbf{N_2O_5.2PbO}$; $\mathbf{2N_2O_5.3PbO.3H_2O}$; and $\mathbf{3N_2O_5.10PbO.4H_2O}$, are formed. The last three are nearly insoluble in water.

Two vanadates of lead are found native, viz., $\mathbf{Pb(VO_3)_2}$, lead metavanadate, or *dechenite*, and $\mathbf{Pb_2V_2O_7}$, lead pyrovanadate, or *descloizite*. Lead orthovanadate, $\mathbf{Pb_3(VO_4)_2}$, has also been prepared; it is a yellow precipitate. An orange-coloured acid salt is also produced on treating one of these vanadates with acetic acid; it has the formula $\mathbf{2V_2O_5.PbO}$. The mineral *vanadinite*, $\mathbf{3Pb_3(VO_4)_2.PbCl_2}$, is a compound of lead orthovanadate and chloride.

Nitrates, vanadates, &c., of members of the vanadium group do not appear to exist. The compound nitric peroxide, N_2O_4, has been viewed as nitrate of nitrosyl, NO, thus, $NO(NO_3)$; but of the justice of this view there is no proof. A nitrate of the oxide V_2O_4 appears to exist; and V_2O_5 is soluble in acids, but the hydrates of tantalum and niobium pentoxides are insoluble in nitric acid.

Similarly, although the oxides of phosphorus and arsenic dissolve in nitric acid, no compound has been isolated. But with antimony, $N_2O_5.Sb_4O_6$, has been prepared; and the pentoxide, Sb_2O_5, is slightly soluble in nitric acid.

Bismuth nitrate, $Bi(NO_3)_3.5H_2O$, is a well crystallised salt. On treatment with water it gives a mixture of three salts, each of which may, however, be prepared fairly pure by careful attention to temperature and dilution. These are $N_2O_5.Bi_2O_3.H_2O = 2\{BiO(NO_3)\}.H_2O$; $2N_2O_5.Bi_2O_3.H_2O = Bi(OH)(NO_3)_2$; and $N_2O_5.2Bi_2O_3.H_2O$ These basic nitrates are insoluble in water.

Molybdenum trioxide is soluble in nitric acid; so too is oxide of tungsten, but no compounds are known. Uranium forms yellow nitrates of uranyl of the formulæ $UO_2(NO_3)_2.3H_2O$ and $6H_2O$.

Samarskite consists chiefly of niobates of uranyl, iron, and yttrium.

A nitrate of tellurium of the formula $N_2O_5.4TeO_2$ is produced on dissolving tellurium in nitric acid, and evaporating.

Rhodium oxide is soluble in nitric acid, but the nitrate is unstable. But on adding sodium nitrate the stable double salt $Rh(NO_3)_3.NaNO_3$ may be obtained in crystals. Palladium nitrate, $Pd(NO_3)_2$ is easily prepared by dissolving palladium monoxide, or the metal, in nitric acid. It is a brown compound; and on evaporation a basic salt is produced.

Osmium oxide is also soluble in nitric acid. Platinic nitrate, $Pt(NO_3)_4$, is unstable, but as with rhodium the addition of potassium nitrate yields a stable double salt of the formula

$$Pt(NO_3)_4.KNO_3.$$

$Cu(NO_3)_2.3H_2O$; $Cu_3(VO_4)_2.H_2O$; $V_2O_5.4CuO.3H_2O$, also H_2O. The latter is possibly $VO.(OCu.OH).(O_2)Cu$. It is found native, and named *volborthite.*

$$Cu(NO_3)_2.NH_4NO_3.$$

Copper nitrate is a soluble blue salt, crystallising well. It is the source of copper oxide for the analysis of organic substances, for, like almost all the nitrates, it yields the oxide on ignition. The vanadates are brown substances.

$AgNO_3$; $AgNO_3.KNO_3$; $AgNO_3.NH_4NO_3$; $2AgNO_3.Pb(NO_3)_2$.
Ag_3VO_4; $Ag_4V_2O_7$; $AgVO_3$.

The first nitrate is an important substance. Great use is made of it in photography, electrotyping, &c., and under its old name "lunar caustic" (*luna* = silver), it is employed as a caustic, being cast into sticks for medical use. It is a white easily fusible salt (m. p. 218°); it is soluble in about its own weight of cold water, and in about four times its weight of alcohol. It crystallises with sodium and lithium, to form double salts like those of potassium and ammonium, but not in molecular proportions. A number of double nitrates and halides are known; *e.g.*,

$$AgNO_3.AgCl; \quad AgNO_3.AgBr; \quad AgNO_3.AgI; \quad 2AgNO_3.AgI;$$
$$4AgNO_3.Pb(NO_3)_2.2AgI; \quad 2AgNO_3.Pb(NO_3)_2.2AgI.$$

These are sparingly soluble salts prepared by mixture.

The mercurous nitrates are numerous, many basic compounds being known. They are as follows:—

$$HgNO_3.H_2O; \; 3N_2O_5.4Hg_2O.H_2O; \; 3N_2O_5.5Hg_2O.2H_2O; \; N_2O_5.2Hg_2O.2H_2O.$$

Others are said to have been obtained, but their existence is questionable. Mercurous nitrate is formed by digesting mercury with cold dilute nitric acid. The basic nitrates are produced by the action of water on the ordinary salt. Double salts with strontium, barium, and lead nitrates are also known, of formulæ such as $3N_2O_5.2PbO.2Hg_2O$. All these salts are crystalline and soluble.

By dissolving mercuric oxide, HgO, in excess of nitric acid and evaporating, crystals of the salt $2Hg(NO_3)_2.H_2O$, are deposited. Crystals with $8H_2O$ may also be produced by cooling the solution. These crystals, when fused, deposit a basic salt, $N_2O_5.2HgO.3H_2O$; and with water they yield $N_2O_5.3HgO.H_2O$. Like silver nitrate, mercuric nitrate combines with mercury halides, forming colourless crystalline compounds, *e.g.*, $Hg(NO_3)_2.HgI_2$; $2Hg(NO_3)_2.HgI_2$; $Hg(NO_3)_2.2HgI_2$; and $2Hg(NO_3)_2.3HgI_2$. These are all decomposed by water. The compound $2Hg(NO_3)_2.4AgI.H_2O$ is also known.

Oxide of gold dissolves in nitric acid, but the solution decomposes spontaneously at the ordinary temperature, again depositing gold oxide.

Compounds of vanadium pentasulphide.—This body is soluble in sulphides of the alkalies. On adding alcohol to its solution in potassium sulphide, a scarlet precipitate is produced, consisting of potassium sulphovanadate; it has probably the formula $V_2S_5.K_2S = KVS_3$, and is a meta-compound. A solution of this substance gives brown precipitates with soluble salts of other elements, but the formulæ of the compounds are unknown.

Compounds containing halogens.

$$VOF_3; \; VOCl_3; \; VOBr_3; \; NbOF_3; \; NbOCl_3; \; NbOBr_3; \; TaOF_3.$$

No corresponding nitrogen compound is known, although a mixture of nitrosyl chloride, $NOCl$, and chlorine reacts with water as if it consisted of $NOCl_3$.

Vanadyl trifluoride is known in combination (see below).

Vanadyl chloride, $VOCl_3$, is produced by heating the oxide, V_2O_2 (or VO ?) in a current of chlorine, when direct union ensues. The higher oxides, mixed with carbon and heated in chlorine, also yield it. The bromide, $VOBr_3$, is similarly prepared; also by passing bromine over the heated trioxide, V_2O_3, but no corresponding iodide seems capable of existence.

Niobium oxyfluoride (**niobyl fluoride**), **oxychloride**, and **oxybromide** are volatile white crystalline bodies. The chloride has a vapour density corresponding to the formula $NbOCl_3$. They are prepared along with the pentahalides by the action of chlorine on a mixture of the pentoxide with charcoal at a bright red heat. The trichloride at a red heat decomposes carbon dioxide, producing carbon monoxide and the oxychloride. Tantalum oxyfluoride is produced by the action of air or water-vapour on the pentafluoride; the oxychloride and oxybromide could probably be similarly produced.

Vanadyl chloride is a golden-yellow liquid, boiling at 127°. Its density leads to the usual formula. The oxybromide forms a dark red liquid boiling at about 130° under a pressure of 100 mms., but it is decomposed at 180° into the dibromide $VOBr_2$.

These oxyhalides form the following compounds with the halides of other elements :—

Fluoxyvanadates :—

$V_2O_5.2VOF_3.6KF.2H_2O$; $V_2O_2.2VOF_3.6NH_4F.2H_2O$; $V_2O_5.2VOF_3.12NH_4F$.

Vanadoxyfluorides :—

$2VOF_3.3KHF_2$; $2VOF_3.3NH_4.HF_2$; $2VOF_3.ZnF_2.ZnO.14H_2O$.

Nioboxyfluorides :—

$NbOF_3.2KF.H_2O$; $NbOF_3.2NH_4F.H_2O$; $NbOF_3.3KF$; $NbOF_3.3NH_4F$; $3NbOF_3.5KF.H_2O$; $3NbOF_3.5NH_4F.H_2O$; $NbOF_3.3KF.HF$; $3NbOF_3.4KF.2H_2O$; $NbOF_3.NH_4F$; $NbOF_3.NbF_5.3NH_4F$; $NbOF_3 ZnF_2.6H_2O$.

Tantaloxyfluoride :—$TaOF_3.3NH_4F$.

These bodies are produced by direct union. They are crystalline salts. The tantaloxyfluorides react with water, forming hydrated tantalum pentoxide and tantalifluorides, such as $TaF_5 2KF$, hence they have been little investigated. The corresponding chlorides and bromides of these elements are also easily decomposed by water, hence their derivatives have not been prepared.

Tetroxides, or dioxides.—These are as follows:—

NO_2 and N_2O_4; **VO_2 or V_2O_4**; **NbO_2 or Nb_2O_4**; **TaO_2 or Ta_2O_4**; **VS_2 or V_2S_4**.

The formula of **nitric peroxide**, as this substance is usually called, depends on the temperature. In the liquid state it is a tetroxide, N_2O_4. The gas, at the lowest possible temperature, also approximates to this formula; but on raising the temperature, dissociation ensues, the extent of dissociation depending on the temperature and pressure, until, at 140°, at atmospheric pressure, the more complex molecules of N_2O_4 are entirely resolved into molecules of NO_2. At higher temperatures the compound NO dissociates in its turn into NO and O, and at 620° the gas contains no molecules of peroxide. On cooling, recombination takes place, and the phenomena are reversed. It is possible to trace these changes by the alteration of colour of the gas; N_2O_4 is an almost colourless substance when solid; NO_2 is dark reddish-black; and a *mixture* of NO and O is also colourless. On heating a tube of hard glass filled with the gas, it turns dark at first, and then lightens in colour, turning nearly colourless at the temperature at which the glass begins to soften. As we have the two substances, one of which is a *polymeride* of the other, it is convenient to give them different names. The first, NO_2, we shall call *nitric peroxide*, reserving the name *tetroxide* for the compound N_2O_4.

Alternative formulæ have been ascribed to the oxides of vanadium, niobium, and tantalum. They are non-volatile solids, and nothing is known regarding their molecular complexity.

Preparation.—1. **By the union of the lower oxides with oxygen.**—Nitrous oxide, N_2O, does *not* combine directly with oxygen; but nitric oxide, NO, mixed with half its volume of oxygen, at once combines, forming a mixture of peroxide and tetroxide. Nitrogen trioxide, N_2O_3, which is a blue liquid, is also slowly converted into peroxide and tetroxide when kept in presence of oxygen or air.

Vanadium tetroxide is formed when the trioxide is heated in air; but on prolonged heating it is oxidised to the pentoxide.

2. **By depriving a higher oxide of oxygen.**—It has been already remarked that nitrogen pentoxide decomposes spontaneously into peroxide and oxygen. Nitric acid is more stable; but when its vapour is led through a red-hot tube, a large proportion is decomposed. It is more convenient, however, to deprive nitric acid of oxygen by distilling it with arsenious anhydride.

The reaction is:—$4N_2O_5.H_2O + As_4O_6 = 4N_2O_4 + 2As_2O_5 + 4H_2O$. The water, however, reacts with the tetroxide, thus:—$3N_2O_4 + 2H_2O = 4HNO_3 + 2NO$; and a mixture of tetroxide, peroxide, and nitric oxide is produced. On condensing the product, these combine to form trioxide, thus, $NO_2 + NO = N_2O_3$. Hence, in order to remove water from the sphere of action, a considerable quantity of strong sulphuric acid or phosphorus pentoxide is added. The product is then pure peroxide and tetroxide. To remove nitric acid, some of which is apt to distil over, the liquid is again distilled, with addition of a little more arsenic trioxide and phosphorus pentoxide.

The tetroxide may also be formed by the action of nitric pentoxide on the trioxide. The blue liquid containing trioxide may be rendered orange by addition of a mixture of nitric acid and phosphoric anhydride, which must contain N_2O_5.

When a nitrate is heated, it decomposes into an oxide and oxides of nitrogen. If the pentoxide were not so unstable, one would expect that it would be formed, but, as a rule, the peroxide resolved by heat into nitric oxide and oxygen is produced by its decomposition. On cooling the resulting gases they re-combine to form tetroxide and peroxide. The most convenient nitrate to employ is that of lead. The equation is:—

$$\mathbf{Pb(NO_3)_2} = \mathbf{PbO} + 2NO_2 + O.$$

Metallic tin may also be used to withdraw oxygen from nitric acid. The equation is:—

$$\mathbf{Sn} + 4HNO_3 = \mathbf{SnO_2} + 2N_2O_4 + 2H_2O.$$

Nitric and nitrous oxides, NO and N_2O, are, however, produced simultaneously. The nitric acid must be strong and somewhat warm. It will be remembered that the tin is oxidised to metastannic acid, $5\mathbf{SnO_2}.5\mathbf{H_2O}$.

The compound, VO_2Cl, decomposes when heated in carbon dioxide into $\mathbf{VO_2}$ and chlorine.

Niobium pentoxide is reduced to tetroxide by heating it to whiteness in hydrogen, and tantalum pentoxide when heated to whiteness in a crucible lined with carbon loses oxygen, leaving the tetroxide.

Properties.—Nitrogen tetroxide is a colourless solid below $-10{\cdot}14°$. At that temperature it melts, but the liquid has a pale straw colour, owing to incipient dissociation. As the temperature rises its colour changes to yellow and then orange-red; it boils at

22°, giving off a brown-red gas, which consists largely of the peroxide. The peroxide is not known in the solid form, but the liquid tetroxide apparently contains some, judging from its colour. The liquid compound is heavier than water (1·45 at 15°). It reacts with ice-cold water, forming nitrous and nitric acids, $N_2O_4 + H_2O = HNO_3 + HNO_2$; and at higher temperatures forming nitric acid and nitric oxide, $3N_2O_4 + 2H_2O = 4HNO_3 + 2NO$. It dissolves in strong nitric acid, forming the red fuming acid often employed for oxidation of sulphides, &c.; and in sulphuric acid, giving salts of nitrosyl, NO (see sulphates). It acts violently on cork and indiarubber, hence, in preparing it, all the joints should be of sealed glass.

Vanadium tetroxide, V_2O_4 or **VO_2**, is a dark green amorphous powder, insoluble in water, but soluble in hydroxides of sodium and potassium, forming hypovanadates, and in acids, forming salts of vanadyl (VO).

Niobium tetroxide is a dense black insoluble powder, which on ignition in air yields the pentoxide; and **tantalum tetroxide** is a dark substance, which acquires metallic lustre under the burnisher.

Tetrasulphides of vanadium, V_2S_4, and tantalum, Ta_2S_4 (?) are known. The first, produced by heating the tetroxide in a stream of hydrogen sulphide, is a black powder, insoluble in water, alkalies, or alkaline sulphides; the second, which may be an oxysulphide, is produced by heating tantalum pentoxide in vapour of carbon disulphide or tantalum pentachloride in hydrogen sulphide. It is a black powder, which when burnished acquires a brass-yellow lustre.

Compounds with oxides and sulphides.—Nitric peroxide does not combine with water, but is decomposed (see above). It combines, however, with lead oxide, producing a compound of the formula **PbN_2O_5**, which may be a salt of the hypothetical acid, $H_2N_2O_5$, or may be a double nitrite and nitrate of lead, $Pb<^{NO_2}_{NO_3}$. Similar compounds, but containing more lead oxide, are produced by heating lead nitrate with metallic lead.

Vanadium tetroxide dissolves in alkalies, forming **hypovanadates.** On addition of a hydroxide to its solution in hydrochloric or sulphuric acids, its hydrate, **$V_2O_4.7H_2O$**, is thrown down as an amorphous black precipitate, which may be viewed as hydrated hydrogen hypovanadate. An arbitrary division is usually drawn between the compounds called hypovanadates and those termed vanadyl salts. They are here considered as chemically similar; both contain vanadium tetroxide in combination with other oxides. They are as follows:—

$$2V_2O_4.K_2O.7H_2O;\ 2V_2O_4.Na_2O.7H_2O;\ 2V_2O_4.(NH_4)_2O.3H_2O;\ 2V_2O_4.BaO;\ 2V_2O_4.PbO;\ 2V_2O_4.Ag_2O.$$

These are termed hypovanadates. There are also **$V_2O_4.3SO_3.4H_2O$** and **$15H_2O$**; **$V_2O_4.2SO_3.7H_2O$** and **$10H_2O$**. These are termed vanadyl sulphates, and will be considered among the sulphates.

Potassium hypovanadate, **$2V_2O_4.K_2O.7H_2O$**, forms dark brown crystals, soluble in water, but nearly insoluble in caustic potash, and quite insoluble in alcohol. The sodium salt is similar. The barium, lead, and silver salts are brown or black, and are produced by precipitation.

Hydrated vanadium tetrasulphide is precipitated on addition of an acid to a solution of the tetroxide in sulphides of the alkalies. It is a brown powder. It dissolves in sulphides of the alkalies, forming the **hyposulphovanadates.** These have been little studied; they are black solids dissolving with a brown colour. Those of the alkalies are soluble, and give precipitates with solutions of the metals.

Compounds with halides.—Compounds of the formulæ $NOCl_3$ and NO_2Cl, though it has been stated that they are formed by various reactions, have been proved to consist of solutions of chlorine in nitrosyl chloride, $NOCl$, or in nitrogen tetroxide. Nor is any compound known of the formula $NOCl_2$.

But vanadium oxytrichloride, $VOCl_3$, when heated to 400° with metallic zinc is converted into **$VOCl_2$**, a light green crystalline solid, deliquescent, and soluble in alkalies. The corresponding bromide is a yellow-brown deliquescent solid, produced by heating the tribromide to 180°. The corresponding fluoride, VOF_2, is known in combination with ammonium fluoride, in the blue monoclinic crystals of **$VOF_2.2NH_4F$**, produced by adding hydrogen ammonium fluoride, HNH_4F_2, to a solution of tetroxide, V_2O_4, in hydrofluoric acid.

Trioxides, N_2O_3; V_2O_3.—Preparation.—Nitrogen trioxide, or nitrous anhydride, is produced by the union of nitric peroxide, NO_2, with nitric oxide, NO. It is apparently formed by all reactions involving these products; but as it cannot exist in the gaseous state, it is formed only on cooling the mixture of its products of decomposition.* Such a gaseous mixture is liberated on treating a nitrite with sulphuric acid, thus:—

* *Chem. Soc.*, **47**, 187.

$$2\mathbf{KNO_2} + H_2SO_4 = \mathbf{K_2SO_4} + H_2O + \mathit{NO_2} + \mathit{NO};$$

or by adding water to hydrogen nitrosyl sulphate:—

$$2H(NO)SO_4 + Aq = 2H_2SO_4.Aq + \mathit{NO_2} + \mathit{NO}.$$

When fairly pure it is a mobile blue liquid, stable only at a very low temperature. It does not solidify even on cooling it to about −90°. If warmed, it decomposes into its constituents; and as more nitric oxide escapes than peroxide, the colour of the remaining portion changes to green, and subsequently to dirty red: for the colour of the remaining peroxide is changed by that of the blue trioxide. It is also formed by the action of a small quantity of water on nitrogen tetroxide, thus:—$2N_2O_4 + H_2O = 2HNO_3 + N_2O_3$; and this is one of the easiest methods of preparing it. A mixture of nitric oxide, *NO*, and oxygen, even if the oxygen be in excess, combines to some extent to form the trioxide when cooled in a freezing mixture.

Vanadium trioxide is produced by heating vanadium pentoxide in a current of hydrogen or with carbon. It is also formed when $\mathbf{V_2O_2}$ is heated gently in air. It is a black insoluble powder, possessing semi-metallic lustre. It is insoluble in acids. When heated to redness in air, it glows and burns to the pentoxide. It is a conductor of electricity. Like nitrogen trioxide, it combines with oxides, forming the vanadites.

No trioxide of niobium or tantalum has been prepared.

Compounds with oxides.—Nitrites and vanadites.—It is probable that two sets of nitrites exist, having the same formulæ but different constitution; these may be regarded as derivatives of two hypothetical nitrous acids, $HN\lessdot^{O}_{O}$, and $HO—N=O$.

It is probable that the silver and mercury salts are derivatives of the first, and the potassium and calcium salts of the second. The reason for this view is as follows:—

The compound of carbon, hydrogen, and iodine, known as methyl iodide, has the formula CH_3I. When heated with silver nitrite in a sealed tube, silver iodide is produced, along with the compound $CH_3.NO_2$, named nitromethane. Now, on exposing this liquid to the action of nascent hydrogen, produced, for example, by the action of tin on hydrochloric acid, the following reaction occurs:—

$$CH_3.NO_2 + 6H = CH_3.NH_2 + 2H_2O;$$

the oxygen is replaced by hydrogen, forming the compound

$(CH_3).NH_2$, analogous to ammonia, NH_3; and it is argued that the nitrogen and the carbon must be combined with each other.

On heating methyl iodide, CH_3I, with potassium nitrite, on the other hand, a compound of the same formula is produced, viz., $CH_3.NO_2$, along with potassium iodide. But this body, which is named methyl nitrite, differs entirely in properties from its isomeride, nitromethane. And on treatment with nascent hydrogen, this reaction takes place:—

$$CH_3.NO_2 + 6H = CH_3.OH + NH_3 + H_2O.$$

The body $CH_3.OH$ is named methyl alcohol, and it is certain that carbon and oxygen are here combined. Hence the formula $CH_3O—NO$ is attributed to it, and $KO—NO$ to the nitrite from which it is derived; whereas silver nitrite has apparently the formula $Ag—NO_2$.

These conclusions are confirmed by a study of the action of caustic potash on these bodies. For while nitromethane reacts thus:—

$$CH_3.NO_2 + KOH = CH_2.KNO_2 + H_2O,$$

methyl nitrite is decomposed, thus:—

$$CH_3.ONO + KOH = CH_3.OH + KONO,$$

the original potassium nitrite being reproduced.

While, therefore, silver nitrite should probably be regarded as a nitride of silver and oxygen, and should be considered among the nitrides, and potassium nitrite as a derivative of nitrous anhydride, yet we do not know which bodies to place in one class and which in the other; and as we are not sure whether some of the compounds named nitrites are not mixtures of both compounds, it is more convenient to include both varieties at present in one class.*

Preparation.—The nitrites are prepared: 1. **By reducing the nitrates.** This is best done by fusing them with metallic lead. For instance, three parts of potassium nitrate fused with two parts of metallic lead with constant stirring yield potassium nitrite and lead monoxide, thus:—

$$KNO_3 + Pb = KNO_2 + PbO.$$

Potassium sulphite may also be employed as a reducing agent.

2. **By the action of a mixture of NO_2 and NO on**

* *Chem. Soc.*, **47**, 203, 205, 631.

hydroxides.—Those reactions which produce such mixtures in correct proportions are to be preferred. An example is—

$$NO + NO_2 + 2KOH.Aq = 2KNO_2.Aq + H_2O.$$

3. **By passing a mixture of oxygen and ammonia over heated platinum black** (finely divided platinum), ammonium nitrite is formed, thus:—

$$2NH_3 + 3O = \mathbf{NH_4NO_2} + H_2O.$$

The nitrites of lead and silver are nearly insoluble, whereas the nitrates are very soluble salts; hence, on adding to a nitrite a soluble salt of one of these metals (nitrates), the respective nitrites are precipitated. They may be converted into other nitrites by digestion with a soluble chloride in the case of silver, or a sulphate in the case of lead.

List of Nitrites.—The following have been prepared:—

$\mathbf{NaNO_2}$; $\mathbf{KNO_2}$; $\mathbf{NH_4NO_2.H_2O}$.

White deliquescent salts. That of sodium is soluble in alcohol. The ammonium salt is produced by addition of nitrous anhydride, N_2O_2, to ammonia, keeping it cold; or by mixing solutions of lead nitrite and ammonium sulphate, filtering off insoluble lead sulphate, and evaporating in a vacuum to crystallisation. When heated, even in solution, it undergoes the curious decomposition $NH_4NO_2 = N_2 + 2H_2O$.

This forms a convenient method of preparing pure nitrogen. It may be carried out more conveniently by heating a mixture of potassium nitrite and ammonium chloride, best after addition of copper sulphate.

The corresponding **vanadites** have not been analysed. They are produced by dissolving vanadium trioxide in alkalies. They are red when hydrated, but green when anhydrous.

$\mathbf{Ca(NO_2)_2.H_2O}$; $\mathbf{Sr(NO_2)_2.H_2O}$; $\mathbf{Ba(NO_2)_2.H_2O}$; $\mathbf{Ba(NO_2)_2.KNO_2.H_2O}$.

These salts may be formed by heating a nitrate of one of these metals, dissolving the product in water, and, in order to separate oxide, passing carbon dioxide to remove it as carbonate. The filtrate is evaporated and crystallised. Calcium nitrite is insoluble in alcohol. These are all soluble white salts.

$\mathbf{Mg(NO_2)_2.3H_2O}$ and $\mathbf{2H_2O}$; $\mathbf{Zn(NO_2)_2.3H_2O}$; $\mathbf{Cd(NO_2)_2.H_2O}$. Also **basic salts**:—$\mathbf{N_2O_3.2ZnO}$, and $\mathbf{N_2O_3.2CdO}$; and **double salts**, $\mathbf{Cd(NO_2)_2.2KNO_2}$, and $\mathbf{Cd(NO_2)_2.4KNO_2}$.

These are all white soluble salts.

Nitrites of chromium and iron have not been investigated.

Manganous nitrite is a pink deliquescent salt; that of **cobalt** is rose-coloured, and of **nickel** green.

The double nitrates of the last two metals are better known. They are as follows:—

$3(Co(NO_2)_2.2KNO_2).H_2O$, also with other amounts of water.
$Ni(NO_2)_2.Ca(NO_2)_2.KNO_2$; also similar strontium and barium salts.

These contain the metals as dyads, and are derivatives of CoO, and NiO.

$2Co(NO_2)_3.4NaNO_2.H_2O$; also $6NaNO_2.H_2O$; $2Co(NO_2)_3.4K_2O$.

These compounds are produced by boiling a cobalt salt with acetic acid and nitrite of sodium or potassium. The cobalt is here triad, as in Co_2O_3. Nickel forms no corresponding compounds, and as the double nitrite of cobalt and potassium is nearly insoluble in water, its formation is used as a means of separating cobalt from nickel. It has a bright yellow colour, and is therefore used as a pigment.

The following compounds of lead are known:—

$Pb(NO_2)_2$; $N_2O_3.2PbO.H_2O$, and $3H_2O$; $N_2O_3.3PbO.H_2O$; $N_2O_3.4PbO.H_2O$.

The last three are yellow bodies, and are made by boiling a solution of lead nitrate with metallic lead; the first, by passing a current of carbon dioxide through one of the latter suspended in water; the excess of lead oxide is removed as carbonate. When lead nitrate solution is boiled with lead, a double nitrate and nitrite is also formed. Its formula is $4Pb{<}^{NO_2}_{NO_3}.H_2O$; a basic salt is also produced, viz., $N_2O_3.N_2O_5.9PbO.3H_2O$. The first of these has been viewed as a salt of the anhydride N_2O_4; as $N_2O_4.PbO$ (see p. 335); but the formula given is more probably correct.

Copper nitrite, $Cu(NO_2)_2$ is an apple-green crystalline salt; and silver nitrite, $AgNO_2$, forms long needle-shaped pale-yellow crystals, sparingly soluble in cold water.

Some interesting double nitrites of platinum have been prepared (see pp. 485 and 544).

Compounds with halides.—NOCl;* VOCl. The first of these bodies has the molecular weight given by the formula. It is prepared (1) by passing a mixture of nitric peroxide and chlorine through a red-hot tube. The nitric peroxide is doubtless dissociated into nitric oxide and oxygen, and the former combines with the chlorine. It is also produced by direct combination of nitric oxide with chlorine at a red heat. (2) By the action of salt (**NaCl**) on hydrogen nitrosyl sulphate, $H(NO)SO_4$, produced by

* *Chem. Soc.*, **27**, 630; **49**, 222.

saturating strong sulphuric acid with NO_2 and NO, thus: $H(NO)SO_4$ + **NaCl** = $HNaSO_4$ + $NOCl$. (3) Along with free chlorine, by heating a mixture of hydrochloric and nitric acids, thus:—$3HCl + HNO_3 = 2H_2O + NOCl + Cl_2$; and probably by the action of hydrogen chloride on nitrogen tetroxide, which may be regarded as nitrate of nitrosyl, $NO(NO_3)$, thus:—

$$HCl + NO(NO_3) = NOCl + HNO_3.$$

A mixture of nitric and hydrochloric acids has been long known under the name "**aqua regia.**" Owing to the nascent chlorine, it has the property of dissolving gold and platinum, converting them into chlorides. It is a powerful oxidising agent, the chlorine reacting with water forming nascent oxygen and hydrogen chloride.

The corresponding vanadosyl chloride, **VOCl**, is a brown powder formed by heating the trichloride, $VOCl_3$, to redness in a current of hydrogen. At the same time the compound $\mathbf{V_2O_2Cl}$ is formed as a heavy shining powder like mosaic gold, and also the oxide $\mathbf{V_2O_2}$ **or VO**. With vanadium we have thus the series, $VOCl_3$, $\mathbf{VOCl_2}$, **VOCl**, and $\mathbf{V_2O_2Cl}$.

Nitric oxide and vanadium dioxide, NO and $\mathbf{V_2O_2}$. The first of these is often erroneously named nitrogen dioxide. Its formula, however, even at $-100°$, is NO, as shown by its vapour-density. No tendency towards increased density has been noticed; the gas contracts *pari passu* with hydrogen. The molecular weight of the vanadium compound is unknown, but as it is derived from $\mathbf{V_2O_2Cl}$, it is possibly $\mathbf{V_2O_2}$.

Nitric oxide is produced in an impure state by the action of nitric acid on certain metals. It is probable that the normal action of nitric acid is similar to that of other acids; that a nitrate is produced with liberation of hydrogen. But nascent hydrogen (*i.e.*, hydrogen in the state of being liberated, when it consists in all probability of single uncombined atoms) cannot exist in presence of nitric acid, but deprives it of oxygen. In theory, the following reactions are possible:—

$$2HNO_3 + M'' = M(NO_3)_2 + 2H.$$
$$N_2O_5 + 2H = N_2O_4 + H_2O;\ N_2O_5 + 4H = N_2O_3 + 2H_2O;$$
$$N_2O_5 + 6H = 2NO + 3H_2O;\ N_2O_5 + 8H = N_2O + 4H_2O;$$
$$N_2O_5 + 10H = N_2 + 5H_2O;\ N_2O_5 + 16H = 2NH_3 + 5H_2O.$$

The conditions determining the prevalence of any one of these reactions are temperature, presence of water, and of the products of reaction. But the oxides of nitrogen produced may themselves react with water or with nitric acid. For example, if N_2O_4 be liberated in presence of water, the reaction described on p. 337 will take place, and a mixture of nitric oxide and nitric acid will be produced. But some peroxide may escape along with nitric oxide. The gases *NO*, N_2O, and nitrogen, not being affected by water, will be liberated as such, if formed.

Nitric acid diluted with its own volume of water, acts on copper at 15° and on aluminium at 60—65°, producing a mixture containing 98 and 97 per cent. respectively of nitric oxide, along with a small amount of nitrous oxide and nitrogen. With silver, acid of the same strength at 15° gives 31 per cent of nitric oxide and 60 per cent. of nitrous oxide, N_2O, while iron with nitric acid of any dilution, gives chiefly nitric oxide (from 86 to 91 per cent.).*

The action of nitric acid on copper therefore forms the most convenient method of preparing nitric oxide. The equation is :— $3\mathbf{Cu} + 8HNO_3.Aq = 3Cu(NO_3)_2.Aq + 4H_2O + 2NO$. To prepare the pure compound, this gas is passed through a strong cold solution of ferrous sulphate, $FeSO_4$, with which nitric oxide combines† (see p. 428). On warming the solution, the compound is decomposed, and pure nitric oxide is liberated. It is a colourless, nearly insoluble gas, which, when mixed with air or oxygen, gives red fumes of nitric peroxide. It condenses to a colourless liquid at −11° under a pressure of 104 atmospheres. Under normal pressure, it boils at −153·6°, and begins to solidify when the pressure is reduced to 138 mms. at −167°. It does not support combustion, but like other gases containing oxygen, it is decomposed at a high temperature, and thus glowing charcoal or phosphorus burn in it. With the vapour of carbon disulphide it forms a mixture which, when set on fire, burns rapidly with a brilliant blue-white flame. When mixed with hydrogen, it can be exploded by a powerful spark.

The corresponding oxide of vanadium, $\mathbf{V_2O_2}$, may be formed by the action of potassium on a higher oxide of vanadium, and used to be considered to be metallic vanadium. It is also produced when a mixture of vanadyl trichloride, $VOCl_3$, and hydrogen are passed through a tube full of red-hot charcoal. It is a light-grey powder with metallic lustre, difficult of fusion, and insoluble in water and acids. When heated in air, it burns to higher oxides.

It may be produced in solution by reducing a solution of vanadium pentoxide in sulphuric acid by means of zinc. Such a solution has a lavender colour, and is one of the most powerful reducing agents known.

* *Chem. Soc.*, **28**, 828; **32**, 52.

† *Compt. rend.*, **89**, 410.

Nitrogen sulphide and selenide, NS and NSe.—The first is produced by the action of ammonia on sulphur chloride dissolved in carbon disulphide, thus:—$8NH_3 + 3S_2Cl_2 = 6NH_4Cl + 2NS + 4S$; the ammonium chloride, being insoluble in carbon disulphide, is removed by filtration, and the carbon disulphide on evaporation deposits nitrogen sulphide in yellow rhombic prisms. The corresponding selenium compound, produced, however, from selenium tetrachloride, is an amorphous, orange-coloured, insoluble substance. Both of these bodies explode by percussion.

When mixed with chloroform and treated with chlorine, sulphur-yellow crystals of the formula **NSCl** are deposited, analogous to nitrosyl chloride, NOCl. A second chloride $(NS)_3Cl$ is also formed; it deposits in copper-coloured needles.

Nitroso-sulphides.—A curious set of compounds of nitric oxide with sulphide of iron and of a metal has been produced* by dropping a solution of ferric chloride into a mixture of solutions of potassium nitrite and ammonium sulphide, when black crystals of $Fe_3S_4(NO)_4.H_2S$ are deposited. When the solution of these crystals is heated with caustic soda, they yield large black crystals of the compound $Fe_2S_3(NO)_2.3Na_2S$; and with an acid, a black precipitate of "nitrososulphide of iron," $Fe_2S_3(NO)_2$ separates. The first compound, heated to 100° with sodium sulphide, deposits red prisms of the body $Fe_2S_3(NO)_2.Na_2S.H_2O$.

The constitution of these bodies is unknown; but they appear to be related to the nitroferricyanides (see p. 566). It is suggested that a corresponding amido-compound has the formula $Fe(NO_2).SNH_2$, and the last nitroso-sulphide may be analogously represented $Fe(SNa).SNO$.

Nitrous oxide, N_2O, is produced (1) by the action of metals on nitric acid. Zinc and pure nitric acid at 15° yield a mixture consisting of 1 per cent. of nitric oxide, 78 per cent. of nitrous oxide, and 21 per cent. of nitrogen. Nickel and cobalt, too, with acid diluted with its own volume of water, yield a mixture containing about 80 per cent.; and tin, at ordinary temperatures, furnishes a mixture containing from 67 to 85 per cent., with acids of all concentrations. (2) The simplest method of preparation is to heat ammonium nitrate to above 185°, when it decomposes like the nitrite, thus:—$NH_4NO_3 = N_2O + 2H_2O$. (3) Nitrous oxide is also formed by the action of an acid or an hyponitrite (see below).

* *Berichte*, **15**, 2600.

Nitrous oxide, or hyponitrous anhydride, as it is sometimes named, is a colourless gas, possessing a faint sweetish smell and taste. It is somewhat soluble in water, and is best collected over hot water, or by downward displacement. When exposed to a sudden shock, as, for instance, the detonation of a fulminate, it explodes into its constituents; this is a property common to bodies produced with absorption of heat. It is condensed by pressure to a liquid, boiling at −88° to −92°, and when the liquid is evaporated by a current of air some of it freezes to a white solid, melting at −99°. Its most striking property is its action on the nervous system when breathed, which has gained for it the name "laughing-gas." When pure, it produces insensibility, and is used as an anæsthetic in minor surgical operations and in dentistry; but when diluted with air it causes excitement and intoxication. It easily decomposes when heated, hence a candle burning brightly continues to burn more brightly in the gas. But if the candle is burning feebly it is extinguished.

Compounds with oxides.—Hyponitrites.*—A solution of potassium nitrate or nitrite, exposed to nascent hydrogen generated from sodium amalgam (an alloy of sodium and mercury) loses oxygen, and potassium hyponitrite, KNO, is produced, about 15 per cent. of the nitrate or nitrite suffering change. The same compound is formed by fusing iron filings with potassium nitrate. The sodium salt forms white, needle-shaped crystals, and has the formula $\mathbf{NaNO.3H_2O}$. With silver nitrate, in presence of acetic acid, the silver salt is precipitated; it is a pale yellow body, of the formula **AgNO**. On addition of hydrochloric acid to the silver salt suspended in water, the acid, presumably HNO, is liberated. It reduces potassium permanganate; and on standing, decomposes into water and nitrous oxide. No other salts have been analysed; but a solution of the sodium salt gives precipitates with soluble salts of most metals, almost all of which are insoluble in acetic acid.

We have thus a series of oxides and acids of nitrogen, vanadium, niobium, and tantalum:—

N_2O, nitrous oxide or hyponitrous anhydride.	HNO acid.
NO, nitric oxide.	
N_2O_3, nitrogen trioxide or nitrous anhydride.	HNO_2 acid.
N_2O_4, NO_2, nitrogen tetroxide and peroxide.	
$\mathbf{N_2O_5}$, nitrogen pentoxide or nitric anhydride.	HNO_3 acid.
	$H_2N_4O_{11}$ acid.
$\mathbf{N_2O_6}$,† nitrogen hexoxide.	

* Divers, *Proc. Roy. Soc.*, **19**, 425; **33**, 401; *Chem. Soc.*, **45**, 78; **47**, 361.

† The hexoxide has been formed by passing sparks through a mixture of

Similarly:—

V_2O_2	—	—	—		
V_2O_3	—	—	HVO_2(?)	—	
V_2O_4	Nb_2O_4	Ta_2O_4	$H_2V_4O_9$(?)	—	
V_2O_5	Nb_2O_5	Ta_2O_5	$\left\{\begin{matrix} H_3VO_4 \\ H_4V_2O_7 \\ HVO_3 \end{matrix}\right\}$	$Nb_2O_5.nH_2O$	$Ta_2O_5.nH_2O$.

Physical Properties.

Mass of 1 cubic centimetre.

	Nitrogen.	Vanadium.	Niobium.	Tantalum.
Monoxides	See below.	—	—	—
Dioxides	—	3·64 at 20°	—	
Trioxides	—	4·72 at 16°	—	—
Tetroxides......	1·49 at 0°	—	—	—
Pentoxides	—	3·5 at 20°	4·37—4·53	7·35—8·01

Mass of 1 c.c. N_2O.

Temp.	−20·6°	−11·6°	−5·5°	−2·2°	+6·6°	+11·7°	+19·8°	+23·7°.
Mass.	1·002	0·952	0·930	0·912	0·849	0·810	0·758	0·698

NS, 2·22 at 15°; VS_2, 4·7 at 21°; V_2S_5, 3·0.

HNO_3, 1·552, at 15°. $2N_2O_5.H_2O$, 1·642 at 18°. $VOCl_2$, 2·88 at 13°; $VOCl_3$, 1·865 at 0°.

Heats of Combination.

$2N + O = N_2O - 180K$. $2N + 3O + Aq = N_2O_3.Aq - 68K$.
$N + O = NO - 215K$. $N + 2O = NO_2 - 77K$.
$2NO_2 = N_2O_4 + 129K$. $2N + 4O = N_2O_4 - 26K$.
$2N + 5O = N_2O_5 + 131K$; $+ Aq = 2HNO_3.Aq + 167K$.
$N_2O_4 = N_2O_4 - 31K$. $N_2O_5 = N_2O_5 - 83K$.

Specific heat of gaseous N_2O_4 *or* NO_2.*

Temp.....	26·5° – 66·7°	27·7° – 103·1°	28·9° – 150·6°	29·0° – 198·5°	29·2° – 253·1°	27·6° – 289·5°
Spec. heat	0·747	0·663	0·513	0·395	0·319	0·298.

---

oxygen and nitrogen, cooled to −23°. From volumetric measurements the compound produced—a volatile crystalline powder—is declared to have the formula NO_3 (*Comptes. rend*, **94**, 1306).

* This great change is due to absorption of heat in the conversion of N_2O_4 into NO_2 (*Compt. rend.*, **64**, 237).

CHAPTER XXIII.

OXIDES, SULPHIDES, SELENIDES, AND TELLURIDES OF ELEMENTS OF THE PHOSPHORUS GROUP.—CONSTITUTION OF PHOSPHORIC ACID, ETC.—THE PHOSPHATES, ARSENATES, ANTIMONATES, SULPHOPHOSPHATES, SULPHARSENATES, AND SULPHANTIMONATES; PYROPHOSPHATES, METAPHOSPHATES, AND ANALOGOUS COMPOUNDS.

Oxides, Sulphides, Selenides, and Tellurides of Phosphorus, Arsenic, Antimony, and Bismuth.

List of Oxides, Sulphides, Selenides, and Tellurides.

P_4O.	—	—	—	P_4S_3.	—	—	—
—	—	—	Bi_2O_2.	—	As_2S_2.	—	Bi_2S_2.
P_4O_6.	As_4O_6.	Sb_4O_6.	Bi_4O_6.	—	As_2S_3.	Sb_2S_3.	Bi_2S_3.
P_2O_4.	—	Sb_2O_4.	Bi_2O_4.	P_2S_4.	—	—	—
P_2O_5.	As_2O_5.	Sb_2O_5.	Bi_2O_5.	P_2S_5.	As_2S_5.	Sb_2S_5.	—

Selenides and tellurides.—P_2Se_5; $AsSeS_2$; As_2Te_2; As_2Te_3; Sb_2Se_3; $SbTe$; Sb_2Te_3; Bi_2Se_3; Bi_3Te; Bi_3Te_2; Bi_2Te_3.

Sources.—**Pentoxide of phosphorus** occurs in combination with oxides of metals, especially calcium and aluminium, as *apatite*, *phosphorite*, *wavellite*, &c. **Arsenious oxide**, As_4O_6, is found as *arsenite*, or *arsenic bloom*; and Sb_4O_6 as *antimony bloom* in trimetric prisms, and as *senarmontite* in regular octahedra. The oxide, Sb_2O_4, is named *antimony ochre*.

The **sulphides** As_2S_2 (*realgar*) As_2S_3 (*orpiment*), Sb_2S_3 (*stibnite*), and Bi_2S_3 (*bismuthine*) also occur native; as well as in combination with many other sulphides.

Preparation.—1. **By direct union.**—When phosphorus is burned in excess of air or oxygen, the pentoxide is formed. Arsenic and bismuth burn to trioxides; and antimony to trioxide and tetroxide. In a limited supply of air, and at moderately high temperature, phosphorus gives P_4O, P_2O_4, and P_2O_5; by careful regulation of air a considerable amount of P_4O_6 is produced, even as much as 50 per cent., the other oxide being mainly P_2O_5.

The process of preparing phosphorus pentoxide is to drop pieces of dry phosphorus through a tube passing through a cork closing the neck of a glass balloon, while a current of air, dried by passing through a U-tube filled with pumice-stone moistened with sulphuric acid, is blown in. The fumes are

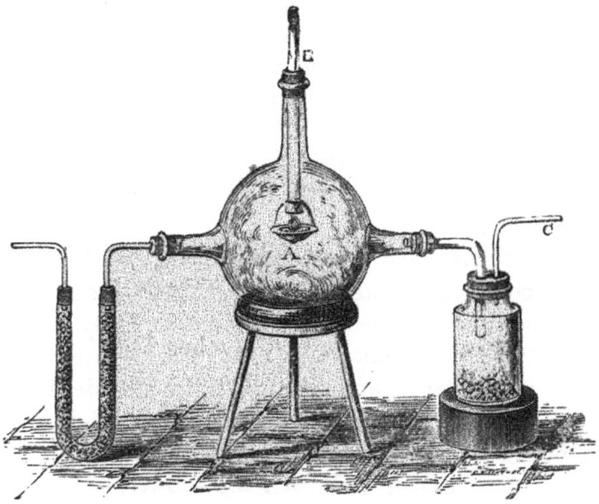

FIG. 40.

condensed partly in the balloon, partly in the bottle communicating with it by a wide-mouthed tube.

By the glowing of phosphorus in dry air the pentoxide is the only product.

Arsenious oxide, As_4O_6, is usually produced by condensing in brick chambers the fumes resulting from the roasting in muffles of arsenical ores of tin, cobalt, and nickel, or arsenical pyrites. To purify it, the condensed product is sublimed in cast-iron pots.

By limiting the supply of air, antimony burns to Sb_4O_6, but with free access of air, to Sb_2O_4.

The sulphides, selenides, and tellurides of all these elements are produced by direct union.

2. **By decomposition of other oxides.**—Phosphorus tetroxide, P_2O_4,* is produced by distilling in a vacuum the product of the combustion of phosphorus in a slow current of air. Bright orthorhombic crystals sublime, of the formula P_2O_4, arising from the decomposition of the phosphorous oxide, thus:—

$$7P_4O_6 = 10P_2O_4 + 2P_4O.$$

Arsenic pentoxide loses oxygen, forming trioxide at a dull red heat; antimony pentoxide yields tetroxide at temperatures above 275°; and bismuth pentoxide, heated to 250°, is converted into

* *Chem. Soc.*, **49**, 833.

tetroxide, and over 305° into trioxide. No known rise of temperature, however great, deprives phosphorus pentoxide of oxygen.

3. **By oxidation in the wet way.**—This method in reality yields hydrates or acids. The usual oxidising agents are a mixture of nitric and hydrochloric acids (*aqua regia*, see p. 341), or caustic potash and chlorine or bromine. Water cannot be expelled by heat from phosphoric acid, $\mathbf{P_2O_5.H_2O} = \mathbf{HPO_3}$; but arsenic acid, $\mathbf{As_2O_5.H_2O}$, is dehydrated at a dull red heat, antimonic acid, $\mathbf{Sb_2O_5.H_2O}$, by heating not above 275°, and hydrated bismuth pentoxide at 120°.

4. **By decomposition of compounds.**—The hydrates, as remarked above, lose water; and the nitrates and sulphates of antimony and bismuth decompose, when strongly heated, leaving trioxides. Phosphoryl chloride, $POCl_3$, when heated with metallic zinc, yields zinc chloride and tetraphosphorus oxide, $\mathbf{P_4O}$; and the same body is formed by heating phosphoryl chloride with phosphorus, thus:—

$$POCl_3 + \mathbf{P_4} = PCl_3 + \mathbf{P_4O}.$$

5. **By double decomposition.**—As a rule, this process yields the hydroxides or acids, for example: $PCl_3 + 3H_2O = H_3PO_3 + 3HCl$; $POCl_3 + 3H_2O = H_3PO_4 + 3HCl$; $2\mathbf{SbOCl} + 2KOH.Aq = \mathbf{Sb_2O_3.Aq} + 2KCl.Aq + H_2O$; $2BiCl_3 + 6KOH.Aq = \mathbf{Bi_2O_3.H_2O} + 6KClAq + 2H_2O$; and, with the exception of the compounds of phosphorus, these yield oxides when heated. It forms, however, the usual method of preparing the sulphides, excepting those of phosphorus: *e.g.*, $2AsCl_3.Aq + 3H_2S = \mathbf{As_2S_3} + 6HCl.Aq$; $2SbCl_3.Aq + 3H_2S = \mathbf{Sb_2S_3} + 6HCl.Aq$, &c.

Properties.—$\mathbf{P_4O}$ is a light red or orange powder resembling red phosphorus, for which it was formerly taken; when prepared by oxidation of phosphorus, it possesses reducing properties; but when by depriving $POCl_3$ of chlorine, it does not reduce salts of mercury, silver, or gold.

Phosphorous oxide or **anhydride**, P_4O_6, forms feathery crystals, melting at 22·5°, and boiling at 173·3°. It is decomposed by heat thus:—

$$2P_4O_6 = 3P_2O_4 + P_2.$$

It is slowly attacked by cold water, with formation of phosphorous acid, H_3PO_3, and immediately and with violence by hot water. It is luminous in the dark in presence of oxygen at a less pressure than that of the air; and when heated gently in air, it burns to P_2O_5. It also burns in chlorine, forming $POCl_3$ and PO_2Cl.

The **tetroxide** forms orthorhombic crystals. It is soluble in water, giving a mixture of phosphorous and phosphoric acids, thus:—$\mathbf{P_2O_4} + 3H_2O = H_3PO_4 + H_3PO_3$. It is, therefore, supposed to have the formula P_2O_4 or $PO(PO_3)$; it would then be named phosphoryl metaphosphate. But of this there is no other proof.

The **pentoxide** or **phosphoric anhydride** is a snow-white powder, volatile below redness. It has a great tendency to combine with water, and is, therefore, used as a dehydrating agent, *e.g.*, in the preparation of nitrogen pentoxide and sulphur trioxide. When heated with carbon, it yields carbon monoxide and phosphorus.

Arsenious oxide or **anhydride,** sometimes called **arsenic trioxide,** exists in three forms. When condensed at high temperatures, it is an amorphous porcelain-like mass; its specific gravity is then 3·74. When cooled quickly, or when it crystallises from solution, it forms colourless regular octahedra, the specific gravity of which is nearly the same, viz., 3·70. But when crystallised at low temperatures, or when it separates from its saturated solution in caustic potash, it forms rhombic crystals of the specific gravity 4·25.

Arsenious oxide is sparingly soluble in water (vitreous, 4 in 100; crystalline, 1·2 or 1·3 parts in 100 of water). It does not combine with water, but crystallises out from its solution in the anhydrous state. It is sparingly soluble in alcohol. Its vapour-density at a white heat corresponds to the formula As_4O_6.* It sublimes without fusion, but when heated under pressure it can be fused.

It is both an oxidising and a reducing agent, tending with certain oxides—nitric acid, chromic acid, &c., to remove their oxygen, while it is itself reduced by carbon, phosphorus, sodium, &c. It is exceedingly poisonous; less than 0·4 gram has been known to cause death; but by continually increasing doses, the system may become inured to as much as 0·2 gram at a time. The antidote is a mixture of hydrated ferric oxide and magnesium chloride, produced by adding magnesium oxide or carbonate in excess to trichloride of iron; such a mixture forms an insoluble arsenite of iron, while the magnesium chloride and oxide act as a purgative.

Arsenic pentoxide is a white mass, dissolving in water to produce arsenic acid. It is poisonous, but is not so deadly as the trioxide.

Antimonious oxide is found native in trimetric prisms as *antimony-bloom*, and in regular octahedra as *senarmontile*. It is a

* *Berichte*, **12**, 1112.

white powder, turning yellow when heated, but white again on cooling. It melts at a red heat, and volatilises at 1550°. Its vapour-density points to the formula Sb_4O_6, like arsenious oxide.* It is insoluble in, and does not combine with water. One of the best solvents is a solution of tartrate of hydrogen and potassium (*cream of tartar*). $HKC_4H_4O_6.Aq$; it forms the potassium salt of the acid $Sb(OH)C_4H_4O_6$, a substituted antimonious acid.

Antimony tetroxide, Sb_2O_4, also occurs native as *antimony ochre*. It is a white powder when cold, and yellow when hot. It has not been melted or volatilised. It is possibly metantimonate of antimonyl, $SbO(SbO_3)$.

The **pentoxide, Sb_2O_5,** is an insoluble lemon-coloured powder.

Bismuth dioxide, Bi_2O_2, is a black crystalline powder, obtained by the reduction with tin dichloride of the trioxide suspended in alkali; it must be dried out of contact with air. On treatment with acid, it gives a salt of the oxide Bi_2O_3, and a precipitate of metallic bismuth. It oxidises at 180°.

The **trioxide, Bi_2O_3,** is a yellow-white solid, which crystallises from fused potassium hydroxide. No compound with an oxide is known, but it is not impossible that such a hot solution contains an easily decomposible bismuthite.

The **tetroxide, Bi_2O_4,** is a brown-yellow solid, produced by treating the trioxide suspended in a cold solution of potash with chlorine; and the **pentoxide, Bi_2O_5,** is a red powder, similarly prepared, the solution of potash being kept boiling during passage of chlorine. The pentoxide combines with water, forming the hydrate $Bi_2O_5.H_2O$.

As hydrogen sulphide has no action on a solution of a phosphate, the **sulphides of phosphorus** are prepared by direct union. There appear to be only three definite compounds.† Phosphorus and sulphur may be melted together, but combination takes place only above 130°. Owing to the great violence of the action and the inflammability of phosphorus in presence of air, a large quantity of sand is added to the melted mixture, and the retort is filled with carbon dioxide. If phosphorus is in excess, the compound produced is P_4S_3. This substance is reddish-yellow, melts at 167°, and boils constantly about 380°. If sulphur is in excess, the pentasulphide, P_2S_5, is formed, melting at 210° and boiling at 519°. Phosphorus and sulphur both dissolve in these compounds, but apparently without altering them. On heating a solution of the body P_4S_3 in carbon disulphide, however, with sulphur, yellow

* *Berichte*, **12**, 1282.

† *Bull. Soc. Chim.*, **41**, 433; *Comptes rend.*, **102**,1386.

crystals of the compound **P_2S_4** are deposited; and intermediate indistinct crystals are said to have been obtained of the formula **$P_8S_{11} = P_4S_3.2P_2S_4$**.

The **selenides of phosphorus** are somewhat doubtful in composition. The bodies **P_4Se, P_2Se, P_2Se_3**, and **P_2Se_5**, are said to have been prepared, but, except perhaps the last, they are probably mixtures of compounds analogous to the sulphides. Phosphorus and tellurium apparently mix in all proportions; no definite compounds have been isolated.

Arsenic disulphide, As_2S_2, is found native as *realgar*, in monoclinic prisms. It is a reddish-orange body, and may be produced by heating arsenic and sulphur together in the right proportions. The **trisulphide, As_2S_3,** similarly produced, occurs native in trimetric prisms as *orpiment;* it forms translucent lemon-yellow crystals. Prepared by double decomposition, it is a yellow powder, which is easily melted and volatilised. When hydrogen sulphide is passed into an aqueous solution of the trioxide no precipitate is produced, but the solution turns yellow. The substance in solution is probably a hydrate and hydrosulphide; on addition of hydrochloric acid, the trisulphide, **As_2S_3**, or more probably its compound with hydrogen sulphide, is thrown down. It is soluble in solutions of hydroxide or hydrosulphide of sodium or potassium, forming oxysulpharsenites and sulpharsenites (see below). The **pentasulphide** is an easily fusible yellow powder; it is formed by direct union; by addition of an acid to a sulpharsenate; and by the action of a *rapid* current of hydrogen sulphide on a solution of arsenic acid. It is easily soluble in solutions of sulphides of the alkalies, forming sulpharsenates (see below). The action of a slow current of hydrogen sulphide on a solution of arsenic pentoxide is first to reduce it, thus:—

$$As_2O_5.Aq + 2H_2S = As_2O_3.Aq + 2H_2O + 2\mathbf{S};$$

and then to precipitate the trisulphide.*

Selenides of arsenic have not been prepared; but two double sulphoselenides have been obtained by direct union, viz., **As_2SeS_2**, and **As_2SSe_2**. They are red bodies; the latter may be distilled unchanged. The tellurides, also directly prepared, have the formulæ **As_2Te_2** and **As_2Te_3**.

Antimony trisulphide, Sb_2S_3, occurs native in trimetric grey metallic-looking or in orange-coloured prisms, as *stibnite*. It can be prepared by direct union, or by the action of hydrogen sulphide on a soluble salt of antimony. The former method yields crystals;

* Bunsen, *Annalen*, **192**, 305; Brauner, *Chem. Soc.*, **53**, 145.

the latter, an orange-red powder, which, until dried, appears to be a hydrosulphide; it dries to a brown powder. It turns grey at 200—220°, and melts easily. The **selenide, Sb_2Se_3**, is a greyish, metallic-looking solid, produced by direct union; the **telluride, SbTe**, is iron-grey; and **Sb_2Te_3**, silver-white. The **pentasulphide** is not produced by direct union, but by decomposition of a sulphantimonate (see below) by an acid. It is a dark orange-coloured powder. The **pentaselenide** is a brown precipitate, similarly prepared.

Bismuth trisulphide, Bi_2S_3, is found in nature as *bismuth-glance*, or *bismuthine*, in rhombic crystals, with steel-grey metallic lustre. A body of similar appearance is prepared by direct union, which becomes crystalline when heated with an alkaline sulphide. The brown-black precipitate, obtained by passing hydrogen sulphide through an acid solution of bismuth nitrate or chloride, is a compound of bismuth sulphide with water and hydrogen sulphide. The action of hydrogen sulphide on an alkaline solution of bismuth trioxide is said to yield the disulphide, Bi_2S_2, in combination with water. The **triselenide** is a black lustrous powder, similarly prepared; and the **telluride** is indefinite. The mineral *telluric bismuth*, **$Bi_2S_3.2Bi_2Te_3$**, occurs native.

Compounds with Water and Oxides; with Hydrogen Sulphide and Sulphides; with Selenides; and with Tellurides.

The constitution of the acids derived from the pentoxides, pentasulphides, &c., of phosphorus, arsenic, and antimony.—Phosphorus, it will be remembered, forms two chlorides, PCl_3 and **PCl_5** (see p. 160). When the pentachloride is treated with a small quantity of water, an oxychloride, of the formula $POCl_3$ is produced (see below). The equation is:—

$$\mathbf{PCl_5} + H_2O = POCl_3 + 2\mathit{HCl}.$$

It is probable that this oxychloride, which corresponds to those of vanadium, $VOCl_3$, and niobium, $NbOCl_3$, and to tantalum oxyfluoride, $TaOF_3$, is in reality the decomposition product of a dihydroxytrichloride, $P(OH)_2Cl_3$, the reaction taking place thus:—

$$PCl_5 + 2H_2O = P(OH)_2Cl_3 + 2HCl;$$

but that body being unstable forms an anhydride, thus:—

$$P(OH)_2Cl_3 = H_2O + POCl_3.$$

The action of water on phosphoryl chloride, $POCl_3$, is to yield orthophosphoric acid, $PO.(OH)_3$, thus:—

$$POCl_3 + 3H_2O = PO(OH)_3 + 3HCl.$$

We have thus the series:—

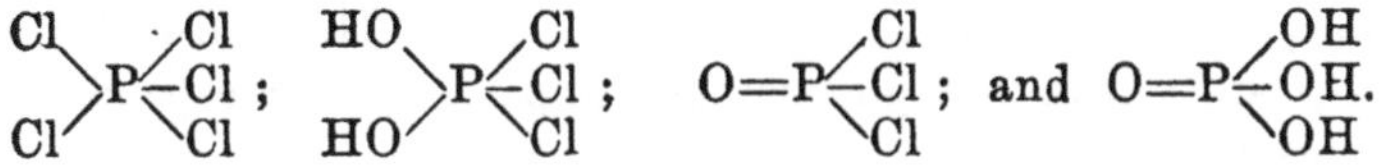

The density of the vapour of phosphoryl chloride, $POCl_3$, shows it to have the molecular weight corresponding to that formula; and the fact that the hydrogen in orthophosphoric acid is replaceable in three stages by such a metal as potassium is a strong argument in favour of the analogy between phosphoryl chloride and phosphoryl hydroxide, or phosphoric acid; such phosphates are:—

$$\mathbf{PO(OH)_2OK};\ \mathbf{PO(OH)(OK)_2};\ \text{and}\ \mathbf{PO(OK)_3}.$$

It would thus appear that phosphoric hydroxide, or the true orthophosphoric acid, should possess the formula $P(OH)_5$; but of this body, the first anhydride, $PO(OH)_3$, is the one to which the name orthophosphoric acid is applied.

By heating the first anhydride, $PO(OH)_3$, the elements of water are expelled, and the second anhydride, metaphosphoric acid, $PO_2(OH)$, is produced, thus:—

$$\mathbf{PO(OH)_3} = H_2O. + \mathbf{PO_2(OH)}.$$

This substance is usually a monobasic acid, that is, its hydrogen is replaceable in one stage; hence its formula (see, however, p. 369). The analogous compound PO_2Cl has also been prepared.

But intermediate between $\mathbf{PO(OH)_3}$ and $\mathbf{PO_2(OH)}$, there exists an acid of the formula $\mathbf{H_4P_2O_7}$, named pyrophosphoric acid. And corresponding to this hydroxide, $\mathbf{P_2O_3(OH)_4}$, a chloride, $P_2O_3Cl_4$, exists, which, however, has not been gasified, inasmuch as it decomposes. But arguing from the relation of the chloride $POCl_3$ to the acid $PO(OH)_3$, the analogy of pyrophosphoric acid to pyrophosphoryl chloride appears justified, for its hydrogen is replaceable in fourths. And just as in the case of the silicic acids, acids are derived from two molecules of the ortho-acid with loss of water, so here. We have therefore the series 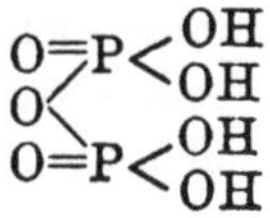 and 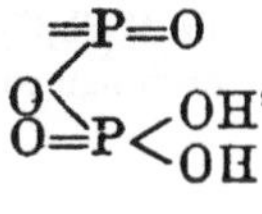, the second member of which has the same composition as metaphosphoric acid, but is a polymeride. Salts of this acid are called dimetaphosphates; the acid is dibasic. Salts of the unknown acid, $H_6P_4O_{13}$, are also known. Such an acid would be the fourth anhydride of tetraphosphoric acid, $H_{14}P_4O_{17}$, also unknown. And salts of the hypothetical acid, $H_{12}P_{10}O_{31}$, are also known, which would be similarly derived. There are also tri-, tetra-, and hexa-metaphosphates, apparently corresponding to condensed acids.

Such compounds can be also represented as formed by union of phosphoric anhydride with oxides. We have, for example, the series:—

$P_2O_5.M_2O = 2PO_2(OM)$, monometaphosphates.
$P_2O_5.2M_2O = P_2O_3(OM)_4$, pyrophosphates.
$P_2O_5.3M_2O = 2PO(OM)_3$, orthophosphates.
$2P_2O_5.2M_2O = 2P_2O_4(OM)_2$, dimetaphosphates.
$2P_2O_5.3M_2O = P_4O_7(OM)_6$, α-phosphates.
$3P_2O_5.3M_2O = 2P_3O_6(OM)_3$, trimetaphosphates.
$4P_2O_5.4M_2O = 2P_4O_8(OM)_4$, tetrametaphosphates.
$5P_2O_5.6M_2O = P_{10}O_{19}(OM)_{12}$, β-phosphates.
$6P_2O_5.6M_2O = P_6O_{12}(OM)_6$, hexametaphosphates.

Such compounds are, as a rule, soluble in water without decomposition. The sodium salts like α- and β-, however, named "Fleitmann and Henneberg's phosphates," are decomposed by much hot water into mixtures of other salts. But the corresponding pyro- and meta-arsenates are converted into ortho-arsenates on treatment with water, unless they happen to be insoluble. For example, the ortho-arsenate, $\mathbf{Na_2HAsO_4}$, is a well-known body; on ignition, it loses water and yields $\mathbf{Na_4As_2O_7}$, corresponding to the pyrophosphate, $\mathbf{Na_4P_2O_7}$; but on treatment with water, while sodium pyrophosphate dissolves as such, sodium pyro-arsenate reacts with the water, thus:—$Na_4As_2O_7 + H_2O = 2Na_2HAsO_4$. No ortho-antimonates are known except that of hydrogen, $\mathbf{SbO(OH)_3}$; some pyroantimonates and many metantimonates have been prepared, and these have the formulæ $\mathbf{M_4Sb_2O_7}$ and $\mathbf{MSbO_3}$.* The hydrate of bismuth, $\mathbf{Bi_2O_5.H_2O}$, is analogous to a meta-acid; it appears to be incapable of combination with other oxides.

Compounds analogous to the orthophosphates have been prepared, in which the oxygen of the phosphate is partially replaced by sulphur, such as $\mathbf{K_3PSO_3}$, $\mathbf{Na_3PS_2O_2}$, and possibly $\mathbf{Na_3PS_3O}$. These bodies are termed **thiophosphates** or **sulphophosphates**. With selenium, compounds analogous to the pyrophosphates have

* It is unreasonable to name compounds of the general formula $M_4Sb_2O_7$ "metantimonates," as is usually done. These bodies have here been named systematically "pyroantimonates."

been prepared, *e.g.*, $\mathbf{K_4P_2Se_7}$. Orthothioarsenic acid, $\mathbf{AsS(SH)_3}$, is said to have been prepared; and ortho-, pyro-, and meta-thioarsenates are known. Similarly, orthothioantimonates are known: but no pyro- or meta-derivatives have been prepared, nor are there any thiobismuthates.

Double compounds of the pentoxides, &c.; phosphates and similar compounds.—Ortho-acids.

Orthophosphoric acid is formed by the oxidation of phosphorus with boiling nitric acid, best in presence of a little iodine; by treating an orthophosphate with some acid which forms an insoluble compound with the metal; and by the action of a pentahalide or an oxytrihalide on water. If the first method be employed, the first product is phosphorous acid. The nitric acid should have the specific gravity 1·2, and should be employed in considerable excess; and at the last, stronger acid may be employed to oxidise the phosphorous to phosphoric acid. The second method is the one employed on a large scale; calcium orthophosphate, $\mathbf{Ca_3(PO_4)_2}$, is mixed with sulphuric acid, and the precipitated calcium sulphate removed by subsidence. The equation is :—

$$\mathbf{Ca_3(PO_4)_2} + 3H_2SO_4.Aq = 3\mathbf{CaSO_4} + 2H_3PO_4.Aq.$$

It is common to use calcined bones or *apatite* (see p. 358) as the source of calcium phosphate. The third method is the most convenient for preparing phosphoric acid in the laboratory, and it may be coupled with the preparation of hydriodic acid. Red phosphorus and iodine in the proportions equivalent to the formula PI_5 are placed in a retort; excess of water is added, and the mixture is distilled. Water distils over first, and then an aqueous solution of hydrogen iodide, while phosphoric acid remains in the retort. It is advisable then to evaporate the viscid residue with nitric acid.

Orthophosphoric acid is also produced by dissolving phosphorus pentoxide in cold water, and boiling the solution of the resulting metaphosphoric acid; and also by oxidation with nitric acid of hypophosphorous, phosphorous, and hypophosphoric acids.

By spontaneous evaporation of its aqueous solution, it crystallises in long colourless prisms, melting at 41·75°, and has the formula $\mathbf{H_3PO_4}$. From the mother liquor of these crystals fresh crystals deposit on cooling, of the formula $\mathbf{2H_3PO_4.H_2O}$; these melt at about 27°. Commercial phosphoric acid is a mixture of these two compounds.

The solution of phosphoric acid is very sour; the acid may be heated to 160° without alteration, but at 212° it is largely converted into pyrophosphoric acid.

By similar processes **orthoarsenic acid** is produced. The most convenient plan is to boil elementary arsenic, or arsenious oxide, with nitric acid, or to pass chlorine through water with which powdered arsenious oxide is mixed. The solution is evaporated to dryness, and heated for some time to 100°; water is then added, and on spontaneous evaporation the hydrated acid $2H_3AsO_4.H_2O$ deposits in small needle-shaped crystals; and on heating to 150° orthoarsenic acid, H_3AsO_4, remains.

Orthoantimonic acid has been produced by treating potassium metantimonate, $KSbO_3$, with nitric acid. It forms an insoluble white precipitate. The usual product of this action, however, is metantimonic acid, $HSbO_3$.

The only corresponding sulphur compound is **orthosulpharsenic acid**, H_3AsS_4, which is precipitated by addition of hydrochloric acid to a solution of sodium sulpharsenate, $Na_3AsS_4.Aq$. Thiophosphates, on similar treatment, give off hydrogen sulphide, and yield phosphates.

List of Orthophosphates and Orthoarsenates.—The following have been prepared:—

Simple salts:—

$2Li_3PO_4.H_2O$; $Na_3PO_4.12H_2O$; K_3PO_4; $(NH_4)_3PO_4$.
$2Li_3AsO_4.H_2O$; $Na_3AsO_4.12H_2O$; K_3AsO_4; $(NH_4)_3AsO_4.3H_2O$.

Mixed salts:—

H_2LiPO_4; $H_2NaPO_4.H_2O$; H_2KPO_4; $H_2(NH_4)PO_4$.
$3H_2LiAsO_4.2H_2O$; $H_2NaAsO_4.H_2O$, and $2H_2O$; H_2KAsO_4; $H_2(NH_4)AsO_4$.
$HNa_2PO_4.12$ and $7H_2O$; $H(NH_4)_2PO_4$.
$HNa_2AsO_4.12$ and $7H_2O$; HK_2AsO_4; $H(NH_4)_2AsO_4$.
$(Li,Na)_3PO_4$; $HNaKPO_4$; $HNa(NH_4)PO_4.4H_2O$; $Na(NH_4)_2PO_4.4H_2O$; $HNaKAsO_4.7H_2O$.
$Na_3PO_4.2NaF$.

These bodies are all white salts. They are prepared by the action of hydroxide or carbonate of lithium, sodium, or potassium, or of ammonia, on phosphoric or arsenic acid. The simple salts are produced only by the action of hydroxide, if in solution, for carbonic acid decomposes them, giving a carbonate and a phosphate or arsenate containing an atom of hydrogen. But the carbonates ignited with the theoretical amount of phosphoric acid yield simple phosphates. The phosphates containing one and two atoms of hydrogen, however, cannot be made by fusion.

Hydrogen di-lithium phosphate has not been obtained pure.

On adding hydrogen disodium phosphate to a concentrated solution of a soluble lithium salt, a precipitate is produced of the formula $(Li_2HPO_4.LiH_2PO_4)H_2O$. It is a sparingly soluble salt (1 in 200 parts of water). The other salts are easily soluble.

Hydrogen disodium phosphate, $HNa_2PO_4.12H_2O$, is the ordinary commercial "phosphate of soda;" the corresponding arsenate is also a commercial product; they crystallise in monoclinic prisms. The salt $HNa(NH_4)PO_4.4H_2O$ is known as "microcosmic salt," because it occurs in urine; the human organism used to be known as the "microcosm." It is used as a blowpipe reagent (see Metaphosphoric Acid).

The following thiophosphates are similar in composition:—*

$$Na_3PO_3S.12H_2O\ ;\ Na_3PO_2S_2.11H_2O\ ;\ (NH_4)_3PO_2S_2.2H_2O.$$

Salts of potassium have been obtained in solution; and also sodium trithiophosphate, Na_3POS_3. These bodies are produced by the action of sodium hydroxide on powdered phosphorus pentasulphide. They are unstable, especially the trithiophosphate, which decomposes when the solution is heated to 50°; a temperature of 90° destroys the dithiophosphates, and they are precipitated by addition of alcohol to their aqueous solutions. They resemble the phosphates in appearance.

Analogous **oxythioarsenates** have been made by dissolving arsenious oxide in a solution of sodium sulphide. They are separated by fractional crystallisation. Their formulæ are:—

$$Na_3AsO_3S.12H_2O\ ;\ Na_2HAsO_3S.8H_2O\ ;\ \text{and}\ Na_3AsO_3S.2H_2O.$$

Analogous to these are the **thioarsenates**, $2Na_3AsS_4.15H_2O$; K_3AsS_4; $(NH_4)_3AsS_4$; and $Na_3(NH_4)_3(AsS_4)_2$. They are produced along with pyro- and meta-thioarsenates by digesting arsenic pentasulphide, As_2S_5, with alkaline sulphides, and evaporating the solution until crystals separate; or by dissolving arsenic trisulphide, As_2S_3, in the solution of a polysulphide. They may also be produced by fusion. If arsenic pentasulphide be dissolved in solution of sodium or potassium hydroxide, a mixture of arsenate and thioarsenate is produced. They form yellowish crystals, and are very soluble in water. The solution of arsenic sulphide in ammonium polysulphide, a process used in qualitative analysis in order to separate sulphide of arsenic from sulphides of copper, lead, bismuth, mercury, and cadmium, depends on the formation of these bodies. The sulphides of antimony and of tin form similar compounds, and may be separated in the same manner from sulphides of lead, copper, &c.

* *J. prakt. Chem.* (2), **31**, 93.

The **thioantimonates** may be classed with the preceding salts. The following are known:—

$Na_3SbS_4.9H_2O$; $2K_3SbS_4.9H_2O$.

They are prepared by boiling a mixture of caustic alkali, sulphur, and antimony trisulphide; they form yellowish crystals. The sodium salt has been long known as "Schlippe's salt." The compound $Na_3SbSe_4.9H_2O$ forms orange-red tetrahedra; it is produced by fusing together sodium carbonate, antimony triselenide, sulphur, and carbon. Trisodium trithioseleno-antimonate,

$$Na_3SbS_3Se.9H_2O,$$

is formed by boiling the tetrathioantimonate with selenium. It forms yellow crystals.

Simple salts:—

$\begin{cases} Be_3(PO_4)_2.7H_2O; & Ca_3(PO_4)_2; & Ba_3(PO_4)_2.H_2O. \\ \quad — & Ca_3(AsO_4)_2; & Ba_3(AsO_4)_2. \end{cases}$

Mixed salts:—

$\begin{cases} HCaPO_4.4, 3, \text{ and } 2H_2O; & HSrPO_4; \\ HCaAsO_4; & HSrAsO_4; \end{cases}$

$Ca_3(PO_4)_2.2CaHPO_4$; $Ba_3(PO_4)_2.2BaHPO_4$.

$\begin{cases} Ca(H_2PO_4)_2; & Ba(H_2PO_4)_2; \\ \quad — & Ba(H_2AsO_4)_2; \end{cases}$

$\begin{cases} LiCaPO_4; KCaPO_4; NaSrPO_4; KSrPO_4; NaBaPO_4; \\ KBaPO_4; NaSrAsO_4; 2NH_4CaAsO_4.H_2O, \text{ also of } Ba; \end{cases}$

$H_2(NH_4)_2Ca(AsO_4)_2$; also of Ba.

$3Ca_3(PO_4)_2.CaF_2$ (*apatite*); $3Ca_3(PO_4)_2.CaCl_2$ (*apatite*).

$7\{Ca(H_2PO_4)_2\}CaCl_2.14H_2O$; $4\{Ca(H_2PO_4)_2\}CaCl_2.8H_2O$;

$Ca(H_2PO_4)_2 CaCl_2.H_2O$.

The simple salts are produced by addition of the chloride of the metal to trisodium phosphate or arsenate. They are insoluble white powders. The salts containing an atom of hydrogen are also insoluble, and are similarly precipitated with hydrogen disodium phosphate or arsenate. By boiling with water these are decomposed, giving the insoluble simple phosphate, while the soluble salt containing one atom of hydrogen goes into solution. The simple salt may also be precipitated by addition of excess of ammonia, or of caustic soda or potash, to the mono- or di-hydrogen salts. These compounds are soluble in acids, the soluble di-hydric salts being formed; but are reprecipitated as simple salts on addition of alkaline hydroxide.

Calcium phosphate is the chief mineral constituent of bones; bone-ash, or calcined bones, contains about 93 per cent. of

$Ca_3(PO_4)_2$. It is also widely distributed in soil. When found native in combination with calcium chloride or fluoride, it is known as *phosphorite*, or *apatite* (see above); the chlorine and fluorine are mutually replaceable. *Coprolites* consist of the remains of the excréta of extinct animals, and are found in the Lias. They contain from 80 to 90 per cent. of phosphates. These bodies are largely used for artificial manure.

To render the tricalcium phosphate soluble, so that its phosphorus may be easily assimilated by plants, it is treated with sulphuric acid in sufficient amount to convert it into monocalcium phosphate, thus:—$Ca_3(PO_4)_2 + 2H_2SO_4 = Ca(H_2PO_4)_2 + 2CaSO_4$.

The mixture of monocalcium phosphate and sulphate is applied to the soil, usually mixed with organic matter containing nitrogen. The old plan of allowing land periodically to lie fallow had the effect of promoting a similar decomposition by aid of the carbon dioxide of the air. It appears that one part of tricalcium phosphate dissolves as monocalcium phosphate in from 12,000 to 100,000 parts of water saturated with carbon dioxide. At the same time the carbon dioxide decomposes silicates, rendering their potash available for the use of plants; and nitrogen in the form of ammonia collects on the soil, being brought down by rain. In the modern system of agriculture, artificial manure is applied to the soil, containing these substances in a soluble form; the phosphorus as monocalcium phosphate, the potash as chloride or carbonate, and the nitrogen as salts of ammonia, or as sodium nitrate; or in the form of animal matter, from which ammonia is formed by putrefaction, such as manure, guano, dried blood, &c.

Calcium arsenate, $\mathbf{CaHAsO_4}$, is found native as *pharmacolite*.

The double phosphates and arsenates are produced by mixture. A arsenato-chloride, corresponding to apatite, has been produced artificially.

The **monothiophosphates** of calcium, strontium, and barium are all insoluble white precipitates; the **dithiophosphates** of strontium and barium, and the **trithiophosphate** of barium, are also insoluble.

Thioarsenates of beryllium and of strontium have been prepared, but not analysed; those of calcium and barium have the formulæ $\mathbf{Ca_3(AsS_4)_2}$, and $\mathbf{Ba_3(AsS_4)_2}$; they are insoluble yellow precipitates, produced by adding alcohol to the product of the action of hydrogen sulphide on $\mathbf{HBaAsO_4}$. The resulting thioarsenate, $HBa(AsS_4)$, decomposes thus:—

$$3HBaAsS_4.Aq = \mathbf{Ba_3(AsS_4)_2} + BaAsS_3.Aq + H_2S;$$

the metathioarsenate remains dissolved. The corresponding **thioantimonate**, $\mathbf{Ba_3(SbS_4)_2}$, has also been obtained from the corresponding sodium salt by precipitation.

Simple salts:—

$Mg_3(PO_4)_2$; $Zn_3(PO_4)_2.5H_4O$; $Cd_3(PO_4)_2$;

$Mg_3(AsO_4)_2$; $Zn_3(AsO_4)_2.3H_2O$; $Cd_2(AsO_4)_2.3H_2O$;

Mixed salts:—

$HMgPO_4.7H_2O$; $HZnPO_4.H_2O$; $Zn(H_2PO_4)_2.2H_2O$.

$HMgAsO_4.7H_2O$; $HZnAsO_4$; $H_2Cd_5(AsO_4)_4 4H_2O$.

$NaMgPO_4$; $KMgPO_4$; $H_2(NH_4)_2Mg(PO_4)_2.3H_2O$;

$NH_4MgPO_4.6H_2O$; $NH_4ZnPO_4.2H_2O$.

$KMgAsO_4$; $NaMgAsO_4.6H_2O$.

$$Mg_3(PO_4)_2.MgF_2 \ (wagnerite) = PO \begin{matrix} O \\ {<} \ O \\ O \end{matrix} \begin{matrix} {>}Mg \\ \ \\ -Mg-F \end{matrix}$$

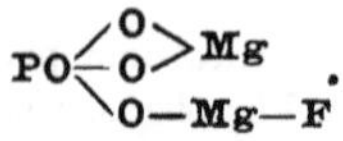

Similar arsenates have been prepared artificially.

Trimagnesium orthophosphate is a constituent of the ash of seeds, especially of wheat. It and the corresponding arsenate are insoluble in water. The other salts are produced by precipitation, and are sparingly soluble. The most important are **ammonium magnesium phosphate** and **arsenate**. The former is a constituent of certain urinary calculi, and is formed by the putrefaction of urine, and separates in crystals. Both of these salts are very sparingly soluble in water (about 1 in 13,000), and are used in the estimation of magnesium, and of phosphoric and arsenic acids. They are produced by adding a solution of magnesium chloride and ammonium chloride, commonly called "magnesia mixture," along with ammonia, to a soluble phosphate or arsenate. On ignition they leave a residue of pyrophosphate or pyroarsenate.

Thioarsenates of magnesium, zinc, and cadmium have also been prepared :- they are soluble crystalline salts.

BPO_4 ; $2YPO_4.5H_2O$ (xenotime) ; $LaPO_4$.

— $YAsO_4$ —

Boron phosphate is an insoluble white substance produced by heating boron hydrate with orthophosphoric acid. Yttrium phosphate and arsenate and lanthanum phosphate are white gelatinous precipitates produced by double decomposition. Yttrium phosphate occurs native, and that of lanthanum occurs in several rare minerals.

$AlPO_4.3$ and $4H_2O$; $AlAsO_4.2H_2O$.

Phosphates of aluminium and hydrogen:—$Al(H_2PO_4)_3$ and $Al_2H_9(PO_4)_5 H_2O$.

Basic phosphates of aluminium:—$6AlPO_4 Al_2O_3.18H_2O$; $4AlPO_4Al_2O_3.12H_2O$; $P_2O_5.2Al_2O_3.8, 6,$ and $5H_2O$.

Thallous phosphates:—Tl_3PO_4 ; $HTl_2PO_4.H_2O$; H_2TlPO_4.

Aluminium phosphate, produced by precipitation, is a white bulky precipitate, closely resembling hydrated alumina, from which it is difficult to distinguish and to separate. The arsenate closely resembles the phosphate. The compound $AlPO_4.4H_2O$ occurs native as *gibbsite;* it is also produced on boiling a solution of hydrogen aluminium phosphate. The first basic phosphate is produced by adding ammonia to a solution of the orthophosphate in hydrochloric acid; the second is *wavellite.* The third, with $5H_2O$, is *turquoise,* which owes its blue colour to a trace of copper; with $6H_2O$ it is *peganite,* and with $8H_2O$ it forms crystals of *fischerite.*

Thallic arsenate is a flocculent, insoluble precipitate; the thallous phosphates are nearly insoluble, and separate from dilute solutions in crystals.

$CrPO_4$.7, 6, 5, and $3H_2O$; $FePO_4$.

The chromic salt exists in two forms: the violet modification, with $7H_2O$, which is soluble and crystalline, and is produced by treating a solution of violet chromic chloride with silver phosphate; and the green modification, precipitated by addition of a soluble phosphate to a green chromium salt. The violet variety, when heated, changes into the green one; and the green precipitate becomes violet and crystalline on standing. Ferric phosphate is a white precipitate produced in a neutral solution of a ferric salt by hydrogen disodium phosphate, or by exposing ferrous phosphate to air. Arsenates give similar precipitates with chromium and iron salts.

Iron also forms the following double phosphates with hydrogen:—

$Fe(H_2PO_4)_3$; $FeH_3(PO_4)_2$; $Fe_6H_3(PO_4)_7$; $Fe_8H_3(PO_4)_9$; and $Fe_4H_3(PO_4)_5$. Also the basic phosphates:—$P_2O_5.2Fe_2O_3.12H_2O$ (*cacoxene*) ; $5H_2O$ (*dufrenite* or *green iron ore*) ; and $12H_2O$ or $18H_2O$ (*delvauxite*).

Basic ferric phosphate is also a frequent constituent of bog-iron ore. Manganic and cobaltic orthophosphates and arsenates are unknown.

Simple salts:—

$Cr_3(PO_4)_2$ (?) ; $Fe_3(PO_4)_2.8H_2O$; $Mn_3(PO_4)_2 7H_2O$; $Co_3(PO_4)_2.8H_2O$; $Ni_3(PO_4)_2.7H_2O$; $Fe_3(AsO_4)_2$; $Co_3(AsO_4)_2.8H_2O$ and $Ni_3(AsO_4)_2.8H_2O$ (*cobalt- and nickel-bloom*).

Mixed salts:—

$Fe(H_2PO_4)_2.2H_2O$; $Mn(H_2PO_4)_2.2H_2O$; $HMnPO_4.3H_2O$; $(NH_4)Fe(PO_4).H_2O$; $(NH_4)Mn(PO_4).H_2O$.

Also the arsenates, $Co(H_2AsO_4)_2$; $Mn(H_2AsO_4)_2$; $MnHAsO_4$.

And the minerals *childrenite,* a phosphate of aluminium, iron, and manganese: *triplite,* $(Fe,Mn)_3(PO_4)_2$, and *triphylline,* $(Li_2,Mg,Fe,Mn)_3(PO_4)_2$.

Chromous phosphate is a blue precipitate; ferrous phosphate is white and insoluble; it occurs native as *vivianite* or *blue iron earth;* the hydrogen manganous salts and the double ammonium salts are obtained by mixture and crystallisation, *e.g.*, $\mathbf{Mn_3(PO_4)_2}$ + $H_3PO_4.Aq = \mathbf{3HMnPO_4} + Aq$; $\mathbf{Mn_3(PO_4)_2} + (NH_4)_3PO_4.Aq = \mathbf{3NH_4MnPO_4} + Aq$. The cobalt salt is reddish-blue, and the nickel salt light-green. Arsenates of cobalt and nickel occur native; *cobalt-bloom* forms red, and *nickel-bloom* green, crystals.

Elements of the carbon-group form no normal phosphates. Carbon phosphate is unknown; titanium forms the compound $\mathbf{Ti_2Na(PO_4)_3}$ when titanium dioxide is fused with hydrogen sodium ammonium phosphate (microcosmic salt); and sodium thorium phosphate, $\mathbf{Th_2Na(PO_4)_3}$, is similarly prepared; zirconium salts by precipitation give the basic phosphate $\mathbf{(ZrO)_3(PO_4)_2}$; thorium phosphate is a white precipitate; cerous phosphate, $\mathbf{CePO_4}$, occurs native as *cryptolite* and *phosphocerite;* prepared artificially, it forms a white precipitate. Arsenates of titanium, zirconium, and thorium have been prepared; also cerous arsenate, $\mathbf{CeAsO_4}$ (?) and sulpharsenate, $\mathbf{CeAsS_4}$(?), which require investigation.

$\mathbf{SiH_2(PO_4)_2.3H_2O}$ is deposited from a solution of silica in phosphoric acid kept at 125° for several days. It is soluble in, and decomposed by contact with, water. Germanium phosphate has not been prepared; a basic phosphate of tin, $\mathbf{P_2O_5.2SnO_2.10H_2O}$, is deposited on treatment of tin dioxide (metastannic acid) with phosphoric acid; this compound is insoluble in nitric acid, and is therefore used in separating phosphoric acid from solutions containing it. Corresponding arsenates are unknown. By fusing stannic oxide with borax and microcosmic salt, crystals of the formula $\mathbf{Na_2Sn(PO_4)_2}$ are produced. With microcosmic salt alone, the body $\mathbf{NaSn_2(PO_4)_3}$ is formed in microscopic crystals.

Stannous phosphato-chloride, $\mathbf{Sn_3(PO_4)_2.SnCl_2}$, is precipitated by adding a solution of ordinary sodium phosphate to excess of tin dichloride; but with excess of sodium phosphate the precipitate has the formula $\mathbf{Sn_3(PO_4)_2.2SnHPO_4.3H_2O}$. The arsenates, similarly produced, are said to have the formulæ $\mathbf{2HSnAsO_4.H_2O}$ and $\mathbf{{ClSn \atop Sn}{>}AsO_4.H_2O}$.

Lead orthophosphate, $\mathbf{Pb_3(PO_4)_2}$, produced by precipitation, is a white amorphous substance, fusible, and crystallising on cooling. By adding phosphoric acid to a dilute boiling solution of lead nitrate, the compound $\mathbf{Pb(H_2PO_4)_2}$ is thrown down in sparkling white laminæ. In the cold, a phosphato-nitrate, of the formula $\mathbf{Pb(NO_3)_2.Pb_3(PO_4)_2.2H_2O}$, is precipitated. It is decomposed by boiling water. By employing a boiling solution of lead chloride and excess of sodium phosphate, the compound

$$Pb_3(PO_4)_2.PbCl_2.H_2O$$

is precipitated. With excess of lead chloride the precipitate consists of $2Pb_3(PO_4)_2.PbCl_2$ (?). *Pyromorphite*, another phosphato-chloride, $3Pb_3(PO_4)_2.PbCl_2$, occurs native in hexagonal prisms, usually of a green colour. The corresponding arsenate,

$$3Pb_3(AsO_4)_2.PbCl_2,$$

is also found in nature, and is named *mimetesite*. Crystals in which arsenic and phosphorus replace each other partially are common. The arsenates $Pb_3(AsO_4)_2$ and $HPbAsO_4$ have been produced by precipitation, and also the sulpharsenate, Pb_3AsS_4.

$(VO)PO_4.7H_2O$; $(VO)AsO_4.7H_2O$; $(VO)_2H_3(PO_4)_3.3H_2O$.

These are the simpler phosphates and arsenates of elements of the nitrogen group. They are brilliant yellow or red crystals. It is to be noticed that these bodies may equally well be conceived as vanadates of phosphoryl and arsenyl, thus:—

$(PO)VO_4.7H_2O$; $(AsO)VO_4.7H_2O$; and $(PO.OH)_3(VO_4)_2.3H_2O$.

Tantalum pentoxide, dissolved in hydrochloric acid, forms a jelly with phosphoric acid, due probably to a combination between them.

A curious compound of the formula $4MgHPO_4.NO_2$ is produced by boiling magnesium pyrophosphate with strong nitric acid, and heating it in a paraffin-bath until it ceases to emit fumes. It is a crystalline whitish-yellow powder, which gives off nitric peroxide when strongly heated.

The vapour-density, and consequently the molecular weight, of phosphorus pentoxide is unknown. If its formula be P_2O_5, it may perhaps be regarded as phosphoryl phosphate, $(PO)PO_4$, $O{=}P{\equiv}O_3{\equiv}P{=}O$; and arsenic pentoxide and the other pentoxides might be similarly regarded.

Many very complicated compounds of the pentoxides with each other have recently been discovered. Among these are

$$P_2O_5.V_2O_5.(NH_4)_2O.H_2O;\ 4P_2O_5.6V_2O_5.3K_2O.21H_2O;$$
$$P_2O_5.20V_2O_5.69H_2O;\ 5As_2O_5.8V_2O_5.27H_2O.$$

Some also contain vanadium dioxide, for example,

$$2P_2O_5.VO_2.18V_2O_5.7(NH_4)_2O.50H_2O.$$

Compounds of arsenious and arsenic oxides are also known; thus:—

$$2As_2O_5.3As_2O_3.H_2O;\ As_2O_5.2As_2O_3.H_2O;\ \text{and}\ As_2O_5.As_2O_3.H_2O.$$

They are produced by partial oxidation of arsenious oxide, As_4O_6, by nitric acid, and are definite crystalline bodies.* The bismuth phosphate corresponding to the last of these, $BiPO_4 = P_2O_5.Bi_2O_3$ is produced by precipitation. The corresponding arsenate, $BiAsO_4.H_2O$ is a yellowish-white precipitate; they may, however, equally well be regarded as metaphosphate and metarsenate of bismuthyl, $(BiO)PO_3$ and $(BiO).AsO_3$.

The compounds with the elements **molybdenum** and **tungsten** are exceedingly complicated. Molybdenum trioxide, MoO_3, and tungsten trioxide, WO_3, combine with phosphorus tri- and pentoxides, and with many other oxides; these compounds will be described among the oxides of molybdenum and tungsten. The only one to be mentioned here is ammonium phosphomolybdate, which is produced by adding ammonium molybdate to any warm solution containing an orthophosphate. It is a bright yellow precipitate, insoluble in nitric acid, and is used as a test for phosphoric acid. Several compounds of **uranyl**, (UO_2), are known. The normal salt has not been prepared, but double salts are known, for example,

$$(UO_2)_3(PO_4)_2.2(UO_2)HPO_4.H_2O,$$

which is formed by precipitation as a light yellow powder. By digestion with phosphoric acid, the salts $(UO_2)HPO_4$, and $(UO_2)(H_2PO_4)_2$ are formed; corresponding arsenates have been prepared. Uranyl sodium salts, $(UO_2)NaPO_4$ and $(UO_2)NaAsO_4$, are produced by addition of sodium phosphate in excess. The calcium salt, $(UO_2)_2Ca(PO_4)_2.8H_2O$, is found native as *uranite;* and a similar copper salt, $(UO_2)_2Cu(PO_4)_2.8H_2O$, occurs as *chalcolite.*

Phosphates and arsenates of the **palladium** and **platinum** groups of metals require investigation. No compound has been analysed (except $H_3Rh(PO_4)_2.H_2O$), although salts of these metals give precipitates with phosphates and arsenates. Compounds of gold are unstable.

Copper orthophosphate, $Cu_3(PO_4)_2$, is a blue-green precipitate; or, when prepared by heating the pyrophosphate with water, yellowish-green crystals with $3H_2O$. The salt $HCuPO_4$ is also a blue-green precipitate. Many basic compounds occur native, *e.g.*,

$P_2O_5.4CuO.H_2O$, $2H_2O$, and $3H_2O$; $P_2O_5.5CuO.2H_2O$ and $3H_2O$; and $P_2O_5.6CuO.3H_2O$.

The last is the most important, and is named *phosphochalcite.*

* *Comptes rend.*, 100, 1221.

The arsenates, $Cu_3(AsO_4)_2$ and $H_2Cu_2(AsO_4)_2.H_2O$, are green and blue powders respectively.

Silver phosphate, Ag_3PO_4, is a yellow precipitate, produced by adding any soluble phosphate to a solution of silver nitrate. It is used as a test for phosphoric acid. Hydrogen disilver phosphate, HAg_2PO_4, produced by digesting the former with phosphoric acid, forms colourless crystals; it is at once decomposed by water into Ag_3PO_4 and H_3PO_4. The arsenate, Ag_3AsO_4, is a red precipitate. It is formed by adding an arsenate to a solution of silver nitrate, and cautiously adding ammonia. It serves as a test for arsenic acid, and distinguishes it from arsenious acid.

Mercurous phosphate, Hg_3PO_4, and mercuric phosphate, $Hg_3(PO_4)_2$, are white crystalline powders. A phosphato-nitrate, $Hg_3PO_4.HgNO_3.H_2O$, is also known. The arsenate Hg_2HAsO_4 is an orange precipitate.

Pyro-compounds. — Pyrophosphoric acid, $H_4P_2O_7 = P_2O_3(OH)_4$, is produced by heating orthophosphoric acid to 215°. The change begins at 160°, but is not complete at 215°, for the mass still contains unchanged orthophosphoric acid. If a higher temperature be employed, metaphosphoric acid begins to be formed. Similarly, **pyroarsenic acid** is formed by heating the ortho-acid to 140—160°. **Pyroantimonic acid**, unlike the corresponding acids of phosphorus and arsenic, is produced by the action of water on the pentachloride. When $SbCl_5$ is mixed with a little water, crystals of the formula $SbCl_5.4H_2O$ are deposited. Addition of more water to the cold solution of this body produces the insoluble oxychloride, $SbOCl_3$; on warming this antimonyl chloride with much water, the sparingly soluble pyroantimonic acid, $H_4Sb_2O_7.2H_2O$ is formed. The water of crystallisation may be expelled at 100°. No corresponding compound of bismuth is known.

These bodies may also be prepared by replacing some metal such as lead, in the pyro-salts, by hydrogen, by the action of hydrogen sulphide, thus:—

$$Pb_2P_2O_7 + 2H_2S + Aq = H_4P_2O_7.Aq + 2PbS.$$

The lead pyrophosphate is insoluble, and is suspended in water. Pyroarsenic acid, however, cannot be thus prepared, for it reacts with hydrogen sulphide, giving arsenic pentasulphide. But as pyrantimonic acid is sparingly soluble, it is precipitated on adding an acid to a solution of a pyroantimonate; *e.g.*,

$$K_4Sb_2O_7.Aq + 4HCl.Aq = H_4Sb_2O_7 + 4KCl.Aq.$$

On standing, even in contact with water, it loses water, changing to $\mathbf{HSbO_3}$, thus :—

$$\mathbf{H_4Sb_2O_7} = \mathbf{2HSbO_3} + H_2O.$$

No pyrosulpho- or pyroselenio-acids are known.

Pyrophosphoric acid is usually a soft colourless glass-like body; it has, however, been obtained in opaque indistinct crystals. Pyroarsenic acid forms hard shining crystals; it unites with water at once, giving out heat, and forming a solution of orthoarsenic acid.

Pyroantimonic acid is a white powder, soluble in a large quantity of water, from which it is precipitated by addition of acids.

Pyrophosphates, &c.—The pyrophosphates and pyroarsenates are produced by heating the mono-hydrogen or mono-ammonium orthophosphates to redness, thus :—

$$\mathbf{2HNa_2PO_4} = \mathbf{Na_4P_2O_7} + \mathit{H_2O};$$
$$\mathbf{2NH_4MgPO_4} = \mathbf{Mg_2P_2O_7} + \mathit{2NH_3} + \mathit{H_2O}.$$

The pyroarsenates require investigation. It is possible that on treatment with water dimetallic orthoarsenates are again formed, but this has not been proved. The pyroantimonates are produced by heating the metantimonates with water, or with an oxide, thus :—

$$2MSbO_3 + M_2O = M_4Sb_2O_7, \text{ and } 2MSbO_3 + H_2O = M_2H_2Sb_2O_7.$$

The pyrophosphates may also be produced by action of pyrophosphoric acid on oxides, hydroxides, or carbonates.

Pyrothioarsenates are the salts usually produced by dissolving arsenic pentasulphide in solutions of soluble sulphides, or hydrosulphides, or the trisulphide in solutions of polysulphides; or by the action of hydrogen sulphide on solutions of the arsenates; or by fusing the sulphides of arsenic and metal together. Many are insoluble, and are precipitated on addition of the sodium salt to a solution of a compound of the element. On treatment with alcohol, they are often decomposed into orthothioarsenates, which are precipitated, while the meta-salts dissolve.

List of Pyrophosphates, &c.

Simple salts :—

$\mathbf{Na_4P_2O_7.10H_2O}$; $\mathbf{K_4P_2O_7.3H_2O}$; $\mathbf{(NH_4)_4P_2O_7}$.—$\mathbf{Li_4As_2S_7}$; $\mathbf{Na_4As_2S_7}$; $\mathbf{K_4As_2S_7}$; $\mathbf{(NH_4)_4As_2S_7}$.—$\mathbf{K_4Sb_2O_7}$.

Mixed salts :—

$\mathbf{H_2Na_2P_2O_7}$; $\mathbf{Na_2(NH_4)_2P_2O_7.5H_2O}$; $\mathbf{H_2K_2P_2O_7}$; $\mathbf{2HK_2(NH_4)P_2O_7.H_2O}$; $\mathbf{Na_2K_2P_2O_7.12H_2O}$; $\mathbf{H_2(NH_4)_2P_2O_7}$.—$\mathbf{H_2Na_2Sb_2O_7}$; $\mathbf{H_2K_2Sb_2O_7}$.

The pyrophosphates are produced by addition of a hydroxide or carbonate to the acid; many of them are precipitated by alcohol. They are white deliquescent salts, and they are not altered by boiling with water; but, on boiling with acids, they combine with water, forming orthophosphates. The double salts are produced by mixture and crystallisation. On heating dihydrogen monosodium orthophosphate, $\mathbf{H_2NaPO_4}$, to 200°, it loses water, giving dihydrogen disodium pyrophosphate, thus:—$2\mathbf{H_2NaPO_4} = \mathbf{H_2Na_2P_2O_7} + H_2O$. Potassium pyroantimonate, $\mathbf{K_4Sb_2O_7}$, is produced by fusing the metantimonate, $\mathbf{KSbO_3}$, with caustic potash, and subsequent crystallisation from water. Dihydrogen dipotassium pyrantimonate is formed, along with potassium hydroxide, by warming the tetrapotassium salt with water. The corresponding sodium salt is very sparingly soluble in water—it is one of the few nearly insoluble salts of sodium—and the formation of a precipitate in a solution free from other metals on addition of a solution of the potassium salt indicates the presence of sodium, owing to the formation of $\mathbf{H_2Na_2Sb_2O_7}$.

Simple salts:—

$\mathbf{Be_2P_2O_7.5H_2O}$; $\mathbf{Ca_2P_2O_7.4H_2O}$; $\mathbf{Sr_2P_2O_7.H_2O}$; $\mathbf{Ba_2P_2O_7.H_2O}$; $\mathbf{Ca_2As_2S_7}$; and others.

Mixed salts:—

$\mathbf{Na_2CaP_2O_7.4H_2O}$; and insoluble white pyrantimonates.

Hydrogen pyrophosphate gives no precipitate with the chlorides of these metals; but with sodium pyrophosphate these pyrophosphates are precipitated. The calcium salt fuses to a transparent glass, which may be substituted for ordinary glass for many purposes.

Simple salts:—

$\mathbf{Mg_2P_2O_7.3H_2O}$; $\mathbf{2Zn_2P_2O_7.H_2O}$ and $\mathbf{10H_2O}$; $\mathbf{Cd_2P_2O_7.2H_2O}$; $\mathbf{Mg_2As_2S_7}$.

Mixed salts:—

$\mathbf{Na_2ZnP_2O_7}$, also with $\mathbf{4H_2O}$; $\mathbf{Na_2CdP_2O_7}$.

The anhydrous magnesium pyrophosphate is left as a white caked mass on igniting ammonium magnesium orthophosphate, $\mathbf{NH_4MgPO_4}$. These anhydrous salts are soluble in sulphurous acid, and crystallise from the solution on evaporation. The double salts crystallise from solutions of oxides in sodium metaphosphate. The sulpharsenate of magnesium is a very soluble yellow salt, also soluble in alcohol.

Pyrophosphates, &c., of the boron group of elements have not been prepared.

$\mathbf{Al_4(P_2O_7)_3.10H_2O}$ is a white precipitate, differing from the orthophosphate by its solubility in ammonia. The salts of gallium, indium, and thallium have not been prepared. The double salt, **$NaAlP_2O_7$**, crystallises from a solution of **Al_2O_3** in fused sodium metaphosphate.

Pyrophosphate of carbon is unknown; titanium, zirconium, and tin pyrophosphates, **TiP_2O_7**, **ZrP_2O_7**, and **SnP_2O_7**, are prepared by dissolving the dioxides in fused orthophosphoric acid.

Silicon pyrophosphate,* **SiP_2O_7**, crystallises in octahedra from a solution of silica in fused metaphosphoric acid, and lead pyrophosphate, **$Pb_2P_2O_7.H_2O$**, produced by precipitation, is a bulky white powder. **$Pb_2As_2S_7$** is also known.

$Cr_4(P_2O_7)_3$; $Fe_4(P_2O_7)_3.9H_2O$.—$2Na_4P_2O_7.Fe_4(P_2O_7)_3.7H_2O$; $Fe_4(As_2S_7)_3$.

These salts are produced by precipitation; that of chromium is green, and those of iron nearly white. They are soluble in excess of sodium pyrophosphate, and doubtless form salts like the double salt of iron of which the formula is given above. Ammonium sulphide does not precipitate chromium or iron from solutions of these double salts. **$NaCrP_2O_7$** crystallises from a solution of chromium sesquioxide in sodium metaphosphate.

$Fe_2P_2O_7$; $Mn_2P_2O_7.3H_2O$; $Co_2P_2O_7$; $Ni_2P_2O_7.6H_2O$; $NaNH_4MnP_2O_7.3H_2O$; $Fe_2As_2S_7$; $Mn_2As_2S_7$; and **$Co_2As_2S_7$** are produced by precipitation.

Of the nitrogen and phosphorus groups, the only pyrophosphate known is that of bismuth, **$Bi_4(P_2O_7)_3$**, which is a white precipitate. It crystallises from a solution of bismuth trioxide in fused sodium metaphosphate. But hydrogen sodium pyrophosphate dissolves antimony trioxide. The pyrophosphates of elements of the palladium and platinum groups have not been prepared.

Cupric pyrophosphate, **$Cu_2P_2O_7.H_2O$**, is a greenish-white powder produced by precipitation. Silver pyrophosphate, **$Ag_4P_2O_7$**, is a white curdy precipitate. Its formation serves to distinguish pyrophosphates from orthophosphates, which give a yellow precipitate of **Ag_3PO_4** with silver nitrate. A double pyrophosphate of gold and sodium, of the formula **$2Na_4P_2O_7.Au_4(P_2O_7)_3.H_2O$**, is formed by mixing gold trichloride with sodium pyrophosphate and evaporation; the sodium chloride separates in crystals, leaving the above salt. Mercuric pyrophosphate, **$Hg_2P_2O_7$**, and mercurous pyrophosphate, **$Hg_4P_2O_7$**, are white precipitates.

It is to be noticed that while there are many double pyrophosphates in which the two atoms of hydrogen of pyrophosphoric acid are replaced by one metal, and two by another, such as **$H_2Na_2P_2O_7$**,

* *Comptes rend.*, **96**, 1052; **99**, 789; **102**, 1017.

$\mathbf{Na_2CaP_2O_7}$, &c., there are few in which the hydrogen is replaced in fourths. Yet instances are known, for example, $\mathbf{NaNH_4MnP_2O_7}$, $\mathbf{HK_2NH_4P_2O_7}$, and one or two others. The conclusion is therefore justified that, inasmuch as such compounds are known, there are four atoms of hydrogen in hydrogen pyrophosphate. With the pyrothioarsenates and pyroantimonates, such double compounds are unknown: the only double salts being those of the pyroantimonates of hydrogen and a metal such as $\mathbf{H_2Na_2Sb_2O_7}$.

Meta-compounds.—Metaphosphates, etc.—It cannot be said with certainty that more than one **metaphosphoric acid** is known, although, as mentioned on p. 354, there are grounds for inferring the existence of at least five sets of metaphosphates: mono-, di-, tri-, tetra-, and hexa-metaphosphates, derived from condensed acids.* When phosphoric anhydride is dissolved in cold water, and the resulting solution evaporated, or when orthophosphoric acid is heated above 213°, a transparent glassy soluble substance remains, the simplest formula of which is $\mathbf{HPO_3}$. The same body is produced by (1) heating microcosmic salt to redness, when sodium metaphosphate is produced, thus:—$\mathbf{HNaNH_4PO_4} = \mathbf{NaPO_3} + H_2O + NH_3$; (2) dissolving this metaphosphate in water, and adding lead nitrate, when lead metaphosphate is formed, thus:—$2NaPO_3.Aq + Pb(NO_3)_2.Aq = 2NaNO_3.Aq + \mathbf{Pb(PO_3)_2}$; and (3) suspending the insoluble lead metaphosphate in water, and passing through the liquid a current of hydrogen sulphide, when lead sulphide and hydrogen metaphosphate are produced, thus:—$\mathbf{Pb(PO_3)_2} + Aq + H_2S = 2HPO_3.Aq + \mathbf{PbS}$. On evaporating the filtered liquid to dryness, the same glassy soluble body is obtained. It is probably a hexametaphosphoric acid, for it forms salts in which one-sixth of the hydrogen is replaceable.

But it has been noticed that during the preliminary stage of phosphorus manufacture, in evaporating orthophosphoric acid with charcoal or coke, and igniting the residue, the black powder of carbon and metaphosphoric acid gives up nothing to water; an insoluble variety is in fact produced. This variety differs therefore from the other, and is possibly monometaphosphoric acid, for that body gives insoluble salts.

On boiling metaphosphoric acid with water, orthophosphoric acid is formed, thus:—$HPO_3.Aq + H_2O = H_3PO_4.Aq$. The meta-acid, when added to a solution of albumen (white of egg) in water, coagulates it, producing a curdy precipitate; the silver salt is white, and is not produced on adding silver nitrate to a solution of

* See also *Zeitschr. f. physik. Chem.*, 6, 122.

2 B

metaphosphoric acid; and it gives no yellow precipitate when warmed with ammonium molybdate and nitric acid. But, after boiling with water, the resulting orthophosphoric acid does not coagulate albumen, gives a yellow precipitate of silver orthophosphate, $\mathbf{Ag_3PO_4}$, with silver nitrate, and a bright yellow precipitate with ammonium molybdate. The two acids are therefore obviously distinct bodies. They are distinguished from pyrophosphoric acid by the fact that silver pyrophosphate is white and curdy.

Metaphosphoric acid is volatile at a high temperature, but it does not lose water to give phosphorus pentoxide.

Metarsenic acid, $\mathbf{HAsO_3}$, is likewise produced by heating ortho- or pyroarsenic acid to 200—206°. It is a white nacreous substance sparingly soluble in cold water; but its solution exhibits no properties differing from those of a solution of orthoarsenic acid, and it appears, therefore, to combine with water to form the latter body. The metarsenates, too, are only known as solids; they may be obtained from the appropriate hydrogen or ammonium orthoarsenates, *e.g.*, $\mathbf{HNaNH_4AsO_4} = \mathbf{NaAsO_3} + H_2O + NH_3$; but on treatment with water they combine, forming dihydrogen metallic ortho-arsenates.

The **metathioarsenates** are produced by the action of alcohol on solutions of the pyrothioarsenates, thus:—

$$K_4As_2S_7.Aq + Alc = \mathbf{K_3AsS_4} + KAsS_3.Aq.Alc.$$

The orthosulpharsenate is precipitated, while the metasulpharsenate remains in solution. The acid is unknown. They have been little investigated.

Metantimonic acid, $\mathbf{HSbO_3}$, results from the spontaneous decomposition of $H_4Sb_2O_7$ dissolved in water; it is also produced when the pyro-acid is heated, or when a metantimonate is treated with an acid. It is also formed by the action of nitric acid on antimony. It is a soft white sparingly soluble powder. This compound and its salts are usually inconsistently named "antimonic acid" and "antimonates." Hydrated pentoxide of bismuth, $\mathbf{Bi_2O_5.H_2O}$ (see p. 350) may be classed here.

(*a.*) **Hexametaphosphates.**—These are the salts prepared by the usual methods from ordinary metaphosphoric acid: $\mathbf{Na_6P_6O_{18}}$; $\mathbf{(NH_4)_6P_6O_{18}}$; $\mathbf{Na_2Ca_5(P_6O_{18})_2}$; $\mathbf{Ag_6P_6O_{18}}$; and others.

The sodium salt is produced by strongly igniting dihydrogen sodium orthophosphate until it fuses, and then rapidly cooling the fused mass. It is an amorphous colourless deliquescent glass, easily soluble in water and in alcohol. It gives gelatinous precipitates with salts of most metals; its hexa-basic character is deduced from the formulæ of double salts such as the one given above, $\mathbf{Na_2Ca_5P_6O_{18}}$. The ammonium salt is produced by saturating ordinary metaphosphoric acid with ammonia, and evaporating.

(*b.*) **Tetrametaphosphates.**—Lead oxide, heated with excess of phosphoric acid, yields large transparent prisms of an insoluble salt. The salt is powdered, and digested with sodium sulphide; lead sulphide and sodium tetrametaphosphate are formed. It is diluted with much water, and filtered. On adding alcohol, an elastic ropy mass, like caoutchouc, is precipitated. Its solution in water gives ropy precipitates with salts of other metals. Its tetra-basicity is inferred from the existence of double salts such as $\mathbf{Na_2Cu''P_4O_{12}}$.

(*c.*) **Trimetaphosphates.**—When a considerable mass of sodium metaphosphate is slowly cooled, the mass acquires a beautiful crystalline structure; and on treatment with warm water the solution separates into two layers, the larger stratum containing the crystalline, and the smaller the ordinary vitreous, salt. The solution of the crystalline variety gives crystalline precipitates with salts of many metals, the silver salt, for example, depositing in crystals of the formula $\mathbf{Ag_3P_3O_9.H_2O}$. The sodium salt deposits in large crystals of the formula $\mathbf{Na_3P_3O_9.6H_2O}$. Its tri-basicity is inferred from formulæ such as $\mathbf{2NaBaP_3O_9.H_2O}$. The salts of this acid uniformly crystallise well.

(*d.*) **Dimetaphosphates.**—By heating copper oxide, $\mathbf{CuO}$, with a slight excess of phosphoric acid to 350°, an insoluble crystalline powder is formed. On digestion with sulphides of sodium, potassium, &c., the corresponding dimetaphosphates are formed, and separate in crystals on addition of alcohol. Double salts are produced by mixture, such as $\mathbf{NaNH_4P_2O_6.H_2O}$; $\mathbf{NaKP_2O_6.H_2O}$; $\mathbf{NaAgP_2O_6}$, &c. These salts, like the trimetaphosphates, are crystalline bodies sparingly soluble in water.

(*e.*) **Monometaphosphates.**—These bodies are insoluble in water. They are produced by igniting together the oxides and phosphoric acid in molecular proportions; or, by adding excess of phosphoric acid to solutions of nitrates or sulphates, evaporating, and heating the residues to 350° or upwards. They are crystalline and anhydrous, and form no double salts; even the salts of the alkalies are nearly insoluble in water. The solution of the potassium salt in acetic acid gives precipitates with salts of barium, lead, and silver.

Metantimonates.—These salts are produced by fusing antimony or its trioxide with nitrates, or the acid $\mathbf{HSbO_3}$ with carbonates; or by double decomposition from the potassium salt, $KSbO_3.Aq$. The chief compounds are:—

$LiSbO_3$; $2NaSbO_3.7H_2O$; $NaSbO_3.3H_2O$; $KSbO_3$; also $2KSbO_3.5H_2O$ and $3H_2O$; $NH_4SbO_3.2H_2O$; $Ca(SbO_3)_2$; $Sr(SbO_3)_2.6H_2O$; $Ba(SbO_3)_2.5H_2O$; $Mg(SbO_3)_2.12H_2O$; $Zn(SbO_3)_2$; $Co(SbO_3)_2$; $Ni(SbO_3)_2.6H_2O$; $Sn(SbO_3)_2.2H_2O$; $Pb(SbO_3)_2$; $Cu(SbO_3)_2$; $Hg(SbO_3)_2$.

All these salts, with exception of the lithium, sodium, potassium, and ammonium salts, are sparingly soluble in water, and crystalline. The compounds $2NaSbO_3.7H_2O$ and $2KSbO_3.5$ and $3H_2O$ are gummy, and may possibly be derived from a poly-metantimonic acid. When boiled with water they are decomposed, giving a residue of $3Sb_2O_5.K_2O.10H_2O$.

"Naples yellow" is a basic antimonate of lead, produced by heating 2 parts of lead nitrate, 1 part of tartar-emetic, and 4 parts of common salt to such a temperature that the salt fuses; the mass is then treated with water, which dissolves the salt, leaving the "Naples yellow" in the form of a fine yellow powder. Another basic antimonate of lead occurs native as *bleinerite*, $Sb_2O_5.3PbO.4H_2O$.

Certain complex phosphates have been prepared by fusing tetrasodium pyrophosphate with metaphosphate in the proportion $Na_4P_2O_7$ to $2NaPO_3$. The product is soluble without decomposition in a small quantity of hot water, and crystallises from the solution; but it is decomposed by much water. With solutions of salts of the metals, it gives precipitates; the silver salt, for example, has the formula $Ag_6P_4O_{13}$. Another salt has been produced by fusing together the same constituents in the proportion $Na_4P_2O_7$ to $8NaPO_3$. The resulting salt is very sparingly soluble; the silver salt derived from it has the formula $Ag_{12}P_{10}O_{31}$. These phosphates go by the name of Fleitmann and Henneberg, their discoverers.

CHAPTER XXIV.

OXIDES, ETC., OF ELEMENTS OF THE PHOSPHORUS GROUP.—COMPOUNDS OF TETROXIDES; HYPOPHOSPHATES AND HYPOANTIMONATES.—CONSTITUTION OF PHOSPHITES, ETC.—THE PHOSPHITES, ARSENITES, AND ANTIMONITES, THIOARSENITES AND THIOANTIMONITES.—THE HYPOPHOSPHITES.—OXYHALIDES AND SULPHOHALIDES.

Oxides, Sulphides, Selenides, and Tellurides of Phosphorus, Arsenic, Antimony, and Bismuth, continued.

Compounds of tetroxides.—It has been already stated that the oxide $\mathbf{P_2O_4}$, when treated with water, gives a mixture of phosphorous and phosphoric acids, thus:—$\mathbf{P_2O_4} + 3H_2O + Aq = H_3PO_4.Aq + H_3PO_3.Aq$. It is therefore concluded to be a phosphite of phosphoryl, thus:—$\mathbf{(PO)'''(PO_3)}$. But a tetrabasic acid is known, of the formula $\mathbf{P_2O_4.2H_2O = P_2O_2(OH)_4}$, which forms distinct salts, and possesses properties differing from those of such a mixture. The sodium salt, $\mathbf{P_2O_2(ONa)_4}$, is converted by bromine and water into dihydrogen disodium pyrophosphate, and, as the acid has no marked reducing properties, it may possibly have the constitution—

$$\begin{array}{l} O{=}P(OH)_2 \\ \quad | \\ O{=}P(OH)_2 \end{array}, \text{ that of pyrophosphoric acid being } \begin{array}{l} O{=}P(OH)_2 \\ \quad >O \\ O{=}P(OH)_2. \end{array}$$

Hypophosphoric acid,* as the acid $\mathbf{P_2O_2(OH)_4}$ is called, is produced along with orthophosphoric and phosphorous acids, by the oxidation of phosphorus exposed to water and air. About one-sixteenth of the phosphorus is converted into hypophosphoric acid. On addition of sodium carbonate, the dihydrogen disodium salt separates out, owing to its sparing solubility in water. To prepare the pure acid, the barium salt is treated with the theoretical

* *Annalen,* **87**, 322; **194**, 28; *Berichte,* **16**, 749; *Comptes rend.*, **101**, 1058; **102**, 110.

amount of sulphuric acid; insoluble barium sulphate is formed, and the acid remains in solution. On evaporation in a vacuum, the acid $\mathbf{H_4P_2O_6.2H_2O}$ separates out in large rectangular tables, melting at about 62°. On standing in a dry vacuum, these crystals lose water, and gradually change to needles of the pure acid $\mathbf{H_4P_2O_6}$. This body, at 70°, suddenly decomposes into phosphorous and metaphosphoric acids:—$\mathbf{H_4P_2O_6 = H_3PO_3 + HPO_3}$.

The following salts are known:—

$\mathbf{Na_4P_2O_6.10H_2O}$; $\mathbf{K_4P_2O_6.5H_2O}$; $\mathbf{(NH_4)_4P_2O_6.H_2O}$; $\mathbf{Mg_2P_2O_6.12H_2O}$; $\mathbf{Ca_2P_2O_6.2H_2O}$; $\mathbf{Ba_2P_2O_6}$; $\mathbf{Pb_2P_2O_6}$; and $\mathbf{Ag_4P_2O_6}$; and the double salts $\mathbf{H_3NaP_2O_6.2H_2O}$; $\mathbf{H_2Na_2P_2O_6}$; $\mathbf{HNa_3P_2O_6.9H_2O}$; $\mathbf{H_3Na_5(P_2O_6)_2.20H_2O}$; $\mathbf{H_3KP_2O_6}$; $\mathbf{H_2K_2P_2O_6\,3H_2O}$; $\mathbf{HK_3P_2O_6.3H_2O}$; $\mathbf{H_3(NH_4)P_2O_6}$; $\mathbf{H_2(NH_4)_2P_2O_6}$; $\mathbf{H_2MgP_2O_6.4H_2O}$; $\mathbf{H_2CaP_2O_6.6H_2O}$; $\mathbf{H_2BaP_2O_6.2H_2O}$.

With lithium salts, sodium hypophosphate gives a white precipitate.

The tetra-metallic salts of the alkalis are easily soluble in water; the dihydrogen disodium salt is sparingly soluble, and is used to separate the acid from its mixture with phosphorous and orthophosphoric acid. The dibarium salt is produced by precipitation; it is nearly insoluble in water, as are most of the other salts. When the salts are heated they give products of decomposition of phosphorous acid (hydrogen phosphide and metaphosphate) and metaphosphate of the metal.

The silver salt may be prepared directly by dissolving 6 grams of silver nitrate in 100 grams of nitric acid diluted with 100 grams of water, and while it is kept hot on a water-bath adding 8 or 9 grams of phosphorus. The mixture must be cooled as soon as the violent evolution of gas ceases, and, on standing, tetrargentic hypophosphate crystallises out. The silver salt is not reduced to metallic silver on boiling, as is silver phosphite; and the sodium salt does not reduce salts of mercury, gold, or platinum.

No similar compounds of arsenic are known; but **antimony tetroxide**, when fused with potassium hydroxide or carbonate, yields a mass from which cold water extracts excess of alkali; the residue, dissolved in boiling water and evaporated to dryness gives a yellow non-crystalline mass which has the composition $\mathbf{Sb_2O_4.K_2O}$. On treatment with hydrochloric acid, it is converted into $\mathbf{2Sb_2O_4.K_2O}$; and excess of acid liberates the compound $\mathbf{Sb_2O_4.H_2O}$.

Compounds of trioxides and trisulphides:—Constitution of the acids, hydroxides, and salts derived from the trioxides and trisulphides of phosphorus, arsenic, antimony, and bis-

muth.—It will be remembered that phosphoryl chloride, $POCl_3$, on treatment with water, yields orthophosphoric acid, $PO(OH)_3$, and it may be supposed that phosphorus trichloride, PCl_3, yields a similar acid, $P(OH)_3$. Such an acid ought to be tribasic, like orthophosphoric acid, and should yield three double salts, *e.g.*, $P(OH)_2(ONa)$, $P(OH)(ONa)_2$, and $PO(ONa)_3$. But the last of these is formed only when the second is mixed with great excess of a strong solution of sodium hydroxide, and left for some time; it is then thrown down on addition of alcohol. It appears not improbable, therefore, that a change has taken place during this time, and that the compound $O{=}P{<}^{H}_{(OH)_2}$ has changed to $PO(H)_3$. And it is also to be noticed that when water acts on phosphorus trichloride, some orthophosporic acid and free phosphorus are formed; this might take place during the change of $P(OH)_3$ to its isomeride $O{=}P(OH)_2H$. Moreover, an acid is known, named ethyl-phosphinic acid (produced by the oxidation of the compound ethyl-phosphine), analogous to hydrogen phosphide (see p. 532), which is certainly dibasic, and in which the phosphorus is doubtless in direct union with carbon. The formulæ are:—

Hydrogen phosphide.	Ethyl phosphine.	Ethyl-phosphinic acid.
$P{\lt}\begin{smallmatrix}H\\H\\H\end{smallmatrix}$	$P{\lt}\begin{smallmatrix}H\\H\\C_2H_5\end{smallmatrix}$	$O{=}P{\lt}\begin{smallmatrix}OH\\OH\\C_2H_5\end{smallmatrix}$.

There are therefore good reasons for believing that, although two phosphorous acids might exist, the one known is $O{=}P(OH)_2H$, and not $P(OH)_3$. The isomerism is analogous to that of the two nitrous acids (see p. 337), $O{=}N{-}OH$, and $O_2{\equiv}N{-}H$. The anhydride of the acid $O{=}P(OH)_2H$ would be therefore not P_2O_3, but O_2PH, an unknown substance. As with orthophosphoric acid pyrophosphates are known, so pyrophosphites exist, *e.g.*, $Na_4P_2O_5$. Such substances also find representatives among the arsenites and thioarsenites, all these series of salts being known, viz., $MAsO_2$, and $MAsS_2$, metarsenites and thioarsenites; $M_4As_2O_5$ and $M_4As_2S_5$, pyroarsenites and thioarsenites; and M_3AsO_3 and M_3AsS_3, ortho-arsenites and thioarsenites. The corresponding metaphosphites are unknown. A few antimonites and sulphantimonites have also been prepared.

Phosphorous acid, etc.

$$H_3PO_3;\ H_4Sb_2O_5 = Sb_2O_3.2H_2O;\ H_3SbO_3 = Sb_2O_3.3H_2O.$$

To prepare crystalline **phosphorous acid, H_3PO_3,** a current of dry air is passed through phosphorus trichloride heated to 60° and passed into water cooled to 0°. When the water is saturated, the crystals which separate are washed with ice-cold water, and dried in a vacuum. It is also slowly formed by union of the anhydride with water; or along with orthophosphoric acid by the action of water on the tetroxide; or along with phosphoric and hypophosphoric acids by the oxidation of phosphorus in air, in contact with water. Phosphorus also abstracts oxygen from a solution of copper sulphate, depositing copper, thus:—$3CuSO_4.Aq + 6H_2O + 2\mathbf{P} = 3H_2SO_4.Aq + 2H_3PO_3.Aq + 3\mathbf{Cu}$. The sulphuric acid may be removed as barium sulphate by cautious addition of solution of barium hydroxide.

Pyroantimonious acid, $\mathbf{H_4Sb_2O_5}$, is produced by addition of copper sulphate to a solution of antimony trisulphide in caustic potash. Copper sulphide is formed, and potassium antimonite; and on addition of an acid to the filtered liquid, the antimonite is decomposed, pyroantimonious acid being precipitated.

Orthoantimonious acid is formed by the spontaneous decomposition of the peculiar compound acid of which tartar-emetic is the potassium salt. This acid is liberated from the barium salt corresponding to tartar-emetic, by the action of sulphuric acid, and has the formula $\mathbf{(C_4H_4O_6)''Sb.OH}$. With water it yields $\mathbf{Sb(OH)_3}$, and tartaric acid, $C_4H_6O_6.Aq$. From this it would appear that tartar-emetic is not, as hitherto supposed, a tartrate of potassium and antimonyl, $K(SbO)C_4H_4O_6$, but a tartaro-antimonite $(C_4H_4O_6)''Sb.OK$, two hydroxyl groups of antimonious acid, $Sb(OH)_3$, being replaced by the dyad group $(C_4H_4O_6)$.

Phosphorous acid forms deliquescent white crystals, melting at 74°. When heated it decomposes into hydrogen phosphide and phosphate:—

$$\mathbf{4H_3PO_3 = 3H_3PO_4} + \mathit{PH_3}.$$

Zinc and iron dissolve in it, and the liberated gas is hydrogen phosphide; this action is somewhat similar to that of nitric acid on certain metals, whereby ammonia is produced. It is a powerful reducing agent, tending to combine with oxygen to form orthophosphoric acid; hence, when added to solutions of salts of silver, gold, and mercury, the metals are deposited. It also reduces sulphurous acid to hydrogen sulphide, thus:—

$$3H_3PO_3.Aq + H_2SO_3.Aq = 3H_2PO_4.Aq + \mathit{H_2S}.$$

The **antimonious acids** are white powders, insoluble in water, but soluble in hot solutions of hydroxides of sodium and potassium, forming antimonites. The corresponding hydroxides of bismuth have no acid properties. The three hydrates, $Bi(OH)_3$, $Bi_2O(OH)_4$, and $BiO(OH)$, are all known. They are produced by heating solutions of bismuth salts with potash or ammonia.

Phosphites.—Na_3PO_3 is the only trimetallic phosphite known. It is produced by addition of a large excess of a strong solution of sodium hydroxide to disodium phosphite, $HNa_2PO_3.Aq$, and after two hours adding alcohol. The trisodium salt settles down as a viscid syrup, which is stirred with alcohol, and finally dried in a vacuum over sulphuric acid.

$Na_2HPO_3.5H_2O$; K_2HPO_3 ; $(NH_4)H_2PO_3.H_2O$; $2HNa(HPO_3).5H_2O$; $HK(HPO_3)$; and $2H_4Na_2(HPO_3)_3H_2O$; and $H_4K_2(HPO_3)_3$.

These bodies form soluble crystals, and are produced by addition of phosphorous acid to hydroxides or carbonates.

$Ca(HPO_3).H_2O$; $2Sr(HPO_3).H_2O$; $2Ba(HPO_3).H_2O$; and $H_2Ca(HPO_3)_2.H_2O$; $H_2Ba(HPO_3)_2.H_2O$.

White sparingly soluble salts.

$Mg(HPO_3)$; $Cd(HPO_3)$ (?) ; $2Zn(HPO_3).5H_2O$.

These and an ammonium magnesium phosphite are produced by precipitation. They are white, crystalline, and sparingly soluble.

Phosphites of aluminium, chromium, and iron have been prepared, but not analysed. They are sparingly soluble precipitates.

$Mn(HPO_2)$; $Co(HPO_2).2H_2O$, and $Ni(HPO_3).3H_2O$.

Coloured precipitates.

$Sn(HPO_3)$ and phosphites of tin dioxide and of titanium have also been prepared; they are white precipitates. $Pb(HPO_3)$ is also white, and is formed by precipitation. It is nearly insoluble. When digested with ammonia, the basic phosphate, $P_2O_3.4PbO.2H_2O$, is produced. Bismuth phosphite is a white precipitate; and copper phosphite, $Cu(HPO_3).2H_2O$, forms sparingly soluble blue crystals; when boiled, metallic copper is precipitated.

All these phosphates decompose when heated, evolving hydrogen and a little hydrogen phosphide, and leaving a phosphate.

It is stated that when the compound $2HNa(HPO_3).5H_2O$ is heated to 160° it loses six molecules of water, forming a pyrophosphite, $Na_2H_2P_2O_5$.* Data concerning the phosphites are exceedingly meagre, and the whole series of salts requires reinvestigation.

* *Comptes rend.*, **106**, 1400.

Some oxythiophosphites* have been prepared by the action of a solution of sodium hydroxide on phosphorus trisulphide (presumably $\mathbf{P_4S_3}$). Hydrogen, mixed with hydrogen phosphide, is evolved, and on evaporation crystals are deposited of the composition $\mathbf{Na_4P_2O_3S_2.6H_2O}$, analogous to a pyrophosphite. With sodium hydrosulphide, NaSH, hydrogen phosphide and sulphide are evolved, and the solution, evaporated in a vacuum, deposits crystals of $\mathbf{Na_4P_2OS_4.6H_2O}$. These crystals lose hydrogen sulphide at the ordinary temperature, probably forming the salt previously mentioned. With ammonium hydrosulphide, crystals of the formula $\mathbf{(NH_4)_4P_2S_5.3H_2O}$ are deposited, which, when dried at 100° in a current of hydrogen sulphide lose hydrogen sulphide, giving the compound $\mathbf{(NH_4)_4P_2O_2S_3.2H_2O}$.

From the mother liquor of these crystals the compound $\mathbf{(NH_4)_4P_2O_3S_2.2H_2O}$, analogous to the potassium salt has been obtained. Solutions of these salts when boiled lose hydrogen sulphide, and yield phosphites.

Arsenites and thioarsenites.—$\mathbf{KAsO_2}$ and $\mathbf{NH_4AsO_2}$ are white soluble salts, produced by dissolving arsenious oxide (As_4O_6) in caustic potash or ammonia. They are apparently metarsenites. By similarly treating arsenic trisulphide with potassium sulphide, either by solution or by fusion, the corresponding thioarsenite, $\mathbf{KAsS_2}$, is produced. It decomposes when treated with warm water.

By adding alcohol to a solution of a large amount of arsenic trioxide in caustic potash, the pyroarsenite, $\mathbf{H_3KAs_2O_5}$, is produced. When digested with caustic potash, the salt $\mathbf{K_4As_2O_5}$ is formed, and may be precipitated with alcohol. A similar ammonium salt is produced by direct addition, $\mathbf{(NH_4)_4As_2O_5}$. The sodium salts are all very soluble, and have not been isolated. The corresponding pyrothioarsenites are unknown; but orthothioarsenites of potassium and ammonium, $\mathbf{K_3AsS_3}$ and $\mathbf{(NH_4)_3AsS_3}$, are precipitated on adding alcohol to a solution of arsenic trisulphide in excess of colourless ammonium sulphide.

$Ca(AsO_2)_2$; $Ca_2As_2O_5$; $Ca_3(AsO_3)_2$; $Sr(AsO_2)_2$; $Ba(AsO_2)_2$; $Ba_2As_2O_5.4H_2O$; $H_4Ba(AsO_3)_2$.

These are white sparingly soluble salts, produced by addition of arsenious oxide to the hydroxides, or arsenites of potassium or ammonium to salts of the metals.

Corresponding to these are $\mathbf{Ca_3(AsS_3)_2.15H_2O}$; $\mathbf{Ba_2As_2S_5}$; and $\mathbf{Ba_3(AsS_3)_2}$; they are soluble substances precipitated by alcohol.

* *Comptes rend.*, **93**, 489; **98**, 45.

$\mathbf{Mg_3(AsO_3)_2}$; $\mathbf{MgHAsO_3}$; $\mathbf{Mg_2As_2O_5}$; $\mathbf{Mg_2As_2S_5}$; and $\mathbf{Zn_2As_2S_5}$ are produced by double decomposition.

Arsenites of the boron and aluminium groups have not been prepared.

Various basic arsenites of iron are known. These are insoluble, and are produced by addition of a ferric salt and an alkali to solutions of arsenious oxide, and for this reason a mixture of ferric hydrate and magnesia is employed as an antidote in cases of arsenical poisoning. Among these are $\mathbf{FeAsO_3.Fe_2O_3}$; $\mathbf{2FeAsO_3.Fe_2O_3.7H_2O}$, and $\mathbf{5H_2O}$.

Ferrous pyroarsenite, $\mathbf{Fe_2As_2O_5}$, is a greenish precipitate; $\mathbf{Mn_3H_6(AsO_3)_4.H_2O}$ and $\mathbf{Co_3H_6(AsO_3)_4.H_2O}$ are rose-red precipitates; the corresponding nickel salt, $\mathbf{Ni_3H_6(AsO_3)_4.H_2O}$, is a greenish-white precipitate which yields $\mathbf{Ni_3(AsO_4)_2}$ on ignition. The sulpharsenites of these metals are all pyro-derivatives, viz., $\mathbf{Fe_2As_2S_5}$, $\mathbf{Mn_2As_2S_5}$, $\mathbf{Co_2As_2S_5}$, and $\mathbf{Ni_2As_2S_5}$.

Stannous and stannic arsenites and sulpharsenites have been prepared, but not analysed. The three lead arsenites, $\mathbf{Pb(AsO_2)_2}$, $\mathbf{Pb_2As_2O_5}$, and $\mathbf{Pb_3(AsO_3)_2}$, are all white precipitates. The compound $\mathbf{Pb(AsS_2)_2}$ is a mineral named *sartorite;* $\mathbf{Pb_2As_2S_5}$ is named *dufrenoysite*, and $\mathbf{Pb_3(AsS_3)_2}$, *guittermannite.* All these are crystals with metallic lustre, and occur native.

The arsenite of hydrogen and copper, $\mathbf{HCuAsO_3}$, is obtained by adding to a solution of copper sulphate a solution of potassium arsenite, a solution of arsenious oxide, and a small amount of ammonia. It is a fine green powder, and is named, from its discoverer, "Scheele's green." The arsenite $\mathbf{Cu(AsO_2)_2}$ is produced by digesting copper carbonate with arsenious oxide and water.

Copper sulpharsenite, $\mathbf{Cu_2As_2S_5}$, is formed by precipitation; and some minerals exist which appear to be compounds of copper sulpharsenite and sulphide, *e.g.*, $\mathbf{Cu_4AsS_4}$, *julianite*, $\mathbf{Cu_6As_4S_9}$, *binnite*, and $\mathbf{Cu_8As_2S_7}$, *tennantite.*

Silver arsenite, $\mathbf{Ag_3AsO_3}$, is a yellow precipitate produced by adding to silver nitrate a solution of arsenious oxide in ammonia. It is soluble in excess of ammonia. It serves, along with Scheele's green, as a distinctive test between arsenious and arsenic oxides; it will be remembered that copper arsenate is blue, and silver arsenate red. The corresponding sulpharsenite, $\mathbf{Ag_3AsS_3}$, occurs native as *proustite;* and the mineral *xanthoconate*, $\mathbf{Ag_9As_3S_{10}}$ appears to be a double sulpharsenite and sulpharsenate of silver.

Only two **antimonites** are known, viz., $\mathbf{NaSbO_2.3H_2O}$, which forms octahedra, and is obtained by dissolving antimonious oxide, ($\mathbf{Sb_4O_6}$) in caustic soda; and an acid compound, $\mathbf{NaSbO_2.2HSbO_2}$, similarly prepared. The corresponding **thioantimonite,*** $\mathbf{NaSbS_2}$, separates on addition of alcohol to a solution of $\mathbf{Sb_2S_3}$ in sodium hydroxide; and copper-coloured crystals of $\mathbf{2NaSbS_2.H_2O}$ deposit from a concentrated solution of the same substances. Many sulphantimonites occur native; among them are $\mathbf{Fe(SbS_2)_2}$, *berthierite;* $\mathbf{Pb(SbS_2)_2}$, *zinkenite;* $\mathbf{Pb_2Sb_2S_5}$, *jamesonite;* $\mathbf{Pb_3Sb_2S_6}$, *boulangerite;* $\mathbf{Pb_4Sb_2S_7}$, *meneghinite;* $\mathbf{Pb_5Sb_2S_8}$, *geocronite;* $\mathbf{CuSbS_2}$, *chalcostibite;* $\mathbf{Cu_2Sb_4S_7}$, *guejarite;* $\mathbf{CuPbSbS_3}$, *bournonite;* $\mathbf{Ag_3SbS_3}$, *pyrargyrite;* $\mathbf{AgSbS_2}$, *miargyrite;* $\mathbf{Ag_5SbS_4}$, *stephanite;* $\mathbf{Ag_9SbS_6}$, *polybasite;* and $\mathbf{Hg(SbS_2)_2}$, *livingstonite.* Besides these, similar compounds of bismuth are known, *e.g.*, $\mathbf{AgBiS_2}$, *silver bismuth glance;* $\mathbf{Pb(BiS_2)_2}$, *galenobismuthite;* $\mathbf{Pb_2Bi_2S_5}$, *cosalite;* $\mathbf{Pb_6Bi_2S_9}$, *beegerite;* $\mathbf{CuBiS_2}$, *emplectite;* $\mathbf{Cu_3BiS_3}$, *wittichenite;* and others. These double sulphides of bismuth have not been made artificially; but the compound $\mathbf{KBiS_2}$, produced by fusing bismuth with sulphur and sodium carbonate, forms steel-grey shining needles.

Hypophosphites.†—Hydrogen hypophosphite, H_3PO_2, is a monobasic acid; and it is therefore concluded that its constitution is somewhat analogous to that of phosphorous acid, inasmuch as it may be regarded as a hydroxyl-derivative of an oxidised hydrogen phosphide, thus, $O{=}P(OH)H_2$. It is only the hydrogen of the hydroxyl which can be replaced by metals. The anhydride of such an acid would not be the oxide P_2O, but the unknown compound $O{=}PH_2{-}O{-}PH_2{=}O = H_4P_2O_3$. Such a body might be expected to be devoid of acid properties.

Hypophosphorous acid, $\mathbf{H_3PO_2}$, is produced by decomposing a solution of the barium salt, $Ba(H_2PO_2)_2$, with its equivalent of sulphuric acid. The dilute solution is boiled down, and finally evaporated at 105°, the temperature being gradually raised to 130°. It is then cooled to 0°, and on shaking it crystallises. It melts at 17·4°. When heated, it decomposes into phosphoric acid and hydrogen phosphide, thus:—

$$\mathbf{2H_3PO_2 = H_3PO_4} + \mathit{PH_3}.$$

It yields salts on neutralisation with hydroxides or oxides. But sodium, potassium, and barium hypophosphites are easily prepared by boiling phosphorus with their hydroxides. The hydrogen phosphide which is evolved is spontaneously inflammable, owing to its containing a trace of liquid hydride, P_2H_4. The reaction is:—

$$\mathbf{4P} + 3KOH.Aq + 3H_2O = \mathit{PH_3} + 3KH_2PO_2.Aq.$$

* See also Ditte, *Comptes rend.*, **102**, 168, for pyrothioantimonites.

† Rammelsberg, *Chem. Soc.*, **26**, 1.

It is from the barium salt, thus prepared, that the acid is obtained.

Hypophosphites.

$\mathbf{LiH_2PO_2.H_2O}$; $\mathbf{NaH_2PO_2.H_2O}$; $\mathbf{KH_2PO_2}$; $\mathbf{(NH_4)H_2PO_2}$.

These salts are white crystalline bodies, produced as described. Those containing water may be rendered anhydrous at 200°. They decompose, when more strongly heated, as follows:—

$$\mathbf{5NaH_2PO_2 = Na_4P_2O_7 + NaPO_3} + 2PH_3 + 2H_2.$$

The ammonium salt undergoes a different change, thus:—

$$\mathbf{7NH_4H_2PO_2 = H_4P_2O_7 + 2HPO_3} + H_2O + 7NH_3 + 3PH_3 + 2H_2.$$

$\mathbf{Ca(H_2PO_2)_2}$; $\mathbf{Sr(H_2PO_2)_2.H_2O}$; $\mathbf{Ba(H_2PO_2)_2.H_2O}$.

White soluble salts. When heated they decompose, thus:—

$$\mathbf{7Sr(H_2PO_2)_2 = 3Sr_2P_2O_7 + Sr(PO_3)_2} + 6PH_3 + H_2O + 4H_2.$$

$\mathbf{Mg(H_2PO_2)_2.6H_2O}$; $\mathbf{Zn(H_2PO_2)_2.6H_2O}$; and $\mathbf{Cd(H_2PO_2)_2}$.

These are also soluble crystalline salts, which can be dried at 200°. When heated they decompose, thus:—

$$\mathbf{5Zn(H_2PO_2)_2 = 2Zn_2P_2O_7 + Zn(PO_3)_2} + 4PH_3 + 4H_2.$$

Aluminium and chromium hypophosphites are gummy solids; the ferric salt is a white sparingly soluble powder (?).

$\mathbf{Fe(H_2PO_2).6H_2O}$; $\mathbf{Mn(H_2PO_2)_2.H_2O}$; $\mathbf{Co(H_2PO_2)_2.6H_2O}$; $\mathbf{Ni(H_2PO_2)_2.6H_2O}$.

The ferrous salt has been prepared by dissolving iron in the acid; the others by neutralisation. They can be dried at 200°. They are all crystalline and soluble. They change thus, when heated:—

$$\mathbf{6Co(H_2PO_2)_2 = 4Co(PO_3)_2 + 2CoP} + 2PH_3 + 9H_2.$$

$\mathbf{Pb(H_2PO_2)_2}$ is crystalline and sparingly soluble; when heated, it decomposes, thus:—

$$\mathbf{9Pb(H_2PO_2)_2 = 4Pb_2P_2O_7 + Pb(PO_3)_2} + 8PH_3 + 2H_2O + 4H_2.$$

Thallous hypophosphite, $\mathbf{TlH_2PO_2}$, forms soluble white crystals. It decomposes like the sodium salt when heated. The uranyl salt, $\mathbf{UO_2(H_2PO_2)_2.H_2O}$, is a sparingly soluble yellow crystalline salt.

Like the phosphites, the hypophosphites possess great power of reduction; the reaction, for example, with silver nitrate, is $Ba(H_2PO_2)_2.Aq + 6AgNO_3.Aq + 4H_2O = 2H_3PO_4.Aq + 4HNO_3.Aq + Ba(NO_3)_2.Aq + 6\mathbf{Ag} + H_2$. The free hydrogen further reduces the nitric acid. With solutions of cupric salts, cuprous salts are first produced, and then a reddish precipitate of copper hydride, **CuH**, is formed. Hypophosphorous acid also withdraws oxygen from sulphur dioxide, liberating sulphur.

Double compounds with halogens.—With phosphorus, compounds of the type $POCl_3$ are best known.

Arsenic forms only one compound of this nature, viz., $\mathbf{AsOF_3.KF.H_2O}$; its characteristic compound is **AsOCl**; and antimony and bismuth resemble arsenic; the compound $\mathbf{SbOCl_3}$ is, however, known.

POF_3; PSF_3; $POCl_3$; $PSCl_3$; $\mathbf{POBr_3}$; $\mathbf{PSBr_3}$; $POCl_2Br$; $PSCl_2Br$.
$\mathbf{SbOCl_3}$; $\mathbf{SbSCl_3}$.

These compounds have the formulæ assigned, inasmuch as their vapour-densities have, in almost all cases, been determined. Their constitution is, without doubt, analogous to $O{=}P{\equiv}Cl_3$; and it will be remembered that when treated with water or alkalies they give rise to orthophosphoric or orthothiophosphoric acid. The corresponding antimony compound, $\mathbf{SbOCl_3}$, on treatment with water, yields the more stable pyrantimonic acid, $\mathbf{H_4Sb_2O_7.2H_2O}$.

POF_3, **phosphoryl trifluoride**, is a colourless gas, liquid at $-50°$, or at $16°$ under a pressure of 15 atmospheres. By evaporation of the liquid a portion solidifies to a snow-like solid. It is produced by direct combination of phosphorus trifluoride and oxygen, which takes place with explosion on passing a spark through the mixed gases. It is more easily produced by distilling powdered *cryolite*, $\mathbf{AlF_3.3NaF}$, with P_2O_5.

PSF_3, the corresponding sulphur compound, is a spontaneously inflammable gas, liquefying under pressure of 10·3 atmospheres at 13·8°. It is best prepared by heating a mixture of phosphorus pentasulphide and lead fluoride, thus:—

$$P_2S_5 + 5\mathbf{PbF_2} = 5\mathbf{PbS} + 2PF_5\text{; and } 3PF_5 + P_2S_5 = 5PSF_3.$$

The density of this gas shows that, like the other similar compounds, it cannot be regarded as the compound $3PF_5.P_2S_5$, but as the simpler body PSF_3.

$POCl_3$, **phosphoryl trichloride**, is produced by the action of water on the pentachloride, thus:—

$$\mathbf{PCl_5} + H_2O = POCl_3 + 2HCl \text{ (see p. 353).}$$

But, as it quickly reacts with water, it is convenient to use combined water, in the form of boracic acid, $\mathbf{H_3BO_3}$, in its formation. It is easily obtained by distilling a mixture of phosphorus pentachloride and boracic acid in theoretical proportions. It can also be produced by heating together the pentachloride and pentoxide of phosphorus, thus :—

$$3PCl_5 + P_2O_5 = 5POCl_3.$$

It is a colourless liquid, heavier than water (1·7), boiling at 110°. It fumes in the air, forming phosphoric and hydrochloric acids. It may be solidified, and melts at 2·5°. It combines with some other chlorides, forming, for example :—

$POCl_3.BCl_3$; a white solid, melting at 11°.
$POCl_3.AlCl_3$; a white solid, melting and boiling without decomposition (?).
$POCl_3.MgCl_2$; a white solid, decomposed when heated.
$POCl_3.ZnCl_2$; white rhombic crystals.
$POCl_3.SnCl_4$; a liquid, boiling at 180°.

It also forms gelatinous compounds with sodium and potassium chlorides.

$PSCl_3$, **sulphophosphoryl trichloride**, is also a colourless fuming liquid, heavier than water, boiling at 124·25°. It could doubtless be prepared by heating phosphorus pentachloride with pentasulphide; but it is more readily obtained by the action of hydrogen sulphide on phosphoryl trichloride, thus :—$POCl_3 + H_2S = PSCl_3 + H_2O$; or by distilling phosphoric chloride with antimony trisulphide, thus :—

$$6PCl_5 + 5Sb_2S_3 = 3P_2S_5 + 10SbCl_3 ; \text{ and}$$
$$3PCl_5 + P_2S_5 = 5PSCl_3.$$

The easiest method of preparation is to distil phosphorus with sulphur chloride, S_2Cl_2 ; the reaction is :—

$$2P + 3S_2Cl_2 = 4S + 2PSCl_3.$$

A compound of this body, with sulphur dichloride, $PSCl_3.SCl_2$, is produced by the action of sulphur on phosphoric pentachloride. It is a colourless liquid, boiling at 100°. Its molecular weight has not been determined.

$POBr_3$ and $PSBr_3$ are crystalline solids, similarly prepared. The former melts at 45° and boils at 195° ; the latter is yellow, and cannot be distilled without partial decomposition. $POCl_2Br$ and $PSCl_2Br$ are also known.

Analogous compounds of arsenic are unknown ; but two compounds of arsenyl fluoride, of the formulæ $AsOF_3.KF.H_2O$ and $AsOF_3.AsF_5.4KF.3H_2O$, have been prepared by treating arsenate of potassium with much hydrofluoric acid. They are colourless crystalline bodies.

Antimonyl trichloride, $SbOCl_3$, is produced by the action of a trace of water on the pentachloride, $SbCl_5$. Another statement is that this body has the formula $SbCl_5.H_2O$; it might well be $SbOCl_3.2HCl$. It crystallises from chloroform. The corresponding compound **$SbSCl_3$** forms white crystals; it is produced by the action of hydrogen sulphide on the pentachloride. It is said to decompose when heated into $SbCl_3$ and S_2Cl_2 (?).

Pyrophosphoryl chloride, $P_2O_3Cl_4$, corresponding to pyrophosphoric acid, $P_2O_3(OH)_4$, has been produced by the action of nitrogen tetroxide, N_2O_4, on phosphorus trichloride. It is a colourless liquid, boiling between 210° and 215°. On treatment with water it yields orthophosphoric acid. Pyrosulphophosphoryl bromide is produced by dissolving phosphorus trisulphide, P_2S_3 (a mixture ?), in carbon disulphide, and adding the necessary quantity of bromine. The carbon disulphide is distilled off, and the residue is extracted with ether, which dissolves the compound $P_2S_3Br_4$. A light yellow oily liquid remains on evaporating the ether. When heated with phosphorus pentabromide, this substance yields the orthosulphophosphoryl bromide, $PSBr_3$; and when distilled alone, the compound P_2SBr_6, which may be regarded as corresponding to the unknown thiodiphosphoric acid, $P_2S(OH)_6$, with hydroxyl replaced by bromine (see p. 353).

This substance is a white solid, melting at −5° to a yellow liquid.

Metaphosphoryl chloride, PO_2Cl, is said to have been obtained by heating in a sealed tube for six hours a mixture of phosphorus pentoxide and phosphoryl trichloride, thus :—

$$P_2O_5 + POCl_3 = 3PO_2Cl.$$

It is a viscid colourless substance. The corresponding sulphophosphoryl bromide, **PS_2Br,** is the insoluble residue after dissolving out the compound $P_2S_3Br_4$ with ether (see above). The analogous metantimonyl chloride, **SbO_2Cl,** is produced by the action of much water on $SbCl_5$.

No chlorine derivatives of the oxide or sulphide, P_2O_3 or P_2S_3, are known. But the bismuth haloid compounds and almost all those of arsenic and antimony are thus composed.

Arsenosyl chloride, or **arsenyl monochloride, $AsOCl$,** is a hard white translucent fuming solid, formed by the action of a small amount of water on arsenic trichloride. It forms the following compounds :—

$$AsOCl.As_2O_3;\ AsOCl.NH_4Cl;\ \text{and}\ AsOCl.H_2O.$$

The last of these may be viewed as orthoarsenious acid, with one hydroxyl group replaced by chlorine, thus:—$As(OH)_2Cl$. Arsenosyl bromide, **AsOBr**, is a brown waxy solid similarly prepared. It forms the compound $2AsOBr.3H_2O$, which may perhaps be conceived as $As_2O(OH)_2Br_2.2H_2O$, a derivative of pyroarsenious acid, two hydroxyl groups being replaced by bromine.

By similarly treating arsenic tri-iodide with water, the compound $AsOI.As_4O_6$ crystallises out in thin plates.

Sulpharsenosyl iodide, **AsSI**, is said to be formed by the action of iodine on arsenic trisulphide; and on addition of powdered arsenic to a solution of sulphur and bromine in carbon disulphide, the compound $AsSBr.SBr_2$ separates in dark-red crystals.

Antimony trifluoride, SbF_3, deliquesces on exposure to moist air, forming the compound $3SbOF.SbF_3$; and bismuth oxyfluoride, **BiOF**, remains as a white powder on heating the crystalline compound **BiOF.2HF**, obtained by the action of concentrated hydrofluoric acid on bismuthous oxide, Bi_2O_3.

Antimonosyl chloride, SbOCl, and **bismuth oxychloride, BiOCl**, are produced by the action of water on the trichlorides $SbCl_3$ and $BiCl_3$. The former is obtained in crystals by mixing 10 parts of the trichloride with 17 of water, and, after allowing to stand for some days, filtering, and washing the precipitate with ether. The corresponding bismuth compound is used as a pigment and cosmetic under the name "pearl-white." When heated in air, it changes, giving the body $3BiOCl.2Bi_2O_3$. Many other compounds are produced by the action of water on antimony trichloride; among these are—

$SbOCl.7SbCl_3$; $2SbOCl\ Sb_2O_3$; $20SbOCl.10Sb_2O_3\ SbCl_3$, &c.

These bodies all dissolve in concentrated hydrochloric acid, giving the trichloride.

Similar bromides are known, similarly prepared; for instance—

$2SbOBr.Sb_2O_3$; $20SbOBr.10Sb_2O_3\ SbBr_3$; $BiOBr$; $7BiOBr.2Bi_2O_3$.
$SbOI$; $2SbOI.Sb_2O_3$; $BiOI$; and $3BiOI.4Bi_2O_3$.

Compounds of sulphantimonosyl chloride are also known, produced by the action of the trichloride on trisulphide of antimony. Crystals of $SbSCl.SbCl_3$ are produced, and on washing them with alcohol, $2SbSCl.3Sb_2S_3$ remains.

Sulphantimonosyl iodide, **SbSI**, is the product of the action of antimony tri-iodide on the trisulphide; it is a brown-red powder; when boiled with zinc oxide and water, the oxysulphide, Sb_2OS_2, is formed.

The compounds **BiSCl** and **BiSI** are similarly produced; and a selenochloride, **BiSeCl**, is formed as steel-grey needle-shaped crystals on adding bismuth triselenide, $\mathbf{Bi_2Se_3}$, to molten $BiCl_3.2NH_4Cl$.

No halogen compounds derivable from or connected with hypophosphorous acid are known; dry hydrogen iodide acts on that acid violently, producing phosphorous acid and phosphonium iodide (see p. 517), thus:—

$$3\mathbf{H_3PO_2} + HI = 2\mathbf{H_3PO_3} + \mathbf{PH_4I}.$$

Physical Properties.

Mass of 1 cubic centimetre—

$\mathbf{P_2O_5}$, 2·387; $\mathbf{As_4O_6}$ (amorphous), 3·74; (crystalline) 3·70.
$\mathbf{As_2O_5}$, 4·0; $\mathbf{Sb_4O_6}$ (octahedral), 5·11; (prismatic) 5 72.
$\mathbf{Sb_2O_5}$, 3·78; $\mathbf{Sb_2O_4}$, 4·07; $\mathbf{Bi_2O_5}$, 5·1; $\mathbf{Bi_2O_4}$, 5·6; $\mathbf{Bi_2O_3}$, 8·08.
$\mathbf{P_4S_3}$, 2·00; $\mathbf{As_2S_2}$, 3·55; $\mathbf{As_2S_3}$, 3·45; $\mathbf{Sb_2S_3}$, 4·22; (*stibnite*) 4·6. $\mathbf{Bi_2S_3}$, 7·00.
$\mathbf{As_2Se_3}$, 4·75; $\mathbf{Bi_2Se_3}$, 6·25; $\mathbf{Sb_2Te_3}$, 6 5; $\mathbf{Bi_2Te_3}$, 7·23—7·94.
$POCl_3$, 1·711 at 0°: $P_2O_3Cl_4$, 1·58 at 7°; $\mathbf{Sb_4O_5Cl_2}$, 5·0; **BiOCl**, 7·2.
$\mathbf{POBr_3}$, 2·82; $\mathbf{PSBr_3}$, 2·87; $\mathbf{AsSBr_3}$, 2·79; **BiOBr**, 6·70 at 20°.

Heats of formation—

$\mathbf{P} = P - 1{\cdot}51K$. $2\mathbf{P} + 5O = \mathbf{P_2O_5} + 3700K + Aq = 2H_3PO_4.Aq + 360K$.
$2\mathbf{P} + 3O + Aq = 2H_3PO_3.Aq + 2 \times 1252K$.
$2\mathbf{P} + O + Aq = 2H_3PO_2.Aq + 2 \times 373K$.
$\mathbf{P} + O + 3Cl = POCl_3 + 1460K$; $\mathbf{P} + O + 3Br = POBr_3 + 1056K$.
$2\mathbf{As} + 5O = \mathbf{As_2O_5} + 2194K$; $+ Aq = 60K$.
$2\mathbf{As} + 3O = \mathbf{As_2O_3} + 1547K$; $+ Aq = -76K$.
$2\mathbf{Sb} + 5O + 3H_2O = 2\mathbf{H_3SbO_4} + 2 \times 1144K$.
$2\mathbf{Sb} + 3O = \mathbf{Sb_2O_3} + 1660K$. $\mathbf{Sb} + O + Cl = \mathbf{SbOCl} + 897K$:
$2\mathbf{Bi} + 3O + 3H_2O = 2\mathbf{H_3BiO_3} + 2 \times 691K$; $\mathbf{Bi} + O + Cl = \mathbf{BiOCl} + 882K$.

CHAPTER XXV.

OZONE (OXIDE OF OXYGEN).—OXIDES, SULPHIDES, SELENIDES, AND TELLURIDES OF MOLYBDENUM, TUNGSTEN, AND URANIUM.—MOLYBDATES, TUNGSTATES, AND URANATES.—SULPHOMOLYBDATES, ETC.—OXYHALIDES.

Ozone.

Ozone.—It has long been known that oxygen through which electric sparks have been passed acquires a peculiar smell, and acts rapidly on mercury. This behaviour is due to the conversion of the oxygen into an allotropic modification, to which the name *ozone* (from ὄζειν, to smell) has been given.* In this instance the molecular weight is known, and consequently the formula of ozone, O_3; and it appears advisable, therefore, to regard it as an oxide of oxygen. It is true that ordinary oxygen, which possesses the molecular formula O_2, might also thus be regarded; but, inasmuch as ozone is the only allotropic modification of an element (except perhaps sulphur gas at a low temperature, which may possess the formula S_8) which is fairly stable and possesses a known molecular weight at the ordinary temperature, it has been given a prominent position.

Sources.—Ozone occurs in small amount in the atmosphere, especially of the country. It may be recognised by its power of turning red litmus paper soaked in a solution of potassium iodide blue, owing to the liberation of potassium hydroxide (see below). Country air contains at most about one seven-hundred-thousandth of its volume of ozone. It appears to contain more ozone in spring than in summer, and more in summer than in autumn or winter; and it is more abundant on rainy than on fine days. Its presence appears also to be favoured (in the northern hemisphere) by west or south-west winds; and its existence has been shown to be largely dependent on the prevalence of atmospheric electricity, for its amount is greatly increased during and after thunderstorms. Its presence in country air in greater amount than in town air may

* Schönbein, *Pogg. Ann*, 50, 616; Andrews, *Chem. Soc. J.*, 9, 168.

be due to the fact that the oxygen evolved from plants contains small traces of ozone, and that in the neighbourhood of towns the ozone is destroyed by its action on organic particles, and on the sulphurous acid produced during the combustion of coal.

Preparation.—Ozone is formed, as mentioned above, by the passage of electric sparks through oxygen; and it is also produced during the oxidation by free oxygen of various substances, such as phosphorus in contact with water, ether vapour, benzene and other hydrocarbons, and also by the combustion of hydrogen; by the action of sulphuric acid on barium dioxide, potassium permanganate, and other substances which evolve oxygen in the cold on being thus treated; and, lastly, by the electrolysis of dilute sulphuric acid. It is never obtained pure. By the first process a quarter of the oxygen present has been converted into ozone, but by the other processes a much smaller proportion undergoes change.

These processes of formation may be illustrated as follows:—

1. **By slow oxidation.**—(*a.*) A few sticks of phosphorus are placed in a large bottle and partly covered with water. After standing for about an hour, the air in the bottle is aspirated through a U-tube containing a solution of potassium iodide mixed with a little boiled starch. The solution will turn blue owing to the liberation of iodine and the formation of blue iodide of starch.—(*b.*) A few drops of ether are poured into a large dry beaker, covered with a plate, and the beaker is shaken so as to mix the ether vapour with the air. The gaseous mixture is then stirred with a glass rod heated over a flame till too hot to touch. On pouring a little solution of potassium iodide and starch into the beaker and shaking, a blue colour will be produced. It has been observed that this reaction is also shown by the air in a bottle containing a little petroleum, frequently opened and shaken, especially if it has been exposed to sunshine for a few days.

2. **During combustion.**—If a small jet of hydrogen be burned below a funnel, and the products be drawn through a solution of potassium iodide and starch, the blue colour is produced; but, besides ozone, hydrogen peroxide and ammonium nitrite are produced, both of which have the property of liberating iodine from such a solution. The method of distinguishing these bodies from ozone is described below.

3. **Dilute sulphuric acid, electrolysed** by eight Grove cells, each electrode consisting of six thin platinum wires, yields oxygen rich in ozone. In one experiment at 9°, about one quarter of the oxygen collected was in the form of ozone. Persulphuric acid is also formed in the liquid by this process.

4. The most convenient method of producing ozone is by the **passage of the "silent discharge" through oxygen.** This "silent discharge" appears to consist of a rain of small sparks, and is best produced between two surfaces of glass placed very near each other, with conducting coatings on their exterior surfaces. A Ruhmkorff coil or an electrical machine may be used as a source of the electricity of the high potential required. The apparatus, of which a

cut is given in fig. 41, serves well for the purpose, and by its means the relation between the alteration in volume of the oxygen during conversion into ozone and the volume of the ozone produced may be shown. It consists of a

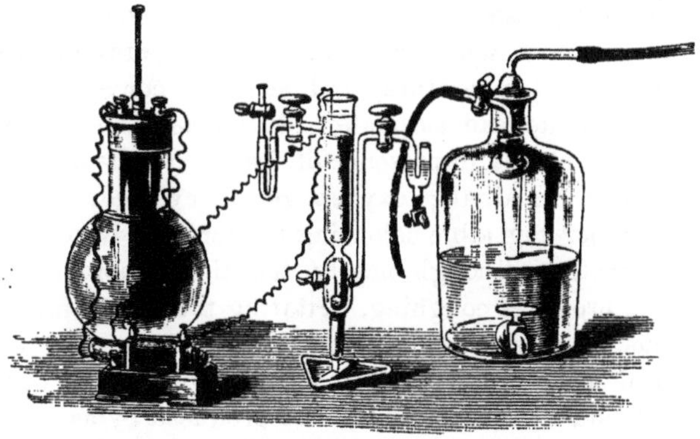

FIG. 41.

wide glass tube, standing on a foot, and constricted about 2 inches from its lower end. At its lower end a paraffined cork, or a glass stopper lubricated with vaseline, is inserted in the tubulus, and on the opposite side to the tubulus a vertical tube, provided with a stopcock, ending in a U-tube, is sealed on. A narrower tube, which should fit the wider one very closely, but without touching, is sealed through its upper end. At the top of the wide tube a gauge, like that shown in the figure, is attached by sealing. The outer tube is covered with tinfoil where it surrounds the inner tube.

As ozone is destroyed by grease, the stopcock should be lubricated with vaseline, and no india-rubber connections should be placed in contact with ozone, for it at once attacks india-rubber.

To illustrate the formation of ozone with this apparatus, a slow current of oxygen is passed through the tube, entering through the gauge-tube, which should contain no liquid. A platinum wire connected with one pole of a coil is dipped in dilute sulphuric acid contained in the inner tube, while the outer coating of tinfoil is connected with the other pole of the coil. The U-tube having been filled with a solution of potassium iodide and starch, a blue colour is produced as soon as the current passes. The characteristic smell may be noticed before the solution is poured into the U-tube.

Properties.—At ordinary pressure and temperature, ozone is a gas, colourless in thin layers; but by looking through a tube several metres in length, filled with ozonised oxygen, it is seen to have a blue colour. When compressed, this blue colour becomes more apparent, and at low temperatures it increases in intensity. A current of ozonised oxygen, cooled to −180° by liquid oxygen boiling at atmospheric pressure, deposits its ozone as a dark-blue liquid, while the oxygen passes on. The blue liquid boils at −106°.*

* *Comptes rend.*, **94**, 1249; *Monatshcft Chem.*, **8**, 69.

When heated to 250—300°, ozone is reconverted into ordinary oxygen, and the volume of the gas is found to have permanently increased. Ozone liberates iodine from a solution of potassium iodide, forming potassium hydroxide and free iodine; it oxidises silver and mercury, which are unaffected by ordinary oxygen at the atmospheric temperature, and at once converts black lead sulphide into white lead sulphate. It reacts with hydrogen peroxide, but only slowly in presence of free acid; the action is rapid in presence of an alkali; oxygen gas is evolved. It bleaches indigo and other vegetable colouring matters. It is very sparingly soluble in water, although nearly ten times more soluble than oxygen. It provokes coughing, irritating the bronchial tubes. It is poisonous when breathed in a concentrated form; and, curiously, the blood of animals killed by it is found to have the dark colour of venous blood; death appears to be produced by asphyxia.

Proof of the formula of ozone.—That ozone has the formula O_3 is rendered probable by the following experiments:—1. Oxygen, when ozonised, undergoes contraction. This may be proved by placing the apparatus shown in fig. 41, filled with oxygen, in water so as to maintain a constant temperature, for on passing the discharge the gas would become heated, and changes of volume, not dependent on the conversion of oxgyen into ozone, would then occur. Some strong sulphuric acid, coloured with indigo, is introduced into the gauge, and the stopcock connecting the apparatus with the U-tube is shut. On passing a current, a momentary expansion will take place at first, due for the most part to heating of the gas; it is, however, followed by a contraction shown by the rise of the liquid in the gauge. Its level is observed.

2. The apparatus is then removed from the water, and by a rapid shake, a small thin bulb filled with oil of turpentine contained in the lower part of the tube is broken. The apparatus is again placed in water, and allowed to stand for a few minutes, so as to regain its original temperature. A further contraction will have taken place, amounting to twice that originally observed. The U-tube is washed out and filled with fresh iodide of potassium and starch; the contents of the apparatus are then expelled through the U-tube by a current of oxygen; no coloration is produced, showing that the ozone has been completely removed.

The relation of the volume of the ozone to that of the oxygen from which it has been produced, can be inferred from these experiments. To take a suppositious case:—Suppose the total volume of the oxygen before electrification is 100 c.c. After partial conversion into ozone, the volume may be imagined to be reduced to

99 c.c.; and after absorption of the ozone by turpentine, the contraction is twice as great as the original one, and the volume is further reduced to 97 c.c. We have thus:—

Volume of original gas	100 c.c.
Volume of oxygen plus ozone	99 c.c.
Volume of oxygen after removal of ozone.....	97 c.c.
Hence oxygen converted into ozone	100−97 c.c. = 3 c.c.
Volume of that ozone	99−97 c.c. = 2 c.c.

We see, therefore, that three volumes of oxygen are converted into two volumes of ozone.

The density of ozone should, therefore, be that of $O_3/2 = 24$.

No direct experiments on its density have been made; but corroborative evidence is furnished by experiments in which the rate of diffusion of ozone mixed with oxygen was compared with that of chlorine mixed with oxygen.* The rate of diffusion of two gases, as shown by Graham, is inversely as the square roots of their respective densities. Now, the density of chlorine is 35·5, the square root of which is 5·96; and that of ozone is presumably 24, the square root of which is 4·90; hence, for every 4·90 grams of chlorine escaping into air by diffusion, 5·96 grams of ozone should escape. The ratio between these numbers is

$$5{\cdot}96 : 4{\cdot}90 : 100 : 82{\cdot}2;$$

the rate of diffusion should be, therefore, the 100/82 of that of chlorine; it was experimentally found to be the 100/84th part, a sufficiently close approximation. A similar set of experiments proved it to have nearly the same rate of diffusion as carbon dioxide, CO_2, the density of which is 22. It may be regarded, therefore, as a compound of three atoms of oxygen, and it possesses the formula O_3, with the molecular weight 48.

Many substances, such as potassium iodide, mercury, and silver, convert ozone into ordinary oxygen, but remove only the third atom of the oxygen of ozone. The equations are:—

$$2KI\,Aq + O_3 + H_2O = 2KOH.Aq + I_2.Aq + O_2;$$
$$2\mathbf{Ag} + O_3 = \mathbf{Ag_2O} + O_2;\ \text{and}\ \mathbf{Hg} + O_3 = \mathbf{HgO} + O_2.$$

Other bodies, such as the *dry* dioxides of lead and manganese, and copper oxide, decompose it without themselves suffering any permanent change. And in some cases it has a reducing action; for example, silver oxide is reduced to metallic silver, thus:—$\mathbf{Ag_2O} + O_3 = 2\mathbf{Ag} + 2O_2$; moist dioxide of lead is also reduced:—

* Soret, *Annales* (4), **7**, 113; **13**, 2.7.

$PbO_2 + O_3 = PbO + 2O_2$. And in alkaline or neutral solution it reduces hydrogen dioxide, $H_2O_2.Aq + O_3 = H_2O.Aq + 2O_2$. It is probable that the decomposing action which silver, mercury, &c., have on ozone is due to a double change, for instance: $2Ag + O_3 = Ag_2O + O_2$, and $Ag_2O + O_3 = 2Ag + 2O_2$. These changes may be regarded as due to the action of *atomic* oxygen, a body incapable of more than a momentary existence at the ordinary temperature, but one which we should suspect to display great chemical activity. In this connection it may be noted that at the moment when the electric discharge begins to pass through pure dry oxygen, a sudden expansion occurs, too sudden to be regarded as due to rise of temperature; an equally sudden contraction ensues. It may be supposed that the first action of the discharge is to partly dissociate the ordinary oxygen, O_2, into atoms, many of which then combine in groups of three, forming ozone, O_3.

Tests for Ozone.—Ozone liberates iodine from potassium iodide, with formation of potassium hydroxide. If, therefore, half of a strip of red litmus paper be moistened with a solution of potassium iodide and starch, the moist portion will become blue, owing to the liberated alkali. This effect is not produced by nitrous acid, hydrogen peroxide, chlorine, or other substances which have also the power of liberating iodine from potassium iodide.

Another test is the power which ozone possesses of oxidising a thallous salt to hydrated thallic oxide. Paper moistened with a solution of colourless thallous hydroxide therefore changes to the brown tint of thallic hydrate on exposure to ozonised oxygen.

Oxides and Sulphides of Molybdenum, Tungsten, and Uranium.

No selenides or tellurides of the elements of this group have been prepared. The following is a list of the oxides and sulphides:—

List.	Molybdenum.	Tungsten.	Uranium.
Oxygen	Mo_2O_3; MoO_2; MoO_3.	WO_2; WO_3.	UO_2; UO_3; UO_4. U_2O_9(?)*; UO_6.*
Sulphur	— MoS_2; MoS_3; MoS_4.	WS_2; WS_3.	US_2; US_3 (?).*

Besides these, the following oxides are known; they may be regarded as compounds of the simpler ones with each other:—

* Possibly exist in combination with other oxides or sulphides.

Mo_2O_5 (in combination with water) = $MoO_2.MoO_3$; U_2O_5; Mo_3O_8 = $MoO_2.2MoO_3$; U_3O_8 = $UO_2.2UO_3$.

Sources.—Molybdenum and tungsten trioxides occur native as *molybdic ochre* and *tungstic* or *wolfram ochre;* the former often coats the surface of the native sulphide as an earthy powder; and the latter forms a bright yellow or yellowish-green powder, sometimes occurring crystallised in cubes. The oxide U_3O_8 is the chief constituent of *pitchblende*, the chief source of uranium. It is a hard greyish, greenish, or reddish-black mineral, sometimes crystallising in regular octahedra. It usually accompanies lead and silver ores. Molybdenum, tungsten, and uranium also occur as **trioxides**, in combination with other oxides. Among such compounds are *wulfenite*, or *yellow lead ore*, lead molybdate, $PbMoO_4$; calcium tungstate, $CaWO_4$, named *scheelite* or *tungsten* (from the Swedish words *tung*, heavy, and *sten*, stone; its specific gravity is 6); ferrous-manganous tungstate, $(Fe,Mn)WO_4$, or *wolfram*, the chief ore of tungsten; and *scheeletine* or lead tungstate, $PbWO_4$. Uranium occurs as carbonate in *liebigite*, $Ca(UO_2)(CO_3)_2$; as phosphate in *uranium vitriol;* also in *uranite*,

$$2(UO_2).Ca(PO_4)_2.8H_2O,$$

and in *chalcolite*, in which copper replaces calcium.

Uranium also occurs in the rare minerals *samarskite*, *fergusonite*, *pyrochlore*, *euxenite*, &c., in combination with oxides of niobium, tantalum, yttrium, and other elements.

Disulphide of molybdenum, MoS_2, occurs native as *molybdenite*, or *molybdenum glance*, in soft, grey, elastic, flexible laminæ, resembling lead in colour and lustre, and graphite in its touch.

Preparation.—1. **By direct union.**—Molybdenum and tungsten, when finely divided, burn when heated in air to the trioxides, MoO_3 and WO_3; uranium yields the oxide U_3O_8. The trioxide of molybdenum is also obtained by heating the metal in water-vapour, or with potassium hydroxide. The sulphides, MoS_2, WS_2, and US_2, are also produced directly, by heating the finely-divided metals with sulphur.

2. **By heating double compounds.**—The oxides Mo_2O_3, MoO_2, MoO_3, WO_3, UO_2, UO_3, and UO_4 are left anhydrous when their hydrates are heated. With Mo_2O_3 and MoO_2, air must be excluded, else oxidation occurs. The hydrate of the oxide UO_2 becomes anhydrous when boiled with water. To dehydrate $UO_4.2H_2O$, the temperature must not be allowed to rise much above 100°, else loss of oxygen ensues.

Molybdates, tungstates, and uranates of ammonium and of mercury leave the trioxides when heated to redness, the volatile bases being expelled. Uranyl nitrate, $\mathbf{UO_2(NO_3)_2}$, at 250° yields the trioxide. When more strongly ignited, $\mathbf{U_3O_8}$ is produced; and at an intense heat, $\mathbf{U_2O_5}$.

3. **By reducing a higher oxide or sulphide.**—Molybdenum sesquioxide is produced by treating the trioxide with nascent hydrogen from zinc and hydrochloric acid. Molybdenum and tungsten dioxides are produced when the trioxides are heated to low redness in hydrogen; at high temperatures the oxides are reduced to metal. Uranium dioxide is produced by heating the complex oxide $\mathbf{U_3O_8}$ to whiteness with carbon, or in a current of hydrogen. It has also been prepared by heating the oxalate, $\mathbf{(UO_2)C_2O_4}$, or the double oxychloride, $\mathbf{UO_2Cl_2.2KCl.2H_2O}$, to redness in a current of hydrogen. Molybdenum dioxide is obtained in a crystalline form, by fusing sodium molybdate, $\mathbf{Na_2MoO_4}$, with metallic zinc, which deprives the trioxide, $\mathbf{MoO_3}$, of its oxygen. Uranium tetroxide, $\mathbf{UO_4}$, loses oxygen when heated to 250°, leaving the trioxide, and at higher temperatures gives the oxide $\mathbf{U_3O_8}$, which loses more oxygen on intense ignition, leaving $\mathbf{U_2O_5}$.

The higher sulphides of molybdenum and tungsten, $\mathbf{US_4}$, $\mathbf{US_3}$, and $\mathbf{WS_3}$, likewise lose sulphur at a red heat, yielding the disulphides.

4. **By oxidation of a lower oxide.**—The oxides $\mathbf{MoO_3}$, $\mathbf{WO_3}$, and $\mathbf{U_3O_8}$ are produced when the lower oxides are ignited in air. The higher sulphides, however, are not formed by heating the lower ones with sulphur.

5. **By replacement, or by double decomposition.**—The only oxide produced by this method is uranium tetroxide; it is formed when a mixture of solutions of uranyl nitrate and hydrogen dioxide, in presence of a large excess of sulphuric acid, are allowed to stand for some weeks, thus:—

$$UO_2(NO_3)_2.Aq + H_2O_2.Aq = \mathbf{UO_4} + 2HNO_3.Aq.$$

This is, however, a method of preparing the sulphides. Molybdenum or tungsten trioxide, heated with sulphur, yields sulphur dioxide, and the disulphide. Disulphide of tungsten is also produced by heating any oxide in a current of hydrogen sulphide or carbon disulphide; and uranium disulphide has been obtained by heating uranium tetrachloride, $\mathbf{UCl_4}$, in a current of hydrogen sulphide.

The higher sulphides are also prepared by this method. Molybdenum and tungsten trisulphides are formed by addition of hydrogen sulphide or ammonium sulphide to the solution of a molybdate or a tungstate, and subsequent addition of an acid. They are then precipitated. Uranium trisulphide is produced by heating the tribromide in a current of hydrogen sulphide. Molybdenum tetrasulphide is precipitated on addition of an acid to a solution of a sulphopermolybdate, such as $Na_2MoS_5.Aq$.

Properties.—Molybdenum sesquioxide was believed by Berzelius to be the monoxide, MoO. It is a black powder, when obtained by igniting the hydrate; but when produced by the action of nascent hydrogen from zinc and hydrochloric acid on the trioxide, a dark brass-yellow precipitate. **Molybdenum dioxide,** from the trioxide with hydrogen, is a dark-brown powder; when prepared from sodium molybdate by fusion with zinc, it forms blue-violet prisms. **Dioxide of tungsten** forms brilliant copper-red plates, insoluble in water and acids; and **uranium dioxide** also possesses metallic lustre; prepared from the oxalate, it is a cinnamon-brown powder; but when obtained from the double chloride, it crystallises in lustrous octahedra. The amorphous form glows when heated in air, burning to $\mathbf{U_3O_8}$.

Molybdenum trioxide forms a light porous white mass of silky scales; it melts at a red heat to a dark-yellow liquid, which, when cooled slowly, solidifies in needles. It is volatile in a current of air, but not alone; this is perhaps due to the transient formation of a dissociable and more volatile higher oxide. It is insoluble in water, but dissolves in acids. **Trioxide of tungsten** is a lemon- or sulphur-yellow powder, turning darker when heated. It may be obtained in transparent trimetric tables by crystallisation from fused borax, or in octahedra by heating it in a current of hydrogen chloride. It melts at a bright-red heat. **Uranium trioxide** is a buff-coloured powder. **Uranium tetroxide** forms a heavy, white, crystalline precipitate. The more complex oxide, $\mathbf{Mo_3O_8}$ is a blue insoluble powder; $\mathbf{U_3O_8}$, when prepared artificially, is a dark-green velvety powder; and $\mathbf{U_2O_5}$ a black powder. When the glaze on porcelain is mixed with the oxide $\mathbf{U_3O_8}$, and baked, an intense black colour is produced, and it is conjectured that during firing, the oxide $\mathbf{U_3O_8}$ is converted into $\mathbf{U_2O_5}$.

Molybdenum disulphide, prepared artificially, is a black lustrous powder; **disulphide of tungsten** forms slender black needles; and that of uranium is a greyish-black amorphous body, becoming crystalline at a white heat. **Molybdenum trisulphide** is a blackish-brown powder; and that of **tungsten** forms black lumps which yield a liver-coloured powder. Both of these bodies dissolve in solutions of sulphides of the alkalis; that of tungsten is slightly soluble in water, but is precipitated on addition of ammonium chloride. It becomes denser when boiled with hydrochloric acid. **Uranium trisulphide** is a black powder. **Molybdenum tetrasulphide** forms dark-red flocks, drying to a dark-green mass with metallic lustre; when triturated, it gives a red powder.

Double compounds.—(*a.*) **With water.**—The hydrates are mostly prepared by double decomposition; those of the trioxides, and of uranium tetroxide, may be termed acids, inasmuch as they correspond in formula to numerous double oxides. Double oxides corresponding to the other oxides of these elements have not been prepared.

Hydrated molybdenum sesquioxide is produced by adding potassium hydroxide to a solution of a molybdate previously exposed for some time to the action of nascent hydrogen from zinc and hydrochloric acid, or better, sodium amalgam. It is a black precipitate, soluble in acids, forming dark-coloured or purple molybdous salts; in dilute solution they have a brownish-red colour.

Hydrated molybdenum dioxide is produced by adding ammonia solution to a solution of molybdenum tetrachloride. It is a rusty-coloured precipitate, sparingly soluble in water, in which it dissolves to a dark red solution; hence it should be washed with alcohol. Its solution gelatinises on standing. It is soluble in acids, forming reddish-brown solutions.

Hydrated dioxide of tungsten is unknown; that of **uranium** is precipitated in red-brown flocks from uranous salts by addition of an alkali. It is soluble in acids, forming green solutions.

Molybdenum trioxide is sparingly soluble in water (1 in 500). By dialysing it, Graham prepared a stronger solution, possessing a yellow colour and an acid taste. This soluble modification has also been prepared by addition of the theoretical amount of sulphuric acid to barium molybdate suspended in water, and filtering off the precipitated barium sulphate. On evaporation it forms a transparent blue-green mass, which slowly dries to the anhydrous oxide. The hydrate, or acid, $\mathbf{2MoO_3.H_2O = H_2Mo_2O_7}$, has been prepared by drying the residue for two months over strong sulphuric acid; and a hydrate, or acid, $\mathbf{MoO_3.H_2O}$, once separated in prisms, after long standing, from a solution of the magnesium salt which had been mixed with nitric acid equivalent to the magnesium.

The hydrated double oxide, $\mathbf{Mo_2O_5.3H_2O}$, forms a blue precipitate on mixing a solution of the dioxide with a solution of the trioxide in hydrochloric acid. It may be regarded as $\mathbf{MoO_2.MoO_3}$, molybdyl molybdate. Similarly, the oxide $\mathbf{Mo_3O_8}$ may be viewed as $\mathbf{MoO_2.2MoO_3}$, or molybdyl dimolybdate.

Hydrated tungsten trioxide, or **tungstic acid,** $\mathbf{WO_3.H_2O}$, $= \mathbf{H_2WO_4}$, forms a yellow precipitate on adding an acid to a *hot* solution of a tungstate. It crystallises without alteration of composition from hydrofluoric acid. By similar treatment of a *cold* solution, a white gelatinous precipitate of $\mathbf{WO_3.2H_2O}$ is formed. It is also produced when water is added to a solution of tungsten chloride or oxychloride.

A soluble modification of tungstic acid, named metatungstic acid, is produced by action of sulphuric acid on barium tungstate suspended in water. On evaporation, the solution deposits yellow crystals of $4(\mathbf{WO_3.H_2O}).31\mathbf{H_2O} = \mathbf{H_2W_4O_{13}.31H_2O}$. This hydrate is easily soluble in water and forms soluble salts. On heating its concentrated solution, ordinary tungstic acid separates out.

Hydrated uranium trioxide, $\mathbf{UO_3.2H_2O}$, has not been obtained pure by precipitation, for alkali is always carried down. But by heating a weak alcoholic solution of the nitrate, oxidation products of alcohol are suddenly evolved; the hydrated oxide remains as a buff-coloured mass. It is also formed by exposing moist $\mathbf{U_3O_8}$ to air. When dried *in vacuo* it forms $\mathbf{UO_3.H_2O} = \mathbf{H_2UO_4}$, a lemon-yellow powder.

The **hydrated tetroxide,** $\mathbf{UO_4.2H_2O}$, is a yellow-white powder obtained by mixing a solution of a uranyl salt with hydrogen dioxide. On treatment with potash, uranic hydrate, $\mathbf{UO_3.2H_2O}$, is precipitated, and the potassium salt of an acid dissolves which may be conceived to have the formula H_8UO_{10}. The hydrated tetroxide may therefore be viewed as $\mathbf{UO_6.2UO_3.6H_2O}$. Attempts to prepare the hydrated oxide UO_6 were, however, unsuccessful; on addition of hydrogen dioxide to a nitric acid solution of uranium nitrate, the ratio of uranium to oxygen in the precipitate corresponded approximately to the formula $\mathbf{U_2O_9}$.

(*b.*) No **hydrosulphides** are known.

(*c.*) **Double oxides and sulphides; salts of molybdic, tungstic, and uranic acids; also of corresponding sulpho-acids.**—A compound of tungsten dioxide and sodium oxide has been prepared, by dissolving in fused sodium tungstate as much trioxide as it will take up, and heating the mixture to redness in hydrogen. On treatment with water, the compound $\mathbf{2WO_2.Na_2O} = \mathbf{Na_2W_2O_5}$ remains in golden-yellow scales and cubes possessing metallic lustre. It cannot be prepared by direct union of the oxides. No similar compounds of molybdenum or uranium are known.

Molybdates, tungstates, and uranates.—These are among the most complex of compounds known. Owing to their complexity, the formulæ of many are somewhat uncertain, different investigators drawing different conclusions from their analytical data. There can be no doubt that, of all oxides, these show most tendency to polymerise, especially when in union with others. It will be convenient, in writing the formulæ of these complex bodies, to represent them as compounds of oxides with oxides, *e.g.*, $mM_2O.nRO_3$, where M stands for any element, and R for molybdenum, tungsten, or uranium. As nothing is known regarding the constitution of these bodies, the water frequently contained in them will be written separately.

$5(MoO_3.Li_2O).3H_2O$; $3(MoO_3.Li_2O).8H_2O$; $MoO_3.Na_2O.H_2O$, and $2H_2O$; $2(MoO_3.K_2O).H_2O$; $MoO_3.K_2O.5H_2O$; $MoO_3.(NH_4)_2O$.—$WO_3.Li_2O$; $WO_3.Na_2O$; $WO_3.K_2O.H_2O$, $2H_2O$, and $5H_2O$; $WO_3.(NH_4)_2O$.—$UO_3.Na_2O.6H_2O$; $UO_3.(NH_4)_2O$.

The lithium, sodium and potassium salts are obtained by fusing the respective trioxides with the carbonates. The molybdates and tungstates are colourless; the uranates yellow. The ammonium salts crystallise from solutions of trioxides in ammonia; they are precipitated by alcohol; the neutral ammonium tungstate is, however, unstable, and yields crystals of $3WO_3.2(NH_4)_2O.3H_2O$. Ammonium molybdate is employed as a reagent for orthophosphoric acid.

The double salt $3MoO_3.K_2O.2Na_2O.14H_2O$ is also known, and is produced by mixture. Sodium tungstate, Na_2WO_4, is used as a mordant in dyeing, and, as it fuses at a red heat, it is employed to render linen and cotton cloth uninflammable.

$2MoO_3.Na_2O$; $2MoO_3.(NH_4)_2O$.—$2WO_3.Na_2O.2H_2O$ and $6H_2O$; $2WO_3.K_2O.2H_2O$ and $3H_2O$.—$2UO_3.Na_2O$; $2UO_3.K_2O$.

These are crystalline salts obtained by acidifying the former, or by adding trioxide in theoretical proportion to their solutions. The uranium salts are also produced by addition of excess of solution of potassium hydroxide to a uranyl salt, such as the nitrate, $UO_2(NO_3)_2$.Aq. They are light orange powders.

$7MoO_3.3Na_2O.22H_2O$; $7MoO.3K_2O.4H_2O$; $7MoO_3.3(NH_4)_2O.22H_2O$.— $7WO_3.3LiO.19H_2O$; $7WO_3.K_2O.6H_2O$.

These molybdates are produced by evaporation to dryness of solutions of the trioxide in solutions of carbonates. From the potassium salt the curious compound $16MoO_3.6K_2O.4H_2O_2$ has been obtained with hydrogen dioxide. The tungstates are obtained by the action of carbonic acid on the former salts.

$5MoO_3.2(NH_4)_2O.3H_2O$.—$5WO_3.2Na_2O.11H_2O$; $5WO_3.2K_2O.2H_2O$.— $9MoO_3.4K_2O.6H_2O$.

These salts, and those which follow, are produced by acidifying those in which the number of molecules of the two oxides are more nearly equal.

$12WO_3.5Na_2O.28H_2O$; $12WO_3.5K_2O.11H_2O$; $12WO_3.5(NH_4)_2O.5H_2O$ and $11H_2O$.
$12WO_3.Na_2O.4K_2O.15H_2O$; $12WO_3.Na_2O.4(NH_4)_2O.12H_2O$; and others.
$3MoO_3.Na_2O.4H_2O$ and $7H_2O$; $3MoO_3.K_2O.3H_2O$; $3MoO_3.Rb_2O.2H_2O$.
$3WO_3.Na_2O.4H_2O$; $3WO_3.K_2O.2H_2O$.

Molybdates of the following types are also known; they are all produced by addition of acid to those containing less trioxide :—

$4MoO_3.Na_2O.6H_2O$; $8MoO_3.Na_2O.17H_2O$; $8MoO_3.K_2O.13H_2O$; $10MoO_3.Na_2O.21H_2O$; $16MoO_3.Na_2O$.

A corresponding tetratungstate, $4WO_3.Na_2O$, remains insoluble on digesting the fused salt, $12WO_3.5Na_2O$, with water; the salt $3WO_3.2Na_2O$ dissolves. A hexuranate, $6UO_3.K_2O$, is also known; it is produced by fusion of uranyl sulphate, $UO_2.SO_4$, with potassium chloride. All these bodies are crystalline, and apparently definite chemical compounds.

The **tetratungstates**, or, as they have been termed, the **metatungstates**, form a separate class, inasmuch as the tungstic acid produced from them is soluble. They are produced by boiling solutions of ordinary tungstates with hydrated tungstic acid, $WO_3.2H_2O$, or by adding phosphoric acid to a solution of a tungstate until the precipitate at first formed redissolves. They are also obtained by adding carbonates to metatungstic acid, produced from the barium salt with sulphuric acid. They form well-defined colourless crystals. Those of the first group have the formulæ $4WO_3.Na_2O.4H_2O$ and $10H_2O$; $4WO_3.K_2O.8H_2O$; and $4WO_3.(NH_4)_2O.8H_2O$.

The sulpho-compounds are less complicated. They are as follows :—

$MoS_3.Na_2S$; $MoS_3.K_2S$; $MoS_3.(NH_4)_2S$.—$WS_3.Na_2S$; $WS_3.K_2S$; $WS_3(NH_4)_2S$.—Also **$2MoS_3.Na_2S$; $2MoS_3.K_2S$.—$2WS_3.Na_2S$;* $2WS_3.K_2S$.** The analogous oxysulphides have also been prepared:—**$MoO_2S.(NH_4)_2S$.—$WO_2S.K_2S$.—$2MoO_2S.Na_2S$.—$WO_3.K_2S.H.O$.**

No similar uranium compounds are known.

The sulphomolybdates and sulphotungstates are prepared by dissolving the trisulphides of these elements in sulphides of the alkalis, and crystallising; or those of potassium by fusing together potassium carbonate, sulphur, carbon, and molybdenum or tungsten trisulphide. Potassium sulphomolybdate forms deep-red prisms, which reflect green light. The sulphomolybdates yield deep-red solutions; the sulphotungstates are yellowish-red. The disulphomolybdates and tungstates are produced by adding acetic acid to the mono-salts; they are precipitated by alcohol. The oxysulphomolybdates and tungstates are produced by mixture, or by adding the hydrosulphide of the metal to a solution of a molybdate or tungstate, and evaporating to crystallisation. They form golden-red or yellow needles.

These bodies form well crystallised double salts with potassium nitrate, *e.g.*, **$MoS_3.K_2S.KNO_3$** and **$WS_3.K_2S.KNO_3$.**

$MoO_3.2BeO.3H_2O$; $MoO_3.CaO$; $MoO_3.SrO$; $MoO_3.BaO$.—$WO_3.CaO$; $WO_3.SrO$; $WO_3.BaO$.—$2UO_3.CaO$; $2UO_3.SrO$; $2UO_3.BaO$.—$7WO_3.3BaO.8H_2O$; $12WO_3.2BaO.3Na_2O.24H_2O$.

These molybdates and tungstates are prepared by fusing the chloride of the metal with molybdenum or tungsten trioxide and sodium chloride, or by precipitation. They are sparingly soluble white crystalline bodies. The uranates are reddish-yellow, and are produced by precipitation.

Calcium tungstate occurs native as *scheelite* or *tungsten* in white quadratic pyramids, associated with tin-stone and apatite. The mineral is insoluble in water.

Metatungstates:—$4WO_3.CaO$; $4WO_3.SrO.8H_2O$; $4WO_3.BaO.9H_2O$,

are soluble salts, prepared by dissolving the carbonate in the acid.

$MoS_3.CaS$; $MoS_3.SrS$; $MoS_3.BaS$.—$WS_3.CaS$: $WS_3.SrS$; $WS_3.BaS$. $3MoS_3.CaS$; $3MoS_3.SrS$; $3MoS_3.BaS$.

The trisulphomolybdates are produced by boiling the trisulphide with solutions of the sulphides; and the monosulphomolybdates deposit from the mother liquor. They are dark-red substances. The sulphotungstates are produced by treating the tungstates with hydrogen sulphide.

* *Annalen*, 232, 244.

$MoO_3.MgO.5H_2O$; $MoO_3.ZnO$; $MoO_3.CdO$.—$WO_3.MgO$; $WO_3.ZnO$; $WO_3.CdO$.—$2UO_3.MgO$; $2UO_3.ZnO$.—Also $7WO_3.2MgO.(NH_4)_2O.10H_2O$; $12WO_3.3MgO.2(NH_4)_2O.24H_2O$; $7WO_3.ZnO.(NH_4)_2O.3H_2O$.

These molybdates and tungstates are produced by fusing together the chloride of the metal with sodium molybdate and chloride. They form colourless crystals. The uranates are produced by igniting the double acetate of uranyl and the metal. They are not crystalline. The double ammonium salts are obtained by mixture.

Metatungstates:—$4WO_3.MgO.8H_2O$; $4WO_3.ZnO.10H_2O$; $4WO_3.CdO.10H_2O$.

These are all colourless crystalline salts, and are prepared from the carbonates.

The sulphomolybdates of zinc and cadmium are dark-brown precipitates. The neutral magnesium salt is soluble, as is also the yellow sulphotungstate. The sulphotungstates of zinc and cadmium are sparingly soluble yellow bodies.

Simple boron and yttrium molybdates have not been prepared. Boron tungstate is also unknown; but double compounds of WO_3, B_2O_3, an oxide, and water are very numerous. They are soluble colourless salts, crystallising well. Owing to their high molecular weights, too great confidence must not be placed in the formulæ given; but they appear to belong to the following classes:—

$10WO_3.B_2O_3.2BaO.20H_2O$.
$9WO_3.B_2O_3.Na_2O.3H_2O$.
$9WO_3.B_2O_3.2BaO.20H_2O$; $9WO_3.B_2O_3.2CdO.15H_2O$.
$14WO_3.B_2O_3.3K_2O.22H_2O$; (the barium and silver salts are also known)—
$12WO_3.B_2O_3.4K_2O.21H_2O$.
$7WO_3.B_2O_3.Na_2O.11H_2O$.

The solution of the cadmium salt has the exceedingly high specific gravity 3·6. The acid corresponding to the nonotungstate has been prepared from the barium salt. It is a syrup, and gives insoluble precipitates with solutions of alkaloids, and may be used to separate quinine, strychnine, &c., from solutions. It may be regarded as boron tungstate. These bodies are all prepared by mixture.

Aluminium tungstate has the formula $7WO_3.Al_2O_3.9H_2O$; it is obtained by precipitation. $3WO_3.Y_2O_3.6H_2O$ is also known. Salts of gallium and indium have not been prepared.—$MoO_3.Tl_2O$ is a crystalline powder.

$MoO_3.FeO$; $MoO_3.MnO.H_2O$; $MoO_3.CoO$; $MoO_3.NiO$.—
$WO_3.FeO$; $WO_3.MnO$; $WO_3.CoO$; WO_3NiO ; $WO_3.(Fe,Mn)O$.

Sulphomolybdates and sulphotungstates of similar formula

have also been prepared. They are produced by precipitation, or by fusion of the trioxide with chloride of the metal and common salt. Ferrous manganous tungstate is *wolfram*, the chief ore of tungsten. It is a hard dark-grey or brownish mineral, associated with tin-ores and galena. $7WO_3.3MnO.11H_2O$ and $7WO_3.3NiO.14H_2O$ are produced by precipitation.

The metatungstates known are $4WO_3.MnO.10H_2O$; $4WO_3.CoO.9H_2O$; and $4WO_3.NiO.8H_2O$. They are soluble. $4MoO_3.Fe_2O_3.7H_2O$; also double salts of ammonium with chromium and with ferric iron of the general formulæ $10MoO_3.M_2O_3.3K_2O.6H_2O$, where M may be aluminium, chromium, ferric iron, or triad manganese. A manganic salt is also known of the formula

$$16MoO_3.Mn_2O_3.5K_2O.12H_2O.$$

$3WO_3.Cr_2O_3.13H_2O$; $7WO_3.Cr_2O_3.9H_2O$; $5WO_3.Fe_2O_3.5(NH_4)_2O.H_2O$; the double salt is soluble.

The uranates have been little studied. Double molybdates and chromates have been prepared, of which an example is $MoO_3.CrO_3.K_2O.MgO.2H_2O$.

These oxides do not combine with oxides of carbon; but with titanium dioxide, compounds similar to these with boron oxide have been prepared. Among them are $12WO_3.TiO_2.4K_2O$ and $10WO_3.TiO_2.4K_2O$. Zirconium, cerium, and thorium compounds appear to exist, but have not been investigated.

The **silicomolybdates** and **tungstates** are also numerous. Silicomolybdic acid has the formula $12MoO_3.SiO_2.13H_2O$; it forms fine yellow crystals. It gives precipitates with salts of rubidium and cæsium, affording a means of separating these metals from sodium and potassium. The corresponding tungstates of silicon and their derivatives have the formulæ

$$12WO_3.SiO_2.4H_2O.nAq \text{ and } 10WO_3.SiO_2.4H_2O.nAq.$$

There are two isomerides having the first formula. The potassium salt of the first is produced by boiling gelatinous silica ($SiO.(OH)_2$) with ditungstate of potassium. This yields a precipitate with mercurous nitrate, from which the acid may be liberated with hydrochloric acid. The salts are produced by its action on carbonates. The ammonium salt of silicodecitungstic acid,

$$10WO_3.SiO_2.4(NH_4)_2O,$$

is produced similarly from ammonium ditungstate. This acid also yields numerous salts. It is unstable, and, on evaporation, is converted into the isomeric acid of the first formula, which has been named decitungstosilicic acid. These acids and their salts as a rule crystallise well.

The salts of tin have not been carefully examined.

Lead molybdate, $MoO_3.PbO$, occurs native as *wulfenite* or yellow lead ore. It is a heavy orange-yellow mineral, occurring in veins of limestone. It may also be obtained as a white precipitate, or in crystals by fusing sodium molybdate with lead chloride and common salt.

Lead tungstate, $WO_3.PbO$, also occurs native as *scheeletine*, in quadratic crystals isomorphous with the molybdate. It has a greenish or brown colour. It can be prepared artificially like the molybdate, and is then white. The salt $7WO_3.3PbO.10H_2O$ is produced by precipitation. The metatungstate, $4WO_3.PbO.6H_2O$, crystallises in needles. $2UO_3.PbO$ is yellowish-red and insoluble. $MoS_3.PbS$ and $WS_3.PbS$ are dark coloured precipitates.

The oxides of the elements of the vanadium and phosphorus groups form exceedingly complex compounds with the trioxides of molybdenum and tungsten, and with the oxides of other elements.* To these names vanadimolybdates, vanaditungstates, &c., are applied, the number of molecules of trioxide being denoted by a numerical prefix. The chief compounds are as follows; they are produced by mixture, and are well-crystallised bodies :—

$5WO_3.V_2O_5.4(NH_4)_2O.13H_2O$.
$6MoO_3.V_2O_5.2(NH_4)_2O.5H_2O$.
$6MoO_3.P_2O_5.Aq$.
$6WO_3.As_2O_5.3K_2O.3H_2O$.
$6WO_3.As_2O_5.4K_2O.2H_2O$.
$10WO_3.V_2O_5.22H_2O$.
$14WO_3.2P_2O_5.6(NH_4)_2O.42H_2O$.
$16MoO_3.P_2O_5.3(NH_4)_2O.14H_2O$.
$16MoO_3.2V_2O_5.5BaO.28H_2O$.
$16WO_3.P_2O_5.CaO.5H_2O$.
$16WO_3.P_2O_5.4K_2O.2H_2O$.
$16WO_3.P_2O_5.6(NH_4)_2O.1? H_2O$.
$16WO_3.As_2O_5.6Ag_2O.11H_2O$.
$18MoO_3.V_2O_5.8(NH_4)_2O.15H_2O$.
$18WO_3.P_2O_5.K_2O.19H_2O$.
$18WO_3.V_2O_5.36H_2O$.
$18WO_3.P_2O_5.6K_2O.23H_2O$.
$18MoO_3.As_2O_5.30H_2O$.
$20MoO_3.P_2O_5.Na_2O.23H_2O$.
$20MoO_3.P_2O_5.2BaO.24H_2O$.
$20MoO_3.P_2O_5.6BaO.42H_2O$.
$20MoO_3.P_2O_5.7K_2O.28H_2O$.
$20MoO_3.P_2O_5.8K_2O.18H_2O$.
$20WO_3.P_2O_5.6BaO.48H_2O$.
$22MoO_3.P_2O_5.3(NH_4)_2O.?H_2O$.
$22WO_3.P_2O_5.2K_2O.6H_2O$.
$22WO_3.P_2O_5.3(NH_4)_2O.21H_2O$.
$22WO_3.P_2O_5.4BaO.32H_2O$.
$24MoO_3.P_2O_5.62H_2O$.
$24WO_3.P_2O_5.53H_2O$.
$24MoO_3.P_2O_5.2K_2O.4H_2O$.
$24WoO_3.P_2O_5.2K_2O.6H_2O$.
$24WO_3.P_2O_5.3K_2O.21H_2O$.

$24MoO_3.P_2O_3.5(NH_4)_2O.20H_2O$.
$24MoO_3.6P_2O.6(NH_4)_2O.7H_2O$.

It is to be noticed that the ratios of the molecules of trioxide

* Wolcott Gibbs, *Amer. Jour. Sci.* (3), **14**, 61; *Amer. Chem. Jour.*, **2**, 217, 281; **5**, , 361, 391; **7**, 209, 313, 392; *Chem. News*, **45**, 29, 50, 60; **48**, 135.

to pentoxide are 5 : 1, 6 : 1, 7 : 1, 8 : 1, 10 : 1, 16 : 1, 18 : 1, 20 : 1, 22 : 1, and 24 : 1; that the number of molecules of the alkaline oxide varies from 1 to 8; and that phosphorus trioxide and monoxide (the hypothetical anhydride of hypophosphorous acid) appear also to be capable of union with these trioxides.

Still more complex bodies have been prepared, containing two or more pentoxides of different elements; or a pentoxide of one and a trioxide of another element, for example:—

$$14MoO_3.8V_2O_5.P_2O_5.8(NH_4)_2O.50H_2O;\ 48MoO_3.V_2O_5.2P_2O_5.7(NH_4)_2O.30H_2O;$$
$$60WO_3.V_2O_5.3P_2O_5.10(NH_4)_2O.6H_2O;\ 16WO_3.3V_2O_5.P_2O_5.5(NH_4)_2O.37H_2O.$$

Apparently arsenic, antimonic, niobic, and tantalic oxides, and the trioxides of boron, phosphorus, vanadium, arsenic, and antimony, are capable of forming similar compounds. A quaternary compound has even been obtained of the formula

$$60WO_3.8P_2O_5.V_2O_5.VO_2.18BaO.150H_2O.$$

Some complex uranium compounds occur native, resembling to some extent those mentioned above. They are:—

Frögerite	$3UO_3.As_2O_5.12H_2O$;
Walpurgin	$3UO_3.5Bi_2O_3.2As_2O_5.10H_2O$;
Zeunerite	$2UO_3.As_2O_5.CuO.8H_2O$;
Uranospinite	$2UO_3.As_2O_5.BaO.8H_2O$, and
Uranosphærite	$UO_3.Bi_2O_3.H_2O$.

They are yellowish or green crystalline minerals.

Indications also appear to exist of complex molybdotungstates, but they have not been investigated.

Uranyl tungstate, $WO_3.UO_2.H_2O$, is a brown precipitate. Triple compounds have also been prepared of the trioxides of molybdenum or tungsten, one of the oxides of sulphur, and the oxide of an alkaline metal, but at present there are no precise data as to their formulæ.

Oxides of the platinum group of metals also form similar compounds. Among the few which have been prepared are:—

$$10MoO_3.PtO_2.4Na_2O.29H_2O \text{ and } 10WO_3.PtO_2.4K_2O.9H_2O.$$

They are analogous to the titani- and silici-decimolybdates and tungstates.

The compounds of copper are:—

$3MoO_3.4CuO.5H_2O$; the metatungstate, $4WO_3.CuO.11H_2O$; the sulphomolybdate, $MoS_3.CuS$; and the sulphotungstate, $WS_3.CuS$.

They are obtained by precipitation.

The silver salts are :—

$MoO_3.Ag_2O$; $2WO_3.Ag_2O$; the metatungstate, $4WO_3.Ag_2O.3H_2O$; the uranate, $2UO_3.Ag_2O$; and the sulphomolybdate and sulphotungstate, $MoS_3.Ag_2S$, and $WS_3.Ag_2S$.

They are all insoluble, except the metatungstate. The action of hydrogen at the ordinary temperature on silver molybdate or tungstate is said to produce sub-argentous salts, containing the oxide Ag_4O, but in the light of recent researches this action is improbable.

The following mercury compounds have been prepared :—

$MoO_3.Hg_2O$; $2MoO_3.Hg_2O$; $WO_3.Hg_2O$; $2WO_3.3HgO$; $3WO_3.2HgO$; the metatungstate, $4WO_3.Hg_2O.25H_2O$; and the sulpho-compounds, $MoS_3.Hg_2S$, $MoS_3.HgS$, $WS_3.Hg_2S$ and $WS_3.HgS$.

These are all produced by precipitation ; even the metatungstate is insoluble. Mercurous tungstate, $WO_3.Hg_2O$, is completely insoluble in water, and on ignition, leaves tungsten trioxide ; hence tungsten trioxide is usually separated from other metals and estimated by precipitation with mercurous nitrate.

The tungstates and molybdates generally resemble the sulphates in their formulæ; and these might with reason be written from analogy M_2MoO_4 and M_2WO_4 ; and a few uranates appear also to possess similar formulæ. Salts analogous to anhydro- or di-sulphates are also known, such as $M_2Mo_2O_7 = 2MoO_3.M_2O$ and $M_2W_2O_7 = 2WO_3.M_2O$; the uranates, as a rule, are thus constituted. But as nothing is known of the constitution of the more complex salts, which, as has been seen, are very numerous, the provisional method of writing the formulæ of the oxides separately has uniformly been adopted.

Peruranates.—It has been stated that the solution of a uranyl salt yields a white compound of the formula $UO_4.2H_2O$, on treatment with hydrogen dioxide. This compound when mixed with solution of potassium hydroxide gives a precipitate of the hydrated trioxide, $UO_3.2H_2O$, while the salt $UO_6.2K_2O.10H_2O$, goes into solution, and may be separated as a yellow or orange precipitate on addition of alcohol. It may also be produced by adding hydrogen dioxide to a solution of hydrated uranium trioxide in caustic potash. The sodium salt, similarly prepared, has the formula, $UO_6.2Na_2O.8H_2O$; and by using a smaller amount of alkaline hydroxide, the compound $UO_6.UO_3.Na_2O.6H_2O$ is formed, and separates on addition of alcohol. The analogous ammonium compound has also been prepared. These compounds would lead to the inference that an oxide of the formula UO_6 is capable of existence; but it has been suggested, apparently on insufficient evidence, that they are in reality compounds of uranium tetroxide with peroxides of the metals, thus :—$\mathbf{UO_4.2K_2O_2.10H_2O}$; $\mathbf{UO_4.2Na_2O_2.8H_2O}$; and $\mathbf{2UO_4.Na_2O_2.6H_2O}$. They readily part with oxygen, forming uranates. Similar permolybdates and pertungstates are said to be capable of existence at low temperatures.

Persulphomolybdates.—Potassium dimolybdate on treatment in solution with hydrogen sulphide yields a mixture of potassium sulphomolybdate, $\mathbf{MoS_3.K_2S}$, and molybdenum trisulphide, $\mathbf{MoS_3}$. Such a mixture, when boiled with water for some hours, gives off hydrogen sulphide, and forms a copious precipitate; it is collected and washed with water until the washings give a red precipitate of $\mathbf{MoS_4}$ with hydrochloric acid. Water extracts potassium persulphomolybdate, $MoS_4.K_2S$ from the residue, leaving the disulphide, $\mathbf{MoS_2}$. On treatment with hydrochloric acid, the tetrasulphide is precipitated, and from it the salts may be obtained by treatment with sulphides. The alkali and ammonium salts are soluble with a red colour; they yield precipitates, usually red or reddish-brown, with soluble salts of the metals. The magnesium salt is an insoluble red precipitate.

(*d.*) **Compounds with halides.**—No simple oxyfluorides of molybdenum are known. But by dissolving molybdates in hydrofluoric acid and evaporating the solutions, compounds isomorphous with stannifluorides, $\mathbf{SnF_4.2MF.H_2O}$, and titani- and zirconifluorides of corresponding formulæ are produced. **Molybdoxyfluorides** of the general formula $\mathbf{MoO_2F_2.2MF.H_2O}$ have been prepared with potassium, sodium, ammonium, and thallium; of the formula $\mathbf{MoO_2F_2.2MF.2H_2O}$ with rubidium and ammonium; and with $\mathbf{6H_2O}$ with zinc, cadmium, cobalt, and nickel.

Tungstoxyfluorides have been similarly prepared; also one of the formula $\mathbf{WO_3.3NH_4F}$. They are isomorphous with the former salts. The oxyfluoride itself is known with uranium, $\mathbf{UO_2F_2}$. It is a white substance produced by evaporating a solution of the trioxide in hydrofluoric acid; and has been obtained in crystals by subliming the tetrafluoride, $\mathbf{UF_4}$ in air. It also forms double salts on mixture; for example, $\mathbf{UO_2F_2.NaF.4H_2O}$; $\mathbf{UO_2F_2.3KF}$; $\mathbf{UO_2F_2.5KF}$; and $\mathbf{2UO_2F_2.3KF.2H_2O}$. They are crystalline yellow bodies. The salt $\mathbf{UO_2F_2.KF}$ is a yellow crystalline precipitate obtained by adding a solution of potassium fluoride to uranyl nitrate, $UO_2(NO_3)_2.Aq$.

Oxychlorides and oxybromides of all these elements are known, viz.: $\mathbf{MoO_2Cl_2}$, $\mathbf{WO_2Cl_2}$, $\mathbf{UO_2Cl_2}$; and $\mathbf{MoO_2Br_2}$, $\mathbf{WO_2Br_2}$, and $\mathbf{UO_2Br_2}$. They are all produced by the action of the halogen on the heated dioxides or by heating the trioxides in a current of hydrogen chloride or bromide. They may also be formed by passing the halogen over a hot mixture of the trioxide with charcoal; and one, $\mathbf{MoO_2Br_2}$, has been prepared by heating a mixture of the trioxide with boron trioxide and potassium bromide :—$\mathbf{MoO_3} + \mathbf{B_2O_3} + \mathbf{2KBr} = \mathbf{2KBO_2} + \mathbf{MoO_2Br_2}$. Molybdyl dichloride, $\mathbf{MoO_2Cl_2}$, forms square reddish-yellow plates; it volatilises without fusion. The bromide also volatilises in crystalline yellow scales. They are soluble in water, alcohol, and ether. Tungstyl dichloride, $\mathbf{WO_2Cl_2}$ forms lemon-yellow scales; and the bromide consists of scales like mosaic gold. They decompose when heated. Uranyl dichloride, $\mathbf{UO_2Cl_2}$, is a yellow crystalline fusible body, volatile with difficulty; the bromide forms yellow needles. An oxyiodide is said to have been made.

Molybdyl and uranyl dichlorides form compounds with water, $\mathbf{MoO_2Cl_2.H_2O}$ and $\mathbf{UO_2Cl_2.H_2O}$. The first of these is a white crystalline substance, very volatile in a current of hydrogen chloride; it is produced, along with the anhydrous body, by passing hydrogen chloride over molybdenum trioxide at 150—200°. Uranyl dichloride unites with chlorides of the alkalies, forming bodies, such as $\mathbf{UO_2Cl_2.2KCl.2H_2O}$, similar to the fluorides; the corresponding bromide also gives salts, *e.g.*, $\mathbf{UO_2Br_2.2KBr.7H_2O}$.

Molybdenum and tungsten also form other oxyhalides, $\mathbf{MoOCl_4}$, $\mathbf{WOCl_4}$, and $\mathbf{WOBr_4}$. These may be named molybdanosyl and tungstosyl tetrachlorides, respectively. The first is produced, along with molybdyl dichloride, by the action of chlorine on a heated mixture of the trioxide with charcoal. It forms green easily fusible crystals, which melt and sublime below 100°; it is soluble in alcohol and in ether. The corresponding tungsten compound is produced when tungstyl chloride is quickly heated above 140°. It forms red transparent needles; it melts at 210·4°, and boils at 227·5°. Its vapour density corresponds with the formula $WOCl_4$. The bromide is similarly prepared by heating tungstyl dibromide; it forms light-brown woolly needles.

Molybdenum forms some other oxyhalides. The action of chlorine on a mixture of molybdenum trioxide and carbon gives, besides the compounds already mentioned, two others: $\mathbf{Mo_2O_3Cl_6}$, which forms dark violet crystals, ruby-red by reflected light, and volatile without decomposition; and $\mathbf{Mo_4O_5Cl_{10}}$, forming large blackish-brown crystals, volatile in a current of hydrogen. The first points to a dimolybdic acid, $Mo_2O_3(OH)_6$, but the second is a derivative of a lower oxide.

Molybdous bromide, $\mathbf{MoBr_2}$, on treatment with alkali, yields a solution from which carbonic or acetic acid throws down the hydroxybromide, $Mo_3Br_4(OH)_2$, as a yellow sparingly soluble precipitate. This body acts as a base, yielding a crystalline sulphate, $Mo_3Br_4.SO_4$, chromate, $Mo_3Br_4.CrO_4$, molybdate, $Mo_3Br_4.MoO_4$, oxalate, $Mo_3Br_4.C_2O_4$, and phosphate, $Mo_3Br_4.2PO(OH)_2$.

The oxychlorides, such as MoO_2Cl_2 and $MoOCl_4$, point to hydroxides like $MoO_2(OH)_2$ and $MoO(OH)_4$; these are known with all three elements and are the respective acids.

No sulphohalides are known.

Physical Properties.

Mass of one cubic centimetre :—

MoO_2, 6·44 grams at 16°. MoO_3, 4·39 grams at 21°. MoS_2, 4·44—4·59 grams. WO_2, 12·11 grams. WO_3, 7·23 grams at 17°. WS_2, 6 26 grams at 20°. UO_2, 10·15 grams; U_3O_8, 7·19 grams; UO_3, 5·02 – 5·26 grams.

The heats of formation are unknown.

CHAPTER XXVI.

COMPOUNDS OF OXYGEN, SULPHUR, SELENIUM, AND TELLURIUM WITH EACH OTHER.—ACIDS AND SALTS OF SULPHUR, SELENIUM, AND TELLURIUM; SULPHATES, SELENATES, AND TELLURATES.

These compounds are most conveniently divided into the two classes:—(1) **the oxides and their compounds**; and (2) **the compounds of sulphur, selenium, and tellurium with each other.**

The following is a list of the oxides:—

Sulphur.	Selenium.	Tellurium.
S_2O_3*; *SO_2*; SO_3; *S_2O_7*†.	SeO_2.	TeO_2; TeO_3.

Besides these, the double oxides $SSeO_3$, $STeO_3$, and $SeTeO_3$ are known, analogous to the oxide S_2O_3.

Sources.—Sulphur dioxide is the only one of these compounds occurring native. It is present in the air in the neighbourhood of volcanoes, being produced by the combustion of sulphur, and also in the air of towns, where its presence is due to the combustion of coal, which almost always contains small quantities of iron pyrites. Air in the neighbourhood of furnaces where sulphides are roasted also contains this gas. It is very injurious to vegetation, and the prevention of its presence in the atmosphere in large quantities should engage the attention of manufacturers.

Sulphur trioxide exists in abundance in combination with other oxides in sea-water, or on the earth's surface, as sulphates, and selenium trioxide has been found native in combination with lead oxide. The more important of the natural sulphates are *Glauber's salt*, or sodium sulphate, $Na_2SO_4.10H_2O$, which is contained in sea-water and in many mineral waters, and when solid, in efflorescent crusts, is named *thenardite; glaserite*, K_2SO_4, in sea-water and spring-water; *schönite*, $K_2Mg(SO_4)_2.6H_2O$, *anhydrite*, $CaSO_4$, and *gypsum*, $CaSO_4.2H_2O$; *celestin*, $SrSO_4$; *heavy spar, or barytes*, $BaSO_4$; *Epsom salt*, $MgSO_4.7H_2O$; *feather alum*,

* *Pogg. Ann.*, **156**, 531.

† *Comptes rend.*, **86**, 20, 277; **90**, 269.

$Al_2(SO_4)_3.18H_2O$; *alum stone*, **$AlKSO_4.2Al(OH)_3$**; *copperas, or green vitriol*, **$FeSO_4.7H_2O$**; *cobalt vitriol*, **$CoSO_4.7H_2O$**; *anglesite*, **$PbSO_4$**; *lanarkite*, a double carbonate and sulphate of lead, and *leadhillite*, **$PbSO_4.PbO$**; and *blue vitriol*, **$CuSO_4.5H_2O$**. Lead selenate, **$PbSeO_4$**, has also been found native.

Preparation.—1. By direct union.—Sulphur, selenium, and tellurium burn with a faint blue flame when heated in air, forming the dioxides. Heated in oxygen, the flame of burning sulphur is much more brilliant, and of a fine lilac colour. Its combustion forms a telling experiment. About 3 or 4 per cent. of the product of the combustion consists of sulphur trioxide, SO_3. The sulphides of many metals, when roasted in air, give the oxide of the metal and sulphur dioxide. Iron pyrites containing from 2 to 4 per cent. of copper is made use of in its commercial preparation, the copper being extracted from the residue. The sulphur dioxide is employed directly in the preparation of sulphuric acid. It is also a by-product in the roasting of zinc sulphide, in the smelting of lead ores (see p. 429), and in various other metallurgical processes.

2. **By oxidation of a lower oxide.**—Sulphur trioxide is thus prepared on a commercial scale. In the laboratory it may be prepared by the following method:—

A dry mixture of gaseous sulphur dioxide and oxygen, the dioxide being made to bubble through the wash-bottle containing strong sulphuric acid twice as quickly as the oxygen, is led through a tube of hard glass, heated to redness, filled with asbestos, previously coated with metallic platinum by moistening it with platinum tetrachloride, and igniting it. Under the influence of the finely-divided platinum, the sulphur dioxide and the oxygen combine, and the sulphur trioxide produced is condensed in a flask. To obtain the pure trioxide, water must be rigorously excluded, and corks should not be exposed to its action, for they are at once attacked.

By passing an electric discharge of high potential through a mixture of perfectly dry sulphur dioxide and oxygen, combination takes place between 4 vols. of sulphur dioxide and 3 vols. of oxygen to form persulphuric anhydride or disulphur heptoxide, S_2O_7.

3. **By reducing a higher oxide.**—The trioxides of sulphur and of tellurium, at a red heat, decompose into the dioxides and oxygen. The vapour of sulphuric or selenic acid also, at a red heat, gives water, sulphur or selenium dioxide, and oxygen, and it is by this method that a mixture in the requisite proportion of

sulphur dioxide and oxygen is obtained on a large scale for the manufacture of sulphur trioxide. The sulphuric acid is decomposed by causing it to flow on to red-hot bricks; and the mixed gases are dried by passage upwards through a tower filled with coke, kept moist by strong sulphuric acid.* The mixture is then passed over asbestos coated with platinum, as on the small scale (see previous page).

The reduction may also be effected by chemical agency. On heating sulphur and sulphuric acid, the dioxide and water are the sole products, thus :—$2(SO_3.H_2O) + \mathbf{S} = 3SO_2 + 2H_2O$. Carbon may be used in the form of charcoal; in this case a mixture of carbon dioxide and sulphur dioxide is produced, from which it is not easy to separate the carbon dioxide:—$2(SO_3.H_2O) + \mathbf{C} = 2SO_2 + CO_2 + 2H_2O$. Almost all metals, when heated with strong sulphuric acid, yield sulphur dioxide, a sulphate and sulphide of the metal, hydrogen sulphide, free sulphur, and water. For example, with copper, the metal most frequently employed in the form of foil or turnings in the ordinary laboratory process for preparing sulphur dioxide :†—

(1.) $\mathbf{Cu} + 2H_2SO_4 = \mathbf{CuSO_4} + SO_2 + 2H_2O$
(2.) $4\mathbf{Cu} + 5H_2SO_4 = 4\mathbf{CuSO_4} + H_2S + 4H_2O$;
(3.) $3\mathbf{Cu} + 4H_2SO_4 = 3\mathbf{CuSO_4} + \mathbf{CuS} + 4H_2O$;
(4.) $4\mathbf{Cu} + 4H_2SO_4 = 3\mathbf{CuSO_4} + \mathbf{Cu_2S} + 4H_2O$; and
(5.) $SO_2 + 2H_2S = 2H_2O + 3\mathbf{S}$.

Reaction (1) is that which predominates; but the other reactions doubtless take place, for the products are found in the residue.

It is probable that these reactions are due to the action of hot nascent or atomic hydrogen on sulphuric acid. Thus, the equations may also be written :—

(1.) $\mathbf{Cu} + H_2SO_4 = \mathbf{CuSO_4} + 2H$; $H_2SO_4 + 2H = 2H_2O + SO_2$;
(2.) $H_2SO_4 + 8H = 4H_2O + H_2S$;
(3.) $\mathbf{CuSO_4} + H_2S = \mathbf{CuS} + H_2SO_4$; and
(4.) $2\mathbf{CuSO_4} + 10H = \mathbf{Cu_2S} + H_2SO_4 + 4H_2O$.

The metals osmium, iridium, platinum, and gold are the only ones which withstand the action of boiling sulphuric acid; but strong acid may be evaporated in iron pans, for the iron becomes protected by a coating of sulphate, which is insoluble in oil of vitriol.

Gold is, however, attacked by selenic acid; the acid is reduced by it and other metals to the dioxide. Selenic acid is also converted

* *Dingl. polyt. J.*, **218**, 128.
† *Chem. Soc.*, **33**, 112.

into selenious acid, with evolution of chlorine, by boiling it with hydrochloric acid, thus:—

$$H_2SeO_4 + 2HCl.Aq = H_2SeO_3.Aq + H_2O + Cl_2.$$

The oxides S_2O_3, $SSeO_3$, $STeO_3$, and $SeTeO_3$ are also formed by reduction. They are produced by dissolving sulphur in fused sulphur trioxide; selenium in sulphur trioxide; and tellurium in strong selenic acid.

4. **By heating compounds.**—Both sulphites and sulphates, and probably also selenites, selenates, tellurites, and tellurates, when heated to a high temperature decompose, leaving the oxide of the metal with which the oxide of sulphur, selenium, or tellurium was combined. But the dioxides are usually produced, for the temperature at which decomposition occurs is almost always so high as to partially, at least, decompose the trioxides. Anhydrosulphates, such as $\mathbf{Na_2S_2O_7}$, however, give off half their trioxide when heated, leaving the monosulphate, $\mathbf{Na_2SO_4}$. The compounds with water, however, in every case, except that of selenic acid, are decomposed by heat, yielding the respective oxide.

Thus a solution of sulphur dioxide in water, presumably containing sulphurous acid, H_2SO_3, loses the oxide when boiled; selenious and tellurous oxides remain on evaporating their aqueous solutions; the latter, indeed, separates out on warming its solution to 40°; sulphuric acid, H_2SO_4, when gasified has the density 44·5, proving it to have split into its constituent oxides, which, however, recombine on cooling; the trioxide is prepared, moreover, by distilling anhydrosulphuric acid, $H_2S_2O_7$, which decomposes thus:— $H_2S_2O_7 = H_2SO_4 + SO_3$; and tellurium trioxide is produced by heating the hydrate to a temperature below redness.

5. **By double decomposition.**—As this process is usually carried out in the presence of water, the hydrates (acids) are the usual products.

6. **By displacement.**—This is a convenient method of preparing sulphur trioxide. Strong sulphuric acid is mixed with phosphoric anhydride, care being taken to keep the acid cold during mixing. It is then distilled, when the trioxide passes over, the phosphoric anhydride having abstracted water from the sulphuric acid, thus:—

$$H_2SO_4 + P_2O_5 = SO_3 + 2HPO_3.$$

A sulphate also, when strongly ignited with silicon dioxide, or with phosphorus pentoxide, yields sulphur trioxide, or its products of decomposition, the dioxide and oxygen. This process finds practical application in the manufacture of glass, where silica in the form of sand is heated with sodium sulphate, lime, and carbon. The addition of carbon causes the conversion of the sulphate into sulphite; the silica replaces the sulphur dioxide at a lower temperature than it would replace the trioxide of the sulphate. A double silicate of sodium and calcium is thus formed, which constitutes one variety of glass. The method, it will be seen, is not available for the preparation of the oxides of sulphur.

Properties.—**Sulphur dioxide** is a gas at the ordinary temperature, but it may be easily condensed to a liquid by passing it first through a tube filled with calcium chloride, to dry it, and then through a leaden worm cooled by a mixture of salt and crushed ice.

It boils at −8° under normal pressure, and melts at about −79°. The liquid oxide is mobile and colourless, and heavier than water (1·45). It forms a white crystalline solid when sufficiently cooled by its own evaporation. The gas has the familiar smell of burning sulphur; it is irrespirable; it supports the combustion of potassium, tin, and iron, which combine both with its oxygen and its sulphur. It is readily soluble in water; one volume of water absorbs about fifty times its volume of the gas at the ordinary temperature, probably with formation of sulphurous acid, H_2SO_3. Hence it cannot be collected over water; but, as its density is high (32), it is easy to collect it in a jar by downward displacement.

Selenium dioxide is a white solid, volatilising to a yellow vapour without melting, at a heat somewhat below redness, and condensing in white quadrangular needles. Its vapour has a sharp acid odour. It is soluble in water, producing selenious acid.

Tellurium dioxide is a white solid, sometimes crystallising in octahedra. It melts to a deep-yellow liquid, and at a high temperature it volatilises. It is sparingly soluble in water, and does not appear to form the acid.

Sulphur trioxide crystallises in long colourless prisms, arranged in feathery groups; it somewhat resembles asbestos. It melts at 15°, and boils at 46°, producing dense white fumes with the moisture of the air. Its molecular weight, as shown by its vapour density, is 80. It unites with water with great violence, hissing like a red-hot iron. It is made in considerable quantity, being used in the manufacture of alizarine or turkey-red, and other artificial dyes.

When this body is kept for some time at a temperature below 25°, it changes into another modification which crystallises in thin needles. When heated above 50° it gradually liquefies, and changes into the first modification. It is distinguished from the first modification by the difficulty with which it dissolves in H_2SO_4, and by its crystallising out of the solution unchanged.

There appear to be indications of the existence of an oxide S_2O_5; for sulphur trioxide dissolves the dioxide in large amount, and the solution is stable up to 5°.

No attempts to prepare selenium trioxide have succeeded. The acid, when heated, decomposes into selenium dioxide, oxygen, and water. Selenious anhydride is the only product of the action of oxygen, even in the state of ozone, on selenium.

Tellurium trioxide is an orange-yellow insoluble substance, which does not dissolve even in hydrochloric or nitric acid. When strongly heated, it loses oxygen, producing tellurium dioxide.

Sulphur, selenium, and tellurium dissolved in pure melted sulphur trioxide give respectively blue, green, and red substances. The sulphur and tellurium compounds have been isolated, and have been shown to have the formulæ $\mathbf{S_2O_3}$ and $\mathbf{STeO_3}$; it is presumed that the others are similarly constituted. Selenium and tellurium also dissolve in concentrated selenic acid, doubtless forming similar compounds. The sulphur compound is insoluble in perfectly pure sulphuric anhydride, and may be separated from it by decantation. It decomposes, on exposure to air at ordinary temperatures, into sulphur dioxide and sulphur. It dissolves in strong sulphuric acid, and, on diluting the acid, it is decomposed. The tellurium compound appears to exist in two modifications, a red one, and a buff-coloured, obtained by heating the red variety to 90°.

Persulphuric anhydride, S_2O_7, at the ordinary temperature forms an oily liquid; when cooled to 0°, it solidifies in long thin transparent flexible needles. It sublimes easily, and decomposes spontaneously on standing for a few days. It dissolves in strong sulphuric acid; it is immediately decomposed by heat.

A. **Compounds with water and oxides; acids and salts of sulphur, selenium, and tellurium. 1.—Compounds of the trioxides; sulphuric, selenic, and telluric acids; sulphates, selenates, and tellurates.**

The trioxide of sulphur dissolves in water with evolution of great heat, forming various hydrates, according to the relative proportion of oxide and water. The following have been isolated:—

$$\mathbf{SO_3.5H_2O};\ SO_3.3H_2O;\ \mathbf{SO_3.2H_2O};\ SO_3.H_2O = H_2SO_4;$$
$$\mathbf{2SO_3.H_2O} = \mathbf{H_2S_2O_7}.$$

Those containing less water than ordinary sulphuric acid are more conveniently produced by dissolving sulphur trioxide in the ordinary acid; those containing more, by pouring ordinary sulphuric acid into water. Salts have been produced corresponding to the acids H_2SO_4 and $H_2S_2O_7$; they are named sulphates, and pyrosulphates or anhydrosulphates respectively.

On boiling a solution of sulphuric acid in water, the water evaporates, and the acid becomes more and more concentrated, until it acquires nearly the composition expressed by the formula H_2SO_4; on further heating, this compound dissociates into trioxide, or anhydride, SO_3, and water, both of which evaporate together.

Some of these hydrates may be dismissed in a few words. The hydrate, $SO_3.5H_2O = H_2SO_4.4H_2O$, crystallises out on cooling sulphuric acid containing the correct amount of water to a very low temperature. It melts at $-25°$. $H_2SO_4.2H_2O$ is the point of maximum contraction of sulphuric acid and water, but has not been obtained in a solid state; and $\mathbf{H_2SO_4.H_2O}$ is also obtained by cooling a mixture in the correct proportion. It melts at 8°. The corresponding selenic acid, $\mathbf{H_2SeO_4.H_2O}$, melts at 25°. The monohydrate requires particular attention.

Sulphuric acid, "oil of vitriol," H_2SO_4.—Sulphur trioxide, as has been mentioned, is decomposed by heat, and hence it cannot be produced in quantity by the combustion of sulphur in air or oxygen, for the temperature of burning sulphur is higher than that at which the trioxide decomposes. Hence an indirect method of preparation must be chosen. It can be prepared in aqueous solution by oxidising sulphur; for example, when boiled with nitric acid, that acid parts with its oxygen, oxidising the sulphur to sulphuric acid, while oxides of nitrogen are liberated. Sulphur may also be oxidised on treatment with chlorine and water, thus:—

$$S + 3Cl_2 + 4H_2O = H_2SO_4 + 6HCl.$$

It will be remembered that the halogen acids may be prepared by the action of the halogens in presence of water on hydrogen sulphide (see p. 106); and, similarly, an aqueous solution of sulphur dioxide is oxidised to sulphuric acid by their action. Chromic acid and other oxidising agents also effect such oxidation.

But such processes are too expensive to be used in manufacture. The main outlines of the process actually in use are given here; the details and the connection of this with other manufactures will be described later (see p. 667).

Sulphur dioxide at once attacks nitrogen peroxide, NO_2. Without discussing intermediate products, which will be afterwards considered, the final reaction, in presence of water at least, is $SO_2 + NO_2 + H_2O = SO_3.H_2O + NO$. In presence of air, as has been seen on p. 333, nitric oxide is oxidised to a mixture of peroxide and tetroxide, NO_2 and N_2O_4. These gases again part with their oxygen when brought in contact with a fresh supply of sulphur dioxide. In theory, then, a small amount of nitrogen dioxide is capable of converting an indefinite amount of sulphur dioxide, in presence of oxygen and water, into sulphuric acid. The nitrogen dioxide required for this process is derived from nitric acid, prepared in the usual manner, *i.e.*, from sodium nitrate and sulphuric acid. On bringing it into contact with sulphur dioxide, it is reduced, and gives an effective mixture of oxides of nitrogen. This process may be illustrated by the following experiment:—D is a flask containing copper turnings and strong sulphuric

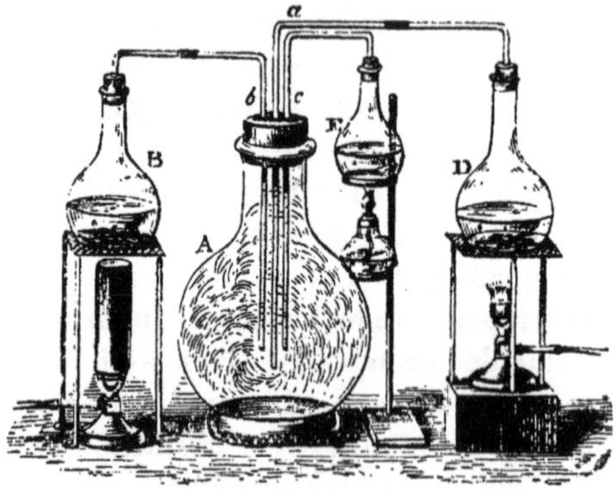

FIG. 42.

acid, from which, on applying heat, sulphur dioxide is generated. B is a similar flask containing copper turnings and dilute nitric acid, and yields a supply of nitric oxide when warmed. E is a flask containing water. The delivery tubes of these flasks all enter the large balloon, A, through a large perforated cork; a glass tube passes to the bottom of the globe through a fourth hole in the cork, and serves as an exit tube for any excess of gas. Nitric oxide is first passed into the globe. It unites with the oxygen of the air, forming a mixture of the dioxide and peroxide, which are at once noticeable as red fumes. Sulphur dioxide is passed in next, and reacts with the peroxide; it will be noticed that the sides of the globe soon become covered with radiating crystals. These are described later; they consist of hydrogen nitrosyl sulphate, $\mathbf{SO_2(OH)(ONO)}$, and are known as "chamber crystals." Steam is then passed into the globe by boiling the water in the flask, E. The crystals disappear and the liquid which collects in the globe is dilute sulphuric acid. It may be concentrated by evaporation in a porcelain or platinum basin, till its strength is little below that indicated by the formula H_2SO_4.

Selenic acid* may be prepared, like sulphuric acid, by the action of chlorine water on selenium, or, better, on selenious acid; but on concentration, the selenic acid is reduced by the hydrochloric acid with evolution of chlorine. A better plan is to saturate a solution of selenious acid with chlorine gas, thereby converting that acid into selenic acid; to saturate the mixed selenic and hydrochloric acids with copper carbonate, forming a mixture of copper selenate and chloride; to evaporate to dryness, and extract with alcohol, which dissolves the copper chloride, leaving the selenate; and, finally, to dissolve the selenate in water, and liberate the selenic acid by precipitating the copper as sulphide by a current of hydrogen sulphide. After filtering off the copper sulphide, the selenic acid is concentrated by evaporation. It can be obtained nearly anhydrous by evaporation in a vacuum at 180°. The acid has then the formula H_2SeO_4. A higher temperature decomposes it into selenium dioxide, water, and oxygen. One other hydrate of selenium trioxide has been prepared by cooling a solution of the acid of the requisite strength to $-32°$. It has the formula $H_2SeO_4.H_2O$, and melts at 25°. Attempts to prepare other hydrates in the solid state have not been successful.

Telluric acid is produced in solution by treating the barium salt (obtained by heating tellurium with barium nitrate) suspended in water, with the requisite amount of sulphuric acid, and, after filtration, concentrating the acid by evaporation. Colourless hexagonal prisms of the formula $\mathbf{H_2TeO_4.2H_2O}$ separate out on cooling. It loses its water a little above 100°, leaving the acid $\mathbf{H_2TeO_4}$ as a white solid.

These acids are also produced by the action of water on the chlorides, SO_2Cl_2, SeO_2Cl_2, and TeO_2Cl_2.

Sulphuric and selenic acids are dense, viscid, colourless

* *Proc. Roy. Soc.*, **46**, 13.

liquids, exceedingly corrosive, inasmuch as they abstract the elements of water from many organic substances containing carbon, hydrogen, and oxygen. A piece of wood placed in strong sulphuric acid is blackened and charred, and sugar placed in contact with it is converted into a tumefied mass of impure carbon. Pure anhydrous sulphuric acid, $\mathbf{H_2SO_4}$, is, however, a solid, melting at 10·5°, and selenic acid, $\mathbf{H_2SeO_4}$, melts at 58°. The presence of a mere trace of water greatly lowers the melting points of these bodies.

The hydrate of telluric acid, $\mathbf{H_2TeO_4.2H_2O}$, dissolves slowly, but to a considerable extent in water. The anhydrous acid, $\mathbf{H_2TeO_4}$, can be dissolved only by prolonged boiling with water.

These acids cannot be said to boil, in the purely physical sense of the word. At the ordinary temperature, sulphuric acid, if perfectly pure, gives off sulphur trioxide, hence the only method of obtaining an acid precisely corresponding to the formula H_2SO_4, is to add sulphuric anhydride to ordinary oil of vitriol. When concentrated by evaporation as far as possible, the acid contains about 98 per cent. of H_2SO_4. On further heating to 327°, this acid dissociates with apparent ebullition into water and trioxide, which recombine on cooling, forming an acid of the same composition. By taking advantage of the different rates of diffusion of water-gas and sulphuric anhydride, which possess respectively the densities 9 and 40, and whose ratio of diffusion is therefore as $\sqrt{40} : \sqrt{9}$, a much stronger acid has been obtained. Acid con-

FIG. 43.

taining 5 per cent. of water was boiled in a flask, while a gentle current of air passed downwards through a tube, sealed on to the bottom of the other flask; after an hour, the composition of the remaining acid was approximately 60 per cent. of H_2SO_4, and 40 per cent. of SO_3. This process of concentration is not applied on a large scale.

Pure selenic acid begins to decompose into dioxide, oxygen, and water at about 200°. On distilling dilute selenic acid, water passes over up to 205°; a little dilute acid then begins to distil over, and, at 260°, white fumes appear, containing a little trioxide, but for the most part consisting of selenium dioxide.

Telluric acid is non-volatile, and parts with its water below a red heat, leaving the anhydride, $\mathbf{TeO_3}$.

A great rise of temperature is produced by the action of water on sulphuric and selenic acids, due to their combination with it to form hydrates.

The specific gravity of ordinary sulphuric acid is approximately 1·84 at 15°; that of selenic acid, 2·61 at 15°; and of telluric acid, 3·42 at 18·8°.

Sulphates, selenates, and tellurates.—These salts are obtained by the action of the acids on aqueous solutions of the hydroxides or carbonates of the metals; by the action of the concentrated acids at a high temperature on most metals, with evolution of the dioxides; by the action of aqueous solutions of the acids on many of the metals themselves, on the oxides, or hydroxides, and on some of the sulphides; and by heating a mixture of the acid and a halide, nitrate, or acetate of a metal, or, in short, with any salt containing a volatile or decomposable oxide. Thus, for example:—

$$H_2SO_4.Aq + 2KOH.Aq = K_2SO_4.Aq + 2H_2O;$$
$$\mathbf{TeO_3} + Na_2CO_3.Aq = Na_2TeO_4.Aq + \mathit{CO_2};$$
$$H_2SO_4.Aq + \mathbf{Zn} = ZnSO_4.Aq + \mathit{H_2};$$
$$H_2SeO_4.Aq + \mathbf{CuO} = CuSeO_4.Aq + H_2O;$$
$$H_2SO_4.Aq + \mathbf{FeS} = FeSO_4.Aq + \mathit{H_2S};$$
$$H_2SO_4 + 2\mathbf{NaCl} = Na_2SO_4 + 2\mathit{HCl};$$
$$H_2SO_4.Aq + \mathbf{CaSO_3} = \mathbf{CaSO_4} + \mathit{SO_2} + Aq;$$
$$H_2SO_4 + \mathbf{CaSiO_3} = \mathbf{CaSO_4} + \mathbf{H_2SiO_3}.$$

The salts of calcium, strontium, barium, and lead are insoluble, or nearly insoluble, and may therefore be produced by addition of a soluble sulphate, selenate, or tellurate, to the solution of a soluble salt of one of these metals, thus:—

$$CaCl_2.Aq + Na_2SO_4.Aq = \mathbf{CaSO_4} + 2NaCl.Aq;$$
$$Pb(NO_3)_2.Aq + K_2SeO_4.Aq = \mathbf{PbSO_4} + 2KNO_3.Aq.$$

The other sulphates and selenates are soluble in water. Many of the tellurates are insoluble, and may be produced by precipitation. The sulphates are also formed by the oxidation of sulphides by boiling with nitric acid; by the action of chlorine water; or by the action of air.

$Li_2SO_4.H_2O$; $Na_2SO_4.7$, and $10H_2O$; K_2SO_4; Rb_2SO_4; Cs_2SO_4; $(NH_4)_2SO_4$.—$Na_2SeO_4.10H_2O$; K_2SeO_4; $(NH_4)_2SeO_4$.—$K_2TeO_4.5H_2O$; $(NH_4)_2TeO_4$.

Lithium sulphate crystallises in flat tables, easily soluble in water and alcohol. Sodium sulphate occurs anhydrous as *thénardite;* and when crystallised with $10H_2O$ it is known as *Glauber's* salt. It is prepared in immense quantity from common salt and sulphuric acid, as a preliminary to the manufacture of sodium carbonate, and is then termed "salt-cake." It is also produced by passing a mixture of steam, air, and sulphur dioxide through sodium chloride, heated to dull redness (Hargreave's process). It is obtained as a residue in the preparation of nitric and of acetic acid; of ammonium chloride; and of common salt by the evaporation of sea-water. It crystallises in an anhydrous state from water at 40° in rhombic octahedra. It is insoluble in alcohol, but very soluble in water; 100 parts of water dissolve 12 parts at 0°, and 48 parts at 18°. It crystallises with $10H_2O$ in large, colourless, monoclinic prisms. Crystals with $7H_2O$ are deposited below 18°. On raising the temperature of a saturated solution above 33°, the anhydrous salt deposits, hence it appears to possess a lower solubility at high than at low temperatures. This apparent abnormality is doubtless explained by the dissociation of the solution of the decahydrate, $Na_2SO_4.10H_2O$, as the temperature rises. Sodium selenate is isomorphous with and closely resembles the sulphate. The tellurate has not been carefully examined.

Potassium sulphate crystallises from the aqueous extract of kelp (burned seaweed), in trimetric prisms or pyramids. It is among the least soluble of the potassium salts, 100 parts of water dissolving 8·36 parts at 0°. It is insoluble in alcohol. Both sodium and potassium sulphates have a saline bitter taste, and a purgative action. Potassium selenate is produced by fusing selenium or potassium selenite with nitre, and crystallisation from water; it resembles the sulphate. The tellurate forms rhombic crystals; they deliquesce in air, becoming converted by carbon dioxide and water into carbonate and ditellurate. Rubidium and cæsium sulphates resemble that of potassium, but are much more soluble in water. Ammonium sulphate, selenate, and tellurate are isomorphous with potassium sulphate, but are more soluble. The sulphate, when heated, decomposes above 280°, yielding ammonia water, and nitrogen, and a sublimate of hydrogen ammonium sulphate; the selenate gives, first, hydrogen ammonium selenate, and then selenium, its dioxide, water, and nitrogen. These salts, with the exception of lithium sulphate, are all insoluble in alcohol.

Double salts:—$HLiSO_4$; $HNaSO_4$; $HKSO_4$; $H(NH_4)SO_4$.—$HKSeO_4$; $H(NH_4)SeO_4$.—$HNaTeO_4$; $2HKTeO_4.3H_2O$.—$HNa_3(SO_4)_2$; $H_3Na(SO_4)_2$; $HK_3(SO_4)_2$; $H_3K(SO_4)_2$; $LiKSO_4$; $NaK_3(SO_4)_2$; $H_2K_4(SO_4)_3$; $Li_4(NH_4)_2(SO_4)_3$; $Li_2K_4(SO_4)_3$; $NaK_5(SO_4)_3$; $HK_3(TeO_4)_2$.

These substances are white crystalline bodies, very soluble in water, and also, as a rule, in alcohol. They are produced by mixture and crystallisation. Bisulphate of potassium, as hydrogen potassium sulphate is generally named, is used in decomposing various minerals, which are for that purpose reduced to fine powder, mixed with the salt, and fused. When carefully heated it loses water and yields the anhydrosulphate, or true disulphate, $K_2S_2O_7$. Sodium tripotassium sulphate is technically named *plate-salt*, from its crystallising in hexagonal plates; it deposits on cooling an aqueous extract of kelp.

The existence of the more complex double sulphates leads to the conclusion that the molecular formulæ of the ordinary sulphates are not so simple as they are usually written. Such formulæ as $H_2K_4(SO_4)_3$ and $NaK_5(SO_4)_3$, lead to the conclusion that the formula of potassium sulphate is probably at least $K_6(SO_4)_3$. Double salts with other acids are also known; *e.g.*, $K_2SO_4.HNO_3$ and $K_2SO_4.H_3PO_4$ separate from solutions of potassium sulphate in nitric or phosphoric acid. They are, however, decomposed by water. The existence of such salts would also favour the supposition of greater complexity of molecule.

$BeSO_4$.2, 4, and $6H_2O$; $CaSO_4.2H_2O$; $2CaSO_4.H_2O$; $SrSO_4$. $BaSO_4$.—
$BeSeO_4.4H_2O$; $CaSeO_4.2H_2O$; $SrSeO_4$; $BaSeO_4$.—
$CaTeO_4$; $SrTeO_4$; $BaTeO_4.3H_2O$.

With the exception of beryllium sulphate, which is soluble, all these compounds may be prepared by precipitation. Beryllium sulphate forms quadratic octahedra; it is insoluble in alcohol but very soluble in water. On evaporation with beryllium carbonate, it yields gummy basic salts of the formulæ

$$SO_3.2BeO.3H_2O;\ SO_3.3BeO.4H_2O;\ SO_3.6BeO.3H_2O.$$

Calcium sulphate occurs abundantly in the native form in salt mines. When anhydrous, it forms trimetric prisms, and is named *anhydrite;* and with two molecules of water it is *gypsum;* individual varieties of gypsum are named *selenite*, *alabaster*, and *satin-spar; selenite* forms transparent colourless monoclinic crystals; the massive variety is *alabaster;* and *satin-spar* is fibrous. When gypsum is heated and ground it forms "plaster of Paris," a material much employed in taking casts, and as a cement. The dihydrated calcium sulphate becomes anhydrous and falls to a powder when heated; on mixing the powder with water, a pasty mass is produced with which casts may be taken. After a few minutes it hardens, expanding slightly at the same time, and forms a fine white material. Plaster of Paris, mixed with a saturated solution of potassium sulphate, gives a paste which solidifies more rapidly than ordinary plaster of Paris, and has a nacreous lustre; for certain purposes this mixture is to be preferred to the ordinary one. A double salt $\mathbf{K_2Ca(SO_4)_2.H_2O}$, is produced. Hydrated calcium sulphate is very sparingly soluble in water, and is more soluble in cold than in hot water (1 in 420 at 20°). This is probably due to the solution containing the dihydrated compound, which loses water, becoming insoluble as the temperature rises. It is much more soluble in weak hydrochloric or nitric acid; or in presence of common salt, or of sodium thiosulphate. Its solubility in the last affords a method of separating calcium from barium. Calcium sulphate melts at a red heat. The selenate closely resembles the sulphate in preparation and properties, and is isomorphous with it. It is reduced to the selenite, however, when boiled with hydrochloric acid, chlorine being evolved.

Calcium tellurate is a white precipitate, soluble in hot water.

Strontium sulphate, $SrSO_4$, occurs native as *cœlestin* in trimetric crystals. It is soluble in about 7,000 parts of cold water; it fuses at a bright red heat. The selenate resembles it.

Barium sulphate occurs as *heavy-spar* or *barytes*, in large quantity; it forms trimetric crystals. A solution of barium chloride or nitrate is the common reagent for sulphuric acid. On adding it to a sulphate, a dense white precipitate is produced, practically insoluble in water and acids. Its insolubility serves to distinguish it from most other bodies of similar appearance. In estimating sulphuric acid, it is always weighed in the form of barium sulphate. It is unaltered by ignition; when heated with charcoal or coke, however, it yields barium sulphide; and this is the usual process of preparing compounds of barium, since the sulphide dissolves in acids. Barium sulphate reacts to a limited extent when boiled with a solution of sodium carbonate; a portion is converted into carbonate, thus:—

$$BaSO_4 + Na_2CO_3.Aq = BaCO_3 + Na_2SO_4.Aq.$$

But the reaction is incomplete. It is only after removal of the sodium sulphate and replacement by fresh sodium carbonate that further decomposition takes place. On fusion with excess of sodium or potassium carbonate, however, it is completely converted into carbonate. Barium sulphate has been used as a pigment under the name *permanent white;* it has too little body, and hence it is generally mixed with white-lead or zinc-white (ZnS). Barium selenate closely resembles the sulphate, but it is decomposed on boiling with hydrochloric acid, selenious acid and chlorine being formed. This serves to distinguish it, and to separate it from the sulphate. The tellurate is fairly soluble in warm water.

Double sulphates, selenates, and tellurates.—$K_2Be(SO_4)_2$; $H_2Ca(SO_4)_2$; $H_2Sr(SO_4)_2$; $H_2Ba(SO_4)_2$, also with $2H_2O$; $H_6Ca(SO_4)_4$; $Na_2Ca(SO_4)_2$; $Na_2Ba(SO_4)_2$; $Na_4Ca(SO_4)_3.2H_2O$; $K_2Ca_2(SO_4)_3.3H_2O$; $H_2Ba(TeO_4)_2.2H_2O$.

Hydrogen calcium and barium sulphates are crystalline bodies produced by dissolving the ordinary sulphates in strong sulphuric acid and crystallising. They are decomposed by water. The tellurate is soluble in water. The double salts are prepared by digesting the simple salts with sodium or potassium sulphate; that of calcium crystallises out with $2H_2O$, but at a higher temperature loses water, and is then identical with the mineral *glauberite*, crystallising in rhombic prisms.

$MgSO_4.7$, 6, and $1H_2O$; $ZnSO_4.7$, 6, 5, 2, and $1H_2O$; $CdSO_4.4H_2O$, also $1H_2O$, and $3CdSO_4.8H_2O$.—$MgSeO_4.7H_2O$; $ZnSeO_4.7$, 6, and $2H_2O$; $CdSeO_4.2H_2O$.—$MgTeO_4$; $CdTeO_4$.

These salts are all easily soluble in water, except magnesium and cadmium tellurates, which are produced by precipitation from concentrated solutions. Magnesium sulphate, as *Epsom salt*, $MgSO_4.7H_2O$, and as *kieserite*, $MgSO_4.H_2O$, occurs native in caves in magnesium limestone and in the salt-beds of Stassfurth. It is a frequent constituent of mineral waters. It has a bitter taste and a purgative action. It is made on the large scale by treating *dolomite*, a carbonate of magnesia and calcium, or *serpentine*, a hydrated silicate, with sulphuric acid. The hepta-hydrated sulphates and selenates of magnesium and zinc are isomorphous, and crystallise in four-sided right rhombic prisms. When heated to 150° they lose $6H_2O$, but retain the seventh molecule even at 200°. Anhydrous magnesium sulphate melts at a red heat; the cadmium and zinc salts lose trioxide, leaving the oxides. These salts are insoluble in alcohol. Zinc sulphate, digested with hydroxide, yields several basic sulphates: $SO_3.2ZnO$; $SO_3.4ZnO.10$ and $2H_2O$; $SO_3.6ZnO.10H_2O$; and $SO_3.8ZnO.2H_2O$. With exception of the first, they are crystalline bodies.

Mixed salts.—

$H_2Mg(SO_4)_2$; $H_6Mg(SO_4)_4$; $Na_2Mg(SO_4)_2.6H_2O$;
$K_2Mg(SO_4)_2.6H_2O$; $K_2Ca_2Mg(SO_4)_4.2H_2O$;
$Na_2Zn(SO_4)_2.4H_2O$; $(NH_4)_2Zn(SO_4)_2.6H_2O$;
$MgZn(SO_4)_2 14H_2O$; $Na_2Cd(SO_4)_2.6H_2O$,

and other similar salts. These are all soluble, and are prepared by mixture.

$B_2(SO_4)_3.H_2O$: $Sc_2(SO_4)_3.6H_2O$; $Y_2(SO_4)_3.8H_2O$; $La_2(SO_4)_3.9H_2O$.— Selenates and tellurates have not been prepared. Boron sulphate is a white mass, produced by evaporating boron trioxide with sulphuric acid.* It is decomposed by water. The other salts of the group are white and crystalline.

Double salts.—

$H_3B(SO_4)_3$; $(NH_4)Sc(SO_4)_2$; $K_4Sc_2(SO_4)_5$; $Na_3Sc(SO_4)_3.6H_2O$;
$K_3Y(SO_4)_3.nH_2O$; $Na_3Y(SO_4)_3.2H_2O$; $(NH_4)La(SO_4)_2.4H_2O$; $K_3La(SO_4)_3$.

These salts are sparingly soluble, and are produced by mixture.

$Al_2(SO_4)_3.18H_2O$; $Ga_2(SO_4)_3$; $In_2(SO_4)_3.9H_2O$; $Tl_2(SO_4)_3.7H_2O$.— $Al_2(SeO_4)_3.nH_2O$; and the thallous salts Tl_2SO_4; $HTlSO_4$; and Tl_2SeO_4.

Sulphate of aluminium, containing $18H_2O$, occurs native as *alunogen*, or *feather alum*; it forms delicate fibrous masses or crusts. It is known in commerce as "concentrated alum," and is prepared by heating finely ground clay with strong sulphuric acid until the latter begins to volatilise. After lying some days, it is treated with water; the solution is freed from iron by precipitating it as ferrocyanide, or by addition of certain peroxides, such as those of lead or manganese; it is then evaporated to dryness and fused. It crystallises with difficulty, being exceedingly soluble (1 in 2 parts of water); its crystallisation may be furthered by addition of alcohol, in which it is insoluble. Basic salts are known, produced by the action of hydrated alumina on the ordinary sulphate, by incomplete precipitation with ammonia, or by the action of zinc on a solution of ordinary sulphate. These are said to have the formulæ $3SO_3 2Al_2O_3$; $3SO_3.3Al_2O_3.9H_2O$ (occurring native as *aluminite*); $3SO_3.4Al_2O_3.36H_2O$; and $3SO_3.5Al_2O_3.20H_2O$. The selenate closely resembles the sulphate,

* *J. Prakt. Chem.* (2), **38**, 118.

and yields corresponding basic salts. The tellurate is a white precipitate. Gallium sulphate, $Ga_2(SO_4)_3$, is very soluble, and crystallises in nacreous scales; indium sulphate has been obtained only as a gummy mass; and thallic sulphate forms thin colourless laminæ, which are decomposed by water into the hydrated trioxide and sulphuric acid.

Thallous sulphate and selenate crystallise in anhydrous rhombic prisms isomorphous with potassium sulphate. They are soluble in water. They establish a link between the aluminium and the potassium groups.

Double salts.—The alums.—These bodies are very numerous. They all crystallise in regular octahedra, are soluble in water, and have the general formula $M'M'''RO_4.12H_2O$, where M' stands for lithium, sodium, potassium, rubidium, cæsium, ammonium, thallium (as a thallous compound), or silver; M''' for aluminium, gallium, indium, chromium, ferric iron, manganic manganese, or cobaltic cobalt;* and **R** for sulphur or selenium. Tellurium alums do not seem to have been prepared. The number of possible different alums is therefore 96; of these some 25 have been prepared. Alums containing aluminium, gallium, and indium are colourless; chromium alums are very deep purple; iron alums, pink; and manganese alums, brownish-red. As they are all isomorphous, they crystallise together. For example, an alum containing aluminium and potassium placed as a nucleus in a solution of chromium alum becomes covered with a regular deposit of the latter, and a coating of iron alum may be deposited on the exterior.

Alums are prepared by mixing solutions of the sulphates or selenates of the two metals, and crystallising. The most important are **potassium aluminium sulphate,** and **ammonium aluminium sulphate,** $KAl(SO_4)_2.12H_2O$, and $NH_4Al(SO_4)_2\,12H_2O$. Ammonium alum, which also occurs native as *tchermigite*, is prepared by mixture; 100 parts of water dissolve 5·22 parts at 0°, and 421·9 parts at 100°. Potassium alum is prepared on a very large scale by calcining aluminous schists, which are essentially impure silicate of aluminium containing quantities of iron pyrites and carbonaceous matter. The pyrites on ignition forms ferrous sulphate, $FeSO_4$, and free sulphuric acid. The ignited mineral is methodically extracted with water, and the liquors are concentrated in leaden pans, giving an acid solution of aluminium sulphate containing ferrous or ferric sulphates. To this liquor, a concentrated

* *Proc. Roy. Soc. Ed.*, **123**, 203.

solution of potassium chloride is added. It is preferable to the sulphate, for it forms, with the iron sulphate, uncrystallisable ferric chloride along with potassium sulphate. After settling, it is run into coolers to crystallise. The confused crystals which separate are washed, drained, dissolved in fresh water, and re-crystallised in casks. It is sometimes freed from iron before the second crystallisation by one of the methods already described (p. 424).

The chief use of alum is as a mordant in dyeing; the sulphate and acetate of aluminium are used for the same purpose. When cloth or any mineral or vegetable fibre is boiled in such a solution, it becomes impregnated with hydrated alumina; and when treated with a dye, a triple combination appears to take place between the fibre, the alumina, and the colouring matter.

Some basic sulphates of aluminium occur native. These are *alunite*, $\mathbf{4SO_3.K_2O.3Al_2O_3.3H_2O}$, found at Tolfa, near Civita Vecchia, at Solfatara, near Naples, and at Puy de Garcy, in the Auvergne. It forms rhombohedral crystals, and is used for the manufacture of Roman alum, which has been prepared from it from very early times. When it is calcined at a moderate heat, the hydrated alumina loses water, and on lixiviation, alum dissolves, and may be crystallised as usual. The basic sulphate, *löwigite*, $\mathbf{4SO_3.K_2O.3Al_2O_3.H_2O}$, is also a natural product.

The difference of solubility of potassium alum from that of rubidium and cæsium alums has afforded a means of separating from each other these elements, which almost always occur together. Rubidium and cæsium alums are insoluble in a cold saturated solution of potassium alum; hence, on concentrating such a mixture, the first portions of the crystals consist chiefly of the former. Cæsium alum is likewise insoluble in a saturated solution of rubidium alum, and may be separated from the latter in a similar manner.

$\mathbf{Mn(SO_4)_2}$.—Produced by dissolving potassium permanganate, $\mathbf{KMnO_4}$, in a mixture of 500 grams of sulphuric acid and 150 of water. It is a yellow substance, which deposits a basic sulphate as a black powder of the formula $\mathbf{MnO.SO_4}$.

$\mathbf{Cr_2(SO_4)_3}$.15 and 5 $\mathbf{H_2O}$; $\mathbf{Fe_2(SO_4)_3.9H_2O}$; $\mathbf{Mn_2(SO_4)_3}$.—There are two hydrated varieties of chromium sulphate, a green and a violet. The green salt is produced when the sulphate is produced by the ordinary methods above 50°, or by heating the violet variety to that temperature; it is soluble in alcohol. The violet variety is produced in the cold; it is also formed when the green modification is allowed to stand. It is precipitated by alcohol, and crystallises best from a mixture of alcohol and water. On heating either variety with excess of sulphuric acid to above 190°, a light yellow mass of anhydrous sulphate is obtained, insoluble in water, and with difficulty in acids. Several basic salts are known, produced by digesting a solution of the ordinary salt with chromium hydrate, or by incomplete precipitation. Among these are $\mathbf{2SO_3.Cr_2O_3}$; $\mathbf{2SO_3.3Cr_2O_3}$; and $\mathbf{3SO_3.2Cr_2O_3}$. They are insoluble and amorphous. Ferric sulphate, $\mathbf{Fe_2(SO_4)_3.9H_2O}$, seems native as *coquimbite*. It is produced by oxidising ferrous sulphate with nitric acid in presence of strong sulphuric acid: $2\mathbf{FeSO_4} + H_2SO_4 + O = \mathbf{Fe_2(SO_4)_3} + H_2O$. It forms small pink scales, and is very difficult to dissolve in water. Manganic sulphate is a non-crystalline green substance produced by heating the hydrated dioxide with sulphuric acid. Many basic sulphates of iron and manganese are known, which resemble those of chromium. The double salts of these oxides, or alums, have already been noticed. A sulphato-nitrate of chromium, $\mathbf{Cr_2(SO_4)(NO_3)_4}$ is produced by dissolving the hydrated basic sulphate, $\mathbf{Cr_2(SO_4)(OH)_4}$. in strong nitric acid. The salt $\mathbf{Cr_2(SO_4)_2(NO_3)_2}$ is also known.

$CrSO_4.Aq$; $\mathbf{FeSO_4}$.7, 5, 3, 2, and $\mathbf{1H_2O}$; $\mathbf{MnSO_4}$.7, 6, 5, 4, and $\mathbf{2H_2O}$; $\mathbf{CoSO_4}$.7, 6, and $\mathbf{4H_2O}$; $\mathbf{NiSO_4}$.7 and $\mathbf{6H_2O}$.
$\mathbf{FeSeO_4}$.7 and $\mathbf{5H_2O}$; $\mathbf{CoSeO_4.7H_2O}$; $\mathbf{Ni_2SeO_4}$.7 and $\mathbf{6H_2O}$.
$\mathbf{FeTeO_4}$; $\mathbf{MnTeO_4}$; $\mathbf{CoTeO_4}$; $\mathbf{NiTeO_4}$.

Chromous sulphate has been obtained as a blue solution, by dissolving the metal in dilute acid. Like all chromous salts, it has powerful reducing properties. Ferrous sulphate occurs native as *green vitriol* or *copperas*, produced by the atmospheric oxidation of iron pyrites. It usually crystallises with $7H_2O$, in light-green monoclinic crystals, which absorb oxygen slowly in moist air, forming a basic ferric sulphate (said to be $2(\mathbf{SO_3.Fe_2O_3}).\mathbf{H_2O}$), but in dry air they are permanent. When heated to redness it evolves sulphur dioxide, and a basic sulphate remains, which, on further heating, leaves a residue of ferric oxide, and yields a distillate of sulphur trioxide. This residue is named *rouge*, and used to be known as "*colcothar vitrioli*," or "*caput mortuum*"; it is used as a pigment. Ferrous sulphate has been obtained with different amounts of water, according to the temperature at which it is crystalised; the hydrates with 3 and $2H_2O$ are formed in presence of sulphuric acid. That with $1H_2O$ is produced by drying the salt at 114°; the last molecule is retained at 280°, and is sometimes termed "water of constitution." Ferrous sulphate absorbs nitric oxide (see p. 342); but the composition of the resulting compound depends on the pressure and temperature, varying from $3FeSO_4.2NO$ to $6FeSO_4.2NO$. Manganous sulphate is a pink salt; cobalt sulphate rose-red, and nickel sulphate grass-green. The anhydrous salts are colourless. The hydrated sulphates of these metals, containing the same number of molecules of water of crystallisation are isomorphous with each other; those with $5H_2O$ resemble copper sulphate, $\mathbf{CuSO_4.5H_2O}$, in crystalline form.

The selenates closely resemble the sulphates; the tellurates are insoluble precipitates. $\mathbf{FeTeO_4}$ occurs native, and has been named *ferrotellurite.*

A large number of **double salts** of the general formula, $M'_2RO_4.M''RO_4.6H_2O$, are known, where **M'** stands for Li, Na, K, Rb, Cs, Tl' and NH_4; M'', for Mg, Zn, Cd, Cr'', Fe'', Mn'', Co'', Ni'', and Cu''; and R for S or Se. They all crystallise in monoclinic crystals, and are isomorphous with each other. They are produced by mixture. The double salts of hydrogen,

$$\mathbf{H_2Mn(SO_4)_2}, \text{ and } \mathbf{H_6Mn(SO_4)_4}$$

have also been prepared.

Sulphate of carbon is unknown. Both the monoxide and dioxide of carbon are insoluble in sulphuric acid.

$\mathbf{Ti_2(SO_4)_3}$; $\mathbf{Ce_2(SO_4)_3}$.5, 6, 8, 9, and $12H_2O$.

Double salts:—$\mathbf{Ce_2(SO_4)_3.2K_2SO_4.2H_2O}$; $\mathbf{Ce_2(SO_4)_3.5K_2SeO_4}$, and others.

The titanous sulphate is violet; the cerous salts colourless.

$\mathbf{Ti(SO_4)_2}$; $\mathbf{Zr(SO_4)_2}$; $\mathbf{Ce(SO_4)_2.4H_2O}$; $\mathbf{Th(SO_4)_2.4H_2O}$.—Also double salts, such as $\mathbf{K_2Ti(SO_4)_3}$; $\mathbf{(NH_4)_6Ce(SO_4)_5.4H_2O}$; $\mathbf{K_4Th(SO_4)_4.2H_2O}$.

The cerium salt is yellow; the others colourless. Cerium also forms a double salt, containing the metal in two states of oxidation; it is called ceroso-ceric sulphate. It has a brown-red colour and the formula $\mathbf{2Ce(SO_4)_2.Ce_2(SO_4)_3.25H_2O}$. These bodies, especially titanium, zirconium, and cerium, also yield basic sulphates. The formation of titanium sulphate serves as a means of separating titanium from silica. The mixture is fused with hydrogen potassium sulphate, dissolved in water, and filtered from silica; on boiling with water the titanium sulphate is decomposed into hydrate and sulphuric acid.

Silica is insoluble in sulphuric acid; and germanium does not appear to form a sulphate.

$SnSO_4$; $PbSO_4$; $PbSeO_4$; $PbTeO_4$.—Double salts:—$K_2Sn(SO_4)_2$; $4K_2Sn(SO_4)_2.SnCl_2$; $(NH_4)_2Pb(SO_4)_2$.

Stannous sulphate is colourless and crystalline. The double salts are obtained by mixture. Lead sulphate occurs native in trimetric crystals as *anglesite*, isomorphous with those of *heavy spar* (barium sulphate). The crystalline variety may be obtained by fusing lead chloride with potassium sulphate. The selenate has also been found native. As lead sulphate and selenate are nearly insoluble, they may be produced by precipitation; they form dense white powders, more easily dissolved by water than by the dilute acid; but they are soluble to a small extent in strong acids. They dissolve in larger quantity in solutions of sulphate, nitrate, acetate, or tartrate of ammonium, and easily in caustic alkali, and in thiosulphates. Lead sulphate also dissolves in sulphuric acid; the solution deposits crystals of $\mathbf{H_2Pb(SO_4)_2.H_2O}$. These bodies melt at a red heat.

Lead sulphate, heated with the sulphide, as in lead smelting, yields metallic lead and sulphur dioxide, thus:—$\mathbf{PbSO_4 + PbS} = 2Pb + 2SO_2$; or the oxide and metal:—$\mathbf{2PbSO_4 + PbS} = 3SO_2 + \mathbf{2PbO} + Pb$.

Lead tellurate is also a white precipitate, but is more easily soluble in water. Basic sulphates and selenates of tin and lead have also been prepared; stannic hydrate dissolves in sulphuric acid, but stannic sulphate is an indefinite non-crystalline body.

Compounds of nitrogen and vanadium usually contain the nitrosyl, or vanadyl groups. Compounds of the pentoxides with sulphuric anhydride are, however, known. The compound, $\mathbf{SO_3.N_2O_5.4H_2SO_4}$, a white crystalline body, is produced by cooling a mixture of sulphur trioxide and nitric acid; it is at once decomposed by water, and, when heated, evolves red fumes, yielding a sublimate supposed to be $\mathbf{SO_3.N_2O_3}$. This would be nitrosyl sulphate, $\mathbf{SO_2(ONO)_2}$, to be alluded to later. The first may be viewed as a compound of nitryl sulphate, $SO_4(NO_2)_2$ with sulphuric acid. The compound, $2(\mathbf{SO_3}).\mathbf{N_2O_5}$, is also known. It is a snowy crystalline mass, produced by the action of induction sparks on a mixture of sulphur dioxide, oxygen and nitrogen; it may be regarded as nitryl anhydrosulphate, $S_2O_7(NO_2)_2$.

Vanadyl sulphate, $\mathbf{3SO_3.V_2O_5}$, is prepared by dissolving vanadium pentoxide in cold sulphuric acid, and expelling excess of sulphuric acid by heat. It may be regarded as $\mathbf{(VO)'''_2(SO_4)_3}$.

It is red and crystalline. During evaporation, the green compound of V_2O_4, $2SO_3.V_2O_4 = (VO)''_2(SO_4)_2$ separates as a crust. By heating the first compound to the temperature of melting lead, the basic sulphate, $(VO)_2O.(SO_4)_2$, is obtained as a red crystalline mass. A double sulphate of the formula, $2SO_3.K_2O.V_2O_5.6H_2O$, is also known.

These bodies are mostly derivatives of the pentoxides of nitrogen and vanadium. Niobium and tantalum are said also to form sulphates, but these compounds have not been investigated.

Nitrosyl sulphate, $(NO)_2SO_4$, may be the substance alluded to on the previous page. Hydrogen nitrosyl sulphate, $H(NO)SO_4$, is better known, and is produced by the action of nitrogen trioxide on sulphuric acid, thus:—$N_2O_3+2H_2SO_4 = 2H(NO)SO_4+H_2O$. Excess of sulphuric acid must be present to combine with the water. The same substance is produced by the action of sulphur dioxide on nitric acid, or by passing the vapours from a heated mixture of nitric and hydrochloric acids (nitrosyl chloride and chlorine, see p. 341) through strong sulphuric acid. It forms long, thin, transparent crystals melting at 85-87°. It is the substance known as "chamber crystals," and its solution in sulphuric acid is produced in the "Gay-Lussac tower," in which the escaping gases from the vitriol chambers are brought in contact with strong sulphuric acid. On treatment with water, it is at once decomposed into oxides of nitrogen (NO and $NO_2+N_2O_4$); this change takes place in the "Glover tower," where the sulphuric acid containing hydrogen nitrosyl sulphate is diluted; the oxides of nitrogen are liberated, and again pass into the chambers (see p. 416). (See also nitrosyl anhydrosulphate, p. 434).

$(PO)'''_2(SO_4)_3$, phosphoryl sulphate, is produced by mixture; it forms thin transparent scales, and is decomposed at 30°, and by water; the corresponding compounds of arsenic, antimony, and bismuth are unknown, the groups $(AsO)'$, $(SbO)'$ and $(BiO)'$ tending, as a rule, to replace only one atom of hydrogen.

By dissolving arsenious oxide, As_4O_6, in sulphuric acid of different concentrations, which must not, however, be more dilute than corresponds with the formula, $H_2SO_4.H_2O$, various white crystalline sulphates of arsenic have been obtained. They appear to have the formulæ $8(SO_3).As_2O_3$; $4(SO_3).As_2O_3$; $3(SO_3).As_2O_3$(?); $2(SO_3).As_2O_3$; and $SO_3.As_2O_3$. The body, $3(SO_3).As_2O_3$, would correspond to $As_2(SO_4)_3$; $2(SO_3)As_2O_3$ may be written $SO_4—As—O—As—SO_4$; and $SO_3.As_2O_3$ may represent arsenosyl sulphate, $(AsO)'_2SO_4$, corresponding in formula to nitrosyl sulphate. These bodies are all decomposed by water, and are all very unstable.

The sulphates of antimony are similar but more stable. The compounds, $4(SO_3).Sb_2O_3$, $3(SO_3).Sb_2O_3$, $2(SO_3).Sb_2O_3$, and $SO_3.Sb_2O_3$, have been prepared. The normal salt, $Sb_2(SO_4)_3 = 3(SO_3).Sb_2O_3$, is produced by boiling antimony with strong sulphuric acid. It crystallises in needles.

With bismuth, the compounds, $3(SO_3).Bi_2O_3$, $2(SO_3).Bi_2O_3$, and $SO_3.Bi_2O_3$, are known. Bismuth dissolves in hot, strong sulphuric acid, with evolution of sulphur dioxide forming the first; it is decomposed by water, giving the third. **Double salts** with hydrogen, $HBi(SO_4)_2.H_2O$; with ammonium, $NH_4Bi(SO_4)_2.4H_2O$; and with potassium, $K_3Bi(SO_4)_3$ are also known. The selenates and tellurates have scarcely been examined. Bismuth tellurate, however, has been found native. Its formula is $TeO_3.Bi_2O_3$; it has been named *montanite.*

Hydrated molybdenum sesquioxide forms a dark-coloured solution with sulphuric acid, which may contain $Mo_2(SO_4)_3$. The dioxide gives a red solution, supposed to contain $Mo(SO_4)_2$.

Uranous sulphate, $U(SO_4)_2.4$ and $8H_2O$, forms green crystals, and is produced by dissolving hydrated uranium dioxide in sulphuric acid. A basic sulphate, $SO_3.UO_2.3H_2O$, is also known; and also the double salt $K_2U(SO_4)_3.H_2O$. They are green, soluble bodies.

$MoO_2(SO_4)$ and $UO_2(SO_4)$, molybdyl and uranyl sulphates, are yellow crystalline bodies, obtained from the hydrated trioxides. This sulphate of molybdenum, when boiled with water, decomposes, depositing the hydrated oxide, $5(MoO_3).H_2O$. Double salts of uranyl sulphate are known, *e.g.*, $H_2(UO_2)(SO_4)_2$, and

$$K_2(UO_2)(SO_4)_2.2H_2O.$$

The selenates and tellurates are little known.

Tellurium dioxide dissolves in hot dilute sulphuric acid, and deposits crystals of $SO_3.2TeO_2$. It is decomposed by warm water.

$Ru(SO_4)_2$; $Rh_2(SO_4)_3.12H_2O$; $PdSO_4.2H_2O$.—Also $KRh(SO_4)_2$.

Ruthenium and rhodium sulphates are orange-brown and red solutions, drying respectively to a yellow-brown amorphous mass, and to a brick-red powder; they are produced by oxidation of the sulphide. Palladium dissolves in sulphuric acid, mixed with a little nitric acid; the solution, when evaporated, deposits brown crystals.

$OsSO_4$; $Os(SO_4)_2$; $IrSO_4$; $IrO.SO_4$; $PtSO_4$; $Pt(SO_4)_2$.

These are all yellow syrups, drying to brown non-crystalline masses; they are all produced by oxidising the respective sulphides with nitric acid, with the exception of platinous sulphate, $PtSO_4$, which is produced when the chloride, $PtCl_2$, is dissolved in sulphuric acid.

Ag_2SO_4; $HAgSO_4$; $H_3Ag(SO_4)_2.H_2O$; $H_6Ag_2(SO_4)_4$; Hg_2SO_4; Hg_2SeO_4; Ag_2TeO_4; cuprous and aurous sulphates are unknown. Auric sulphate, however, can be prepared in solution by dissolving auric oxide in dilute acid. It decomposes on standing.

Sulphates of silver and mercury are sparingly soluble white salts, produced by precipitation, or by dissolving the metals in sulphuric acid. The silver salt is isomorphous with anhydrous sodium sulphate. The tellurate is a dark-yellow powder. It has been found native, and named *magnolite*.

$CuSO_4.5H_2O$; $CuSeO_4.5H_2O$; $HgSO_4$.—Basic salts:—$SO_3.2CuO.H_2O$; $SO_3.3CuO.3H_2O$; $SO_3.4CuO.3H_2O$; $SO_3.3HgO$. Double salts:—Those of copper belong to the class **$M_2'M''(SO_4)_2.6H_2O$**; those of mercury resemble **$3K_2Hg(SO_4)_2.2H_2O$**. Also **$HgSO_4.HgI_2$; $2HgSO_4.HgS$**.

Copper sulphate, or *blue vitriol*, is produced on a large scale by the spontaneous oxidation of copper pyrites, or by the action of air on ignited cuprous sulphide, Cu_2S, whereby cupric oxide is produced at the same time. It crystallises with water in large blue monoclinic prisms, isomorphous with ferrous sulphate of the same degree of hydration. Indeed, copper sulphate, if present in excess in a solution containing ferrous sulphate, induces the latter to adopt its crystalline form; and, similarly, ferrous, zinc, magnesium, or nickel sulphate in excess, causes copper sulphate to assume their special form. When heated to 100°, **$CuSO_4.5H_2O$** loses four molecules of water; the last molecule is retained up to 200°, and is regarded as "water of constitution." It is easily soluble in water, but insoluble in alcohol. The tetrabasic salt occurs native as *brochantite*. The selenate closely resembles the sulphate. Mercuric sulphate is decomposed by water into a soluble acid salt, **$3SO_3.HgO.nH_2O$**, and the basic salt, **$SO_3.3HgO$**, a lemon-yellow powder, which used to be called *turpeth mineral*. The compound, **$2HgSO_4.HgS$**, is precipitated by the action of a moderate quantity of hydrogen sulphide on a solution of the sulphate. It is a white precipitate.

Anhydro- or pyrosulphuric acid, $H_2S_2O_7$.—This substance is, as will appear hereafter, an analogue of pyrophosphoric acid, inasmuch as it may be regarded as constituted of two molecules of sulphuric acid, minus a molecule of water, thus.—

$$HO—(SO_2)—O—(SO_2)—OH.$$

But it cannot be prepared by heating ordinary sulphuric acid, for that acid, as already remarked, distils as a whole. It may be obtained by dissolving sulphur trioxide in ordinary sulphuric acid, thus:—$H_2SO_4 + \mathbf{SO_3} = \mathbf{H_2S_2O_7}$. The old method of preparation, which gained for this acid the name "Nordhausen sulphuric acid," is still carried out at Nordhausen in Saxony; it consists in distilling partially dried ferrous sulphate from tube-shaped retorts of very refractory fire-clay. The products are sulphur dioxide and anhydrosulphuric acid, while ferric oxide of a fine red colour remains in the retort, and is made use of as a pigment under the name of "Venetian red" or "rouge." This method of manufacture is a very ancient one. When ferrous sulphate, $\mathbf{FeSO_4.7H_2O}$, is dried, it loses six molecules of water, retaining the seventh. On distilling the monohydrated salt, sulphur dioxide and water are evolved first, leaving basic ferric sulphate, thus:—

$$\mathbf{2FeSO_4.H_2O} = SO_2 + H_2O + \mathbf{Fe_2O_2(SO_4)}(= \mathbf{SO_3.Fe_2O_3}).$$

The sulphur dioxide escapes; the temperature is then raised, when sulphur trioxide distils over, and combines with the water, leaving iron sesquioxide in the retort.

Anhydrosulphuric acid is a white solid, crystallising in needles, and melting at 35°. It gives off sulphur trioxide when heated. It hisses when dropped into water, evolving great heat.

It is probable that still more condensed sulphuric acids are formed when more sulphur trioxide is added to sulphuric acid; but they have been little investigated. Corresponding compounds of selenium and tellurium are unknown.

Pyrosulphates and polytellurates.—The pyrosulphates are produced (1) by the action of pyrosulphuric acid on the oxides; (2) in a few cases by heating the double salts of hydrogen and a metal; and (3) by the action of sulphur trioxide on the normal sulphate. The following salts are known:—

$\mathbf{Na_2S_2O_7}$; $\mathbf{K_2S_2O_7}$; $\mathbf{Ba_2S_2O_7}$; $\mathbf{Ag_2S_2O_7}$; also the double salt $\mathbf{HKS_2O_7}$.

The sodium and potassium salts may be prepared by all these methods. They are crystalline salts, which combine with water, forming hydrogen metallic sulphates. Hydrogen potassium pyrosulphate crystallises from a solution of the anhydrosulphate in strong sulphuric acid; the other salts are best prepared by method (3).

Nitrosyl anhydrosulphate, $S_2O_7(NO)_2$, is produced as a white crystalline substance by the action of sulphur dioxide on nitric peroxide. It is at once decomposed by water into sulphuric acid and the products of decomposition of nitrous anhydride.

Several polysulphates of arsenic, antimony, &c., have already been described among the sulphates.

Di- and tetra-tellurates are also known. The ditellurates probably correspond to the anhydrosulphates; and the tetratellurates are produced by the action of water on the monotellurates. The formulæ of the following have been ascertained:—

$K_2Te_2O_7$; $(NH_4)_2Te_2O_7$; $PbTe_2O_7$; $Ag_2Te_2O_7$; also $4TeO_3.K_2O$; $4TeO_3.(NH_4)_2O$; $4TeO_3.BaO$; $4TeO_3.PbO$; and $4TeO_3.Ag_2O$.

These bodies are more soluble than the ordinary tellurates.

CHAPTER XXVII.

COMPOUNDS OF OXYGEN, SULPHUR, SELENIUM, AND TELLURIUM WITH EACH OTHER (CONTINUED).—SULPHITES, SELENITES, AND TELLURITES; HYPOSULPHITES, THIONATES, THIOSULPHATES, ETC.—OXYHALIDES.

Compounds with Water and with Oxides (continued):—(2) Compounds of the Dioxides; Sulphurous, Selenious, and Tellurous Acids; Sulphites, Selenites, and Tellurites.

Sulphurous, selenious, and **tellurous acids,** in aqueous solution, are produced either by direct combination of the anhydrides with water, or by displacement.

Water absorbs at 15° about 45 times its volume of sulphur dioxide; and on cooling the solution several definite hydrates have been obtained.

By passing a current of the gas through a solution cooled to −6°, white crystals, fusing at 4°, of the formula $\mathbf{H_2SO_3.8H_2O}$, were produced. By a similar process, crystals melting at 14°, of the formula $\mathbf{H_2SO_3.6H_2O}$, were obtained; and it is also stated that the compound $\mathbf{H_2SO_3}$ has been thus isolated in cubical crystals. A solution of sulphurous acid may also be produced by adding almost any acid to a dilute solution of a sulphite; if the solution be strong, the anhydride, SO_2, is evolved. On boiling a solution of sulphur dioxide the gas is evolved; but it does not wholly escape, except the boiling be considerably prolonged. The solution possesses the smell and taste of the gas, and, like many other similar solutions, it doubtless contains the free anhydride as well as the acid. When heated to 180—200° in a sealed tube, an aqueous solution of sulphur dioxide yields sulphuric acid and free sulphur.

It shows a great tendency to absorb oxygen. On exposure to air, it is gradually converted into sulphuric acid; and this conversion may be effected by the addition of a solution of a halogen, or of a chromate, of a manganate or permanganate, &c., which readily yields oxygen. A convenient test for a sulphite consists in boiling it with a solution of potassium dichromate, acidified with hydrochloric acid; the orange colour of the dichromate changes to the green colour of a chromic salt, and the solution then contains a sulphate, which yields a precipitate with barium chloride.

This power of reduction has led to the employment of sulphur dioxide in bleaching animal fibres, such as silk and wool. The colouring matters, which are insoluble, are converted into colourless substances by exposure in a moist state to the fumes of burning sulphur It also finds use as a disinfectant, and in the form of sulphites is used in brewing for checking fermentation.

Sulphurous acid at once reacts with hydrogen sulphide, giving a deposit of sulphur:—$2H_2S + H_2SO_3.Aq = 2H_2O + Aq + 3\mathbf{S}$ (see Pentathionic Acid, p. 451).

Selenious acid, H_2SeO_3, is produced by direct union of the dioxide with water, or by boiling selenium with nitric acid. It deposits in colourless prismatic crystals when its solution is cooled. The crystals lose water on exposure to air, and when gently heated they yield the dioxide. When a current of sulphur dioxide is passed through its solution, it is decomposed, depositing selenium, thus: $H_2SeO_3.Aq + H_2O + 2SO_2 = 2H_2SO_4.Aq + \mathbf{Se}$. This is the usual method of separating selenium from its compounds. The solution of selenious acid has a very acid taste. It is not altered by boiling with hydrochloric acid, but may be oxidised to selenic acid by the usual oxidising agents; not, however, by nitric or nitrohydrochloric acid (see p. 417).

Tellurous acid, H_2TeO_3, is precipitated by pouring strong nitric acid, in which tellurium has been boiled, at once into water, or by the action of water on tellurium tetrachloride. It is a white bulky precipitate, drying to a white powder, only sparingly soluble in water. It dissolves in acids, but is reprecipitated on dilution.

The **sulphites, selenites,*** and **tellurites** are prepared by the usual methods of preparing salts. They are all decomposed by such acids as sulphuric or phosphoric, with liberation of the respective acid. They are also all decomposed by heat. Like the sulphates, they form two main classes, the normal salts, such as M_2SO_3, and the anhydro- or pyro-salts, such as $M_2S_2O_5$ (compare also Phosphates). The latter are known only in a few instances.

$Li_2SO_3.6H_2O$; $Na_2SO_3.8$, and $7H_2O$; $K_2SO_3.2H_2O$; $(NH_4)_2SO_3$.—$Li_2SeO_3.H_2O$; Na_2SeO_3; K_2SeO_3; Rb_2SeO_3; Cs_2SeO_3; $(NH_4)_2SeO_3$.—Li_2TeO_3; $Na_2TeO_3.nH_2O$; K_2TeO_3.

These are all white soluble salts. At a dull red heat potassium

* *Bull. Soc. Chim.* (5), **23**, 260, 335.

sulphite gives sulphate and sulphide, thus:—$4K_2SO_3 = 3K_2SO_4 + K_2S$

Double salts.—$HNaSO_3$; $HKSO_3$; $H(NH_4)SO_3$; $NaKSO_3.2H_2O$;* $KNaSO_3.H_2O$.*—$HLiSeO_3$; $HNaSeO_3$; $HKSeO_3$; also $H_3Li(SeO_3)_2$; $H_3Na(SeO_3)_2$; $H_3K(SeO_3)_2$; and $H_2(NH_4)_4(SeO_3)_3$.—$H_3Na(TeO_3)_2.H_2O$; $H_3K(TeO_3)_2.H_2O$; $H_3(NH_4)(TeO_3)_2.H_2O$.

These are produced by mixture. When heated, the acid sulphites give off water, and leave a residue of sulphate and thiosulphate.

They are all white and soluble, and smell of sulphur dioxide.

The normal sulphites of potassium and ammonium form compounds with nitric oxide, NO, named *nitrososulphites* (see p. 455).

$BeSO_3$; $CaSO_3.2H_2O$; $SrSO_3$; $BaSO_3$.—$BeSeO_3.2H_2O$; $CaSeO_3.2H_2O$; $SrSeO_3.2H_2O$; $BaSeO_3.H_2O$.—$CaTeO_3$; $SrTeO_3$; $BaTeO_3$.

With the exception of beryllium sulphite, these salts are sparingly soluble in water, and may be produced by precipitation, when they come down in small crystals. The tellurites are also produced by fusion of the respective carbonate with tellurous anhydride. They are all white bodies, and the sulphites decompose, when heated, into sulphate and sulphide.

Little is known of the double sulphites of these metals. Solutions of the neutral salts absorb sulphur dioxide, but the neutral salts again crystallise out on evaporation over sulphuric acid. Such a solution of calcium sulphite is made use of in sugar refining to prevent fermentation.

Double salts.—$H_2Be(SeO_3)_2$; $H_2Ca(SeO_3)_2.H_2O$; $H_2Sr(SeO_3)_2$.

These are white soluble salts, obtained by mixture and crystallisation. The existence of corresponding tellurites is doubtful.

$MgSO_3$.6 and $3H_2O$: $ZnSO_3.5H_2O$; and $CdSO_3.H_2O$.

These are sparingly soluble white salts.

$MgSeO_3.6H_2O$; $ZnSeO_3.H_2O$; and $CdSeO_3$.

These are sparingly soluble salts, which dissolves in selenious acid, forming double salts with hydrogen, which have the formulæ $H_2Mg(SeO_3)_2$; $H_4Mg(SeO_3)_3.3H_2O$; $H_2Cd_2(SO_3)_3.H_2O$; and $H_2Cd_3(SeO_3)_4$.

$MgTeO_3$; $ZnTeO_3$; and $CdTeO_3$.

These are also obtained by precipitation.

No sulphite, selenite, or tellurite of boron is known. Scandium

* These salts are isomeric (see p. 453). They are formed respectively thus:—$2HNaSO_3.Aq + K_2CO_3.Aq = 2KNaSO_3.Aq + CO_2 + H_2O$; and $2HKSO_3.Aq + Na_2CO_3.Aq = 2NaKSO_3.Aq + CO_2 + H_2O$.

forms a selenite of the formula $10SeO_2.3Sc_2O_3.4H_2O$. Yttrium sulphite, $Y_2(SO_3)_3$, and selenite, $Y_2(SeO_3)_3.12H_2O$, are white insoluble powders. Yttrium tellurite is a white precipitate. The compounds of this group have scarcely been examined.

Aluminium sulphite is basic: $SO_2.Al_2O_3.2H_2O$. The selenite, however, $Al_2(SeO_3)_3$, is normal. These salts, and the tellurite, are white and insoluble; but the selenite dissolves in selenious acid.

Indium sulphite, $In_2(SO_3)_3.8H_2O$, is a white insoluble powder. Its formation is made use of in separating indium from traces of copper, lead, zinc, and iron. The gallium salts have not been prepared. The compounds $9SeO_2.4Al_2O_3.36H_2O$ and the double salts $H_3Al(SeO_3)_3.4H_2O$ and $H_3In(SeO_3)_3.6H_2O$ have been prepared.

Thallous sulphite, selenite, and tellurite are nearly insoluble.

$Cr_2(SO_3)_3.16H_2O$ is a yellow precipitate, thrown down by alcohol. $Fe_2(SO_3)_3$ may exist as a red solution, but is rapidly changed by reduction into ferrous sulphate; but if alcohol be added at once, a yellow-brown basic salt, $3SO_2.2Fe_2O_3$, is precipitated. On treatment with water it decomposes, yielding the salt $SO_2.Fe_2O_3.6H_2O$. On addition of caustic potash to the original red solution, the basic double salt $3SO_2.2K_2O.Fe_2O_3.5H_2O$ is precipitated. A double salt of cobalt, $KCo^{III}(SO_3)_2$, is produced by digesting cobaltic hydrate with hydrogen potassium sulphite.

$Cr_2(SeO_3)_3$ and $Fe_2(SeO_3)_3$ are insoluble powders. The tellurates are also insoluble.

$FeSO_3.3H_2O$; $MnSO_3.2H_2O$; $CoSO_3$; $NiSO_3.6H_2O$.
$FeSeO_3$; $MnSeO_3.2H_2O$; $CoSeO_3.2H_2O$; $NiSeO_3.2H_2O$.

These salts are sparingly soluble, but crystallise from dilute solutions. The tellurites are insoluble. A double selenite, of the formula $H_2Ni(SeO_3)_2.2H_2O$, has been prepared.

The sulphites and tellurites of cerium, zirconium, and thorium are said to be white insoluble powders. The selenites, $Ce_2(SeO_3)_3.12H_2O$; $Th(SeO_3)_2.8H_2O$; and the acid salts $H_2Ce_2(SeO_3)_4.5H_2O$, $H_2Th(SeO_3)_3.6H_2O$ and $H_6Th(SeO_3)_5.5H_2O$ have been prepared.

Salts of silicon and germanium are unknown. Stannic selenite is said to be an insoluble precipitate. $PbSO_3$, $PbSeO_3$, and $PbTeO_3$ are nearly insoluble white precipitates.

Compounds of nitrogen, vanadium, niobium, tantalum, phosphorus, arsenic, and antimony are unknown. Bismuth sulphite, $SO_2.Bi_2O_3$, is basic, and sparingly soluble.

Compounds of molybdenum and tungsten have not been prepared. But uranyl sulphite, $(UO_2)(SO_3).3H_2O$, is known; and double sulphites, of the general formula $(UO_2)HM'(SO_3)_2$, where M′ is Na, K, or NH_4, are produced by mixture. They are yellow, sparingly soluble, crystalline precipitates. Osmous sulphite, $OsSO_3$, is produced by dissolving the tetroxide, OsO_4, in sulphurous acid. It forms a double sulphite with potassium, $K_6Os(SO_3)_4.5H_2O$.

Double sulphites of palladium, rhodium, iridium, and platinum with the alkalies are known. That of palladium, $Na_6Pd(SO_3)_4.2H_2O$, is produced by adding sulphurous acid to palladium dichloride, and precipitating with caustic soda. The precipitate gradually becomes yellow and crystalline. The iridium compound has a similar formula. Other double salts are also formed at the same time, viz., $H_2Na_6Ir(SO_3)_5.4H_2O$ and $10H_2O$. They form whitish-yellow scales. A double salt, which crystallises well, is produced by the action of sulphurous acid on ammonium iridichloride, $Ir_2Cl_6.6NH_4Cl$, viz., $IrCl_2.H_2SO_3.4NH_4Cl$, which reacts with carbonates, yielding salts, such as $IrCl_2.K_2SO_3.2NH_4Cl.4H_2O$.

Platinous compounds are also known. Sulphur dioxide, passed through water in which platinic hydrate is suspended, reduces and dissolves it; and on addition of a sodium salt, a precipitate of the formula $Na_6Pt(SO_3)_4.3H_2O$ is produced. The action of sulphurous acid on ammonium platinochloride is to form the compound $H(PtCl)SO_3.2NH_4Cl$, in which the group (PtCl) functions as a monad. The hydrogen in this body may be replaced by metals. Substituting potassium platinochloride for the ammonium salt, the corresponding compound $H(PtCl)SO_3.2KCl$ is obtained. And by the action of excess of a sulphite on such compounds, bodies such as $H(PtCl)SO_3.K_2SO_3.3H_2O$ are formed. Lastly, by the action of hydrogen ammonium sulphite on ammonium platinochloride, $PtCl_2.2NH_4Cl$, both atoms of chlorine are replaced, and the compound $Pt(SO_3H)_2.2NH_4Cl.H_2O$ is obtained in crystals. Possibly selenious acid would form similar combinations.

Cu_2SO_3; $Cu_2SO_3.H_2O$; Ag_2SO_3; Hg_2SO_3.—Cu_2SeO_3; Ag_2SeO_3; Hg_2SeO_3.

These salts are insoluble, and are produced by precipitation, or by the action of sulphurous acid on the hydrates. Cuprous sulphite forms red microscopic quadratic prisms; silver sulphite is white.

Double Salts.—$NaCuSO_3.H_2O$; $(NH_4)Cu(SO_3)$; $K_4Cu_2(SO_3)_3$, and perhaps more complex salts, *e.g.*, $8K_2SO_3.CuSO_3.16H_2O$; $5Na_2SO_3.CuSO_3.38H_2O$, &c.

These are produced by mixture. Double sulphites of gold are also known, such as $3Na_2SO_3.Au_2SO_3.3H_2O$; this compound has a purple colour, and from it other double salts may be prepared. Normal cupric sulphite is unknown; $\mathbf{CuSeO_3.2H_2O}$ forms blue needles. The tellurite is green.

A basic cupric sulphite, $\mathbf{SO_2.4CuO.7H_2O}$, is precipitated on addition of cupric hydrate to a solution of sulphur dioxide in absolute alcohol; and also several cuprous-cupric sulphites, *e.g.*, $\mathbf{Cu_2SO_3.CuSO_3.2H_2O}$, produced by warming a solution of cupric sulphite with hydrogen potassium sulphite. They are red crystalline powders.

Hydrogen cupric selenite, $\mathbf{H_2Cu(SeO_3)_2.2H_2O}$, is prepared by mixture. $HgSO_3$ does not exist. The selenite, $\mathbf{HgSeO_3}$, is a white precipitate, and the tellurite, a brown precipitate. A basic salt, $\mathbf{SO_2.2HgO}$, is produced by precipitation. It is a heavy white crystalline body. $\mathbf{H_2Hg(SO_3)_2}$, $\mathbf{Na_2Hg(SO_3)_2}$, $\mathbf{K_2Hg(SO_3)_2}$, and $\mathbf{(NH_4)_2Hg(SO_3)_2}$ have also been prepared. They are soluble. Ammonium sulphite unites with mercuric chloride, forming the salt $\mathbf{2(NH_4)_2SO_3.3HgCl_2}$.

A double auric sulphite, $\mathbf{5K_2SO_3.Au_2(SO_3)_3.5H_2O}$, is produced by adding potassium sulphite to a solution of potassium aurate. It forms yellow needles.

Polysulphites, selenites, and tellurites.—The compounds analogous to the anhydrosulphates, and to the pyrophosphates, are not very numerous. Those which have been prepared are as follows:—

$\mathbf{Na_2S_2O_5}$; $\mathbf{K_2S_2O_5}$; $\mathbf{(NH_4)_2S_2O_5}$.—$\mathbf{CaSe_2O_5}$; $\mathbf{BaSe_2O_5}$; $\mathbf{CdSe_2O_5}$; $\mathbf{MnSe_2O_5}$; $\mathbf{CoSe_2O_5}$; $\mathbf{PbSe_2O_5}$.—$\mathbf{Li_2Te_2O_5}$; $\mathbf{Na_2Te_2O_5}$; $\mathbf{K_2Te_2O_5}$; $\mathbf{CaTe_2O_5}$.

Sodium and potassium anhydrosulphites are produced by passing a current of sulphur dioxide through hot solutions of the respective carbonates; they separate in crystals. The ammonium compound is formed when the normal sulphite is heated. The corresponding selenites and tellurites are formed by warming solutions of the normal salts with the requisite excess of acid or anhydride; and some of the tellurites have been prepared by fusing the dioxide with the required amount of the carbonate. They are almost all soluble.

Three salts are known which contain a smaller excess of dioxide over the normal salt, viz., $\mathbf{4SO_2.3HgO}$; $\mathbf{4SeO_2.3HgO}$; and $\mathbf{3SeO_2.CoO.H_2O}$.

One tetraselenite, $\mathbf{4SeO_2.NiO.H_2O}$, which may be regarded as hydrogen nickel anhydroselenite, $\mathbf{H_2Ni(Se_2O_5)_2}$, and the following tetratellurites, $\mathbf{4TeO_2.Li_2O}$, $\mathbf{4TeO_2.K_2O}$, $\mathbf{4TeO_2.CaO}$, and $\mathbf{4TeO_2.BaO}$ have been prepared; the tellurites are formed when excess of anhydride is added to the normal salts.

Compounds of oxides and halides.—As these compounds

are related solely to the dioxides and trioxides of sulphur, selenium, and tellurium, it appears advisable to consider them here, before treating of the other compounds of these elements.

Sulphuryl, selenyl, and telluryl compounds.—These contain the groups $(SO_2)''$, $(SeO_2)''$, and $(TeO_2)''$. They are as follows:—SO_2Cl_2; SeO_2Cl_2 (?); $\mathbf{SO_2Br_2}$.

Sulphuryl chloride is produced by the direct combination of sulphur dioxide and chlorine in sunlight, or in presence of charcoal at a moderate temperature. It is more easily prepared by passing a current of sulphur dioxide through hot antimony pentachloride, which parts with two atoms of chlorine; and it is likewise obtained by distilling sulphuric acid with phosphorus pentachloride, thus:—$4SO_2(OH)_2 + 2\mathbf{PCl_5} = 2PO(OH)_3 + 4SO_2Cl_2 + 2HCl$. This action, however, yields other products. It is a colourless liquid, boiling at 77°; its vapour-density is normal; but it decomposes at 440° into sulphur dioxide and chlorine. The corresponding bromide forms white crystals, volatile at the ordinary temperature. These bodies rapidly react with water, forming sulphuric acid and the halogen acid, *e.g.*,

$$SO_2Cl_2 + 2H.OH = SO_2(OH)_2 + 2\mathit{HCl}.$$

It is this, and analogous actions, which lead to the conclusion that sulphuric acid may be regarded as analogous to sulphuryl chloride, and that their formulæ are comparable:—

$$SO_2{<}^{Cl}_{Cl}, \qquad SO_2{<}^{OH}_{OH}$$

(see p. 268). But we are ignorant of the molecular weight of sulphuric acid; and the existence of double sulphates such as those mentioned on p. 421 would lead to the belief that the molecular weight is higher than that expressed by the formula H_2SO_4.

Chlorosulphonic acid.—The existence of two bodies related like SO_2Cl_2 and $SO_2(OH)_2$ would lead to the inference that an intermediate compound is possible; a body containing one atom of chlorine and one hydroxyl group. This body is chlorosulphonic acid, $SO_2(OH)Cl$. It is produced by the action of dry hydrogen chloride on sulphur trioxide, thus:—$\mathbf{SO_3} + \mathit{HCl} = SO_2(OH)Cl$; or on anhydrosulphuric acid, $H_2S_2O_7$. It is also formed when sulphuric acid is distilled with phosphorus pentachloride, thus:—$SO_2(OH)_2 + \mathbf{PCl_5} = POCl_3 + SO_2(OH)Cl + \mathit{HCl}$; or with phosphoryl chloride:—$4SO_2(OH)_2 + 2POCl_3 = 4SO_2(OH)Cl + 2HPO_3 + 2\mathit{HCl}$. It is a fuming colourless liquid, boiling at

158·4° (another statement gives 151°); its density is 1·78. At 200°, it decomposes into sulphuric acid, sulphuryl chloride, and other products. Near its boiling point its density* corresponds with the formula $SO_2(OH)Cl$; but at higher temperatures it is lower, owing to decomposition.

A few salts of this acid are known.† Dry nitrosyl chloride, *NOCl*, acts on sulphur trioxide, giving nitrosyl chlorosulphonate, $\mathbf{SO_2(O.NO)Cl}$; it is a white crystalline mass which can be melted, but which decomposes on raising the temperature. A salt derived from sulphur tetrachloride, SCl_4, is produced by its action on sulphur chloride, S_2Cl_2, in presence of chlorine, thus: - $S_2Cl_2 + 2SO_2(OH)Cl + 3Cl_2 = 2\mathbf{SO_2(OSCl_3)Cl} + 2HCl$. The group $(SCl_3)'$ behaves in this case like a monad metal. This chlorosulphonate is a white crystalline substance, subliming at 57°; it is converted by heating in a sealed tube into a mixture of sulphuryl and sulphurosyl chlorides, $\mathbf{SO_2(OSCl_3)Cl} = SO_2Cl_2 + SOCl_2$. Similar bodies are produced by the action of chlorosulphonic acid on selenium and on titanium tetrachlorides at 100°. The first, $\mathbf{SO_2(OSeCl_3)Cl}$, is a yellow amorphous powder; the second, $\mathbf{SO_2(O.TiCl_3)Cl}$, melts at 165° and boils at 183°. Both of these bodies decompose when heated. Salts of the ordinary kind are unknown, because the acid is at once energetically attacked by water, yielding sulphuric and hydrochloric acids:—

$$SO_2(OH)Cl + H_2O = SO_2(OH)_2 + HCl.$$

(compare Chlorochromates, p. 268).

Anhydrosulphuryl chloride, $S_2O_5Cl_2$, corresponding to anhydrosulphuric acid, $S_2O_5(OH)_2$, is also known. It is produced by distilling phosphorus pentachloride with sulphur trioxide:— $\mathbf{PCl_5} + 2\mathbf{SO_3} = POCl_3 + S_2O_5Cl_2$; or with chlorosulphuric acid, $\mathbf{PCl_5} + 2SO_2(OH)Cl = S_2O_5Cl_2 + POCl_3 + 2HCl$. It is also formed when phosphoryl chloride and sulphur trioxide are heated in a sealed tube to 160°:—$2POCl_3 + 6\mathbf{SO_3} = 3S_2O_5Cl_2 + \mathbf{P_2O_5}$. It is a colourless liquid, of density 1·82, boiling at 153°. Its density is normal at 184°, but at 250° it decomposes, giving chlorine, and sulphur dioxide and trioxide.

No direct compounds of sulphur dioxide with halogen acids are known. But selenium and tellurium dioxides form the following:—$SeO_2.HCl$, an amber-coloured liquid, stable below 26°; $SeO_2.2HCl$, stable at −20°; $SeO_2.2HBr$, stable below 55°; $2SeO_2.5HBr$, stable at −25°; $TeO_2.HBr$, a brown solid, stable

* *Comptes rend.*, **96**, 646; *Berichte*, **16**, 479, 602.

† *Annalen*, **196**, 265; *Chem. Soc.*, **41**, 297.

at 15°; $TeO_2.2HBr$, stable at 14°, and decomposed at 40° into $TeO_2.2HBr$, which on further heating yields water and black needles of tellurosyl bromide, $TeOBr_2$. The compound $2TeO_2.3HCl$ is stable at −10°; on rise of temperature it yields $TeO_2.HCl$, which at 110° gives a white mass of $\mathbf{TeOCl_2}$. There is some reason to doubt the definite nature of these so-called compounds.

Sulphurosyl (thionyl), selenosyl, and tellurosyl halides, $SOCl_2$; $\mathbf{SeOCl_2}$; $\mathbf{TeOCl_2}$; $\mathbf{SeOBr_2}$; $\mathbf{TeOBr_2}$.—No fluorides or iodides are known.

Sulphurosyl chloride is prepared by passing sulphur dioxide over heated phosphorus pentachloride :—$SO_2 + \mathbf{PCl_5} = POCl_3 + SOCl_2$; or by distilling calcium sulphite, $\mathbf{CaSO_3}$, with phosphoryl chloride. It is also obtained by distilling a mixture of sulphur chloride, S_2Cl_2, and sulphur trioxide, through which chlorine is being passed :—$SCl_4 + \mathbf{SO_3} = SOCl_2 + SO_2 + Cl_2$. It is a colourless liquid, boiling at 82°; it bears to sulphurous acid the same relation as sulphuryl chloride to sulphuric acid, as is shown by its action on water :—$SOCl_2 + 2H.OH = SO(OH)_2 + 2HCl$. The sulphurous acid, however, decomposes into water and the dioxide.

Selenosyl chloride, $SeOCl_2$, is produced by heating together selenium tetrachloride and selenium dioxide :—$\mathbf{SeCl_4} + \mathbf{SeO_2} = \mathbf{2SeOCl_2}$; by the action of water on selenium tetrachloride; or, most readily, by distilling selenium dioxide with sodium chloride, thus :—$\mathbf{2SeO_2} + \mathbf{2NaCl} = \mathbf{Na_2SeO_3} + \mathbf{SeOCl_2}$. It is a yellowish substance, melting at 10° and boiling at 179·5°. Its specific gravity is 2·44. The corresponding tellurium compounds have been obtained, as described, by heating the compounds of the dioxide with the halides of hydrogen.

Other acids of sulphur and selenium.—The following is a list :—

(1.) Thiosulphuric acid,*† $H_2S_2O_3$.
(2.) Seleniosulphuric acid,* H_2SSeO_3.
(3.) Hyposulphurous acid,* $H_2S_2O_4$.
(4.) Dithionic acid,‡ $H_2S_2O_6$.
(5.) Trithionic acid, $H_2S_3O_6$.
(6.) Seleniotrithionic acid,* $H_2S_2SeO_6$.
(7.) Tetrathionic acid, $H_2S_4O_6$.
(8.) Pentathionic acid, $H_2S_5O_6$.
(9.) Hexathionic acid (?), $H_2S_6O_6$.
(10.) Persulphuric acid (?), $H_2S_2O_8$.
(11.) Dithiopersulphuric acid,*§ $H_2S_4O_8$.

* These acids are unknown in the free state; but their salts have been prepared.

† Formerly named hyposulphurous acid.

‡ Also named hyposulphuric acid.

§ This name is a provisional one.

1. **Thiosulphuric acid,** $H_2S_2O_3$.—On adding dilute hydrochloric or sulphuric acid to a weak solution of the sodium salt, the acid appears to be liberated; but it decomposes almost immediately. Sulphur separates and sulphurous acid is formed, thus:—$H_2S_2O_3 = H_2SO_3 + \mathbf{S}$. But a secondary reaction appears to take place at the same time, for hydrogen sulphide may be recognised at first by its smell:—$H_2S_2O_3 = \mathit{H_2S} + \mathit{O} + \mathit{SO_2}$. This may indeed be the first stage of the decomposition, the nascent oxygen reacting with the hydrogen sulphide, giving water and sulphur.

$\mathbf{Na_2S_2O_3.5H_2O}$; $\mathbf{K_2S_2O_3.2H_2O}$; $\mathbf{3\{(NH_4)_2S_2O_3\}H_2O}$.

Sodium thiosulphate is produced by boiling a solution of sodium sulphite with sulphur, thus:—

$$Na_2SO_3.Aq + \mathbf{S} = Na_2S_2O_3.Aq,$$

or by boiling sulphur in a solution of sodium hydroxide:—$6NaOH.Aq + 12\mathbf{S} = 2Na_2S_5.Aq + Na_2S_2O_3.Aq + 3H_2O$; the potassium salt is prepared similarly; and they may be obtained by adding a solution of the respective carbonate to a solution of calcium thiosulphate, $CaS_2O_3.Aq + M_2CO_3.Aq = M_2S_2O_3.Aq + \mathbf{CaCO_3}$; insoluble calcium carbonate is precipitated, and the soluble thiosulphate remains dissolved.

These are very soluble white salts. The **sodium salt** forms large monoclinic crystals. It is made use of as an "antichlore;" cloth bleached with chloride of lime its dipped in its solution to remove adhering chlorine, which might attack the fibre. It reacts with the halogens, thus:—$2Na_2S_2O_3.Aq + \mathit{Cl_2} = 2NaCl.Aq + Na_2S_4O_6.Aq$; sodium tetrathionate is formed. It is also used in ' fixing" photographic negatives or prints (see Silver Thiosulphate).

When heated, sodium thiosulphate yields sulphate and pentasulphide, $\mathbf{4Na_2S_2O_3 = 3Na_2SO_4 + Na_2S_5}$.

$\mathbf{CaS_2O_3.6H_2O}$; $\mathbf{SrS_2O_3.6H_2O}$; $\mathbf{BaS_2O_3.H_2O}$.

Calcium thiosulphate is prepared on a large scale from "soda-waste," which is essentially a sulphide of calcium. It is exposed to the air in a moist state for some days, when the sulphide is partially converted into sulphite, $CaSO_3$. At the same time sulphur is liberated, probably by the action of atmospheric carbonic acid and oxygen, and it reacts with the sulphite, forming thiosulphate. On treating the oxidised waste with water, a solution of sulphite and thiosulphite is obtained. The sulphite deposits in crystals on evaporation; they are removed, and the thiosulphate crystallises from the mother liquor. Calcium thiosulphate is the usual source of the thiosulphates generally. It crystallises in large clear triclinic prisms.

Strontium and barium thiosulphates are precipitated on mixing solutions of the respective chlorides with sodium thiosulphate. The precipitation is completed by adding alcohol. They are white, sparingly soluble salts.

The double salt $\mathbf{CaNa_2(S_2O_3)_2.}n\mathbf{H_2O}$ is produced by treating calcium sulphate with a solution of sodium thiosulphate. It is a soluble salt. Barium and strontium sulphates do not give this reaction, and it therefore affords a means of separating calcium from the sulphates of these metals.

$\mathbf{MgS_2O_3.6H_2O}$; $\mathbf{ZnS_2O_3.}n\mathbf{H_2O}$; $\mathbf{CdS_2O_3.}n\mathbf{H_2O}$.

These are very soluble salts. The two last may be produced along with sulphite by passing sulphur dioxide through water in which the sulphides are suspended:—$\mathbf{ZnS} + H_2SO_3.Aq = \mathbf{ZnSO_3} + H_2S.Aq$; $2H_2S + SO_2 = 2H_2O.Aq + 3\mathbf{S}$; $\mathbf{ZnSO_3} + Aq + \mathbf{S} = ZnS_2O_3.Aq$. Solutions of the zinc and cadmium salts are decomposed by heat into sulphuric and sulphurous acids and zinc sulphide and sulphate. The double salt, $\mathbf{K_2Mg(S_2O_3)_2.6H_2O}$, is prepared by mixture.

The thiosulphates of the boron group of elements are unknown, as are also those of the aluminium group, with one exception; a double thiosulphate of thallium and sodium, $\mathbf{Na_6Tl_4(S_2O_3)_5.10H_2O}$, is produced by mixture; it forms fine silky needles.

Chromic and ferric thiosulphates are unknown.

$\mathbf{2FeS_2O_3.5H_2O}$; $\mathbf{CoS_2O_3.6H_2O}$; $\mathbf{NiS_2O_3.6H_2O}$.

The manganese salt has not been obtained solid; it decomposes on concentration into sulphur, sulphur dioxide, and manganous sulphide. The ferrous salt is formed, along with sulphite, by dissolving iron in sulphurous acid. On evaporation, the less soluble sulphite crystallises first; the thiosulphate separates on concentrating the mother liquor. These salts are all soluble and unstable.

The only known thiosulphates of an element of the carbon group are those of zirconium and thorium. The former is precipitated as thiosulphate, $\mathbf{Zr(S_2O_3)_2}$ (?), by boiling a solution of its chloride with sodium thiosulphate. This precipitation serves as a means of separating zirconium from yttrium, the cerium metals, and iron. On ignition of the white precipitate, pure zirconia is left. Thorium thiosulphate is also a white precipitate.

Thiosulphates of the elements of the silicon group are unknown, with the exception of that of lead, $\mathbf{PbS_2O_3}$. It is a white precipitate, very sparingly soluble in water, but dissolving in solutions of thiosulphates of other metals, forming, for example, $\mathbf{K_2Pb(S_2O_3)_2}$, $\mathbf{BaPb(S_2O_3)_2}$, &c.

Thiosulphates of elements of the nitrogen and phosphorus groups have not been prepared, with the exception of the double salts with bismuth, which have formulæ such as $\mathbf{K_3Bi(S_2O_3)_3.H_2O}$. They are very soluble in water and also in alcohol, in which the simple salts are nearly insoluble; they are thrown down on adding a solution of potassium chloride.

No thiosulphate of molybdenum, tungsten, or uranium is known.

Platinum forms a double thiosulphate with sodium, of the formula $\mathbf{Na_6Pt''(S_2O_3)_4.10H_2O}$. It is precipitated by alcohol from a mixture of ammonium platinochloride and sodium thiosulphate. It forms yellow crystals.

Double salts of copper, silver, and gold with sodium thiosulphate are also known. $\mathbf{KCu'S_2O_3.H_2O}$ precipitates on adding potassium thiosulphate to *cupric* sulphate, as a yellow precipitate, which rapidly changes to cuprous sulphide. With more potassium thiosulphate, the salt $\mathbf{K_3Cu(S_2O_3)_2}$ is precipitated on addition of alcohol. Similar sodium salts are known. Silver thiosulphate is exceedingly unstable, giving sulphide; but two varieties of double salt are known, produced by dissolving silver oxide in a solution of a thiosulphate, or by dissolving silver chloride, nitrate, &c., in a solution of an alkaline thiosulphate. These are $\mathbf{R_4Ag_2(S_2O_3)_3}$ and $\mathbf{RAgS_2O_3}$, R standing for a monad metal. Such a double salt is formed during the "fixing" of photographic negatives and prints. That portion of the silver bromide or iodide not exposed to light, and not reduced to the metallic state by treatment with the "developer," is removed from the plate or paper by immersion in a solution of sodium thiosulphate, or, as it is familiarly termed, "hypo." Salts of the first series are easily soluble in water; hence the necessity of using excess of thiosulphate in fixing, else a salt of the second series is formed, which is insoluble.

A double thiosulphate of gold and sodium is prepared by mixing solutions of auric chloride and sodium thiosulphate and adding alcohol; the barium salt, $\mathbf{BaAu_2(S_2O_3)_4}$, is insoluble, and is formed from the sodium salt, $\mathbf{NaAu(S_2O_3)_2}$, by double decomposition.

A double mercuric salt, $\mathbf{K_{10}Hg''_3(S_2O_3)_8}$, is produced by dissolving mercuric oxide in a solution of potassium thiosulphate. It forms sparingly soluble white prisms.

The chief insoluble thiosulphates are those of barium and lead.

The tendency of copper, silver, gold, and mercury to form double thiosulphates should be remarked.

The estimation of a thiosulphate depends on its action on free iodine dissolved in a solution of potassium iodide, whereby an iodide and a tetrathionate are produced, thus:—$2Na_2S_2O_3.Aq + I_2 = Na_2S_4O_6.Aq + 2NaI.Aq$. The amount of thiosulphate is easily calculated from the amount of iodine employed.

2. Closely allied to the thiosulphates are the **selenio-sulphates**, formed by boiling a sulphite with selenium. The potassium and sodium salts are crystalline bodies thus prepared; on addition of a cadmium salt insoluble cadmium selenio-sulphate, $\mathbf{CdSSeO_3}$, is precipitated; but most selenio-sulphates of metals decompose, a selenide or selenium being precipitated. Thioselenates, which it might be supposed would be formed on boiling sulphur with a solution of a selenite, do not appear to exist (see below, *Constitution of Sulphur Acids*).

3. **Hyposulphurous acid,** $H_2S_2O_4$.—This acid is said to be produced by the action of zinc on an aqueous solution of sulphurous acid. No hydrogen is evolved, and the solution acquires a brownish-yellow colour and great reducing power. Its tendency to unite with free oxygen is such that it turns warm on exposure to air. The sodium salt is better known. It is prepared by digesting in a closed vessel in the cold finely-divided zinc with a concentrated solution of hydrogen sodium sulphite, $HNaSO_3$. Its formation is expressed by the equation:—

$$\mathbf{Zn} + 4HNaSO_3.Aq = \mathbf{Na_2Zn(SO_3)_2} + Na_2S_2O_4.Aq + 2H_2O.$$

The zinc-sodium sulphite separates out on addition of alcohol, and on cooling the remaining liquid slender needles separate, still containing a little zinc, from which they may be freed by dissolving in water and again mixing the solution with alcohol.

M. Schützenberger, the discoverer of these compounds, believed them to have the formulæ H_2SO_2 and $HNaSO_2$ respectively.* But the formulæ appear to be $H_2S_2O_4$ and $Na_2S_2O_4$, for the following reasons:—The action of a dilute solution of the sodium salt on a solution of copper sulphate in ammonia is to yield sodium sulphite and a cuprous compound. Now it has been shown that for every two atoms of sulphur contained in sodium hyposulphite, two molecules of copper sulphate are reduced. This implies the gain of *one*, not two, atoms of oxygen for every two atoms of sulphur. If, for instance, the formula were $HNaSO_2$, the oxidation by means

* *Bull. Soc. Chim.*, **12**, 121; **19**, 152; **20**, 145; Bernthsen, *Berichte*, **14**, 438; *Annalen*, **211**, 285.

of cupric oxide would be expressed by the equation $HNaSO_2 + 2CuO = HNaSO_3 + Cu_2O$. But then two atoms of copper would be equivalent to one atom of sulphur. The reaction is in fact $Na_2S_2O_4 + 2CuO + H_2O = Cu_2O + 2NaHSO_3$.

Again, iodine in presence of water converts this salt into sulphate; and if its formula were $HNaSO_2$ four atoms of iodine would be required for each atom of sulphur, thus:—$HNaSO_2 + 2H_2O + 2I_2 = HNaSO_4 + 4HI$. But it is found that in actual fact three atoms of iodine for each atom of sulphur are necessary; hence the equation $Na_2S_2O_4 + 3I_2 + 4H_2O = 2NaI + 4HI + 2H_2SO_4$. The formula of the acid must therefore be $H_2S_2O_4$, and not H_2SO_2. Other salts have not been investigated.

The sodium salt when added in excess to copper sulphate gives a precipitate of copper hydride, $\mathbf{Cu_2H_2}$ (see pp. 382 and 577). A solution of the sodium salt, as has been remarked, absorbs free oxygen, and on this fact is founded a method of estimating oxygen dissolved in water, the end point of the reaction being denoted by the action of the sodium salt on indigo, used as an indicator. A quantity of indigo solution is decolorised by addition of hyposulphite solution of known reducing power, ascertained by its reaction with ammoniacal copper sulphate; the solution of free oxygen is then added, the indigo-white being thereby partially reconverted into indigo-blue; and the unoxidised indigo is again decolorised by addition of hyposulphite solution.

The impure calcium salt in aqueous solution is employed in the arts, in dyeing with indigo. Insoluble indigo-blue is by its means combined with hydrogen, and thereby converted into soluble indigo-white. The goods are then dyed, and, on exposure to air, the indigo is again oxidised, and changes to insoluble blue. The dye-bath is named the "hyposulphite vat."

4. **Dithionic or hyposulphuric acid,** $H_2S_2O_6$.—The manganous salt of this acid, $\mathbf{MnS_2O_6.6H_2O}$, is produced by passing a current of sulphur dioxide through water, kept cold, and containing manganese dioxide in suspension. Dithionate and sulphate of manganous are produced, thus:—

$$\mathbf{MnO_2} + 2SO_2.Aq = MnS_2O_6.Aq; \quad \mathbf{MnO_2} + SO_2.Aq = MnSO_4.Aq.$$

To separate the dithionate and sulphate, a solution of barium hydroxide is added to the solution. White barium sulphate and manganese hydrate are precipitated as insoluble powders, while barium dithionate goes into solution:—$MnS_2O_6.Aq + Ba(OH)_2 = BaS_2O_6.Aq + \mathbf{Mn(OH)_2}$. From the barium salt a solution of the acid may be obtained by careful addition of dilute sulphuric acid.

It may be concentrated in a vacuum over sulphuric acid till it acquires the specific gravity 1·35; if an attempt be made to concentrate further, it decomposes into sulphuric acid and sulphur dioxide.

Dithionic acid is a syrupy strongly acid liquid. Its salts are prepared by addition of the required sulphate to the barium salt. They are as follows:—

$\mathbf{Li_2S_2O_6.2H_2O}$; $\mathbf{Na_2S_2O_6.2H_2O}$; $\mathbf{K_2S_2O_6}$; $\mathbf{Rb_2S_2O_6}$; $\mathbf{(NH_4)_2S_2O_6.H_2O}$.

These are all colourless soluble crystals, insoluble in alcohol.

$\mathbf{CaS_2O_6.4H_2O}$; $\mathbf{SrS_2O_6.H_2O}$; $\mathbf{BaS_2O_6.2}$ and $\mathbf{4H_2O}$; and the double salts $\mathbf{Na_2Ba(S_2O_6)_2.4}$ and $\mathbf{6H_2O}$.

These salts are all soluble.

$\mathbf{MgS_2O_6.6H_2O}$; $\mathbf{ZnS_2O_6.6H_2O}$; $\mathbf{CdS_2O_6}$.—$\mathbf{MgBa(S_2O_6)_2.4H_2O}$.

Yttrium dithionate has been prepared; and also the aluminium salt, $\mathbf{Al_2(S_2O_6)_3.18H_2O}$.

$\mathbf{Tl_2S_2O_6}$ forms crystals isomorphous with $\mathbf{K_2S_2O_6}$.

$\mathbf{Cr_2(S_2O_6)_3.18H_2O}$; the ferric salt is basic.
$\mathbf{FeS_2O_6.5}$ and $\mathbf{7H_2O}$; $\mathbf{MnS_2O_6.6H_2O}$; $\mathbf{CoS_2O_6.6}$ and $\mathbf{8H_2O}$; $\mathbf{NiS_2O_6.6H_2O}$.

Ceric hydrate is insoluble in dithionic acid. $\mathbf{Th(S_2O_6)_2.4H_2O}$ is very unstable. Lead dithionate, $\mathbf{PbS_2O_6.4H_2O}$, and the basic salt, $\mathbf{(Pb_2O)S_2O_6}$, are crystalline.

No compounds of the elements of the nitrogen group are known; and bismuthyl dithionate, $\mathbf{(BiO)_2S_2O_8}$, is the only representative of the phosphorus group.

Three basic uranyl salts, $\mathbf{6UO_2.S_2O_5.10H_2O}$; $\mathbf{7UO_2.S_2O_5.8H_2O}$; and $\mathbf{8UO_2.S_2O_5.21H_2O}$, have also been made. No salts of metals of the palladium or platinum groups are known.

$\mathbf{Ag_2S_2O_6.2H_2O}$; $\mathbf{NaAgS_2O_6.2H_2O}$ and $\mathbf{Tl_4Ag_2(S_2O_6)_3}$, are all soluble. $\mathbf{CuS_2O_6.4H_2O}$; $\mathbf{HgS_2O_6.4H_2O}$; and basic cupric and mercuric salts exist.

The dithionates are almost all crystalline and soluble in water.*

5. **Trithionic acid.**—The potassium salt of this acid is produced by digesting at a gentle heat a solution of hydrogen potassium sulphite with sulphur:—$6HKSO_3.Aq + \mathbf{S_2} = 2K_2S_3O_6.Aq + K_2S_2O_3.Aq + 3H_2O$. The warm filtered solution deposits the trithionate in crystals. Also, by passing sulphur dioxide through a saturated solution of potassium thiosulphate mixed with alcohol:—

$$2K_2S_2O_3.Aq + 3SO_2 = 2K_2S_3O_6.Aq + \mathbf{S}.$$

* *Annalen*, **246**, 179 and 284.

It is crystallised from dilute alcohol. Also by passing sulphur dioxide through a mixture of solution of hydrogen potassium sulphite and potassium sulphide:—$4HKSO_3.Aq + K_2S.Aq + 4SO_2 = 3K_2S_3O_6.Aq + 2H_2O$.

The acid is produced by substituting hydrogen for the potassium in the potassium salt by addition of hydrosilicifluoric acid, $H_2SiF_6.Aq$, which forms an insoluble salt of potassium, $\mathbf{K_2SiF_6}$. When dilute the acid is stable; but on attempting to concentrate it, even at 0°, it evolves sulphur dioxide, deposits sulphur, and sulphuric acid remains in solution:—$H_2S_3O_6 = H_2SO_4 + SO_2 + \mathbf{S}$.

Potassium trithionate, in aqueous solution, soon decomposes into pentathionate, sulphate, and sulphur dioxide.

The trithionates are very unstable, and appear to be all soluble in water. The following have been prepared:—

$\mathbf{K_2S_3O_6}$; $\mathbf{(NH_4)_2S_3O_6}$; $\mathbf{BaS_3O_6.2H_2O}$; $\mathbf{ZnS_3O_6}$; $\mathbf{Tl_2S_3O_6}$, and $\mathbf{KCuS_3O_6}$.

The sodium salt cannot be prepared like that of potassium.

6. **Seleniotrithionic acid,** $H_2S_2SeO_6$. The potassium salt of this acid is formed by digesting selenium with potassium sulphite, or with hydrogen potassium sulphite; or by evaporating together a mixture of solutions of hydrogen potassium sulphite and potassium seleniosulphate, K_2SSeO_3. It is most easily obtained by mixing a solution containing hydrogen potassium sulphite and thiosulphate with selenious acid. The liquid becomes warm, and the potassium salt then crystallises out in needles on cooling. The salt $K_2S_2SeO_6$ is stable in solution for some time, but gradually decomposes, forming partly potassium dithionate and free selenium, and partly selenium, potassium sulphate, and sulphurous acid.

7. **Tetrathionic acid,** $H_2S_4O_6$.—Tetrathionates are produced by adding iodine in small successive portions to the solution of thiosulphates, thus:—$2Na_2S_2O_3.Aq + \mathbf{I_2} = 2NaI.Aq + Na_2S_4O_6.Aq$. They are precipitated by addition of alcohol. In this manner tetrathionates of sodium, potassium, strontium, and barium have been prepared. Also, by adding dilute sulphuric acid to a mixture of lead thiosulphate and lead peroxide:—$2PbS_2O_3.Aq + \mathbf{PbO_2} + 2H_2SO_4.Aq = PbS_4O_6.Aq + \mathbf{2PbSO_4} + 2H_2O$. The acid is obtained by treating the solution of the lead salt with dilute sulphuric acid, filtering from the precipitated lead sulphate, and evaporating in a vacuum. When heated, it decomposes into sulphuric and sulphurous acids and free sulphur. Its solution is colourless, and has a strong acid taste. When heated with hydrochloric acid, it gives off hydrogen sulphide.

The tetrathionates are all soluble, but are precipitated by alcohol from their aqueous solutions. They crystallise well; and when heated they give a sulphate or a sulphide, sulphur dioxide, and free sulphur. The solution of the potassium salt, on standing, contains a mixture of trithionate and pentathionate. Those prepared are as follows:

$\mathbf{Na_2S_4O_6}.n\mathbf{H_2O}$; $\mathbf{K_2S_4O_6}$; $\mathbf{SrS_4O_6.6H_2O}$; $\mathbf{BaS_2O_6.2H_2O}$; $\mathbf{CdS_4O_6}$; $\mathbf{FeS_4O_6}$; $\mathbf{NiS_4O_6}$; $\mathbf{PbS_4O_5}$; and $\mathbf{Cu_2S_4O_6}$.

The last is obtained when a solution of barium thiosulphate is digested with copper sulphide.

8. **Pentathionic acid,** $H_2S_5O_6$.*—On passing a slow current of hydrogen sulphide through a weak solution of sulphurous acid at 0°, sulphur is deposited, and a solution is obtained containing *liquid* sulphur, sulphuric acid, a trace of trithionic acid, tetra-, penta-, and an acid containing still more sulphur, probably hexathionic acid. It is also said to contain *dissolved sulphur*, which can be precipitated by a solution of potassium nitrate. Such a solution is known as "Wackenroder's solution." It may be concentrated over sulphuric acid. To prepare a pentathionate from it, very dilute potash is added with constant stirring; potassium pentathionate is at once decomposed by excess of alkali; but it is stable in, and may be recrystallised from, acid solution. Salts may also be prepared by mixing the concentrated Wackenroder's solution with acetates, and leaving to evaporate. The acetic acid evaporates away, leaving the thionate in a crystalline condition. The potassium and the copper salts have been analysed; the former has the formula $\mathbf{2K_2S_5O_6.3H_2O}$; the latter $\mathbf{CuS_5O_6.4H_2O}$. They crystallise well.

The first action between hydrogen sulphide and sulphurous acid appears to result in the formation of tetrathionic acid:—$3SO_2.Aq + \mathit{H_2S} = H_2S_4O_6.Aq$. Tetrathionic acid and free sulphurous acid form trithionic and thiosulphuric acids, thus:—$H_2S_4O_6.Aq + H_2SO_3.Aq = H_2S_3O_6.Aq + H_2S_2O_3.Aq$. The thiosulphuric acid, being unstable, gives up its sulphur; the nascent sulphur adds itself to the trithionic acid, forming pentathionic acid, while much of the sulphur separates in the free state. By the action of excess of hydrogen sulphide for a long time, the equation $2\mathit{H_2S} + \mathit{SO_2} = 2\mathit{H_2O} + 3\mathbf{S}$ is realised. The action of the nascent sulphur from the decomposing thiosulphuric acid appears also to give rise to

* Debus, *Chem. Soc.*, **53**, 278.

9. **Hexathionic acid,** $H_2S_6O_6$, the potassium salt of which separates from the mother liquors of the pentathionate in a nearly pure state, in white wart-like masses.

10. The sodium salt of **dithiopersulphuric acid,** $H_2S_4O_8$,* is said to be produced by saturating a solution of sodium thiosulphate, containing more sodium thiosulphate than it can dissolve, with sulphur dioxide. The process is repeated for several days, the solution being occasionally allowed to stand at rest. On evaporation over sulphuric acid, anhydrous crystals of $\mathbf{Na_2S_4O_8}$ separate out. They crystallise from water with $2H_2O$. The equation suggested is $2Na_2S_2O_3 + 5SO_2 = 2Na_2S_4O_8 + S$. It may be noticed that such a body is the analogue of hexathionic acid, two atoms of sulphur being replaced by oxygen.

Constitution of the acids of sulphur and selenium.—The constitution of the sulphates and pyrosulphates has already been discussed. It is probable that that of the selenates and tellurates is similar; and it now remains to discuss the other compounds which have not yet been considered.

Just as there are probably two nitrous acids (see p. 337) and two phosphorous acids (see p. 375), so theory leads to the suggestion that two sulphurous acids are also capable of existence.†

Their formulæ should be—

$$O{=}S{<}^{OH}_{OH} \quad \text{and} \quad {}^{O}_{O}{>}S{<}^{OH}_{H}.$$

Now sulphurosyl chloride, $SOCl_2$, cannot be conceived other than $O{=}S{=}Cl_2$; on treating it with water, it would naturally follow that $O{=}S(OH)_2$ should be formed. And if, instead of acting on it with water, alcohol or ethyl hydroxide, $(C_2H_5)OH$ be chosen, the corresponding sulphite of ethyl $(C_2H_5)'$, viz., $O{=}S(OC_2H_5)_2$, should result.‡ This is, in fact, the case. And, moreover, ethyl sulphite reacts with boiling caustic potash, producing potassium sulphite and ethyl hydroxide, thus:—

$$O{=}S(OC_2H_5)_2 + 2HOK = 2HOC_2H_5 + K_2SO_3.$$

It might be expected that the same compound, ethyl sulphite, would be produced by heating a sulphite with iodide of ethyl, $(C_2H_5)I$, thus:—

$$O{=}S(ONa)_2 + 2I(C_2H_5) = 2NaI + O{=}S(OC_2H_5)_2.$$

* *Compt. rend.*, **106**, 851, 1354.

† Divers, *Chem. Soc.*, **47**, 205; **49**, 533. Röhrig, *J. prakt. Chem.* (2), **37**, 217.

‡ The compound actually used is sodium ethoxide, C_2H_5ONa.

But the product is a different body. It has a higher boiling point than ethyl sulphite, and, moreover, on boiling with an alkali, a different change occurs; only one ethyl group is replaced by the alkaline metal, and a salt termed an ethyl-sulphonate is produced.

The conclusion follows, therefore, that sodium sulphite has a constitution different from that of ethyl sulphite. The alternative formula, $(O_2)^{IV}S(OH)H$, is therefore adopted, and the formulæ for these bodies are, therefore:—

$$\begin{matrix}O\\O\end{matrix}\!\!>\!S\!<\!\begin{matrix}O(C_2H_5)\\(C_2H_5)\end{matrix};\quad \begin{matrix}O\\O\end{matrix}\!\!>\!S\!<\!\begin{matrix}ONa\\(C_2H_5)\end{matrix};\ \text{and}\ \begin{matrix}O\\O\end{matrix}\!\!>\!S\!<\!\begin{matrix}OH\\H\end{matrix}.$$

Ethyl-sulphonate of ethyl. Ethyl-sulphonate of sodium. Sulphurous acid.

This view of the constitution of ethyl-sulphonate of sodium is confirmed by the relation of this body to ethyl hydrosulphide, $(C_2H_5)SH$, a compound analogous to alcohol, which is ethyl hydroxide, $(C_2H_5)OH$. On oxidising ethyl hydrosulphide, ethyl-sulphonic acid is produced:—

$$(C_2H_5)SH + 3O = (C_2H_5)SO_3H.$$

And, conversely, by reducing ethyl sulphonyl chloride, $(C_2H_5)SO_2Cl$, with nascent hydrogen, ethyl hydrosulphide is formed, thus:—

$$(C_2H_5)S(O_2)^{IV}Cl + 6H = (C_2H_5)SH + 2H_2O + HCl.$$

Other considerations lead to the same conclusion, but the proof given is the most important, because it is the most direct. It must then be concluded that, when sulphurosyl chloride is decomposed by water, the sulphurous acid originally formed, $O{=}S(OH)_2$, undergoes molecular rearrangement, and changes into sulphonic acid, $(O_2)S(OH)H$. It has also been ascertained that two sodium potassium sulphites, $NaKSO_3$, exist. These are $KO.SO_2Na$ and $NaO.SO_2K$, and they differ in properties.*

Selenious acid, on the contrary, appears to have the formula $O{=}Se(OH)_2$;† for by acting on selenosyl chloride, $O{=}SeCl_2$, with ethyl hydroxide, or by heating together sodium selenite and ethyl iodide, the same product is obtained, viz., $O{=}Se(OC_2H_5)_2$; this is known because it reacts with caustic soda, forming selenite again, thus: $O{=}Se(OC_2H_5)_2 + 2HONa = O{=}Se(ONa)_2 + 2HO(C_2H_5)$. There appears, therefore, to be only one selenious acid, $O{=}Se(OH)_2$.

The constitution of tellurous acid has not been investigated.

* *Berichte*, **22**, 1729.

† *Ibid.*, **13**, 656.

Thiosulphuric acid is thus named because it may be regarded as sulphuric acid, H_2SO_4, of which one atom of oxygen has been replaced by an atom of sulphur, *θεῖον*. Here again alternative formulæ are possible. The oxygen may be replaced in the group $(SO_2)''$, or in one of the hydroxyl groups. The alternative formulæ are :—

$$\begin{matrix}S\\O\end{matrix}\!\!>\!S\!<\!\begin{matrix}OH\\OH\end{matrix},\ \text{and}\ \begin{matrix}O\\O\end{matrix}\!\!>\!S\!<\!\begin{matrix}OH\\SH\end{matrix}.$$

Preference is given to the latter formula, for this among other reasons: an aqueous solution of sodium thiosulphate, when boiled with ethyl bromide, forms sodium ethyl thiosulphate,* thus :—

$$Na_2S_2O_3.Aq + (C_2H_5)Br = Na(C_2H_5)S_2O_3.Aq + NaBr.Aq.$$

On mixing this salt with barium chloride, it is to be presumed that barium ethyl thiosulphate is produced. This compound is unstable, and in a few hours decomposes into barium dithionate, BaS_2O_6, and ethyl disulphide, thus showing that the ethyl group, (C_2H_5), was attached to sulphur, not to oxygen, thus :—

$$O_2S\!<\!\begin{matrix}S(C_2H_5)\ \ (C_2H_5)S\\O\text{———}Ba\text{———}O\end{matrix}\!>\!SO_2 = BaS_2O_6 + (C_2H_5)_2S_2.\dagger$$

As regards hyposulphurous acid, too little is known regarding it to establish any formula as probable; the formula $O_2S\!\diagup\!\overline{K\ K}\!\diagdown\!SO_2$ has been suggested.

The constitution of dithionic acid follows from the decomposition of barium ethyl-thiosulphate. It may be regarded as certain that ethyl, once attached to sulphur, will not readily leave it; the constitution of barium dithionate is therefore,

$$Ba\!<\!\begin{matrix}O-SO_2\\O-SO_2\end{matrix}\!>,\ \text{and probably}\ Ba\!\begin{matrix}\diagup O-S(O_2)\\ \quad\quad\ \ | \\ \diagdown O-S(O_2)\end{matrix};$$

although the smooth decomposition of this body into barium sulphate and sulphur dioxide might lead to the conjecture that the two sulphuryl groups are united by virtue of their oxygen atoms. But this argument may have little value.

Seleniosulphuric acid, H_2SSeO_3, has doubtless the constitution $O_2S\!<\!\begin{matrix}OH\\SeH\end{matrix}$, for the isomeric acid, $O_2Se\!<\!\begin{matrix}OH\\SH\end{matrix}$, cannot be prepared by

* *Berichte*, **7**, 646.
† *Chem. Soc.*, **28**, 687.

the action of sulphur on sodium selenite, which, as has been pointed out, has the constitution $SeO(OH)_2$.

It is useless, in the present state of our knowledge, to construct constitutional formulæ for the remaining thionic acids. The possible formulæ are discussed by Debus (*Transactions of the Chemical Society*, 1888, p. 354); but no decided reason has yet been adduced for giving preference to one formula over others.

The relation between the so-called dithiopersulphuric acid and hexathionic acid has already been suggested (p. 452).

Nitrososulphates.—By passing a current of nitric oxide, *NO*, through a cooled solution of ammonium or potassium sulphites, two molecules of the oxide unite with each molecule of the sulphite, forming crystalline compounds, possessing respectively the formulæ $\mathbf{(NH_4)_2SO_3(NO)_2}$ and $\mathbf{K_2SO_3(NO)_2}$. The method of formation would suggest an analogy with the thiosulphates. It has been suggested that because, when exposed to the action of nascent hydrogen from sodium amalgam in strong alkaline solution, these salts yield a hyponitrite and a sulphite, they are constituted similarly to thiosulphates, viz., $O{=}S{<}^{OK}_{(NO)_2K}$. The equation representing their reduction would therefore be:—

$$(KSO_3)(NO)_2K + 2Na = (SO_3K)Na + NaNO + KNO.$$

But that the reduction of nitric oxide in alkaline solution should yield a hyponitrite is to be expected, and the argument is of little value. These compounds would repay further investigation.

Compounds of sulphur, selenium, and tellurium with each other.—It is questionable whether the bodies produced by fusing sulphur and selenium together are mixtures or compounds. A yellow compound or mixture is produced by passing a current of hydrogen sulphide through a solution of selenious acid. It is probable that this reaction is analogous to that of hydrogen sulphide on sulphurous acid, and if so it must be a very complicated one. A brick-red solid is formed when selenium and sulphur are fused together in the proportion of SeS_3. Sulphur and selenium crystallise together from hot benzene in orange crystals, but the ratio between the two is indefinite.

Similarly, an iron-grey mass is formed when selenium and tellurium are melted together.

The compounds of sulphur with tellurium are more definite. Hydrogen sulphide produces in acidified solutions of tellurites a dark-brown precipitate of $\mathbf{TeS_2}$, soluble in solutions of sulphides of the alkalies, forming sulphotellurites. These bodies are also formed when hydrogen sulphide is passed through a solution of a tellurite. The sodium and lithium salts are amorphous pale-yellow masses. The potassium salt, $\mathbf{K_6TeS_5}$, separates in pale yellow needles when its solution is evaporated in a vacuum. The ammonium salt, $\mathbf{(NH_4)_6TeS_5}$, crystallises in pale yellow quadratic prisms. The salts of calcium, strontium, and barium are prepared by boiling solutions of the corresponding sulphides with tellurium disulphide. The barium salt crystallises well.

The other sulphotellurites are obtained by precipitation.

The following have been analysed :—

$\mathbf{Zn_3TeS_5}$; $\mathbf{Pt_3(TeS_5)_2}$; $\mathbf{Ag_6TeS_5}$; $\mathbf{Au_2TeS_5}$; $\mathbf{Hg_6TeS_5}$; and $\mathbf{Hg^{II}_3TeS_5}$.

They are brown or black insoluble bodies.

Tellurium trisulphide, $\mathbf{TeS_3}$, is deposited in lustrous dark-grey spangles from telluric acid saturated with hydrogen sulphide. The sulphotellurates are produced by substituting a tellurate for telluric acid, and filtering off the precipitated trisulphide. The salts of sodium and potassium are yellow and crystalline. Their formulæ are unknown, but it is probable that their investigation would throw light on the constitution of sulphuric, selenic, and telluric acids.

Ortho-acids.—Analogy with orthocarbonic acid (see p. 292), orthosilicic acid (see p. 306), and phosphoric acid (see p. 353) would lead to the supposition that orthosulphuric acid should possess the formula $S(OH)_6$, corresponding to the theoretical $P(OH)_5$ and $C(OH)_4$ (of which, however, the ethyl salt is known), and the known $Si(OH)_4$. It is, indeed, possible that the acid containing two molecules of water, $H_2SO_4.2H_2O$, may be hydrogen orthosulphate. But no other salts are known. The first anhydride of such an acid would be the crystalline monohydrated acid, $H_2SO_4.H_2O = SO(OH)_4$; but, again, metals do not appear to replace hydrogen. The second anhydride, $SO_2(OH)_2$, is the ordinary acid, which forms numerous salts.

Similarly, orthosulphurous acid would correspond with the unknown orthosilicoformic acid, $H.Si(OH)_3$, and the likewise unknown orthoformic acid, $H.C(OH)_3$, of which, however, the ethyl salts are in both cases known. The formula of orthosulphurous acid would, therefore, be $H.S(OH)_5$. It is unknown, nor have any derivatives been prepared. But the sulphotellurites may be its sulphur-tellurium analogues, and have the constitution $M.Te(SM)_5$. It is not improbable that the sulphotellurates, on further investigation, should supply the link missing in orthosulphates.

It is thus evident that a systematic study of the rarer elements is greatly to be desired, inasmuch as light is thereby thrown on the relations of atoms with each other; in other words, on the *structure* of compounds.

Physical Properties.

Mass of one cubic centimetre:—

SO_2.	Temp.	−20·5°	−9·9°	−2·1°	0°	21·7°	35·2°	52°	62°	82·4°
	Mass	1·491	1·461	1·438	1·434	1·376	1·337	1·287	1·252	1·184

SO_2.	Temp.	102·4°	120·4°	130·3°	140·8°	146·6°	151·7°	154·3°
	Mass	1·104	1·017	0·956	0·869	0·806	0·732	0·671
	Temp.	155·05°	156·0°.					
	Mass	0·637	0·52.	Critical temp., 156°.				

SO_3, 1·936 gram at 20°. **SeO_2**, 3·954. **TeO_2**, 5·784 at 14°. **TeO_3**, 5·112 at 11°. H_2SO_4, 1·839 at 15°; 1·836 at 20°; 1·833 at 25°.—$H_2SO_4.H_2O$, 1·778 at 15°.—$H_2SO_4.2H_2O$, 1·651 at 15°.—$H_2SO_4.3H_2O$, 1·551 at 15.°—$H_2S_2O_7$, 1·9.—H_2SeO_4, 2·62.

	Li_2SO_4.	Na_2SO_4.	K_2SO_4.	Rb_2SO_4.	Cs_2SO_4.	$(NH_4)_2SO_4$.
	2·21	2·68	2·66	3·64	4·10	1·76
Se	—	3·21	3·08	3·90	4·34	2·20

	$BeSO_4$.	$CaSO_4$.	$SrSO_4$.	$BaSO_4$.	$MgSO_4$.	$ZnSO_4$.	$CdSO_4$.
	2·44	2·97	3·97	4·48	2·71	3·62	4·45
Se	—	2·93	4·23	4·75	—	—	—

	$Sc_2(SO_4)_3$.	$Y_2(SO_4)_3$.	$Al_2(SO_4)_3$.	$In_2(SO_4)_3$.	$Ce_2(SO_4)_3$.	(Tl_2SO_4).
	2·58	2·61	2·71	3·44	3·91	6·80
Se	—	—	—	—	—	—

	$Cr_2(SO_4)_3$.	$Fe_2(SO_4)_3$.	$FeSO_4$.	$MnSO_4$.	$CoSO_4$.	$NiSO_4$.
	3·01	3·01	3·35	3·28	3·47	3·42
Se	—	—	—	—	4·03	—

	$PbSO_4$.	$CuSO_4$.	$HgSO_4$.	Ag_2SO_4.	Hg_2SO_4.	H_2TeO_4.	$(NH_4)_2TeO_4$.
	6·00	3·61	6·47	5·49	7·56	3·42	3·00
Se	6·22	—	—	5·92	—		

Tl_2TeO_4.	$BaTeO_4$.
6·75	4·54

Heat of combination:—

*$\mathbf{S} + O + Aq = SOAq + 109K$.
$\mathbf{S} + 2O = SO_2 + 710K$; $+ Aq = 77K$. $SO_2 = SO_2 + 62K$.
$\mathbf{S} + 3O = \mathbf{SO_3} + 1033K$; $+ Aq = H_2SO_4.Aq + 392K$.
$2H_2SO_4.Aq + O = H_2S_2O_8.Aq - 283K$.
$2\mathbf{S} + 2O + Aq = H_2S_2O_3.Aq + 689K$.
$2\mathbf{S} + 2H + 6O + Aq = H_2S_2O_6.Aq + 2796K$.

* Rhombic; monoclinic 23K more.

$4\mathbf{S} + 5O + Aq = H_2S_4O_6.Aq + 1928K.$
$\mathbf{S} + O + 2Cl = SOCl_2 + 498K;\ SO_2 + 2Cl = SO_2Cl_2 + 187K.$
$2\mathbf{S} + 5O + 2Cl = S_2O_5Cl_2 + 1630K.$

$\mathbf{Se} + 2O = \mathbf{SeO_2} + 572K;\ + Aq = H_2SeO_3.Aq - 9K.$
$\mathbf{Se} + 3O + Aq = H_2SeO_4.Aq + 768K.$
$\mathbf{Te} + 2O + H_2O = \mathbf{H_2TeO_3} + 773K.$
$\mathbf{Te} + 3O + Aq = H_2TeO_4.Aq + 985K.$
$2NaOH.Aq + H_2SO_4.Aq = Na_2SO_4.Aq + 314K.$ Similarly for—

Li_2SO_4.	K_2SO_4.	Tl_2SO_4.	$CaSO_4$.	$SrSO_4$.	$BaSO_4$.	$(NH_4)_2SO_4$ all with Aq.
313K.	313K.	311K.	311K.	307K.	369K.	282K.

$MgSO_4.Aq$.	$ZnSO_4.Aq$.	$CdSO_4.Aq$.	$FeSO_4.Aq$.	$MnSO_4.Aq$.	$CoSO_4.Aq$.	$NiSO_4.Aq$.
311K.	235K.	238K.	249K.	266K.	247K.	263K.

$CuSO_4.Aq$.	$Al_2(SO_4)_3.Aq$.	$Cr_2(SO_4)_3Aq$.	$Fe_2(SO_4)_3.Aq$.
184K.	632K.	493K.	338K.

CHAPTER XXVIII.

COMPOUNDS OF THE HALOGENS WITH OXYGEN, SULPHUR, SELENIUM, AND TELLURIUM.—OXY-ACIDS OF THE HALOGENS; HYPOCHLORITES, CHLORITES, CHLORATES, AND PERCHLORATES; BROMATES, IODATES, AND PERIODATES.

In this group, as in the preceding, the compounds with oxygen present marked difference in most points from those with sulphur, selenium, and tellurium, which have already been described as halides on p. 167.

While fluorine forms no compound with oxygen, those of chlorine, bromine, and iodine are numerous; and the compounds of their oxides with other oxides are well defined, and have long been known. The following is a list of the oxides:—

Chlorine.	Bromine.	Iodine.
Cl_2O; ClO_2.*	(?)	I_2O_3(?)†; I_2O_5.

Sources.—Iodine pentoxide occurs in combination with sodium oxide as sodium iodate in *caliche*, the crude sodium nitrate found in Peru (see p. 325).

Preparation.—Chlorine monoxide is produced by passing a current of dry chlorine over dry precipitated mercuric oxide, contained in a tube cooled with ice. The chlorine combines with the mercury, forming an oxychloride, and with the oxygen, forming chlorine monoxide, thus:—$2\mathbf{HgO} + 2Cl_2 = \mathbf{Hg_2Cl_2O} + Cl_2O$. The gaseous monoxide is condensed in a freezing mixture of finely-powdered ice and salt. If a lower temperature can be obtained for condensation it should be employed, for the yield is thereby much increased.

Ordinary red oxide of mercury, owing to its compact nature, cannot be used in this preparation; the yellow variety, produced by addition of caustic soda to mercuric chloride, and dried at 400°, must be employed.

Compounds of the formulæ ClO and Cl_2O_3 are unknown.

Chlorine peroxide, ClO_2, is formed by heating chloric acid, $HClO_3$ (see p. 464).

* *Annalen*, **177**, 1; **213**, 113. *Berichte*, **14**, 28.

† *Compt. rend.*, **85**, 957.

The reaction is expressed by the equation :—$3HClO_3 = HClO_4 + 2ClO_2 + H_2O$; perchloric acid being formed simultaneously. It is more convenient to prepare it from potassium chlorate and concentrated sulphuric acid, yielding chloric acid, which decomposes as above. The temperature should not exceed 40°, else a dangerous explosion will result.

No oxides of bromine have been isolated.

Iodine trioxide, I_2O_3, is said to be produced by the action of ozone on iodine.

Iodine pentoxide, $\mathbf{I_2O_5}$, is produced by heating iodic acid, $\mathbf{HIO_3}$, to 170°. It is the anhydride of this acid :—$\mathbf{2HIO_3} = \mathbf{I_2O_5} + H_2O$.

Properties.—Chlorine monoxide, Cl_2O, is a yellowish-brown gas, condensing to a dark brown liquid,* which boils at 5° under a pressure of 738 mm. (about 6° under normal pressure). It is soluble in water, forming a yellow solution of hypochlorous acid; hence it is sometimes named **hypochlorous anhydride.** It is inadvisable to collect more than a drop or two in a test-tube, for it is exceedingly explosive. The gas can be exploded by gentle heat, by throwing into it a pinch of flowers of sulphur, or by contact with organic matter. Its density at 10° is normal.

Chlorine trioxide does not exist. The gas, formerly believed to be this substance, produced by the mutual action of nitric acid, potassium chlorate, and arsenic trioxide, has been shown to consist of the peroxide mixed with variable amounts of free chlorine.

Chlorine peroxide, ClO_2 (comp. nitric peroxide, NO_2), is a dark red liquid, boiling at 9° under a pressure of 730 mm. (about 10·6° at 760 mm.). It forms a reddish-brown gas, which explodes when heated, often, indeed, at the atmospheric temperature. Its density at 10·7° and 718 mm. was found to correspond with the formula ClO_2; it does not appear, therefore, to resemble nitric peroxide in forming a polymeride. With water it forms a mixture of chlorous and chloric acids.

Chloric acid and hydrogen chloride decompose one another to a great extent, giving a mixture of chlorine peroxide and free chlorine. This mixture, which is evolved by the action of hydrochloric acid diluted with its own volume of water on potassium chlorate, or by distilling a mixture of potassium chlorate, salt, and dilute sulphuric acid, was long believed to be a definite oxide of chlorine, and was named by Sir Humphrey Davy, its discoverer,

* *Berichte*, **16**, 2998; **17**, 157. *Annalen*, **230**, 273.

euchlorine. The equation expressing complete decomposition would be $HClO_3 + 5HCl = 3H_2O + 3Cl_2$; but the reaction is only a partial one, the chloric acid yielding perchloric acid and chlorine peroxide, as already described, mixed with variable quantities of chlorine.

Iodine pentoxide, I_2O_5, is a white solid, crystallising in the trimetric system. When heated to 180—200° it decomposes, without explosion, into iodine and oxygen. It combines with water, forming iodic acid.

Iodine forms no other oxides capable of free existence.

Compounds with water and other oxides.—The oxides described combine with water, forming compounds termed acids. They are as follows:—

(1.) $HClO$,* hypochlorous acid.	$MBrO$, hypobromite.	MIO, hypoiodite.
(2.) $HClO_2$,* chlorous acid.	—	—
(3.) $HClO_3$, chloric acid.	$MBrO_3$, bromate.	$\mathbf{HIO_3}$, iodic acid.
(4.) $HClO_4$, perchloric acid.	—	$\mathbf{H_5IO_6}$, periodic acid.

It is to be noticed that the perchloric and hypobromous acids and salts of bromic and hypoiodous acids are known, whereas the free oxides have not been prepared, owing to their instability.

1. **Hypochlorites, hypobromites,** and **hypoiodites** of the metals of the sodium and calcium groups are produced by the action of the halogen on solutions of the respective hydroxides, thus:—$Cl_2 + 2KOH.Aq = KCl.Aq + KOCl.Aq + H_2O$; $2Cl_2 + 2Ca(OH)_2.Aq = CaCl_2.Aq + Ca(OCl)_2.Aq + 2H_2O$; and the hypochlorites are also formed by acting on the hydroxides with hypochlorous acid.

Hypochlorous acid is easily prepared in dilute solution by shaking precipitated mercuric oxide with chlorine-water, and filtering from the precipitated mercuric oxychloride, thus:— $2\mathbf{HgO} + 2Cl_2.Aq = \mathbf{HgCl_2.HgO} + 2HClO.Aq$. It forms a pale yellow solution, with a pleasant smell of seaweed; it possesses very powerful oxidising and bleaching properties. It reacts at once with hydrochloric acid, forming chlorine and water, thus:— $HClO.Aq + HCl.Aq = H_2O + Cl_2 + Aq$. It cannot be obtained in concentrated solution, for it decomposes into chlorine and oxygen. It can also be produced by passing chlorine through water containing calcium carbonate in suspension, thus:—$\mathbf{CaCO_3} + Aq + Cl_2 = CaCl_2.Aq + CO_2 + 2HClO.Aq$. By distillation the hypochlorous acid may be separated from the calcium chloride;

* Known only in solution.

it comes over, along with water, in the first portion of the distillate. A similar action takes place with solution of sodium carbonate. When a current of chlorine is passed through it, a mixture of chloride and hypochlorite is formed at first, thus :—

$$Na_2CO_3.Aq + Cl_2 = NaCl.Aq + NaClO.Aq + CO_2;$$

further action of chlorine liberates hypochlorous acid, thus :—

$$NaClO.Aq + Cl_2 + H_2O = NaCl.Aq + 2HClO.Aq.$$

Hypobromous acid may be obtained in dilute aqueous solution, by shaking precipitated mercuric oxide with bromine-water. It is a yellow liquid, with a smell closely resembling that of hypochlorous acid.

Hypoiodous acid has not been obtained in the free state.

Hypochlorites, hypobromites, and hypoiodites.—Only one salt, viz., calcium hypochlorite, $Ca(OCl)_2.4H_2O$, has been prepared in an approximately pure state.

It has been stated that when a hydroxide, dissolved in water, is saturated with chlorine in the cold, a mixture of a hypochlorite and chloride is formed. Thus with sodium hydroxide :—

$$2NaOH.Aq + Cl_2 = NaCl.Aq + NaOCl.Aq + H_2O.$$

But sodium hypochlorite, owing to its instability, has never been isolated. Such a solution has, however, great oxidising power, and is named "Labarroque's disinfecting liquid." A similar mixture of potassium chloride and hypochlorite used to be known as "Eau de Javelle," and was formerly used for bleaching.

The most important compound of this acid is a double chloride and hypochlorite of calcium, known commercially as **bleaching-powder** or **"chloride of lime."*** It is produced on the large scale by passing chlorine over slaked lime (calcium hydroxide), spread in thin layers on slate shelves, in a building specially constructed for the purpose (see Alkali-manufacture, p. 670). The reaction which takes place is :—

$$Ca(OH)_2 + Cl_2 = Ca(OCl)Cl + H_2O.$$

That this body really is a definite compound of the formula $Ca{<}{}^{OCl}_{Cl}$, and not a mixture of chloride and hypochlorite of calcium, is shown by the fact that calcium chloride is deli-

* Gay-Lussac, *Annales*, **26**, 163; Odling, *Manual*, 1861, I, 56; Kopfer, *Chem. Soc.*, **28**, 713; Kingzett, *ibid.*, **28**, 404; Stahlschmidt, *Dingl. polyt. J.*, **221**, 243, 335.

quescent, soon liquefying by attracting moisture from air; but bleaching-powder does not deliquesce. Moreover, calcium chloride and hypochlorite are both exceedingly soluble in water; but 1 part of bleaching-powder requires about 20 parts of water to effect solution; it usually leaves a slight residue consisting of calcium hydroxide, some of which it almost always retains in the uncombined state. But this compound, **Ca(OCl)Cl**, is decomposed by water. For on cooling a saturated solution, or on evaporating it over sulphuric acid, calcium hypochlorite, **$Ca(OCl)_2$**, crystallises out, in transparent, very unstable crystals, while calcium chloride, which is more soluble, remains in solution.

Bleaching-powder is a white non-crystalline powder, smelling of hypochlorous acid. Its solution bleaches, owing to its parting with oxygen, thus:—$Ca(OCl)_2.Aq = CaCl_2.Aq + 2\mathit{O}$. This change is greatly facilitated by addition of an acid, whereby hypochlorous acid is liberated, which gives up its oxygen, being converted into hydrochloric acid. Hence, goods to be bleached are first run through an aqueous solution of bleaching-powder, and then through a bath of dilute sulphuric acid.

Calcium hypochlorite gives no precipitate with silver nitrate, for silver hypochlorite is very soluble. Metallic mercury is converted by hypochlorous acid into oxychloride, but by chlorine into chloride; hence it is possible to distinguish the one from the other in aqueous solution.

When distilled with dilute sulphuric, nitric, phosphoric, or even hydrochloric acid, if the last is not in excess, hypochlorous acid is found in the distillate. Excess of hydrochloric acid produces the decomposition:—$HClO.Aq + HCl.Aq = H_2O + \mathit{Cl_2} + Aq$; but, if only enough hydrochloric acid is used to liberate hypochlorous acid, the latter distils over.

The action of a cobalt salt on a solution of chloride of lime is to form hydrated cobalt sesquioxide. On boiling, even a minute proportion of this oxide causes evolution of oxygen from the solution of bleaching-powder. It is supposed that the black hydrated oxide of cobalt is further oxidised to an oxide, the formula of which is unknown, and that this higher oxide is simultaneously decomposed, liberating oxygen. Such an action is termed "catalytic." The final reaction is $2CaCl(OCl).Aq = 2CaCl_2.Aq + \mathit{O_2}$.

Chlorine acts on silver hydroxide suspended in water, forming silver chloride and hypochlorous acid. If the oxide be present in large excess, and if the solution be shaken, the odour of hypochlorous acid disappears, and the solution contains the very soluble silver hypochlorite. But on standing, decomposition soon ensues, chlorate and chloride of silver being produced:—

$$6AgClO.Aq = 5\mathbf{AgCl} + AgClO_3.Aq.$$

The only other known compound of chlorine monoxide is a red body, crystallising in needles, produced by its action on sulphur trioxide. It has the formula $\mathbf{Cl_2O.4SO_3}$. It melts at about 50°, and is at once resolved by water into sulphuric and hypochlorous acids.

Hypobromites and **hypoiodites** have not been obtained free from admixture with bromides and iodides. They are even less stable than hypochlorites, and are similarly produced. Both bromine and iodine dissolve in caustic soda or potash solution, forming yellowish liquids, which possess a fragrant chlorous smell. They presumably contain in solution the respective hypobromite or hypoiodite. They rapidly decompose on standing into bromide or iodide, and bromate or iodate, thus:—

$$6KBrO.Aq = 5KBr.Aq + KBrO_3.Aq;$$
$$6KIO.Aq = 5KI.Aq + KIO_3.Aq.$$

Chlorous acid and chlorites.—When chlorine peroxide, ClO_2, is added to water, it yields a mixture of chlorous acid, presumably $HClO_2$, and chloric acid, $HClO_3$, thus:—$2ClO_2 + H_2O + Aq = HClO_2.Aq + HClO_3.Aq$. But chlorous acid has not been examined. The reaction is strictly analogous to that between nitric peroxide and water (see p. 334).

Similarly, the addition of chlorine peroxide to an aqueous solution of a hydroxide yields a mixture of a chlorite and a chlorate. Potassium chlorite is more soluble than the chlorate, and may be obtained crystallised in thin needles. It has the formula $\mathbf{KClO_2}$. The lead and silver salts are sparingly soluble, and may be precipitated. They crystallise from a warm solution in thin yellow plates.

Chloric, bromic, and iodic acids:—chlorates, bromates, and iodates.—These compounds may be viewed as combinations of the unknown chlorine and bromine pentoxides, and of the known iodine pentoxide, with water and oxides.

Chloric acid, $HClO_3$, is best prepared by adding the requisite amount of dilute sulphuric acid to barium chlorate (see below), filtering from the precipitated barium sulphate, and concentrating by evaporation *in vacuo* over sulphuric acid. It is a colourless syrupy liquid, which is at once decomposed at 100° into perchloric acid, water, and chlorine peroxide, which itself explodes into chlorine and oxygen. It oxidises organic matter with great energy, often igniting it.

Bromic acid, $HBrO_3.Aq$, is similarly prepared. It is even more unstable than chloric acid, and decomposes, giving off bromine and oxygen, before it can be rendered syrupy by evaporation.

Iodic acid may be similarly prepared from barium iodate. But it is more conveniently prepared by direct oxidation of iodine by means of strong nitric acid. Convenient proportions are 5 grams of iodine and 200 grams of strong nitric acid; the mixture is kept at 60° for an hour. Iodic acid separates out; and a further quantity may be obtained by distilling off the nitric acid.

The oxidation may also be effected by means of chlorine and water, or of potassium chlorate and hydrochloric acid, which yield nascent oxygen. Iodine suspended in about ten times its weight of water is treated with a current of chlorine till the iodine is completely dissolved. Sodium carbonate is then added, which throws down a portion of the iodine, to be collected and treated as before. The liquid is then mixed with barium chloride, which throws down barium iodate, which is collected and decomposed by boiling with the requisite quantity of sulphuric acid. The solution is filtered from the insoluble barium sulphate, and boiled down, when the iodic acid separates in crystals.

The acid $\mathbf{HIO_3}$ is a white, easily soluble substance, crystallising in hexagonal tables. At 130°, or when digested with absolute alcohol, it loses water, forming the less hydrated compound, $\mathbf{HI_3O_8 = 3I_2O_5.H_2O}$. At 170°, this body forms the anhydride, $\mathbf{I_2O_5}$; and the acid may again be produced by dissolving the anhydride in water.

Another hydrate, $\mathbf{I_2O_5.5H_2O = 2H_5IO_6}$, has also been obtained, crystallising in hexagonal tables.

Iodic and hydriodic acids cannot exist in the same solution; they react forming iodine and water, thus:—

$$5HI.Aq + HIO_3.Aq = 3I_2 + 3H_2O + Aq.$$

Chlorates, bromates, and iodates.—These bodies are produced (1) by heating the hypochlorites, hypobromites, or hypoiodites, thus:—$3MXO = MXO_3 + 2MX$; (2) by treating the acids with hydroxides or carbonates; or (3) by acting on barium chlorate, bromate, or iodate with the solution of a sulphate. Some are produced by precipitation, *e.g.*, lead, mercurous, and silver bromates, and many iodates.

$2LiClO_3.H_2O$; $NaClO_3$; $KClO_3$; $RbClO_3$; NH_4ClO_3.—$LiBrO_3$; $NaBrO_3$; $KBrO_3$; NH_4BrO_3.—$LiIO_3$; $NaIO_3.H_2O$ and $5H_2O$; KIO_3; NH_4IO_3.

These salts are white, soluble crystalline bodies, all of which are decomposed by heat, giving off oxygen, and leaving the halide, thus:—$2MXO_3 = 2MX + 3O_2$. If a chlorate be employed, part of the oxygen converts chlorate into perchlorate, thus:—$MClO_3 + O = MClO_4$. But there appears to be no definite ratio between the amount of perchlorate formed and the amount of oxygen evolved. The application of heat should be continued until a sample of the residue gives no yellow coloration on addition of hydrochloric acid. Perbromates and periodates do not appear to be thus produced. The ammonium salts decompose with explosive violence, giving nitrogen, water, and the halide of hydrogen.

Potassium chlorate is the most important of these salts. It is formed by boiling a solution of the hypochlorite, thus:—$3KClO.Aq = KClO_3.Aq + 2KCl.Aq$. As the chlorate is much less soluble than the chloride, it crystallises out on evaporation. The preparation of the chlorate may, however, be carried out at one operation. Chlorine passed into a cold solution of potassium hydroxide yields chloride and hypochlorite; if the solution be heated during passage of the chlorine, the chlorate is produced. The complete reaction is:—$6KOH.Aq + 3Cl_2 = 5KCl.Aq + KClO_3.Aq + 3H_2O$ (see also p. 462). Potassium carbonate may be substituted for the hydroxide; in this case carbon dioxide is evolved.

Potassium chlorate crystallises in monoclinic six-sided plates often of considerable size. It is insoluble in alcohol, and sparingly soluble in water, 1 part of the salt requiring at 15° about 15 parts of water for solution. When heated, it fuses at 388°, and at a some-higher temperature begins to evolve oxygen. If manganese dioxide be mixed with the chlorate, a much lower temperature suffices to cause evolution of oxygen, while a little chlorine is also evolved. It is suggested that the nature of the change which takes place is the temporary formation of potassium permanganate, according to the equation $2MnO_2 + 2KClO_3 = 2KMnO_4 + Cl_2 + O_2$, and that the permanganate is further decomposed into oxygen and peroxide of manganese, ready to undergo further oxidation. The reaction would then to some sense be analogous to that of cobalt sesquioxide on a hot solution of bleaching-powder (see p. 463). As has been already mentioned, the decomposition of potassium and other chlorates probably occurs in two stages:—*

* Spring and Prost, *Bull. Soc. Chim.* (3), 1, 340.

(1.) The chlorine pentoxide of the chlorate $\mathbf{K_2O.Cl_2O_5}$ is decomposed into chlorine and oxygen; and (2) the nascent chlorine displaces oxygen from the potassium oxide, K_2O. That this is the case, appears to follow from the behaviour of other chlorates, in which the oxygen of the metallic oxide is only partially displaced by chlorine; such chlorates yield a mixture of oxygen and free chlorine. For example, 100 grams of barium chlorate yield 0·28 gram of free chlorine; of mercuric chlorate, 3·7; of lead chlorate, 8·0; of copper chlorate, 12·5; and of zinc chlorate, 14·4 grams. In such cases the oxygen of the metallic oxide remains behind in part, while chlorine is evolved in greater or less quantity, according to the conditions of the reaction. If potassium chlorate be perfectly pure, no chlorine is evolved; the displacement of oxygen is perfect.

The bromates, when heated, yield up oxygen, but no perbromate is formed. Experiments as regards free bromine have not been made. The iodates likewise decompose into iodides and oxygen, no periodates being formed; but iodine is liberated along with oxygen from sodium iodate.

Double salts.—$\mathbf{HNa(IO_3)_2}$; $\mathbf{H_2Na(IO_3)_3}$; $\mathbf{HK(IO_3)_2}$, $\mathbf{H_2K(IO_3)_3}$.

These salts are prepared by acidifying the ordinary salts with hydrochloric, nitric, or iodic acid, which practically amounts to mixture of the constituents. They form white soluble crystals. Their existence would lead to the conjecture that the formula of iodic acid is a multiple of HIO_3.

$\mathbf{NaIO_3.2NaBr.9H_2O}$; $\mathbf{2NaIO_3.3NaCl.9H_2O}$; $\mathbf{NaIO_3.NaI.10H_2O}$;
$\mathbf{HK(IO_3)_2.KCl}$.—$\mathbf{KIO_3.HKSO_4}$.

These soluble crystalline salts are obtained by mixture.

$\mathbf{Ca(ClO_3)_2.2H_2O}$; $\mathbf{Sr(ClO_3)_2.5H_2O}$; $\mathbf{Ba(ClO_3)_2}$.—
$\mathbf{Ca(BrO_3)_2.H_2O}$; $\mathbf{Sr(BrO_3)_2.H_2O}$; $\mathbf{Ba(BrO_3)_2.H_2O}$.—
$\mathbf{Ca(IO_3)_2.5H_2O}$; $\mathbf{Sr(IO_3)_2.\mathit{n}H_2O}$; $\mathbf{Ba(IO_3)_2.H_2O}$.

These are sparingly soluble white crystalline salts, best produced by mixing potassium chlorate with the acetate or chloride of calcium, strontium, or barium, and evaporating to crystallise. The more soluble acetate or chloride of potassium remains dissolved, while the halate crystallises out. Beryllium iodate is said to be a gummy mass.

$\mathbf{Mg(ClO_3)_2.6H_2O}$; $\mathbf{Zn(ClO_3)_2.6H_2O}$.—$\mathbf{Mg(BrO_3)_2.6H_2O}$; $\mathbf{Zn(BrO_3)_2.6H_2O}$;
$\mathbf{Cd(BrO_3)_2.H_2O}$.—$\mathbf{Mg(IO_3)_2.4H_2O}$; $\mathbf{Zn(IO_3)_2.2H_2O}$.

The chlorates and bromates and the iodate of magnesium are easily soluble in water; zinc iodate dissolves sparingly. They are all white and crystalline.

$Y(ClO_3)_3$.—$Y(BrO_3)_3$; $La(BrO_3)_3.9H_2O$.—$Y(IO_3)_3$; $2La(IO_3)_3.3H_2O$.

These are sparingly soluble crystalline white salts.

$Al(ClO_3)_3$ (?), $Al(BrO_3)_3$ (?), $Al(IO_3)_3$ (?).

These are deliquescent syrups; the iodate appears to crystallise. The gallium and indium salts have not been prepared. $TlClO_3$, $TlBrO_3$, and $TlIO_3$ are white sparingly soluble crystals.

The salts of chromium and ferric iron are indefinite. A basic iodate of the formula $2I_2O_5.Fe_2O_3.8H_2O$, or $4Fe(IO_3)_3Fe_2O_3.24H_2O$, has been prepared.

$Fe(BrO_3)_2$; $Fe(IO_3)_2$; $Co(ClO_3)_2.6H_2O$; $Ni(ClO_3)_2.6H_2O$.—$Co(BrO_3)_2.6H_2O$; $Ni(BrO_3)_2.6H_2O$.—$Mn(IO_3)_2.H_2O$; $Co(IO_3)_2.H_2O$; $Ni(IO_3)_2.H_2O$.

These are coloured crystalline salts. The chlorate and bromate of manganese are known in solution. They all readily decompose, the metal being oxidised to a higher oxide.

$Ce(ClO_3)_3.nH_2O$; $Ce(BrO_3)_3.nH_2O$; $Ce(IO_3)_3.nH_2O$.

White salts; the iodate is sparingly soluble.

$Pb(ClO_3)_2.H_2O$; $Pb(BrO_3)_2.H_2O$ and $Pb(IO_3)_2$.

Sparingly soluble salts. No compounds of the other elements of this group have been prepared.

The compound $SO_3.I_2O_5$ is said to be obtained in granular crystals by the action of sulphur trioxide on iodine pentoxide at 100°.*

$2Bi(IO_3)_3.H_2O$, and a basic bromate, $2Br_2O_5.3Bi_2O_3.6H_2O$, have been prepared. They are insoluble. $(VO_2)(IO_3)_2.5H_2O$ is a yellow precipitate. It is the only known compound of this group.

No compounds of the palladium or platinum groups, or of gold, have been prepared.

$AgClO_3$; $AgBrO_3$; $AgIO_3$; $HgClO_3$; $HgBrO_3$; $HgIO_3$.

Chlorate of silver is soluble; the other salts given above are sparingly soluble white bodies, produced by precipitation.

* *J. prakt. Chem.*, **82**, 72.

$Cu(ClO_3)_2.6H_2O$; $Cu(BrO_3)_2.5H_2O$; $2CuIO_3.3H_2O$.—
$Hg(ClO_3)_2$; $Hg(BrO_3)_2.2H_2O$; $Hg(IO_3)_2$.

Chlorate and bromate of copper and mercuric chlorate are easily soluble; the remaining bodies are nearly insoluble.

It will be noticed that as a rule the chlorates become more soluble, the bromates and iodates less soluble, as the elements follow in the periodic order. It would be advisable to attempt to prepare double salts of the chlorates and bromates, and also of such iodates as those of calcium and magnesium.

Double iodates.—Chromiodates.*—By dissolving chromium trioxide in iodic acid, and evaporating over sulphuric acid, ruby-red rhombic crystals of chromiodic acid, $\mathbf{IO_2.O.CrO_2.OH}$, are deposited. With iodates, corresponding chromiodates have been prepared, viz., $\mathbf{IO_2.O.CrO_2.OLi.H_2O}$, $\mathbf{IO_2.O.CrO_2.ONa.H_2O}$, $\mathbf{IO_2.O.CrO_2.OK}$, and $\mathbf{IO_2.O.CrO_2.ONH_4}$. Manganese, cobalt, and nickel salts have also been prepared. These compounds have a brilliant red colour, and are decomposed by water into chromates and iodates.

Perchloric and periodic acids, perchlorates, and periodates.—Perbromic acid and perbromates are unknown. These acids and salts may be regarded as compounds of the unknown heptoxides, Cl_2O_7 and I_2O_7, with water and oxides. Perchlorate of potassium is the starting point for the perchlorates. It is produced by heating the chlorate, some of the nascent oxygen combining with the chlorate and oxidising it; or by heating the chlorate with nitric acid, thus:—

$$3\mathbf{KClO_3} + 2HNO_3 = \mathbf{KClO_4} + 2\mathbf{KNO_3} + H_2O + \mathit{Cl_2} + \mathit{2O_2}.$$

By the first method, a mixture of chloride and perchlorate of potassium is produced; by the second, a mixture of perchlorate and nitrate. They are separated by crystallisation, the perchlorate being much less soluble than the chloride or nitrate. From potassium perchlorate, **perchloric acid** is produced by distillation with sulphuric acid; it comes over at 203° as an oily liquid containing 70·3 per cent. of $HClO_4$. On mixing this hydrate with twice its volume of oil of vitriol and again distilling, anhydrous perchloric acid distils as a yellowish strongly fuming liquid. On further distillation, the oily hydrate passes over, and when it comes in contact with the anhydrous acid they combine to form a hydrate, $\mathbf{HClO_4.H_2O}$; a little sulphuric acid also distils over. The

* *Compt. rend.*, **104**, 1514.

crystals, collected and distilled alone, yield pure perchloric acid in the first portions of the distillate.

The anhydrous acid, $HClO_4$, is a colourless very volatile liquid Its specific gravity is 1·782 at 15·5°. It explodes violently when brought in contact with any oxidisable matter, and hisses when dropped into water. It decomposes when boiled, leaving a black explosive residue. It also decomposes, frequently with explosion, at the ordinary temperature. The monohydrate, $\mathbf{HClO_4.H_2O}$, produced by addition, is a white solid, melting at 50°, and decomposing at 110° into pure acid and the oily hydrate mentioned above, which resembles sulphuric acid in appearance, and distils unchanged at 203°. An aqueous solution of perchloric acid does not bleach, and reddens litmus. It is also not reduced by hydrogen sulphide or sulphur dioxide.

Periodic acid, $\mathbf{H_5IO_6}$ (= $\mathbf{HIO_4.2H_2O}$), is prepared from a periodate. The starting point is sodium iodate, $\mathbf{NaIO_3}$, which is oxidised by sodium hypochlorite. A mixture of sodium iodate and caustic soda is saturated with chlorine, when the reaction occurs:—$NaIO_3.Aq + 3NaOH.Aq + Cl_2 = H_3Na_2.IO_6.Aq + 2NaCl.Aq$; or 1 part of iodine and 7 parts of sodium carbonate are dissolved in 100 parts of water and saturated with chlorine. The iodate at first formed is converted into periodate, which crystallises out, being sparingly soluble in water. This periodate is dissolved in nitric acid free from nitrous acid, and silver nitrate is added; the precipitated trihydrogen diargentic periodate is dissolved in hot dilute nitric acid and evaporated until monoargentic periodate, $\mathbf{AgIO_4}$, crystallises out. On treatment with water, this salt undergoes the change $2\mathbf{AgIO_4} + 4H_2O = \mathbf{H_3Ag_2IO_6} + H_5IO_6$. The silver salt is removed by filtration, and the filtrate on evaporation deposits crystals of periodic acid.

Periodic acid, $\mathbf{H_5IO_6}$, forms white, oblique, rhombic prisms which melt between 130° and 136° with decomposition into iodine pentoxide, water, and oxygen. It is easily soluble in water, and sparingly in alcohol and in ether. Unlike perchloric acid, it is at once reduced by hydrochloric or sulphurous acids and by hydrogen sulphide.

The perchlorates and periodates are produced in the usual manner.

$\mathbf{NaClO_4}$; $\mathbf{KClO_4}$; $\mathbf{NH_4ClO_4}$; $\mathbf{LiIO_4}$; $\mathbf{NaIO_4}$; $\mathbf{KIO_4}$.

The sodium salts are very soluble; potassium perchlorate is one of the least soluble of potassium salts; hence perchloric acid may be used as a means of precipitating potassium. It is almost insoluble in alcohol. The iodate is also sparingly soluble.

$Ca(ClO_4)_2$; $Ba(ClO_4)_2$.

These are very soluble. No periodates are known, except **$Ca(IO_4)_2$**.

$Zn(ClO_4)_2$; $Cd(ClO_4)_2$.—$Mg(IO_4)_2.10H_2O$; $Cd(IO_4)_2$.

These are white soluble salts.

The remaining perchlorates which have been prepared are:—

$Fe(ClO_4)_2.6H_2O$; $Mn(ClO_4)_2$; $Pb(ClO_4)_2.3H_2O$; $(Pb_2O)(ClO_4)_2.H_2O$; $Cu(ClO_4)_2$; $AgClO_4$; $HgClO_4.3H_2O$; $Hg(ClO_4)_2$.

They are all soluble in water and crystalline, except the silver salt, which is a white powder.

Only a few corresponding periodates are known, viz., **$Fe(IO_4)_3$**, a bright yellow powder; **$AgIO_4$**, crystallising in orange-yellow crystals; and **$Pb(IO_4)_2$**, an amorphous red salt.

Many **complex periodates** are known, which may be best explained by reference to the conception of a normal hydroxide, as follows:—Sodium and elements of that group tend to form only a monohydroxide, M.OH; those of the magnesium and calcium groups, dihydroxides, $M^{II}(OH)_2$; those of the boron and aluminium groups, trihydroxides, $M^{III}(OH)_3$; silicon, and possibly other members of the carbon and silicon groups, tetrahydroxides, $M^{IV}(OH)_4$; but here we notice instability, so that the first anhydrides of such bodies, $M^{IV}O(OH)_2$, are more stable; the pentahydroxides of elements of the nitrogen and phosphorus groups are non-existent; but their first anhydrides are known with phosphorus, arsenic, &c., in phosphoric and arsenic acids, $P^{V}O(OH)_3$ and $As^{V}O(OH)_3$. Hexahydroxides may be inferred in the case of normal sulphuric, selenic, and telluric acids; again, their existence is doubtful in any definite cases; but their second anhydrides, such as $S^{VI}O_2(OH)_2$, are well-known bodies; and it would follow that the elements of the chlorine group should produce heptahydroxides, $M^{VII}(OH)_7$. Such substances are unknown with chlorine, but the assumption of their existence affords a means of representing systematically many of the compounds of periodic acid.

The perchlorates, which possess the general formula $MClO_4$, may be regarded as the metallic derivatives of the third anhydrides of the hypothetical heptahydroxides, thus:—

Normal salt......	$Cl(OM)_7$;	
First derivative..	$ClO(OM)_5$;	
Second derivative.	$ClO_2(OM)_3$;	
Third derivative..	$ClO_3(OM)$.	(Ordinary perchlorate.)

Such derivatives are known with the periodates. Those of the last class have already been considered; it remains to describe and classify the derivatives of the first and second anhydrides of the normal acid, H_7IO_7, which, however, is unknown as such. The acid HIO_4 is unknown; the known acid $\mathbf{H_5IO_6}$, which has been described, is the first anhydride of the theoretical normal acid, H_7IO_7. It has been suggested that derivatives of the type $M'IO_4$ should be named **meta-periodates**, as the phosphates of the type MPO_3 are termed meta-phosphates; derivatives of the formula M_3IO_5 have been termed **meso-** (middle) **periodates**; of the type M_5IO_6, **para-periodates**; and of the type M_7IO_7, **ortho-periodates**. Besides these derivatives, some of a still more condensed type are known.

Single Salts, containing only one Metal.

Para-periodates.—$\mathbf{H_5IO_6}$; $\mathbf{Li_5IO_6}$; $\mathbf{Ba_5(IO_6)_2}$; $\mathbf{Fe^{II}_5(IO_6)_2}$; $\mathbf{Ag_5IO_6}$; $\mathbf{Hg_5IO_6}$; $\mathbf{Cu_5(IO_6)_2.5H_2O}$; $\mathbf{Hg_5(IO_6)_2}$.

Meso-periodates.—$\mathbf{Ni_3(IO_5)_2}$; $\mathbf{Pb_3(IO_5)_2}$; $\mathbf{Ag_3IO_5}$.

Double Salts.

Ortho-periodates.—(M_7IO_7). $\mathbf{H_6NaIO_7}$; $\mathbf{H_5K_2IO_7.2H_2O}$; $\mathbf{H_4(NH_4)MgIO_7.H_2O}$; $\mathbf{2H_5ZnIO_7.H_2O}$; $\mathbf{H_5CaIO_7.2H_2O}$; $\mathbf{H_5SrIO_7.H_2O}$; $\mathbf{H_5BaIO_7.H_2O}$; $\mathbf{HFe^{III}_2IO_7.20H_2O}$.

Para-periodates.—(M_5IO_6). $\mathbf{H_3Li_2IO_6}$; $\mathbf{H_3Na_2IO_6}$; $\mathbf{H_2Na_3IO_6}$; $\mathbf{H_3(NH_4)_2IO_6}$; $\mathbf{H_3MgIO_6.9H_2O}$; $\mathbf{HZn_2IO_6}$; $\mathbf{HCd_2IO_6.H_2O}$; $\mathbf{H_3CaIO_6}$; $\mathbf{2H_3SrIO_6.H_2O}$; $\mathbf{H_3BaIO_6}$, anhydrous, and with $\mathbf{4H_2O}$; $\mathbf{H_3FeIO_6}$; $\mathbf{H_4Pb_3(IO_6)_2}$; $\mathbf{H_3Ag_2IO_6}$; $\mathbf{H_2Ag_3IO_6}$; $\mathbf{HCu^{II}_2IO_6}$, anhydrous, and with $\mathbf{3H_2O}$.

Meso-periodates.—(M_3IO_5). $\mathbf{HCdIO_5}$, anhydrous and with $\mathbf{4H_2O}$; $\mathbf{H_2AgIO_5}$; $\mathbf{HAg_2IO_5}$.

This classification must be regarded as merely provisional; it is, as a rule, impossible to decide whether hydrogen should be included in the formula, or should belong to water of crystallisation. For example, the salt $HCdIO_5$ is a meso-periodate when thus written; but it has also been prepared of the formula $HCdIO_5.4H_2O$. If one molecule of this water be included, it becomes a para-periodate, thus:—$H_3CdIO_6.3H_2O$; if two molecules of water be included, its formula is that of an ortho-periodate, thus:—$H_5CdIO_7.2H_2O$. Lastly, the salt $HCdIO_5$ still contains hydrogen; by doubling its formula and subtracting the elements of water, we have $2HCdIO_5 - H_2O = Cd_2I_2O_9$, a diperiodate. To form any definite conclusion is difficult, for the individuality of each of the four classes of compounds is not well marked, as in the case of the analogous phosphates.

Derivatives of condensed periodic acids are also known. These may be viewed as derived from a hypothetical diperiodic acid, analogous to pyrophosphoric and phosphoric acid and to anhydro-sulphuric acid, of the formula $I_2O(OH)_{12}$, from which there are deducible the five anhydro-acids, (1) $I_2O_2(OH)_{10}$; (2) $I_2O_3(OH)_8$; (3) $I_2O_4(OH)_6$; (4) $I_2O_5(OH)_4$; and (5) $I_2O_6(OH)_2$. Derivatives of the dodeca-, of the deca-, and of the hexa-hydroxyl acids, $I_2O(OH)_{12}$, $I_2O_2(OH)_{10}$, and $I_2O_4(OH)_6$, are unknown; the others may be classified as follow:—

Octohydroxyl acid.—$Mg_4I_2O_{11}$; $Ag_8I_2O_{11}$; $Hg_8I_2O_{11}$.
Tetrahydroxyl acid.—$K_4I_2O_9$; $HK_3I_2O_9$; $Mg_2I_2O_9$; $HFeI_2O_9$; $Ni_2I_2O_9$; $Ag_4I_2O_9$, anhydrous, also with H_2O and $3H_2O$.

The salt $Ag_4I_2O_9.H_2O$ is not identical with the para-periodate of similar percentage composition, HAg_2IO_5; they differ in appearance, and while the molecule of water in the first salt is lost at 100°, no change occurs on heating the second salt until the temperature rises to 300°. And the two compounds $H_3Ag_2IO_6$ and $Ag_4I_2O_9.3H_2O$ are also quite different from each other in chemical and physical properties.

Formation.—To describe the method of formation of each individual salt would occupy too much space. The general methods are:—

Meta-periodates (MIO_4) are produced by boiling di-, meso-, or para-periodates with nitric acid; thus, for example:—

$$H_3Ag_2IO_6 + HNO_3 = AgIO_4.H_2O + H_2O + AgNO_3.$$

Di-periodates are changed to para-periodates by treatment with silver nitrate:—

$$K_4I_2O_9.Aq + 4AgNO_3.Aq + 3H_2O = 2H_3Ag_2IO_6 + 4KNO_3.Aq.$$

Para-periodates yield meso-periodates when their double salts with hydrogen are heated:—

$$H_4Pb_3(IO_6)_2 = Pb_3(IO_5)_2 + 2H_2O.$$

Meso-periodates may in analogous manner yield di-periodates:—

$$2HAg_2IO_5 = Ag_4I_2O_9 + H_2O.$$

Meta-periodates are often decomposed by water, yielding para-periodates, *e.g.*:—

$$2AgIO_4 + 4H_2O = H_3Ag_2IO_6 + H_5IO_6.$$

Derivatives of a few still more condensed types are known, *e.g.*:—

$$Ba_5I_4O_{19}:\ Ag_{10}I_4O_{19};\ Cd_{10}I_6O_{31},15H_2O.$$

A remarkable compound of the formula $N_2O_6.Cl_2O_7$ = $N_2O_5(ClO_4)_2$, or $Cl_2O_7(NO_3)_2$, has been produced by passing a silent electric discharge through a mixture of chlorine, nitrogen, and oxygen.* It is a white solid, easily volatile in a vacuum; it deliquesces in air, giving a mixture of nitric and perchloric acids.

Physical Properties.

Mass of one cubic centimetre:—

I_2O_5, 4·8 grams at 9°. HIO_3, 4·87 grams at 0°.

Heats of combination:—

$2Cl + O = Cl_2O - 178K;\ + Aq = 2HOCl.Aq + 94K.$
$2Cl + 5O + Aq = 2HClO_3.Aq - 204K.$
$2Cl + 7O + Aq = 2HClO_4.Aq + 42K.$
$2Br + O + Aq = 2HOBr.Aq - 162K.$
$2Br + 5O + Aq = 2HBrO_3.Aq - 436K.$
$2I + 5O = I_2O_5 + 453K;\ + H_2O = 2HIO_3 + 26K.$
$2I + 7O + Aq = 2H_5IO_6.Aq + 268K.$

* *Comptes rend.*, **98**, 626.

CHAPTER XXIX.

OXIDES, SULPHIDES, AND SELENIDES OF RHODIUM, RUTHENIUM, AND PALLADIUM:—OF OSMIUM, IRIDIUM, AND PLATINUM; PLATINONITRITES, AND PLATINOCHLOROSULPHITES; CARBONYL COMPOUNDS, AND PLATINOPHOSPHITES; OXIDES, SULPHIDES, SELENIDES, AND TELLURIDES OF COPPER, SILVER, MERCURY, AND GOLD.—OXYHALIDES.—CONCLUDING REMARKS ON OXIDES AND SULPHIDES.

No tellurides of metals of the palladium or platinum groups are known. The following is a list of the compounds which have been prepared:—

Note.—The following method may be advantageously employed in separating the metals of the palladium and platinum groups from each other:—The ore is treated with *aqua regia* under pressure. The solution contains platinum, palladium, rhodium, ruthenium; the residue, osmium, iridium, and some rhodium and ruthenium. The solution is boiled with caustic soda, and mixed with a solution of potassium chloride and alcohol; the sparingly soluble platinichloride of potassium separates out. On ignition, it is converted into metallic platinum. A sheet of zinc is placed in the solution from which the platinum has been removed; the remaining metals are precipitated.

The insoluble residue is heated in a platinum retort in a current of oxygen, when osmium tetroxide volatilises. The residue in the retort, and the metals precipitated with zinc are melted with four times their weight of zinc under a layer of zinc chloride. The alloy is heated with hydrochloric acid until palladium begins to dissolve with a brown colour. The black residue is boiled with aqua regia; a residue of a portion of the rhodium and ruthenium still remains. From the solution, palladium di-iodide is precipitated with potassium iodide. The solution is treated with a current of hydrogen, which precipitates all the remaining metals except iridium. The precipitate is mixed with the rhodium and ruthenium, and heated with barium chloride in a current of chlorine to volatilise any remaining osmium as dichloride. The residue, consisting of barium rhodiochloride, is dissolved in water, and the barium removed with sulphuric acid. The rhodium is then thrown down with sodium hydrogen sulphite, which precipitates the insoluble compound $3Na_2O.Rh_2O_3.6SO_2$. The ruthenium, again precipitated from the filtrate with zinc, is boiled with potassium hydroxide and chromate, treated with excess of potash, and boiled with sodium sulphate. The precipitate contains only ruthenium.—See also *Iron*, 1879, **13**, 654.

		Oxygen.				
Rhodium	—	RhO;	Rh_2O_3;	RhO_2;	RhO_3;	—
Ruthenium	—	RuO;	Ru_2O_3;	RuO_2;	RuO_3;*	RuO_4
Palladium	Pd_2O;	PdO	—	PdO_2	—	—

		Sulphur.			Selenium.
Rhodium	—	RhS;	Rh_2S_3	—	—
Ruthenium	—	—	Ru_2S_3;	RuS_2.	—
Palladium	Pd_2S;	PdS;	—	PdS_2.	$PdSe$.

None of these compounds occurs native.

Preparation.—1. **By direct union.**—Finely-divided rhodium, prepared by heating the double chloride $RhCl_3.3NH_4Cl$ to redness, when heated to dull redness in oxygen, yields the monoxide, RhO; powdered ruthenium is oxidised to Ru_2O_3; palladium yields dark-grey Pd_2O. Ruthenium tetroxide, RuO_4, is formed by direct union at about 1000°, but decomposes on cooling. This is a most curious result, for the tetroxide is decomposed violently when heated to 108°; and it is only by rapid cooling, by means of an inside tube through which cold water circulates, which passes through the centre of the outside tube, heated to bright redness, that it is possible to isolate some of the tetroxide without decomposition. If allowed to cool slowly, the body dissociates again into ruthenium and oxygen.†

Rhodium and palladium monosulphides, and palladium monoselenide are formed with incandescence when the metals are heated with sulphur or selenium.

2. **By decomposing a higher compound by heat.**—Rhodium sesquioxide, when heated, yields the monoxide; and, similarly, the sesquisulphide yields the monosulphide.

3. **By the action of heat on a double compound.**—Rhodium nitrate, when heated to dull redness, leaves a residue of Rh_2O_3; ruthenic sulphate, $Ru(SO_4)_2$, yields the dioxide RuO_2 on ignition; and palladous nitrate, $Pd(NO_3)_2$, moderately heated, yields the monoxide, PdO. The hydrates, when they exist, yield the oxides when heated.

4. **By replacement.**—Sulphur displaces chlorine when heated with the compound $RhCl_3.3NH_4Cl$; and oxygen, when a mixture of the sesquioxide and sulphur are heated in a current of carbon dioxide. In each case the monosulphide, RhS, is formed.

* Known only in combination.

† Debray and Joly, *Comptes rend.*, **106**, 100, and 328. See also *ibid.*, **84**, 946.

Conversely, the sulphides, roasted in air, are converted into the more stable of the oxides.

5. **By double decomposition.**—Ruthenium dichloride, $RuCl_2$, calcined with sodium carbonate in an atmosphere of carbon dioxide, yields the monoxide; the excess of sodium carbonate is removed by washing with water. Palladium monoxide, PdO, is similarly prepared. On boiling a solution of palladium tetrachloride with a solution of sodium carbonate, the anhydrous dioxide is precipitated.

Rhodium sesquisulphide is produced by heating the trichloride, $RhCl_3$, in a current of hydrogen sulphide to 360°; ruthenium sesquisulphide and disulphide are produced by passing hydrogen sulphide through solutions of the respective chlorides. Palladium monosulphide, PdS, is formed by the action of hydrogen sulphide on the dichloride, and the disulphide by the action of hydrochloric acid on sodium sulphopalladate, Na_2PdS_3.

6. **By oxidation of the metal or of a lower oxide** by nascent oxygen. This process yields the higher oxides. When rhodium or an oxide is fused with a mixture of potassium hydroxide and nitrate, the dioxide, RhO_2, is formed, and may be separated from the excess of soluble salts by boiling with dilute nitric acid. Chlorine acts on rhodium sesquioxide in presence of caustic potash (forming hypochlorite of potassium) giving a green precipitate of the hydrated dioxide, and a violet-blue solution, from which a green powder deposits on standing. On warming with nitric acid, this powder leaves the anhydrous trioxide, RhO_3. Similarly, when ruthenium is heated with nitre and potassium hydroxide, a soluble yellow mass is obtained, believed to contain the trioxide in combination with potassium oxide. On saturating its hot solution in aqueous caustic potash with chlorine, a sublimate is formed of RuO_4.

Palladium hemisulphide, Pd_2S, is formed by a method which cannot easily be classified. A mixture of the monosulphide, PdS, sodium carbonate, ammonium chloride, and sulphur is heated to redness for twenty minutes. On digesting with water, sodium sulphopalladite, $NaPdS_3$ (see below), goes into solution, while the hemisulphide remains as a fused mass.

Properties.—The **monoxides** are dark-grey, insoluble powders, with semi-metallic lustre. Those of ruthenium and rhodium are not attacked by acids; palladium monoxide dissolves. Monoxides of rhodium and palladium are reduced to metal by hydrogen at a dull-red heat; that of ruthenium at the ordinary temperature. The **monosulphides** of rhodium and palladium are bluish-white substances with metallic lustre. The latter melts at about 1000°. Palladium **selenide** resembles the sulphide, but has not been fused.

Rhodium **sesquioxide** is a grey porous mass, with metallic iridescence; and that of ruthenium, a dark blue, insoluble powder. Rhodium **sesquisulphide** forms brownish-black crystalline plates; the ruthenium compound is a brown powder.

Rhodium **dioxide** is a dark-brown insoluble powder; ruthenium dioxide prepared by roasting the disulphide in a blackish-blue powder; prepared by heating the sulphate, it is a greyish metallic-looking substance, showing a blue iridescence. When heated in oxygen to the melting-point of copper, it crystallises in quadratic prisms, isomorphous with *cassiterite*, SnO_2, and with *rutile*, TiO_2. Anhydrous palladium dioxide is a black powder, soluble in acids, even in strong hydrochloric acid; but, curiously, with dilute hydrochloric acid, chlorine is evolved.

Ruthenium **disulphide** is a brownish-yellow precipitate; and palladium disulphide a blackish-brown crystalline powder.

Rhodium **trioxide** is a blue flocculent precipitate.

Ruthenium **tetroxide** forms volatile golden-yellow rhomboidal prisms, melting at 25·5° when pure, and decomposing rapidly at 106°. It is sparingly soluble in water. Its vapour density corresponds with the formula RuO_4. Its aqueous solution is said to yield an oxysulphide with hydrogen sulphide.

Compounds with water and oxides, and with hydrogen sulphide and sulphides.—The following have been prepared:—

$PdO.nH_2O$.—$Rh_2O_3.3$ and $5H_2O$; $Ru_2O_3.3H_2O$; $Rh_2S_3.3H_2S$; $Rh_2S_3.3Na_2S$.—$RhO_2.2H_2O$; $RuO_2.2H_2O$; $PdO_2.nH_2O$.—$PdS_2.Na_2S$; $PdS_2.2K_2S$; $PdS_2.Ag_2S$; $PdS_2.Pd_2S.K_2S$; $PdS_2.2PdS$; $PdS_2.PdS.Ag_2S$; $RuO_3.Na_2O$; $RuO_3.K_2O$; $RuO_3.MgO$; $RuO_3.CaO$; $RuO_3.SrO$; $RuO_3.BaO$; $RuO_3.Ag_2O$.

Complex oxides.—$Ru_2O_5.2H_2O$; $Ru_2O_5.K_2O$; and $Ru_4O_9.2H_2O$.—$Ru_2O_7.K_2O$.

$PdO.nH_2O$.—A dark-brown precipitate, produced by addition of solution of sodium carbonate to solution of palladous chloride, $PdCl_2.Aq$. It reacts with acids, forming palladous salts.

$Rh_2O_3.3H_2O$.—A gelatinous precipitate, produced by adding an alcoholic solution of potash to a solution of the double chloride $RhCl_3.3NaCl$. It is nearly insoluble in acids, though a red solution is formed when it is digested with hydrochloric acid. By use of aqueous solution of potash, the pentahydrate is thrown down as a somewhat soluble yellow precipitate.

$Rh_2S_3.3H_2S$ and **$Rh_2S_3.3Na_2S$** are produced by addition of hydrogen or sodium sulphide respectively to a solution of the trichloride, $RhCl_3.Aq$. The first is a brownish-black insoluble precipitate; the second a dark-brown crystalline body. The former compound is noticeable as being one of the very few hydrosulphides known.

$Ru_2O_3.3H_2O$ is a dark precipitate produced by sodium carbonate in a solution of ruthenium trichloride. It is soluble in acids, forming ruthenic salts.

$RhO_2.2H_2O$ is a green precipitate formed by the action of chlorine on a solution of the hydrate, **$Rh_2O_3.5H_2O$**. It is somewhat soluble in water, with a violet-blue colour; and with hydrochloric acid it evolves chlorine, dissolving to $RhCl_3$.

$RuO_2.2H_2O$ is a yellow precipitate produced by treating a solution of potassium ruthenichloride, $RuCl_4.2KCl.Aq$, with sodium carbonate. It dissolves in acids with a yellow colour, forming ruthenic salts; and in alkalies with a light-yellow colour.

$PdO_2.nH_2O$ is similarly prepared, but it appears to be impossible to obtain it free from admixed alkali.

Sulphopalladites.—By fusing together palladium monosulphide, sodium carbonate, ammonium chloride, and sulphur, a whitish-grey metallic-looking button of **Pd_2S**, is formed. The fused mass of salts covering this button, when washed with alcohol, gives a residue of sodium sulphate, and the compound **Na_2PdS_3**, forming reddish-grey metallic-looking needles. With silver nitrate, a blackish-brown precipitate of **Ag_2PdS_3** is formed. If potassium carbonate be employed in the above fusion, the residue on treatment with alcohol contains the sub-palladous salt **$K_2PdS_3.Pd_2S$**, which forms blue metallic-looking hexagonal laminæ. It is insoluble; when heated in hydrogen, palladium is produced, along with the soluble salt **K_4PdS_4**, thus :—$2K_2Pd_3S_4 + 4H_2 = 4H_2S + 5Pd + PdS_2.2K_2S$. The salt **$K_2PdS_3.Pd_2S$** when treated with hydrochloric acid yields the hydrogen salt **$H_2PdS_3.Pd_2S$**, which, on oxidation in air, yields the compound **Pd_3S_4**, thus :—$H_2PdS_3.Pd_2S + O = H_2O + Pd_3S_4$. When heated in air, **$Pd_3S_4$** is converted into **PdS**. The silver compound, **$Ag_2PdS_3.Pd_2S$**, forms whitish-grey plates.

Ruthenates and perruthenates.—Ruthenium tetroxide fused under water and added to strong potash solution at 60° evolves oxygen, and on cooling deposits blackish-brown quadratic octahedra of potassium perruthenate, **$KRuO_4$**. The mother liquor from this salt on evaporation gives crystals of potassium ruthenate, **$K_2RuO_4.H_2O$**, crystallising in rhombic prisms, and soluble in a little water, with a yellow colour. On diluting its solution it decomposes, thus:—$4K_2RuO_4.Aq + 5H_2O = 2KRuO_4.Aq + \mathbf{Ru_2O_5.2H_2O} + 6KOH.Aq$. Diruthenium pentoxide, $\mathbf{Ru_2O_5.2H_2O}$, is a crystalline body; the substance $\mathbf{KRuO_3}$, which may be named hyporuthenite of potassium, is its compound with potassium oxide, $\mathbf{Ru_2O_5.K_2O}$. This oxide, $\mathbf{Ru_2O_5}$, heated in a vacuum to 260°, loses oxygen, depositing black scales of $\mathbf{Ru_4O_9}$; the same substance is produced when a solution of the tetroxide is heated with water to 100°.

From potassium ruthenate the following bodies have been prepared:—$\mathbf{Na_2RuO_4.2H_2O}$, $\mathbf{MgRuO_4}$, and $\mathbf{CaRuO_4}$, black precipitates; $\mathbf{BaRuO_4}$, a vermilion precipitate; and $\mathbf{Ag_2RuO_4}$, a dense black precipitate.

The formulæ of potassium hyporuthenite, $\mathbf{KRuO_3}$, is analogous to that of potassium chlorate; and those of potassium ruthenate, $\mathbf{K_2RuO_4}$, and of potassium perruthenate, $\mathbf{KRuO_4}$, to potassium manganate and permanganate respectively; but the crystalline forms are not the same. We have at present no certain knowledge regarding the constitution of these bodies.

Oxides, Sulphides, and Selenides of Osmium, Iridium, and Platinum.

List:—

			Oxygen.		
Osmium	OsO;	Os_2O_3;	OsO_2;	OsO_3;*	OsO_4.
Iridium..........	IrO;	Ir_2O_3;	IrO_2;	IrO_3	—
Platinum	PtO	—	PtO_2	—	—

			Sulphur.		
Osmium	OsS, &c. (?)	—	—	OsS_4.	
Iridium..........	IrS;	Ir_2S_3;	IrS_2;	IrS_3.	—
Platinum	PtS	—	PtS_2	—	—

The oxide $\mathbf{Pt_3O_4}$ has also been prepared. The selenides and tellurides require investigation; they have not been analysed.

None of these compounds occurs in nature.

Preparation.—1. By direct union.—Osmium tetroxide is formed when finely-divided osmium is heated to bright redness in oxygen or in air. Platinum monosulphide is also directly formed.

2. **By decomposing a higher compound by heat.**—Platinum dioxide, gently heated, is converted into the oxide $\mathbf{Pt_3O_4}$; and platinum disulphide yields the monosulphide at a low red heat. Iridium monosulphide is also produced when higher sulphides are

* Known only in combination.

heated. As a rule, however, the oxides or sulphides, when heated, give off oxygen or sulphur, leaving the metals. The compound OsO_3 appears to be incapable of separate existence. When liberated from its compound with potassium oxide, $\mathbf{K_2OsO_4}$, by dilute nitric acid, it decomposes into dioxide and tetroxide, thus:—$\mathbf{2OsO_3 = OsO_4 + OsO_2}$.

3. **By the action of heat on a double compound.**—Osmium dioxide, iridium dioxide, and platinum mono- and di-oxides are produced by gently heating the hydrates. Osmium monoxide is formed when the sulphite, $\mathbf{OsSO_3}$ (see p. 439), is heated in hydrogen.

4. **By double decomposition.**—This method serves for the preparation of most of these compounds. Osmium sesquioxide, $\mathbf{Os_2O_3}$, is produced from potassium osmochloride, $\mathbf{OsCl_3.3KCl}$; the dioxide, $\mathbf{OsO_2}$, from potassium osmichloride, $\mathbf{OsCl_4.2KCl}$; iridium monoxide, $\mathbf{IrO}$, from the compound $\mathbf{IrS_2O_5.6KCl}$; and iridium sesquioxide from potassium iridochloride, $\mathbf{IrCl_3.3KCl}$; by gently heating these salts with potassium carbonate, in a current of carbon dioxide. In aqueous solution, with caustic potash, the hydrates are generally formed, which, on heating, leave the oxides.

The sulphides of osmium are said to be produced from solutions of the corresponding compounds by the action of hydrogen sulphide, and those of iridium and platinum have been similarly obtained. As a rule, sulphides of the alkalies may also be used as precipitants, but the resulting sulphides are soluble in excess. Iridium disulphide, $\mathbf{IrS_2}$, has been prepared by igniting ammonium iridichloride, $\mathbf{IrCl_4.2NH_4Cl}$, with sulphur.

5. **By oxidation of the metal by means of nascent oxygen.**—Finely-divided osmium distilled with nitrohydrochloric acid is oxidised to the tetroxide, which volatilises. Iridium fused with potassium and barium nitrate is oxidised to the trioxide $\mathbf{IrO_3}$, which to some extent remains combined with the potassium or barium. Compounds of platinum dioxide (platinates) are similarly formed.

Properties.—Osmium monoxide, sesquioxide, and dioxide, iridium monoxide and sesquioxide, and platinum monoxide and dioxides are black amorphous powders, insoluble in water and in acids. Osmium dioxide, prepared by heating the hydrate, forms copper-red lumps. Iridium trioxide is black and crystalline. Osmium trioxide is said to be formed as a waxy substance, in combination with the tetroxide when the latter is distilled from strong sulphuric acid. Osmium tetroxide forms large crystals; it is very volatile; it melts below 100°, and volatilises at a somewhat higher temperature. It has an exceedingly pungent, disagreeable smell, strongly irritating the eyes, and is very poisonous when its vapour is inhaled. An antidote is said to be hydrogen sulphide well diluted with water. It is soluble in water, and also in alcohol and in ether: the solution in the latter two menstrua deposit metallic osmium on standing. It also dissolves in alkalies to red or yellow solutions; these, when heated, part with it to some extent, and to some extent lose oxygen, retaining an osmite in solution. It is best obtained pure by saturating the first third of the distillate obtained from osmiridium (a native alloy of iridium and osmium) and nitrohydrochloric acid with caustic potash, and again distilling. Part of the tetroxide volatilises over in crystals, and a portion dissolves in the distillate. It is made use of as a means of hardening animal preparations for the microscope.

The sulphides of osmium are black insoluble substances. The one best known is $\mathbf{OsS_4}$, which is a black precipitate obtained in saturating a solution of the tetroxide in hydrochloric acid with hydrogen sulphide.

Iridium monosulphide is a blackish-blue insoluble substance; the sesquisulphide, brownish-black, and sparingly soluble; the disulphide a yellow-brown powder, and the trisulphide has a dark yellow-brown colour; it is difficult to precipitate. All the sulphides of iridium dissolve in solutions of alkaline sulphide, no doubt forming compounds; none of which, however, have been investigated.

Platinum mono- and disulphides are also black and insoluble. The disulphide is soluble in alkaline sulphides, forming double compounds, regarding which there are no data. Platinum selenide, directly prepared, is a grey infusible mass. Its formula is unknown.

Double compounds.—1. With water: hydrates.

Hydrated osmium monoxide, $\mathbf{OsO.nH_2O}$; sesquioxide, $\mathbf{Os_2O_3.nH_2O}$; dioxide, $\mathbf{OsO_2.2H_2O}$; iridium sesquioxide, $\mathbf{Ir_2O_3.2H_2O}$, and $\mathbf{5H_2O}$; dioxide, $\mathbf{IrO_2.2H_2O}$; platinum monoxide, $\mathbf{PtO.H_2O}$; and dioxide, $\mathbf{PtO_2.nH_2O}$.

These are all prepared by acting on solutions of corresponding compounds with sodium or potassium hydroxide. Thus:— $\mathbf{OsO.nH_2O}$, from $OsSO_3$ and KOH in a closed vessel; $\mathbf{Os_2O_3.nH_2O}$, from $OsCl_3.3KCl$; $\mathbf{OsO_2.2H_2O}$, from $OsCl_4.2NaCl$; $\mathbf{Ir_2O_3.2H_2O}$ from $IrCl_3$, with KOH and alcohol; $\mathbf{Ir_2O_3.5H_2O}$, from $IrCl_3.3KCl$ and a little potash; $\mathbf{IrO_2.2H_2O}$, from $IrCl_4.2KCl$ with boiling potash; or from $IrCl_3.3KCl$ and potash, with subsequent exposure to oxygen; **$PtO.H_2O$**, from $PtCl_2$ and warm KOH; some remains dissolved, but is precipitated on addition of dilute sulphuric acid; and **$PtO_2.nH_2O$**, by the action of potassium hydroxide, or finely divided calcium carbonate, or a solution of platinic nitrate or sulphate, $Pt(NO_3)_4$, or $Pt(SO_4)_2$.

Hydrated osmium monoxide is a blue-black powder, soluble with a blue colour in hydrochloric acid; it rapidly oxidises on exposure to air. The hydrated sesquioxide is a brown-red precipitate, soluble in acids; the hydrated dioxide is a gummy-black precipitate, insoluble in acids.

Hydrated iridium sesquioxide, **$Ir_2O_3.2H_2O$**, is a black precipitate; the compound **$Ir_2O_3.5H_2O$** is yellow, and easily oxidised. It dissolves in excess of potassium hydroxide. The hydrated dioxide, **$IrO_2.2H_2O$** is an indigo-coloured precipitate soluble in acids with a dark-brown colour.

Hydrated platinum monoxide is a black powder, soluble in acids, forming unstable salts; the hydrated dioxide, **$PtO_2.nH_2O$**, is also a black powder, soluble in acids, forming platinic salts.

2. **Compounds with other oxides:**—Of these only the **osmites**, and **platinates** have been investigated.

Potassium osmite, **$K_2OsO_4.2H_2O$**, is prepared by dissolving the tetroxide in potassium hydroxide, and adding alcohol, which reduces the tetroxide to trioxide in presence of the potash. Thus prepared, it is a brick-red powder. If the reduction be effected more slowly by potassium nitrite, KNO_2, the osmite crystallises in octahedra. The sodium salt is more soluble. No other salts have been investigated. These salts are composed of the oxide OsO_3, unknown in the free state. Although osmium tetroxide dissolves in alkalies, yet the solution, when distilled, yields free tetroxide again; hence the combination, if there is one, must be very unstable.

Hydrated iridium sesquioxide, **$Ir_2O_3.5H_2O$**, dissolves in alkalies, possibly forming compounds; but none have been isolated. The trioxide, prepared by fusing finely divided iridium with potassium nitrate, is mixed or combined with a variable amount of potassium oxide, from which it cannot be freed by washing.

Platinites, compounds of platinum monoxide with oxides of the alkaline metals, appear to be formed when platinum is heated with caustic alkalies. They are uninvestigated. **Platinates** are produced by the action of excess of alkali on a solution of platinum tetrachloride. The following have been analysed:— **$3PtO_2.Na_2O.6H_2O$**; a reddish-yellow crystalline precipitate formed by exposing a solution containing platinum tetrachloride and sodium carbonate to sunlight; **$3PtO_2.K_2O$**, produced by heating potassium platinichloride with caustic potash to dull redness, and exhausting with water; it is a rust-coloured substance; and **$3PtO.BaO$**, formed by exposing a mixture of platinum tetrachloride and barium hydroxide to sunshine. These compounds all require investigation. The sulphides of osmium, iridium, and platinum dissolve in solutions of sulphides of the alkalies, no doubt forming compounds, none of which, however, have been investigated.

Oxysulphides.—Osmium tetrasulphide on oxidation yields a body of the formula, $Os_3S_7O_5.2H_2O$; and on further oxidation, $OsSO_3.3H_2O$. The latter is not a sulphite, but hydrated osmium tetroxide, in which one atom of oxygen has been replaced by sulphur; it differs from osmous sulphite, $OsSO_3$, produced by dissolving the tetroxide in sulphurous acid. A somewhat similar body is said to be formed when platinum disulphide is boiled with nitric acid.

Double compounds of platinum.—Platinonitrites.—Solutions of potassium nitrite and potassium platinichloride mixed, deposit crystals of potassium platinonitrite, the empirical formula of which is $K_2Pt(NO_2)_4$. This compound contains dyad platinum, for on treatment with chlorine, two atoms add themselves on, converting the compound into one containing tetrad platinum. The potassium salt gives with silver nitrate a precipitate of silver platinonitrite, $Ag_2Pt(NO_2)_4$, from which the barium salt, $BaPt(NO_2)_4$, is produced by the action of barium chloride. From the barium salt many others have been produced by the action of solutions of sulphates of the metals. The platinonitrites are uniformly soluble, and crystallise well. The following have been prepared :—

$Li_2Pt(NO_2)_4.3H_2O$; $Na_2Pt(NO_2)_4$; $K_2Pt(NO_2)_4.2H_2O$; $Rb_2Pt(NO_2)_4.2H_2O$; $Cs_2Pt(NO_2)_4$; $(NH_4)_2Pt(NO_2)_4.2H_2O$.
$MgPt(NO_2)_4.5H_2O$; $ZnPt(NO_2)_4.8H_2O$; $CdPt(NO_2)_4.3H_2O$.
$CaPt(NO_2)_4.5H_2O$; $SrPt(NO_2)_4.3H_2O$; $BaPt(NO_3)_4.3H_2O$.
$Y_2\{Pt(NO_2)_4\}_3.9H_2O$; $Al_2\{Pt(NO_2)_4\}_3.14H_2O$; $Tl_2Pt(NO_2)_4$.
$MnPt(NO_2)_4.8H_2O$; $CoPt(NO_2)_4.8H_2O$; $NiPt(NO_2)_4.8H_2O$.
$Ce_2\{Pt(NO_2)_4\}_3.18H_2O$; $PbPt(NO_2)_4.3H_2O$.
$CuPt(NO_2)_4.3H_2O$; $Ag_2Pt(NO_3)_4$; and $Hg_2Pt(NO_3)_4.Hg_2O.H_2O$.

Salts of erbium, lanthanum, and didymium are also said to have been prepared.

These salts, treated with an alcoholic solution of iodine, form **iodoplatininitrites,** containing two atoms of iodine in excess of the above formulæ, *e.g.*, $K_2(PtI_2).(NO_2)_4$. The hydrogen, lead, and silver salts are insoluble, and are thrown down from the potassium salt by adding nitrate of hydrogen, lead, or silver to its solution. These salts are, as a rule, amber coloured, and crystallise well. The platinum is not thrown down by hydrogen sulphide, nor is the iodine removed by silver nitrate, but on addition of a mercuric salt, mercuric iodide separates.

Attempts to prepare platinonitrites of beryllium, iron, or indium, by addition of the sulphates of these metals to the barium salt result in the formation of diplatinonitrites, or more correctly **diplatinoxynitrites**, the products of decomposition of nitrogen trioxide being evolved, thus :—

$$2BePt(NO_2)_4.Aq = Be(Pt_2O)(NO_2)_4.Aq + Be(NO_2)_2.Aq + \mathit{NO} + \mathit{NO_2}.$$

The beryllium, aluminium, indium, chromic, ferric, and silver salts have been prepared.

A third compound, still more condensed, named **triplatinonitrous acid**, is produced when a solution of platinonitrous acid is allowed to evaporate spontaneously, thus :—

$$3H_2Pt(NO_2)_4.Aq = H_4(Pt_3O)(NO_2)_8.Aq + 2\mathit{NO} + 2\mathit{NO_2} + H_2O.$$

The potassium salt has been prepared; it is a yellow, well-crystallised substance which may be heated to 130° without change.

It is suggested that these compounds have formulæ such as

$$Pt{<}\begin{matrix}O{-}NO{=}NO{-}OK\\ O{-}NO{=}NO{-}OK\end{matrix}, \quad \begin{matrix}I\\ I\end{matrix}{>}Pt{<}\begin{matrix}O{-}NO{=}NO{-}OK\\ O{-}NO{=}NO{-}OK\end{matrix};$$

$$O{<}\begin{matrix}Pt{-}O{-}NO{=}NO{-}OK\\ Pt{-}O{-}NO{=}NO{-}OK\end{matrix}.$$

It is possible that these bodies may be in some measure analogues of the nitrosulphides, described on p. 343; but too little is still known of these compounds.

Platinimolybdates and platinitungstates, analogous to the silicitungstates, have already been briefly described on p. 404.

Chloroplatinosulphites, of the general formula $\mathbf{ClPtSO_3M}$, are produced by the action of sulphurous acid on ammonium platinichloride. These compounds have been already described on p. 439.

Carbonyl, the group CO, also enters into combination with platinum and halogens, to form **platinicarbonyl compounds.** They are produced by direct union of platinum dichloride with carbon monoxide. Three compounds are thus formed,

$$Cl_2{=}Pt{=}CO; \; Cl_2{=}Pt{=}(CO)_2; \text{ and } Cl_2{=}Pt{<}\begin{matrix}CO{-}PtCl_2\\ |\\ CO{-}CO\end{matrix}.$$

On heating the mixture to 150°, the second and third give off carbon monoxide; and on raising the temperature to 240° in a stream of carbon dioxide, platinicarbonyl chloride, $Cl_2 = Pt = CO$, sublimes in golden-yellow needles, melting at 195°. It also crystallises from carbon tetrachloride, CCl_4. The original crude product heated to 150° in a current of carbon monoxide yields a sublimate of platinidicarbonyl chloride, $Cl_2Pt(CO)_2$: it forms white needles melting at 142°. The third compound, diplatinitricarbonyl tetrachloride is extracted from the crude product by boiling carbon tetrachloride, from which it crystallises in slender yellow needles, melting at 130°. The two latter compounds sublime if heated in a current of carbon monoxide, whereas they decompose if heated alone.

Platinous chloride also forms double compounds with phosphorous acid. The combination does not take place directly, but by the action of water on monophosphoplatinic chloride (see p. 174), thus:—$Cl_2Pt=PCl_3 + 3H_2O = 3HCl + Cl_2=Pt=P(OH)_3$.

Dichloroplatini-phosphonic acid forms deliquescent orange-red prisms. The silver and lead salts have been prepared; the acid is decomposed by alkalies.

Diphosphoplatinic chloride, produced by dissolving monophosphoplatinic chloride in phosphorus trichloride, forms yellow crystals. It yields with water cooled with ice, **dichloro-platinidiphosphonic acid,** thus:—

$$Cl_2Pt=P_2Cl_6 + 6H_2O = 6HCl + Cl_2Pt{<}\genfrac{}{}{0pt}{}{P(OH)_3}{P(OH)_3},$$

which consists of yellow, very deliquescent needles. If the temperature rises to 10° or 12°, the body $ClPt{=}P_2\overset{\displaystyle O}{\overset{/\backslash}{}}(OH)_5$ is formed; it is a colourless crystalline acid, which at 150° loses water, leaving $ClPt{=}P_2\overset{\displaystyle O}{\overset{/\backslash}{}}O(OH)_3$, a light yellow powder.

Platinum alloys easily with tin, forming a compound of the formula Pt_2Sn_3. This substance burns when heated in air, forming an oxide, $Pt_2Sn_3O_3$. It also forms a black compound when treated with hydrochloric acid, which appears to be the corresponding chloride. This body with dilute ammonia yields a hydroxide, $Pt_2Sn_3O_2(OH)_2$, as a brownish-black insoluble body. When it is gently heated in a current of dry oxygen, the oxide $Pt_2Sn_3O_4$ is formed.*

Oxides, Sulphides, Selenides, and Tellurides of Copper, Silver, Gold, and Mercury.

List:—	Oxygen.	Sulphur.
Copper....	Cu_2O; CuO; Cu_2O_3;† CuO_2.	Cu_2S; CuS; CuS_2†.
Silver	Ag_2O; AgO; — ; — ;	Ag_2S; — ; — .
Gold	Au_2O; AuO(?); Au_2O_3§; — .	Au_2S; — ; Au_2S_3.
Mercury ..	Hg_2O; HgO; — ; — .	— ; HgS; —

List:—	Selenium.	Tellurium.
Copper	Cu_2Se; $CuSe$.	?
Silver...............	Ag_2Se; $AgSe$.	Ag_2Te.
Gold	— —	Au_2Te.
Mercury	— $HgSe$.	—

Cu_4O is also said to have been prepared, and there is also some doubtful evidence of the existence of a similar suboxide of silver, Ag_4O.‡

Sources.—Many of these bodies occur native. Cuprous oxide, Cu_2O, occurs as *red copper ore* in regular octahedra, and as *copper bloom* in trimetric needles. Cu_2S is known as *copper glance;* it forms trimetric hexagonal prisms; it is also a constituent of *copper pyrites* and of *purple copper ore* (see p. 257). Silver sulphide, Ag_2S, occurs as *silver glance*, or *argyrose*, in dark grey masses with dull metallic lustre. *Argentiferous copper glance*, or *stromeyerite*, has the formula $Cu_2S.Ag_2S$. Cuprous selenide, Cu_2Se, forms the rare mineral, *berzelianite*, occurring in silver white crusts. *Hessite*, or *telluric silver*, Ag_2Te, and the double telluride, $Au_2Te.4Ag_2Te$, also occur native.

Cupric oxide, CuO, forms the important *black copper ore*, or *melaconite;* it crystallises in cubes. The sulphide, CuS, is known as *indigo copper* or *corellin*, crystallising in hexagonal plates. Mercuric sulphide, HgS, when found native is named *cinnabar;* it has a dull red colour; it is the chief ore of mercury; it usually occurs in heavy earthy lumps, but is occasionally found crystallised in acute hexagonal rhombohedra. The crystals are sometimes bright red and transparent. Mercuric selenide, $HgSe$, also occurs native.

* *Comptes rend.*, **98**, 985.

† Known only in combination.

‡ As regards the existence of Ag_4O, see *Chem. Soc.*, **51**, 416; *Berichte*, **20**, 1458; 2554.

§ *Berichte*, **19**, 2541.

One oxyhalide also occurs in nature. *Atacamite* is a native oxychloride of copper, $3CuO.CuCl_2.5H_2O$; it crystallises in green rhombic crystals.

Preparation.—1. By direct union.—Copper heated in air becomes covered with scales; these consist on the exterior of black cupric oxide, **CuO**, and on the interior of red cuprous oxide, Cu_2O. Silver does not combine with oxygen at the ordinary pressure, but under an increased pressure of 15 atmospheres, a portion of the silver oxidises at 300°, forming argentous oxide, Ag_2O; mercury slowly oxidises when kept boiling in an atmosphere of oxygen or air for several weeks. This fact was discovered by Boyle, and it will be remembered that by means of the oxidation of mercury Lavoisier made his all-important discovery of the nature of oxygen (see p. 11). The red powder which slowly gathers on the surface of the boiling mercury used to be termed "*mercurius præcipitatus per se.*"

Silver (argentic) oxide, **AgO**, is produced by the direct oxidation of silver by means of ozone, or by nascent oxygen, when a solution of silver nitrate is electrolysed with silver poles. Gold does not directly unite with oxygen.

The sulphides may all be prepared by direct union. Cuprous sulphide, Cu_2S, is formed when finely divided copper and sulphur are rubbed together in a mortar. The mass grows red hot, great heat being evolved during the union. Red hot copper burns in sulphur-gas, yielding the same compound. Silver and sulphur also unite directly to form argentous sulphide, Ag_2S. Gold and sulphur do not unite when heated together, because the sulphides easily decompose by heat; but on heating a mixture of gold and silver with sulphur, dark grey crystals of the formula $2Au_2S_3.5Ag_2S$ are produced. Mercuric sulphide, **HgS**, is formed as a black amorphous mass by rubbing together mercury (200 parts) and sulphur (32 parts). After sublimation it is brilliant red, and is known as *vermilion*, and used as a paint.

Cuprous and cupric selenides, Cu_2Se and **CuSe**; argentous and argentic selenides, Ag_2Se and **AgSe**, and mercuric selenide, **HgSe**; also argentous and aurous tellurides, Ag_2Te and Au_2Te, and copper telluride, are formed by heating the elements together in the required proportions.

2. **By reducing a higher compound.**—Cuprous oxide is produced by heating a mixture of cupric oxide, **CuO**, or better copper sulphate, $CuSO_4$, with metallic copper to an intense red heat. The sulphate loses SO_3, forming oxide, which is reduced by the metallic copper. Cuprous oxide is also produced by boiling cupric hydroxide with grape sugar and caustic soda,

better in presence of tartaric acid, which keeps the hydroxide in solution as double tartrate. The grape sugar is oxidised, while the copper oxide loses oxygen. Five molecules of grape sugar, $C_6H_{12}O_6$, are capable of reducing one molecule of cupric oxide. This is the basis of **Fehling's process** for estimating sugar. Copper dioxide, $\mathbf{CuO_2}$, and argentic oxide, **AgO**, are very unstable bodies, yielding cupric oxide, **CuO**, and argentous oxide, $\mathbf{Ag_2O}$, on gentle heating.

3. **By decomposing a double compound.**—The hydroxides yield the oxides when gently heated. Silver oxide is formed from the carbonate, $\mathbf{Ag_2CO_3}$, at 200°. Cupric oxide is produced from cupric sulphate, $\mathbf{CuSO_4}$, at a white heat, and from cupric nitrate or carbonate at a red heat. Gold sesquioxide, $\mathbf{Au_2O_3}$, is produced by addition of an acid to an aurate, *e.g.*, $\mathbf{Au_2O_3.K_2O}$, with sulphuric acid. Compounds of gold trioxide with oxides are, as a rule, decomposed by water. The oxide dissolves in strong nitric acid, doubtless forming auric nitrate, but on addition of water the oxide is again deposited. The same change takes place with argentic oxide. It dissolves in moderately strong nitric acid, but the nitrate decomposes on dilution, the oxide being precipitated. Mercuric oxide, **HgO**, like cupric oxide, is usually produced by heating the nitrate; mercurous nitrate, $\mathbf{HgNO_3}$, also leaves mercuric oxide when heated.

Cuprous sulphide, $\mathbf{Cu_2S}$, is formed when cupric sulphate is heated to whiteness in a crucible lined with carbon; and aurous telluride remains on heating sulphotellurate of gold, $\mathbf{TeS_2.Au_2S_3}$.

4. **By double decomposition.**—This process, as a rule, yields hydroxides, but as these bodies are unstable in this group of elements, the oxides are formed.

$\mathbf{Cu_2O}$.—Heating together cuprous chloride, $\mathbf{Cu_2Cl_2}$, and sodium carbonate, or boiling together cuprous chloride and solution of caustic soda.

$\mathbf{Ag_2O}$.—Solution of silver nitrate, $AgNO_3$, and hot barium hydroxide (used because commercial sodium or potassium hydroxide almost always contains chloride); boiling together silver chloride, **AgCl**, and strong caustic potash.

$\mathbf{Au_2O}$.—Aurous chloride, **AuCl**, and cold potash solution.

$\mathbf{Hg_2O}$.—Mercurous chloride or nitrate, and cold caustic potash in the dark.

CuO.—Cupric nitrate, or sulphate, and hot caustic soda or potash solution. In the cold the hydrate is precipitated.

AuO.—Adding solution of hydrogen potassium carbonate to a solution of gold in aqua regia, gold being present in excess; it precipitates when the temperature is raised to 50°.

The sulphides are generally prepared by passing hydrogen sulphide through a solution of a suitable salt of the metal, thus:— **Cu_2S** from **Cu_2Cl_2** suspended in water; **Ag_2S** from $AgNO_3$.Aq; **Au_2S** from $KAu(CN)_2$.Aq (see p. 572); Hg_2S appears not to be formed from a mercurous salt and hydrogen sulphide; the precipitate produced consists of mercuric sulphide, **HgS**, mixed with free mercury; **Cu_2Se** and **Ag_2Se** have been similarly formed with aid of H_2Se; **CuS** from $CuSO_4$.Aq, &c.; **AuS** appears to be the formula of the precipitate produced in a cold dilute solution of auric chloride (?); **HgS** from $HgCl_2$.Aq, &c. With mercury, intermediate sulphochlorides are formed (see below). Cupric selenide, **CuSe**, has been similarly obtained.

Properties.—Cuprous oxide, Cu_2O, is a bright red, or a yellow red, powder, according to the method of preparation; it can be fused at a very high temperature. **Argentous oxide, Ag_2O,** is a dense black powder, which decomposes above 200° into silver and oxygen. **Aurous oxide, Au_2O,** is also a black powder, soluble in alkalies. On standing it changes to auric oxide, **Au_2O_3**, and metallic gold. **Mercurous oxide** is also black, and is even more easily decomposed into mercuric oxide, **HgO**, and mercury; the change is brought about by sunlight, or even by trituration in a mortar.

Cuprous sulphide, Cu_2S, is a black substance. When heated to redness in air, it burns to cupric oxide and sulphur dioxide. It undergoes a reaction when heated with cupric oxide, whereby metallic copper and sulphur dioxide are formed:—**Cu_2S + 2CuO = 4Cu** + SO_2. This reaction takes place during the preparation of metallic copper (compare the action of lead sulphate and oxide on the sulphide, p. 429). **Argentous sulphide** is a leaden grey body with dull metallic lustre; when produced by precipitation it is black. When heated in air it is oxidised to sulphate, **Ag_2SO_4**. **Aurous sulphide** is a dark brown precipitate, which loses its sulphur when strongly ignited.

Cuprous selenide, prepared by fusion, is a silver white substance; by precipitation it is a black powder. **Argentous selenide** is a black precipitate, grey when dry, and melting at a red heat to a silver-white button. **Argentous telluride** forms leaden grey granules, and **aurous telluride** is a grey brittle substance.

Cupric oxide is black. It melts at a bright red heat and crystallises from fused potassium hydroxide in tetrahedra. **Argentic oxide, AgO,** is a white substance; it dissolves in cold nitric acid with a deep brown colour (forming argentic nitrate $Ag(NO_3)_2$?); but on dilution it is precipitated unchanged.

Auric monoxide, AuO, is a black substance, soluble in hydrochloric acid to a dark green solution; and **mercuric oxide,** produced in the dry way, is a brownish-red or red crystalline powder; when heated it becomes bright red, and then turns black and begins to decompose.

Auric sesquioxide, Au_2O_3, prepared by decomposing an aurate with an acid, still retains alkali. To purify it, it is dissolved in strong nitric acid, and on dilution the oxide is precipitated pure. It is a brownish-black powder, soluble in nitric or sulphuric acids, but the nitrate and sulphate are decomposed by addition of water. It is very unstable, being decomposed by light.

Copper dioxide, CuO_2, produced by adding dilute hydrogen dioxide at 0° to cupric hydroxide, **$Cu(OH)_2$,** is a yellowish-brown substance, very unstable, yielding oxygen and cupric oxide.

Cupric sulphide, CuS, produced by precipitation, is black.

Auric monosulphide, AuS, is yellow, and loses sulphur when gently heated, giving aurous sulphide. **Mercuric sulphide** is a velvety-black powder, when produced by direct union or by precipitation. When sublimed, or when warmed in contact with an alkaline sulphide, or when heated with excess of sulphur and solution of potassium hydroxide to 45—50° for 10 hours, it acquires a brilliant red colour; it is by one or other of these methods that vermilion is prepared.

Cupric selenide, CuSe, is a black precipitate, acquiring metallic lustre when rubbed in a mortar; it loses half its selenium by heat. **Argentic selenide, AgSe,** is a white lustrous substance, and **mercuric selenide, HgSe,** is also white, with metallic lustre.

Auric sesquisulphide, Au_2S_3, is a black precipitate.*

Copper disulphide is known only in combination.

Double compounds. 1. With water.—Cuprous, mercurous, and gold oxides do not form hydrates: **silver hydroxide, AgOH,** produced by precipitation in the cold, is a grey flocculent substance, losing water at 60°, and leaving the oxide. It is sparingly soluble in water.

Cupric hydroxide, $Cu(OH)_2$, is a pale blue precipitate, drying to greenish-blue lumps. It has a metallic taste, hence it must be slightly soluble in water. It loses water below 100°, even in presence of water, leaving the black oxide. **Hydrated auric dioxide** is an olive-green precipitate, which cannot be dried without loss of water and conversion into the black oxide, **AuO**.

Mercuric hydrate, $Hg(OH)_2$, produced by precipitation, is a

* See *Berichte*, **20**, 2369, and **2704**.

yellow powder, which may be heated to 100° without decomposition. At a higher temperature it loses water, leaving the *yellow* oxide. It has also a strong metallic taste, hence it must be somewhat soluble.

Hydrated auric sesquioxide, $Au_2O_3.H_2O$, is a dark-brown powder, thrown down from a solution of auric chloride, $AuCl_3.Aq$, mixed with a solution of hydrogen potassium carbonate, on addition of sodium sulphate. If dilute potash be used for precipitation, the **trihydrate, $Au_2O_3.3H_2O = Au(OH)_3$,** is produced. Both of these substances loses water with great readiness.

2. **With oxides.**—These bodies are produced by direct union. They are as follows:—**$HgO.K_2O$.**—Formed by heating mercuric oxide with fused potash. It consists of white crystals.

$Cu_2O_3.nCaO$. This substance forms rose coloured crystals, and is produced by the action of a solution of calcium hypochlorite (bleaching powder, Cl—Ca—OCl) on cupric nitrate. This substance is so unstable that the ratio between sesquioxide of copper and oxide of calcium has not been ascertained; but it appears to be proved that the copper and oxygen bear to one another the proportion indicated by the formula Cu_2O_3.

3. **Aurates.**—These are analogous compounds of gold sesquioxide. Only one has been carefully investigated, viz. :— **$Au_2O_3.K_2O.6H_2O = KAuO_2.3H_2O$.** It crystallises on evaporation from a solution of auric hydrate in caustic potash. It forms yellow crystals, and its solution, mixed with solutions of the salts of other metals gives precipitates, no doubt of analogous composition. On mixing its solution with hydrogen potassium sulphite, yellow needles are formed, of the formula **$KAuO_2.4HKSO_3$**, which are nearly insoluble in dilute alkali, but dissolve in water with decomposition.*

4. **Double sulphides.**—**$4Cu_2S.K_2S$** and **$Au_2S.K_2S$**, are soluble substances crystallising from solutions of the respective sulphides in a strong solution of potassium sulphide.

$(CuS_2)_2(NH_4)_2S$ crystallises in soluble red needles from a solution of cupric sulphide, **CuS**, in yellow polysulphide of ammonium. Mercuric sulphide, precipitated, also dissolves in a mixture of solutions of potassium monosulphide and hydroxide, giving crystals of **$HgS.K_2S.5H_2O$**, which are decomposed by water. The compound **$5HgS.K_2S$**, is also produced by mixture; it crystallises in golden-yellow plates, with one molecule of water; in colourless crystals with **$7H_2O$**; and when anhydrous in black needles.

* As regards the "Purple of Cassius," see *J. prakt. Chem.*, **121**, 30, 252. This substance, used as a test for gold, and produced by addition of stannous chloride to a compound of gold, owes its colour to finely divided metallic gold.

5. **With halides.**—$CuO.CuF_2.H_2O$, and $HgO.HgF_2$ are formed by heating solutions of the respective fluorides. The first is green and insoluble; the second, an orange-yellow powder.

$2CuO.CuCl_2.4H_2O$ is produced by addition of a small amount of potash to $CuCl_2.Aq$; $2HgO.HgCl_2$ is a brick-red powder.

$3CuO.CuCl_2.5H_2O$ occurs native in green rhombic crystals as *atacamite;* prepared by the action of ammonium chloride on metallic copper in presence of air, it forms the pigment *Brunswick green,* with 4 or $6H_2O$. $3HgO.HgCl_2$, produced by heating mercuric chloride with solution of hydrogen potassium carbonate, forms yellow crystals; $4HgO.HgCl_2$ crystallises in brown crusts from the filtrate. $3CuO.CuCl_2$ is a green hydrated substance produced by the action of water on the compound $CuCl_2.N_2H_4$ (see p. 545). The following oxyhalides of mercury are produced by mixture:—$HgO.2HgCl_2$, white and soluble; $2HgO.HgBr_2$, yellow soluble needles; $3HgO.HgI_2$, a yellow brown powder.

Similar **sulphohalides of mercury** are known, viz.:—$2HgS.HgCl_2$; $HgS.HgBr_2$; $2HgS.HgI_2$, all yellowish-white substances, produced by the action of a limited amount of hydrogen sulphide on the respective halide of mercury. From the nitrate and the sulphate, corresponding compounds, $2HgS.Hg(NO_3)_2$ and $2HgS.HgSO_4$ are produced. The compounds $3HgS.HgCl_2$ and $4HgS.HgCl_2$ have also been prepared. Lastly, by boiling mercuric sulphide with a solution of cupric chloride, a brilliant orange-coloured substance is formed, viz., $HgS.CuCl$, sulphur separating at the same time.

Physical Properties.

Mass of one cubic centimetre.

Cu_2O, 6·13 grams; CuO, 6·40; Ag_2O, 7·52; Hg_2O, 10·7; HgO, 11·30.
Pd_2S, 7·30; PtS, 8·85; PtS_2, 7·22.—Cu_2S, 5·79; CuS, 4·64; Ag_2S, 7·36; HgS, 8·10 (cinnabar).

Heats of formation:—

$Pd + O + H_2O = Pd(OH)_2 + 227K$.—$Pd + 2O + 2H_2O = Pd(OH)_4 + 304K$.
$Pt + O + H_2O = Pt(OH)_2 + 179K$.
$2Cu + O = Cu_2O + 408K$.—$Cu + O = CuO + 372K$.—$2Cu + S = Cu_2S + 183K$.
$Cu + S = CuS + 81K$.—$2Ag + O = Ag_2O + 59K$.—$2Ag + S = Ag_2S + 33K$.
$2Au + 3O + 3H_2O = 2Au(OH)_3 - 132K$.—$2Hg + O = Hg_2O + 422K$.
$Hg + O = HgO + 302K$.—$Hg + S = HgS + 149K$.

Concluding Remarks on the Oxides, Sulphides, &c.

In concluding the description of oxides, sulphides, selenides, and tellurides it may be pointed out that the available data as regards compounds of selenium and tellurium are very scanty. A complete theory of chemistry can only be constructed by supplementing deficiencies in one set of compounds by examples from others; and it would follow that in spite of their want of commercial importance, the selenides and tellurides greatly require exhaustive study. The existence of hydrosulphides, analogous to the hydroxides, for example, is possible. But few of these appear to be stable, at least at the ordinary temperature, although the precipitated sulphides usually contain a small amount of sulphur in excess of that required by their formulæ, denoting the presence of a trace of undecomposed hydrosulphide. In this connection it may be noted that the sulphides of many elements, such as arsenic, copper, lead, silver, gold, &c., when produced in dilute neutral solution are soluble in water, and are precipitated only on addition of an acid or salt. Such solutions, however, do not contain appreciable amounts of hydrosulphides; and it is probable that they are either hydroxy-hydrosulphides, or colloidal and soluble varieties of sulphides.

The physical properties of the oxides, sulphides, &c., still require investigation; our knowledge in this respect greatly falls short of our acquaintance with the halides.*

Classification of the oxides.—It has been customary to divide the oxides into three classes:—**basic oxides, acid-forming oxides**, and **peroxides**, to which may perhaps be added a fourth class, **suboxides**. This classification is founded partly on the behaviour of these oxides towards water, towards each other, and when exposed to heat. It cannot be strictly maintained, and indeed it has tended to obscure the relations between different families of oxides. Yet as this nomenclature is still in vogue, it is advisable to insert a short sketch here of the nature of the division.

A **suboxide** is one which shows no tendency towards the formation of double compounds; and which, when treated with acids, is either indifferent to their action, or if attacked, decomposes into element, and a higher oxide, which combines with the distinctive oxide of the acid. Thus, suboxide of lead, or lead dross, of some-

* For a list of double sulphides, see *Pogg. Ann.*, **149**, 381, and **153**, 588. The individual compounds have been described in their place.

what indefinite composition, when treated with acetic acid, for example, yields metallic lead, while lead acetate passes into solution.

A **basic oxide** is one which, when treated with an acid, combines with the distinctive oxide of the acid, forming a *salt*, with liberation of water. Thus calcium oxide and nitric acid give calcium nitrate and water, and so with a multitude of instances. Such bodies are often soluble in water, and it is at present a disputed point whether they are resolved by water into their constituents, basic oxide and acid oxide. This, however, is certain, that in most cases, on evaporating the water, they remain, as a rule, in an anhydrous state. We have seen, however, that many so-called basic oxides are capable of entering into combination with each other; and in such cases, it is a matter of opinion which to term the basic and which the acid oxide. An **acid oxide**, conversely, is defined as one which combines with a basic oxide, forming a salt. It is noticeable that, as a rule, such acid oxides contain more than one atom of oxygen. But, again, many examples of compounds of acid oxides with each other have been noticed, and, as with basic oxides, it is impossible to ascribe to each its peculiar function.

A **peroxide** is defined as one which on treatment with certain acids (especially with strong sulphuric acid) gives off oxygen; or, which on treatment with an aqueous solution of a halogen acid evolves halogen. Such bodies, as a rule, are also decomposed by moderate heat into a lower oxide and oxygen. Here again, however, we notice that almost all so-called peroxides, when suitably treated, yield compounds both with basic and with acid-forming compounds. Such compounds, however, are not usually stable, and lose oxygen readily when heated.

Constitutional formulæ.—In the foregoing chapters on the oxides, **constitutional formulæ** have been adopted, when the molecular weight of the compound has been determined from its vapour-density; as, for example, SO_2Cl_2; or, where the formula can be directly deduced from such compounds by simple and direct connection, as, for instance, $SO_2(OH)_2$. But the latter formulæ are by no meant so well vouched for as the former, and we have seen (p. 421) reason to believe that the simple formula of sulphuric acid does not, in all probability, represent its true molecular weight. Where such evidence is not at hand, the double compounds have been classified as addition products. This, however, by no means precludes the ascribing to them constitutional formulæ, when data sufficient to warrant this course have been accumulated; and, in some cases, we have been guided to such formulæ by analogy with compounds, the proof of whose constitutional formulæ is fairly satisfactory.

Constitution of the double halides and oxyhalides.—But having in many cases seen reasons for giving constitutional formulæ to certain double oxides, it may not be amiss to inquire whether the double halides, which have uniformly been treated as additive compounds, should not also have constitutional formulæ ascribed to them.

It has been suggested that such combination occurs by virtue of the halogen elements, which in such compounds function as *triads* towards each other. We are acquainted, for example, with the compound **ICl_3**, in which iodine acts as a triad. Now, if this supposition be granted, it becomes possible to represent any double halides whatever constitutionally.

For example, the compounds KF.HF, KF.2HF, and KF.3HF, are known. They may be written:—

$$K{-}F{=}F{-}H;\quad K{-}F\left<\begin{matrix}F{-}H\\ |\\ F{-}H\end{matrix}\right.;\quad \text{and}\quad K{-}F\left<\begin{matrix}FH\\ \\ FH\end{matrix}\right>F{-}H.$$

We have also: $Be\left<\begin{matrix}Cl{=}Cl{-}K\\ Cl{=}Cl{-}K\end{matrix}\right.;\quad Ba\left<\begin{matrix}F{=}Cl\\ F{=}Cl\end{matrix}\right>Ba;$

$$Zn\left<\begin{matrix}Cl & Cl{=}Cl.H\\ Cl{=}Cl & \end{matrix}\right>Zn;\quad B\left<\begin{matrix}F{=}KF\\ F\\ F\end{matrix}\right.;$$

and so on. Given the hypothesis, all double halides may be thus represented by a little ingenuity.

It is an ascertained fact that the vapour-densities of many simple halides increase with rise of temperature. For example, the formula of gaseous stannous chloride at temperatures not far above its boiling point, 601°, is Sn_2Cl_4, but with rise of temperature it falls, until at a high temperature its formula is $SnCl_2$. It might be conceived that such bodies are analogously constituted, thus:— $Sn\left<\begin{matrix}Cl{=}Cl\\ Cl{=}Cl\end{matrix}\right>Sn$; but we shall see reason to doubt this explanation in considering compounds of such elements with hydrocarbon radicles (see p. 506). And if such an explanation is unsatisfactory for bodies in the gaseous state, it appears inadvisable to apply it to solid and liquid substances, about whose molecular weights we can only speculate.

PART V.—THE BORIDES; THE CARBIDES AND SILICIDES.

CHAPTER XXX.

BORIDES.—CARBIDES AND SILICIDES.—ORGANO-METALLIC COMPOUNDS.—CONSTITUTION OF DOUBLE COMPOUNDS.

Borides.

Very few of these compounds are known. The list which follows comprises all that have been investigated:—*H_3B*; **Mg_3B_2**; **AlB**; **AlB_6**; **Mn_3B_2**; **B_3C**; **BN**; **Ag_3B**; also borides of iron, palladium, iridium, and platinum. The compound of boron with nitrogen will be considered under the heading "Nitrides."

Hydrogen boride, *H_3B*,* is produced by gradual addition of strong hydrochloric acid to magnesium boride, **Mg_3B_2**, contained in a flask. A colourless gas, consisting of more than 99 per cent. by weight of hydrogen and less than 1 per cent. of hydrogen boride is evolved. It has a disagreeable smell, and produces headache and nausea. It is sparingly soluble in water, but its solution is permanent, retaining its smell for several years. The gas is decomposed into boron and hydrogen at a red heat. Hydrogen boride communicates a brilliant green colour to the flame of the burning hydrogen.

Its formula was determined by comparing the weight of water obtained by passing a known volume of the gas over red-hot copper oxide with that obtained from an equal volume of hydrogen. Had its formula been H_2B, occupying the same volume as the hydrogen it contains, the weight of water should have been identical in both cases; had its formula been HB, a less amount of water would have been produced. Actual experiment showed that a somewhat greater amount was formed; hence the conclusion that it contains more hydrogen than H_2B; and the probable conjecture that its formula is H_3B.

* Jones, *Chem. Soc.*, **35**, 41; **39**, 215.

Magnesium boride, Mg_3B_2, is a grey semifused mass, produced by direct union; or when formed by the action of boron chloride on hot magnesium, an almost black substance. It is most easily obtained, although in an impure state, by heating together to redness in a covered crucible a mixture of boron trioxide and magnesium dust. It is insoluble in water.

Aluminium borides, AlB, and **AlB_6,*** are produced when aluminium and boron trioxide are heated together. The first forms golden-yellow hexagonal laminæ; the latter large black laminæ. At the same time boron carbide is formed, due to the carbon of the furnace, probably of the formula **B_3C**, in hard black crystals with metallic lustre, insoluble in nitric acid.

Manganese boride, Mn_3B_2, forms grey-violet crystals, produced by heating manganese carbide to redness with boron trioxide. They are decomposed by water at 100°, and by acids, hydrogen alone being evolved.

Silver boride, possibly **Ag_3B,** is a black precipitate produced when hydrogen, containing hydrogen boride, is passed through a solution of silver nitrate. It reacts with hot water, evolving hydrogen boride.—**Boride of iron,** a substance with white metallic lustre, is formed by heating ferrous borate in a stream of hydrogen, or by the action of boron trichloride on red-hot iron.

The remaining borides have been little investigated.

Carbides and Silicides.

Compounds of carbon with hydrogen have been very fully investigated, and those containing hydrogen and other elements (the "organo-metallic" compounds) are also, in many instances, known. To describe the "hydrocarbons" is beyond the province of this book, as the subject is fully treated in text-books of organic chemistry. Some of the more important, however, will be considered here.

Methane or **marsh-gas,** CH_4.—**Sources.**—As its name implies, this compound is formed in marshes, where it is produced by the decomposition of carbon compounds such as woody fibre, out of contact with air, at the bottom of stagnant pools. When the mud is stirred, the marsh-gas comes to the surface in large bubbles mixed with carbon dioxide, also arising from the decay of the organic matter, and with nitrogen dissolved in the water. Methane is also known as "fire-damp" by miners, and occurs in coal mines. When the pressure of the atmosphere diminishes, as shown by a

* *Comptes rend.*, **97**, 456.

falling barometer, methane issues from cracks in the coal, mixing with the air of the mine, and forming an exceedingly dangerous explosive mixture, by the ignition of which many lives are annually lost. Methane requires for complete combustion to carbon dioxide and water twice its volume of oxygen, corresponding to approximately ten times its volume of air. But, in order to fire such a mixture, the temperature must be high. Sir Humphrey Davy took advantage of this fact in his invention of the "Davy Safety Lamp," an oil lamp completely surrounded with copper gauze; the copper conducts away the heat, so that, even if a mixture of methane and air penetrates to the interior of the lamp, the heat it evolves is distributed over a considerable mass of copper, and the temperature is thereby lowered below the point of ignition of the mixture of methane and air in the mine. When such combustion takes place in the interior of the lamp, however, the miner should withdraw.

Methane issues from borings in the ground, in the vicinity of the oil-springs of Pennsylvania, in enormous quantity, and it is utilised as a fuel.

Methane is also one of the chief constituents of coal-gas, formed by the destructive distillation of coal; London gas contains from 35 to 42 per cent. by volume.

Preparation.—When carbon and hydrogen combine directly in the intense heat of the arc-light, they form not methane, but acetylene, C_2H_2. From this compound, however, methane can be produced by the action of nascent hydrogen.

Methane is also obtainable by double decomposition. 1. When a mixture of hydrogen sulphide and carbon disulphide vapour, produced by passing a current of the former through a wash-bottle containing liquid carbon disulphide, is led through a red-hot tube of hard glass filled with copper turnings, the copper withdraws sulphur, and hydrogen and carbon combine, thus:—

$$2H_2S + CS_2 + 8\mathbf{Cu} = 4\mathbf{Cu_2S} + CH_4.$$

2. Pure methane is formed by the action of water or ammonia on zinc-methyl (see p. 503); the zinc is converted into hydroxide or into zinc-amide, and its place is filled by hydrogen, thus:—

$$\mathrm{Zn(CH_3)_2} + \mathrm{2H_2O} = \mathbf{Zn(OH)_2} + 2CH_4;$$
$$\mathrm{Zn(CH_3)_2} + 2NH_3 = \mathbf{Zn(NH_2)_2} + 2CH_4.$$

The decomposition of a mixture of sodium acetate and hydroxide by heat also yields impure methane. It is better to substitute for pure caustic soda, "soda-lime," produced by slaking lime with caustic soda solution. Sodium acetate may be regarded as sodium carbonate, one sodoxyl group (—ONa) of which is replaced by methyl, *i.e.*, methane minus an atom of hydrogen. When heated with sodium hydroxide, its sodoxyl replaces the methyl-group, which combines with the hydrogen of the caustic soda, forming methane, thus:—

$$CH_3\text{—}CO.ONa + H\text{—}ONa = CH_4 + CO{<}^{ONa}_{ONa}.$$

A flask of hard glass should be used, and the mixture of acetate and soda-lime should be thoroughly dried before it is placed in the flask.

Properties.—Methane is a colourless gas, without smell, almost insoluble in water. When pure it burns with a non-luminous flame. It decomposes, when strongly heated, into other hydrocarbons and hydrogen. It is unattacked by acids or alkalies; with chlorine in direct sunlight, however, it gives a series of substitution-products, in which one, two, three, or four atoms of hydrogen are successively replaced by chlorine. The names and formulæ of these bodies are as given below:—

Chloromethane, or methyl chloride, CH_3Cl;
Dichloromethane, or methylene dichloride, CH_2Cl_2;
Trichloromethane, or chloroform, $CHCl_3$;
Tetrachloromethane, or carbon tetrachloride, CCl_4.

These bodies may have their chlorine replaced, atom by atom, by the action of nascent hydrogen, yielding methane as a final product. The existence of this series of compounds strongly corroborates the conclusions drawn from the vapour-density of methane, that it contains four atoms of hydrogen, inasmuch as its hydrogen is replaceable in fourths.

Methane boils at −164°, and solidifies to a white snow-like mass about −185·8°.

Hydrogen silicide, SiH_4, the corresponding compound to methane, is also a colourless gas. It does not occur free in nature.

Preparation.—1. By direct combination.—Nascent hydrogen, evolved by electrolysing a solution of common salt by a battery, the negative pole of which consists of aluminium containing silicon, unites with the silicon, and spontaneously inflammable bubbles of hydrogen containing hydrogen silicide are evolved 2. **By double decomposition.**—Magnesium silicide, produced by heating to redness magnesium powder and sand in a small crucible, is treated with dilute hydrochloric acid. The magnesium silicide yields magnesium chloride and siliciuretted hydrogen, as hydrogen silicide is often called, thus:—

$$\mathbf{Mg_2Si} + 4HCl.Aq = \mathit{H_4Si} + 2MgCl_2.Aq. \text{ (Compare } BH_3, \text{ p. 497.)}$$

This forms a convenient lecture experiment.

A third method is to heat the compound $SiH(OC_2H_5)_3$ with sodium, which is itself unchanged, but which induces the following decomposition:—$4SiH(OC_2H_5)_3 = \mathit{SiH_4} + 3Si(OC_2H_5)_3$.

Properties.—Hydrogen silicide is a colourless, spontaneously inflammable gas, insoluble in water, burning to water and silica. It also burns in chlorine, producing hydrogen and silicon chlorides. It is decomposed at 400° into amorphous silicon and hydrogen. It liquefies under a pressure of 50 atmospheres at −11°, 70 atmospheres at −5°, and 100 atmospheres at −1°.

A liquid compound analogous to chloroform, of the formula $SiHCl_3$, is produced by heating silicon to redness in a current of hydrogen chloride. It boils at 55—60°. With water, it yields SiH_2O_3, or silicoformic anhydride. The corresponding iodide is also known; it boils at 220°.

Ethane, $\mathit{C_2H_6}$.—When methyl iodide, CH_3I, is treated with sodium, sodium iodide and di-methyl or ethane is produced, thus:—

$$CH_3\overline{\underline{I \quad Na}}\,|\overline{\underline{Na \quad I}}|CH_3, \text{ or } 2CH_3I + 2\mathbf{Na} = 2\mathbf{NaI} + \mathit{C_2H_6}.$$

From this synthesis it is argued that the constitution of ethane is represented by the formula $H_3C—CH_3$. This compound differs in formula from methane by containing the group (CH_2) in addition; and many other hydrocarbons are known of the general formula C_nH_{2n+2}. To treat of such compounds is the province of Organic Chemistry; but it may be here stated that such a series of compounds is termed a homologous series. Thus we have:—**Methane,** $\mathit{CH_4}$; **ethane,** $\mathit{C_2H_6}$; **propane,** $\mathit{C_3H_8}$; **butane,** $\mathit{C_4H_{10}}$; **pentane,** $\mathit{C_5H_{12}}$, and so on. Like methane, ethane is a colourless insoluble gas, devoid of taste or smell; it issues, along with methane, from the soil in parts of Pennsylvania.

Hexahydrogen disilicide, or **silicon-ethane,** $\mathbf{Si_2H_6}$.—When an electric discharge is passed through hydrogen silicide, a yellow compound encrusts the tube, having the composition of silicon-ethane. It burns in the air, and ignites on percussion; like the tetrahydride, it burns in chlorine.* This compound is analogous to the hexachloride described on p. 151, and its formula is deduced by analogy, for its molecular weight is unknown. No other mem-

* *Comptes rend.*, **89**, 1068.

ber of the homologous series is known with silcon, but a compound containing both carbon and silicon, named **silico-nonane**, has been produced by treating zinc ethyl (see p. 503), $Zn(C_2H_5)_2$, with silicon tetrachloride. Its formula is $Si(C_2H_5)_4$, and it is noteworthy as the analogue of the corresponding carbon compound nonane, C_9H_{20}.

For other compounds displaying analogy between carbon and silicon, a text-book of Organic Chemistry must be consulted.

Double compounds of methyl and ethyl; "Organo-metallic compounds."—This name is given to compounds in which the metallic elements replace the hydrogen of methane, ethane, and similar hydrocarbons; or they may be regarded as compounds of groups such as $(CH_3)'$, methyl, or $(C_2H_5)'$, ethyl, with the elements. As these compounds are usually capable of existence in the gaseous state, and as their vapour-densities are, as a rule, known, they are well adapted to throw light on the formulæ of other compounds of the elements. They will be considered in their order.

Preparation.—1. By the action of methyl or ethyl iodide (iodomethane, or iodoethane), CH_3I, or C_2H_5I, **on the elements,** or **on their alloys with potassium or sodium,** thus:—

$$\mathbf{ZnNa_2} + 2C_2H_5I = Zn(C_2H_5)_2 + \mathbf{2NaI};$$
$$\mathbf{SbK_3} + 3C_2H_5I = Sb(C_2H_5)_3 + \mathbf{3KI}.$$

2. **By heating the halides of some elements with zinc-ethyl,** $Zn(C_2H_5)_2$, thus:—

$$\mathbf{HgCl_2} + Zn(C_2H_5)_2 = Hg(C_2H_5)_2 + \mathbf{ZnCl_2};$$
$$\mathbf{Sn(C_2H_5)_2I_2} + Zn(C_2H_5)_2 = Sn(C_2H_5)_4 + \mathbf{ZnI_2}.$$

3. **By replacing zinc in zinc-ethyl by another metal by direct action,** *e.g.*:—

$$Zn(C_2H_5)_2 + \mathbf{2Na} = 2Na(C_2H_5) + \mathbf{Zn}.$$

Properties.—These compounds are colourless or yellow liquids with nauseous smell, heavier than water, and with few exceptions ($Bi(C_2H_5)_3$, $Pb_2(C_2H_5)_6$, and $Sn_2(C_2H_5)_4$), volatile without decomposition at comparatively low temperatures. Hence their vapour-densities have in most cases been determined. They are almost all decomposed by water, yielding the hydroxide of the element and ethane, thus:—$Zn(C_2H_5)_2 + 2H.OH = \mathbf{Zn(OH)_2} + 2C_2H_6$. Most of them react energetically with oxygen, and inflame spontaneously in air. By cautious oxidation they yield either organo-metallic oxides or double oxides of metal and ethyl or methyl, for example:—$Sb(C_2H_5)_3 + O = \mathbf{O{=}Sb(C_2H_5)_3}$; $Sn_2(C_2H_5)_6 + O = \mathbf{O{<}^{Sn(C_2H_5)_3}_{Sn(C_2H_5)_3}}$; zinc-ethyl gives a double oxide, named zinc-ethylate or ethoxide, thus:—$Zn(C_2H_5)_2 + 2O = Zn(OC_2H_5)_2$.

In many cases they form addition-products with halogens, as for example, $Sb(C_2H_5)_3 + \mathbf{I_2} = \mathbf{I_2Sb(C_2H_5)_3}$; in other instances, the compound is split, forming an iodide of the organo-metal and ethyl iodide, thus:—

$$Hg(C_2H_5)_2 + \mathbf{I_2} = \mathbf{I.Hg.C_2H_5} + C_2H_5I;$$
$$(C_2H_5)_3Sn{-}Sn(C_2H_5)_3 + \mathbf{I_2} = \mathbf{2ISn(C_2H_5)_3}.$$

But halogen compounds, if they contain two atoms of halogen, yield oxides on treatment with hydroxides of the alkalies; for example, $\mathbf{I_2Sn(C_2H_5)_2}$ yields $\mathbf{O{=}Sn(C_2H_5)_2}$; and the compounds $\mathbf{OSn_2(C_2H_5)_6}$, and $\mathbf{O{=}Sb(C_2H_5)_3}$, are produced by direct oxidation.

Hydroxides are similarly produced from the monohalides, thus:—$\mathbf{C_2H_5HgI} + KOH.Aq = KI.Aq + \mathbf{C_2H_5.Hg.OH}$; and, similarly, $\mathbf{(C_2H_5)_2Tl.OH}$, $\mathbf{(C_2H_5)_3Sn.OH}$, and $\mathbf{(C_2H_5)_4Sb.OH}$ have been prepared. These are soluble compounds, which, on treatment with acids form definite salts, such as nitrate, $C_2H_5.Hg.ONO_2$; sulphate, $\{(C_2H_5)_2TlO\}_2SO_2$, &c.

List.—$Na(C_2H_5)$; $K(C_2H_5)$; these bodies have been produced only in combination with zinc ethyl, by acting on that substance with metallic sodium or potassium. The compounds have the formulæ $\mathbf{Na(C_2H_5).Zn(C_2H_5)_2}$, and $\mathbf{K(C_2H_5).Zn(C_2H_5)_2}$, and consist of large clear crystals.

$Be(C_2H_5)_2$; from beryllium and mercury ethyl, $Hg(C_2H_5)_2$, b. p. 287°.

$Mg(C_2H_5)_2$; from magnesium and ethyl iodide. Not volatile.

$Zn(CH_3)_2$ (b. p. 46°), and $Zn(C_2H_5)_2$ (b. p. 118°); from an alloy of zinc and sodium and the respective iodide.

$B(C_2H_5)_3$ (b. p. 95°); from boron trichloride and zinc ethyl. This body is analogous to hydrogen boride; with oxygen it yields the ethyl salts of ethyl boracic acid, $C_2H_5.B(OC_2H_5)_2$, and of diethyl boracic acid, $(C_2H_5)_2.B(OC_2H_5)$. These substances are more easily produced by the action of zinc ethyl on ethyl borate, $B(OC_2H_5)_3$, formed by the action of ethyl hydroxide (ordinary alcohol) on boron trichloride, thus:—

$$\mathit{BCl_3} + 3C_2H_5OH = B(OC_2H_5)_3 + 3\mathit{HCl}.$$

The reactions of this substance with zinc ethyl are as follows:—

$$B(OC_2H_5)_3 + Zn(C_2H_5)_2 = B{<}^{C_2H_5}_{(OC_2H_5)_2} + Zn{<}^{C_2H_5}_{OC_2H_5} \text{; and}$$

$$B{<}^{C_2H_5}_{(OC_2H_5)_2} + Zn(C_2H_5)_2 = B{<}^{(C_2H_5)_2}_{(OC_2H_5)} + Zn{<}^{C_2H_5}_{OC_2H_5}.$$

The analogy of the compound $Zn(OC_2H_5)_2$, also known, with the hydroxides, will be perceived.

On treatment of ethyl borate of ethyl, and of diethyl borate of ethyl with water, ethyl and diethyl boracic acids are produced, thus:—

$$B{<}^{C_2H_5}_{(OC_2H_5)_2} + 2H_2O = B{<}^{C_2H_5}_{(OH)_2} + 2C_2H_5OH\text{; and}$$

$$B{<}^{(C_2H_5)_2}_{(OC_2H_5)} + H_2O = B{<}^{(C_2H_5)_2}_{OH} + C_2H_5OH.$$

These compounds, it will be noticed, are the analogues of compounds occupying an intermediate position between boron hydride and hydroxide, such compounds being incapable of existence, owing to their instability, thus:—$B{<}^{H}_{(OH)_2}$ and $B{<}^{H_2}_{OH}$; and, similarly, the compound $Zn{<}^{C_2H_5}_{(OC_2H_5)}$ is analogous to a mixed hydride and hydroxide, H—Zn—OH.

$Al(CH_3)_3$; $Al(C_2H_5)_3$; $Al_2(CH_3)_6$; $Al_2(C_2H_5)_6$.—A great deal of interest attaches to these compounds. They are produced by the action of metallic aluminium on mercury methide or ethide. The methide boils at 130°; the ethide at 194°.

The first point which a study of their vapour-densities is able to decide is the precise constitutional formulæ of the halides of such elements as aluminium, chromium, iron, &c. We have seen, on page 143, that while dichloride of chromium, in the state of gas, appears to exist partly as $CrCl_2$, at 1600° its high vapour-density shows the presence of molecular aggregates, probably of the complexity Cr_2Cl_4. Iron dichloride, at 1400—1500°, possesses the simple formula $FeCl_2$. Chromic chloride, about 1060°, has the simple formula $CrCl_3$, while ferric chloride, below 620°, has a complex formula, probably Fe_2Cl_6, but at higher temperatures (750° and upwards) it has the simpler formula $FeCl_3$. The densities of aluminium halides, and their bearing on the molecular weights of these compounds, has been discussed on p. 135, with similar conclusions.

Precisely similar results are obtained with the methide and ethide of aluminium. The following tables give a summary of the results obtained:—

Ethide.—Temperatures :—	234°	235°	258°	310°	350°
Densities :—	65·1(?)	121 (?)	87·4	36·2	36·2

The theoretical density of the compound $Al(C_2H_5)_3$ is 57, and of $Al_2(C_2H_5)_6$ is 114.

Methide.—Temperatures :—	130°	163°	220°	240°
Densities :—	63·1	59·3	40·7	40·5

Theoretical density of $Al(CH_3)_3$, 36·0; of $Al_2(CH_3)_6$, 72.

Two different sets of observers are responsible for the densities of the ethide at 234°, viz., Gay-Lussac, and Buckton and Odling; and at 235°, Victor Meyer, and Louise and Roux. At 310° and 350° the substance is wholly decomposed. All that can be gathered from these results is that, at low temperatures, the formula appears to be $Al_2(C_2H_5)_6$, and at higher temperatures, the molecular formula is a simpler one. With the methide the results show more symmetry. The density at 130°, 63·1, approaches the theoretical density of $Al_2(CH_3)_6$, and at 240°, the number 40·5 is not far removed from 36, the density of $Al(CH_3)_3$. These results generally confirm those obtained with the chloride, given on p. 135; and it may, consequently, be concluded, that at high temperatures all these bodies have the simpler formulæ, and at lower temperatures, more complex formulæ, probably due to association of two simpler molecules.

But a second question arises, which involves a point of extreme theoretical importance. It is: What is the nature of the combination which exists in the double molecular formulæ? To this question there are three possible answers.

First, from analogy with the carbon compounds C_2H_6 and C_2Cl_6, the former of which betrays its constitution by its synthesis from methyl iodide and sodium (p. 501), such compounds as Al_2Cl_6, $Al_2(C_2H_5)_6$, &c., may be regarded as thus constituted :—

$$Cl_3Al—AlCl_3.$$

Second. The complex halides, Al_2Cl_6, &c., may be regarded as analogous to double halides, such as have been discussed on p. 496, constituted thus:

$$Al \begin{array}{l} \diagup Cl=Cl \diagdown \\ —Cl=Cl— \\ \diagdown Cl=Cl \diagup \end{array} Al.$$

Third. They may be regarded as purely additive compounds, groups such as $AlCl_3$ being capable of associating in twos, threes, &c.

The second hypothesis may be at once disposed of by noticing that such a constitution as $(AlCl_3)(Cl_3Al)$ is completely excluded, from the fact that no such formula can be applied to the methides and ethides, for the ethyl and methyl groups have no further power of combination; no addition compounds of methane and ethane are known. This of itself forms a strong argument against the proposed graphic formulæ for the halides, mentioned on p. 496. Hence we are confined to the first and third hypotheses. Against the first, which looks plausible from the analogy between Al_2Cl_6 and C_2Cl_6, it may be urged that while compounds of carbon, in which it functions as a tetrad, as in CH_4, CCl_4, &c., are the rule, such compounds of aluminium are wholly unknown. Moreover, none of the elements of the aluminium group form such compounds; those of manganese, which presents some slight analogy with aluminium, such as MnF_4, are exceedingly unstable.

We are, therefore, obliged to accept the third hypothesis that two classes of chemical compounds can exist; the substitutive, of which we have had many examples, and the additive. It is impossible to class compounds containing water of crystallisation otherwise than as additive compounds, and there appears no reason to believe that double halides are otherwise constituted. Moreover, the uncertainty attaching to molecular formulæ, such as Al_2Cl_6, which appear to be constant over a very small range of temperature, and the consideration of the molecular weight of hydrogen fluoride, and similar bodies, would lead to the supposition that the molecular aggregation is not necessarily restricted to that of twice the simpler molecule. But these considerations should not lead us to exclude substitutive formulæ, for which, as has been shown repeatedly, we have abundant evidence; it would merely lead to the conclusion that it is impossible to represent all forms of chemical combination by their aid.

$Si(C_2H_5)_4$, from $SiCl_4$ and $Zn(C_2H_5)_2$, and the derivatives, $ClSi(C_2H_5)_3$, $Cl_2Si(C_2H_5)_2$, $Cl_3Si(C_2H_5)$, confirm and exemplify the compounds SiH_4,$SiHCl_3$, and $SiCl_4$. Besides these, the existence of compounds

$$Si(OC_2H_5)_4;\ O{=}Si{<}^{C_2H_5}_{OH};\ O{=}Si(C_2H_5)_2;\ \text{and}\ HO{-}Si(C_2H_5)_3,$$

justify the views expressed on p. 306, regarding the constitution of orthosilicic acid.

$Si_2(C_2H_5)_6$, from disilicon hexiodide $\mathbf{Si_2I_6}$, and zinc ethyl, confirms the formula of the hexahydride, $\mathbf{Si_2H_6}$, and renders likely the suggested formulæ for the polysilicic acids, described on p. 307.

The compounds of tin, $Sn(C_2H_5)_4$, from $SnCl_4$ and $Zn(C_2H_5)_2$, or from $\mathbf{Sn(C_2H_5)_2I}$ and $Zn(C_2H_5)_2$; $Sn_2(C_2H_5)_6$, from an alloy of tin and sodium in the proportion expressed by the formula $SnNa_3$, and zinc ethyl; and of $Sn_2(C_2H_5)_4$, from an appropriate alloy by the same reaction, confirm the relationship between silicon and tin. The last compound is analogous to the chloride,

$$Cl_2{=}Sn{=}Sn{=}Cl_2,$$

in the state of vapour at low temperatures.

The relations of lead to silicon and tin are exemplified by the existence of the compounds $Pb(C_2H_5)_4$, and $Pb_2(C_2H_5)_6$.

It will be remembered that lead tetrachloride is a very unstable compound, and that diplumbic hexachloride is unknown (see p. 153).

The similar compounds of the nitrogen and phosphorus groups will be alluded to later, in discussing the nitrides and phosphides of hydrogen.

Lastly, dyad compounds of mercury, $Hg(CH_3)_2$, $Hg(C_2H_5)_2$, and others, are known, the vapour-densities of which establish the formulæ of the dihalides of mercury, although it is not improbable that the latter may have more complex formulæ in the solid state.

Ethylene, C_2H_4.—Like methane, this hydrocarbon forms the first member of a homologous series; it is termed **the olefine series**, from the old name for ethylene, "olefiant gas," due to the fact that ethylene and chlorine combine directly to form a dichloride, $C_2H_4Cl_2$, which is an oily body.

Preparation.—Ethylene is one of the products of the distillation of wood and coal, being probably formed by the action of heat on methane, thus:—$2CH_4 = C_2H_4 + 2H_2$. It is usually prepared by the action of sulphuric acid on alcohol (ethyl hydroxide), $C_2H_5.OH$. The first action is the formation of hydrogen ethyl sulphate, $HO{-}SO_2{-}OC_2H_5$, thus:—$C_2H_5OH + HO{-}SO_2{-}OH = HO{-}SO_2{-}OC_2H_5 + H_2O$. On raising the temperature, the sulphate is decomposed, thus:—$HO{-}SO_2{-}OC_2H_5 = HO{-}SO_2{-}OH + C_2H_4$. The operation should be performed with a large excess of sulphuric acid in a large flask, for the mixture is very apt to froth up. Ethylene and homologous hydrocarbons are also formed by the action of acids on carbide of iron, or cast iron; this mode of formation is analogous to that by which hydrogen boride and silicide are prepared.

Properties.—Ethylene is a colourless gas, without odour, almost insoluble in water. It may be condensed to a liquid boiling at $-102{\cdot}3°$, and when frozen it is a solid melting at $-169°$. It combines directly with the halogens, forming oily bodies, which may be regarded as substitution products of ethane, C_2H_6. From this and other reactions it is assumed that the formula of ethylene is $H_2C{=}CH_2$, analogous to the $Cl_2Sn{=}SnCl_2$, and other similar compounds.

Ethylene burns with a luminous flame, and is one of the constituents of coal-gas which cause it to burn brightly. The luminosity is due to the presence of solid white-hot particles of carbon. That this is the case is proved by the fact that solar light reflected from a candle or coal-gas flame shows when viewed through the spectroscope its characteristic vertical black lines; now gases cannot reflect light, hence the presence of a solid in the flame is demonstrated. By mixing with excess of air, so as to supply sufficient oxygen to wholly consume the carbon within the flame, a non-luminous and hotter flame is obtained; this is the principle of the Bunsen's burner, so necessary in our laboratories.

Acetylene, C_2H_2.—This hydrocarbon is formed along with ethylene during the distillation of wood, coal, &c., and hence it forms one of the constituents of coal-gas. It has been prepared by exposing methane or ethylene to a red heat, probably according to the equations:—$2CH_4 = C_2H_2 + 3H_2$; $C_2H_4 = C_2H_2 + H_2$. One of the easiest methods of preparing acetylene is to partially burn methane; or coal-gas, the methane in which is sufficient for the purpose. When a Bunsen's burner "burns below," a disagreeable smell is perceived, due to this gas. It may be still more conveniently prepared by burning air in coal-gas, by means of the arrangement shown in Fig. 44. The air enters by a small tube into an atmosphere of coal-gas, where it burns in excess of the latter. The acetylene is drawn off by means of an aspirator and after being cooled it is passed through an ammoniacal solution of cuprous chloride, with which it reacts, forming a red insoluble compound.

It is also formed by the direct union of carbon with hydrogen at an intensely high temperature, produced by the electric arc in an atmosphere of hydrogen.

Acetylene is a colourless gas with a disagreeable smell; it liquefies at 1° under a pressure of 48 atmospheres. It unites directly with chlorine, &c., forming a tetrachloride, and it is therefore concluded that it possesses the constitutional formula $HC{\equiv}CH$. When passed over heated sodium, one or both atoms of hydrogen are replaced by the metal, forming **HC≡CNa** and **NaC≡CNa**, solid bodies which on treatment with water at once yield sodium hydroxide and acetylene, thus:—$\mathbf{HC{\equiv}CNa} + HOH = \mathit{HC{\equiv}CH} + NaOH$. With ammoniacal solutions of cuprous or argentous chloride it yields red or yellowish-white precipitates

FIG. 44.

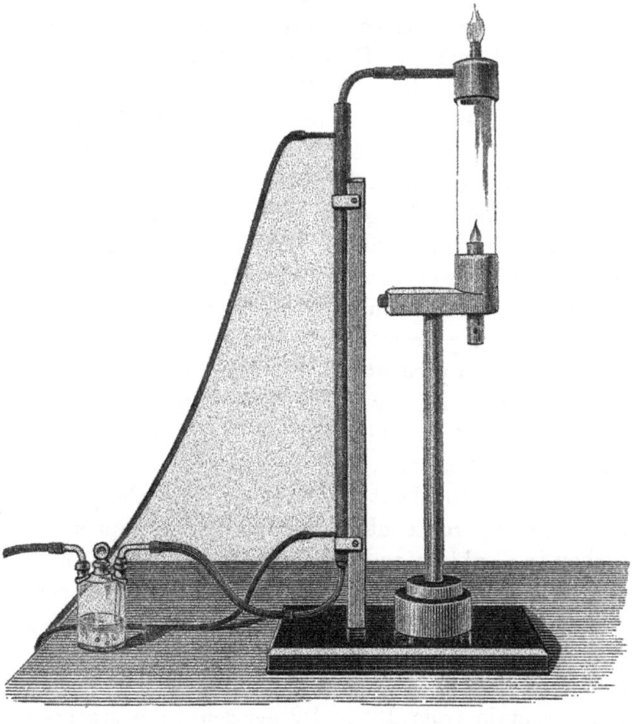

of $\mathbf{H{-}C{\equiv}C{-}Cu{-}CuOH}$ and $\mathbf{H{-}C{\equiv}C{-}Ag}$, which are exceedingly explosive when dry, and which, on treatment with acids, yield acetylene, *e.g.*, $\mathbf{H{-}C{\equiv}C{-}Ag} + ClH.Aq = \mathit{H{-}C{\equiv}C{-}H} + \mathbf{AgCl} + \mathbf{Aq}$. Homologues of acetylene are also known which give similar metallic compounds.

This scanty notice of the hydrocarbons must here suffice. Enough has been said to show the relations of these bodies to the silicon compounds, and to throw some light on the nature of the compounds of other elements. The hydrocarbons may be termed the "elements" of Organic Chemistry, and their derivatives are as numerous as, and more complex than, those of almost all other elements together.

The remaining carbides and silicides have been little investigated, but appear worthy of more careful study.

Iron carbide exists in pig-iron, the crude iron produced by smelting iron ore in a blast furnace with coal and lime. The greater part of the carbon is, however, uncombined. If finely-divided iron be kept fused with charcoal in a crucible until it has united with its maximum of carbon, a dark-grey mass is obtained, exceedingly brittle, and containing 94·36 per cent. of iron and 5·64 per cent. of carbon. Dividing 94·36 by the atomic weight of iron, 56·02, and 5·64 by 12·0, the atomic weight of carbon, the ratio of the number of atoms of iron to that of carbon is ascertained; it is 1·68 Fe to 0·47 C., or approximately Fe_7C_2, which would require 5·7 per cent. of carbon.

Some specimens of pig-iron when broken are seen to have a grey, and some a white colour. The grey specimens contain carbide of iron and free carbon, for when treated with acid, carbon is left in the form of graphite, while hydrocarbons like ethylene, C_2H_4, are evolved. From this it may perhaps be concluded that the carbide of iron present has the formula Fe_2C_2, corresponding to C_2H_4. But such a body has not been obtained. The amount of "combined carbon" in white pig-iron is about 2·5 per cent., and of graphite 0·9 per cent., and in grey pig-iron the "combined carbon" amounts to about 1·0 per cent., and the graphite to 2·6 per cent. If the iron contains manganese, its capacity for retaining carbon in combination is much increased. An iron containing 10 per cent. of manganese retains as much as 4 per cent. of carbon in chemical combination, and is known as "spiegel iron." Steel is also a mixture of iron with its carbide. If the proportion of carbon does not exceed 0·3 per cent., the steel is comparatively soft; if it contain from 1·0 to 1·2 per cent., the steel is hard, and is employed for cutting instruments; 1·4 cent. of combined carbon renders it like white cast iron, more fusible, and very brittle.

The effect of adding **silicon** to **iron** is to modify its properties very considerably. Samples have been obtained containing as much as 10 per cent. of silicon, but the iron has at the same time contained 1·12 per cent. of free carbon and 0·69 per cent. of combined carbon, besides phosphorus, manganese, and sulphur. "Silicon pig," as this mixture is termed, forms better castings than ordinary cast iron. The best results, most free from air-holes, are obtained with from 1·5 to 3 per cent. of silicon. The iron is usually bluish, and has a close-grained fracture, but with 10 per cent. of silicon the colour is nearly white, and the fracture shows large silvery facets. No definite compound has, however, been isolated.

Like iron, **nickel** unites with carbon, forming a brittle carbide of unknown formula.

Compounds of **tin** and **lead** with silicon appear to be formed y direct union.

On dissolving a large amount of pig-iron containing **titanium** in dilute hydrochloric acid, a number of minute cubes with metallic lustre were obtained, having the composition **TiC**.

A carbide of **palladium** is produced by fusing palladium in a crucible filled with lampblack. It is so brittle that if struck with a hammer when red-hot it falls to powder and gives off a white fume. A piece of palladium heated in an alcohol flame unites with carbon (?) before it becomes red-hot, and when removed from the flame it glows until the carbon is consumed. **Iridium** behaves similarly.

Certain carbon compounds containing platinum are said to leave a carbide on gentle ignition. It may be a mixture, however, for on treatment with nitro-hydrochloric acid the platinum dissolves, leaving carbon.

Platinum and **silicon** readily combine, forming a white metallic looking mass.

Copper and **silver** also unite with **silicon**, and three **carbides** of **silver** are said to exist, viz., Ag_4C, Ag_2C, and **AgC**, produced by heating certain compounds of carbon containing silver. Their existence, however, is doubtful.

PART VI.—THE NITRIDES, &c.

CHAPTER XXXI.

NITRIDES, PHOSPHIDES, ARSENIDES, AND ANTIMONIDES OF HYDROGEN: DOUBLE COMPOUNDS: AMINES AND AMIDES.

THE compounds of nitrogen, phosphorus, arsenic, and antimony with hydrogen are best known. Many combinations containing nitrogen and hydrogen have been prepared; the nitrides of the other elements are but little investigated. The phosphides, generally produced by direct union, require investigation. A considerable number of arsenides and antimonides are found native.

1. **Compounds with hydrogen; hydrogen nitrides (ammonia and hydrazine), phosphides (phosphines), arsenides (arsines), and antimonide (stibine).**

List:—

Nitrogen.	Phosphorus.	Arsenic.	Antimony.
NH_3; N_2H_4.	PH_3; P_2H_4; $\mathbf{P_4H_2}$.	AsH_3; $(\mathbf{AsH})_n$.	SbH_3.

Sources.—Ammonia occurs in the atmosphere in very small proportion (3 or 4 parts per million). But its presence is essential to the life of plants, and indirectly of animals. It is washed down by the rain into the soil, whence it is absorbed as food by vegetation, probably after oxidation to nitrates. It is produced by the putrefaction of nitrogenous organic matter, especially by the decomposition of a constituent of urine, urea, $\mathbf{CON_2H_4}$, under the influence of a ferment named *bacillus ureæ*. The change produced is represented thus:—$CON_2H_4 + H_2O = CO_2 + 2NH_3$. Although it is exceedingly soluble in water, yet it is retained by soil, and is available for plant-food. Some ammonia, however, is washed down by streams, and hence natural water always contains traces.

Ammonium chloride is found encrusting the soil in the neighbourhood of volcanoes. The ammonia prepared from this source is named "volcanic ammonia."

The remaining hydrides do not occur free in nature.

Preparation.—By direct union.—Although the decomposition of ammonia is attended by absorption of heat, showing that its formation, as usual in the case of stable compounds, should take place with evolution of heat, yet there is insufficient evidence that ammonia has been produced by direct union of its elements. But if one of the two constituent elements is in the nascent state, union occurs. This may be effected (1) by the action of the induction discharge on a mixture of the gases; and (2) by leading a mixture of moist hydrogen and nitrogen over red-hot iron filings. It is to be assumed in the first case that the induction discharge dissociates molecules of nitrogen and of hydrogen into atoms, which then combine; and in the second that a hydride or nitride (most probably the latter) of iron is formed, which is then attacked by the nitrogen or hydrogen. In both cases, however, mere traces are produced. (3.) Moist nitric oxide passed over hot iron filings yields ammonia; here both hydrogen and nitrogen are nascent. (4.) By the action of nascent hydrogen from zinc and sodium hydroxide (see p. 229) on oxygen compounds of nitrogen, such as nitric oxide, nitrites, or nitrates.

Hydrogen arsenide and antimonide are also obtained by this process. A solution of chloride of arsenic or antimony is placed in a flask containing hydrochloric acid and pure zinc. The nascent arsenic or antimony, liberated from the chloride, unites with the nascent hydrogen, forming arsenide or antimonide of hydrogen.

2. **By the decomposition of their compounds by heat.**—All ammonium and phosphonium salts dissociate when heated, and with the exception of the nitrite, nitrate, chlorate, and a few others, they yield acid and ammonia. Thus ammonium chloride yields ammonia and hydrogen chloride; phosphonium iodide, phosphine and hydrogen iodide, thus:—

$$\mathbf{NH_4Cl} = NH_3 + HCl;\ \mathbf{PH_4I} = PH_3 + HI.$$

Recombination takes place on cooling; hence this process cannot be practically applied.

3. **By double decomposition.**—Nitrides of boron, silicon, magnesium, titanium, &c., when heated in a red-hot tube in a current of steam, or with an alkali, yield ammonia, thus:—

$$\mathbf{2BN} + 3H_2O = \mathbf{B_2O_3} + 2NH_3.$$

Attempts have been made to utilise this reaction, but without commercial success.

Hydrogen phosphides, arsenides, and antimonide are also similarly produced by the action of hydrochloric acid or of water on phosphides, arsenides, or antimonide of sodium or calcium. These bodies, prepared by direct union of the elements in the desired proportion, undergo a change such as this:—$\mathbf{Na_3As} + 3HCl.Aq = 3NaCl.Aq + \mathit{AsH_3}$.

As ammonia does not combine with hydroxides or with warm chlorides of sodium, potassium, calcium, &c., it is usually prepared by heating its hydrochloride (ammonium chloride) with excess of calcium oxide, or with caustic soda, thus:—

$$\mathbf{NH_4Cl} + \mathbf{NaOH} = \mathit{NH_3} + \mathbf{NaCl} + \mathit{H_2O};$$
$$\mathbf{2NH_4Cl} + \mathbf{CaO} = \mathit{2NH_3} + \mathbf{CaCl_2} + \mathit{H_2}O.$$

Similarly, phosphine, PH_3, may be produced from phosphonium iodide and a caustic alkali.

4. **Hydrogen phosphide,** PH_3, is formed when phosphorus is boiled with a solution of potassium or sodium hydroxide. This may be regarded as a union of the phosphorus both with the hydrogen of the water, forming PH_3, and with oxygen, the hydroxyl and hydrogen forming hypophosphorous acid; the latter afterwards may be supposed to react with the alkali, forming water and a hypophosphite. The complete reaction is—

$$\mathbf{P_4} + 3H_2O + 3NaHO.Aq = \mathit{PH_3} + 3NaH_2PO_2.Aq.$$

It may be regarded as occurring in the two stages:—

$$\mathbf{P_4} + 6H.OH = 3H_2PO(OH) + \mathit{H_3P};$$
and $$3H_2PO(OH) + 3NaOH.Aq = 3H_2PO(ONa).Aq + 3H_2O.$$

(see Hypophosphorous Acid, p. 380). This reaction is essentially analogous to that of sulphur on sodium hydroxide; only in this case a further change occurs, whereby sulphur dioxide and hydrogen sulphide mutually decompose each other, yielding sulphur, which further reacts on the undecomposed sulphite.

Hypophosphorous acid, H_3PO_2, and phosphorous acid, H_3PO_3, when heated, yield phosphoric acid and hydrogen phosphide. It will be remembered that hypophosphorous acid is probably two-thirds hydrogen phosphide, and phosphorous acid one-third hydrogen phosphide, thus:—

$$H_2{=}\underset{\overset{\|}{O}}{P}(OH) \quad \text{and} \quad H{-}\underset{\overset{\|}{O}}{P}(OH)_2.$$

When heated, these bodies decompose, thus:—

$$2H_3PO_2 = H_3PO_4 + \mathit{PH_3}; \; 4H_3PO_3 = 3H_3PO_4 + \mathit{PH_3}.$$

5. The **usual source of ammonia** is the "gas-liquor," produced by causing coal gas to pass through water in the "scrubbers." The nitrogen and hydrogen of the coal unite,

forming ammonia, which escapes with the coal gas, but is retained in the water, owing to its high solubility. The gas-liquor is neutralised with hydrochloric acid, the water expelled by evaporation, and the ammonium chloride is then sublimed in hemispherical iron pots, covered by hemispherical lids. Ammonia is produced from the chloride by the action of quicklime.

6. **Hydrazine,** N_2H_4.*—The method of producing this substance involves the use of carbon compounds, and can hardly be understood without a knowledge of their nature and reactions. But, for completeness' sake, the method will be indicated here.

The hydrochloride of ethyl amidoacetate,

$$NH_2.CH_2.COOC_2H_5.HCl,$$

is treated with a solution of sodium nitrite. The following change takes place:—

$$NH_2—CH_2—COOC_2H_5.HCl + NaNO_2.Aq = NaCl.Aq + 2H_2O + \begin{matrix} N \\ \| \\ N \end{matrix}\!\!>\!CH.COOC_2H_5.$$

The last compound, diazoacetate of ethyl, when heated with caustic soda, polymerises, forming a triple group, while the ethyl group is replaced by hydrogen. This group, on treatment with acids, reacts with water, forming oxalic acid and hydrazine, thus:—

$$\{(N_2){=}CH.COOH\}_3 + 6H_2O = 3(COOH)_2 + 3N_2H_4.$$

An acid, named hydrazoic acid, has been prepared by Curtius, of the formula $H(N_3)$. It is a soluble gas, with a penetrating smell, forming salts. The silver salt, AgN_3, is an insoluble, explosive powder; the barium salt, BaN_6, forms large, transparent crystals. It is derived from benzoyl-azo-imide,

$$C_6H_5—CO—N\!<\!\begin{matrix} N \\ \| \\ N \end{matrix}.$$

Properties.—Ammonia, NH_3, is a colourless gas, with a pungent odour, very soluble in water, 1 volume of water dissolving more than 800 volumes of the gas. This solution is the *liquor ammoniæ*, or "spirit of hartshorn" of the shops, so called because it used to be obtained by distilling stags' or harts' horns. In reality, this yields the carbamate, which, however, is easily converted into ammonia by quicklime.

Liquid ammonia boils at $-40°$, and solidifies to a white crystalline solid at about $-80°$.

Owing to its solubility in water, the gas must be collected over mercury, or, as it is very light (8·5 times as heavy as hydrogen),

* Curtius, *Berichte*, **20**, 1062.

by upward displacement. As it is absorbed by the usual drying agents, sulphuric acid or calcium chloride, it must be dried by passage through a tube filled with calcium or barium oxide.

A considerable rise of temperature occurs when ammonia gas is passed into water, owing partly to the liquefaction of the gas, and partly, no doubt, to chemical combination with the water. It appears probable that a solution of ammonia contains, besides liquid ammonia mixed with water, a small amount of the compound NH_4OH. But of this there is no satisfactory proof as yet. The solution, however, reacts as if it contained such a body; like caustic soda and potash, it has a strong alkaline reaction and a caustic taste. But the ammonia is easily expelled by boiling the solution; hence ammonium hydroxide, if it exists, must be very unstable.

When heated to a few degrees above 500°, ammonia decomposes; but at that temperature the rate of decomposition is exceedingly slow. As there is no recombination between its constituents, a sufficiently long exposure to that temperature ultimately completely decomposes it. Decomposition takes place more rapidly the higher the temperature, and is aided by porous surfaces.*

Ammonia does not evolve sufficient heat by burning to continue ignited in air, for a considerable amount of heat is absorbed to effect its decomposition before free hydrogen is produced, which will unite with oxygen. But it burns in oxygen with a yellowish flame, giving nitrogen and water. It instantly reacts with halogens, forming halogen substitution products if halogen is in excess, and nitrogen if excess of ammonia be present (see pp. 54 and 158).

It unites with very many compounds; these substances will be considered later. Its heat of formation is $N + 3H = NH_3 + 120K + Aq = 204K$.

Phosphoretted hydrogen, as **trihydrogen phosphide,** PH_3, is usually called, is also a colourless gas, possessing a disagreeable smell of garlic. Unlike ammonia, it is nearly insoluble in water. The liquid boils at $-85°$, and solidifies at $-132{\cdot}5°$.† It is exceedingly poisonous, air containing one ten-thousandth of its volume of the gas speedily producing death. When contaminated with the liquid phosphide, P_2H_4, it is spontaneously inflammable; such a mixture is produced by every method of preparation except that of decomposing phosphonium iodide, $\mathbf{PH_4I}$, with caustic

* *Chem. Soc.*, **45**, 92.

† *Monatsh. Chem.*, **7**, 371.

alkali, or by boiling phosphorus with an alcoholic solution of soda or potash. In preparing such an inflammable compound, care must be taken to expel air from the flask in which it is generated by means of a current of coal-gas, or of carbon dioxide. When it is allowed to bubble through water, each bubble takes fire

FIG. 45.

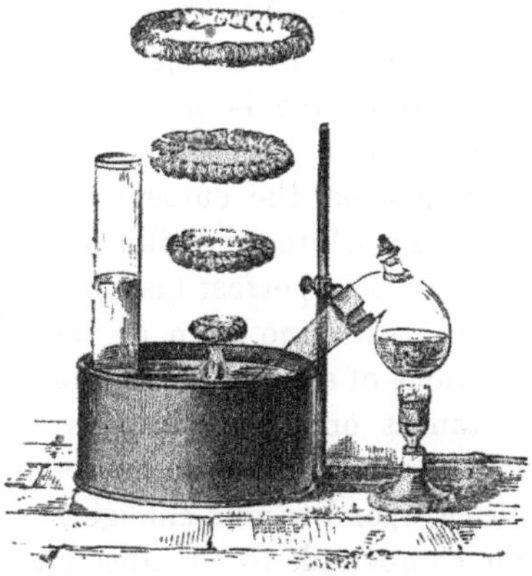

spontaneously as it bursts, and produces a beautiful vortex ring of finely divided phosphoric acid.

The heat of formation of PH_3 is $\mathbf{P} + 3H = PH_3 + 43\text{K}$.

Like ammonia, hydrogen phosphide unites directly with hydrogen chloride, bromide, iodide, and sulphate, but compounds with other acids have not been prepared. The "**phosphonium**" **salts,** as these compounds have been named, from their analogy with ammonium compounds, have the formulæ $\mathbf{PH_4Cl}$, $\mathbf{PH_4Br}$, and $\mathbf{PH_4I}$; the formula of the sulphate has not been ascertained.

Phosphonium chloride, PH_4Cl, is produced by mixing equal volumes of phosphuretted hydrogen and hydrogen chloride, and compressing the mixture. At 20 atmospheres, small white crystals deposit on the side of the tube. The same substance is produced by cooling the mixture to —30° or —35°.

Phosphonium bromide, PH_4Br, is produced when the gases are mixed and cooled; or by the action of a strong solution of hydrobromic acid on phosphorus at 100—120° in a sealed tube. It forms white crystals resembling the chloride.

Phosphonium iodide, PH_4I, is produced by mixing hydrogen phosphide and hydrogen iodide at the ordinary temperature. A more convenient method of preparation is to dissolve 400 grams of yellow phosphorus in its own weight of carbon disulphide, and to add very gradually 680 grams of iodine, keeping the solution cold. The carbon disulphide is then completely distilled off by means of a water-bath. The product is a mixture of iodides of phosphorus with free phosphorus. While carbon dioxide is passed through the retort, 240 grams of water are slowly added, the temperature still being kept low. The following reaction takes place:—$13\mathbf{P} + 9\mathbf{I} + 21H_2O = 3H_4P_2O_7 + 7\mathbf{PH_4I} + 2HI$. The condenser is then removed and a long wide tube adapted to the neck of the retort, closed at its further end by a perforated cork, through which a narrow tube is inserted leading to a draught. On careful heating over a sand-bath, the phosphonium iodide sublimes into the wide tube, the current of carbon dioxide being maintained. It forms a white crystalline crust, which on careful resublimation crystallises in perfect lustrous cubes. It is interesting to note that the crystalline form of these bodies is identical with that of the halides of sodium, potassium, &c.

All these substances, on passing into vapour, decompose into their constituents, thus resembling ammonium chloride. Their vapour densities correspond to this change, and are half what they would be were the compounds to volatilise unchanged.

Phosphonium sulphate is formed by passing hydrogen phosphide into strong sulphuric acid at $-35°$. It forms a white crystalline mass, which decomposes as temperature rises, the sulphuric acid being reduced to hydrogen sulphide, sulphur, and sulphur dioxide, while acids of phosphorus are produced. No nitrate is formed under similar circumstances. The nitric acid is reduced, and inflammable hydrogen phosphide is formed.

Hydrogen arsenide, arsine, or arseniuretted hydrogen, H_3As, and **hydrogen antimonide, stibine, or antimoniuretted hydrogen,** H_3Sb, usually written AsH_3 and SbH_3, are colourless gases, exceedingly poisonous. They have very disagreeable smells; that of AsH_3 resembling garlic: liquid AsH_3 boils at $-54{\cdot}8°$, and the solid melts at $-113{\cdot}5°$; and solid $\mathbf{SbH_3}$ melts at $-91{\cdot}5°$, but decomposes before its boiling point is reached.* They are very sparingly soluble in water. As ordinarily prepared, they are mixed with large quantities of hydrogen. Stibine, indeed, cannot be obtained pure, except at a very low temperature; even at $-60°$ a tube containing liquid stibine becomes coated with metallic antimony. They do not unite with acids.

This means of recognising arsenic and antimony is taken advantage of in "Marsh's test." Compounds of arsenic or antimony, placed in a flask contain-

* Olzewski, *Monatsh. Chem.*, 5, 127; 7, 371.

ing zinc and acid, which yield nascent hydrogen, unite with the hydrogen, producing arsine or stibine. As commercial zinc often contains arsenic and antimony, specially purified zinc must be employed. The gas, after being dried by passage through a tube contaiuing calcium chloride, is set on fire at the exit tube, which should be drawn out into a jet, as shown in the figure. On holding

FIG. 46.

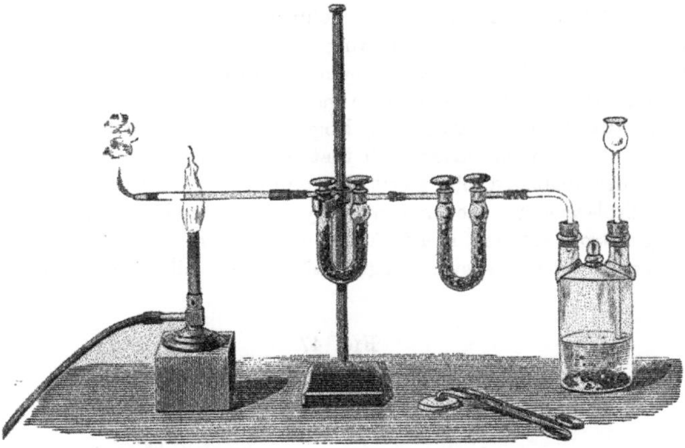

the lid of a porcelain crucible in the flame, arsenic or antimony is deposited, the former with a grey, and the latter with a black, colour. These deposits may be distinguished from each other by their behaviour with a solution of calcium hypochlorite. While the grey deposit of arsenic is easily oxidised and dissolved, the black stain of antimony remains unaffected.

If the exit tube be heated to redness, the arsine or stibine is decomposed, and deposits of arsenic and antimony are obtained, which may be dissolved and tested by the usual means. This process is well adapted for testing for these poisons in complex organic mixtures, such as the contents of the stomach, &c. For further details concerning this process, a work on analytical chemistry must be consulted.

Hydrazine, N_2H_4 is a gas, with an exceedingly sharp pungent smell, somewhat resembling that of ammonia. It is very hygroscopic and difficult to free from water. Like ammonia, it unites with acids to form salts; its hydrochloride, for example, having the formula $\mathbf{N_2H_4.HCl}$. Its name is derived from the French term for nitrogen, *azote*.

Tetrahydrogen diphosphide, P_2H_4, commonly termed liquid phosphoretted hydrogen, is produced along with the gaseous phosphine by most of the reactions which serve to prepare the latter. It may be separated by passing the gaseous spontaneously inflammable product through a U-tube cooled by a freezing mixture. It is a colourless mobile refractive liquid, which, on standing, decomposes into phosphine and dihydrogen tetraphosphide, $\mathbf{P_4H_2}$, a red solid. Liquid phosphoretted hydrogen is spontaneously inflammable. No compounds with acids are known.

A velvety brown substance, said to have the empirical formula **AsH**, is produced when sodium or potassium arsenide is treated with water.

Composition of ammonia.—The volume relations of the constituents of ammonia may be shown by the following experiments:—

1. To prove that ammonia gas contains half its volume of nitrogen.—The principle of the operation is to place gaseous ammonia in contact with some substance capable of removing its hydrogen by oxidation, and to compare the volume of the ammonia taken with that of the residual nitrogen. For this purpose a dry graduated tube, about 40 cm. in length, is filled with ammonia by upward displacement; the ammonia may be prepared for this purpose by warming a strong solution, in a flask, through a cork in the neck of which issues a long vertical tube, as shown in figure 47. When full, the graduated tube is

FIG. 47.

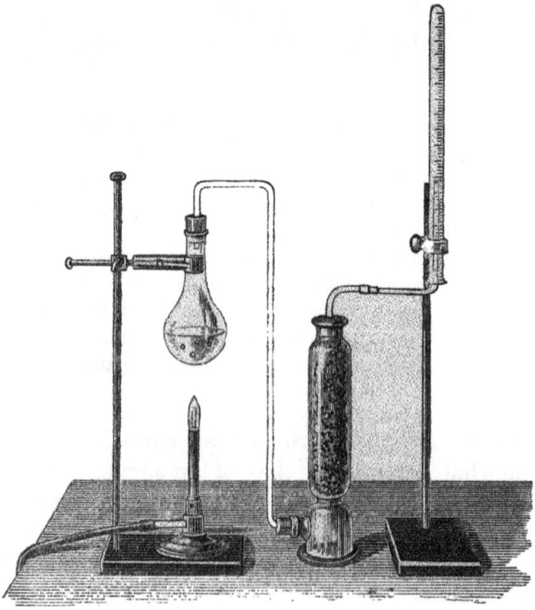

slowly raised, and when free from the vertical tube conveying the ammonia, it is closed with the thumb. It is then transferred to a basin containing a strong solution of sodium hypobromite, NaBrO (Fig. 48). Some of the solution will enter: the tube is now shaken, and its open end is again dipped into the solution of hypobromite. The reaction $3NaBrO.Aq + 2\mathit{NH}_3 = 3NaBr.Aq + 3H_2O + \mathit{N}_2$ takes place. On removing the graduated tube to a jar of water, and equalising the level of the liquid inside the tube with that of the water in the jar, it will be found that the nitrogen occupies half the space originally occupied by the ammonia. It is thus seen that *two volumes of ammonia yield one volume of nitrogen.*

FIG. 48.

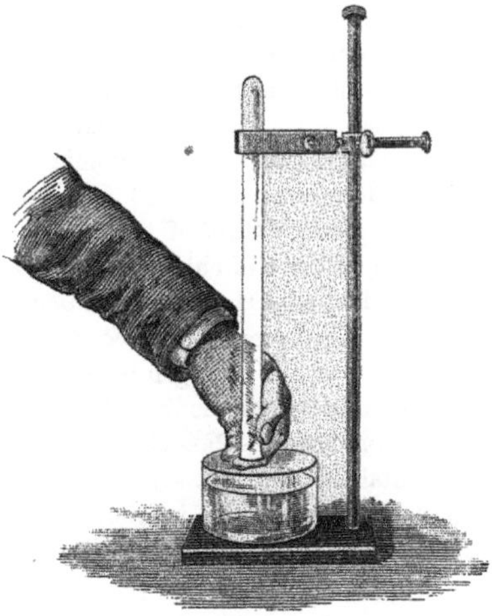

FIG. 49.

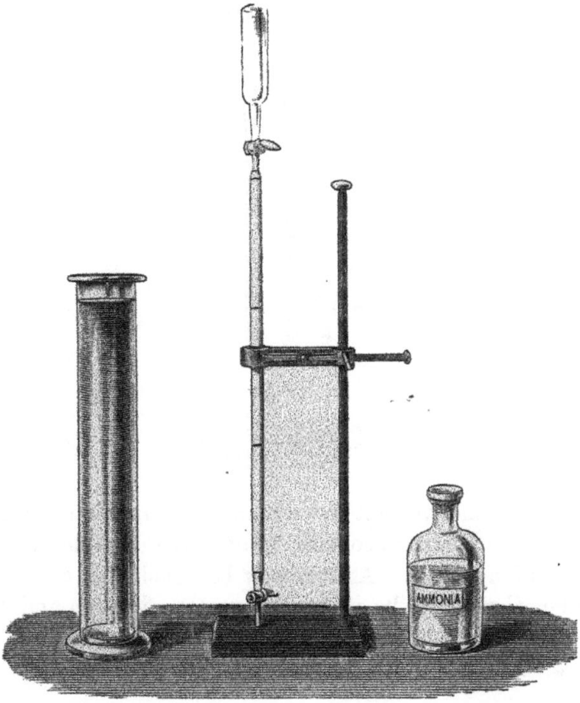

2. To prove that for every three volumes of hydrogen contained in ammonia, it contains one volume of nitrogen.—A tube, provided with a stopcock at each end, is filled with chlorine (Fig. 49). It is divided into three equal parts by two indiarubber rings. A solution of ammonia is then poured into the funnel at one end, and the upper stopcock is opened, when some ammonia solution enters the tube. A flame is seen to run down the tube. It is now shaken, when dense white fumes of ammonium chloride are formed. More ammonia solution is passed in, and the tube is again shaken. Finally the funnel is rinsed out, and some weak sulphuric acid is passed into the tube, to combine with the excess of ammonia. On placing the tube in a jar of water, opening the lower stopcock, and equalising levels, it is seen that the remaining nitrogen occupies one-third of the volume originally occupied by the chlorine. But, as equal volumes of chlorine and hydrogen combine to form hydrogen chloride, it is evident that the three volumes of chlorine must correspond to three volumes of hydrogen: hence, for every three volumes of hydrogen in ammonia, one volume of nitrogen is present.

3. To show that, on decomposing ammonia by heat, the resultant gases occupy twice the volume of their compound.—Pass electric sparks from a Ruhmkorff's coil through ammonia contained in a tube standing in a mercury trough. The ammonia will be completely decomposed in about three-quarters of an hour, and it will be seen that its volume has doubled (Fig. 50).

FIG. 50.

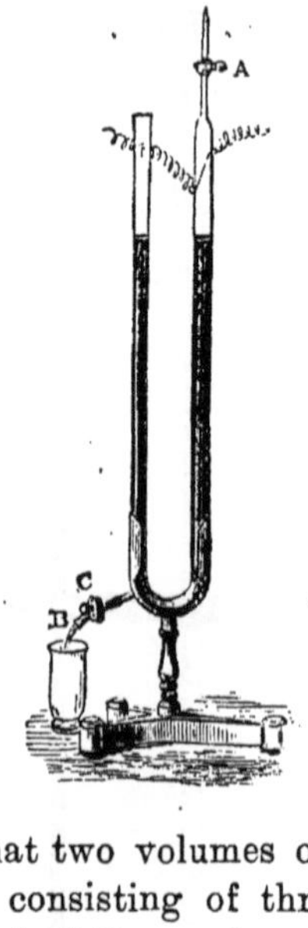

It is thus shown that two volumes of ammonia when decomposed yield a mixture consisting of three volumes of hydrogen with one of nitrogen. And it may be concluded, conversely, that if combination could be induced between nitrogen and hydrogen, one volume of the former would unite with three of the latter, to produce two volumes of ammonia.

It may also be shown by weighing a vacuous flask of known volume, filling it with ammonia, and weighing again, that ammonia is 8·5 times as heavy as hydrogen. This corresponds to a molecular weight of 17, implying the formula NH_3.

The formulæ of phosphine has been deduced from analysis, and from its density; that of arsine from analogy, and that of stibine from the formula of the compound it forms, **Ag_3Sb**, when passed into a solution of silver nitrate.

Compounds of hydrogen nitride, phosphide, arsenide, and antimonide. The halogen substitution compounds of ammonia have already been described on p. 158. Analogous to NH_2Cl, which may be named monochloramine, is the compound—

Hydroxylamine, *NH_2OH*.*—It is produced by the reduction of nitric oxide or nitric acid by means of nascent hydrogen, generated by the action of hydrochloric acid on tin, zinc, cadmium, aluminium, or magnesium. To prepare it, nitric oxide is passed through a mixture of tin and hydrochloric acid, to which a few drops of chloride of platinum have been added, to promote galvanic action and facilitate the evolution of the hydrogen. The solution then contains stannous chloride, $SnCl_2$, and hydroxylamine hydrochloride, $NH_2.OH.HCl$. The tin is removed as sulphide, **SnS**, by the passage of hydrogen sulphide through the solution. The filtrate from the sulphide is evaporated to dryness, and extracted with absolute alcohol, in which ammonium chloride, also produced by reduction of the nitric oxide, is insoluble, but in which hydroxylamine hydrochloride dissolves. On filtering from the undissolved ammonium chloride, and again evaporating to dryness, hydroxylamine hydrochloride remains as a white crystalline mass.

From the hydrochloride, the sulphate is produced by evaporation with weak sulphuric acid; and from a solution of the sulphate hydroxylamine may be liberated by addition of the requisite amount of baryta-water.

If the solution is distilled, a considerable portion of the hydroxylamine passes over with the steam, but most of it is decomposed thus:—$3NH_2OH = NH_3 + N_2 + 3H_2O$. The heat of formation of hydroxylamine is:—$3N + H + O + Aq = NH_3O.Aq + 181$ K (Thomsen gives + 243 K).

Hydroxylamine is a powerful reducing agent. When added to solutions of salts of silver or mercury, the metals are precipitated; and when boiled with copper sulphate, cuprous oxide, **Cu_2O**, is thrown down.

* *Chem. Soc.*, **43**, 443; **47**, 71; **51**, 50, 659.

The following salts of hydroxylamine have been prepared:—$NH_2OH.HCl$; $2NH_2OH.HCl$; $3NH_2OH.HCl$; $NH_2OH.HNO_3$; $2NH_2OH.H_2SO_4$; $3NH_2OH.H_3PO_4$; $2NH_2OH.H_2C_2O_4$ (oxalate). Some double salts have also been prepared, which in crystalline form resemble those of ammonium, *e.g.*, $(NH_2OH).HAl(SO_4)_2.12H_2O$; $(NH_2OH).HCr(SO_4)_2.12H_2O$; and $(NH_2OH).HFe(SO_4).12H_2O$; corresponding to the alums; and $(NH_2OH)_2 H_2SO_4.MgSO_4.6H_2O$, corresponding to the double sulphates of dyad metals (see pp. 425 and 428).

The constitution of hydroxylamine is doubtless $H_2N—OH$.

No similar compound of phosphorus is known; but attention may be directed to hypophosphorous acid, the oxidised analogue of hydroxylamine, $O{=}P\begin{cases}H\\H\\OH\end{cases}$ (p. 380); and the somewhat analogous constitution of hyponitrous acid, $O{=}N—H$ (see p. 344). Phosphorous and nitrous acids may also be compared (see pp. 337 and 345). Arsenic and antimony do not form similar combinations; but these bodies may be compared with $O{=}As—Cl$, $O{=}Bi—Cl$, described on pp. 384, 385.

Amido-compounds or amines.—As the group named hydroxyl, —OH, may be regarded as capable of entering into combination with the elements, forming hydroxides and acids; so the group $—NH_2$, named the "**amido-group**" or "**amine**" enters into similar combinations. And such compounds may be regarded as substituted ammonia, just as the hydroxides and acids may be viewed as substituted water. Thus we have Na—OH, sodium hydroxide, to which corresponds $Na—NH_2$, sodamide; and $Zn{<}^{OH}_{OH}$, zinc hydroxide, with its analogue $Zn{<}^{NH_2}_{NH_2}$, zinc-amide. But few of these simpler compounds are known; because the nitrogen still retains its power of combining with haloid and other acids to form salts. Such bodies are so numerous that only an incomplete sketch can be given here. We shall begin with the simpler compounds, considering the salts subsequently.

Simple compounds:—

$NaNH_2$; KNH_2; $Zn(NH_2)_2$; $ZnPH$; $P(NH_2)_3$.

Sodamide and **potassamide*** are produced by passing ammonia over gently heated sodium or potassium. They are olive-green substances, transmitting brown light when in thin scales. They melt a little above 100°, and when heated to dull redness

* *Annalen*, **108**, 88.

give nitride of potassium or sodium, $\mathbf{K_3N}$ or $\mathbf{Na_3N}$, and ammonia. With water they yield hydroxide and ammonia, thus establishing their constitution; thus: $\mathbf{KNH_2} + \mathrm{H.OH} = \mathbf{KOH} + H.NH_2$.

Zinc-amide,* $\mathbf{Zn(NH_2)_2}$, is a white powder, insoluble in ether, produced along with methane or ethane by treating zinc-methyl or zinc-ethyl with ammonia. When heated to redness it yields the nitride, $\mathbf{Zn_3N_2}$. The compound **ZnPH** forms a yellow mass; it is produced by passing a current of phosphine into zinc-ethyl, $Zn(C_2H_5)_2$.

Phosphorosamide, $\mathbf{P(NH_2)_3}$, appears to be produced by the action of ammonia on phosphorus trichloride, PCl_3, thus:—$PCl_3 + 3HNH_2 = \mathbf{P(NH_2)_3} + 3HCl$. The hydrogen chloride combines with the excess of ammonia, forming ammonium chloride, from which the phosphorosamide has not been separated. The mixture is a white crystalline mass.

The carbon compound, $C(NH_2)_4$, appears incapable of existence; a body differing from it by the elements of ammonia is however known; it is named **guanidine**, and its formula is $\mathbf{HN{=}C(NH_2)_2}$.

Double compounds.—Halides, and, generally speaking, salts of such amides, are formed by the action of ammonia on most compounds of the metals; but here a difficulty in classification meets us, for a considerable number of molecules of ammonia very frequently add themselves to such compounds, and it is at present as impossible in many cases to assign reasonable constitutional formulæ to such bodies, as it is to understand in what manner of combination water of crystallisation exists in salts which contain it. We shall, therefore, assume that, where it appears reasonable to suppose so, an amide is formed; and any further molecules of ammonia which add themselves on to such compounds will often be represented as if they were merely additive molecules. At the same time there are compounds, such as those of cobalt and of platinum, where such additive molecules of ammonia appear to form an essential portion of the total molecule. Where such is the case, attention will be drawn to the fact.

The elements will, as usual, be considered in their periodic order.

$\mathbf{NH_4Cl.3NH_3}$; $\mathbf{NH_4Cl.6NH_3}$; $\mathbf{NH_4Br.3NH_3}$.

The first of these melts at 7°, the second at −18°. They are produced by heating ammonium chloride with excess of ammonia, and allowing it to cool in contact with the gas.† The compound

* *Annalen*, **134**, 52. † *Comptes rend.*, **88**, 578.

$NH_4NO_3.3NH_3$ is also formed by direct union, ammonium nitrate liquefying in contact with dry ammonia.*

$Ca(NH_2)_2.2HCl.6NH_3$ and $Sr(NH_2)_2.2HCl.6NH_3$.

White substances produced by saturating calcium or strontium chloride with ammonia. When warmed the original constituents are re-formed. The corresponding barium compound is unknown.

$ClZn(NH_2).HCl$; $HO.CO.O.Zn(NH_2)$; $HO.SO_2Zn(NH_2)$.— $Zn(NH_2)_2.2HCl$; $Cd(NH_2)_2.2HCl$; $Zn(NH_2)_2.2HCl.2$ and $3NH_3$; $Zn(NH_2)_2.2HI.2$ and $3NH_3$; $Zn(NH_2)_2.H_2SO_4.H_2O$; $Zn(NH_2)_2.H_2S_2O_3$; $Zn(NH_2)_2.H_2SO_4.2$ and $3NH_3.4H_2O$; $Cd(NH_2)_2.H_2CrO_4.2NH_3$; $3(Zn(NH_2)_2.2HIO_3).2NH_3$; $2Zn(NH_2)_2.Zn(OH)_2.12H_2O$; $2(Zn(NH_2)_2.H_4P_2O_7)Zn(OH)_2.8H_2O$.

These compounds are all made by treating the respective salts with ammonia.

$BF_3.3NH_3$; $BF_3.2NH_3$; $BF_3.NH_3$.

Produced by the action of dry ammonia on boron trifluoride, BF_3. The last of these might be written $BN.3HF$. But boron nitride, BN, when treated with aqueous hydrofluoric acid, yields $BF_3.NH_4F$, ammonium borofluoride, which, it may be supposed, is not $BN.4HF$. The formula is more probably $F_2B(NH_2)HF$. It may be volatilised without decomposition. The formula of the second may be written $FB(NH_2)_2.2HF$, and of the first $B(NH_2)_3.3HF$. The first and second, when heated, lose ammonia, leaving the third. Boron chloride, BCl_3, is said to yield **$2BCl_3.3NH_3$**, which may possibly be a mixture of $ClB(NH_2)_2.2HCl$ and $Cl_2B(NH_2).HCl$.

Compounds of scandium, yttrium, and lanthanum have not been examined.

The **compounds of aluminium** are similar to those of boron, **$Al(NH_2)_3.3HCl$** and **$ClAl(NH_2)_2.2HCl$** having been prepared. It is interesting to note a similar compound of aluminium chloride with phosphuretted hydrogen, **$3AlCl_3.PH_3$**. Compounds of gallium and indium have not been examined. But thallium trichloride reacts with ammonia in presence of ammonium chloride, giving a dense white precipitate of the trihydrochloride of thallamide, **$Tl(NH_2)_3.3HCl$**.

The **chromium compounds** are somewhat complex. Chromium hydrate, digested with ammonium chloride and ammonia, dissolves with a deep red colour; and on exposing the solution to air, a violet powder precipitates; this powder dissolves in hydro-

* *Proc. Roy. Soc.*, **21**, 1091; *Comptes rend.*, **94**, 1117.

chloric acid, forming the salt $\mathbf{CrCl_3.4NH_3.H_2O}$. This compound might be regarded as $Cr(NH_2)_3.3HCl.NH_3.H_2O$; but while it loses its water of crystallisation at 100°, ammonia is retained up to 200°, which would lead to the conclusion that even the fourth molecule is in intimate relation to the chromium. These compounds may be supposed to contain the group —(N_2H_5)—, or —NH_3—NH_2—, a group which may be named the diamido-group; it might, perhaps, preferably be termed the ammonium-amido-group.

Such a supposition would bring such compounds into conformity with those of cobalt and of other elements; but the heptamines cannot be classified thus, unless a further condensation of the ammonia molecule is supposed possible.

There are five series of these compounds: the **triamines**, the **tetramines**, the **pentamines**, the **hexamines**, and the **heptamines**.

Of the **triamines**, the oxalate, $\mathbf{Cr_2O_3.3C_2O_3.6NH_3.3H_2O}$, has been prepared. Of the **tetramines**, the compound $\mathbf{CrCl_3.4NH_3.H_2O}$ is an example; the bromide, $\mathbf{CrBr_3.4NH_3.H_2O}$; the iodide, $\mathbf{CrI_3.4NH_3.H_2O}$; the chlorodibromide, $\mathbf{CrClBr_2.4NH_3.H_2O}$; the dichlorobromide, $\mathbf{CrBr_2Cl.4NH_3.H_2O}$; the chlorodi-iodide, $\mathbf{CrClI_2.4NH_3.H_2O}$; the bromosulphate, $\mathbf{CrBr(SO_4).4NH_3.H_2O}$; the chlorochromate, $\mathbf{CrCl(CrO_4).4NH_3.H_2O}$, and the chloronitrate,

$$\mathbf{CrCl(NO_3)_2.4NH_3.H_2O},$$

have been prepared. They have a deep red colour.

The starting point for the **pentamines** is chromous chloride, $\mathbf{CrCl_2}$, produced by the action of hydrogen on red-hot chromic chloride, $\mathbf{CrCl_3}$ (see p. 138). It is added to a solution of ammonium chloride in strong ammonia, in which it dissolves with a blue colour. Air is then passed through the liquid until oxidation is complete. Excess of hydrochloric acid is added, and the mixture is boiled, when the hydrochloride of the pentamine is precipitated. It is purified by solution in weak sulphuric acid, and filtering into a large excess of strong hydrochloric acid, washing with water and alcohol, and drying in air. Its formula is $\mathbf{CrCl_3.5NH_3}$. It is a red crystalline powder. Numerous salts have been obtained, the composition of all of which is analogous to that of the chloride. These bodies have been named **purpureo-chromium** compounds.

If, instead of treating with hydrochloric acid, dilute hydrobromic acid be used, the hydroxybromide of the dipentamine, $\mathbf{HO{-}Cr_2Br_5.10NH_3.H_2O}$, crystallises out in carmine needles. On digestion with hydrochloric acid, the chloride is formed, from which, by suitable means, other salts can be prepared. They all crystallise well, and have a carmine-red colour. On treatment with silver hydroxide, the chloride yields the hydroxide, a blue solution, which rapidly changes to red. These hydroxy-derivatives have been named **rhodochromium salts.** Isomeric with these are the **erythrochromium** compounds, produced by digesting the former with dilute ammonia. While solutions of the former have a blue colour, the latter are red.

The **roseochromium** compounds, containing two hydroxyl-groups, are produced by precipitating purpureo-compounds with sodium dithionate after boiling with dilute ammonia.

By digesting roseo- or purpureo-salts with ammonia in a close vessel, **luteo**-salts are produced, in which six molecules of ammonia are in combination with chromium trichloride.

Two **heptamines** have also been prepared as double salts. It should be stated that these formulæ are usually doubled, chromium trichloride being assumed to have a molecular weight corresponding to the formula Cr_2Cl_6; but, with respect to this view, see the statements on p. 505.

Supposing the halogen compounds of all these amines to exist, and the atom of halogen to be represented by X, the chromamines may be classified as follows:—

Triamines: $CrX_3.3NH_3$, possibly $Cr(NH_2)_3.3HX$.
Tetramines: $CrX_3.4NH_3$, possibly $Cr(NH_2)_2.(N_2H_5).3HX$.
Pentamines: $CrX_3.5NH_3$ (purpureo-chromic compounds); possibly $Cr(NH_2)(N_2H_5)_2.3HX$.
Hydroxydipentamines: $Cr_2X_5(OH).10NH_3$ (rhodo- and erythro-chromic compounds).
Hydroxypentamines: $CrX_2(OH).5NH_3$.
Hexamines: $CrX_3.6NH_3$; possibly $Cr(N_2H_5)_3.3HX$.
Heptamines: $CrX_3.7NH_3$.

Ferric chloride combines with dry ammonia to form **$FeCl_3.NH_3$**, or **$Cl_2{=}Fe(NH_2).HCl$**, a red mass, volatilising when heated, leaving a residue of ferrous chloride.

The manganamines have not been examined.

Cobaltamines. — These compounds closely resemble the chromamines. They fall into the following classes:—

Diamines: $CoX_3.2NH_3$.
Triamines: $CoX_3.3NH_3$; also $CoCl_3.2NH_3.NO_2H$.
Tetramines: $CoX_3.4NH_3$; also $CoX(NO_2)_2.4NH_3$, and $Co_2O(NO_2)_4.8NH_3$.
Pentamines: $CoX_3.5NH_3$; also $CoX(NO_2)_2.5NH_3$ and $CoX_2(NO_2).5NH_3$.
Hexamines: $CoX_3.6NH_3$.

Diamines.—These are prepared from the pentamines (purpureo-cobaltamines (see below) by adding a solution of hydrogen ammonium sulphite, HNH_4SO_3, until the liquid smells of sulphur dioxide. Sparingly soluble brown octahedra of the sulphite, $Co_2(SO_3)_3.4NH_3.5H_2O$, are deposited. In this case the sulphite is the only salt known.

Triamines.—If more free ammonia be present, and if the addition of hydrogen ammonium sulphite is stopped as soon as the smell of ammonia disappears, insoluble yellow needles of the triamine sulphite, $Co_2(SO_3)_3.6NH_3.H_2O$, are deposited. Here, again, other salts have not been prepared.

A series of compounds, in which one molecule of ammonia of the triamines is replaced by the group HNO_2, nitrous acid, are formed by the action of ammonium nitrite on neutral ammoniacal solutions of cobalt salts. The salt $Co(NO_2)_3.2NH_3.NH_4NO_2$ crystallises out; and the ammonium group is replaceable by other metals. Thus the salts $Co(NO_2)_3.2NH_3.TlNO_2$ and $Co(NO_2)_3.2NH_3.Hg'NO_2$ and others have been prepared by precipitation, the groups NH_4NO_2, $TlNO_2$, $HgNO_2$, &c., being substituted for one molecule of ammonia, while the three groups of NO_2 are in combination with the cobalt. These may also be regarded as ammoniacal double nitrites of cobalt and ammonium, &c.

Tetramines.—These substances are known as **fuscocobaltamines**. They are produced by the action of water on the oxycobaltamines, which are also tetramines. The nitrate has the formula $Co_2O(NO_3)_4.8NH_3$; the chloride, $Co_2OCl_4.8NH_3$, &c. The **croceocobaltamines** are closely allied to the fuscocobaltamines; they are produced by the action of nitrites on ammoniacal solutions of cobalt salts. The nitrate, $Co(NO_2)_2.(NO_3).4NH_3$, forms sparingly soluble sherry-coloured crystals. It is produced by mixing a solution of cobalt chloride with ammonium nitrite, and then adding a solution of ammonium nitrate containing much ammonia; the equation showing its formation is

$$2CoCl_2.Aq + 2NH_4NO_3.Aq + 6NH_3.Aq + 4NH_4NO_2.Aq + O = 4NH_4Cl.Aq + H_2O + 2\{Co(NO_2)_2NO_3.4NH_3\}.$$

The sulphate is similarly prepared from cobalt sulphate, ammonia, and potassium nitrate. The chloride, iodide, chromate, and dichromate have been prepared. A tri-iodide is also known:

$$Co(NO_2)_2.I_3.4NH_3.$$

It appears possible for the ammonia in the tetramine to exchange with the group NO_2. By acting on cobalt chloride with potassium nitrite in presence of a large excess of ammonium chloride, the body $Co(NH_3)_2Cl.4NO_2$ is produced. It will be observed that this formula is strictly analogous to that of the croceocobaltamines, $\mathbf{Co(NO_2)_2Cl.4NH_3}$.

Pentamines.—There are two isomeric series of pentamines: the **roseocobaltamines** and the **purpureocobaltamines**. The formula of both is $CoX_3.5NH_3$. The first are produced by exposing a brown ammoniacal solution of cobalt sulphate, $CoSO_4$, to air, when it turns cherry-red, and deposits a brownish-black powder. On addition of hydrochloric acid, care being taken to keep the mixture cold, a brick-red powder is precipitated, which is collected, and washed, first with strong hydrochloric acid, then with ice-cold water. The formula of this substance is $\mathbf{CoCl_3.5NH_3.H_2O}$. The nitrate is similarly prepared, and is a yellow precipitate. The sulphate and other salts have been obtained. From the sulphate, by addition of solution of barium hydroxide, an alkaline liquid is produced, probably containing the hydroxide; it absorbs carbon dioxide from the air, and from it other salts may be produced. The roseocobaltamines all contain water of crystallisation.

By allowing the temperature to rise during the neutralisation of an ammoniacal solution of cobalt sulphate with hydrochloric acid, violet-red anhydrous prisms of **purpureocobaltamine** chloride, $\mathbf{CoCl_3.5NH_3}$, are deposited. The same compound is produced by heating fuscocobaltamine chloride, $\mathbf{CoCl_3.4NH_3}$, in a sealed tube with aqueous ammonia. The nitrate, acid sulphate, chromate, and pyrophosphate, and other salts have been prepared; the hydroxide and sulphite are also known. The purpureocobaltamines are all anhydrous.

Closely connected with these are the **xanthocobaltamines**, in which one atom of chlorine is replaced by one molecule of the nitro-group, NO_2, thus:—$\mathbf{Co(NO_2)Cl_2.5NH_3}$. They are produced by mixing cobalt nitrate with excess of an alcoholic solution of ammonia, and passing in a mixture of nitric oxide and peroxide, produced by the action of nitric acid on starch, care being taken to keep the mixture cold. Yellow-brown prisms of

$$\mathbf{Co(NO_2)(NO_3)_2.5NH_3}$$

are deposited. The sulphate is similarly prepared, and from it the chloride, hydroxide, and carbonate have been made. The xanthocobaltamines, when digested with hydrochloric acid and ammonium chloride, lose the nitro-group, which is replaced by chlorine; the purpureocobaltamine is formed, $\mathbf{CoCl_3.5NH_3}$.

The **flavocobaltamines** are similar to the xanthocobaltamines, but in these two atoms of chlorine are replaced by two nitro-groups. They are produced by treating a purpureocobaltamine, *e.g.*, $CoCl_3.5NH_3$, with potassium nitrite, and adding a little acetic acid. The formula of the chloride is $Co(NO_2)_2Cl.5NH_3$. The nitrate, iodide, and other salts have been prepared.

The **trinitropentamine,** $Co(NO_2)_3.5NH_3$, is produced by treating the trichloropentamine, $CoCl_3.5NH_3$ (purpureocobaltamine) with silver nitrite. It forms brown-orange octahedra.

Hexamines.—These are named **luteocobaltamines.** The chloride, $CoCl_3.6NH_3$, is formed by digesting cobalt chloride with solid ammonium chloride and ammonia. On shaking, the liquid turns brown. Lead or manganese dioxide is then added, and after heating, the liquid is filtered and saturated with hydrogen chloride; yellow-brown crystals are deposited. The bromide, iodide, carbonate, nitrate, pyrophosphate (insoluble), phosphate, and sulphate are among the salts which have been prepared.

Derivatives of the unknown chloride $CoCl_4$ have also been obtained as pentamines. The general formula of these bodies is $Co^{IV}OX_2.5NH_3$. They are formed by direct oxidation of ammoniacal cobalt solutions. They are decomposed by water, with evolution of oxygen, and formation of the usual pentamines (purpureocobaltamines). They form olive-brown crystals.

Here, again, the usual formulæ have been halved; for there appears to be no valid reason for supposing that the formula of cobalt trichloride is Co_2Cl_6 in preference to $CoCl_3$. Where the actual molecular weight is unknown, preference is always given to the simplest formulæ. The additive formulæ given, however, certainly do not express the constitution of these bodies. But it is possible to represent them all as derivatives of ammonia, NH_3, if it can be supposed that a di-ammonia is capable of existence, $—NH_3—H_3N—$, a reasonable enough supposition, inasmuch as ammonia can combine directly with hydrogen halides to form bodies such as $H—NH_3—Cl$. If this be granted, then, the formulæ of the cobaltamines may be thus represented :—

Diamines :—$Cl—Co(NH_2)_2.2HCl$.
Triamines :—$Co(NH_2)_3.3HCl$.
Tetramines :—$Cl—Co(NH_3—NH_2)_2.2HCl$.
Pentamines :—$NH_2—Co(NH_3.NH_2)_2.3HCl$.
Hexamines :—$Co(NH_3.NH_2)_3.3HCl$.

Or, again, it may be supposed that they are thus constituted :—

$$Cl—Co \left< \begin{matrix} NH_2 < {Cl \atop NH_4} \\ NH_2 < {NH_4 \atop Cl} \end{matrix} \right.$$

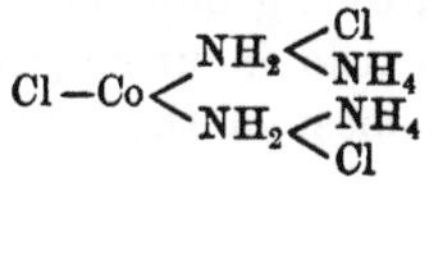

&c., the group $-NH_2<$ having the power of combination with the monad element chlorine, and with the monad group (NH_4). But these formulæ are speculative and have little to support them.

Chromosamines, derivatives of CrX_2, are unknown.

Ferrosamines, manganosamines, cobaltosamines, and nickelosamines have been prepared. They are as follows :—

> **$FeCl_2.6NH_3$; $Fe_2P_2O_7.NH_3$.—$MnCl_2.6NH_3$(?); $MnSO_4.4NH_3$ (?).**
> **$NiCl_2.6NH_3$; $NiBr_2.6NH_3$; $NiI_2.4NH_3$; $NiI_2.6NH_3$.**
> **$Ni(NO_3)_2.4NH_3.H_2O$; $Ni(NO_2)_2.4$ and $6NH_3$; $Ni(BrO_3)_2.2NH_3$;**
> **$Ni(IO_3)_2.4NH_3$; $NiSO_4.4NH_3.2H_2O$; $2NiSO_4.10NH_3.7H_2O$;**
> **$NiSO_4.6NH_3$; $NiS_2O_3.4NH_3.6H_2O$; $NiS_2O_6.NH_3$.**

These salts are all crystalline, and are produced by direct addition.

One of the **amido-compounds of carbon** has already been mentioned (see *Guanidine*, p. 525). Others are known in which one or more of the hydrogen atoms of the hydrocarbons is replaced by the amido-group. Thus we have **methylamine,** $CH_3.NH_2$, **dimethylamine,** $(CH_3)_2NH$, and **trimethylamine,** $(CH_3)_3N$, all forming salts resembling those of ammonium. Here, too, similar **phosphines** are met with, viz., $CH_3.PH_2$, $(CH_3)_2PH$, and $(CH_3)_3P$, **monomethyl, dimethyl,** and **trimethyl phosphines,** respectively, as well as many others, in which ethyl, propyl, and other paraffin radicles replace the hydrogen of phosphoretted hydrogen. Similar derivatives are known of **arsine,** but in this case hydrogen is no longer in combination with the arsenic, but chlorine, oxygen, sulphur, &c. For example, we know the compounds :—CH_3AsCl_2; $(CH_3)_2AsCl$; $(CH_3)_3As$; and these bodies, and similar compounds containing oxygen, such as CH_3AsO, or sulphur, CH_3AsS, &c., have the power of combining with other two atoms of chlorine, or with another atom of oxygen forming such compounds as CH_3AsCl_4, $(CH_3)_2AsCl_3$, $(CH_3)_3AsCl_2$, $(CH)_4AsCl$, and even $(CH_3)_5As$. The last of these compounds is specially interesting as a representative of the unknown NH_5.

The **stibines,** derivatives of SbH_3, are similarly constituted; but for detailed accounts of these bodies a treatise on carbon compounds must be consulted.

Corresponding to guanidine, $C(NH)''(NH_2)_2$, is—

Carbamide, or **urea, $CO(NH_2)_2$,** which may be regarded as carbonic acid, the hydroxyl-groups of which have been replaced by amido-groups. It exists in urine (from 2 to 3 per cent.), and is the form in which most of the nitrogen consumed in food is eliminated from the organism. It may be separated from urine after evaporation to about one quarter of its original volume by addition of nitric acid, which precipitates the sparingly soluble nitrate. From the nitrate, carbamide may be separated by addition of the requisite amount of potassium hydroxide, evaporation to dryness, and extraction with alcohol, from which it crystallises on cooling in white prisms.

It may also be produced by treating carbonyl chloride, $COCl_2$, with ammonia, thus:—$COCl_2 + 2HNH_2 = \mathbf{CO(NH_2)_2} + 2HCl$; also by heating solutions of ammonium carbonate or carbamate to 140—150° in sealed tubes:—$\mathbf{CO(ONH_4)_2} = \mathbf{CO(NH_2)_2} + 2H_2O$; $\mathbf{NH_2{-}CO(ONH_4)} = \mathbf{CO(NH_2)_2} + H_2O$. Urea unites with acids; thus the hydrochloride has the formula $\mathbf{CO(NH_2)_2.HCl}$; the nitrate $\mathbf{CO(NH_2)_2.HNO_3}$. It is decomposed by a solution of potassium hypochlorite or hypobromite, thus:—$CO(NH_2)_2.Aq + 3KBrO.Aq = 3KBr.Aq + CO_2 + 2H_2O + N_2$. By measuring the volume of nitrogen liberated, the amount of urea in such a liquid as urine may be estimated.

Sulphocarbamide, $\mathbf{CS(NH_2)_2}$, resembles carbamide.

$HO{-}CO{-}NH_2$, or **carbamic acid,** is unknown in the free state. Its **ammonium salt** is produced by the union of one volume of carbon dioxide with two volumes of ammonia, thus:—$CO_2 + 2NH_3 = \mathbf{NH_4O.CO.NH_2}$, or by digesting a strong solution of ammonium carbonate with saturated aqueous ammonia. It is completely decomposed at 60° into its constituents. The following salts of the acid have been prepared:—

$$\mathbf{NaO.CO.NH_2;\ KO.CO.NH_2;\ Ca(O.CO.NH_2)_2;\ Sr(O.CO.NH_2)_2;\ Ba(O.CO.NH_2)_2.}$$

Those of calcium, strontium, and barium are soluble, thus differing from the carbonates.

Carbamates are attacked by hypobromites, but not by hypochlorites; they may be thus distinguished from ammonium carbonate, which is decomposed by both.

The hydrochlorides of **titanamine,** $\mathbf{Ti(NH_2)_4.4HCl}$, and of **zirconamine,** $\mathbf{Zr(NH_2)_4.4HCl}$, are produced by heating the chlorides in a current of ammonia gas. They form white deliquescent crystals.

Difluosilicamine dihydrofluoride, $\mathbf{F_2{=}Si(NH_2)_2.2HF}$; **iodostannamine trihydriodide,** $\mathbf{I{-}Sn(NH_2)_3.3HI}$, and the compounds $\mathbf{Sn(NH_2)_4.4HI}$ and $\mathbf{SnI_4.6NH_3}$, are all produced by direct addition. The monophosphine of stannic chloride, $\mathbf{Cl_3Sn(PH_2).HCl}$, and probably the monamine are also known.

The known stannous compounds are **$SnCl_2.3NH_3$**; **$SnI_2.4NH_3$**; and also the plumbous compound, **$PbI_2.2NH_3$**. The hydroxide, **$Pb(OH)_2.2NH_3$**, has also been prepared, and a chloride of the formula **$2PbCl_2.3NH_3$**.

Vanadium trichloride combines with ammonia; niobium and tantalum chlorides have not been examined in this respect.

Phosphorus triamide has already been described.

The acids of phosphorus, also, form compounds in which the amido-group replaces hydroxyl, more or less completely. They are as follows:—

Orthophosphamide, $PO(NH_2)_3$, is produced by the action of dry ammonia gas on phosphoryl chloride, thus:—$POCl_3 + 3HNH_2 = \mathbf{PO(NH_2)_3} + 3HCl$. The excess of ammonia combines with the hydrogen chloride, forming ammonium chloride, which is removed by washing with water, in which phosphamide is insoluble. It is a white powder, not acted on by boiling water, but decomposed by hot sulphuric acid into ammonium sulphate and phosphoric acid. A similar body, **sulphophosphamide, $PS(NH_2)_3$**, is obtainable from sulphophosphoryl chloride, $PSCl_3$.

When heated, phosphamide gives rise to substances containing less nitrogen, and corresponding to the anhydrophosphoric acids, so far as the analogy between the amido-group and hydroxyl will permit. **Phosphorylamide-imide** (the group $=NH$ is named the "imido-group") and **phosphoryl nitride**, thus produced, are white insoluble powders. These three bodies have the formulæ **$O{=}P.(NH_2)_3$**; **$NH{=}(OP){-}NH_2$**; and **$N{\equiv}(OP)$**, the first two of which correspond to **$O{=}P(OH)_3$**, and **$O{=}(OP){-}OH$**.

Orthophosphamic acid is unknown; but its **sulphur** analogue, $PS(NH_2)(OH)_2$, is produced by treating sulphophosphoryl chloride, $PSCl_3$, with strong aqueous ammonia. It forms soluble salts. The anhydride of phosphamic acid, in which two hydroxyl-groups are replaced by the imido-group, NH, may be termed **phosphimic acid**. Its formula is **$O(NH)''P(OH)$**. It is a pasty soluble mass, produced by the action of dry ammonia on phosphoric anhydride, thus:—

$$P_2O_5 + 2NH_3 = 2O(NH)P(OH) + H_2O.$$

It is monobasic, and forms white crystalline salts, of which those of sodium, potassium, ammonium, calcium, ferrous, manganous, nickel, cadmium, zinc, lead, silver, and mercury have been prepared. This acid may be regarded as metaphosphoric acid, $O_2P(OH)$, with one atom of oxygen replaced by the imido-group, $(NH)''$.

Phosphodiamic acid, $OP(NH_2)_2.OH$, is also unknown, but its sulphur analogue, **$SP(NH_2)_2.OH$**, is produced by treating sulpho-chloride of phosphorus with ammonia, and digesting the resulting solid product with water. It may be supposed that the intermediate body $S.PCl(NH_2)_2$ is formed. The free acid has not been obtained, but its zinc, lead, copper, and silver salts are insoluble precipitates, and establish its formula.

Pyrophosphamic acids are also known, and correspond to pyrophosphoric acid, $P_2O_3(OH)_4$, in which one, two, three, or four hydroxyl-groups are replaced by the amido-group.

Pyrophosphamic acid, $P_2O_3(NH_2)(OH)_3$, is produced by heating a solution of pyrophosphodiamic acid (see below); the amido-group of the latter is exchanged for hydroxyl, thus:— $P_2O_3(NH_2)_2(OH)_2 + H.OH = P_2O_3(NH_2)(OH)_3 + HNH_2$.

Pyrophosphodiamic acid, $P_2O_3(NH_2)_2(OH)_2$, is produced by adding phosphoryl chloride, $POCl_3$, to a cold saturated solution of ammonia, thus:—$2POCl_3 + 2NH_3.Aq + 3H_2O = P_2O_3(NH_2)_2(OH)_2.Aq + 6HCl.Aq$. It is also formed by boiling phosphoryl amide-imide, **$PO(NH)(NH_2)$**, with dilute sulphuric acid, thus:— $2PO(NH)(NH_2) + H_2SO_4.Aq + 3H_2O = P_2O_3(NH_2)_2(OH)_2.Aq + (NH_4)_2SO_4.Aq$. It is soluble in water, yielding salts.

Pyrophosphotriamic acid should have the formula $P_2O_3(NH_2)_3(OH)$, and should, therefore, be a monobasic acid. But the body of that formula is tetrabasic, and hence is more probably **tri-imido-pyrophosphoric acid, $P_2(NH)''_3(OH)_4$**. It is produced by the action of ammonia on phosphoryl chloride, with subsequent treatment with water, thus:—$2POCl_3 + 9NH_3.Aq + 2H_2O = P_2(NH)_3(OH)_4 + 6NH_4Cl.Aq$. It is a white, amorphous, sparingly-soluble powder, forming salts; those of potassium and ammonium are monobasic, thus:—**$P_2(NH)_3(OH)_3(OK)$**; they are white and insoluble. Mono-, di-, and tri-basic lead salts have been prepared, and a tetrabasic mercuric salt.

Other more complex compounds, probably amido-derivatives of still more condensed phosphoric acids, are produced by the action of ammonia on phosphoryl chloride. Among them are **$P_4N_5O_{11}H_{17}$**; **$P_4N_4O_9H_{10}$**; and **$P_4N_5O_7H_9$**.

Closely connected with these bodies is **phospham, PN_2H**, produced by the action of ammonia on phosphoric chloride, **PCl_5**. The intermediate product is phosphorus trichlorodiamide, **$PCl_3(NH_2)_2$**, which when heated evolves hydrogen chloride, thus:— **$PCl_3(NH_2)_2 = PN_2H$** $+ 3HCl$. The constitution of phospham may be taken as **$N{\equiv}P{=}(NH)$**. It is a white insoluble powder, unattacked by chlorine or by sulphur gas. If boiled with alkalies, it yields a phosphate and ammonia, thus :—$\mathbf{N\equiv P=(NH)}$ + OH_2 + $3H—OK.Aq = O=P(OK)_3.Aq + 2NH_3$.

Phosphorus trichloride, similarly treated with ammonia, yields, as has been mentioned on p. 525, phosphorus triamide, $\mathbf{P(NH_2)_3}$, as a white mass. When heated out of contact with air, a whitish-yellow residue is left, probably containing $\mathbf{(HN)=P(NH_2)}$ and $\mathbf{P\equiv N}$.

Compounds are produced by treating halides of arsenic, antimony, and bismuth with ammonia. They have not been exhaustively studied; but the following are known :—

$\mathbf{2AsCl_3.7NH_3}$; $\mathbf{2AsI_3.9NH_3}$; $\mathbf{As_2ClNH_6O_4}$; $\mathbf{SbCl_3.NH_3}$; $\mathbf{SbCl_5.6NH_3}$; $\mathbf{2BiCl_3.NH_3}$; $\mathbf{BiCl_3.2NH_3}$; $\mathbf{BiCl_3.3NH_3}$; $\mathbf{BiBr_3.2NH_3}$; $\mathbf{2BiBr_3.5NH_3}$; $\mathbf{BiBr_3.2NH_3}$.

These compounds are all formed by direct addition. The body $\mathbf{As_2ClNH_6O_4}$ is produced by the action of water on the amine $\mathbf{2AsCl_3.7NH_3}$, and it appears possible that their constitution is closely related. Perhaps the formlæ may be adopted :—

$$As\begin{cases}(NH_2)_2.2HCl \\ NH \\ (NH_2)_2.2HCl\end{cases} .2NH_4Cl \quad \text{and} \quad As\begin{cases}NH_2.HCl \\ OH \\ O \\ (OH)_2\end{cases}$$

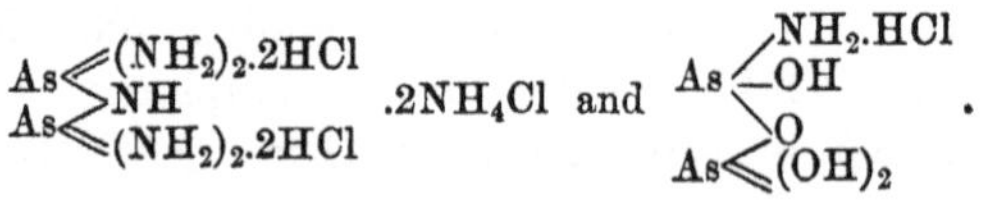

The formulæ of the remaining compounds are easily represented as chloramine hydrochlorides, with the exception of $\mathbf{2AsI_3.9NH_3}$, which requires further investigation.

Amido-compounds of **molybdenum** are unknown. The action of ammonia at a high temperature on the chloride yields the nitride, $\mathbf{Mo_3N_2}$, as a grey powder.

Complicated compounds are produced by the action of ammonia on **tungsten** hexachloride. One of these compounds is said to have the formula $\mathbf{W_7N_8O_4H_4}$; another, $\mathbf{W_5N_6O_5H_3}$; a third, $\mathbf{W_2N_2O_3}$; while $\mathbf{W_2N_3}$ is formed from WO_2Cl_2, and also from WCl_6. As regards the constitution of these bodies, no conclusions can be suggested until they are further investigated.

Uranium tetrachloride absorbs ammonia, producing

$$\mathbf{3UCl_4.4NH_3}.$$

Sulphur dichloride, treated with ammonia gas, yields the compounds $\mathbf{S(NH_2)_2.2HCl}$ and $\mathbf{SCl_2.4NH_3}$; disulphur-dichloride, S_2Cl_2, yields $\mathbf{S_2Cl_2.4NH_3}$. They are fairly stable crystalline bodies. No similar compounds of selenium have been produced; but tellurium tetrachloride, $TeCl_4$, with ammonia yields $\mathbf{Te(NH_2)_4.4HCl}$.

Compounds containing oxygen have been more closely examined. Sulphurosyl chloride, $SOCl_2$, with ammonia yields sulphurosamine, **$SO(NH_2)_2$**, and hydrogen chloride. It is noticeable that with compounds of phosphorus and sulphur the basic character of the amido-group appears to be neutralised by the acid functions of the groups $(SO)''$ or $(PO)'''$; hence the hydrochlorides are not formed, but hydrogen chloride is liberated. Such alteration in the functions of an amide, produced by the entry of groups, which, when combined with water, give rise to acids, is of common occurrence among the amido-derivatives of the carbon compounds.

Sulphur dioxide mixed with its own volume of ammonia yields **sulphurosamic acid, $HO—(SO)—NH_2$**; and with twice its volume of ammonia, ammonium sulphurosamate, **$NH_4O—(SO)—NH_2$**, is produced. No other compounds of this acid have been prepared.

Sulphuryl chloride, SO_2Cl_2, with ammonia, does not yield sulphuramine, $SO_2(NH_2)_2$, as might be expected, but the reaction takes place with formation of ammonium sulphamate, **$NH_2—(SO_2)—ONH_4$**, and ammonium chloride; but it is impossible to represent such a change without the interposition of a molecule of water. The reaction requires further investigation. Ammonium sulphamate is however easily obtained by the action of ammonia on sulphur trioxide, thus:—**SO_3** + $2NH_3$ = **$NH_4O—(SO_2)—NH_2$**; if less ammonia be used, **sulphamic acid** is formed, thus:— **SO_3** + NH_3 = **$HO—(SO_2)—NH_2$**. Ammonium sulphamate is crystalline, and is sparingly soluble in water, and yields, with barium chloride and ammonia, a white precipitate of a basic barium sulphamate, from which the neutral salt may be prepared by cautious addition of sulphuric acid. Its formula is **$Ba(OSO_2NH_2)_2$**; and from it the potassium and other salts have been produced by treatment with sulphates.

Similar compounds of selenium and tellurium are unknown.

Compounds with oxides of chlorine, bromine, and iodine are unknown. The amido-derivatives of these elements have been described among the halogen compounds of nitrogen (see p. 158).

The **amines of ruthenium, rhodium,** and **palladium** differ in character from each other. Those of ruthenium occupy a unique position; those of rhodium closely resemble the chromamines and cobaltamines; while those of palladium are more closely related to the nickelamines and similar bodies. They are all specially stable.

The compounds of **ruthenium** are produced by the action of solution of ammonia on ammonium ruthenichloride, $(NH_4)_2RuCl_6$.

The mixture is heated, evaporated to dryness, and extracted with alcohol, which leaves a pale yellow crystalline powder of the formula **$RuCl_2.4NH_3.3H_2O$.** It is to be observed that the ruthenium is not said to be reduced by this action; hence the formula usually assigned, $Ru''(N_2H_5)_2.2HCl$, is obviously inapplicable. On treatment with silver hydroxide, the chloride yields a hydroxide in solution, from which several salts—the carbonate, nitrate, sulphate, &c.—have been prepared.

When the hydroxide is heated it loses two molecules of ammonia, leaving the hydrate of the diamine, **$Ru(NH_3)_2(OH)_2.4H_2O$**, in the solid state. From this body salts have been prepared, which have a darker yellow colour than those of the diamine. As ruthenium dichloride is blue, compounds derived from it would almost certainly not have the colour of the ammonium rutheni-chloride, in which the tetrachloride, $RuCl_4$, is contained, unless there were some close connection between them.

The **rhodium amines** are similar to the purpureo- and roseocobaltamines. The compounds may be thus classified:—

$Rh(NH_3)_5.X_3$; roseorhodamines, or rhodamines;
$Cl—Rh(NH_3)_5.X_2$; purpureorhodamines, or chlororhodamines;
$(NO_3)—Rh(NH_3)_5.X_2$; nitratorhodamines;
$(NO_2)—Rh(NH_3)_5.X_2$; nitrorhodamines, or xanthorhodamines.

In the first set of compounds X_3 is replaceable; in the remaining three sets, X_2. Hence these bodies confirm the suggested formulæ for the cobaltamines. For the first set we may suggest the formula:—

$$Rh\begin{cases}NH_2 & HX\\ (N_2H_5)_2.2HX\end{cases}, \text{ and for the second, } X—Rh(NH_3)_5.2HX.$$

The roseorhodamines are obtained by digesting rhodochloride of ammonium, **$RhCl_3.2NH_4Cl$**, with ammonia. This yields the chloride, **$RhCl_3.5NH_3$**, from which the hydrate is obtainable by the action of silver hydroxide; and from it the carbonate, oxalate, sulphate, &c.

The **purpureorhodamines** are formed from rhodium trichloride and ammonia; that they contain one atom of chlorine in more intimate combination than the other two is proved by the fact that on treatment with silver nitrate, sulphate, &c., only two atoms of chlorine are exchanged for the groups $(NO_3)_2$, (SO_4), &c. Corresponding bromine and iodine compounds are produced from the roseorhodamines, in which all the halogen atoms are replaceable, by keeping the bromides and iodides at a high temperature for some time. The nitrato- and nitro-compounds may be similarly produced. All these bodies form series of salts, such as hydroxide, nitrate, sulphate, &c.

Three series of **palladium** compounds are known, two derived from the dichloride, $PdCl_2$, and the third from the trichloride, $PdCl_3$. They are as follows:—

$PdX_2.2NH_3 = Pd(NH_2)_2.2HX$, palladamine dihydrohalide;
$PdX_2.4NH_3 = Pd(N_2H_5)_2.2HX$, palladodiamine dihydrohalide;
$PdX_3.2NH_3 = X—Pd(NH_2)_2.2HX$, halopalladamine dihydrohalide.

The hydrochloride of the **palladamine** is produced by digesting palladium dichloride with a slight excess of ammonia solution. It has a red colour, but at 100° it turns yellow, possibly owing to isomeric change. The fluoride, the bromide, the iodide, the hydroxide (which is crystalline and a strong base), and several other salts, such as the carbonate, nitrite, sulphite, and sulphate, have been prepared by the usual methods. All crystallise well.

By digesting palladium dichloride with a greater excess of ammonia solution, the hydrochloride of **palladodiamine** is produced. Of this series, the fluoride, bromide, iodide, hydroxide, silicifluoride, carbonate, nitrate, sulphite, and sulphate are known. When heated to 100°, these salts lose ammonia, being converted into compounds of the first series.

The hydrochloride of the third series is produced by oxidising that of the first with nitrohydrochloric acid, or by exposing its solution to the action of chlorine. It and the other salts are dark red crystalline compounds.

Osmamines, Iridamines, and Platinamines.

The **osmamines** are derived from tetrad osmium, as in $OsCl_4$. They are as follows:—

$Cl_2{=}Os(NH_2)_2.2HCl.nH_2O$; $O{=}Os(NH_2)_2.(OH)_2$; and $OsCl_2(NH_3)_4$.

The second of these is formed by the action of excess of ammonia on osmium tetroxide, OsO_4; it is a blackish-brown powder, soluble in hydrochloric acid, yielding the first.

The third, supposed by its discoverer, Claus, to be analogous to the ruthenamines of similar formula, is produced by the action of ammonium chloride on potassium osmate, K_2OsO_4. On treatment with silver hydroxide, it yields a hydroxide. This compound evidently also contains tetrad osmium; but, like the corresponding compounds of ruthenium, it requires reinvestigation, for its formula implies that it is a derivative of dyad osmium. It is probable that there is error in supposing it to contain twelve atoms of hydrogen.

The compound $\mathbf{Os_2N_2O_5K_2}$ is produced by the action of ammonia and potassium hydroxide on osmium tetroxide, thus:— $6\mathbf{OsO_4} + 8NH_3 + 6KOH = 3\mathbf{Os_2N_2O_5K_2} + N_2 + 15H_2O$. From the barium salt the acid has been obtained. Many salts are known, of which those of sodium, potassium, ammonium, barium, and zinc are soluble; and those of lead, silver, and mercury insoluble. They explode when heated.

The **iridamines** are derived, first, from iridium dichloride. Such are **iridosamine** hydrochloride, $\mathbf{Ir(NH_2)_2.2HCl}$, and its derivatives, and **iridosodiamine** hydrochloride,

$$\mathbf{Ir(NH_2.NH_3)_2.2HCl};$$

the nitrate and sulphate have also been prepared. The first is produced by cautiously heating iridium trichloride till it changes to the dichloride, dissolving the brown residue in solution of ammonium carbonate, and adding a slight excess of hydrochloric acid. It is a yellow-orange substance. The sulphate is produced by evaporating the hydrochloride with the requisite amount of sulphuric acid. When the hydrochloride is boiled with excess of ammonia, the diamine hydrochloride, $\mathbf{Ir(N_2H_6)_2.2HCl}$, is produced as a whitish precipitate. The nitrate and sulphate are crystalline.

Second, from iridium trichloride, $\mathbf{IrCl_3}$. The hydrochloride, $\mathbf{IrCl_3.5NH_3}$, is analogous to that of the purpureocobaltamines and the rhodamines. It is produced by gently heating, for some weeks, ammonium iridichloride, $\mathbf{IrCl_3.3NH_4Cl}$, with excess of ammonia solution, then neutralising with hydrochloric acid, and washing the flesh-coloured precipitate with water. With silver hydroxide it yields the hydroxide, from which the nitrate, sulphate, &c., may be prepared by neutralisation.

Third, from iridium tetrachloride, $\mathbf{IrCl_4}$. By heating iridosamine hydrochloride, $\mathbf{Ir(NH_2)_2.2HCl}$, with strong nitric acid, the nitrate of dichloroiridodiamine is produced. When evaporated with hydrochloric acid, violet crystals of the hydrochloride are formed. This compound is known to have the constitution $\mathbf{Cl_2{=}Ir(N_2H_5)_2.2HCl}$, because silver nitrate removes only half its chlorine, forming the dinitrate. If its formula were $Ir(NH_2)_4.4HCl$, all the chlorine would be then removed. The sulphate, produced by evaporating the hydrochloride with the requisite amount of sulphuric acid, forms greenish needles.

The **amido-derivatives of platinum** have been very thoroughly investigated, and are very numerous. They are divisible into three main groups, according as they are derived from PtX_2, PtX_3, or PtX_4, where X represents a halogen, &c.

1. Platinous derivatives (from PtX_2).

1. **$Pt''(NH_2)_2.2HX$**;* salts of **platinosamine.**
2. **$X—Pt''(N_2H_5).HX$**;† salts of **haloplatinosodiamine** (the name diamine may be given to the group ($—NH_2—NH_3—$).

These compounds are isomeric with those of group 1.

3. $Pt''<^{(N_2H_5)HX}_{(NH_2)HX}$‡; salts of **platinosomonamine-diamine.**
4. **$Pt''(N_2H_5)_2.2HX$**;§ salts of **platinosodiamine.**

1. The hydrochloride of **platinosamine, $Pt(NH_2)_2.2HCl$**, is produced by heating the hydrochloride of platinosodiamine, **$Pt(N_2H_5)_2.2HCl$**, to 220—270°; or by boiling platinum dichloride with solution of ammonium carbonate, filtering, and crystallising. It forms yellow rhombohedra. Silver salts remove all the chlorine; and in this way many of the salts have been prepared. Among these are the bromide, iodide, oxide, oxalate, nitrite, nitrate, sulphite, sulphate, chlorosulphite, $Pt<^{NH_2Cl}_{NH_2.SO_3H}$, and sulphite, $Pt<^{NH_2—SO_3H}_{NH_2—SO_3H}$, which have replaceable hydrogen, and furnish series of metallic derivatives.

2. By the action of ammonia on a solution of platinous chloride in hydrochloric acid, a precipitate (the "green salt of Magnus") is formed, and the filtrate, on cooling, deposits yellow prisms of **chloroplatinosodiamine** hydrochloride,

$$ClPt(N_2H_5).HCl.$$

From its solution silver salts remove only half the total chlorine, yielding corresponding salts, among which are the bromide, iodide, cyanide, nitrite, nitrate, sulphate, chlorosulphite, and sulphite. The two last retain hydrogen, like the compounds of group 1, and yield metallic derivatives. With caustic soda the chloride yields hydroxyplatinosodiamine hydroxide, **$HO—Pt—(N_2H_6).OH$**.

3. The action of a small amount of ammonia on platinum dichloride yields a double compound of the dichloride with **platinosomonamine-diamine** hydrochloride, of the formula $Pt<^{NH_2.HCl}_{(N_2H_5).HCl}.PtCl_2$. When treated with nitric acid, the nitrate

* Reiset, *Annales* [3], **11**, 426; Peyronne, *ibid.*, **16**, 462; Cléve, *Bull. Soc. Chim.*, **16**, 203.

† Cleve, *ibid.*, **16**, 207.

‡ Cleve, *ibid.*, **16**, 21.

§ Peyronne, *Annales* [3], **12**, 193; Cléve, *Bull. Soc. Chim.*, **7**, 12.

is produced, $\mathbf{Pt}<{}^{\mathbf{NH_2.HNO_3}}_{\mathbf{(N_2H_5).HNO_3}}$; and from it the chloride, sulphate, &c.

4. When the hydrochloride of the monamine, $\mathbf{Pt(NH_2)_2.2HCl}$, is digested with ammonia, or when its isomeride (the "green salt of Magnus," see below) is similarly treated, the compound $\mathbf{Pt(N_2H_5)_2.2HCl}$, is formed, and deposits in large yellow crystals.

Its platinochloride, $\mathbf{Pt(N_2H_5)_2.2HCl.PtCl_2}$, is the green salt of Magnus* (the first discovered of these amido-derivatives), as is proved by its formation by direct addition of platinum dichloride to the hydrochloride of platinosodiamine. This platinochloride forms insoluble needles of a deep green colour. It was originally produced by addition of ammonia to a mixture of platinum tetrachloride with sulphur dioxide; a mixture containing platinum dichloride, owing to the reduction of the tetrachloride by the sulphur dioxide.

Of the diamine, the bromide, iodide, hydroxide, carbonate, oxalate, nitrate, chromate, dichromate, sulphate, hydrogen sulphate, hydrogen phosphate, and other salts have been prepared.

Two series of **double salts of the di-diamine** are known, produced by addition. The first of these (Buckton's) have the formula $\mathbf{Pt(N_2H_5)_2 2HCl.MCl_2}$, where M stands for a dyad metal; the second (Thomsen's), $\mathbf{M(N_2H_5)_2.2HCl.PtCl_2}$. This second series really forms the platinochlorides of the diamines of other metals. The salt of Magnus is the platinochloride of this series, where M stands for platinum, thus:—$\mathbf{Pt(N_2H_5)_2.2HCl.PtCl_2}$.

II. Compounds of platinum trichloride, $PtCl_3$ (or, as $PtCl_3$ is unknown, possibly of Pt_2Cl_6; but, as there is as little reason to suppose the existence of the latter body, the simpler formulæ are adopted). Derivatives of the following series are known.

1. **X—$\mathbf{Pt(NH_2)_2.2HX}$;† haloplatinodimonamine.**
2. **X—$\mathbf{Pt(N_2H_5)_2 2HX}$;‡ haloplatinodi-diamine.**
3. **$\mathbf{X_2}$—$\mathbf{Pt(N_2H_5)}$;§ dihaloplatinodiamine.**

1. The first of these is produced as **iodoplatinodimonamine** hydriodide, by the action of ammonia on di-iodoplatinidiamine hydriodide (see below), $I_2Pt^{IV}(NH_2)_2.2HI$. It is a crystalline powder, and is the only compound of the series known.

2. The starting point for salts of the second series is the nitrate of platinosodiamine (I 4). On treatment with iodine,

* Magnus, *Pogg. Ann.*, **14**, 242.
† Cléve, *Bull. Soc. Chim.*, **17**, 100.
‡ Cléve, *ibid.*, **15**, 168.
§ Cláve, *ibid.*, **17**, 100; Blomstrand, *Berichte*, 1871, 639.

it yields the nitrate of di-iodoplatini-di-diamine (see below), $I_2{=}Pt(N_2H_5)_2.2HNO_3$. This substance, with ammonia, loses iodine, and is converted into the oxide of **iodoplatino-di-diamine** nitrate, $I{-}Pt{<}^{N_2H_5.HNO_3}_{N_2H_5 - - -} \overset{HNO_3.N_2H_5}{O - - - N_2H_5}{>}Pt{-}I$, which forms microscopic yellow needles. With nitric acid, it yields the nitrate of iodoplatinodi-diamine, $I{-}Pt(N_2H_5)_2.2HNO_3$, a soluble orange powder, crystallising in small prisms. From the nitrate, the sulphate, phosphate, and oxalate have been made.

By boiling the oxide of iodoplatino-di-diamine nitrate with silver nitrate, **hydroxyplatino-di-diamine** nitrate is produced, $HO{-}Pt(N_2H_5)_2.2HNO_3$, and from this salt the chloride, sulphate, orthophosphate, dichromate, and oxalate have been prepared, the hydroxyl-group retaining its position.

By further treatment with nitric acid, the hydroxyl-group is replaced by the group $(NO_3)'$, yielding $NO_3.Pt(N_2H_5)_2.2HNO_3$, a body resolved by water into the hydroxy-compound.

The bromochloride, bromonitrate, bromosulphate, and bromoxalate have also been prepared, of the formula

$$Br{-}Pt(N_2H_5)_2.2HX.$$

3. The chloride alone is known, $Cl_2{=}Pt(N_2H_5)$; it is an amorphous yellow powder, produced by the action of nitrohydrochloric acid on a base of the formula $HO{-}Pt(N_2H_5)$, which, nevertheless, differs from the salt I 2. Opinions differ (Blomstrand, Cléve) as to the composition of the compound

$$HO{-}Pt{-}(N_2H_5);$$

it and its derivatives are amorphous explosive black substances, produced by boiling chloroplatinosodiamine hydrochloride with soda.

III. **Platinic amines**, derived from PtX_4. These compounds fall into the following classes:—

1. $X_2{=}Pt^{iv}(NH_2)_2.2HX.$* **Dihaloplatinamine** dihydrohalide.
2. $X_3{\equiv}Pt^{iv}(NH_2).HX.$† **Trihaloplatinamine** hydrohalide.
3. $X_2{=}Pt^{iv}{<}^{NH_2.HX}_{N_2H_5.HX}$.‡ **Dihaloplatinimonodiamine** hydrohalide.
4. $X_2{=}Pt(N_2H_5)_2.2HX.$§ **Dihaloplatinidi-diamine** hydrohalide.

---

* Gerhardt, *Rev. Scient.*, **28**, 273; Cléve, *Bull. Soc. Chim.*, **17**, 100.

† *Ibid.*, **17**, 105.

‡ *Ibid.*, **17**, 107.

§ Gros, *Annales* [2], **69**, 204; Raewsky, *ibid.* [3], **22**, 273; Cléve, *Bull. Soc. Chim.*, **7**, 19; and **15**, 162.

1. These compounds are prepared from the chloride, which is formed by the action of chlorine on platosamine hydrochloride (I 1), suspended in water. It forms small yellow quadratic octahedra, and is sparingly soluble in water. Its formula is $\mathbf{Cl_2{=}Pt(NH_2)_2.2HCl}$. From it many other salts have been prepared, among which are the dibromodihydrobromides, the di-iodo-dihydriodide, the dihydroxydihydroxide, the dinitrato-dinitrates, the dihydroxydinitrate, the dinitrato-dinitrite, the nitrato-chloro-dinitrite, the dichlorodinitrite, &c. The formula of the last may be given to illustrate the nomenclature. It is

$$\mathbf{Cl_2{=}Pt(NH_2)_2.2HNO_2.}$$

2. Compounds containing the groups $\equiv Pt(N_2H_5)$—are similarly formed by the action of chlorine on chloroplatinosodiamine (I 2). They are isomeric with the former. The chloride forms sparingly soluble yellow laminæ, and has the formula $\mathbf{Cl_3{\equiv}Pt(N_2H_5).HCl}$. From it the tribromohydrobromide, the trihydroxy-nitrate, the dichloro-nitro-nitrite,

$$(NO_2)Cl_2{\equiv}Pt(N_2H_5).HNO_2,$$

the dihydroxy-sulphate, $(HO)_2{=}Pt{<}^{N_2H_5H}_{\_\_\_\_}SO_4$, and other compounds have been prepared by the usual methods.

3. Derivatives of $=Pt{<}^{N_2H_5}_{NH_2}$ are prepared, like those of the former two groups, by the action of chlorine on the hydrochloride of platinoso-mono-diamine, $\mathbf{Pt(NH_2)(N_2H_5).2HCl}$. Its hydrochloride forms yellow rhombic scales. The dihydroxy-dinitrate, the dibromo-dinitrate, and the dibromo-sulphate have been prepared.

4. Derivatives of $=Pt(N_2H_5)_2$, are similarly prepared by the action of chlorine on a solution of platinosodi-diamine hydrochloride (I 4); or by dissolving the hydrochloride of dichloroplatinamine (IV 1) in ammonia. The hydrochloride of the dichloride forms sparingly soluble yellow transparent regular octahedra. Many derivatives of this amine are known; among others, the dibromo-hydrobromide, the hydroxybromo-dibromide, the di-iodo-hydriodide, the chlorobromo-hydrochloro-hydrobromide ($\mathbf{ClBr{=}Pt(N_2H_5)_2.HCl.HBr}$), &c., &c.

Some similar compounds with hydroxylamine have been prepared.*

* *Chem. Centralbl.*, 1887, 1254.

Cuprosamines and Cupramines; Argentamines, Auramines, Mercurosamines, and Mercuramines.

The **amido-derivatives of copper** are divisible into two classes:—

I. Those containing **cuprous** copper, as in cuprous chloride, Cu_2Cl_2.

II. Those containing **cupric** copper, as in cupric chloride, $CuCl_2$.

But this difference is to be noted between these compounds and those of the previous palladium and platinum groups, viz., that on treatment with halogen acids they give double halides, such as $CuCl.NH_4Cl$; $CuCl_2.2NH_4Cl$, &c. They are not so stable as the compounds of the preceding groups, but rather resemble the zinc and cadmium compounds.

I. **Cuprous compounds.***†—

1. **$Cu_2Cl_2.NH_3 = Cu_2{=}NH.2HCl$, dicuprosamine dihydrochloride**, produced by direct action in a gentle heat, is a black powder.

2. **$CuCl.NH_3 = Cu{-}NH_2.HCl$**, cuprosamine hydrochloride, is a non-crystalline substance, produced by the action of ammonia on copper monochloride, in the cold. After long continued action of ammonia, it appears probable that the compound $CuCl.2NH_3$ is formed.

3. **$CuI.2NH_3 = CuNH_2.NH_4I$** forms white crystalline plates. It is produced by digesting copper with cupric chloride, and then adding a solution of potassium iodide.

II. **Cupric compounds.**‡—1. **Cupramine** hydrochloride, **$Cu(NH_2)_2.2HCl$**, is produced by the action of ammonia on cupric chloride, $CuCl_2$, at 140°. The corresponding carbonate, **$Cu(NH_2)_2.H_2CO_3$**, and sulphate, **$Cu(NH_2)_2.H_2SO_4$**, have been prepared by direct addition. They are apple-green compounds.

2. **$Cu<^{NH_2}_{(N_2H_5)}.2HBr$** is precipitated by alcohol from a mixture of ammonia with cupric bromide.

3. **$Cu(N_2H_5)_2.2HI.H_2O$** is formed by the action of atmospheric oxygen on a solution of cuprous iodide, CuI, in ammonia. This body is of special interest, for cupric iodide, CuI_2, is unstable.

* Dehérain, *Compt. rend.*, **55**, 807; Leval, *J. Pharm.* (3), **4**, 328.

† As we are ignorant of the molecular weight of combined cuprous chloride, the simple formula is given.

‡ Rammelsberg, *Pogg. Ann.*, **48**, 162; **55**, 246.

The hydroxide is also known, $Cu(N_2H_5)_2H_2.(OH)_2.2H_2O$, forming blue octahedra; also the sulphate, dithionate, and iodate.

4. $Cu(N_2H_5)_2.(HBr)(NH_4Br)$ is produced by direct union.

5. $Cu(N_2H_5)_2.2NH_4Cl$ is similarly produced. It is blue and amorphous.

Argentamines.—These appear to be very numerous, inasmuch as almost all salts of silver are soluble in ammonia, and presumably unite with it. But only a few have been isolated. These are all argentous compounds.

1. **Argentamine**, $Ag(NH_2)$ (?), is a black explosive powder, commonly called "fulminating silver," produced by the action of ammonia solution on silver hydroxide

2. $(AgNH_2.HI).AgI$. A double salt, formed by the action of gaseous ammonia on dry silver iodide which has not been fused.

3. $Ag(NH_2).HNO_3$ and the corresponding chlorate, sulphate, and chromate, are soluble crystalline bodies, formed by direct union.

4. $(AgNH_2.HCl).2NH_4Cl$, is a double salt, formed by the action of ammonia on silver chloride, suspended in water. It is very soluble, and crystallises out on concentration. Silver bromide does not absorb ammonia. The corresponding nitrate has been prepared.

Auramines.*—Aurous oxide, Au_2O, dissolves in strong ammonia, giving $NAu_3.NH_3$. When boiled with water, ammonia is evolved, and gold nitride, Au_3N, remains. Gold monoxide, AuO, on similar treatment, yields a similar compound, but the gold is present as hydroxide; its formula is $N(AuOH)_3.NH_3$. It is a very explosive substance, which, when boiled with water, undergoes a similar decomposition, the product being $N(AuOH)_3$. Auric chloride, $AuCl_3$, digested with ammonia, yields "fulminating gold," a mixture of $HN{=}AuCl$, and $HN{=}Au{-}NH_2$. The latter, digested with sulphuric acid, yields a salt of the formula $Au(N_2H_5)_2.H_2SO_4$. These compounds are all very unstable.

Mercurosamines and **mercuramines.**†—Of these many are known; and a few examples of corresponding phosphorus and arsenic compounds have also been prepared, which will be considered along with their analogues.

I. **Mercurosamines.**—Of these there are three classes; the

* Dumas, *Annales* (2), **44**, 167; Raschig, *Annalen*, **235**, 341.

† The chief references on this subject are:—Mitscherlich, *Pogg. Ann.*, **9**, 387; **16**, 41; **55**, 248; Kane, *Annales* (2), **72**, 215; Millon, *ibid.* (3), **18**, 392; Plantancour, *Annalen*, **40**, 115; Hirzel, *ibid.*, **84**, 258; Schmieder, *J. prakt. Chem.*, **75**, 128.

first contains the group $(NH)_2'$; the second the group $(NH)''$, and the third the triad atom $(N)'''$. These are analogous to the monamimes, diamines, and triamines, where the hydrogen of a molecule of ammonia is replaced successively by one, two, and three hydrocarbon groups, such as methyl, ethyl, &c., as NH_2CH_3, $NH(CH_3)_2$, and $N(CH_3)_3$.

1. **$HgNH_2.HF$**, a black substance, decomposed by water, produced by the action of gaseous ammonia on mercurous fluoride. A double salt of the chloride, with ammonium chloride, **$HgNH_2.NH_4Cl$** (or possibly $Hg(N_2H_5)HCl$), is formed when dry ammonia is absorbed by dry mercurous chloride. When warmed this body loses ammonia, leaving the hydrochloride, **$HgNH_2.HCl$**. The iodide is unstable.

2. The compounds **$Hg_2NH.HCl$** and **$Hg_2NH.HBr$** are black substances, produced by treating mercurous chloride or bromide with solution of ammonia. The formation of this precipitate serves to distinguish mercurous chloride from that of silver or lead, both of which are also insoluble. Mercurous nitrate, with ammonia solution, gives a corresponding compound,

$$2(Hg_2NH.HNO_3).H_2O,$$

as a greyish-black powder, decomposed by light.

The action of hydrogen arsenide on mercuric chloride is to produce an analogous compound, in the form of a double salt, **$Hg_2AsCl.HgCl_2$**, as a brownish-yellow precipitate. Here hydrogen is replaced by chlorine; and it may be remembered that compound arsines containing hydrogen in union with arsenic are also unknown.

3. **$2(Hg_3N.HNO_3).3H_2O$**, **trimercurosamine** nitrate, is also formed by treating mercurous nitrate with ammonia solution.

II. **Mercuramines.**—These may be divided into four classes:—

1. Those in which the mercury-atom shares its two powers of combination with the amido-group, and with some other group.

$Cl—Hg—NH_2$, **chloromercuramine**, is a white precipitate, formed by adding a slight excess of ammonia to a cold solution of mercuric chloride, $HgCl_2.Aq$. When gently heated, its hydrochloride, **$Cl—Hg—NH_2.HCl$**, sublimes, leaving the compound **$ClHg—N{=}Hg.HgCl_2$** (see below). Chloromercuramine is named "infusible white precipitate." The hydrochloride may also be produced by acting on dry mercuric chloride with gaseous ammonia, or by digesting mercuric oxide, **HgO**, with ammonium chloride. It resembles mercuric chloride in appearance, and may be sublimed without sensible decomposition.

Bromomercuramine, $BrHgNH_2$, and its hydrobromide resemble the chloro-compounds in properties and reactions.

Iodomercuramine hydriodide, produced by evaporating a solution of mercuric iodide in ammonia, forms small white needles. A double salt of oxymercuramine nitrate with ammonium nitrate, $O<\begin{smallmatrix}Hg-NH_2.HNO_3\\Hg-NH_2.HNO_3\end{smallmatrix}.NH_4NO_3$, is formed by mixing solutions of mercuric nitrate and ammonia, and evaporating the filtered liquid. When boiled with water, its solution deposits yellow needles of the simple nitrate.

2. **Mercuramines.**—Mercuramine hydrochloride,

$$Hg(NH_2)_2.2HCl,$$

or "fusible white precipitate," is produced by adding a solution of mercuric chloride to a boiling mixture of solutions of ammonium chloride and ammonia as long as the precipitate at first formed redissolves; also by boiling chloromercuramine, $ClHgNH_2$, with solution of ammonium chloride. It forms white rhomboidal dodecahedra. The hydriodide is a substance of a dull-white colour, produced by the action of ammonia gas on mercuric iodide. In presence of water, crystals of $Hg(NH_2)_2.2HI.3H_2O$ are formed. A double salt of the oxide with mercuric nitrate,

$$O<\begin{smallmatrix}Hg.NH_2\\Hg.NH_2\end{smallmatrix}.Hg(NO_3)_2,$$

is produced by adding ammonia in small quantity to a slightly acid solution of mercuric nitrate. It deposits slowly as a white precipitate.

3. The third group resembles the first, in having halogen or similar groups directly connected with the mercury; but two hydrogen-atoms of the ammonia are thereby replaced.

The fluoride, $(FHg)_2.NH.H_2O$, is a gelatinous precipitate produced by treating mercuric fluoride with ammonia.—**Dichloromercuramine** hydrochloride, $(ClHg)_2.NH.HCl$, is formed as a white insoluble precipitate by the action of great excess of ammonia on mercuric chloride.—Hydroxychloromercuramine, $\begin{smallmatrix}HOHg\\ClHg\end{smallmatrix}>NH$, is produced when chloromercuramine hydrochloride, $Cl-Hg-NH_2.HCl$, is boiled with water, thus:—$2ClHgNH_2.HCl + H_2O + Aq = NH_4Cl.Aq + 2HCl.Aq + (HOHg).(ClHg).NH$. It is a dense yellow powder. With solution of potassium iodide it gives the corresponding hydroxyiodide, which is also formed as a brown precipitate by the action of ammonia on mercuric iodide.—

The hydroxyoxide, $O<\begin{matrix}Hg(NH)-Hg.OH\\Hg(NH)-Hg.OH\end{matrix}$, is produced by the action of strong ammonia solution on yellow mercuric oxide. It is a brown insoluble substance, turning white on exposure to air owing to absorption of carbon dioxide. Many salts of this body are known.

Oxydimercuramine nitrate, $O<\begin{matrix}Hg\\Hg\end{matrix}>NH.HNO_3$, is a granular white powder, formed by boiling oxymercuramine mercuric nitrate, $O(HgNH_2)_2.HgNO_3$ (see 2), with water; ammonium nitrate is also formed. The mercuro-hydroxynitrate of oxydimercuramine, $O<\begin{matrix}Hg\\Hg\end{matrix}>NH.Hg<\begin{matrix}OH\\NO_3\end{matrix}$, is produced by adding a great excess of ammonia to mercuric nitrate; it is a whitish-yellow precipitate.

Compounds such as these are very numerous. Carbonates, chromates, sulphites, phosphates, arsenates, iodates, and other compounds have been prepared. Their methods of preparation, constitution, and properties may, however, be inferred from those of the halides and nitrates described above.

4. The last series of compounds is one in which the hydrogen of ammonia is entirely replaced by mercury. **Trimercuramine,** $N_2(Hg'')_3$, or more correctly mercuric nitride, is a dark-brown powder, produced by passing ammonia over hot mercuric oxide at 130°. It is exceedingly explosive. The action of liquefied ammonia on mercuric iodide yields the compound $IHg-N=Hg$; and the hydrate $HO.Hg-N=Hg$ is formed by digesting mercuric oxide with aqueous ammonia. It may be heated to 100°, and yields the oxide $Hg=N-Hg-O-Hg-N=Hg$ as a deep-brown explosive powder. From the oxide the chloride, $ClHg-N=Hg$, is produced by treatment with hydrochloric acid. A compound of this chloride with mercuric chloride, $ClHg-N=Hg.HgCl_2$, forms small red crystalline laminæ, and is left as a residue when the hydrochloride of chloromercuramide, $ClHgNH_2.HCl$, is sublimed from chloromercuramide, $ClHgNH_2$. Similar to this last compound is one produced by the action of hydrogen phosphide on a solution of mercuric chloride; its formula is $2(ClHg-P=Hg).HgCl_2$. It is a yellow powder. The corresponding bromide has also been prepared.*

* Rose, *Pogg. Ann.*, 40, 75.

CHAPTER XXXII.

NITRIDES, PHOSPHIDES, ARSENIDES, AND ANTIMONIDES (CONTINUED); CYANIDES AND DOUBLE CYANIDES.

Na_3N;* Na_3P;† Na_3As;‡ Na_3Sb.‡—K_3N;* K_3P† (?); K_3As;‡ K_3Sb.‡

Sodium nitride is a greenish mass, produced by heating sodamide (see p. 524) to redness, thus:—$3NaNH_2 = Na_3N + 2NH_3$. **Potassium nitride** is similarly produced. These compounds burn brilliantly when heated in air, and are decomposed by water. —**Sodium** and **potassium phosphides** are produced by direct union, best under a layer of xylene, C_8H_{10}; the union is exceedingly energetic, and is accompanied by evolution of heat and light. Excess of phosphorus is dissolved out by treatment with carbon disulphide, and the blackish powder remaining is dried in a current of dry carbon dioxide.—**Arsenide** and **antimonide of sodium** and **potassium** are metallic-looking substances, of crystalline fracture, produced by direct union at a red heat; the union takes place with incandescence. With water, these bodies yield hydrogen arsenide or antimonide.

No nitrides of beryllium, calcium, strontium, or barium have been prepared; beryllium is said to combine directly with phosphorus: but the compound obtained was impure. **Calcium** and **barium phosphides** have been produced mixed with pyrophosphates§ by the action of phosphorus gas on the oxides, thus: $7BaO + 12P = 5BaP_2 + Ba_2P_2O_7$. The mixture is a brownish-black lustrous substance, giving with water phosphine and barium hypophosphite. Arsenides appear to be similarly produced. No antimonides have been prepared.

* Gay-Lussac and Thenard, *Rech. physico-chim.*, 1, 354; Beilstein and Geuther, *Annalen*, **108**, 88.

† Vigier, *Bull. Soc. Chim.*, 1861, 6.

‡ Landolt, *Annalen*, **89**, 201.

§ Dumas, *Annales*, **32**, 364.

Mg_3N_2;* **Mg_3P_2**;‡ **Mg_3As_2**.‡—**Zn_3N_2**;† **Zn_3P_2**;§ **ZnP**; **ZnP_2**; (?) **ZnP_6**; (?) **Zn_3As_2**; **Cd_3As_2**‖.—Nitrides of cadmium have not been prepared; but that metal unites directly with **phosphorus**, forming **Cd_3P_2** and **Cd_2P**. The **arsenide** is said to have the formula Cd_2As.

Magnesium nitride is a greenish-yellow amorphous mass, produced by direct union of nitrogen with red-hot magnesium, or even by burning magnesium in a limited supply of air; it reacts with water, forming ammonia.—**Zinc nitride** is produced by heating zincamine, **$Zn(NH_2)_2$**, to redness (see sodium nitride); it is a grey powder, reacting violently with water, yielding ammonia and zinc hydroxide. **Magnesium phosphide**, produced by direct union at a red heat, is a steel-grey, crystalline substance with metallic lustre. The compound **Zn_3P_2** is produced by direct union of the vapours of zinc and phosphorus, either directly or when zinc phosphate is strongly heated with charcoal. It forms iridescent prismatic crystals, or a grey mass. It volatilises at a higher temperature than zinc. The phosphide, **ZnP**, is said to form brilliant needles; it is probable that the compounds ZnP_2 and ZnP_6 are mixtures of amorphous Zn_3P_2 and red phosphorus.¶ **Cadmium phosphide, Cd_3P_2**, is a crystalline body with grey metallic lustre. The **phosphide, Cd_2P**, is said to be formed at the same time.¶ These bodies require further study.—**Magnesium arsenide** is a brown slightly lustrous substance, produced by direct union; zinc and arsenic combine with incandescence, giving brilliant grey octahedra, which, when heated, yield a brittle grey button of **Zn_3As**; and **cadmium arsenide**, a bright metallic-looking substance with a reddish tinge, is produced by the action of potassium cyanide, KCN, at a red heat on cadmium arsenate. Compounds of these elements with antimony may be prepared by fusion; two crystalline **antimonides of zinc**, of the formulæ **ZnSb** and **Zn_3Sb_2**, are known; they appear to be definite compounds. They decompose water at the ordinary temperature.

BN.**—No other compounds of the elements of this group have been prepared.—**Boron nitride** is produced by direct union; by the action of ammonium chloride on boron oxide at a red heat; and by passing the product of the action of boron chloride on

* Brieglieb and Geuther, *Chem. News*, **38**, 39; *Annalen*, **123**, 228.

† *Phil. Mag.* (4), **15**, 149.

‡ Parkinson, *Chem. Soc.* (2), **5**, 117.

§ Vigier, *Bull. Soc. Chim.*, 1861, 5.

‖ *Compt. rend.*, **86**, 1022, 1065.

¶ Renault, *Annales* (4), 9, 162.

** Wöhler, *Annalen*, **74**, 70; Martius, *ibid.*, **109**, 80.

ammonia through a red hot tube. It is a soft white amorphous infusible powder. It is very stable; but when heated in steam it yields boron oxide and ammonia (see p. 513).

AlN;* the phosphide and arsenide have been prepared, but not analysed. The only other compound of the group which has been prepared in a definite form is **TlSb**.—**Aluminium nitride** was prepared by heating aluminium with sodium carbonate to an exceedingly high temperature (the nitrogen is evidently derived from air); it forms yellow, lustrous crystals, becoming dull on exposure to moist air, and finally evolving ammonia, leaving a residue of aluminium hydroxide.—The **phosphide** and **arsenide** are grey masses, produced by direct union.—**Thallium antimonide**, also produced directly, is brittle, and possesses metallic lustre.

Cr_3N_4; CrP.—Fe_3N_2(?);† Fe_3P_4; FeP; Fe_2P;†—Fe_3As_2; FeAs; Fe_3As;‡ Fe_2As_3; $FeAs_2$; $FeAs_4$.—Mn_3P_2(?); MnAs.—Co_3P_2; $CoAs_2$; $CoAs_3$.—Ni_2P; Ni_3P_2; Ni_3As; Ni_2As; Ni_3As_2; NiAs; $NiAs_2$; NiSb.

Many of these compounds occur native; among them are *leucopyrite* or *arsenosiderite*, **Fe_2As_3**; *lölingite*, **$FeAs_2$**; *kaneite*, **MnAs**; *smaltite*, **$CoAs_2$**; *skutterudite*, **$CoAs_3$**; *kupfernickel*, or *niccolite*, **NiAs**; *rammelsbergite*, **$NiAs_2$**; and *breithauptite*, **NiSb**.

Chromium nitride has been obtained by heating anhydrous chromium trichloride in ammonia. It is an insoluble brown powder, burning in air to chromium sesquioxide and nitrogen. The **phosphide**, similarly prepared, is a black powder, insoluble in water, and not attacked by acids. **Phosphides of cobalt and nickel, Co_3P_2**, and **Ni_3P_2**, are grey powders, produced by heating the dichlorides in a current of phosphine. **Iron** at a red heat is hardly attacked by molecular nitrogen. But if the nitrogen is nascent, as, for instance, if ammonia be passed over red-hot iron, a white brittle substance is formed, and the gain in weight corresponds to the formula **Fe_2N**. A similar substance is formed by heating ferrous chloride, **$FeCl_2$**, in a current of ammonia; but its composition appears to correspond with **Fe_3N_2**. At a higher temperature it loses nitrogen, and is converted into **Fe_3N**. Iron nitrides burn in air, and when heated in hydrogen yield ammonia and metallic iron; with steam, iron oxide and ammonia are the products.

Many **phosphides of iron** have been obtained,§ but the

* Mallet, *Chem. Soc.*, **30**, 349.

† Stahlschmidt, *Pogg. Ann.*, **125**, 37.

‡ *Compt. rend.*, **86**, 1022 and 1065.

§ Freese, *Pogg. Ann.*, **132**, 225.

separate existence of many of them as distinct chemical individuals is doubtful. Those which appear best established are Fe_3P_4, produced as a black powder by heating the disulphide, FeS_2, in a current of phosphine; **FeP**, a grey tumefied mass, obtained by the action of phosphorus vapour on finely divided iron, reduced from its oxide by hydrogen; and Fe_2P, a hard brittle mass with metallic lustre, produced by throwing phosphorus on to red-hot iron filings. These substances are insoluble, and are attacked with difficulty by acids. A **phosphide of manganese**, which appears to approximate in composition to the last mentioned phosphide of iron, is produced similarly, or by reducing manganous pyrophosphate, $Mn_2P_2O_7$, by charcoal at an intense heat. Corresponding **phosphides of nickel** and **cobalt** are similarly prepared, and have similar properties. The **arsenides of iron** are whitish-grey brittle substances with metallic lustre, which are either found native, or have been prepared by direct union. A white hard magnetic alloy is also formed when **antimony** and **iron** are heated together. The native **arsenide of manganese** is a hard grey substance, approximating in composition to the formula **MnAs**. **Cobalt diarsenide** or *smaltine* is the most abundant of cobalt ores, and is found native in silver-white regular crystals. When heated, a portion of the arsenic is evolved, and a fusible brittle metallic-looking mass remains. *Skutterudite*, or *modumite*, $CoAs_3$, forms regular crystals of a grey-white metallic appearance, which evolve arsenic when heated. *Chloranthite*, or *white nickel*, $NiAs_2$, forms tin-white regular crystals, or as *rammelsburgite*, trimetric prisms, which oxidise in moist air to arsenate of nickel. *Copper nickel* or *niccolite*, **NiAs**, usually forms compact masses of a copper-red colour, and sometimes hexagonal prisms; it is one of the chief ores of nickel. The rare mineral *breithauptite*, **NiSb**, occurs in thin, copper-coloured, hexagonal plates. **Speiss** is a deposit formed in the pots when roasted arsenide of cobalt, mixed with copper nickel, is fused with potassium carbonate and silica in the preparation of **smalt**, a blue glass containing cobalt. It contains cobalt, manganese, iron, antimony, bismuth, and sulphur, but consists mainly of nickel arsenide; and the proportions of these constituents correspond best with the formula Ni_3As_2. It is sometimes found in dimetric crystals, but is generally a white metallic-looking substance with a reddish tinge. The arsenide, Ni_2As, has been produced by direct union.

Double compounds.—These are found native, and uniformly contain sulphur; the most important of the double sulphides and arsenides are :—

Mispickel, or *arsenopyrites*, **$FeSAs$**; *pacite*, **$Fe_5S_2As_8$**; *glaucopyrites*, **$Fe_{13}S_2As_{24}$**; *glaucodot* **$(Fe,Co)SAs$**; *cobaltite*, **$CoSAs$**; *gersdorffite*, **$NiSAs$**; *ullmannite*, **$NiSSb$**; *corynite*, **$NiS(As,Sb)$**; and *ulloclasite*, **$Co_3S_4As_6Bi_4$**.

These have a yellow or grey colour, and possess metallic lustre.

Nitride of carbon or **cyanogen**, C_2N_2, is such a remarkable substance, and forms so many double compounds, that it is preferable to consider it apart. It will therefore be treated of after the other nitrides have been described.

Titanium forms two well-defined nitrides, which for long were mistaken, owing to their metallic lustre, for titanium itself. Wöhler was the first to show their true nature.* Their formulæ are **TiN** and **Ti_3N_4**, the first corresponding to the oxide, **Ti_2O_3**, and the latter to **TiO_2**. **TiN** is produced by heating titanium dioxide in a current of ammonia; **Ti_3N_4**, by similarly treating $TiCl_4$; on heating it to a high temperature for a sufficiently long time, it loses nitrogen, yielding **TiN**. Titanium mononitride forms golden-yellow crystals, and trititanic tetranitride forms crystals of a copper-red colour with metallic lustre. The existence of other nitrides, described by Wöhler, appears to be disproved. When heated in steam these nitrides yield titanium oxide and ammonia. Titanium easily unites directly with nitrogen, forming a mixture of these compounds.

Zirconium also forms nitrides when the element or its tetrachloride is heated in ammonia; yellow crystals have been obtained also by the action of atmospheric nitrogen on zirconium at an intense heat.† Their composition is unknown, but they are decomposed by steam, yielding ammonia. Cerium and thorium compounds are unknown.

No phosphide of carbon is known. **Titanium phosphide** has been prepared by heating the phosphate with carbon. Its formula is unknown; but it is said to form white brittle fragments. Phosphides of cerium, zirconium, and thorium have not been prepared; nor are arsenides or antimonides of these elements known.

Nitride of silicon is a white amorphous mass, of the formula **Si_2N_3**, infusible, and unoxidisable by heating in air, and insoluble in all acids but hydrofluoric. It is produced by heating silicon in a current of nitrogen; the action of ammonia on silicon tetrachloride yields a chloride, **$Si_5N_6Cl_2$**, a white powder, which when

* *Annales* (3), **29**, 175; and **52**, 92; also Friedel and Guérin, *Compt. rend.*, **82**, 972, and *Annales* (5), **7**, 24.

† Mallet, *Sill. Amer. J.*, **28**, 346.

heated in ammonia loses hydrogen chloride, leaving $\mathbf{Si_2N_3H}$.* It slowly evolves ammonia on exposure to moist air. Compounds of silicon with phosphorus, arsenic, and antimony are unknown. The germanium compounds have not been investigated; and nitrides of tin and lead are unknown. **Tin** combines with phosphorus directly, forming a brilliant crystalline mass which appears to have the formula $\mathbf{Sn_3P_2}$. It is less fusible than tin, but white, softer, and more malleable. Another phosphide is produced by the action of phosphine on stannic chloride; on treatment with water, a yellow powder remains which has the formula $\mathbf{SnP_3}$. Phosphides of tin, heated in a current of hydrogen, leave a residue of tin, while phosphorus sublimes. **Lead** dissolves about 15 per cent. of phosphorus. The product is like lead, and may be cut with a knife; but it breaks when hammered. Excess of phosphorus crystallises from the lead in the form of "red" (black metallic) phosphorus (see p. 59). Phosphine is said to throw down a brown precipitate of lead phosphide from a solution of the acetate.

Arsenides of tin and lead appear to be of the nature of alloys. They form metallic-looking masses, and lose arsenic on distillation. An arsenide of tin containing 1 part of arsenic to 15 parts of tin crystallises in large leaves; it is less easily fused than tin. The compound $\mathbf{Sn_2As_3}$ has also been prepared. The alloy of lead and arsenic is also crystalline and brittle; **PbAs**, $\mathbf{Pb_3As_4}$, and $\mathbf{Pb_2As}$ are known.† Lead shot contains 0·1 to 0·2 per cent. of arsenic; its presence makes the lead assume the form of drops, and renders it harder. The antimonides will be treated of in the next chapter, for they are of the nature of alloys.

PN (?); VN; $\mathbf{VN_2}$.—AsP; SbP.—$\mathbf{Sb_2As}$; $\mathbf{Sb_2As_3}$;—$\mathbf{Bi_3As_4}$.

The product of the action of ammonia on phosphorus trichloride is probably $P(NH_2)_3$ (see p. 525). When heated, a residue is left which may contain phosphorus nitride, **PN**, but this subject requires further investigation. Vanadyl trichloride, $VOCl_3$, or vanadium trioxide, $\mathbf{V_2O_3}$,‡ when heated to a high temperature in a current of ammonia, yield a greyish-brown powder mixed with small plates with metallic lustre, possessing the formula **VN**. The first product of the action of ammonia on vanadyl trichloride is the **dinitride, $\mathbf{VN_2}$**, a brown powder, which loses nitrogen at a white heat, leaving the **mononitride, VN**. The **phosphide,**

* *Compt. rend.*, **93**, 1508.
† *Ibid.*, **86**, 1022 and 1068.
‡ Roscoe, *Annalen, Suppl.*, **6**, 114, and **7**, 70.

which has not been analysed, is said to be formed by the action of carbon at a white heat on vanadyl phosphate, and to form a grey porous mass. Similar compounds of niobium and tantalum have not been prepared, nor have arsenides and antimonides of vanadium.

Nitrides of arsenic, antimony, and bismuth are unknown. It would be advisable to investigate the action of a high temperature on the compounds of the trichlorides with ammonia.

The action of arsine on phosphorus trichloride, or of phosphine on arsenic trichloride, yields **phosphide of arsenic**, as a red-brown solid, soluble in carbon disulphide, of the formula **AsP**. It is changed by water into an oxide, $\mathbf{As_3P_2O_2}$. When heated in carbon dioxide phosphorus sublimes, and then arsenic. A similar red **phosphide of antimony, SbP**, is produced by the action of phosphorus, dissolved in carbon disulphide, on a solution of antimony bromide in carbon disulphide. Antimony and arsenic combine when heated together, forming a crystalline substance of the formula $\mathbf{Sb_2As}$; and the mineral *allamontite* has the formula $\mathbf{Sb_2As_3}$.

$\mathbf{Mo_3N_2}$; $\mathbf{W_2N_3}$;* $\mathbf{U_3N_2}$ (?).—**MoP**; $\mathbf{W_3P_4}$; $\mathbf{W_2P}$.—Arsenides and antimonides unknown.

$\mathbf{Mo_3N_2}$ is produced by the action of ammonia at a red heat on molybdenum chloride; it is a grey powder. **Tungsten nitride** is a black powder, produced by the action of ammonia at a red heat on WO_2Cl_2, or on WCl_6 (see p. 536). Similarly, uranium pentachloride, heated in a current of ammonia, yields a black powder of doubtful formula. **Molybdenum phosphide** is a grey metallic-looking mass, formed by heating a mixture of molybdenum pentoxide, metaphosphoric acid, and charcoal to whiteness; one **phosphide of tungsten,** $\mathbf{W_3P_4}$, is produced by direct union at a red heat, and is a dark-grey powder; and the other phosphide, $\mathbf{W_2P}$, forms fine hexagonal steel-grey crystals with metallic lustre, produced by reducing with charcoal a mixture of metaphosphoric acid and tungsten pentoxide. Phosphides of uranium have not been prepared.

Although ruthenium, rhodium, and palladium combine with phosphorus, arsenic, and antimony, no compounds, except PdP_2, have been investigated. No simple nitrides of these metals are known. The same remark may be made of osmium. The phosphides, arsenides, and antimonides are much more easily fusible than the metals themselves.

* Wöhler, *Annalen*, **108**, 258.

Platinum nitride, Pt_3N_2, is produced by heating the oxide of platinosodiamine, $Pt^{II}(NH_2)_2.H_2O$ to 280°, thus: $3Pt(NH_2)_2.H_2O = 3H_2O + 4NH_3 + Pt_3N_2$. It is a greyish substance, decomposing suddenly at 290° into platinum and nitrogen.

The phosphide, Pt_3P_5, is a white substance, with metallic lustre, much more easily fusible than platinum, produced by direct union, and crystallising in cubes. When heated in a muffle, the residue Pt_2P is left. A corresponding iridium compound is similarly formed,* and is known as "cast iridium." The **arsenide, $PtAs_2$,** also formed by direct union, resembles the phosphide, and the **antimonide** is also white, brittle, and easily fused. A hydroxy-arsenide, **$Pt.AsOH$**, is formed by passing a current of arsine through a solution of platinic chloride; it forms black scales.

Cu_3N;† Cu_3P;‡ Cu_3P_2; CuP; Cu_9As; Cu_6As; Cu_3As; Cu_2As; Cu_3As_2.—Ag_3N (?); Ag_3P; Ag_3P_2; AgP_2; Ag_3As; Ag_3As_2; $AgAs$; Ag_3Sb; Ag_3Sb_2.—Au_3N; Au_2P_3; Au_4As_3.—Hg_3N_2; Hg_3P_2; Hg_3As_2; $HgAs$.

The **nitride, Cu_3N,** is produced by passing ammonia over cuprous oxide heated to 250°; it is a brown substance, decomposing about 360°. It has been suggested that fulminating silver (see p. 546) is in reality a nitride, but it appears more probable that it is silver amide, **$AgNH_2$**. Gold nitride has already been mentioned (p. 546). **Mercuric nitride, Hg_3N_2,** is a black substance, produced by the action of ammonia on mercuric oxide at 130°, which detonates when heated or struck.

Cuprous phosphide, Cu_3P, is a grey powder, produced by heating cuprous chloride, **Cu_2Cl_2**, in a current of phosphine. **Cupric phosphide, Cu_3P_2,** is similarly prepared from cupric chloride, **$CuCl_2$**, and forms a black powder. It is attacked by hydrogen chloride, yielding spontaneously inflammable phosphine. At a high temperature, in a current of hydrogen, it yields the phosphide, **CuP_2**, as a grey crystalline powder. **Phosphides of silver** are formed by direct union. The formulæ given above have been ascribed to them, but are not certain. A compound of the formula **$Ag_3P.3AgNO_3$**, and a similar compound, **$Ag_3As,3HNO_3$**, are produced in yellow crystals by saturating a strong solution of silver nitrate with phosphine or arsine at 0°. They are very unstable, almost at once depositing metallic silver. There appears also to be a similar compound of antimony.

Gold phosphide, AuP, is produced by direct union between

* *Chem. News*, **48**, 285.
† Schrötter, *Annalen*, **37**, 131.
‡ Rose, *Pogg. Ann.*, **4**, 110; **6**, 209; **14**, 188; **24**, 328.

spongy gold and phosphorus; it is a grey mass, more fusible than gold, with metallic lustre. The phosphide, $\mathbf{Au_3P_2}$, is produced by precipitation with phosphine, and is a black powder; it is mixed with metallic gold. **Mercury phosphide,** probably $\mathbf{Hg_3P_2}$, is a black compound, formed when mercuric oxide or chloride is heated with phosphorus; it is also formed in brown flakes when a mercuric salt is treated with phosphine, or as a yellow sublimate when mercuric chloride is heated in a current of phosphine. Along with this phosphide, a yellow powder of the formula $\mathbf{Hg_3P_2.3HgCl_2}$ is produced, which decomposes thus when boiled with water:— $\mathbf{Hg_3P_2.3HgCl_2} + 6H_2O + Aq = 6HCl.Aq + \mathbf{2H_3PO_3} + 6Hg$.

Mercuric phosphide also forms double compounds with basic mercuric nitrate and sulphate, $\mathbf{Hg_3P_2.3Hg_2O(NO_3)_2}$ and $\mathbf{Hg_3P_2.3Hg_3O(SO_4)_2.4H_2O}$, formed by he action of phosphine on the nitrate or sulphate.

The **arsenides,** $\mathbf{Cu_3As}$ and $\mathbf{Ag_3As}$, occur native as *arsenical* copper, or *domeykite*, and *arsenical silver*, or *huntilite*. The other arsenides are formed directly, as is also arsenide of gold. Mercury and arsenic do not easily combine directly, but when arsine is passed into a solution of mercuric chloride, the compound $\mathbf{Hg_3As_2.3HgCl_2}$ is precipitated. It is a yellowish-brown powder, and when in contact with water slowly decomposes into arsenious oxide, $\mathbf{As_4O_6}$, mercury, and hydrogen chloride.

Further investigation of all these compounds is much to be desired. Data concerning most of them are very meagre, and many have not been examined since the time of Berzelius.

Nitride of Carbon, or Cyanogen, C_2N_2, and its Compounds.

Cyanogen, $(CN)_2$,* is not formed by direct union. It is best prepared by heating cyanide of silver, gold, or mercury, preferably the last, thus:—$\mathbf{Hg(CN)_2} = Hg + (CN)_2$. It may be more conveniently prepared by heating a mixture of mercuric chloride with dry potassium ferrocyanide, or better with potassium cyanide (see below).

Cyanogen is a colourless gas with a sharp smell, resembling that of bitter almonds. It is exceedingly poisonous. It burns with a blue-purple flame, forming carbon dioxide and nitrogen. Water at the ordinary temperature dissolves about four and a half times, and alcohol twenty-three times, its volume of cyanogen. In

* Gay-Lussac, *Annales*, **77**, 128; **95**, 136.

the liquid state it is colourless, and boils at $-20°$, and at a lower temperature it freezes to a white solid melting at $-34·4°$.

Its formula is shown to be $(CN)_2$ by its vapour density, and it may be regarded as similar to molecular chlorine, hydrogen, or oxygen, Cl_2, H_2, or O_2, or to ethane (dimethyl), $(CH_3)_2$, (see p. 501); and it has been shown to contain its own volume of nitrogen by decomposing a known volume by means of an electric spark. Its heat of formation is: $2\mathbf{C} + 2N = C_2N_2 - 657K$.

Cyanides.—The starting point for preparing the cyanides is potassium cyanide, produced by heating the ferrocyanide, $\mathbf{K_4Fe(CN)_6}$ (see p. 562).

Hydrogen cyanide, hydrocyanic acid, or **prussic acid,** HCN.*—Hydrogen and cyanogen do not combine directly, but hydrogen cyanide is produced when the electric arc passes through moist air, by the union of carbon, hydrogen, and nitrogen. Anhydrous hydrogen cyanide may be prepared by heating mercuric cyanide, better mixed with ammonium chloride, with strong hydrochloric acid, passing the vapours over powdered marble to remove excess of hydrogen chloride, and through a tube filled with ignited calcium chloride, to dry the gas. Or by decomposing mercuric cyanide at 30° or 40° in a tube with hydrogen sulphide, and causing the resulting gases to pass through a layer of lead carbonate to remove excess of hydrogen sulphide. The pure compound should never be prepared without the utmost precautions being taken against its escape into the air of the laboratory, as it is an intense poison. The anhydrous compound may also be prepared by distilling its strong aqueous solution with fused calcium chloride dropped into the acid in small pieces at a time, to abstract water. It must be condensed in a receiver, best in a U-tube, cooled by a freezing mixture, and the exit from the receiver should lead away to a good draught.

An aqueous solution of the acid may be prepared by distilling potassium cyanide with dilute sulphuric acid:—$KCN.Aq + H_2SO_4.Aq = HCN.Aq + KHSO_4.Aq$. Or ferrocyanide of potassium may be employed (10 parts, water 30 parts, sulphuric acid 6 parts), thus:—

$$2K_4Fe(CN)_6.Aq + 3H_2SO_4.Aq = 3K_2SO_4.Aq + \mathbf{K_2Fe_2(CN)_6} + 6HCN.Aq.$$

Hydrogen cyanide is a colourless liquid, boiling at 27°; it freezes to a solid, which melts at $-15°$. It has a strong odour to

* Gay-Lussac, *Annales*, **77**, 128; **95**, 136.

those who can smell it, but about one person out of every five is incapable of perceiving it. It can always be detected by the choking sensation which it produces in the glands of the throat. It is miscible with water in all proportions. It burns, forming water, carbon dioxide, and nitrogen.

It is exceedingly poisonous; a few drops of the strong aqueous solution cause immediate death. It is employed medicinally in a 2 per cent. solution. It may be produced by distilling crushed peach-stones or laurel leaves with water, and it is known that such preparations were used in the middle ages by professional poisoners.

The heat of formation of hydrogen cyanide is:—H + **C** + N = HCN − 275K.

The analogy between chlorine and cyanogen, and between hydrogen chloride and cyanide, is a striking one. The cyanides in many respects resemble the chlorides, but while hydrogen chloride is not easily produced from its salts except by the action of acids like sulphuric, phosphoric, &c., even carbonic acid expels hydrogen cyanide from some cyanides. Hence, solid potassium cyanide always smells of hydrogen cyanide.

The cyanides and double cyanides are very numerous. It is only possible here to give a partial sketch of these compounds.

LiCN; NaCN; KCN; RbCN; CsCN; NH_4CN.

These salts are produced by the action of hydrocyanic acid on the hydroxides of the metals, or by direct combination of cyanogen with the metals; that of ammonium by direct combination of equal volumes of hydrogen cyanide and ammonia, or by distilling a mixture of potassium cyanide and ammonium chloride in requisite proportions. Cyanides are also produced by passing cyanogen into solutions of the hydroxides; a cyanate and cyanide are formed thus:—2KOH.Aq + $(CN)_2$ = KCN.Aq + KCNO.Aq. This reaction is exactly analogous to that which takes place between chlorine and caustic alkali (see p. 462). An interesting synthesis of potassium cyanide is carried out by passing nitrogen over a red-hot mixture of carbon and potassium carbonate, produced by igniting the tartrate, citrate, or some similar salt; it may be formulated:—$\mathbf{K_2CO_3}$ + **4C** + N_2 = **2KCN** + $3CO$. Sodium cyanide is also formed when any nitrogenous carbon compound is heated with sodium; this affords a means of testing for nitrogen in carbon compounds. It is also produced in a blast furnace, where iron ores are smelted with coal and lime;

the sodium is contained in the coal ash and the limestone; the carbon is derived from the coal, and the nitrogen from the air. It may be separated from the escaping gases by passing it through scrubbers filled with water, as in the extraction of ammonia from coal-gas.

Potassium cyanide is most conveniently prepared by heating the ferrocyanide, $\mathbf{K_4Fe(CN)_6}$, previously dried, in an iron crucible. It decomposes, giving an indefinite carbide of iron and the cyanide. Sometimes potassium carbonate is added to increase the yield. It may be purified by crystallisation from alcohol. If required perfectly free from cyanate, **KCNO**, it is best produced by passing the vapour of hydrogen cyanide into an alcoholic solution of potassium hydroxide, when it is precipitated.

These cyanides are all white deliquescent solids, crystallising in the regular system; they smell of prussic acid. They are very soluble in water, and somewhat soluble in alcohol. They are all poisonous.

Although cyanogen and hydrogen cyanide are produced with absorption of heat, potassium cyanide is formed with heat evolution:—$\mathbf{K} + \mathbf{C} + N = \mathbf{KCN} + 325\text{K}$.

$\mathbf{Ca(CN)_2}$; $\mathbf{Sr(CN)_2}$, and $\mathbf{Ba(CN)_2}$.

White deliquescent solids. Barium cyanide may be prepared by the action of the nitrogen of the air on a red hot mixture of barium carbonate and carbon; when heated to 300° in a current of water-vapour, it yields its nitrogen in the form of ammonia, thus:—$\mathbf{Ba(CN)_2} + 4H_2O = \mathbf{BaCO_3} + 2NH_3 + CO + H_2$. This process has been proposed as a method of producing ammonia from atmospheric nitrogen, but is not commercially successful.

$\mathbf{Mg(CN)_2}$; $\mathbf{Zn(CN)_2}$; $\mathbf{Cd(CN)_2}$.
Double compounds:—$\mathbf{2(Zn(CN)_2).NaCN.5H_2O}$; $\mathbf{Zn(CN)_2.KCN}$; $\mathbf{Zn(CN)_2.2NH_4CN}$; $\mathbf{Zn(CN)_2.Ba(CN)_2}$.

Magnesium cyanide is soluble; the cyanides of zinc and cadmium are white precipitates, thrown down from solutions of their soluble salts by addition of potassium cyanide. The double cyanides are soluble, and are obtained by mixture.

Yttrium cyanide is soluble. Aluminium cyanide appears to be incapable of existence. Gallium and indium cyanides have not been prepared. **Thallous cyanide, TlCN,** is thrown down from a solution of thallous hydroxide in hydrocyanic acid by addition of alcohol and ether as a white precipitate. It crystallises from a hot solution, and is readily soluble in water. It forms the double cyanides **$2TlCN.Zn(CN)_2$**; also **$Tl(CN)_3.TlCN$**, produced by the action of hydrocyanic acid on moist thallic oxide; the latter crystallises from strong hydrocyanic acid, but is decomposed by water.

$Cr(CN)_3$. Ferric cyanide is unknown in the solid state; nor are simple manganic or cobaltic cyanides known. Nickelicyanides are unknown even in combination.

Chromic cyanide, $Cr(CN)_3$, is said to be the formula of the bluish-grey precipitate, produced on adding a solution of chromium trichloride to a solution of potassium cyanide. Its formula is doubtful.

$Cr(CN)_2$; $Fe(CN)_2$; $Mn(CN)_2$ (?); $Co(CN)_2$, and $Ni(CN)_2$.

Prepared by addition of solution of potassium cyanide to solutions of chromous, ferrous, manganous, cobaltous, or nickelous salts. Chromous cyanide, prepared from chromous chloride, is white; ferrous cyanide, the formula of which is doubtful, is yellowish-red; that of manganese is reddish-white; of cobalt, flesh-coloured; and of nickel, apple-green.

Double cyanides.—The double cyanides of this group of elements are very numerous. They may be divided into three classes:—1, those containing the elements in dyad forms of combination; 2, those containing the elements as triads; and 3, those in which the iron, &c., exists in both dyad and triad states.

1. Chromocyanides are unknown; they are probably capable of existence, for chromium dicyanide dissolves in excess of solution of potassium cyanide. So also do cyanides of manganese and cobalt.

Ferrocyanides are compounds containing ferrous cyanide, **$Fe(CN)_2$**, in combination with four molecules of an alkaline cyanide, as in **$K_4Fe(CN)_6 = Fe(CN)_2.4KCN$**; or with two molecules of the cyanide of a dyad metal, as in

$$Ba_2Fe(CN)_6 = Fe(CN)_2 2Ba(CN)_2.$$

The starting point for the ferrocyanides is the **potassium salt, $K_4Fe(CN)_6$**. It is produced on the large scale, by heating together in a shallow iron pan nitrogenous animal matter, such as chips of horn, hair, fragments of skin, woollen rags, &c., with crude potassium carbonate and iron filings. Cyanide of potassium and ferrous sulphide are produced, the latter deriving its sulphur partly from the organic matter, partly from the sulphate present as impurity in the crude carbonate of potassium.* Only one-sixth to one-tenth of the nitrogen present in the animal matter is utilised. On treatment with water, the potassium cyanide and ferrous sulphide react, thus:—$6KCN.Aq + \mathbf{FeS} = K_2S + K_4Fe(CN)_6.Aq$. The liquors are then evaporated, and the impure crystals which separate out are recrystallised.

The formula of the crystals is $\mathbf{K_4Fe(CN)_6.3H_2O}$; they are truncated dimetric pyramids of a lemon-yellow colour, easily soluble in water, and not poisonous. When heated, iron carbide and potassium cyanide are produced (see p. 560). When distilled with strong sulphuric acid carbon monoxide is evolved, thus:—

$$\mathbf{K_4Fe(CN)_6} + 6H_2SO_4 + 6H_2O = \mathbf{2K_2SO_4} + \mathbf{FeSO_4} + 3\mathbf{(NH_4)_2SO_4} + 6CO.$$

It may be supposed that the hydrogen cyanide at first liberated combines with water, forming ammonium formate, $HCO.ONH_4$, which is decomposed by the sulphuric acid, liberating carbon monoxide. But such stages cannot be recognised in the decomposition.

On adding to a strong solution of potassium ferrocyanide, previously boiled to expel air, strong hydrochloric acid, also boiled and cooled, and a little ether, thin white scales of **hydrogen ferrocyanide, $H_4Fe(CN)_6$,** separate out. They may be collected on a filter, washed with a mixture of alcohol and ether, and dried over sulphuric acid in a vacuum. Hydroferrocyanic acid is easily soluble in water and alcohol, but insoluble in ether.

Barium ferrocyanide, $Ba_2Fe(CN)_6$, may be produced by action on barium cyanide (obtained from the carbonate, carbon, and nitrogen) of ferrous sulphate, thus:—$3Ba(CN)_2.Aq + FeSO_4.Aq = \mathbf{BaSO_4} + Ba_2Fe(CN)_6.Aq$; also by precipitating a boiling solution of potassium ferrocyanide with great excess of barium chloride, and boiling the resulting precipitate with solution of barium chloride. It crystallises in flattened yellow monoclinic prisms with six molecules of water.

The other ferrocyanides are prepared either by treating the hydroxide or carbonate of the metal with a solution of hydrogen ferrocyanide; by mixing a solution the sulphate of the metal with solution of barium ferrocyanide; or by precipitation, many ferrocyanides being insoluble.

The following is a list of the more important ferrocyanides:—

* Liebig, *Annalen*, **38**, 20.

$Li_4Fe(CN)_6$; $Na_4Fe(CN)_6.12H_2O$; $K_4Fe(CN)_6.3H_2O$; $(NH_4)_4Fe(CN_6).3H_2O$; also **$Li_2K_2Fe(CN)_6.6H_2O$; $NaK_3Fe(CN)_6$; $K_2(NH_4)_2Fe(CN)_6$; $K_3(NH_4)Fe(CN)_6$;** and the double salts **$(NH_4)_4Fe(CN)_6.2NH_4Cl.3H_2O$,** and **$(NH_4)_4Fe(CN)_6.2NH_4Br.3H_2O$.**
$Ca_2Fe(CN)_6.12H_2O$; $Ba_2Fe(CN)_6.6H_2O$; also **$K_2CaFe(CN)_6.3H_2O$;** and **$K_2BaFe(CN)_6.3H_2O$.**

(The last two double salts are produced by precipitation.)

$Mg_2Fe(CN)_6.12H_2O$; $Zn_2Fe(CN)_6.3H_2O$; $K_2MgFe(CN)_6$.
$Al_4\{Fe(CN)_6\}_3$; $Fe_4'''\{Fe''(CN)_6\}_3.18H_2O$ (Prussian blue); also **$KFe'''Fe(CN)_6$.**

The aluminium compound and the ferric compound are produced by precipitation. The latter is prepared industrially as a blue pigment, by precipitation, thus :—

$$4FeCl_3.Aq + 3K_4Fe(CN)_6.Aq = \mathbf{Fe_4'''\{Fe(CN)_6\}_3} + 12KCl.Aq;$$

or by the oxidation by air, or other oxidising agents of potassium ferrous ferrocyanide, probably thus :—$6K_2Fe\{Fe(CN)_6\}.Aq + 3O = Fe_2O_3n.H_2O + 3K_4Fe(CN)_6.Aq + \mathbf{Fe_4\{Fe(CN)_6\}_3}$. It is by this last method that it is usually prepared commercially. At the same time, potassium ferric ferrocyanide, $\mathbf{KFe'''Fe''(CN)_6}$, is produced, which is soluble in water. It appears to be formed, if the ferrocyanide of potassium is present in insufficient quantity, thus :—

$$K_4Fe(CN)_6.Aq + FeCl_3.Aq = KFe'''Fe''(CN)_6.Aq + 3KCl.Aq.$$

When digested with more ferrocyanide, it is converted into Prussian blue. This compound may also be regarded as potassium ferrous ferricyanide, $KFe''Fe'''(CN)_6$ (which see).

Ferrous ferrocyanide, $Fe_2''Fe''(CN)_6$ (white), has the same percentage composition as ferrous cyanide, $Fe(CN)_2$. But as ferrocyanides of manganese (white), cobalt (pale blue), and nickel (light green), with corresponding formulæ, are known, it is probable that the formula is the more complex one. By addition of solution of potassium ferrocyanide to a solution of iron wire in aqueous sulphurous acid, the potassium ferrous salt, $\mathbf{K_2Fe''Fe''(CN)_6}$, is thrown down as a white precipitate. This compound is also produced when potassium ferrocyanide is distilled with dilute sulphuric acid, as in the preparation of hydrocyanic acid, thus :—$2K_4Fe''(CN)_6.Aq + 3H_2SO_4.Aq = 3K_2SO_4.Aq + 6HCN + \mathbf{K_2Fe''Fe''(CN)_6}$. With a ferrous salt containing, as it usually does, a little ferric salt, this precipitate is light blue, and serves for the detection of ferrous iron. It also rapidly turns blue on exposure to air, owing to oxidation.

Lead ferrocyanide is white; that of bismuth also white; of monad copper, white, $\mathbf{Cu_4Fe(CN)_6}$; potassium cuprous ferrocyanide, $\mathbf{K_2Cu_2Fe(CN)_6}$, forms deep brown crystals; potassium cupric, $\mathbf{K_2CuFe(CN)_6}$, is the brown-red precipitate produced by a solution of potassium ferrocyanide in solutions of copper salts; with great excess of ferrocyanide of potassium a reddish-purple precipitate of $\mathbf{Cu_2Fe(CN)_6}$ is produced. The silver salt is white, and is not acted on by hydrochloric acid; the mercuric salt is also white. Ferrocyanides of cupramine and of mercuramine are also known.

The **manganocyanides** are analogous to the ferrocyanides, and are also produced by dissolving manganous cyanide in excess of an alkaline cyanide. The potassium salt forms deep violet tabular crystals of the formula $\mathbf{K_4Mn(CN)_6}$. It would be interesting to compare the salts $\mathbf{K_2MnFe(CN)_6}$ and $\mathbf{K_2FeMn(CN)_6}$ with a view of seeing whether or not they are identical.

The **double cyanides of nickel** have formulæ differing from the ferro- and manganocyanides. The potassium salt, $\mathbf{K_2Ni(CN)_4}$, produced by mixture, forms yellow oblique rhomboidal prisms. Ammonium, calcium, and barium compounds have also been prepared.

Chromicyanides, ferricyanides, manganicyanides, and **cobalticyanides.—Chromicyanide of potassium,** $\mathbf{K_3Cr'''(CN)_6}$, produced by dissolving chromium hydrate in solution of potassium cyanide in presence of potassium hydroxide, forms brown crystals, from which the red silver salt may be produced by precipitation. The silver salt with hydrogen sulphide gives the hydrogen salt $\mathbf{H_3Cr'''(CN)_6}$, which is a crystalline body. Ferrous chromicyanide is a brick-red powder.

The starting point for **ferricyanides** is potassium ferrocyanide. When a current of chlorine is passed through its solution, the following reaction takes place:—$2K_4Fe''(CN)_6.Aq + Cl_2 = 2KCl.Aq + 2K_3Fe'''(CN)_6.Aq$.* A still better method† is to digest potassium ferric ferrocyanide with a solution of potassium ferrocyanide, thus:—$KFe'''Fe(CN)_6.Aq + K_4Fe(CN)_6.Aq = K_3Fe'''(CN)_6.Aq + \mathbf{K_2Fe''\{Fe''(CN)_6\}}$. The insoluble potassium ferrous ferrocyanide is removed by filtration, and may be reconverted into potassium ferric ferrocyanide by digestion with nitric acid, and thus rendered available for a second operation. The filtrate on evaporation yields dark orange-red crystals of ferricyanide. The sodium salt is similarly prepared. The double salt of sodium and

* Gmelin, *Handbook*, **7**, 468.

† Williamson, *Annalen*, **57**, 237.

potassium has the formula $\mathbf{Na_3K_3\{Fe(CN)_6\}_2}$; hence the formula of the simple salts is often written $\mathbf{Na_6Fe_2'''(CN)_{12}}$. The lead salt is sparingly soluble in water, and crystallises in brown-red plates. From it the hydrogen salt is produced by the action of the requisite amount of dilute sulphuric acid; the filtrate from the lead sulphate is evaporated, and deposits brownish needles of $\mathbf{H_3Fe(CN)_6}$.

The iron salts are specially interesting. **Ferrous ferricyanide,** $\mathbf{Fe''_3\{Fe'''(CN)_6\}_2}$, is a deep-blue precipitate, known as *Turnbull's blue*, which shows on its fractured surfaces a copper-red lustre. It is extensively used in calico-printing, and is produced by addition of solution of ferrous sulphate to potassium ferricyanide, thus:—

$$3Fe''SO_4.Aq + 2K_3Fe'''(CN)_6.Aq = \mathbf{Fe_3''\{Fe'''(CN)_6\}_2} + 6KCl.Aq.$$

Potassioferrous ferricyanide, $\mathbf{KFe''Fe'''(CN)_6}$, is a blue-violet compound, produced by boiling white potassium ferrous ferrocyanide, $\mathbf{K_2Fe''Fe''(CN)_6}$, with dilute nitric acid. When digested with a ferrous salt, it yields Turnbull's blue, thus:—

$$\mathbf{2KFe''Fe'''(CN)_6} + FeSO_4.Aq = \mathbf{Fe_3''\{Fe'''(CN)_6\}_2} + K_2SO_4.Aq;$$

with a ferric salt Prussian blue is formed:—

$$\mathbf{3KFe''Fe'''(CN)_6} + FeCl_3.Aq = 3KCl.Aq + \mathbf{Fe'''Fe_3''\{Fe'''(CN)_6\}_3} = \mathbf{Fe_4'''\{Fe''(CN)_6\}_3}.$$

By the action of excess of chlorine on a solution of potassium ferro- or ferricyanide, **Prussian green** is formed. It is a green ferricyanide of the formula $\mathbf{Fe_3''Fe_4'''\}Fe(CN)_6\}_6}$.

With ferric chloride, potassium ferricyanide gives a brown solution, which may contain ferric ferricyanide, $Fe'''Fe(CN)_6$, or perhaps ferric cyanide, $Fe(CN)_3$.

Nitro-prussides.—A class of compounds containing nitric oxide is produced by the action of nitric acid mixed with its own volume of water on ferro- or ferricyanides;* the mixture after standing is heated in a water-bath, when gases are evolved. When it no longer gives a blue precipitate with ferrous sulphate it is cooled, when nitre and oxamide crystallise out. The mother liquor is neutralised with sodium carbonate and again filtered. The filtrate on evaporation deposits first crystals of nitre, and afterwards deep-red crystals of **sodium nitroferricyanide,** $\mathbf{Na_2Fe'''(CN)_5.NO}$. Many salts have been prepared. The most striking reaction of the nitroprussides is that with a soluble

* Playfair, *Phil. Mag.* (3), **36**, 197, 271, 348. Roussin, *Annales* (3), **52**, 285. Pavel, *Berichte*, **15**, 2600.

sulphide; a splendid purple colour is produced, which, however, is transient.

It is suggested that these compounds are closely allied to the nitrosulphides of iron (see p. 343); for on adding mercuric cyanide to sodium ferrinitrosulphide, hydrogen sodium nitroferricyanide is formed, thus :—

$$2NaFeS_2.NO.Aq + H_2O + 5Hg(CN)_2.Aq = 2HNaFe(CN)_5.NO.Aq + 4\mathbf{HgS} + \mathbf{HgO};$$

and, conversely,

$$2Na_2Fe(CN)_5.NO.Aq + Na_2S.Aq = 2NaFeS_2.NO.Aq + 10NaCN.Aq.$$

At present, however, there are not data sufficient to make it possible to suggest constitutional formulæ for these compounds.

Manganicyanide of potassium,* $\mathbf{K_3Mn'''(CN)_6}$, is formed by exposing the manganocyanide to air. It is amorphous with the ferricyanide, and forms reddish-brown crystals. The ferrous salt is cobalt-blue, but is unstable.

Cobalticyanide of potassium† is similarly prepared. The hydrogen salt, produced from the lead salt with sulphuretted hydrogen, $\mathbf{H_3Co(CN)_6}$, forms colourless needles. Ferrous cobalticyanide, $\mathbf{Fe_3''\{Co(CN)_6\}_2}$, is a white precipitate, analogous in formula to Turnbull's blue. The corresponding cobaltous salt is a light-red precipitate.

Nickelicyanides are unknown.

Cyanide of titanium has not been investigated. But a **double nitride and cyanide,** $\mathbf{Ti(CN)_2.3Ti_3N_2}$,‡ occurs in copper-coloured crystals in the beds of blast-furnaces, and was formerly believed to be the element titanium. It may be produced by heating to a high temperature a mixture of titanium dioxide and potassium ferrocyanide. When heated in steam these crystals yield ammonia, hydrogen, and hydrocyanic acid, leaving a residue of titanium dioxide; and in chlorine, titanium chloride and crystals of a double compound of the chlorides of titanium and cyanogen, $TiCl_4.CNCl$.—Cerium cyanide is said to be a white precipitate; cyanides of zirconium and of thorium have not be prepared.

Cyanides of silicon and of germanium are also unknown; tin appears not to form a cyanide; and lead yields only a white precipitate of a hydroxycyanide, **HO—Pb—CN**, in presence of ammonia.

* Eaton and Fittig, *Annalen*, **145**, 157; Descamps, *Bull. Soc. Chim.*, **9**, 443.
† Zwenger, *Annalen*, **62**, 137.
‡ Wöhler, *Chem. Soc.*, **2**, 352.

Cyanides of elements of the nitrogen-group have not been prepared.

Phosphorous cyanide, $P(CN)_3$,* forms long, white needles, which catch fire when touched with a warm glass rod. It is produced by heating to 130° in a sealed tube a mixture of silver cyanide and phosphorus trichloride, and subsequent sublimation in a current of carbon dioxide. It melts at 200°—203°. **Cyanide of arsenic** may be similarly prepared. Cyanides of antimony and bismuth are unknown.

Cyanides of molybdenum, tungsten, and uranium have not been examined.

$(CN)_2O$ is unknown. The corresponding **sulphide, $(CN)_2S$,** is produced by the action of a solution of cyanogen iodide in ether on silver sulphocyanide, thus:—**AgSCN** + ICN. eth. = **AgI** + $(CN)_2S$. eth. It forms volatile, colourless, rhombic tables.

Cyanogen hydroxide, or **cyanic acid, (CN)OH,** and the corresponding **(CN)SH, sulphocyanic acid** form numerous compounds in which the hydrogen is replaced by metals. The potassium salts are produced by oxidation of potassium cyanide, with atmospheric oxygen, or better, by lead oxide or manganese dioxide; and by direct combination of potassium cyanide with sulphur. For an account of their salts, a text-book on the carbon compounds must be consulted. **Ferric sulphocyanide, $Fe(CNS)_3$,** a blood-red, soluble salt, is produced by the action of an aqueous solution of potassium or ammonium sulphocyanide on ferric salts; it is noticeable as a test for iron in the ferric condition; ferrous sulphocyanide being colourless. **Seleniocyanic anhydride†** or **selenium cyanide,** and **seleniocyanides‡** are similarly prepared to the sulphur compounds. They are unstable, decomposing easily with separation of selenium. Tellurium compounds have not been prepared.

Fluoride of cyanogen will no doubt soon be prepared by Moissan. **Cyanogen chloride, CNCl, bromide, CNBr,** and **iodide, CNI,** are produced by the action of the halogen on mercuric cyanide. Chlorine in the dark; bromine on the cyanide, cooled by ice; and iodine at a gentle heat, yield the respective halides. The chloride is a colourless gas, liquefying at −12° to −16°, and solidifying at −18° to long, transparent prisms. It forms a hydrate with water. The bromide forms long, colourless needles which soon change to minute cubes. It melts above 40°, but

* Hübner and Wehrhahne, *Annalen*, **128**, 254; 132, 277.

† Linnemann, *Annalen*, **120**, 36.

‡ Crookes, *Chem. Soc. J.*, **4**, 12.

volatilises rapidly at 15°. The sublimed iodide also forms long, white needles, but crystallises from alcohol or ether in four-sided tables. It boils above 100°, but volatilises at the ordinary temperature. All these bodies are very poisonous.

Some double compounds of the chloride are known, produced by direct union, *e.g.*, **$BCl_3.CNCl$**; **$FeCl_3.CNCl$**; **$TiCl_4.CNCl$**; and **$SbCl_5.CNCl$**. The boron and antimony compounds form white crystals; the titanium compound is a yellow, crystalline mass; and the iron compound is black, and apparently amorphous.

Ruthenium forms **ruthenocyanides**, similar in formula to the ferrocyanides, and isomorphous therewith.* The potassium salt is formed by fusing chloride of ruthenium and ammonium with potassium cyanide; the filtrate deposits it in small, colourless tables of the formula **$K_4Ru(CN)_6$**. On warming this compound with hydrochloric acid, a violet precipitate of ruthenous cyanide, **$Ru(CN)_2$**, is produced. The hydrogen salt is liberated from the potassium salt in presence of ether, like hydrogen ferrocyanide; it forms white laminæ. Ruthenic cyanides have not been isolated.

Rhodocyanides, on the other hand, are unknown. **Rhodium tricyanide, $Rh(CN)_3$**, formed by addition of acetic acid to a solution of rhodicyanide of potassium, is a red powder, soluble in potassium cyanide solution, forming rhodicyanides. The rhodichloride of potassium, **K_3RhCl_6**, fused with potassium cyanide, yields potassium **rhodicyanide, $K_3Rh(CN)_6$**, analogous to ferricyanide. It forms large, anhydrous, easily-soluble crystals.

The cyanides of palladium require investigation. **Palladium dicyanide** is said to be a white precipitate, produced on adding palladium dichloride to mercuric cyanide. It dissolves in a solution of potassium cyanide, giving crystals of **$K_2Pd(CN)_4$**, analogous to the double cyanides of nickel. The tetracyanide, **$Pd(CN)_4$**, is said to be a rose-coloured precipitate produced by mercuric cyanide in a solution of potassium palladichloride, $K_2PdCl_6.Aq$.

Potassium **osmocyanide,† $K_4Os(CN)_6$**, analogous to the ferrocyanide, is produced by adding solution of potassium cyanide to a solution of osmium tetroxide, **OsO_4**, in aqueous caustic potash. The solution is evaporated to dryness, and heated to dull redness. On treatment with water osmocyanide of potassium dissolves, and may be purified by crystallisation. It forms yellow, quadratic

* Claus, *Jahresb.*, 1855, 446.

† Claus, *Beiträge zur Chemie der Platinmetalle*, Dorpat, 1854; Martius, *Jahresb.*, 1860, 233.

crystals isomorphous with the ferrocyanides. Its solution gives a light-blue precipitate with ferrous salts, which is oxidised by nitric acid into a violet compound analogous to Turnbull's blue, probably $\mathbf{Fe_3''\{Os(CN)_6\}_2}$. A violet compound, said to have the same formula, is produced by addition of a ferric salt to potassium osmocyanide. Barium osmocyanide, $\mathbf{Ba_2Os(CN)_6}$, crystallising in reddish-yellow prisms, is produced by treating this compound with baryta-water, which separates ferric hydrate. The acid is also known, and is prepared in the same manner as hydroferrocyanic acid. When boiled, the solution of the acid gives a violet precipitate of osmium dicyanide, $\mathbf{Os(CN)_2}$.

The **iridicyanides,*** of which the potassium salt is produced by fusing ammonium iridichloride, $\mathbf{IrCl_3.3NH_4Cl}$, with potassium cyanide, or metallic iridium with potassium ferrocyanide, are analogous to the ferricyanides and also resemble the rhodicyanides. The hydrogen, potassium, and barium salts are white and crystalline; the zinc, ferrous, lead, and mercurous salts are white and insoluble; the ferric salt yellow; and the cupric salt blue.

Platinum forms two series of cyanides; 1, those analogous to the double cyanides of nickel, for example, $\mathbf{K_2Pt(CN)_4}$; and 2, dihalo-platinocyanides, such as $\mathbf{I_2Pt(CN)_4.K_2}$. The potassium salt of the first series is produced by heating platinum with cyanide or ferrocyanide of potassium, or, better, by dissolving ammonium platinichloride, mixed with caustic potash, in a strong solution of potassium cyanide, boiling until ammonia is expelled, and crystallising. It forms rhombic prisms, yellow by transmitted, and blue by reflected, light. The copper salt is a green precipitate, produced by adding solution of copper sulphate to a solution of potassium platinocyanide; and from it the hydrogen salt may be prepared by the action of hydrogen sulphide; the barium salt, by the action of barium hydroxide; and from the barium salt the platinocyanides of other metals may be produced by adding the requisite amounts of sulphates of other metals. The platinocyanides all exhibit remarkable dichroism; the magnesium salt is one of the most beautiful; it forms square-based prisms, deep red by transmitted light; the sides of the prisms reflect brilliant metallic green, and the extremities are purple-blue.

As regards the products of their oxidation by nitric acid, chlorine and bromine in presence of water, lead dioxide, &c., considerable doubt still exists. On the one hand, they are stated to have formulæ such as $K_2Pt(CN)_5$; and analyses of many such

* Martius, *Annalen*, **117**, 357.

compounds are given by several well-known chemists.* On the other hand, the compounds produced by the action of chlorine are said to have formulæ such as $6(K_2Pt(CN)_4).Cl_2.(H_2O)$.† And recently, the formula $[K_2Pt(CN)_4.3H_2O]_3Cl$ has been ascribed to the potassium compound, and that of $[K_2Pt(CN)_4.3H_2O]_3HCl$‡ to a similar compound produced by the action of hydrogen chloride. Regarding the action of excess of halogen, only one view exists; such compounds have the formulæ like the one already given, viz., $Cl_2.Pt(CN)_4.K_2$. Compounds containing chlorine, bromine, the nitro-group, NO_2, the group SO_4, &c., have been prepared. They all display remarkable dichroism.

CuCN; AgCN; AuCN.—Double cyanides.—**Cuprosocyanides**, such as $KCu(CN)_2$; $K_2Cu_3(CN)_5$; and $K_3Cu(CN)_4$.—**Argentocyanides**, such as $KAg(CN)_2$; $K_2NaAg_3(CN)_6$.—**Aurocyanides**, such as $KAu(CN)_2$.

Cuprous and **argentous cyanides** are white powders. The first is obtained by adding a solution of potassium cyanide to a solution of cuprous chloride, Cu_2Cl_2, in hydrochloric acid. It may be obtained in crystals by treating with hydrogen sulphide lead cuproso-cyanide, $PbCu(CN_3)_3$, suspended in water; the compound $HCu(CN)_2$ appears to be formed, which, when filtered from the lead sulphide and evaporated, decomposes, depositing crystals of cuprous cyanide.

Silver cyanide is easily produced by adding to a solution of silver nitrate a solution of the requisite amount of potassium cyanide; excess of cyanide redissolves the precipitate, producing the double cyanide $KAg(CN)_2$, which separates in crystals on evaporation. **Aurous cyanide** is produced by decomposing potassium aurocyanide, $KAu(CN)_2$, with nitric acid. It is a yellow crystalline powder.

Mercurous cyanide does not exist. On addition of a soluble cyanide to a mercurous salt, metallic mercury is precipitated, and mercuric cyanide goes into solution.

The **double cyanides** are produced by the action of potassium cyanide in excess on the cyanides, or on the chlorides, oxides, &c. From the potassium salts other derivatives may be prepared. Those of silver and of gold are largely used in electro-plating. A double nitrate and cyanide of silver is known, $AgCN.AgNO_3$, crystallising from a solution of silver cyanide in a solution of silver nitrate.

* Knop, *J. prakt. Chem.*, **37**, 461; Wiselsky, *ibid.*, **69**, 276; Martius, *loc. cit.*

† Hadow, *Chem. Soc. J.*, **14**, 106.

‡ Wilm, *Berichte*, **19**, 959.

$Cu(CN)_2$; $Hg(CN)_2$.—Double salts, such as **$K_2Cu(CN)_4$**, and **$K_2Hg(CN)_4$**.—**$Cu(CN)_2.2CuCN$**; **$Cu(CN)_2.4CuCN$**. Mercuric cyanide also forms numerous compounds with other salts, such as **$Hg(CN)_2.KCl$**; **$2Hg(CN)_2.CaCl_2.6H_2O$**; **$Hg(CN)_2.2CoCl_2 4H_2O$**; and similarly with bromides and iodides; also **$Hg(CN)_2.K_2CrO_4$**; **$Hg(CN)_2.Ag_2Cr_2O_7$**; **$2Hg(CN)_2.Ag_2SO_4$**; **$2Hg(CN)_2.K_2S_2O_3$**; **$Hg(CN)_2.AgNO_3$**, and many others.

Cupric cyanide is very unstable, giving off cyanogen. It is a yellow-green precipitate, produced by adding copper sulphate to excess of potassium cyanide. It dissolves rapidly, giving a double salt. When allowed to stand it changes into the double compound, **$Cu(CN)_2.2CuCN$**, which forms green granular crystals, or the other cuprous-cupric cyanide, **$Cu(CN)_2.4CuCN$**. These bodies rapidly decompose, forming cuprous cyanide.

Mercuric cyanide is produced by boiling mercuric sulphate with a solution of potassium ferrocyanide, or by digesting mercuric oxide with hydrocyanic acid. It forms colourless dimetric crystals. The double compounds crystallise well, and are produced by mixture.

$Au(CN)_3$ is unknown in the free state. **Potassium auricyanide, $KAu(CN)_4$**, is produced by crystallising a mixture of auric chloride, $AuCl_3$, with potassium cyanide. It forms large colourless tables. The silver salt, produced by precipitation with silver nitrate, yields, on treatment with hydrochloric acid, the hydrogen salt, which, after evaporation over sulphuric acid, separates from its solution in large colourless tables of the formula

$$2HAu(CN)_4.3H_2O.$$

Constitution of the cyanides.—Two formulæ are possible for hydrogen cyanide; either $H—C\equiv N$, representing it as methane with one atom of nitrogen replacing three atoms of hydrogen; or H—NC, representing it as ammonia, in which carbon replaces two atoms of hydrogen. These are suggested by the following considerations among others:—Compounds are known in which the hydrogen of hydrogen cyanide is replaced by methyl, CH_3. One of those, when treated with nascent hydrogen, yields an ammonia, ethylamine, $C_2H_5.NH_2$, the constitution of which has been proved to be $CH_3.CH_2.NH_2$; this cyanide is not easily attacked by hydrochloric acid, but when boiled with caustic potash it assimilates the elements of water, yielding ammonia and potassium acetate, thus:—$CH_3CN + H_2O + H.OH = CH_3.COOH + NH_3$, the nitrogen being replaced by oxygen plus hydroxyl. The other compound, $CH_3.NC$, is not easily reduced, but at once decomposes on treatment with hydrochloric acid, giving methylamine and formic acid, thus :—$CH_3NC + H_2O + H.OH = CH_3.NH_2 + HCO.OH$. Compounds of the first class are not specially poisonous, and have a not unpleasant smell; compounds of the second class are exceedingly poisonous, and have an unbearable odour. Both are produced together by the reactions:— $CH_3I + Ag(CN) = CH_3(CN) + AgI$; and $CH_3KSO_4 + KCN = K_2SO_4 + CH_3(CN)$. Thus it would appear that either potassium and silver cyanides are mixtures of two such salts as $AgCN$ and $AgNC$; or that the compound CH_3CN, which is a more stable body than CH_3NC, is produced when heated, owing to molecular change during the reaction. The latter view appears, on the whole, the most tenable; and it would therefore follow that the cyanides belong to the class represented by HNC. This may also be inferred from their very poisonous nature.

It is also noticeable that the cyanides show great tendency towards the formation of complex compounds. Cyanogen itself polymerises; the chloride $C_3N_3Cl_3$, is known as cyanuric chloride; and it has been suggested that the grouping of the cyanogen is in

$$\begin{array}{c} Cl-C-N=C-Cl \\ \| \quad\quad | \\ N-C=N \\ | \\ Cl \end{array}$$

such cases . The group C_3N_3, therefore, may be regarded as a triad-group; and potassium ferrocyanide may be thus represented as $Fe''(C_3N_3K_2)_2$. Similarly, dicyanogen compounds are known, in which the grouping has been supposed to be

$$\begin{array}{c} -C=N \\ | \quad | \\ N=C- \end{array}$$

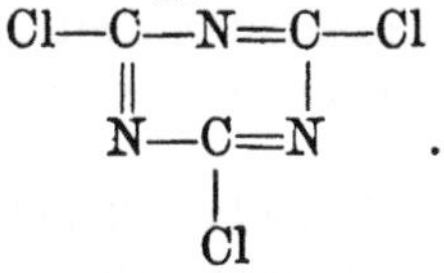

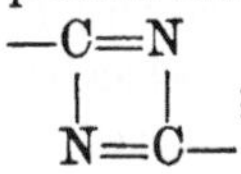

; this is a dyad-group; and ferricyanide of potassium on this theory would be $Fe'''(C_2N_2K)_3$, and potassium nickelocyanide and similar salts $Ni''(C_2N_2K)_2$. But it must be remembered that such methods of representation, however suggestive, have little to recommend them except inasmuch as they may prove to be working theories.

PART VII.—ALLOYS.

CHAPTER XXXIII.

Alloys appear to form three distinct classes: (1.) **Mixtures in which no chemical combination has occurred, and which may be regarded as solidified solutions**; (2.) **Chemical compounds with definite formulæ**; and (3) **substances intermediate between these two classes, which contain the elements partly in a state of mixture, partly as compounds.*** But in the present state of our knowledge, we can seldom discriminate between these three classes, and hence, in the following chapter, attention will be drawn to cases of definite combination, wherever suspected, while the elements will, as usual, be taken mainly in the periodic order.

It frequently happens that two elements will not mix, or that one will mix with a minimal quantity of another. Thus, while copper and silver, or lead and tin, mix in all proportions, iron and silver do not; a little silver, it is true, enters the iron, considerably modifying its properties; and a little iron enters the silver; but mixtures in any desired proportion cannot be prepared. It has been suggested that alloys often contain one or more of their constituent metals in an allotropic condition; the alloy of zinc and rhodium, when freed from zinc by treatment with acid, leaves the rhodium in an allotropic state; and this modification is converted into ordinary rhodium with slight explosion at 300° (see p. 78).

A new method of investigating alloys has recently been devised,† and it has been applied to determining the constitution of alloys of copper and zinc, and of copper and tin. The principle of the method is as follows:—If a battery be constructed containing, instead of a plate of any single metal, a compound plate made of two metals, the electromotive force of the circuit is that of the more positive metal. If an *alloy* of two metals is used as material for a plate, the electromotive force is still that of the more electro-

* Matthiessen, *Brit. Assn. Reports*, 1863, 37; *Chem. Soc. J.*, 1867, 201.

† Laurie, *Chem. Soc. J.*, 1888, 104.

positive metal, if the alloy is a mere mixture; but if a compound, it has its own electromotive force, differing from that of the more electropositive element of the compound. If the alloy consists of both a mixture and a compound, then, as the plate dissolves away, the more electropositive element will disappear first, and the electromotive force will fall suddenly to that due to the compound. In this way the existence of definite compounds of the formulæ $\mathbf{CuZn_2}$ and $\mathbf{Cu_3Sn}$, have been rendered probable.

By extension and development of such experiments as these, we may hope soon to enlarge our knowledge of alloys.

The alloys will now be described in their order.

Hydrides.—Sodium and **potassium** heated in a current of hydrogen yield hydrides, of which the sodium compound has the formula $\mathbf{Na_2H}$. Sodium hydride* is a soft substance like wax, which becomes brittle at a temperature somewhat below its melting point. It is formed at 300°, and dissociates at 421°. Its melting point lies below that of sodium. When treated with mercury, the sodium and mercury unite, and hydrogen is expelled. The potassium compound is white, brittle, and shows crystalline structure. Lithium absorbs only seventeen times its volume of hydrogen.

No other hydrides are met with till the metal iron is reached. The action of zinc ethyl, $Zn(C_2H_5)_2$, on ferrous iodide in presence of ether, is represented by the equation:—$Zn(C_2H_5)_2 + \mathbf{FeI_2} = \mathbf{ZnI_2} + 2C_2H_4 + FeH_2$. The product is a black powder, resembling finely-divided iron, which evolves hydrogen when gently heated, or on treatment with water. Its composition is conjectural, for it has not been analysed, and apparently some hydrogen is evolved along with the ethylene.

Metallic iron absorbs a small quantity of hydrogen gas if exposed to it in a finely-divided condition. Meteoric iron has been found to contain 2·85 times its volume of gas, containing 86 per cent. of hydrogen. Similar results have not been obtained with chromium, manganese, or cobalt; but nickel, in the porous state, if made the negative electrode of a battery, absorbs about 165 times its volume of hydrogen, losing it gradually in the course of a few days. If it be recharged several times, it falls to powder.†

* Troost and Hautefeuille, *Comptes rendus*, **78**, 807.

† Raoult, *Comptes rendus*, **69**, 826.

Vanadium absorbs about 1·3 per cent. of its weight of hydrogen, forming a compound which oxidises easily in air.

A hydride of **niobium**, of the formula **NbH**, is formed by the action of sodium on niobifluoride of potassium. It is a black powder, soluble only in hydrofluoric acid and in fused hot potassium hydroxide.*

Rhodium and ruthenium do not appear to combine with hydrogen; but **palladium** may be made to absorb as much as 936 times its volume, or 4·68 per cent. of hydrogen, when made the negative pole of a battery by which dilute sulphuric acid is being electrolysed. The specific gravity of the palladium is thereby reduced from 12·38 to 11·79; and it becomes magnetic, implying that solid hydrogen is magnetic. This affords a means of determining the specific gravity of solid hydrogen, on the assumption, which is nearly true in most instances, that the specific gravity of an alloy is the mean of that of its constituents. It is 0·62 at 15°. From the sodium alloy, $\mathbf{Na_2H}$, it is 0·63.† The specific heat of solid hydrogen can also be calculated on a similar assumption. The following results were obtained:—

Sp. ht. 4·49; 8·87; 4·99; 8·31; 4·08; 4·96; 5·76; 4·06; 4·46; 9·10; 4·58; 6·02; 4·82; 6·55; 4·34.

The mean result is 5·7.‡ It will be seen that the atomic heat of hydrogen may lie near that of other elements—about 6·2.

This alloy,§ which has been supposed to contain the metal "**hydrogenium,**" as it was termed by Graham, resembles palladium in colour. It loses its hydrogen when heated, best in a vacuum. The hydrogen possesses very active properties, resembling those which it possesses in the nascent state; thus it combines directly in the dark with bromine and with iodine; reduces mercuric chloride to mercurous chloride, ferric to ferrous salts, &c. The alloy approximates in composition to that expressed by the formula $\mathbf{Pd_2H}$, but it appears to absorb hydrogen in excess, which is not chemically combined, but in a state of mixture. To this phenomenon Graham gave the name "occlusion." The metal platinum has also the power of occluding gases. "Platinum black," produced by dissolving platinous chloride, $\mathbf{PtCl_2}$, in caustic potash, adding alcohol gradually to the hot liquid, and washing the precipitated metallic powder successively with alcohol, hydro-

* Krüss, *Berichte*, **20**, 169.

† Troost and Hautefeuille, *ibid.*, **78**, 968; also Dewar, *Phil. Mag.* (4), **47**, 324.

‡ Roberts and Wright, *Chem Soc.*, **26**, 112.

§ Graham, *Roy. Soc. Proc.*, **16**, 422; **17**, 212, 500.

chloric acid, caustic potash, and distilled water, so as to remove all foreign substances, has the power of absorbing hydrogen in greater quantity than compact platinum. It also absorbs about 250 times its volume of oxygen, and when introduced into a mixture of these gases, it causes their combination. "Spongy platinum," prepared by igniting ammonium platinichloride, **$(NH_4)_2PtCl_6$**, has a similar power in less degree. If a jet of hydrogen be directed on it, it grows red hot, and inflames the hydrogen. A recent application of this power of causing the combination of such gases has been applied to the detection of fire-damp (mainly CH_4) in mines. The bulb of a thermometer, coated with finely-divided platinum, or better, palladium, registers a sudden rise of temperature when it is brought into a mixture of marsh gas and air.

CuH, or **Cu_2H_2**, **copper hydride**,* is a brown powder, produced by adding to a strong solution of copper sulphate a solution of hydrogen hypophosphite, H_3PO_2, made by adding sulphuric acid to a solution of barium hypophosphite. It decomposes rapidly between 55° and 60°; it is attacked by hydrochloric acid, giving cuprous chloride and hydrogen, thus :—

$$\mathbf{CuH} + 2HCl.Aq = CuCl.HCl.Aq + H_2.$$

Alloys of lithium, sodium, potassium, and ammonium.—Lithium, sodium, and potassium mix with each other in all proportions. Alloys of sodium and potassium, containing 10 to 30 per cent. of the latter, are liquid at 0°, and have very high surface tension.

Sodium dissolves in liquid ammonia, forming an opaque liquid, with coppery metallic lustre, said to have the formula $Na_2N_2H_6$;† it has been named **sodammonium**, and is supposed to be the analogue of the undiscovered N_2H_8, diammonium. With excess of ammonia a blue liquid is formed. This substance, released from pressure, deposits sodium; it can hardly be a simple solution of that metal in liquid ammonia, for, if so, it is the only instance of a metal dissolving in a compound without chemical action.

The following alloys of the elements have been examined (to economise space, symbols are here employed; where indices are given formulæ have been ascribed):—

Na, Zn.—Zinc is insoluble in sodium; but they may be mixed together.

Na, Cd.—Sodium dissolves about 3 per cent. of cadmium.

Na, In.—Mix easily. **Na, Tl.**—Mix easily.

* Wurtz, *Annales* (3), **11**, 251.

† Weyl, *Pogg. Ann.*, **121**, 607; **123**, 365; also, Seeley, *Chem. News*, **22**, 317.

KFe_2, and KFe_3.—More fusible than iron; produced by heating iron sesquioxide with potassium tartrate. Sodium alloys are similarly made.

Na, Pb; K, Pb.—Similarly produced; **Na, Pb** alloys are malleable and bluish; those of **K, Pb** are white, brittle, and granular. Lead is insoluble in melted sodium.

Na, Sn; K, Sn.—Similarly prepared; white and brittle.

Na, Bi; K, Bi.—Similarly prepared; white and brittle.

Na, Pt; &c.—Metals of the platinum group are attacked at a red heat by sodium or potassium.

Na, Ag.—Silver is insoluble in melted sodium.

Na, Au.—Sodium dissolves about one-third of its weight of spongy gold; the alloy is white, and harder than sodium.

Li, Hg.—Lithium amalgam is produced by mixing the metals, or by electrolysing a concentrated solution of lithium chloride with mercury as the negative pole. It crystallises in needles, and is at once acted on by water.

Na, Hg.—Prepared by mixture; best under a layer of heavy paraffin; much heat is evolved by the union. When it contains under 1·5 per cent. of sodium, it is liquid; over that percentage, solid. It is much employed as a source of nascent hydrogen in alkaline solution. It is slowly attacked by water; but quickly by dilute acids.

K, Hg; Rb, Hg.—Similarly prepared, and similar in properties. Both crystallise in needles. **Rb, Hg** may be prepared like **Li, Hg.**

NH_4, Hg (?).—A buttery metallic mass, produced by the action of Na, Hg on ammonium chloride. Its difference in properties from metallic mercury would lead to the supposition that it contains the ammonium group; but it is very unstable, and splits quickly into mercury, and ammonia and hydrogen, which are evolved. It may be frozen, and then forms a solid bluish-grey brittle metal. A similar spongy bismuth compound may be prepared.

Ca, Zn.—A white alloy, containing from 2·6 to 6·4 per cent. of calcium, produced by heating together calcium chloride, zinc, and sodium. It does not decompose cold water, and is not appreciably tarnished by air.

Ca, Al.—Similarly prepared; white; contained 8·6 per cent. of calcium.

Ba, Al.—Similarly prepared; white; contained 24 to 36 per cent. of barium. It easily decomposes cold water.

Ba, Fe.—Lead-coloured; prepared by direct union; very oxidisable.—**Ba, Pd**; white.

Ba, Pt.—Bronze metal falling to red powder.

Ca, Hg; Sr, Hg; Ba, Hg.—Produced by electrolysis, like lithium amalgam. The first contains very little calcium, the second is somewhat richer, and the third may easily be obtained crystalline. Barium amalgam is also produced by shaking a solution of barium chloride with sodium amalgam; it is pasty and crystalline. Even at a white heat it retains 77 per cent. of mercury. **Ba, Sn** and **Ba, Bi** have been similarly prepared.

Mg, Zn; Mg, Cd; Mg, Al; Mg, Tl; Mg, Pb; Mg, Sn; Mg, Sb; Mg, Bi; Mg, Pt; Mg, Ag; Mg, Au; and **Mg, Pt** have been prepared. They are brittle, harder than the constituent metals, and have a granular fracture. They are all easily oxidised. Iron and cobalt are said not to alloy with magnesium, but the addition of a little magnesium to nickel lowers its melting point, and renders it ductile and malleable. This discovery has greatly increased the uses of nickel. Magnesium dissolves in mercury.

Zn, Cd.—Zinc and cadmium mix in all proportions.

Zn, Tl.—A soft white alloy.

Zn, Fe.—Iron and zinc do not easily unite. But iron may be "galvanized," that is, coated with zinc, by passing it through a bath of zinc kept melted under a coating of ammonium chloride. The iron must be first scrupulously cleaned with acid, and polished first with sand, then with bran. "Galvanized iron" is used for roofing, and for utensils of various kinds. The zinc does not easily oxidise, but if, owing to imperfect coating, oxidation takes place, the zinc oxidises and not the iron.

Some iron dissolves in the zinc, which is then known as "hard spelter;" it is purified by distillation.

Zn, Sn.—An alloy of 91·5 per cent. of tin with 8·3 per cent. of zinc is permanent; other alloys "liquate," that is, when heated on a sloping bed, the more easily fusible tin melts and runs down, leaving the less fusible zinc behind.

Zn, Sn_9, Pb_2.—An alloy made in these proportions melts at 168°.

Zn, Pb.—Harder than lead. Lead dissolves only 1·2 per cent. of zinc; and zinc only 1·6 per cent. of lead. A process for desilverising lead is based on this fact. The lead containing silver is mixed mechanically with zinc, and the mixture is stirred. The silver is carried up by the zinc, which floats. The zinc solidifies at a higher temperature than the lead, and the cake of zinc is removed and distilled; the silver remains behind. The lead is easily freed from zinc by oxidation; the zinc is removed as dross. **(Parke's process** for desilverising lead.)

Zn, Pb, Bi.—An alloy containing equal parts by weight of each melts under boiling water.

Zn, Bi.—Zinc dissolves 2·4 per cent. of bismuth; bismuth from 8·6 to 14·3 per cent. of zinc.

Zn, Ru.—Hexagonal prisms, burning when heated in air.

Zn, Rh.—Rhodium melted, zinc added, and excess of zinc removed by treatment with hydrochloric acid. A white crystalline compound.

Zn, Pt.—Similarly prepared. Zinc and platinum unite with incandescence. The compound is very hard, and bluish-white in colour. It fuses easily.

Zn, Cu.—Brass, pinchbeck, Muntz metal, tombac. Produced by melting copper and adding zinc. Copper may be superficially coated with brass by exposing it to zinc-vapour. The colour of brass and tombac is yellow, resembling gold. Brass tarnishes in air, but it may be protected by "lacquering," that is, coating it with a varnish made of shellac dissolved in alcohol, and coloured with gamboge. It is harder, not so tough, and more easily fusible than copper, and is ductile and malleable. English brass contains usually 70 per cent. of copper and 30 per cent. of zinc; pinchbeck, 86 per cent. Cu and 14 per cent. Zn; Muntz metal, used for castings, axle-bearings, &c., 66 per cent. Cu and 34 per cent. Zn; tombac, 86 per cent. Cu and 14 per cent. Zn; and bronze powder, for imparting the appearance of bronze to castings, &c., 83 per cent. Cu and 17 per cent. Zn.

Zn, Cu, Ni.—"German silver." A white alloy, hard, malleable, and ductile. Contains Cu, 62 per cent., Zn, 23 per cent., Ni, 15 per cent. Used for coins.

Zn, Ag.—A malleable, permanent alloy.

Zn, Au.—Zinc alloys easily with gold. An alloy of 11 parts of gold to 1 part of zinc is greenish-yellow and brittle; of equal parts is white and hard; and of 2 parts of zinc with 1 part of gold, hard, and whiter than zinc.

Zn, Hg.—Easily prepared. It is not attacked by dilute sulphuric acid; hence zinc battery plates are amalgamated. Until they are made the negative pole of the battery, hydrogen is not evolved. The compound **Zn_3Hg** is produced by electrolysing a solution of zinc chloride with a mercury electrode; and by squeezing a saturated zinc amalgam the compound **Zn_2Hg** is said to remain.

Cd, Pb; Cd, Sn.—Malleable white alloys.—**Cd_2Tl**, white and crystalline.

Cd_2Pt.—A white granular very brittle compound, prepared like the compound **ZnPt**.

Cd, Cu.—Brittle and whitish-red. **Cd, Ag.**—An alloy of 2 parts of cadmium with 1 part of silver is malleable and tenacious; with 1 part of cadmium with 2 of silver, is brittle. **Cd, Hg.**—Cadmium amalgamates easily. An amalgam containing 21 per cent of cadmium is brittle; alloys with more cadmium are malleable and very tenacious.

Al, Tl.—A soft dull-coloured alloy.

Al, Fe.—An alloy containing as much as 11·5 per cent. of aluminium is made by the Cowle's process of heating corundum (native oxide of aluminium) by the electric arc in a chamber along with charcoal, previously soaked in lime-water and dried, so as to prevent its conducting; scrap iron is present, which boils at the enormously high temperature of the arc, and washes down the aluminium, forming an alloy. It has been shown that this process does not depend on electrolysis, because an alternating current, which is incapable of electrolysing, yields equally good results. The alloy is named "ferro-aluminium," and is employed as follows: a quantity of nearly pure iron (fine wrought iron, containing but little carbon) is heated to its point of fusion, and enough ferro-aluminium is added to give a percentage of about 0·1 of aluminium to the total mass of iron employed. The melting point of the iron is thereby greatly lowered, it is said as much as 500°, and the iron is then directly employed for castings. Another noteworthy point is that iron containing even such a small proportion of aluminium retains the carbon in combination until it is just on the point of solidifying, and then rejects it almost completely to the outside. Castings made thus are very fine grained, and show no porosity. Such iron is named "mitis-iron." An alloy possessing the approximate composition **$AlFe_4$** is very hard, and may be forged.

Al_3Mn.—This alloy is apparently of definite composition; it forms square-based prisms.

Al Ni_6.—Large shining crystals, also of definite composition.

Al, Ti.—Brilliant brown iridescent crystals.

Al_3Zr; or including silicon **$(Al_3Zr)_2Si$.**—Crystalline laminæ.

Al, Sn.—Aluminium and tin alloy in all proportions; if the alloy contains from 7 to 19 per cent. of aluminium it is ductile, if over 30 per cent., it is brittle.

Al, Pb.—Aluminium and lead do not alloy.

Al_3Nb and **Al_3Ta.**—Crystals with metallic lustre; hard and brittle.

Al_4W.—Hard brittle grey ortho-rhombic crystals.

Al, Cu.—"Aluminium bronze." These alloys are blue-white when they contain little copper, and gold-coloured to red when the copper predominates. That with 60—70 per cent. of aluminium is brittle, very hard, and crystalline; with 30 per cent., soft; with under 30, it again becomes hard; an alloy with 20 per cent. is so brittle that it may be powdered in a mortar. Alloys containing from 11 to $1\frac{1}{2}$ per cent. of aluminium possess great tenacity, malleability, and ductility, and may be easily worked. They resemble gold in colour.

Tl_2Sn.—White, difficult to fuse. **Tl_2Pb.**—Soft and non-crystalline; lead-coloured. **Tl_4Sb.**—Hard, and not permanent in air. **Tl, Bi.**—Greyish-red, soft, and fusible. **Tl, Cu.**—Brass coloured, soft; may be cut with a knife. **Tl_2Hg.**—Of butter-like consistence.

Cr, Fe.—"Ferro-chrome." Produced by simultaneous reduction of a mixture of the oxides, *e.g.*, of a mixture of chrome iron ore and hæmatite. The alloy resembles iron in appearance. It is used for addition to iron, which it renders very hard, and especially adapted for cutting instruments, as it can be easily tempered.

Cr, Mn.—Very hard, and only slowly attacked by nitro-hydrochloric acid.

Mn, Fe.—"Ferro-manganese." Produced by simultaneous reduction of oxides of iron and manganese; resembles iron in appearance. It is used for addition to metallic iron, to which it communicates valuable properties. An iron containing about 10 per cent. of manganese crystallises in large brilliant plates, and is known as "spiegel iron," or "specular iron." The presence of manganese in steel causes it to harden, however slowly it is cooled; and if much manganese be present, the iron loses its magnetic properties. "Hadfield's manganese steel," containing from 7 to 20 per cent. of manganese, is so hard that it cannot be filed; yet it is ductile, and may be drawn into fine wire. It is very tenacious.

Fe, Co.—A hard, very compact alloy.

Fe, Ni.—The presence of 3 to 10 per cent. of nickel in iron renders it harder and more brittle.

Fe, Ti.—Is present in some pig-irons.

Fe, Sn.—A very brittle alloy. Six parts of tin and one of iron forms a hard white metal. Iron is "tinned" by plunging scrupulously clean plates (see Zn, Fe) into molten tin, kept liquid under a layer of grease. It is left for some time, withdrawn, passed through rollers, and passed quickly through a fresh bath of tin, to destroy the crystalline foliated appearance of the first coating. Where such a "moiré" effect is required, it may be produced by brushing the surface of the "tin plate" with weak nitro-hydrochloric acid. Such tinned iron resists the action of moist air; but, as the coating of tin is seldom quite perfect, galvanic action begins after some time, and the iron is oxidised. Hence "tinned iron" does not last so well, and is not so generally applicable as "galvanised iron." An alloy of 5 per cent. of iron, 6 per cent. of nickel, and 89 per cent. of tin is used under the name "polychrome" for tinning copper. It adheres easily to iron.

Fe, Pb.—Iron may be similarly coated with lead. Cubes of the formula $FePb_2$, of a brass yellow colour, have been found in cracks in the hearth of a blast furnace. Iron and lead alloy; the product is very hard; and may be fused at a white heat.

Fe, Ta.—Very hard; scratches glass. Not ductile.

Fe, Sb.—"Réaumur's alloy." Very hard, melting at a white heat. It is produced by melting together under charcoal 70 per cent. of antimony and 30 per cent. of iron.

Fe, Mo.—Greyish-blue, brittle, with granular fracture. Difficult to fuse.

Fe, W.—Whitish-brown, and compact. Alloys with more than 10 per cent. of tungsten do not fuse. An alloy has, however, been prepared with 80 per cent. of tungsten. It is very hard, and contains 5 per cent. of carbon. Tungsten is sometimes added to steel, in order to render it hard.

Fe, Rh.—These metals alloy easily. Steel, containing a little rhodium, is greatly improved in quality. An alloy containing more rhodium takes a high polish.

Fe, Pd.—One per cent. of palladium in steel renders it very brittle.

Fe, Pt.—Easily formed. The alloy takes on a high polish, and is very unalterable. One containing 9 parts of steel to 4 of platinum is ductile and hard.

Fe, Cu.—Iron alloys with copper in all proportions. An alloy of 2 parts of copper and 1 part of iron is of a greyish-red colour and very tenacious. The presence of 1 or 2 per cent. of copper in steel renders it brittle.

Fe, Ag.—A small quantity of silver in steel renders it hard.

Fe, Au.—The metals alloy easily. An alloy containing 1 part of iron and 12 parts of gold has a pale-yellow colour, and is very ductile. One containing 1 part of iron to 6 of gold is known as "grey gold," and is used by jewellers.

Fe, Hg.—Mercury and iron do not unite directly. But in presence of sodium, mercury alloys with iron. The amalgam may also be prepared by electrolysis (see Zn, Hg). It is insoluble in mercury, and soon splits into its constituents. The residue, after squeezing the excess of mercury through chamois leather, is said to have the formula **FeHg**.

Mn, Pb.—Hard and ductile.

Mn, Sn; Mn, Cu; Mn, Ag; Mn, Au.—Manganese easily forms these alloys. They resemble the corresponding alloys of iron. That with copper is reddish-white, and malleable.

Mn, Hg.—Produced by shaking a strong solution of manganese dichloride with sodium amalgam, or by electrolysis. It is grey and crystalline, and soluble in mercury.

Co, Ni.—The metals alloy easily, forming a brittle white metal.

Co, Pb.—This alloy is brittle; when fused it separates into two layers.

Co, Sn.—Bluish-white, somewhat ductile.

Co, Sb.—Easily prepared; grey and brittle.

Co, Pt.—A fusible alloy. **Co, Cu.**—Brittle dull-red alloy.

Co, Ag.—Brittle; when fused, separates into two layers.

Co, Au.—An alloy of 1 part of cobalt and 19 of gold is very brittle, and has a deep-yellow colour; even 1 part in 130 of gold renders it brittle.

Co, Hg.—Prepared like manganese amalgam. White and magnetic.

Ni, Sn.—Hard white brittle alloy. **Ni, Pb.**—Grey brittle laminæ. **Ni, Bi.**—Ditto.

Ni, Pd.—A brilliant alloy, capable of high polish; very malleable. It absorbs 70 times its volume of hydrogen.

Ni, Pt.—Whitish-yellow; as fusible as copper.

Ni, Cu.—An alloy containing 10 parts of copper to 4 of nickel is silver-white. The alloy containing zinc in addition is German silver (see Zn, Ni, Cu).

Ni, Ag.—A malleable alloy. **Ni, Au.**—Hard, very malleable, and ductile, yellowish-white, and capable of a good polish.

Ni, Hg.—Like cobalt amalgam.

Sn, Pb.—"**Solder,**" "**Pewter.**"—Alloys of tin and lead are harder than tin. Plumbers' solder, for soldering lead pipes, &c., contains 66 per cent. of lead and 33 per cent. of tin; tinsmiths' solder consists of equal weights of both. Lead and tin may be mixed in all proportions. Pewter, much used for drinking vessels, taps, &c., consists of 80 per cent. of lead and 20 per cent. of tin. The tin protects the lead from the action of acid liquors. Pewter sometimes consists almost entirely of tin, with a little copper to give it hardness. "**Britannia metal**" consists of equal parts of brass, tin, antimony, and bismuth; "**Queen's metal**" of one part each of antimony, lead, and bismuth, and 9 parts of tin.

Sn, Pb, Bi.—Alloys of these metals, to which cadmium is sometimes added, have very low melting-points, and are hence termed "fusible alloys." The following is a list:—

	Newton's.	Darcet's.	Rose's.	Wood's.	Lipowitz's.
Sn	3	1	207	2	4
Pb..........	5	1	236	2	8
Bi..........	8	2	420	7 to 8	15
Cd..........	—	—	—	1 to 2	3
M. p........	94·5°	93°	80—90°	66—71°	60°

They are used for safety taps, to prevent excess of pressure in boilers, or to melt and allow the escape of water in case of fire.

Sn, Sb.—A hard white sonorous alloy; when composed of 1 part antimony to 3 parts of tin, it is somewhat malleable, but is apt to crack.

Sn, Bi.—A hard, brittle alloy.

Sn_3Ru.—Prepared like the rhodium alloy.

Sn_2Ru.—Produced by direct union; cubical crystals, resembling bismuth.

Sn_3Rh.—Brilliant crystals left on treating tin containing 3 per cent. of rhodium with hydrochloric acid.

SnRh.—Brilliant black crystals of the formula given.

SnPd.—Brilliant scales, corresponding to the formula.

Sn_3Ir.—Similar to the rhodium alloy.

Sn_2Ir.—Cubical crystals.

Sn_4Pt.—Dilute hydrochloric acid on an alloy of tin and platinum, containing not more than 2 per cent. of the latter.

Sn_3Pt_2.—White brilliant fusible laminæ, consisting of small cubes.

Sn, Cu.—Bronze, speculum metal.—This alloy has been known for ages, and was produced by reducing copper and tin ores at one operation. For bells, 8 to 11 parts of tin and 100 parts of copper are employed. With more than 11 per cent. of tin, the alloy is malleable if quickly cooled, and may be fused at a red heat. A common alloy consists of 22 parts of tin and 78 parts of copper. The hardness of bronze is greatly increased by the addition of a small amount of phosphorus, as copper phosphide. Speculum metal for astronomical mirrors consists of 32 per cent. of tin, 67 of copper, and 1 of arsenic. It is susceptible of very high polish. The tin may be "liquated" out of such alloys. Copper cooking vessels are often protected against corrosion by "tinning." The copper is cleaned by scouring with ammonium chloride, or better with the double chloride of zinc and ammonium. The tin is then run over it, and the excess poured out. A thin coherent coating covers the copper.

Sn, Ag.—Tin alloys in all proportions with silver, forming hard white alloys; on liquation, however, the tin is removed.

Sn, Au.—This alloy is whitish-yellow and brittle.

Sn, Hg.—An amalgam of 1 part of tin with 10 of mercury is liquid; one with 1 part of tin and 3 of mercury crystallises in cubes. This alloy is employed in silvering mirrors; also for coating the rubbers of frictional electrical machines.

Pb, Sb.—Type-metal.—Type-metal consists of lead containing 17 to 18 per cent. of antimony. It is a dull-grey alloy, much harder than lead; and it is rendered still harder by addition of 8 to 10 per cent. of tin.

Pb, Bi.—White brittle laminæ.

Pb, W.—An alloy of lead and tungsten is brown, spongy, and ductile.

Pb, Cu.—Lead and copper do not alloy easily; the alloy

separates into two layers when left in repose. When stirred, however, castings can be made. They have a dull reddish-grey colour, and are said to withstand the action of sulphuric acid. Two or three per cent. of lead added to brass makes it less fibrous and more easily worked.

Pb, Ag.—Most commercial lead contains silver. When allowed to cool slowly, nearly pure lead crystallises out, the silver remaining in the melted portion. By a systematic procedure of this nature, the crystals being separated from the still liquid portion, silver is separated from lead (**Pattinson's process** for desilverising lead). The alloy is white, and harder than lead. It may be freed from lead by cupellation; that is, by melting the alloy in a cupel or shallow vessel made of bone-ash in a current of air. The lead is oxidised, and the oxide is absorbed by the porous cupel, leaving metallic silver.

Pb, Au.—Lead makes gold very brittle; even 1 part in 2000 greatly alters its malleability and ductility. This alloy is produced in the process of cupelling gold.

Pb, Hg.—Lead amalgam is easily obtained. From a strong solution of lead in mercury, crystals separate, which are said to possess the formula $\mathbf{Pb_2Hg_3}$.

Sb, Bi.—Antimony and bismuth mix in all proportions. The alloy, like bismuth, expands on solidification.

Sb, Cu.—Antimony renders copper brittle; even 0·00015 of its weight produces the effect. The alloy of the formula $\mathbf{SbCu_2}$ used to be known as "Regulus of Venus," and has a purple metallic lustre. When melted with lead and cooled slowly, the upper layer has approximately the formula $\mathbf{SbCu_4}$, and is nearly white, with vitreous fracture.

Sb, Ag.—Antimony and silver alloy easily, forming a similar alloy. Silver displaces copper from the copper-antimony alloy.

Sb, Au.—One part of antimony to nine parts of gold yields a brittle white alloy; even 0·05 per cent. of antimony renders gold brittle.

Sb, Hg.—Antimony dissolves in boiling mercury, but not to a great extent. Crystals separate on cooling.

Bi, W.—A porous brittle alloy, with dull metallic lustre.

Bi, Rh.—A white brittle alloy, completely dissolved by nitric acid.

Bi, Pd.—Grey, as hard as steel.

Bi, Pt.—Bluish brittle fusible laminæ.

Bi, Cu.—Pale red and brittle.

Bi, Ag.—White, brittle, and crystalline.

Bi, Au.—Greenish-yellow, granular, and brittle; 0·05 per cent. of bismuth renders gold quite brittle.

Bi, Hg.—Bismuth easily dissolves in mercury: a concentrated solution deposits crystals on cooling.

Mo, Pt.—Hard, grey, and brittle.

W, Cu.—Spongy and somewhat ductile.

W, Ag.—Brownish-white, somewhat malleable.

Rh, Pt.—Platinum containing 30 per cent. of rhodium is more fusible than rhodium, and easily worked.

Rh, Cu.—Alloy easily; the alloy is completely soluble in nitric acid.

Rh, Ag.—Very malleable.

Rh, Au.—Rhodium containing 4 or 5 per cent. of gold is gold-coloured, very ductile, and difficult of fusion.

Pd, Pt.—Grey, hard, somewhat ductile.

Pd, Cu.—The metals unite with incandescence. An alloy of 4 parts of copper to 1 of palladium is white and ductile; one with equal parts is pale-yellow and susceptible of high polish.

Pd, Ag.—Used for dentists' enamel; contains 1 part of silver to 9 of palladium. It is grey and harder than iron.

Pd, Au.—The metals unite with incandescence. With the proportion of 1 part of palladium to 6 of gold, the alloy is almost white; with 1 to 4, it is white, hard, and ductile; with equal parts, bright grey.

Os, Ir.—Osmiridium: found native, containing other metals of the platinum-group. It forms white scales, is exceedingly hard, and is used for pointing gold pens, for the bearings of small wheels, &c.

Os, Cu.—**Os, Ag.**—**Os, Au.**—Ductile alloys. **Os, Hg** adheres to glass.

Ir, Pt.—Harder than platinum and less easily fusible. An alloy containing about 10 per cent. of iridium is used for crucibles, &c.

Ir, Ag.—Ductile and white. **Ir, Au.**—Ductile and yellow.

Pt, Cu.—A ductile alloy, with the colour and density of gold. It is used in ornamental jeweller's work; the alloy contains 5 per cent. of copper and 95 per cent. of platinum.

Pt, Ag.—Less ductile and harder than silver, sometimes used for jewellery. Its colour is white.

Pt, Au.—With 2 parts of platinum to 1 of gold, the alloy is brittle; with equal parts, gold-coloured; with 3 of platinum, grey. Gold is used for soldering platinum vessels.

Pt, Hg.—Produced, like iron amalgam, in presence of sodium. If squeezed through leather, the definite compound $PtHg_2$ remains. When treated with nitric acid, it gives a residue retaining 7 to 8 per cent. of mercury, which behaves like platinum-black.

Cu, Ag.—White; if copper predominates, gold-coloured. All alloys liquate, except one of the composition Cu_2Ag_3. The alloys take a much higher polish than pure silver. This alloy is extensively used for **coinage**. English "silver" contains 7·5 per cent. of copper; its specific gravity is 10·20. French money contains copper and zinc. Silver-copper alloys, if cooled slowly, do not remain homogeneous; the metals separate partially. **Silver solder**, used for soldering jewellery, contains 66 per cent. of silver, with copper and zinc.

Cu, Au.—This alloy is used for coins, watches, jewellery, &c., owing to its greater hardness than gold. The English standard is 11 parts of gold to 1 of copper; in France and the United States, 9 parts of gold to 1 of copper. The alloys are more fusible than gold itself. Gold solder consists of 5 parts of gold to 1 part of copper. The richness of a gold alloy is estimated in "carats;" 24-carat gold is pure; 23-carat gold contains $\frac{1}{24}$th of copper.

Cu, Hg.—Prepared by boiling copper in mercury; by triturating finely-divided copper, first with mercurous nitrate, and then with mercury. When heated, it exudes drops of mercury; and if then ground up, it is so soft as to be moulded by the fingers; but it speedily becomes hard. By pressing through leather, the alloy **CuHg** remains.

Ag, Au.—Pale greenish-yellow. Found native, and known as "electrum." Sometimes used for jewellery. It is harder and more fusible than gold; that containing 1 part of silver to 2 of gold is the hardest.

Ag, Hg.—By placing mercury in a mixture of 2 parts of mercuric nitrate with three of silver nitrate, a crystalline growth of silver amalgam takes place, which is sometimes called the "Tree of Diana." Silver amalgamates readily with mercury; the amalgam deprived of excess of mercury by squeezing through leather has the formula $AgHg_2$. Silver amalgam is nearly insoluble in mercury.

Au, Hg.—Gold amalgamates readily. Crystals separate which are said to possess the formula $AuHg_4$. They are white, and dissolve sparingly in mercury.

It is seen that very few alloys have definite formulæ, and some of those which appear to be definite chemical compounds do not possess marked metallic lustre. It is very difficult to determine whether or not an alloy contains a definite compound in solution. In a mixture of two or more metals, that alloy which possesses the lowest melting point has not a definite formula. The lowering of the melting point of a metal appears to be proportional, at all events for dilute solutions, to the absolute amount of the metal present in smallest quantity, and inversely proportional to its molecular weight. Hence, by determining the lowering of the melting point of a metal such as sodium due to the addition of small amounts of other metals, the relative molecular weights of the dissolved metals may be ascertained; and it appears that, assuming the molecule of mercury to be monatomic, an assumption which is justified on other grounds, the molecular weights of most of the other metals are also identical with their atomic weights.

Alloys are not electrolysed into their constituents by the passage of an electric current; they are all good conductors; but the conductivity of those which conduct best, such as silver, copper, and gold, is greatly diminished by the presence of small amounts of other metals.

It has been already remarked that one of the elements of an alloy appears in some cases to be present in an allotropic condition. It is noteworthy that, by dissolving out the zinc from an alloy of zinc and rhodium, the latter metal should be left in an allotropic condition, so unstable, that on rise of temperature a slight explosion takes place, and the allotropic rhodium returns to its usual form. It appears not improbable that metallic iron is capable of existing in two allotropic states: one soft and not capable of permanent magnetisation; while the other form, steel, is hard, can be tempered, and remains magnetic for a long time after magnetisation. This change is apparently induced by the presence of a small amount of carbon.

Altogether, our knowledge of the chemical nature of alloys is very scanty; but the attention of chemists is again turning to this subject, so important both from a scientific as well as from a practical standpoint.

PART VIII.

CHAPTER XXXIV.

THE RARE EARTHS.—ENERGY RADIATED FROM MATTER; SPECTROSCOPY; CONNECTION OF THE SPECTRA OF THE ELEMENTS WITH ATOMIC WEIGHT.—APPLICATION OF SPECTRUM ANALYSIS TO THE ELUCIDATION OF THE RARE EARTHS.—SKETCH OF SOLAR AND STELLAR SPECTRA.

THE elements and their compounds have now been classified; and it has been seen that the arrangement adopted, that of the periodic table, has been fairly justified, inasmuch as elements displaying similarity, although always offering a regular gradation of properties with increase of atomic weight, have fallen into the same groups.

But a number of rare elements, comprising those contained in such scarce minerals as *orthite*, *euxenite*, *cerite*, *samarskite*, and *gadolinite*, have been only cursorily alluded to; these elements are, yttrium, lanthanum, ytterbium, terbium, didymium, erbium, and samarium; and to this list a number of others might be added of even more doubtful individuality, to which the names neodymium, praseodymium, decipium, phillipium, holmium, thulium, dysprosium, and gadolinium have been given. The state of our knowledge of these rare bodies is such that it appears advisable to consider them in a separate chapter; moreover, it is the opinion of Lecoq de Boisbaudran and Crookes, two of the chief authorities on such bodies, that they do not find their place in the periodic system of the elements.

Before they are described, it is necessary to have some acquaintance with the methods of spectroscopy, and with the nature of the vibrations emitted from matter.

Spectrum analysis.*—*General considerations.*—It has already been mentioned (see p. 92) that all matter is in a state of molecular motion. This motion is of two kinds. Gases, which inhabit space great in comparison with the actual volume of their con-

* See Roscoe's *Spectrum Analysis;* Schellen's *Spectral-Analyse*, and the original papers to which reference is made in notes to this chapter.

stituent molecules, have great freedom of motion, or, as it is termed, great "free path;" the duration of time in which their molecules are in a state of unimpeded motion is great in comparison with that in which they are in collision with neighbouring molecules, or with the sides of the containing vessel. But besides this "translatory" motion, or motion through space, the molecules are permanently in a state of vibratory motion. The true nature of this vibrational motion is, however, as yet uncertain. They are capable of communicating this vibratory motion to, and receiving it from, a medium which pervades all space, termed "ether." To discuss the nature of "ether" would lead us beyond our province; it may, however, be stated that no form of matter is impermeable to ether; and that it does not appear to be comparable in nature or properties with the usual forms of matter with which we are acquainted. The necessity for inferring its existence is obvious, however, when we consider that such vibrations are transferred across a vacuum, and that they *spend time in passing*. Light and radiant heat are special kinds of such vibrations; and it is known that light is not instantaneously transmitted across empty space, but travels at the rate of 185,000 miles per second. There must be something to convey this motion—something set into vibration by, and communicating its vibration to, material bodies—and this medium is called ether.

The molecules of a gas, being far apart, do not materially interfere with each other's vibrations; hence each single molecule assumes such rates and modes of vibration as are compatible with its structure; and such modes of vibration are transmitted through the ether to surrounding objects, which in their turn generally take up and exhibit vibrations of similar rates and modes to those of the gaseous molecules which incite them.

Such vibrations are, however, of varying frequency; each different kind of vibrating molecule having its own special rate or rates of vibration. When a vibrating molecule causes waves in the ether which pass any stationary point at a rate greater than 20 million-million per second, it produces effects of sensation, of expansion, &c., which we term heat. If they pass at a rate greater than 392 million-million per second, they affect chemically the compounds composing the lining membrane of the retina of the eye, and we have then light; and it is possible to recognise vibration as frequent as 4000 million-million per second, by their effect on certain other compounds, notably the salts of silver, and to vibration of such wave-lengths is given the name "actinic." To refract such exceedingly rapid vibrations, quartz prisms must be used, for they are absorbed by glass. But it must not be supposed that there is any difference in kind, but only in frequency, between vibrations to which we give these different names.

Our eyes, then, are capable of distinguishing as light vibrations whose number lies between 392 million-million and 757 million-million per second. Now, as we know the velocity of light, and the number of vibrations per second, the length of a wave capable of exciting any definite vibration is easily calculated; it is, expressed in millimetres, the velocity of light in millimetres per second, divided by the number of waves per second. Thus the wave-length of the slowest visible vibration is 766·7 millionths of a millimetre; and of the fastest visible, about 397 millionths of a millimetre. Waves of ether of different lengths produce on us the effect of colour. Red light, for example, is caused by waves of about 686 millionths of a millimetre in length; yellow light, by waves of 589 millionths of a millimetre long; green, by waves of 527 millionths of a millimetre; and blue, by waves of 486 millionths of a millimetre; while waves of 405 millionths of a millimetre produce the impression of violet light. The result of the impinging on the retina of waves of all visible wave-lengths is to produce what we term white light. We can give no corresponding names to waves capable only of inciting a rise of temperature, or of only producing chemical action; they must be distinguished by the number indicating the particular wave-length referred to.

Such waves are propagated more quickly through some media than through others; they pass more rapidly through gases, such as air, than through glass, or quartz, or liquids. If they fall on thick plates with parallel surfaces at right angles to that direction of their propagation, they merely undergo retardation, while they pass through the medium; but if they fall on such plates at any angle (not a right angle) to the surface, the ray is bent or *refracted* during its passage through the plate, returning to its original direction on issuing from its other surface. If, however, they fall obliquely on a prism, that is, a block of glass or other transparent material of triangular section, they are doubly bent, both on entering and on issuing.

But white light, or, to speak more generally, radiant energy, consists of vibrations of all conceivable rates; and it is known that those waves whose frequency is most rapid are more refracted during their passage through a prism than waves of less frequent vibrations; and they are therefore bent through a more acute angle, or, to express it in the usual language, they are more powerfully refracted than those of less frequency. Hence it happens that white light, when passed through a prism, is sorted out into coloured lights; violet light with its rapid vibrations being more bent than red light. Now, if the ray of light passing through the prism comes from a circular aperture, say in a window shutter, and is homogeneous, that is, consists of waves of some definite frequency, the result will be that an image of the circular aperture will be projected on a screen placed to receive it. If white light comes from a circular aperture, then a number of coloured circles must appear on the screen. But the number is practically infinite; hence these circles will overlap each other, except at the ends of the image; there the light will appear coloured, the outside colour being red at one end, and violet at the other (see Fig. 51); but the

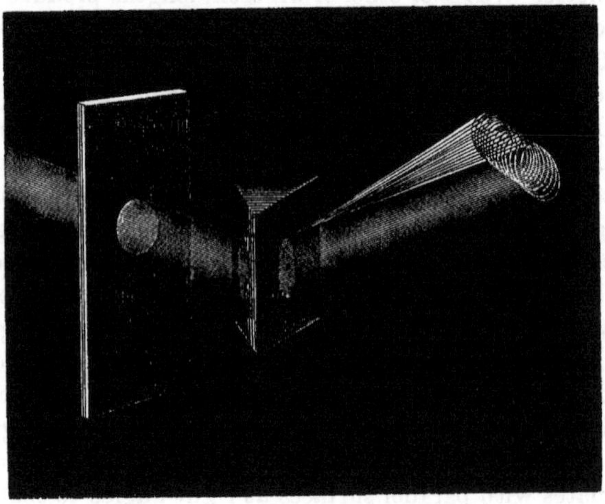

FIG. 51.

major part of the image will appear white, owing to the overlapping of the coloured light. To obviate this, a slit is used as the aperture; and an infinite number of narrow parallelograms are thus thrown on the screen. The finer the slit, the less the mixture of waves of different frequency on the screen; and though overlapping cannot entirely be avoided, it may be greatly reduced. The resulting image is termed a spectrum, and is shown in Fig. 52.

It has been stated that when the molecules of a gas are so hot as to emit radiant energy which can be observed, the vibrations which they perform are of certain definite frequencies, inasmuch as they are seldom interfered with by collision. It is otherwise with a solid, or with a liquid. In them the molecules are so closely packed as to leave little room for independent motion—they possess small free path. It therefore happens that no molecule is free to oscillate or vibrate without interference from its neighbour mole-

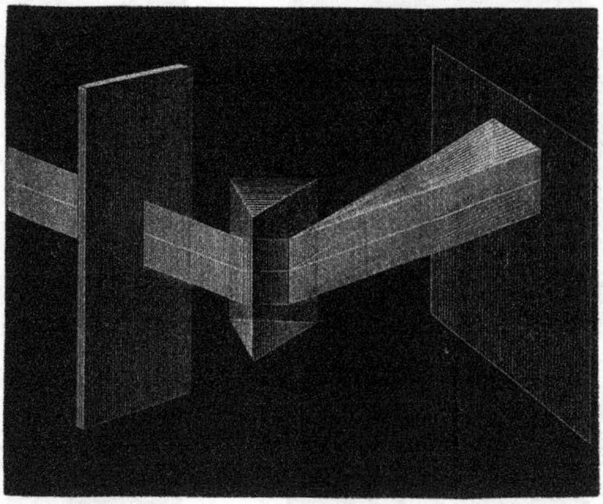

FIG. 52.

cules; hence, while each molecule tends, no doubt, to execute vibrations of such frequency and character as correspond with its individual nature, it is forced to execute vibrations of different periods. Hence (to confine ourselves to light) an incandescent solid or liquid emits light of all visible wave-frequency, that is, of every visible colour. But, at temperatures at which they first become luminous (about 550°), they emit dark-red light, and are said to be "dull red-hot." With rise of temperature, they emit yellow along with red light, and are said to be "bright red-hot;" at still higher temperatures green and blue light is emitted, and they are then "white-hot;" and it may be noticed that the electric arc-light is blue in colour, owing to its intensely high temperature. The spectrum of most radiating solids is a "continuous" one, that is, it is composed of light of all possible wave-lengths. The solar spectrum, due to that immense mass of incandescent matter, the Sun, is mainly of this character, but it is crossed by various dark lines, implying absence of waves where they occur, the nature of which will be afterwards described.

Such spectra appear to depend on the complexity, as well as on the near neighbourhood, of the molecules. It is probable, from what we know of dissociation, that a high temperature will split complex molecules into simpler ones, and that the spectra of the simpler molecules will themselves be simpler. But just as it is

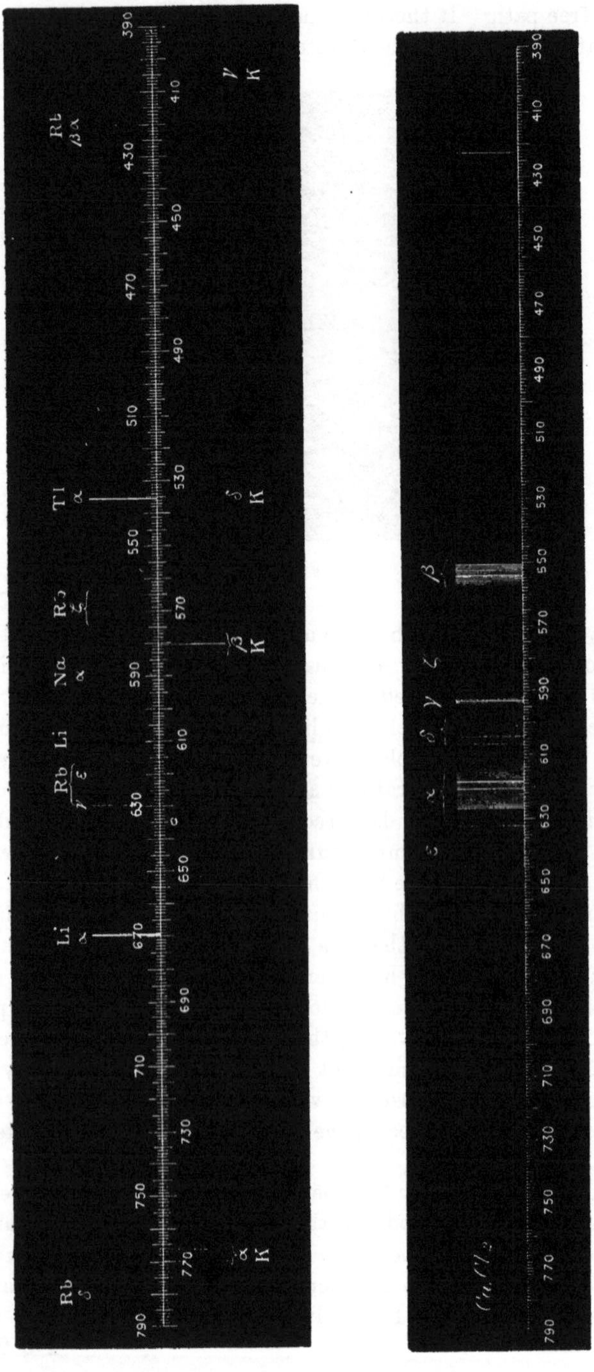
Rb δ
Rb β α
Tl α
Rb ξ
Na α
Li
Rb γ ε
Li α
K γ
K δ
K β
K α
CuCl2
β
ζ
γ
δ
α
ε
390 410 430 450 470 490 510 530 550 570 590 610 630 650 670 690 710 730 750 770 790

Fig. 53.

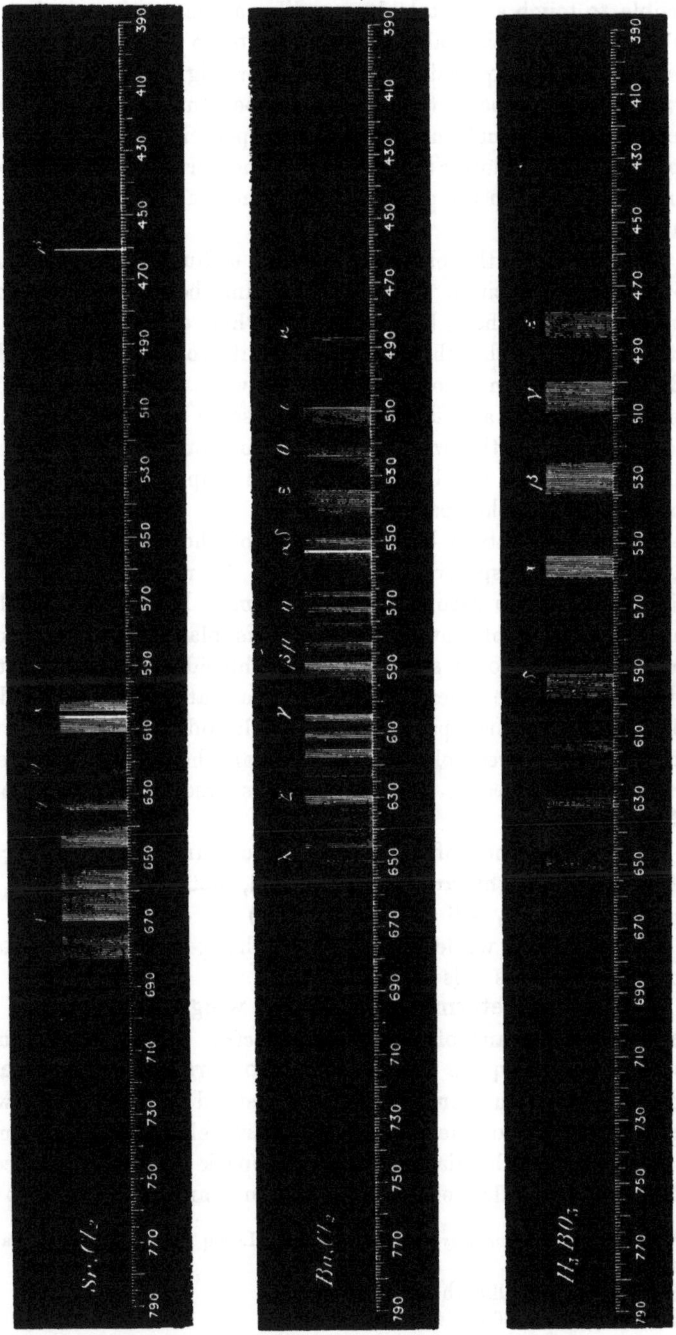
Sr Cl2
Ba Cl2
H3 BO3
390 410 430 450 470 490 510 530 550 570 590 610 630 650 670 690 710 730 750 770 790

FIG. 53.

possible to touch a stretched wire, like a piano-wire, so that its fundamental vibration alone is audible, or to cause it to vibrate strongly, when unpleasing over-tones or higher notes are perceived, so it is probable that a high temperature may cause vibrations in a simple molecule, which are unperceived, owing to their small intensity, at lower temperatures; and certain spectra become more complex when the gaseous bodies emitting them are strongly heated.

The spectra of the elements lithium, sodium, potassium, rubidium, cæsium, calcium, strontium, barium, boron, gallium, thallium, and some others, become visible when a compound of the metal (preferably the chloride, owing to its volatility) is heated in a Bunsen's flame in a loop of platinum wire. The accompanying woodcut* (pp. 596 and 597) reproduces some of these spectra in black and white; the wave-lengths in millionths of a millimetre are given, and also the letters which are employed to denote the principal lines of the spectra.

If the temperature be higher, that of the electric spark for example, different spectra are produced. To render such spectra visible, one of the secondary wires from a Rühmkorff's coil is connected with a platinum wire, which is placed about 0·2 mm. above the surface of a solution of the chloride of the element to be tested, while the other wire, also with a platinum terminal, dips in the liquid. Sparks pass from wire to liquid, and *vice versâ*, and some of the dissolved solid is volatilised and heated to a high temperature. The spectra may then be observed. They are shown in Fig. 54.

(For convenience of reference, the colours corresponding to certain wave-lengths are given†:—Red, 686 μ; yellow, 589 μ; green, 527 μ; blue, 486 μ; violet, 405 μ.)

It is seen that, while in some cases the spectrum is more complex, in other cases it is simpler.

Method of determining atomic weights by means of spectra.—By means of the spark spectra, it has been found possible by Lecoq de Boisbaudran **to predict the atomic weights** of certain elements.‡ The method of calculation will be shown for that of the recently discovered element germanium.

There are two brilliant lines in the spark spectrum of silicon, and also of its fellow elements germanium and tin; and also in

* Copied from *Spectres lumineux*, by M. Lecoq de Boisbaudran, Paris, 1874.

† μ denotes one millionth of a millimetre.

‡ *Chem. News*, 1886 (2), 4.

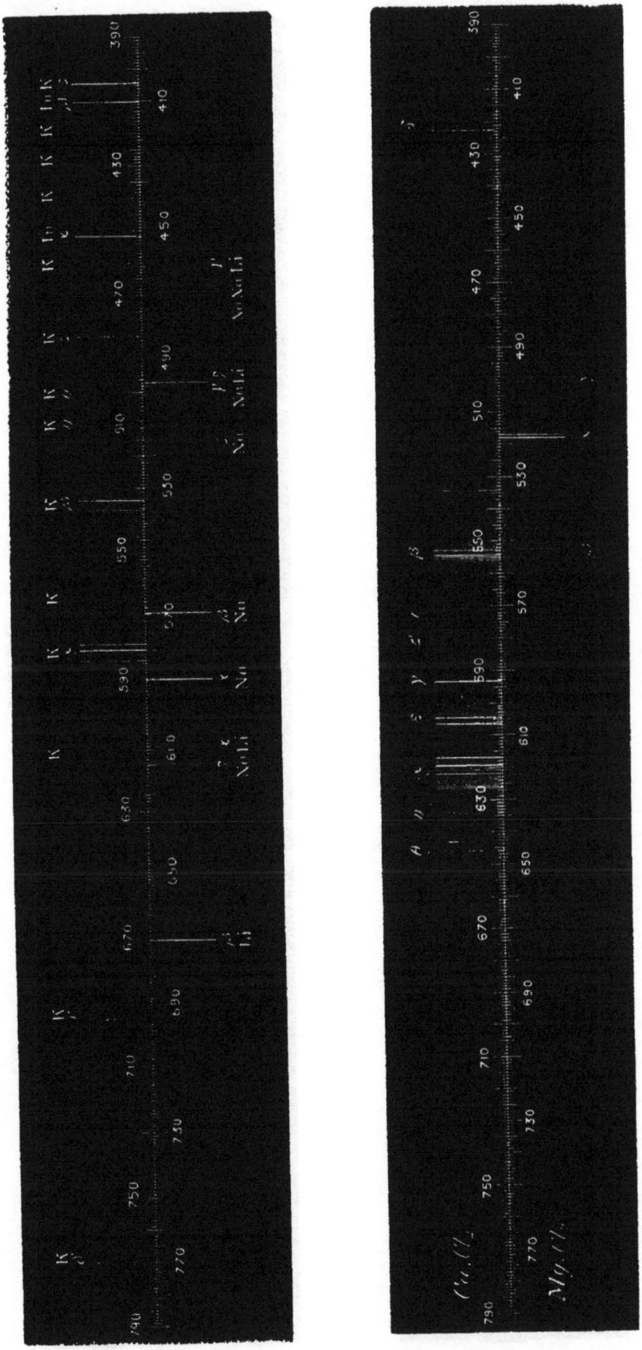

FIG. 54.

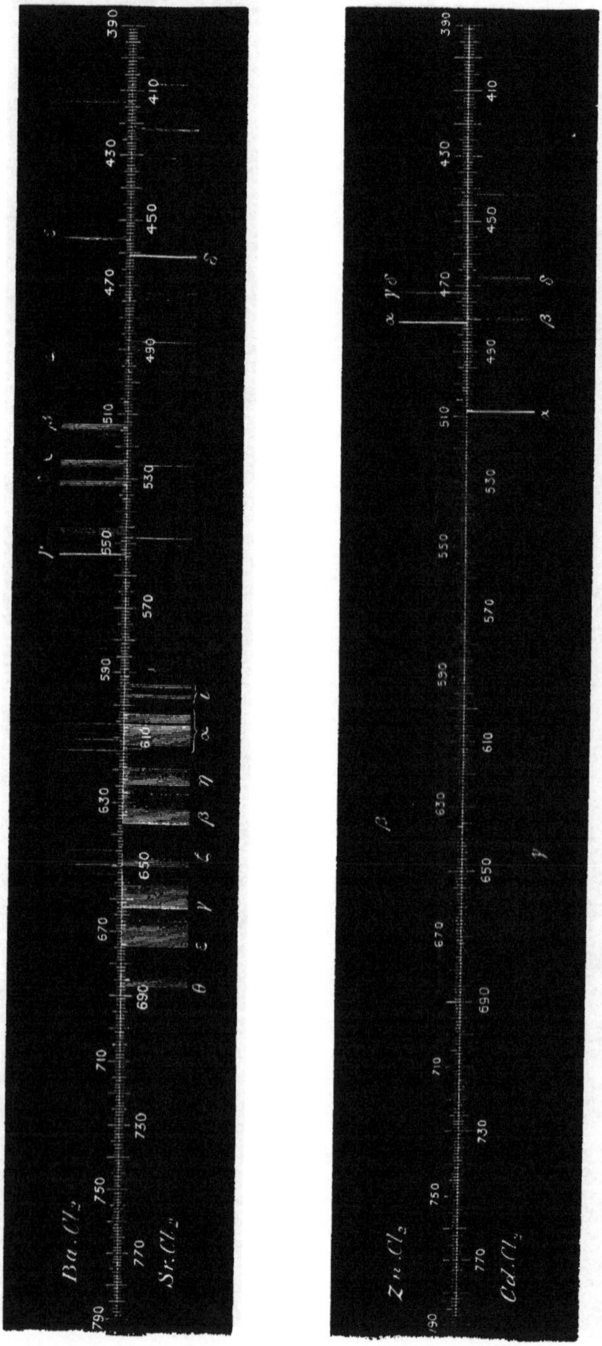
Ba.Cl2
Sr.Cl2
Zn.Cl2
Cd.Cl2
390
410
430
450
470
490
510
530
550
570
590
610
630
650
670
690
710
730
750
770
790

FIG. 54.

those of aluminium, gallium, and indium, three corresponding elements belonging to the previous group of the periodic table. Their wave-lengths are as follows (λ = wave-length in millionths of a millimetre):—

	Si.	Ge.	Sn.	Al.	Ga.	In.
1st line.... λ =	412·9	468·0	563·0	—	—	—
2nd line... λ =	389·0	422·6	452·4	—	—	—
Mean wave-length	401·0	445·3	507·7	395·2	410·1	430·6

The atomic weights of these elements of which that constant is known are, Si = 28; Ge ?; Sn = 118; Al = 27·5; Ga = 69·9; In = 113·5. Comparing the differences between the atomic weights of the members of both series with those between the mean wave-lengths of their two characteristic rays, the following table results:—

Atomic weights.		Δ.	Variations.	λ(mean).	Δ.	Variations.
Si......	28	90	—	401·0	44·3 62·4	$\frac{40·51}{100}$
Ge	?			445·3		
Sn.....	118			507·7		
Al.....	27·5	42·4 43·6	$\frac{2·8302}{100}$	395·2	14·9 20·5	$\frac{37·584}{100}$
Ga.....	69·9			410·1		
In.....	113·5			430·6		

Under the heading "Variations" is stated the percentage of the first difference which must be added to it to obtain a number equal to the second difference; thus, $42·4 + \frac{2·8303 \times 42·4}{100} =$ 43·6; and $44·3 + \frac{40·51 \times 44·3}{100} = 62·4$. The ratio which follows gives a means of determining what the values of Δ for the atomic weights of silicon and germanium, and for germanium and tin, should be:—

Al−2Ga + In (λ)	:	Al−2Ga+In (at. wt.).	::	Si−2Ge+Sn. (λ)	:	Si−2Ge+Sn (at. wt.).
37·584		2·8302	::	40·51	:	3·051

The number 3·051 is the percentage of the difference between the atomic weights of silicon and germanium, by which this difference must be increased to make it equal to the difference between the atomic weights of germanium and tin. The first difference, therefore, is 90/2·03051 = 44·32. Hence we have the series—

$$\begin{array}{l} Si = 28·00 \\ Ge = 72·32 \\ Sn = 118·00 \end{array} \begin{array}{l} < 44·32 \\ < 45·68 \end{array}$$

The atomic weight subsequently found by Winkler, the discoverer of germanium, was 72·3.

It is thus seen that there exists a close relation between the atomic weights of allied elements and certain lines of their spectra. This subject has, however, as yet been very little studied.

To render gases luminous which do not emit light at the temperature of a Bunsen's flame, such as hydrogen, oxygen, &c., a discharge of electricity of high potential is passed through them when rarefied to a pressure of under 5 millimetres. The gases are confined in tubes, generally called "vacuum tubes," through which platinum wires, or sometimes wires of aluminium, are sealed. These wires are connected with the secondary coil of a Rühmkorff's induction coil, and on passing an alternating current of high potential the gas in the tube is raised to a high temperature and emits light. By directing a spectroscope on the narrow capillary portion of the tube, the spectrum may be observed.

Many solid bodies exposed to an electric discharge of high potential in vacuum tubes also emit coloured light. Such substances are said to "phosphoresce." The form of tube employed by Mr. Crookes, the discoverer of this property, is shown in Fig. 55. Among such substances are *phenakite* (beryllium silicate), which emits a blue glow; *spodumene* (lithium aluminium silicate), which shines with a rich golden-yellow light; the ruby, which exhibits a very brilliant crimson phosphorescence; and the diamond, the light of which is exceptionally brilliant and of a greenish-white colour.

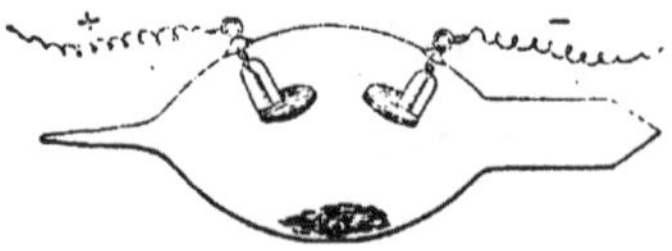

Fig. 55.

The rare earths.—Seen through a spectroscope, such coloured lights resolve themselves into bands of greater or less brilliancy at various parts of the spectrum. The oxides of the rare elements previously mentioned when examined in vacuum tubes by an inductive discharge are particularly rich in such lines and bands, and it is by this means that Crookes has investigated their nature, while Lecoq de Boisbaudran, Clève, Delafontaine,

Marignac, Soret, Nilson, Brauner, and others have, as a rule, employed the spark spectrum.

These elements are divisible into three main groups, the **didymium group**, comprising bodies to which the names neodymium, praseodymium, samarium, and dysprosium have been given; the **erbium group**, members of which have been named scandium, ytterbium, terbium, erbium, holmium, and thulium; and the **yttrium group**, to individual members of which names have not yet been given.*

The main lines of separation are as follows; but it should be mentioned that many other processes have been adopted:—The mineral is finely powdered, and boiled for some hours with hydrochloric acid (gadolinite, thorite), or mixed with strong sulphuric acid, and gently heated (cerite, euxenite); or fused with hydrogen sodium sulphate; or treated with hydrofluoric acid (samarskite, &c.). The product is then treated with cold water and filtered, and the residue is again treated similarly. To this solution ammonium oxalate is added, which precipitates the metals as oxalates. The precipitate is dissolved in hydrochloric acid; then thrown down with ammonia to remove lime, and next ignited, thus leaving a residue of oxides, usually of a reddish-brown colour. The oxides are dissolved by long boiling with nitric acid; the excess of acid is evaporated, and solid potassium sulphate is added until the solution is saturated. Double sulphates of members of the cerium, lanthanum, and didymium groups with potassium separate out, and are removed by filtration, while sulphates of the yttrium group remain in solution. The elements of both groups are then precipitated as oxalates; to separate cerium, the oxalates are dissolved in nitric acid and heated to incipient decomposition, and the solution is poured into a large excess of hot water. Basic cerium nitrate separates as a whitish-yellow precipitate. By two or three repetitions of this process, cerium may finally be obtained free from lanthanum and didymium. The residues are submitted to repeated fractionation, either by addition of an amount of ammonia insufficient to precipitate more than a fraction of the total amount of element present; or by treating the mixed oxides with an amount of nitric acid insufficient for complete solution; or by heating the mixed nitrates cautiously, so as to produce partial conversion into oxides, and treatment with water, in which the undecomposed nitrates alone are soluble; or by other processes comparable with

* Brauner, *Chem. Soc.*, **41**, 68; also numerous papers by Clève, Nilson, and others.

the above. It should be understood that such processes must be repeated methodically thousands of times before any definite elements are isolated. And it is also remarkable that minerals from different sources yield very different results, nature having often performed a partial separation.

Elements of the didymium group have been investigated by means of the absorption spectra seen when a solution of one of the compounds of the elements in water is examined through a spectroscope. The old "didymium" has the spectrum shown below (A).

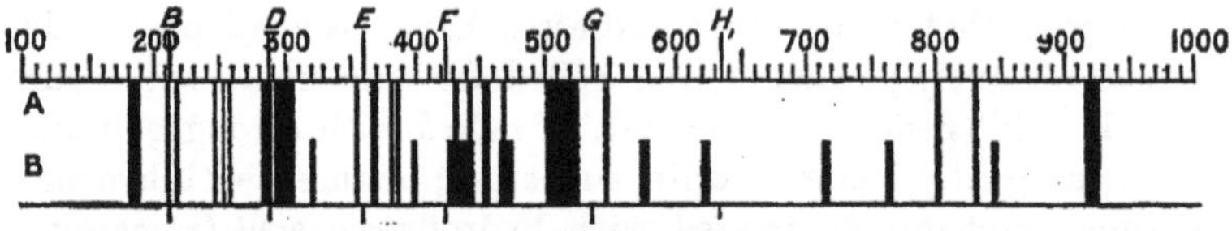

FIG. 56 (A).

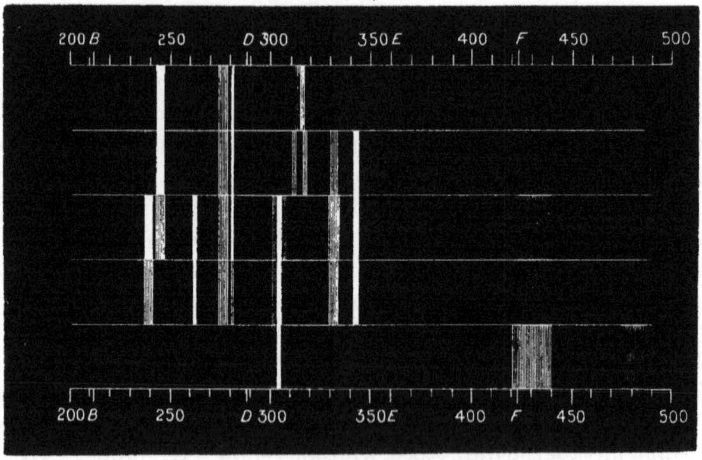

FIG. 56 (B).

The absorption spectrum of bodies separated from didymium by processes of fractional crystallisation, precipitation, &c., are shown at (B). The thin line at 320, the line at 400, the band between 420 and 440, the narrow band between 460 and 470, and lines at 575, 620, 715, 765, and 845 form together the spectrum of the so-called samarium, a pseudo-element separated from what was at one time believed to be the pure element didymium by Delafontaine* in 1878, and by Lecoq de Boisbaudran. In 1885, Carl

* *Chem. News*, **38**, 223; **40**, 99. This description is largely taken from Mr. Crookes' address to the Chemical Society, *Chem. Soc.*, **55**, 256.

Auer* succeeded in isolating two new bodies by fractionally crystallising the double nitrates of elements of this group with ammonium nitrate. Of these, one had pink salts, and he named it *neodymium;* the other green, and to it he gave the name *praseodymium.* The absorption spectrum of the former includes the lines at 187, 207, the three faint lines at 250, the broad band at 300, the thin line at 355, the band at 365, and the two bands about 385; while the exceedingly fine line at 549 is strengthened to a distinct band. The latter has the other part of the thick band at 287, one at 430, the band at 455, and the thick band between 497 and 515. There are still two bands unclaimed, viz., those at 462 and 475, which might lead to the supposition that a third substance is present which has not been identified. But Mr. Crookes states that by other methods of fractionation he has obtained evidence of other cleavages; for it must be noticed that treatment with any one reagent will effect a separation into only two groups; and that the particular results obtained by Auer depend on the nature of the process which he adopted. Krüss and Nilson† believe the old didymium to contain at least nine separate components. But it is dangerous to draw any definite conclusion from such results; for Brauner has shown‡ that on mixing a dilute solution of a salt of samarium with one of didymium, the three bands at 430—460 disappear, while samarium bands do not take their place until a large proportion of a salt of the latter has been added.

In an exactly similar manner, the original "erbium" has been resolved into fractions giving absorption spectra corresponding in part with that of their parent earth. The investigations of Delafontaine, Marignac, Soret, Nilson, Clève, Brauner, and others, have pointed towards at least six different earths, three, scandia, ytterbia, and terbia, having no absorption spectrum, while others, viz., the new erbia, holmia, and thulia give absorption spectra. The earth called phillipia has been proved by Roscoe to be a mixture of yttria and terbia.

Elements of the yttrium group do not give absorption spectra; hence they have been investigated by Lecoq de Boisbaudran chiefly by spark spectra, and by Crookes, by means of phosphorescence spectra. The old "yttrium," by a system of fractionation repeated many thousand times, has been separated into a number of portions. The spectra of the extreme ends of the fractions differ from that of the original yttrium by showing new bands,

* *Monatsh. Chem.*, **6**, 477.
† *Berichte*, **20**, 2134.
‡ *Chem. Soc.*, **43**, 286.

and by the disappearance of some of the old ones. And a gradual transition may be noticed from fraction to fraction, the new band appearing dimly, and gaining strength as the separation proceeds; or the old band becoming fainter, finally to vanish. The spectra of the fraction to which the names "samarium," "yttrium α," "mosandrium," and "yttrium" have been given are shown in Fig. 56 (B), as well as the result of continuing the fractionation to the other side, and separating these substances as far as practicable. It is right to observe, however, that de Boisbaudran does not agree with Crookes in such conclusions; he maintains that there are three perfectly characterised earths, to which the provisional symbols $Z\alpha$, $Z\beta$, and $Z\gamma$ have been given, which Crookes has not succeeded in separating. He regards it as probable also that the oxide of $Z\beta$ is identical with a deep-brown oxide obtained from the so-called terbium. Crookes has also found that the addition of foreign elements, for instance, lime or alumina, has a profound influence in modifying such phosphorescence spectra.

We see therefore that there is great reason to believe that such substances are mixtures. Crookes has revived the bold speculation of Marignac, that not all the molecules of any given element are uniform in mass or other properties, and that what we name an element is simply the mean result of a number of atoms or molecules, closely approximating to each other in properties, but not identical; and he suggests that, provided suitable methods, such as the various plans of fractionation, be made use of, it may be possible to effect a separation of more or less complete nature. Whether this view is a correct one, or whether the rare elements of which he treats are merely mixtures of some eight or ten new bodies, which, owing to their similar behaviour, are very difficult to separate, must be left to the future to determine. He suggests, indeed, that there may be different degrees of elemental rank. But the fact that the presence of one element tends to modify, and sometimes to thoroughly alter, the spectrum of another should lead to great caution in accepting the help of spectroscopy in identifying elements.

It may here be mentioned that Brauner has succeeded in obtaining elements from tellurium, or at least from what usually passes under that name. Further research will doubtless throw light on the question of its elementary nature.

Solar and stellar spectra.—One other application of spectroscopy remains to be mentioned. It has led to the identification of many elements existing in the sun, the fixed stars, nebulæ, and comets with those existing on our earth. The principle underlying this discovery is as follows:—Matter not only radiates energy, it also absorbs energy. We see a wall coloured because the paint with which it is covered absorbs vibrations of certain wave-lengths from the white light which illumines it, while reflecting vibrations of other wave-lengths.

A gaseous molecule can be made to oscillate by the impingement of ether waves of the same period of oscillation, and not, as a rule, by waves of a different period; and if ether vibrations of some definite periods impinge on a large number of molecules all capable of vibrating, to the same period, then it will cause them to vibrate. But it may be that the intensity, that is, the amplitude, of the vibration of the ether is not sufficient to cause so many molecules themselves to vibrate with an amplitude great enough to be perceived by our senses; the ether vibrations are then practically extinguished, because they distribute their energy through such a large number of molecules.

Now the light given out by an incandescent gas is, as we have seen, generally composed of waves of a few definite lengths. And if its waves, propagated through the ether, be caused to impinge on a quantity of the same gas, the molecules of that gas will be set in vibratory motion; but that motion, being distributed over a large number of molecules, will be so weakened in amplitude as not to be perceptible. To confine ourselves to light: if the yellow light emitted from incandescent sodium gas be caused to impinge on a sufficient quantity of sodium gas, which is not at so high a temperature as to incandesce, it will be completely absorbed, and *the sodium light will not pass through the sodium gas;* or, more correctly speaking, the vibrations of the second portion of gas will be too small in amplitude to affect our eyes.

The sun is a vast mass of incandescent matter, sending forth energy of all conceivable wave-lengths. Among these vibrations are some which coincide in period with those given out by incandescent sodium. But the sun is surrounded by an atmosphere of sodium gas; hence these vibrations will not be transmitted through the sodium vapours to the ether with amplitude sufficient to be perceived. It is for this reason that we see in the solar spectrum a dark line, or more correctly, two dark lines very near together, in the yellow part of the spectrum. This double line is named "the D line," and its position is absolutely identical with that of the bright line visible when the light of incandescent sodium gas is viewed through a spectroscope. It is therefore legitimate to conclude that the phenomenon which can be produced on a small scale is identical with that occurring in the sun and its atmosphere;

and it follows that the sun is surrounded by an atmosphere consisting partly of gaseous sodium.

By similar means, by comparing the dark lines seen in the solar spectrum and in the spectrum of the fixed stars with the bright lines produced by incandescent gases, it has been discovered that many of our elements are present in these heavenly bodies. The following is a list of those which have been identified in the sun; with the number of lines which have been observed coinciding with the ordinary spectra:—

Element....	H.	Na.	K.	Ca.	Sr.	Ba.	Mg.	Zn.	Cd.	Al.	Cr.	Fe.	Mn.	Co.
Lines......	4	2	2	75	4	11	4	2	2	2	18	450	57	19

Element..	Ni.	Ti.	Ce.	Pb.	V.	Mo.	U.	Pd.
Lines....	33	118	2	3	4	8	2	5

There are also present, lithium, iridium, and copper.

The following are doubtful:—In, Rb, Cs, Bi, Sn, Ag, Be, La, Y, C, Si, Th, and the halogens, F, Cl, Br, and I are absent.

Oxygen has been lately observed; but it gives bright bands, and not absorption lines.

Prominences are continually observed on the disc of the sun. These appear to be enormous outbursts of the gaseous atmosphere; they have been examined, and appear to consist of hydrogen, and of the vapours of magnesium and iron, besides other elements.

The moon reflects the solar spectrum without alteration, adding nothing and absorbing nothing. This affords an argument for the non-existence of a lunar atmosphere, borne out by other considerations.

The fixed stars may be divided into four classes:—

(1.) White stars, such as Sirius; they have been found to contain hydrogen, sodium, magnesium, and iron.

(2.) Yellowish stars, such as Arcturus; they possess a complex spectrum like the sun, and no doubt our sun belongs to this class.

(3.) Red stars, like α-Orionis and α-Herculis; these show sets of bands resembling those caused by the solar spots. It has been suggested that they are surrounded with a thick atmosphere.

(4.) Stars usually of small magnitude, such as γ-Cassiopeiæ. They show lines of hydrogen and of sodium.

Nebulæ and comets show a faint continuous spectrum, together with certain lines which are identical with those of certain hydrocarbons when exposed to an electric discharge under low pressure in vacuum tubes.

It is probable that some of them, at least, consist of incandescent solid matter, accompanied by a gaseous hydrocarbon.

This short sketch will suffice to give an idea of the manner in which we have acquired a knowledge of the chemical composition of the heavenly bodies. A large number of facts has been accumulated, but it cannot be doubted that improved methods, and the extension of observations to the invisible parts of the spectrum will greatly add to our conceptions of the nature of the galaxy of suns with which we are surrounded.

The nature of the vibrations which are transmitted to us through the ether is not as yet understood, and the sciences of chemistry and spectroscopy touch at only a few points as yet; but it is evident that further research will greatly increase our knowledge of the atomic and molecular constitution of matter. Already it has afforded a means in the hands of Bunsen and Kirchhoff, of Crookes, of Lecoq de Boisbaudran, and of Reich and Richter, of detecting the presence of undiscovered elements in minerals and in waste products; the spectrum lines have been shown to be in close relationship with the atomic weights of the elements; and some success has been met with in applying spectroscopy to quantitative analysis. These are great achievements; but there are undoubtedly greater to follow.

CHAPTER XXXV.

THE ATOMIC AND MOLECULAR WEIGHTS OF ELEMENTS AND COMPOUNDS.—THE KINETIC THEORY OF GASES.—THE STANDARD OF MOLECULAR WEIGHTS.—THE VAPOUR-DENSITY OF ELEMENTS AND COMPOUNDS.—DISSOCIATION OF MOLECULES OF ELEMENTS AND COMPOUNDS.—ATOMIC AND MOLECULAR HEATS.—REPLACEMENT.—ISOMORPHISM.—MOLECULAR COMPLEXITY.—MONATOMIC STATE OF MERCURY GAS.

In stating the objects of the science of chemistry, in the first pages of this book, the composition, nature, synthesis, and classification of different kinds of matter were first noticed; for it is obviously necessary that they should be known in order that the further objects of the science, viz., the nature of the changes which matter undergoes, and the classification of these changes, may be understood. To discuss the nature of such changes from a physical point of view is not within our province here; but it has been seen that, in order to obtain a connected view of the relations between various classes of compounds, certain conceptions must be entertained regarding the ultimate nature and the constitution of matter. These conceptions depend on the behaviour of gaseous matter when exposed to different conditions of pressure, temperature, &c., and, for the most part, our classification is one strictly applicable only to gases. To assign formulæ to liquids and solids, as has been frequently remarked, is usually an extension, not warranted by our knowledge, of the principles which represent our conceptions only of gaseous matter.

It is therefore advisable to consider in detail four classes of constants, all of which are indispensable to correct classification; these are:—

1. **The atomic weights of the elements;**
2. **The molecular weights of the elements;**
3. **The molecular weights of compounds; and**
4. **The structure of compounds.**

The atomic and molecular weights of elements and the molecular weights of compounds.—(*a.*) **Density in the state of gas.**—It has been told in Chapter II what led Dalton to assign certain atomic weights to the elements which he investigated. Having ascertained the equivalents of certain elements, he was guided by the principle of "simplicity" and "similarity;" that is, the atomic weight of an element was taken to be that multiple of its equivalent which gives the simplest proportions between the numbers of atoms contained in all known compounds of the element; and like compounds were assigned like formulæ. This scheme was also followed out by Berzelius. We have seen again in Chapter VIII, p. 109, how the atomic weights of elements may be deduced from the densities of their gaseous compounds; the history of the discovery is as follows:—In 1805, **Gay-Lussac** and **Humboldt**, in investigating the volume composition of water, found that two volumes of hydrogen unite with one volume of oxygen to form two volumes of water-gas (see p. 193). This led Gay-Lussac to make further researches on the relative volumes in which gases combine, and he discovered that two volumes of nitrous oxide consist of one volume of oxygen and two volumes of nitrogen; that one volume of chlorine unites with one volume of hydrogen to form two volumes of hydrogen chloride; that two volumes of ammonia consist of three volumes of hydrogen and one of nitrogen; and that one volume of carbon monoxide unites with one volume of chlorine to form two volumes of carbonyl chloride, or, as it was then termed, "phosgene gas." Towards the end of 1808 he made the important generalisation in the *Mémoires de la Société d'Arçueil*, **2**, 207, that—(1) "**there is a simple relation between the volumes of gases which combine**;" and (2) "**a similar simple relation exists between the volumes of the combining gases and that of the resulting gaseous compound.**" And from these statements it follows:—"**The weights of equal volumes of both simple and compound gases** (or in other words, their densities), **are either proportional to their combining weights, or are a simple multiple thereof.**"

As a sequel to Gay-Lussac's discovery, **Avogadro** announced in 1811 (*Journal de Physique*, **73**, 58) that equal volumes of gases contain equal numbers of particles (*molécules intégrantes*), but that these are not of the nature of atoms, indivisible, but consist of several atoms. Otherwise stated, the molecular weights are proportional to the densities. The relation between the relative masses and rates of motion of gases has been since worked out by **Clerk Maxwell, Sir William Thomson, Clausius**, and others, and starting with the assumption that the expansive tendency of gases is due to the motion of their molecules, they deduced the kinetic theory of gases, for an account of which the reader is referred to Maxwell's *Theory of Heat*, pp. 289, *et seq.*, or to the article on "Heat" in Watt's *Dictionary of Chemistry*, **3**, p. 131.

By means of a mechanical conception, viz., that the pressure of a gas is due to the impacts of its molecules on the walls of the containing vessel; and that its temperature is due to the motion of its molecules; it is shown that Avogadro's law is true, viz., that the number of molecules in equal volumes of all gases is equal, provided the gases be compared under similar conditions of temperature and pressure.

Standard of molecular weights.—We know by experiment the relative weights of many molecules. We know that the relative densities of hydrogen and chlorine are as 1 : 35·4; that is, the molecule of chlorine is 35·4 times as heavy as the molecule of hydrogen. But still the question is not answered, *what are their relative atomic weights?* In other words, how many atoms are contained in a molecule of hydrogen and in a molecule of chlorine? We have seen on pp. 158, 159, and 160 that if hydrogen chloride consist of 1 atom of hydrogen in combination with 1 atom of chlorine, it is a reasonable deduction that a molecule of hydrogen contains also 2 atoms of hydrogen, and a molecule of chlorine 2 atoms of chlorine, although the possibility is not excluded that it may consist of more than 2. This leads us to consider the densities of gaseous elements in so far as these have been determined, so that we may ascertain whether they correspond with hydrogen and with chlorine in the complexity of their molecules.

Density of Elements in the Gaseous State.

Hydrogen....	1 (unit of density).
Sodium......	12·7* (Victor Meyer's method, in platinum vessel).
Potassium....	18·8* (" " ").
Zinc	34·15 at 1400°.†
Cadmium	57·01 at 1040°.‡
Mercury.....	100·94 at 446°;§ 101·3 at about 1730°.‖

* Scott, *Proc. Roy. Soc. Ed.*, **14**.
† Mensching and Meyer, *Berichte*, **19**, 3295.
‡ Deville and Troost, *Annales*, **113**, 46.
§ Dumas, *Annales*, **33**, 337.
‖ Biltz and Meyer, *Zeitschr. für Phys. Chem.*, **4**, 265.

Thallium	206·2 at 1730°.*
Nitrogen	14·08 at atmospheric temperature;† not altered at highest temperatures.
Phosphorus ..	63·96 at 313°;‡ 55·7 at 800—900°; 53·65 at 1200—1300°; 52·45 at 1500°; 46·59 at 1680°; 45·58 at 1708.§
Arsenic	154·2 at 644° and 860°;‖ 79·5 above 1700°.¶
Antimony....	155·5 at 1572°; 141·5 at 1640°.¶
Bismuth	146·5 at 1640°.¶
Oxygen......	15·99 at atmospheric temperature;** not altered at the highest temperature.
Sulphur	114·9 at 468°;†† 94·8 at about 500°;‡‡ 39·1 at 714—743°; 34·7 at 800—900°; 31·8 at 1100—1160°;†† 31·8 at 1719°.¶
Selenium	111·0 at 860°; 92·2 at 1040°; 82·2 at 8420°.§§
Tellurium ...	130·2 at 1390—1439°.§§
Fluorine.....	18·3 at atmospheric temperature.‖‖
Chlorine.....	35·90 at 20°;¶¶ 35·45 at 200°; 35·31 at 630°;*** 31·83 at 800°; 27·06 at 1000°; 24·02 at 1200°; 23·3 at 1560°.
Bromine	80·16†††; 82·77 at 102·6°; 81·03 at 175·6; 79·93 at 228‡‡‡; 52·7 at 1500°.§§§
Iodine	127·66 at 253°; 127·37 at 580°; 98·4 at 840°; 82.5 at 1000°;‖‖‖ 63·7 at 1500° under low pressure.¶¶¶

The list given above does not include all determinations; but the more important researches are referred to in the notes.

These results lead to a division of the elements into two groups:—(1) Those which undergo **no change in density** on

* Biltz and Meyer, *Zeitschr. für Phys. Chem.*, **4**, 265.
† Regnault, corrected by Jolly, *Wied. Ann.*, **6**, 536.
‡ Dumas, *Annales*, **49**, 210.
§ Meyer and Biltz, *Zeitschr. phys. Chem.*, **4**, 259.
‖ Mitscherlich; and Deville and Troost.
¶ Meyer and Biltz, *loc. cit.*, 263.
** Regnault, corrected by Jolly, *loc. cit.*
†† Bineau. See also Biltz, *Zeitschr. phys. Chem.*, **2**, 920, and **3**, 228; Ramsay, *ibid.*, **3**, 67.
‡‡ Dumas, *Annales*, **50**, 178; Mitscherlich, *Pogg. Ann.*, **29**, 217.
§§ Deville and Troost, *Annales*, **58**, 273.
‖‖ Moissan, *Compt. rend.*, 1889, Dec. 2nd.
¶¶ Ludwig, *Berichte*, **1**, 232.
*** Meyer, *Berichte*, **12**, 1428 (the chlorine was produced from platinous chloride, and was nascent; it mixed at once with nitrogen).
††† Mitscherlich.
‡‡‡ Jahn, *Monatsh. Chem.*, 1882, 176.
§§§ Meyer and Züblin, *Berichte*, **13**, 405 (from $PtBr_4$).
‖‖‖ Meyer, *Berichte*, **13**, 394.
¶¶¶ Crafts and Meier, *Compt. rend.*, **92**, 39.

rise of temperature; that is; those of which the co-efficient of expansion remains equal to that of hydrogen. Such are mercury, nitrogen, oxygen; and, although experiments in this direction have not been made, probably sodium, potassium, zinc, cadmium, thallium, and tellurium (?). (2) Those of which the **vapour density decreases with rise of temperature**; among these are phosphorus, arsenic, antimony, bismuth (?), sulphur, selenium, fluorine (?), chlorine, bromine, and iodine.

Now, we can calculate the maximum atomic weights of many of these elements from the vapour densities of their volatile compounds, generally of their halides, or of their hydrides. An example of each will be given.

Compound.	Density × 2.	Mol. wt.	Constituents. Parts by weight of			
Water	9·0	18·00	H =	1·00	O =	16·00
Sodium (forms no sufficiently volatile compound).						
Potassium iodide	168·9	165·99	K =	39·14	I =	126·85
Zinc chloride	133·4	136·22	Zn =	65·3	Cl_2 =	70·92
Cadmium bromide	267·1	272·00	Cd =	112·1	Br =	159·90
Mercuric chloride	283·0	271·12	Hg =	200·2	Cl_2 =	70·92
Thallium monochloride	236·7	239·66	Tl =	204·2	Cl =	35·46
Ammonia	17·2	17·03	N =	14·03	H_3 =	3·00
Phosphine	33·1	34·03	P =	31·03	H_3 =	3·00
Arsine	77·8	78·09	As =	75·09	H_3 =	3·00
Antimony trichloride	224·7	226·68	Sb =	120·30	Cl_3 =	106·38
Bismuth trichloride	327·7	314·48	Bi =	208·10	Cl_3 =	106·38
Nitric oxide	30·0	30·03	O =	16·00	N =	14·03
Hydrogen sulphide	34·4	34·06	S =	32·06	H_2 =	2·00
Selenium dioxide	116·0	111·00	Se =	79·0	O_2 =	32·00
Methyl fluoride	34·76	34·00	F =	19·0	CH_3 =	15·00
Hydrogen chloride	36·0	36·46	Cl =	35·46	H =	1·00
Mercuric bromide	351·0	360·1	Br_2 =	159·9	Hg =	200·2
Hydrogen iodide	128·0	127·85	I =	126·85	H =	1·00

From these examples, it will be seen that the smallest weight of element which enters into the molecule of one of the above-mentioned compounds, and which therefore is the *maximum* atomic weight, is as follows (in whole numbers):—

H.	Na.	K.	Zn.	Cd.	Hg.	Tl.	N.	P.	As.	Sb.	Bi.	O.	S.	Se
1	23(?)	39	65	112	200	206	14	31	79	120	208	16	32	79

Te.	F.	Cl.	Br.	I.
125	19	35·5	79	126

while the vapour densities are sometimes equal to these numbers, as in the case of hydrogen, nitrogen, and oxygen (thallium and fluorine) at all temperatures; sometimes equal at low temperatures, and half at high temperatures, as is the case with chlorine, bromine, and iodine; sometimes half at all available temperatures, as with sodium, potassium, zinc, cadmium, and mercury; sometimes twice as great at low temperatures, and becoming equal with rise of temperature, as with phosphorus, arsenic, and antimony; and sometimes a greater multiple than twice, as with sulphur.

It is necessary to postulate the inviolable nature of Avogadro's law that equal volumes of gases, under similar conditions of temperature and pressure, contain the same number of molecules, and the above apparently capricious data become clear. To take an example from each class :—

1. The atomic weight of hydrogen is accepted as unity. The molecular weight of hydrogen, equal to 2, is the standard of molecular weight; and 2 grams of hydrogen inhabit, at 0° and 760 mm. pressure, 22·32 litres. The same volume is inhabited by 28 grams of nitrogen under similar conditions of temperature and pressure. The molecular weight of nitrogen is therefore 28, and 28 is twice 14, the atomic weight. Hence the molecule of free nitrogen contains 2 atoms, and its molecular formula is N_2. It appears to be unaltered by rise of temperature. The same reasoning holds with oxygen, and also with thallium and fluorine, and with chlorine, bromine, and iodine at low temperatures, also with arsenic at 1700°, with sulphur above 1100°, with selenium at 1420°, and with tellurium at 1439°.

2. Considering mercury as an instance of the second class, we find that 200 grams of its vapour inhabit the same space as 2 grams of hydrogen under similar conditions of temperature and pressure. But from the density of its chloride, 283 (as nearly 271 as can be expected from the experimental error), an atom of mercury is seen to be at most 200 times as heavy as an atom of hydrogen, for, on subtracting (2 × 35·4) from 271, the remainder, 200, represents the maximum atomic weight of mercury. But if the molecule of mercury were represented by the formula Hg_2, it should weigh 400 times as much as an atom of hydrogen, or 200 times as much as a molecule. But it weighs only 100 times as much as an equal volume of hydrogen. Hence, we must conclude either that the formula of mercuric chloride should be written Hg_2Cl_2, mercury having the atomic weight 100, or that the molecule of mercury consists of 1 atom, inhabiting the same space as a molecule of hydrogen, which consists of 2 atoms. We shall see that the specific heat of mercury confirms the latter view. So with zinc, cadmium, sodium, and potassium.

3. The justice of this view is borne out by the behaviour of the elements chlorine, bromine, and iodine at high temperatures. With rise of temperature the density diminishes progressively, until with iodine at 1500° the density is 63·7, that is, equal to half the atomic weight of iodine, as deduced from the formula of hydrogen iodide. The molecules of these elements appear to consist of 2 atoms within a wide range of temperature; but, finally, they gradually dissociate into monatomic molecules; and the gradual decrease of density is caused by the gradual transition from diatomic to monatomic molecules. Bismuth, from the one observation available, appears to be undergoing this transition; but not to have reached the final monatomic stage.

4. Lastly, phosphorus at 313° is 64 times as heavy as hydrogen; its molecules are therefore 128 times as heavy as an equal number of atoms of hydrogen. But its maximum atomic weight, from the density of phosphine, PH_3, is 31; hence we must conclude that its molecules are tetratomic. With rise of temperature they tend to become diatomic; but, even at the very high temperature 1708°, the vapour contains many tetratomic molecules. The molecules of arsenic, tetratomic up to 860°, become diatomic at 1700°; and those of antimony have not become wholly diatomic at 1640°. The molecular weight of sulphur is especially anomalous, as shown by its vapour-density. At 464°, it has been found by Biltz as high as 230, implying a molecular formula of about S_8; it decreases in density without a halt till at 1100° its molecular formula is S_2; but above that temperature, up to 1719°, no further change occurs. We may, therefore, conclude that at high temperatures its vapour-density shows its molecular complexity to be S_2; leaving undecided the precise molecular complexity at low temperatures. With selenium and tellurium there is evidence of similar change, but over a much smaller range of temperature; these substances become gaseous at temperatures so high that the more complex molecules are already decomposed.

In the last two classes, we have phenomena similar to those observed during the dissociation of a compound; but, for example, while phosphorus pentachloride, PCl_5, dissociates into unlike molecules, viz., PCl_3 and Cl_2, iodine, I_2, dissociates into like molecules, viz., I and I. There is no reason to suggest different causes for these similar phenomena; and we are, therefore, justified in regarding the vapours of elements, with the few exceptions of sodium, potassium, zinc, cadmium, and mercury, as consisting of complex molecular groups, which become more simple as temperature rises. It is indeed possible, knowing the complexity of two molecular states, to calculate the proportion of dissociated and undissociated molecular groups by means of the formula

$$p = \frac{100(d - D)}{D},$$

where p represents the number of molecules decomposed per 100 undecomposed molecules originally present, d, the theoretical density of the undecomposed substance; and D, the found density of the partially dissociated gas. The rate of increase of dissociation with rise of temperature and fall of pressure may thus be followed. But the end result alone concerns us here.

(*b.*) **Specific heats of elements and compounds.**—A determination of the specific heats of elements furnishes an arbitrary law, discovered by **Dulong and Petit*** in 1819, which has been already explained on p. 126, viz., that "**the atoms of all elements in the solid state have equal capacity for heat;**" the specific heats of elements are therefore inversely proportional to their atomic weights, and, as the theoretical specific heat of solid hydrogen is apparently 6·0, compared with water as unity (see pp. 128 and 576), the approximate atomic weights of the elements may be ascertained by dividing that number by the found specific heat.

Attention has already been drawn to the exceptions to this law in the case of beryllium, boron, carbon, and silicon (see p. 155), and to the fact that at high temperatures their atomic heats approximate to those of other elements at lower temperatures; but it is not so well known that many other elements show similar, if not so great, deviations. The following table shows the specific heats of some elements at 50° and at 300°.†

Element.	Specific heat at 50°.	Atomic heat.	Specific heat at 300°.	Atomic heat.‡
Cadmium......	0·0551	6·18	0·0617	6·92
Zinc..........	0·0929	6 08	0·1040	6·81
Iron..........	0·1113	6·23	0·1376	7·71
Silver.........	0·0556	6·00	0·0609	6·57
Copper........	0·0932	5·90	0·0985	6·24
Nickel........	0·1090	6·43	0·1327	7·83
Antimony.....	0·0495	5·95	0·0537	6·46
Lead..........	0·0304	6·29	0·0338	7·00
Aluminium....	0·2164	5·87	0·2401	6·51

* *Annales*, **10**, 395.

† *Gazzetta*, **18**, 13; also *Chem. Soc.*, **54**, 1237.

‡ The atomic heat is the product of specific heat into atomic weight, *i.e.*, the heat in calories required to raise the temperature of the atomic weight of an element taken in grams through 1°.

The increase is very remarkable, amounting in the case of iron to nearly 25 per cent. of its amount at 50°; and it is also to be noticed that it is not of equal rate for all the elements investigated. It must be remembered that the specific heat of solids, as we determine it, is the sum of very different actions; first, the temperature of the body is increased; second, internal work, due to the separation of molecules and increasing their rate of motion, is done; and, third, external work, due to the expansion of the body, forms a portion of that for which heat is expended. The last, however, is so small that it may be neglected. We cannot, therefore, attempt to explain the true nature of the specific heat of solids, and we must therefore accept Dulong and Petit's law as an empirical statement of facts, which has, however, proved of great service to chemical theory.

In 1831, **Neumann** extended Dulong and Petit's law to compounds. His statement is*:—"**The specific heats of similar compounds is inversely as their molecular weights**:" or, otherwise expressed, "**the molecules of similar compounds have equal capacities for heat.**" As examples, the following instances may be quoted:—

Compound.	Molecular weight.	Specific heat.	Product.	Calculated specific heat.
$CaCO_3$	100·08	0·2044	20·46	0·2057
$MgCO_3$	94·30	0·2161	20·38	0·2211
$ZnCO_3$	125·3	0·1712	21·45	0·1669
$BaSO_4$	233·0	0·1068	24·88	0·1061
$CaSO_4$	136·14	0·1854	25·24	0·1804
$SrSO_4$	183·5	0·130	23·85	0·1346
MgO	40·3	0·276	11·12	0·270
HgO	216·2	0·049	10·59	0·051
ZnO	81·3	0·132	10·73	0·138
HgS	232·3	0·052	12·08	0·052
PbS	239·0	0·053	12·67	0·051
ZnS	97·36	0·112	10·90	0·125

Neumann's extension of Dulong and Petit's law was confirmed by **Regnault,** in 1840; and **Kopp,** in 1864,† made numerous determinations of the specific heats of compounds, which led to the conclusion that the atomic heat of chlorine in its compounds varies from 6·1 in some double chlorides to 6·4 in chlorides such

* *Pogg. Ann.*, **23**, 1.
† *Annalen*, Suppl., **3**, 1 and 289.

as MCl; of bromine, from 6·5 to 6·9; of iodine from 6·5 to 6·7. In the case of compounds containing oxygen, however, the specific heat of oxygen, deduced by subtracting the known specific heat of the element in the compound from the molecular heat of the compound, is in general about 4; similarly combined, hydrogen has the approximate specific heat 2·3; carbon, 1·8; boron, 2·7; silicon, 3·8; and phosphorus and sulphur, each 5·4.

If it is required to calculate the molecular heat of such a compound as ferric oxide, Fe_2O_3, we have—

Atomic heat of iron, 6·16; mean atomic heat of oxygen, 4·0;

hence $(6{\cdot}16 \times 2) + (4 \times 3) = 24{\cdot}32$; found, 25·1 (Kopp).

It must be noticed here that a determination of the specific heat of a compound throws no light on its molecular weight. For the molecular heat is the product of specific heat and molecular weight, and this product evidently depends on the particular molecular weight chosen. Thus in the above example the molecular weight has been assumed to correspond to the formula Fe_2O_3; had we assumed the formula as Fe_4O_6, which may be true, the molecular heat would have been doubled.

By means of these laws, the atomic weights of elements can be deduced with great probability. First, analysis of a compound furnishes the equivalent of the element; second, the vapour density of a compound of the element reveals the maximum number for its atomic weight; third, the specific heat of the element shows whether its true atomic weight is this maximum number, or some fraction thereof.

(*c.*) **Replacement.**—The law of **replacement** is also adduced as a means of determining formulæ, and, taken in conjunction with the methods previously described, it often furnishes a valuable aid. It may be best understood by a concrete instance. It is argued that ethane, C_2H_6, contains six atoms of hydrogen, because it is possible to replace them by sixths by chlorine; the series of compounds, C_2H_6, C_2H_5Cl, $C_2H_4Cl_2$, $C_2H_3Cl_3$, $C_2H_2Cl_4$, C_2HCl_5, and C_2Cl_6, is known. Selenium tetrachloride is regarded as containing four atoms of chlorine, because they are replaceable by fourths; the series $SeClBr_3$, $SeCl_3Br$, and $SeBr_4$ being known. It is by this means that the basicity of acids is usually ascertained; thus sulphuric acid, H_2SO_4, is generally taken as dibasic, because its hydrogen may be easily replaced by halves; but here the existence of such sulphates as $H_3Na(SO_4)_2$ and $NaK_3(SO_4)_2$ would lead to the inference that the molecule of sulphuric acid is expressible by the more complex formula $H_4S_2O_8$, while the definite crystalline compounds, $H_2K_4(SO_4)_3$ and $NaK_5(SO_4)_3$, would cause us to regard the molecule of sulphuric acid as $H_6S_3O_{12}$. It must be remembered that the molecular weight of gaseous sulphuric acid is unknown, and that to deduce its molecular complexity from the apparently analogous compound, SO_2Cl_2, which has undoubtedly that formula in the gaseous state, is not permissible.

(*d.*) **Isomorphism.**—The law of **isomorphism** also furnishes data whereby atomic weights may sometimes be deduced. As stated by **Mitscherlich**, its discoverer, in 1818, it is, "**substances of similar chemical constitution possess similar crystalline form.**"* This statement is by no means reversible; it is not true that similar crystalline form implies similar chemical constitution. But if two bodies form "mixed crystals;" that is, if a mixture of solutions of two compounds deposits crystals containing both compounds in indeterminate amount, of similar crystalline form to that which either salt assumes when pure, they may be taken to possess similar constitution. The following is a list of elements which, as a rule, replace each other in such a manner, and which are therefore said to form isomorphous compounds.

I. F, Cl, Br, I; Mn (in permanganates).
II. S, Se; Te (in tellurides); Cr, Mn, Fe (in chromates, manganates, and ferrates); As and Sb in arsenides and antimonides of the formula MR_2.
III. As, Sb, Bi; Te (as element); P, V (in salts); N, P in ammonium and phosphonium compounds.
IV. Li, Na, K, Rb, Cs; Tl, Ag.
V. Sr, Ba, Pb, Cu; Mg, Zn, Fe, Mn; Ni, Co, Cu; Ce, La, Di, Er, and Y with Ca; Cu, Hg with Pb; Be, Cd, In with Zn; Tl with Pb.
VI. Al, Cr, Mn, Fe; Ce, U in sesquioxides.
VII. Cu, Ag in cuprous and argentous oxides; Au.
VIII. Rh, Ru, Pd, Os, Ir, Pt; Fe, Ni, Au; Sn, Te.
IX. C, Si, Ti, Zr, Sn, Th; Ti, Fe.
X. Nb, Ta.
XI. Cr, Mo, W.

Those elements separated from the others by a semicolon display only partial isomorphism.

As an application of isomorphism to the determination of an atomic weight, the case of gallium may be cited. Before this constant had been determined by the vapour density of its chloride, or by the specific heat of the element,† it was found that the 69 parts by weight of gallium replaced 27 parts by weight of

* *Annales*, **14**, 172, and **19**, 350.

† Lecoq de Boisbaudran, *Compt. rend.*, **83**, 824; Berthelot, *ibid.*, **86**, 786.

aluminium, in a gallium alum, crystallising in the usual form of alums, the octahedron, with twelve molecules of water. Its equivalent had been found from the analysis of its chloride to be 23, approximately. Knowing that the atomic weight of aluminium is three times its equivalent, the conclusion was drawn that gallium also is a triad element in such compounds, and that its atomic weight is $23 \times 3 = 69$.

Further reference to this principle will be made in treating of the periodic arrangement of the elements.

(*e.*) The method devised by **Lecoq de Boisbaudran**, and described in the previous chapter, should not be omitted in stating the methods of determining atomic weights.

(*f.*) Lastly, the atomic weight may be deduced from the **position of the element in the periodic table,** allocated to it by a consideration of the nature of its compounds. This is discussed on p. 639.

The complexity of the molecules of those substances which cannot be obtained in the gaseous state has been fully considered by **Henry.*** He discusses the chlorides and oxides, but the arguments which he deduces in favour of molecular complexity would apply to other compounds.

Let us contrast first the volatility of chlorides and oxides; some instances are given in the following table :—

Volatile oxides.			Non-volatile oxides.		
Compound.	Mol. wt.	Boiling-points.	Compound.	Mol. wt.	Boiling-points.
O.O	32	−186°	B_2O_3	$n \times 70$	—
OCl_2......	87	+ 6°	BCl_3.....	117	17°
SO_2	64	− 8°	SiO_2.....	$n \times 60$	—
$SOCl_2$.....	119	+ 82°	$SiCl_4$.....	170	59°
SO_3	80	+ 46°	TiO_2	$n \times 82$	—
SO_2Cl_2....	135	+ 77°	$TiCl_4$	192	135°
CO_2.......	44	− 79°	Nb_2O_5 ...	$n \times 174$	—
$COCl_2$.....	99	+ 8°	$NbCl_5$ ...	271·5	240·5°
CCl_4......	154	+ 76°	Cr_2O_3....	$n \times 153$	—
C_2Cl_4O....	182	+118°	$CrCl_3$ ...	159	volatile
C_2Cl_6	237	+182°	As_4O_6 ...	396	200°
PCl_3O.....	153·5	+110°	$AsCl_3$....	181·5	134°
PCl_5......	208·5	+148°†	Sb_4O_6 ...	564	1500°
WCl_4O ...	342	+227°	$SbCl_3$....	228·5	225°
WCl_6	395	+346°			

* *Phil. Mag.*, Aug., 1885.

† Decomposes into $PCl_3 + Cl_2$.

Such a table might be greatly extended; but it suffices to show that while the oxides in the first column have invariably lower boiling-points than the corresponding chlorides, those in the second column are, as a rule, non-volatile, while the chlorides are volatile and have simple molecular formulæ, as shown by their vapour-densities. Now it is noticeable that the volatility of a halide depends on the atomic weight of the halogen it contains. The chlorides volatilise at lower temperatures than the bromides, and the bromides at a lower temperature than the iodides. It may therefore be expected that the oxides, containing the still more volatile element oxygen, should have lower volatilising-points than the chlorides. That this is the case, when the substance exists in a simple molecular state, is proved by the first column. The high volatilising-points of the oxides of arsenic and antimony, the molecular weights of which are known to correspond to the formulæ As_4O_6 and Sb_4O_6, shows that they are connected with increased molecular complexity. It is a fair inference, therefore, that the non-volatility of many of the oxides is due to their complex molecules. Examples of the same kind may be found in great number among the compounds of carbon.

The progessive nature of the dehydration of hydrated oxides points to the same conclusion. Thus boracic acid—

Dried at ordinary temperature, has the formula..	H_3BO_3;
Dried at 100°, ...	$H_2B_2O_4$;
Dried at 160°, ...	$H_2B_4O_7$;
Dried at 270°, ...	$H_2B_{16}O_{25}$.

It appears to be a legitimate conclusion that the anhydrous oxide is $(B_2O_3)_n$, where n is a high number.

Similar arguments may be adduced from the density of the oxides compared with that of the chlorides.

It is also noticeable that one oxide forms double compounds with others; and it is a fair inference to draw that, in default of a different oxide, it combines with itself.

The fluorides occupy an abnornal position as regards the chlorides; here also their volatility is as a rule not so great; and they form more numerous and more stable double compounds than the chlorides do. It may therefore be equally well inferred that the molecular weights of the fluorides of most elements are not represented by the simple formulæ which it is customary to ascribe to them.

It would follow from these arguments that the molecular weight of liquid water is not represented by the simple formula H_2O, for it boils at a higher temperature than hydrogen sulphide; a change in the molecular complexity has not been observed in steam at a low temperature; but such a change undoubtedly takes place with hydrogen fluoride before the liquid state is reached (see p. 115).

Although we may therefore assume the complexity of most of the oxides, no method has yet been devised whereby the precise value of the molecular weight may in all cases be determined.

The operation of solution appears in many instances to exercise a dissociating action on molecular aggregates. It has been proved experimentally, by **Raoult,*** that **the depression of the freezing-point of a solvent, produced by the presence of dissolved substance, is approximately inversely proportional to the molecular weight of the latter, and directly proportional to its absolute weight.** Raoult's experimental proof has been substantiated by a theory of the nature of matter in solution, devised by **Van't Hoff,†** depending on certain thermodynamical relations which cannot be explained here. Hence a measurement of the depression of the freezing-point of a solution containing a known percentage of dissolved substance may be made to yield data regarding its molecular weight. The method has been applied with great success to the determination of the molecular weights of carbon compounds, dissolved in liquids which are also compounds of carbon; but data derived from the lowering of the freezing-point of water, due to the presence of a dissolved salt, point to the dissociation of the salt into the ions of which it is composed, that is, the products which are obtained from it under the influence of an electric current.

The method of application is as follows:—The observed lowering of freezing-point of 100 grams of water or other solvent, caused by the presence of P per cent. of a dissolved compound, is termed C; Raoult terms the ratio C/P *the co-efficient of lowering of freezing-point* for the dissolved compound. If M be the molecular weight of the compound, the ratio MC/P, or the *lowering of freezing-point per molecule of the dissolved substance*, is constant for compounds of similar nature. The value of this number for compounds which do not dissociate, or, what corresponds therewith, which do not conduct electrolytically, is 19, water being the solvent in each case.

Until this method has been more carefully investigated, it is premature to give an extended statement of results; it may, however, be mentioned that Paterno and Nasini‡ have thereby deter-

* *Annales* (6), **8**, 317. † *Phil. Mag.*, August, 1888.

‡ *Berichte*, **21**, 2153.

mined the molecular weights of sulphur, phosphorus, bromine, and iodine. For sulphur dissolved in benzene, the freezing-point of benzene was found to be depressed about 0·26° for each part per cent. of dissolved sulphur. The theoretical molecular depression of its freezing-point, calculated by Van't Hoff on theoretical grounds, is 53; now $0·26 \times 192 = 50$; but $192 = 6 \times 32$; and it would follow that the molecule of sulphur dissolved in benzene has the formula S_6, a probable conclusion from its vapour-density (see p. 614). Similar experiments with bromine dissolved in water and in acetic acid led to the formula Br_2; for iodine, I_2 mixed with I; and for phosphorus, P_2 mixed with P_4.

The depression in the vapour-pressure of a liquid produced by the presence of dissolved substances was also found by Raoult to be approximately inversely proportional to the molecular weight of the latter; and Van't Hoff has also shown that thermodynamical considerations lead to a similar conclusion (*loc. cit.*). This method has not been so widely applied as that depending on the lowering of freezing-point; but by its help Loeb* has determined the molecular weight of iodine dissolved in ether and in carbon disulphide; his conclusion is that the brown solution in ether contains mostly molecules of I_4, while the violet solution in carbon disulphide contains a large proportion of I_2 molecules. This conclusion, it will be observed, does not agree with that of Paterno and Nasini. Ramsay has also applied this method to determine the molecular weights of some metals, the solvent being mercury; and the evidence is in favour of a monatomic molecular state.† Experiments by Heycock and Neville on the depression of the freezing-point of tin and of sodium used as solvents for metals appear to point to a similar conclusion.‡ From the vapour-densities of these metals, which have been volatilised, the conclusion would seem to be justified. But our knowledge is as yet too immature to allow of positive conclusions.

The monatomic nature of mercury gas.—Before dismissing the subject of molecular weights, very valuable experiments of **Kundt and Warburg**§ must be mentioned, which lend great support to the view that the molecules of mercury consist of single atoms; and, consequently, that molecules of hydrogen, nitrogen, oxygen, chlorine, &c., consist each of two atoms. The argument is as follows:—Assuming that the pressure of a gas on the walls of

* *Chem. Soc.*, **53**, 405.
† *Ibid.*, **55**, 251.
‡ *Ibid.*, **55**, 666.
§ *Pogg. Ann.*, **127**, 497; **135**, 337 and 527.

the vessel containing it is due to the impacts of its molecules on the walls; and that the effect of a rise of temperature of a gas is to increase the number of impacts in unit of time, and hence its pressure, it is possible to calculate the increase of kinetic energy given to a gas by raising its temperature through 1°. The amount of heat required to raise through 1° the molecular weight expressed in grams of any gas has been calculated to be 3·00 calories, provided the gas be not allowed to expand; if it be allowed to expand, it must overcome the pressure of the air; or if it be supposed to be confined in a vertical cylinder with a piston it must lift a column of air through some height; and, in order that it may be able to accomplish this work, more heat must be communicated to it than that which produces merely a rise of temperature. This may be shown to amount to 2·00 calories more per gram molecule of the gas. The first of these quantities of heat, 3·00 calories, represents the molecular heat of the gas at constant volume; the second, 5·00 calories, the molecular heat at constant pressure. The ratio between these numbers is 1 : 1·66. The actual specific heat of mercury vapour has not been determined; but it has been found by Kundt and Warburg that this ratio actually exists between the specific heat of mercury vapour at constant volume and at constant pressure. But with oxygen, hydrogen, nitrogen, carbon monoxide, nitric oxide, and other gases, the molecules of which are presumably diatomic, the molecular heat at constant volume is for O_2, 4·96; for H_2, 4·82; for N_2, 4·82; for CO, 4·86; for NO, 4·95; instead of 3·00, the calculated molecular heat. The specific heat at constant pressure is found by adding 2·00 calories to each of these numbers; thus O_2, 6·96; H_2, 6·82; N_2, 6·82; CO, 6·86; NO, 6·95. The ratio between these numbers is as 1 : 1·41 approximately. Why does a gas such as oxygen require more heat to raise its temperature at constant volume than mercury gas? The answer is, that the heat is not wholly utilised in causing molecular motion, but is partly employed in causing the atoms in the molecule to rotate, or oscillate; while with mercury vapour the monatomic nature of its molecules makes such an expenditure of energy impossible.

Granting that mercury gas consists of monatomic molecules, it follows from the density of mercury gas compared with those of hydrogen and oxygen, &c., that the molecules of the latter consist of two atoms; and from this we can deduce the molecular weights of all bodies obtainable in the gaseous state without decomposition.*

* For a detailed account of this subject, see Clerk-Maxwell's *Theory of*

The structure of compounds has been dealt with as opportunity arose during their classification; and little can be added with advantage to what has already been said. Our knowledge of the structure of carbon compounds, which forms the basis of organic chemistry, is in a much more advanced state than that of compounds of other elements; and further investigation of the compounds of carbon containing other elements besides carbon, hydrogen, and oxygen is likely to shed light on the subject.

Heat, Third Edit., chap. XI; also Naumann's *Thermochemie* (1882), **71** *et seq.*; also Ostwalds' *Allgemeines Chemie*, **1**, 266 (1885).

CHAPTER XXXVI.

THE PERIODIC ARRANGEMENT OF THE ELEMENTS.

THE **relation between the atomic weights of the elements** has, almost since the announcement of the atomic hypothesis by Dalton, been a subject of speculation. The first conjecture, published by **Prout** (1815), and shortly afterwards by **Meinecke** (1817), was that, as it was conceivable that the ancient doctrine of the uniformity of matter was true, the primary material must be hydrogen, and the atomic weights of the other elements should therefore be multiples of that of hydrogen. This hypothesis was warmly advocated by **Thomas Thomson**, in whose text book Dalton's discovery was first formally announced; but **Berzelius** pronounced against it, relying on his own determinations of atomic weights. But in 1842 it was discovered by **Liebig** and **Redtenbacher**, and confirmed by **Dumas** and **Stas**, and by **Erdmann** and **Marchand**, that Berzelius had made an error in his determination of the atomic weight of carbon, and that the correct value was 12; and shortly afterwards **Dumas** determined with great precautions the ratio of the weights of hydrogen and oxygen in water, obtaining the value 16 for oxygen, and by similar work the value 14 for nitrogen; and Prout's hypothesis was accordingly resuscitated, not in its original form, however; but it was supposed that the atomic weights of the elements were multiples of 0·5, half the atomic weight of hydrogen. This change was necessitated by Berzelius's determination of the atomic weight of chlorine, which is approximately 35·5, confirmed by Penny, by Marignac, and by Pelouze; it was ultimately disproved, however, by Stas's determinations of the atomic weights of silver, chlorine, bromine, iodine, potassium, sodium, lithium, sulphur, nitrogen, and lead, which were executed with a precision not to be surpassed.

In 1817 **Döbereiner** pointed out that the atomic weight of strontium is the arithmetical mean of those of calcium and barium. This is not actually the case, but the number 87·5 closely approaches 88·5, the true arithmetical mean. Many similar "triads" exist, as will afterwards be shown; **Zeuner**, in 1857, tried to arrange all the atomic weights then known as "triads."

In 1850 **Pettenkofer** suggested that the atomic weights of similar elements formed arithmetical series. This view was adopted and extended by **Bremers, Gladstone,** and **Dumas.**

The first fruitful attempt to introduce order into the seeming chaos of numbers was due to **Newlands** in 1863 and 1864. It has recently been pointed out that **de Chancourtois,*** Professor of Geology at the École des Mines in Paris, had independently anticipated Newlands by about a year; but his suggestions were encumbered with untenable theories, and met with no attention. Newlands arranged all the elements then known in the order of their atomic weights, and observed that every eighth element, as a general, but not absolute, rule, belonged to the same class, manifesting similar properties. He termed this relation the "**law of octaves.**"

In 1869 **D. Mendeléeff,** Professor at St. Petersburg, and **Lothar Meyer,** now Professor at Tübingen, in Würtemburg, simultaneously published on the subject, both pointing out, independently of the other, that "the properties of the elements are periodic functions of their atomic weights." The methods of representation, though the idea was essentially the same, differed slightly from each other. Meyer's scheme was as follows:†—

Li	Be	B	C	N	O	F.	
Na	Mg	Al	Si	P	S	Cl.	
K	Ca	Sc	Ti	V	Cr	Mn	Fe, Co, Ni.
Cu	Zn	Ga	Ge	As	Se	Br.	
Rb	Sr	Y	Zr	Nb	Mo	—	Ru, Rh, Pd.
Ag	Cd	In	Sn	Sb	Te	I	
Cs	Ba	La, Di	Ce	—	—	—	
—	—	—	—	—	—	—	
—	—	Yb	—	Ta	W	—	Os, Ir, Pt.
Au	Hg	Tl	Pb	Bi	—	—	
—	—	—	Th	—	U		

Mendeléeff's table is somewhat differently constructed, although essentially the same. It may be regarded as Meyer's table turned through a right angle:—

* *Nature*, **41**, 986.

† The table has been given as published in the last edition of his "Modernen Theorien;" see also the translation by Bedson and Williams, 1888. The position of cerium has been altered to the carbon group.

R_2O I..........	Li	K	Rb	Cs	—	—	RCl.
RO II..........	Be	Ca	Sr	Ba	—	—	RCl_2.
R_2O_3 III	B	Sc	Y	La	Yb	—	RCl_3.
RO_2 IV	C	Ti	Zr	Ce	—	Th	RCl_4.
R_2O_5 V (III)....	N	V	Nb	Di	Ta	—	RCl_3RCl_5.
RO_3 VI (II)....	O	Cr	Mo	—	W	U	$RCl_2(RCl_6)$.
R_2O_7 VII (I)....	F	Mn	—	—	—	—	$RCl(RCl_7)$.
RO_4 VIII		Fe, Co, Ni	Ru, Rh, Pd		Os, Ir, Pt		
R_2O I..........	Na	Cu	Ag	—	Au	—	RCl.
RO II..........	Mg	Zn	Cd	—	Hg		RCl_2.
R_5O_3 III	Al	Ga	In	—	Tl	—	RCl_3.
RO_2 IV	Si	Ge	Sn	—	Pb	—	RCl_4.
R_2O_5 V (III)....	P	As	Sb	—	Bi	—	RCl_3,RCl_5.
RO_3 VI (II)	S	Se	Te	—	—	—	$RCl_2,(RCl_6)$.
R_2O_7 VIII (I)...	Cl	Br	I	—	—	—	$RCl.(RCl_7)$.

While in L. Meyer's table the alternate elements show analogy with each other, in Mendeléeff's table the elements are divided into two classes.

We shall consider: 1, **the numerical relations** of the numbers expressing atomic weights; 2, some of the **physical properties of the elements and their compounds,** varying with atomic weight; 3, **a comparison of** the formulæ of compounds of the elements; and 4, **the fulfilment of predictions** of the atomic weights and properties of undiscovered elements, and the changes in recognised atomic weights, due to the periodic arrangement.

1. **Numerical relations.**—These are best seen by reference to Lothar Meyer's arrangement.

(*a.*) *Vertical columns.*—It is to be noted that the atomic weights of the elements of the Na, Mg, Al, Si, P, S, and Cl columns are nearly the mean of those of the nearest elements of the Li, Be, B, C, N, O, and F columns. Thus, for example, the mean of the atomic weights of Li and $K = \frac{1}{2}(7{\cdot}03 + 39{\cdot}14) = 23{\cdot}08$; the atomic weight of sodium is 23·06. Calculating in this manner, we have the following numbers:—

	Na.	Mg.	Al.	Si.	P.	S.	Cl.
Calculated	23·08	24·5	27·5	30·0	32·6	34·2	37·0
Found	23·06	24·4	27·1	28·4	31·0	32·1	35·5
Difference	+0·02	+0·1	+0·4	+1·6	+1·6	+2·1	+2·5

Here there is a constantly increasing difference.

	Cu.	Zn.	Ga.	Ge.	As.	Se.	Br.
Calculated	62·3	63·7	66·4	69·4	72·7	74·1	—
Found	63·3	65·5	69·9	72·3	75·0	79·1	—
Difference	−1·0	−1·8	−3·5	−2·9	−2·3	−5·0	—

We note that the negative difference shows signs of increase; but it is irregular.

	Ag.	Cd.	In.	Sn.
Calculated.......	109·15	112·2	113·6	115·4
Found..........	107·94	112·1	113·7	118·1
Difference.......	+1·21	+0·1	−0·1	−2·7

There are no data for further calculations; it is to be noticed, however, that the difference is, to begin with, a positive one, becoming negative. It would be possible, however, to calculate, with some probability of correctness, the atomic weight of the element succeeding niobium, in the nitrogen group, in the following manner:—The difference between the calculated and the found values of silicon and phosphorus is, in each case, + 1·6; that between the calculated and found values of germanium is −2·9, and of arsenic −2·3; it is, therefore, to be expected that the difference will be approximately −2·7 between the calculated and found values of antimony, for it is approximately the same for the members of the fourth and fifth groups. The atomic weight of antimony is 120·3; subtracting from it 2·7, we have 117·6; multiplying by 2, we obtain 235·2, as the sum of the atomic weights of niobium and the element immediately following it, and preceding tantalum. Subtracting from 235·2 the atomic weight of niobium, 94·2, we obtain 141·0 as the probable number for the element "eka-niobium." Whether this is identical with neodymium, one of the products into which the element didymium was split up by Auer von Welsbach in 1885,* and to which he assigns the atomic weight 140·8, time must determine. The properties of its compounds, so far as they have been investigated, hardly lend support to the view; but it is probable that it has not yet been obtained in a state of sufficient purity to allow of definite conclusions.

Another method of averaging may be tried for the atomic weights of elements of the Li, Be, B, C, N, O, and F groups. It is seen at once that the atomic weight of calcium, for example, which is 40·0, differs greatly from the mean atomic weights of magnesium, 24·4, and zinc, 65·5 = 44·9. But a closer approximation may be obtained in some instances by taking the average of the atomic weights of the two nearest elements in the same vertical row. We thus obtain the "triads," to which attention was directed by Döbereiner. This method is not available for

* *Monatsh. f. Chem.*, 6, 477. The author has been verbally informed by C. M. Thompson, that he has confirmed von Welsbach's work.

potassium, calcium, scandium, &c.; for example, Li 7·03 + Rb 85·4 = 92·43, and ½(92·43) = 46·21, a number very far from K = 39·14. But Rb = 85·4 is approximately the mean of K = 39·14 and Cs = 132·9, viz., 86·02; Sr = 87·5 is nearly the mean of Ca = 40·0 and Ba = 137·0, viz., 88·5; Y = 88·7 does not greatly differ from ½(Sc = 44·1 + La = 138·5) = 91·3; and Zr 90·7 may be compared with ½(Ti = 48·1 + Ce = 140·2) = 94·1. The difference, it will be noticed, is an increasing one.

Applying the same method to Cu, Zn, Ga, Ge, As, Se, and Br, *i.e.*, Cu = ½(Na + Ag), Zn = ½(Mg + Cd), &c., we obtain:—

	Cu.	Zn.	Ga.	Ge.	As.	Se.	Br.
Calculated....	65·5	68·2	70·4	73·2	75·7	78·5	81·15
Found.......	63·3	65·5	69·9	72·3	75·0	79·1	79·96
Difference....	+2·2	+2·7	+0·5	+0·9	+0·7	−0·6	+1·19

Here, too, the approximation is fair, but the differences are irregular.

(*b.*) Horizontal rows.—The question here is as regards the position of the atomic weight of an element with respect to those immediately preceding and succeeding it in numerical order. The comparison is as follows:—

	Be.	B.	C.	N.	O.	F.
Calculated......	9·02	10·55	12·52	14·00	16·5	19·5
Found	9·10	11·01	12·00	14·04	16·0	19·0
Difference......	−0·08	−0·46	+0·52	−0·04	+0·5	+0·5
	Mg.	**Al.**	**Si.**	**P.**	**S.**	**Cl.**
Calculated......	25·08	26·39	29·06	30·23	33·24	35·60
Found.........	24·38	27·1	28·4	31·03	32·06	35·45
Difference......	+0·70	−0·72	+0·66	−0·80	+1·18	+0·15
	Ca.	**Sc.**	**Ti.**	**V.**	**Cr.**	
Calculated......	41·6	44·0	47·6	50·2	53·1	
Found.........	40·0	44·1	48·1	51·2	52·3	
Difference......	+1·6	−0·1	−0·5	−1·0	+0·8	
	Zn.	**Ga.**	**Ge.**	**As.**	**Se.**	
Calculated......	66·6	68·9	72·4	75·7	77·5	
Found.........	65·5	69·9	72·3	75·0	79·1	
Difference......	+1·1	−1·0	+0·15	+0·7	−1·6	
	Sr.	**Y.**	**Zr.**	**Nb.**	**Mo.**	**?.***
Calculated......	87·0	89·1	91·4	93·3	—	99·6
Found.........	87·5	88·7	90·7	94·2	—	—
Difference......	−0·5	+0·4	+0·7	−0·9	—	—

* This atomic weight has been calculated on the assumption that the average number would be 1·0 too small.

	Cd.	In.	Sn.	Sb.	Te.
Calculated......	110·8	115·1	117·0	121·5	123·6
Found..........	112·1	113·7	118·1	120·3	125·0
Difference......	−1·32	+1·4	−1·1	+1·2	−1·4

	Ba.	La.	Hg.	Tl.	Pb.
Calculated......	135·7	138·6	200·6	203·6	206·0
Found..........	137·0	138·5	200·4	204·1	206·9
Difference......	−1·3	+0·1	+0·2	−0·5	−0·9

It has not been thought permissible to take the mean of the second last member of one row and the first member of the next, so as to obtain the atomic weight of the last member; hence, the atomic weights of the first and last members of the rows are, as a rule, omitted. The results, however, show that, although it is possible to approximate to the atomic weights of unknown elements, the numbers are too irregular to allow of accurate prediction.

Various attempts have been made to devise some scheme which should reduce these numbers to order. One which holds out hopes of ultimate success is due to G. Johnstone Stoney.* Until more accurate numbers have been obtained in all cases, there is little to be done. It may be stated, however, that the variations from a mean curve passing among points representing atomic weights appear themselves to vary periodically.

To discover the true relations between these numbers must remain one of the problems of chemistry; and its discovery will, in all probability, prove a key to many problems at present unsolved.

As an example of attempts which have frequently been made to assign some cause for the periodic arrangement, a recent note by A. M. Stapley will be here quoted. His table is given on the next page.†

These results are obtained by taking (as a rule) the first member of a column, and adding to it 16, or multiples thereof, for elements on the left, and the product of the first member by 2, adding to it 16, or multiples thereof, for the elements on the right of each column. There is, generally speaking, a certain concordance between the numbers obtained and the true atomic weights; but there are many wide discrepancies, especially in the row containing Os, Ir, Pt, Au, &c. The author

* "The logarithmic law of atomic weights," *Chem. Soc. Abstr.*, 1888–89, 55.

† *Nature*, **41**, 57.

VII.	VIII.			I.
H . 1.				
R^i = 3.	R^{ii} = 4	R^{iii} = 5	R^{iv} = 6	Li 7
3 + 16 F.	—	—	—	7 + 16 Na
Cl 3 + 32	—	—	—	K 7 + 32
6 + 48 Mn.	8 + 48 Fe	10 + 48 Co	12 + 48 Ni	14 + 48 Cu
Br 3 + 80	—	—	—	Rb 7 + 80
6 + 96?	8 + 96 Ru	10 + 96 Rh	12 + 96 Pd	14 + 96 Ag
I 3 + 128	—	—	—	Cs 7 + 128
6 + 144?	—	—	—	—
?3 + 176	—	—	—	—
6 + 192?	8 + 192 Os	10 + 192 Ir	12 + 192 Pt	14 + 192 Au
?3 + 224	—	—	—	—

II.	III.	IV.	V.	VI.
Be 9	B 11	C 12	N 14	O 16
9 + 16 Mg	11 + 16 Al	12 + 16 C	14 + 16 P	16 + 16 S
Ca 9 + 32	Sc 11 + 32	K 12 + 32	V 14 + 32	Cr 16 + 32
18 + 48 Zn	22 + 48 Ga	12 + 48 Ge	14 + 48 As	32 + 48 Se
Sr 9 + 80	Y 11 + 80	Zr 12 + 80	Nb 14 + 80	Mo 16 + 80
18 + 96 Cd	22 + 96 In	24 + 96 Sn	28 + 96 Sb	32 + 96 Te
Ba 9 + 128	La 11 + 128	Ce 12 + 128	Di 14 + 128	—
—	22 + 144 Er	—	—	—
—	—	—	—	W 16 + 176
18 + 192 Hg	22 + 192 Tl	24 + 192 Pb	28 + 92 Bi	—
—	—	Th 12 + 224	—	U 16 + 224

of this suggestion compares these figures with those of the oxides, *i.e.*, MO, MO_2, M_2O_3, MO_5, &c. This has no real significance, but only typifies the fact that 16 is an average difference. The scheme has little to recommend it, and is adduced merely as a sample of numerous attempts of the kind.

2. **Physical properties.**—(*a.*) **Volumes** and their connection with atomic weights.—The specific volume of a substance expressed in units is the volume of one gram; the specific gravity is the weight of one cubic centimetre; and it is obvious that the one is reciprocal to the other. The atomic weight, multiplied by the specific volume, or divided by the specific gravity, gives the atomic volume of an element; that is, the number of cubic centimetres occupied by its atomic weight expressed in grams. If it were certain that the space between the atoms of liquids and solids were equal in all cases, the resulting numbers would give the comparative volumes of the atoms; but, as this is probably not the case, the numbers represent merely the volume

of the atom plus the space it inhabits. That such constants bear definite relations to the periodic arrangement of the elements is seen from the following table:—*

	1.	2.	3.	4.	5.	6.
1	Li = 11·9	K = 45·5	Rb = 56·3	Cs = 70·6	—	—
2	Be = 4·3	Ca = 25·3	Sr = 34·5	Ba = 36·5	—	
3	B = 4·1	Sc = ?	Y = ?	La = 22·9	—	—
4	C = 3·4	Ti = ?	Zr = 21·9	Ce = 21·0	—	Th = 29·9
5	N —	V = 9·3	Nb = 14·5	—	Ta = 17·0	—
6	O —	Cr = 7·7	Mo = 11·1	—	W = 9·6	U = 13·0
7	F —	Mn = 7·4	—		—	—
8	—	Fe = 6·6	Ru = 9·2	—	Os = 8·9	—
		Co = 6·7	Rh = 9·5	—	Ir = 8·6	—
		Ni = 6·7	Pd = 9·3	—	Pt = 9·1	—
1	Na = 23·7	Cu = 7·2	Ag = 10·3	—	Au = 10·2	—
2	Mg = 13·3	Zn = 9 5	Cd = 13·0	—	Hg = 14·8	—
3	Al = 10·1	Ga = 11·8	In = 15·3	—	Tl = 17·2	—
4	Si = 11·3	Ge = 13·2	Sn = 16·1	—	Pb = 18·2	—
5	P = 13·5	As = 13·3	Sb = 17·9	—	Bi = 21·2	—
6	S = 15·7	Se = 18·5	Te = 20·3	—	—	
7	Cl = 25·6	Br = 25·1	I = 25·7	—		

Similar regularities are observable in the molecular volumes of the oxides;† those of the lithium and sodium groups are taken as M_2O; those of the beryllium and magnesium groups as M_2O_2; those of the boron and aluminium groups as M_2O_3; of the carbon and silicon groups as M_2O_4; of the nitrogen and phosphorus groups as M_2O_5; and of the chromium and sulphur groups as M_2O_6. The volumes are those of oxides supposed to contain 1 atom of element.

	1.		2.		3.		4.		5.		6.	
1	Li_2O	7	K_2O	17	Rb_2O	21	Cs_2O	25	—		—	
2	Be_2O_2	7	Ca_2O_2	18	Sr_2O_2	22	Ba_2O_2	28	—		—	
3	B_2O_3	19	Sc_2O_3	18	Y_2O_3	23	La_2O_3	25	—		—	
4	C_2O_4	46	Ti_2O_4	20	Zr_2O_4	25	Cl_2O_4	26	—		Th_2O_4	27
5	N	—	V_2O_5	26	Nb_2O_5	30	—		Ta_2O_5	31	—	
6	O	—	Cr_2O_6	37	Mo_2O_6	33	—		W_2O_6	37	U_2O_6	56
7	F	—	Mn	—	—	—	—		—		—	
1	Na_2O	11	Cu_2O	12	Ag_2O	14	—		Au_2O	18	—	
2	Mg_2O_2	12	Zn_2O_2	14	Cd_2O_2	16	—		Hg_2O_2	19	—	
3	Al_2O_3	13	Ga_2O_3	17	In_2O_3	19	—		Tl_2O_3	23	—	
4	Si_2O_4	23	Ge_2O_4	22	Sn_2O_4	22	—		Pb_2O_4	27	—	
5	P_2O_5	30	As_2O_5	31	Sb_2O_5	42	—		—		—	
6	S_2O_6	41	Se	—	Te	—	—		Bi_2O_5	42	—	
7	—	—	Br	—	I	—	—		—		—	

* Somewhat altered from that given by L. Meyer, *Annalen, Suppl.*, **7**, 354.

† Brauner and Watts, *Berichte*, **14**, 48.

(*b.*) **Melting-points.***—These, so far as they are known, are on the absolute scale as follows:—

	1.	2.	3.	4.	5.	6.
1	Li 453	K 335	Rb 311	Cs 300	—	—
2	Be 1230+	Ca +	Sr +	Ba 748	—	
3	B + +	Se ?	Y ?	La + +	—	—
4	C ∞	Ti ∞	Zr + +	Ce + +	—	Th ?
5	N 59	V ∞	Nb ∞	—	Ta ∞	—
6	O 59—	Cr 2270+	Mo ∞	—	W + +	U 2008?
7	F —	Mn 2170	— —	—	—	—
8	-	Fe 2080 Ni 2070 CO 1870	Ru 2070 Rh 2270 Pd 1775	—	Os 2770 Ir 2220 Pt 2050	—
1	Na 369	Cu 1330	Ag 1230	—	Au 1310	
2	Mg 1023	Zn 690	Cd 590	—	Hg 234	
3	Al 1123	Ga 303	In 449	—	Tl 563	
4	Si + +	Ge 1173	Sn 503	—	Pb 599	
5	P 528	As 773+	Sb 710	—	Bi 540	
6	S 388	Se 490	Tl 728	—	—	
7	Cl 198	Br 266	I 387	—	—	

The signs + and — affixed to a number signify that the melting-point is above or below the number given. The sign + standing alone means that the melting-point is higher than the one immediately above. The sign ∞ means that the element has not been melted. Similar relations have been traced for the halides and ethides by Carnelley† for boiling-points, showing similar regularity.

The periodic arrangement also shows analogies between the refraction-equivalents and the conductivity for heat and electricity of the elements, and also heats of formation of halides, oxides, &c., which lack of space will not allow us to discuss here. Enough has been given, however, to fully justify the statement that the properties of elements are functions of their atomic weights.

3. **Comparison of the elements and their compounds.**—These must be very summarily discussed. Beginning with halides, it is to be noticed that elements of the lithium group form only halides of the formula MX; while sodium resembles them in this respect. The bodies are not soluble in, and do not react with, water; or, to speak more correctly, water may be expelled from their solutions without decomposing them. Of the sodium group, the elements copper, silver, and gold more closely re-

* Lothar Meyer.

† *Phil. Mag.*; July, 1884; Sept., 1885. *Chem News,* Nos. 1375–1378.

semble those of the iron, palladium, and platinum groups than they do sodium. Their monohalides are insoluble; and they form soluble dihalides, which connect them with the elements of the magnesium group; while gold forms trihalides. The halides of the beryllium group are, without exception, dihalides, as are also those of the magnesium group, with exception of mercury, which harks back, as it were, to the sodium group, forming monohalides closely resembling those of copper, silver, and gold. Trihalides of elements of the boron group are the only ones known; the elements of the aluminium group, also, all form trihalides; but a dihalide of aluminium is known in combination; also a dihalide of gallium; and indium forms both a di- and a monohalide. The dihalides are wanting with thallium, but the monohalides are the more stable compounds. The elements of the carbon group form only tetrahalides; the compounds such as M_2X_6, which might pass for trihalides, are, without doubt, combinations in which two atoms of the element M are conjoined; cerium, however, may form a genuine trihalide. The silicon-group of elements also forms tetrahalides, those of lead being unstable; while silicon, germanium, and tin also form dihalides.

Elements of the nitrogen group, nitrogen itself excepted, form more than one halide; those of vanadium being specially numerous; but, while niobium forms two, tantalum forms only one. This order is reversed with the phosphorus group; for, while phosphorus is capable of uniting with as many as eleven atoms of halogen, but not with less than three, bismuth appears incapable of combining with more than three atoms of a halogen, and also forms a dihalide. The elements chromium, manganese, iron, cobalt, and nickel form halides much more closely resembling each other than the halides of chromium and manganese resemble those of the oxygen and fluorine groups; while molybdenum, tungsten, and uranium are again noted for the great number of their halogen compounds. Sulphur, selenium, and tellurium appear to combine with two and with four atoms of halogen; although the compounds are not all represented. Iodine forms a mono- and a tri-chloride. The halides of the iron and of the palladium groups correspond to the formulæ MX_2, MX_3, and MX_4; and those of the platinum group are similar.

The capacity for forming double halides appears to increase, as a rule, with rise in atomic weight; but it must be remarked that the investigation of these bodies is closely connected with their stability in presence of water; few compounds having been isolated which are unable to resist the action of that agent.

Taking next in order the alkides, for example, the ethides, we have as the only representative of the lithium group the double ethide of zinc and potassium, $K(C_2H_5).Zn(C_2H_5)_2$, and of the sodium group, the corresponding compound $Na(C_2H_5).Zn(C_2H_5)_2$, corresponding with the halides of sodium and potassium. The beryllium group is represented by $Be(C_2H_5)_2$, corresponding to the halides; and we have next $Mg(C_2H_5)_2$ and $Zn(C_2H_5)_2$, representing the dihalides. The mercury compound, $Hg(C_2H_5)_2$, also corresponds to its dihalide; of the boron-group, $B(C_2H_5)_3$ is the only ethide known; but with aluminium the two compounds, $Al(C_2H_5)_3$ and $Al_2(C_2H_5)_6$, have been prepared; the theoretical consequences of the existence of these bodies have been alluded to on p. 504. With silicon, $Si(C_2H_5)_4$ and $Si_2(C_2H_5)_6$ are known; with tin, $Sn(C_2H_5)_4$, $Sn_2(C_2H_5)_6$, and $Sn_2(C_2H_5)_4$; and with lead, $Pb(C_2H_5)_4$ and $Pb_2(C_2H_5)_6$.

Of the nitrogen group, we have $N(C_2H_5)_3$ and $N_2(C_2H_5)_4$; and of the phosphorus group, $P(C_2H_5)_3$ and $P_2(C_2H_5)_4$; also $As(C_2H_5)_5$, $As(C_2H_5)_3$, and $As_2(C_2H_5)_4$. Similarly, with antimony, the compound $Sb(C_2H_5)_3$ is known.

The hydroxides in general correspond to the halides; but it may be pointed out again here that $C(OH)_4$, $P(OH)_5$, and $S(OH)_6$ are unstable, and hence unknown, although their derivatives $CO(OH)_2$, $PO(OH)_3$, and $SO_2(OH)_2$ lead us to infer their possibility.

The oxides, sulphides, selenides, and tellurides are much more numerous and complex. For with elements of the lithium and sodium groups we have not merely what we may term normal oxides, such as Li_2O and Na_2O, but also such bodies as K_2O_2, K_2O_4, K_2S_5, &c. If we include copper, silver, and gold, we have also M_2O, MO, M_2O_3, and MO_2; but it is doubtful whether these elements are not really members of the iron, palladium, and platinum groups. With the beryllium group, we have CaO_2, SrO_2, and BaO_2, besides BaS, BaS_2 BaS_3, SrS_4, and BaS_5. In the magnesium group, however, the existence of dioxides is doubtful; but its last member, mercury, forms an oxide, Hg_2O, corresponding to its monochloride; and in the boron group the only known oxides are trioxides, except with yttrium and lanthanum, where evidence of higher oxides, Y_4O_9 and La_4O_9 has been obtained. Here, too, we notice a well-marked tendency to form double oxides with those of other elements. The borates, although indefinite, contain the oxide B_2O_3 as their "acidic" constituent. Elements of the aluminium group, thallium excepted, form exclusively sesquioxides and sulphides; aluminium, too, forms many double oxides, among others, the *spinels*. Both a monoxide and monosulphide and sesquioxide and sesquisulphide of thallium are known, and also a dioxide in combination with other oxides, where it plays an "acidic" part.

With the carbon group, the oxides become more numerous. We know those of the general formulæ MO, M_2O_3, MO_2, M_2O_5, MO_3, and M_2O_7, the dioxides, MO_2, exhibiting definite acid characters, and the greatest stability, except with cerium. Silicon and its "homologues" form monoxides and monosulphides, sesquioxides and sesquisulphides, and dioxides and disulphides. Again, those of the formula MR_2 are most stable, and display "acidic" characters.

The elements of the nitrogen group also form numerous oxides and sulphides. We know M_2O, MO, M_2O_3, MO_2, M_2O_5, and MO_3. Those of the formula M_2O_5 are best characterised, and their compounds are most stable. The closely analogous phosphorus group is characterised by compounds of the formulæ P_4O, P_4S_3, BiO, P_4O_6, P_2O_4, P_2O_5.

The compounds of phosphorus and arsenic of the formulæ P_2O_5 and As_2O_5 are the most stable; those of antimony and bismuth find their most stable stage in Sb_4O_6 and Bi_4O_6.

Of the group of which oxygen is the first member, we have chromium with derivatives such as MO; chromium and molybdenum forming compounds such as Cr_2O_3, Mo_2O_3; chromium, molybdenum, tungsten, and uranium giving dioxides, MO_2, and trioxides, MO_3; while molybdenum forms a tetrasulphide, MoS_4, and uranium a tetroxide, UO_4. Still higher oxides of uranium appear to exist in combination. Those of the type MO_3 appear to be best characterised. Oxides of elements of the sulphur group are not as a rule stable, except in combination; but SO_2, SO_3, SeO_2, TeO_2, and TeO_3 are known. Compounds of S_2O_2, S_2O_3, and S_2O_7, besides others more complex, are known as hyposulphites, thiosulphates, and persulphates. Perhaps the oxides MO_3 may be regarded as most characteristic.

The oxides of manganese are very numerous, resembling those of aluminium and chromium on the one hand, and those of iron, nickel, and cobalt on the other. We are acquainted with MnO, Mn_2O_3, MnO_2, and with MnO_3 and Mn_2O_7 in combination. The most stable oxide is MnO; that characterising the position of manganese in the periodic table is Mn_2O_7. The sulphides are fewer in number.

The oxides of the halogens, taken together, furnish representations of M_2O, M_2O_3, MO_2, M_2O_5, and M_2O_7. The second and fifth are known only in combination. The sulphides range from Cl_4Se to Cl_2S_2. The oxide M_2O_7 is regarded as characteristic.

Proceeding to the iron group, we have iron itself, with oxides FeO, Fe_2O_3, and FeO_3: the last in combination; and the sulphide FeS_2; cobalt, with CoO and Co_2O_3; and nickel, with the sulphide Ni_2S, and the oxides NiO and Ni_2O_3.

The palladium group contains compounds representing MO, M_2O_3, MO_2, and MO_3; and the platinum group MO, M_2O_3, MO_2, MO_3, and MO_4.

The reason for inserting the general formulæ of the oxides and chlorides at the beginning and end of the table on p. 629 will now be seen. These oxides and chlorides certainly exist in most cases, either free or in combination. But the elements do not form such compounds exclusively. It is possible that when we know more about the temperature of maximum stability of such compounds, we shall have light thrown on the subject; for the present, all that can be said is that there is certainly a definite order visible in the periodic arrangement of the elements, in which, in the main, elements possessing similar properties are grouped together; and we must trust to experiment to give us knowledge of the true meaning of all its apparent inconsistencies and anomalies.

The borides, carbides, nitrides, phosphides, &c., do not throw any further light on the periodic arrangement. They have as yet been too little investigated.

4. **Prediction of undiscovered elements.**—When Mendeléeff gave its present form to the periodic table, the element indium was believed to possess the atomic weight 76, twice its equivalent, which is 38; the formula of its most stable chloride was, therefore, accepted as $InCl_2$, and of its corresponding oxide, InO. The properties and reactions of the element and its compounds forbade it having a place in the potassium or calcium groups; moreover, there was no vacant place for it; its atomic weight could not, therefore, be 38. With the then accepted atomic weight 76, it would have fallen between arsenic and selenium, a position for which it showed no analogy, and, again, one occupied fully. With the atomic weight $38 \times 3 = 114$, it falls in the aluminium group; and its specific heat, shortly afterwards determined, confirmed its right to that position.

The element uranium was supposed to possess the atomic weight 120. The formula of the uranyl compounds, which are very characteristic, contained the group $(U_2O_2)^{ii}$, and its stable oxide was regarded as U_2O_3. But with the atomic weight 120, its place is among elements such as tin, antimony, and tellurium, with which it has no connection; and, again, these places were already filled. Hence it was decided to double the atomic weight; assigning the formula $(UO_2)^{ii}$ to the uranyl group, and UO_3 to its oxide. It then fell into its true position as the analogue of molybdenum and tungsten. This conclusion was afterwards ratified by Zimmerman.

The element with the approximate atomic weight 44, in the boron group, was then unknown. Mendeléeff predicted its existence. Belonging to the boron group, "**eka-boron,**" as he named the element, should have an oxide of the formula M_2O_3, without very marked tendencies towards combination, inasmuch as it lies between calcium and titanium; its sulphate should display analogy with that of calcium, and be sparingly soluble; it should accompany the next member of the group, yttrium, and should be difficult to separate from that element. "The oxide will be insoluble in alkalies;" it will give gelatinous precipitates with potassium hydroxide and carbonate, sodium phosphate, &c., &c. Ten years later, "**scandium**" was discovered by Nilson, possessing these identical characteristics.

At the same time, Mendeléeff predicted the existence of two other elements, also then unknown, viz., "**eka-aluminium**" and "**eka-silicon.**" Eka-aluminium should have the atomic weight 68; its compounds should resemble those of aluminium in formula. It should be easily reduced, stable, of sp. gr. about 5·9, and should decompose water at a red heat. All these predictions were subsequently fulfilled by "**gallium,**" discovered in 1875 by Lecoq de Boisbaudran.* Eka-silicon was not discovered till much later. It is the element "**germanium,**" discovered by Winkler in 1886. As with gallium, it fulfils all that Mendeléeff predicted.

Among other points to be mentioned are the correction of the atomic weight of **beryllium,** long maintained to be three times its equivalent 4·6, instead of twice (see p. 128); the placing in their true position the elements **osmium, iridium,** and **platinum,** since confirmed by the re-determination of their atomic weights, by Seubert (Os = 190·8; Ir = 192·64; and Pt = 194·46); also of gold by Thorpe and Laurie (197·34), confirmed by Krüss (197·12); the numbers previously accepted were Os = 200; Ir, 197; and Pt, 197 to 198, while gold was taken as 197·1, differing little from those more recently obtained.

Brauner announced, in 1889,† that he has for six years been

* Mendeléeff, *Compt. rend.*, **81**, 969.

† *Chem. Soc.*, 1889, 382.

engaged in determining the atomic weight of tellurium. At present, the data represent it with the atomic weight 128. But it must obviously lie between antimony, on the one hand, with the atomic weight 120·3, and iodine, 126·86. Mendeléeff assigned it the atomic weight 125 in his earliest table. Brauner finds that the element named tellurium is a mixture of, at least, three elements, and he is at present engaged in their separation. It is not improbable that his work will result in the discovery of elements of higher atomic weight, for which there are vacant places in the antimony group, and also following tellurium in the sulphur group.

The rare elements, described in Chapter XXXIV, now being investigated by Crookes, Cléve, Nilsson, and others, present a problem difficult to solve. Should it appear, as believed by Crookes, that these elements are capable of resolution into an almost infinite variety of others, the conclusion will be difficult to reconcile with the periodic arrangement; but should it finally prove, as the author believes to be more likely, that they supply the places wanting between the atomic weights 141 and 172 inclusive, comprising in all 15 undiscovered, or, at least, unidentified, elements, the periodic table will be nearly completed. Time alone will show which of these surmises is the correct one.

PART IX.

CHAPTER XXXVII.

PROCESSES OF MANUFACTURE.

VARIOUS processes of manufacture will now be discussed, for they cannot be correctly understood in all their bearings without a previous knowledge of scientific chemistry. In a work of this extent, the treatment must necessarily be incomplete; and attention will be drawn to the chemical nature of the changes involved in manufactures, rather than to the mechanical appliances and plant requisite to carry them out. For detailed information on such questions, special treatises must be consulted.

The matter will be arranged under three main heads:—

1. **Combustion**; and means for generating high temperatures.
2. **The metals**: their extraction from their ores.
3. **Other processes of chemical manufacture** arranged, so far as is possible, as the operations are carried out on a large scale, those manufactures which are carried on under one roof being grouped together.

1. **Combustion.**—For practical purposes, combustion is the union of carbon, or of gases containing hydrogen and hydrocarbons, with the oxygen of the air. It is true that other substances burn, and evolve heat, and that combustion may take place in gases other than air; but economy prevents the employment of such processes, except in one or two unimportant instances.

The substance burned is termed "fuel." In using fuel, there are two objects which may be sought for:—(*a*) to obtain the maximum heating effect; and (*b*) to produce the highest possible temperature. The maximum amount of heat obtainable from a fuel is termed its *heating-power*. The highest temperature which can be produced by burning fuel, under favourable circumstances, is termed its *calorific intensity*.

(*a*.) *Heating-power.*—The theoretical heating-power of a fuel is approximately calculable from its elementary composition. It is calculable absolutely in the theoretical case of the fuel consisting of a pure element, or of a mixture of pure elements and pure combustible compounds in known proportions; or if the impurity itself is non-combustible, and of known specific heat.

1. The fuel is pure carbon.—The thermal equation with pure wood charcoal is:—$C + O_2 = CO_2 + 969{\cdot}8K$; whence it follows that 1 gram of carbon in burning to CO_2 evolves 8080 cal. This is its heating-power.

If the carbon contain ash, the heating-power relative to that of pure carbon is represented by the percentage of carbon. Thus, if a specimen of wood charcoal or coke contain 2 per cent. of ash, its heating-power is 98 per cent. of 8080 = 7920 calories.

2. The fuel is pure hydrogen.—Hydrogen burns to form *gaseous* water, and the water is seldom or never restored to the liquid state by condensation, for a temperature of over 100° in the flue is required to keep up the draught. Hence the equation is $H_2 + O = H_2O + 587K$; and multiplying by 100 and dividing by 2, one gram of hydrogen gives 29,350 calories when burned. This is its heating-power.

3. The fuel is carbon monoxide.—The thermal equation for CO, burning to CO_2, is:—$CO + O = CO_2 + 680K$; whence, multiplying by 100 and dividing by 28, 1 gram of carbon monoxide, in burning to dioxide, has a heating power of 2428 calories.

4. A mixture containing hydrogen and carbon monoxide has a heating-power depending on their relative proportion, and on their several heating-powers. Thus, a mixture containing 50 per cent. of each has the heating-power

$$\{(50 \times 29{,}350) + (50 \times 2428)\}/100 = 15{,}890 \text{ calories.}$$

5. The fuel is a hydrocarbon.—Suppose the hydrocarbon to be methane (marsh-gas), CH_4. Its composition is C = 75 per cent.; H = 25 per cent. Calculated as in (4), the heating-power should be, were the carbon and hydrogen free,

$$\{(75 \times 8089) + (25 \times 29{,}000)\}/100 = 13{,}400 \text{ calories.}$$

It actually amounts to only 12,030 calories. The difference, 1370 calories, is absorbed in decomposing methane into its elements, and is lost, so far as heating-power is concerned.

The following table gives the heating-power of the hydrocarbons up to C_6H_{14} in 100 calories = K, for molecular weights:—

	CH_4.		C_2H_3.		C_3H_5.		C_4H_{10}.		C_5H_{12}.		C_6H_{14}.
	1925		3413		4904		6387		7889		9313
Δ =		1488		1491		1483		1502		1424	

The difference is a nearly regular one; hence it is possible to calculate the heat of combustion of any paraffin, of which the molecular weight is known, by adding to 1925 some number approximating to 1480 for every group CH_2 above methane; multiplying by 100, and dividing by its molecular weight. In this manner the heating-power of liquid fuel, which is now coming into use, may be calculated with fair accuracy.

6. The fuel is coal, wood, or peat. — Only roughly approximate results can be calculated from a knowledge of the percentage composition of the fuel, since the data are wanting for the heat of union of the carbon, hydrogen, and oxygen in the fuel. It is customary, but incorrect, to suppose that the oxygen is already combined with hydrogen as water, and to calculate the heat of combustion of the residue as if it consisted of a mixture of free carbon and gaseous hydrogen. Hence it is found by the formula: Heating-power $= 8080C + 29{,}350(H - \frac{1}{8}O)/100$. For such complex fuels a practical essay is best.

(b.) *Calorific intensity.*—The highest temperature theoretically obtainable from a fuel may be calculated; but here the values obtained are usually far from the truth. The calorific intensity depends on the heat of combustion of the fuel, and as the heat is employed in raising the temperature of the products of combustion, it also depends on their specific heat; and as air is almost always used to promote combustion, a quantity of inert gas, viz., nitrogen, has also to be heated; nor is this all; for more air must be admitted than can be wholly utilised; hence the excess of air has also to be heated. We must know, therefore, in order to make an approximate calculation, the heating-power of the fuel; the amount, and the specific heat of the products of combustion; the specific heat of nitrogen and that of air.

1. Suppose the fuel to be pure carbon burned in oxygen. The heating-power of 12 grams of carbon is 97,000 calories. It forms 44 grams of carbon dioxide, the specific heat of which is usually taken as 0·2164. Now, 12 grams of carbon, in burning to carbon dioxide, would raise the temperature of 97,000 grams of water through 1°. It would raise the temperature of 44 grams of water through 97,000/44 = 2204°. But as the specific heat of carbon dioxide is 0·2614, it should raise the temperature of 44 grams of CO_2, its product of combustion, through 2204/0·2164 = 10,184°. Now it has been shown by E. Wiedemann that the specific heat of carbon dioxide is not constant, being at 100° 0·2169, and at 200° 0·2387; hence the assumption that it is a constant is unwarranted, and the temperature calculated above is

certainly too high. But for another reason the result is totally fallacious. Carbon dioxide dissociates long before such a temperature is reached; it begins to dissociate, indeed, at 1200—1300°. It is, therefore, improbable that the temperature as as high as 2000°.

2. As a further example of the method of calculation, an instance is chosen where the fuel is a hydrocarbon, viz., methane, CH_4; it is supposed to be burnt in twice as much air as is necessary for complete combustion. The data are as follows :—

16 grams of methane evolve, on burning,		192,500 calories,
and produce carbon dioxide,	„	44 grams,
and water-gas,	„	36 „
consuming oxygen	„	64 „
equivalent to air, containing nitrogen		256 „
but also part with heat to		320 .

of air. The heat evolved is, therefore, utilised in raising the temperature of a mixture of carbon dioxide, water-gas, nitrogen, and air. Their specific heats are given as $CO_2 = 0{\cdot}2164$; H_2O gas $= 0{\cdot}475$; $N = 0{\cdot}244$; and air $= 0{\cdot}238$.

The temperature is, therefore,

$$T = \frac{192{,}500}{(44 \times 0{\cdot}2164) + (36 \times 0{\cdot}475) + (256 \times 0{\cdot}244) + (320 \times 0{\cdot}238)} = 1167°.$$

This is a probable result, as the amount of dissociation at 1167° can be but small; but it is still inaccurate, owing to the assumption that the specific heats of carbon dioxide and water-gas are constant between 200 and 1200°.

Apparatus for measuring high temperatures: pyrometers. —For detailed description of such instruments, a treatise on Technical Chemistry must be consulted; the principles involved depend on the following considerations :—

1. The expansion of a gas. A cylindrical bulb of porcelain, or of platinum, provided with a long capillary neck, the capacity of which is small in comparison with that of the bulb, is heated in the furnace of which the temperature is to be measured. The escaping air is collected and measured. By this means temperatures as high as 1700° have been measured. A similar method consists in confining the air or other gas at constant volume, and noting the rise of pressure produced by the increased temperatures. Both of these methods involve calculation, but are subject

to errors small in comparison with the total temperature measured.

2. Water is caused to circulate through a spiral tube exposed to the heat. Its temperature on entering the tube is read by means of a thermometer, as also its temperature on leaving. Comparative measurements are thus possible, the rate of flow being maintained constant.

3. Siemens' pyrometer consists of a platinum wire, through which a current is passed, exposed to the high temperature. Its resistance is increased by rise of temperature, and the increase is measured. A formula having been obtained showing the ratio between the rise of resistance and the temperature, the latter can be calculated.

4. The fusing-points of a number of salts have been determined with fair accuracy by Carnelley and Williams up to about 900°. By noting the particular salts which fuse, or which remain unfused, at the temperature to be measured, an estimation may be made to within 10—20°.*

5. For higher temperatures, cones are sold by the Jena Glass Company, composed of various silicates, by means of which approximate estimations may be made in a similar manner.

The first is the standard method, but is sometimes inconvenient of application. Siemens' pyrometer gives good results; and a sixth method, depending on the communication of heat from an iron ball heated in the furnace in a clay tube to water, into which it is dropped, also yields fairly accurate results.

Varieties of fuel.—Coal consists of carbon, hydrogen, oxygen, nitrogen, and ash composed of silicates, sulphates, and phosphates of alumina, iron, lime, and magnesia. It usually contains some iron pyrites. The varieties of coal may be classified as follows:—

	Carbon, p. c.	Hydrogen, p. c.	Oxygen and Nitrogen, p. c.
1. Caking coal......	83·0—88·0	5·0—5·2	3·0—5·5
2. Splint coal.......	75·0—83·0	6·3—6·5	5·0—10·5
3. Cherry, or soft coal	81·0—85·0	5·0—5·5	8·5—12·0
4. Cannel coal......	83·0—86·5	5·4—5·7	8·0—12·5
5. Anthracite.......	90·0—94·0	1·5—4·0	3·0—4·8

The first, as its name implies, "cakes" readily, and undergoes a semifusion when heated, which causes it to become spongy. The second does not cake, but burns brightly. The third does not cake, but is easily broken; it, as well as the second, is much

* For a list of such salts, see *Chem. Soc.*, **33**, 273-281.

used for household purposes. The fourth is used for gas manufacture, as it evolves much more gas and oils when distilled than any of the other varieties. It is hard, and does not soil the fingers. "Jet" is a special variety of cannel coal.

Coke is produced by charring or distilling coal, either by heating the coal in open heaps, covered to prevent too free access of air, or by distilling coal in coke ovens. It consists almost entirely of carbon and ash. During its formation, gases escape, consisting mainly of compounds of carbon and hydrogen, some of which may be liquefied, and of ammonia, most of which is now recovered by "scrubbing" with water, *i.e.*, by causing the gases to pass through water contained in U-shaped iron tubes. The amount of coke produced from different varieties of coal varies within wide limits. From some coals 80 per cent. of coke may be obtained, while others yield as little as 56 per cent. Anthracite furnishes from 85 to 92 per cent. of coke.

Gaseous fuels.—These are produced in one of two ways. Either the coal is distilled in a special form of apparatus termed a "producer," by the combustion of a portion; or steam is led through white-hot coke. If produced by the latter method, the product is termed "water-gas." The Siemens gas producer has been found to yield the following mixture of gases:—

CO_2.	CO.	N.	H.	Hydrocarbons.	
4—6	21·5—24	60--64	5·2—9·5	1·3—2·6	p. c. by volume.

Produced by the latter method, the gases have been found to consist of—

CO.	H.	N.	CO_2.
28·6	14·6	53·0	4·0

The nitrogen is due to the air forced into the fuel along with the steam.

Liquid fuels.—These consist of natural oils, consisting mainly of hydrocarbons. The fuel is burnt by injection by means of compressed air against a plate, or a bed of coke; by percolation upwards through a bed of heated fire bricks; by vaporisation in a separate still, the products of distillation being burned; or by injection by means of superheated steam. The last plan is said to yield the best results.

To ensure complete combustion of solid fuels, such as coal, a regular supply of combustible must be introduced into the furnace. The stoking is now-a-days often performed mechanically, sometimes by travelling fire-bars, sometimes by the introduction at regular intervals of time of known amounts of fuel below the ignited mass, so that the gases distilled off by the heat may travel upwards through the incandescent upper layer, and so be completely burned at the surface, where they mix with air.

Uses of fuel.—The chief uses to which fuel is put is (1) in *evaporation;* the flame and hot gases are either allowed to pass over the surface of the liquid to be evaporated (surface evaporation), which is contained in long tanks; or they impinge at the bottom of flat shallow pans filled with the solution to be evaporated. The "Yaryan" system, which is now coming largely into use, and which is a very economical one, consists in the use of a number of straight tubes passing from end to end of a shell, and coupled to each other at their ends by connecting tubes. The liquid flows through these tubes, which are kept vacuous by a pump and heated with steam. Three or more such sets of tubes are worked in concert, the vacuum being maintained at the exit from the last set. The steam from the first set serves to heat the tubes in the second, and that from the second heats the third set of pipes. The temperature is thus kept low, for the liquid boils under reduced pressure; the surface for evaporation is large and constantly renewed, and the heat is economised, since the steam derived from the first set of pipes is utilised in heating the second, and that from the second goes to heat the third.

2. *Distillation.*—This operation is not frequently practised in the manufacture of compounds other than those of carbon. The apparatus is usually of the simplest description; the heat is applied by an open fire to a retort connected with a condensing worm, as in the manufacture of nitric acid; by the products of combustion of coal, as in distilling zinc, sodium, phosphorus, &c., from clay retorts; or a large Bunsen burner is used as source of heat, as in the evaporation of sulphuric acid in glass vessels, which is in reality effected by distillation. For carbon compounds, such as alcohol, hydrocarbons, &c., more complicated forms of apparatus are used; but the method of heating is of the simplest kind; either a direct flame or a steam jacket is employed.

3. *Reverberatory furnaces.*—In such furnaces the products of combustion come into direct contact with the substances to be heated. Such furnaces serve for the calcination of ores, for firing porcelain and bricks, and in glass making; in the last instance, the glass is contained in fire-clay pots, which are exposed to the products of combustion of coal burned in a separate compartment.

In other operations, the solid fuel comes into direct contact with the object to be heated, as in lime-burning, in lead-smelting, and in iron-smelting. In some cases combustion is furthered by a blast of air; and if the blast be heated, a great saving in fuel is effected, for the temperature in the furnace is higher, less heat being withdrawn from the furnace in heating the air. This process is made use of in iron-smelting; and the Siemens *regenerative furnace* is constructed on a similar principle; but in the latter case it is extended, so that the gaseous fuel employed is also heated. The "regenerators" are large chambers constructed of fire-clay bricks, and filled loosely with bricks. The products of combustion pass from the chamber in which the combustion takes place through these chambers, before escaping into the flues. The gases part with their heat to the bricks; and, after a certain time, a second pair of similar chambers is brought into operation. The current is then reversed, by opening appropriate valves, and the air enters through one of the already hot chambers, while the "producer" gas is heated by the other. As before, the products of combustion pass through the second pair of chambers. When the first pair has grown cool, and the second pair hot, the current is again reversed. By this means a great saving in heat is effected; for the heat of the escaping gases, instead of being dissipated, is to a great extent trapped by the brick chambers and utilised.

Such furnaces are employed in iron-smelting, in glass making, and, indeed, in most chemical operations where economy of fuel is an object. It is also possible, in using such regenerators, to render the flame oxidising or reducing as required, by regulating the relative amounts of air and producer-gas. The air and gas mix and burn at the spot where the high temperature is required.

For certain operations where a sudden intense and uniform rise of temperature is required, the heat radiated from flame is made use of. Before describing the device adopted to secure such flames, the subject of flame must be itself considered.

The cause of the luminosity of flame has long been a question under discussion. It has been urged on the one hand, that the presence of solid particles rendered incandescent by a high temperature is essential to light; and the luminosity of the flame of a candle or of burning hydrocarbons has been ascribed to the presence in the flame of particles of white-hot carbon. In a candle flame there are three separate regions or zones; first the faintly blue interior cone, where the compound of carbon with hydrogen, or of carbon with hydrogen and oxygen, is being partly distilled, partly decomposed into other hydrocarbons and free carbon by the heat radiated towards the wick by the luminous zone. Next follows the luminous zone, to which the oxygen of the air has some access, but is yet not present in quantity sufficient for complete combustion; and last, there is the indefinite hazy bluish-pink zone, surrounding the luminous zone, where the combustion is completed. In a candle flame it has been conclusively proved that incandescent solid particles are present; because if the sun's rays be focussed on the flame by a lens, the light emitted from the brilliant spot is seen to yield the absorption-bands peculiar to the solar spectrum when viewed through a spectroscope. Now, gases cannot reflect light, but only solids and liquids. It is unlikely that liquids are present; but it is exceedingly probable that unburnt, yet white-hot, particles of carbon are present. Of course if a plate be held over the flame of a candle, it will be smoked; this would appear to favour such a conclusion; yet it is not inconceivable that the presence of a cold body, such as a plate, should produce that separation of carbon for which it is intended as a test. Such presence is, however, proved indubitably by the reflection of the solar spectrum. Yet it has been shown that a gas, such as hydrogen, burning under high pressure, gives a luminous flame; and the Bunsen flame, non-luminous because of the complete combustion of the gas, may be rendered luminous if its temperature be raised by heating the tube through which it issues. The luminosity of a coal-gas flame is caused by the presence in it of solid carbon particles, produced by the incomplete combustion of hydrocarbons of the ethylene and acetylene series, and by the vapours of benzene, C_6H_6, and naphthalene, $C_{10}H_8$, the vapour-pressures of which are sufficiently high at the ordinary temperature to permit of their existing as gases, when mixed with such gases as hydrogen, methane, and carbon monoxide, which are the other chief constituents of coal-gas.

By regulating the supply of producer gas and air, and by a special construction of furnace, whereby the flame is allowed to pass without striking the arch of the combustion chamber, Siemens has succeeded in producing a luminous flame, the radiating power of which is very great. While non-luminous flames give up their heat by contact, luminous particles lose heat chiefly by radiation. Hence the temperature of such a flame is very intense; it may be so adjusted as to be evenly distributed; and by its means large sheets of cold plate glass may be heated to the softening point without cracking in less than two minutes. Such flames are capable of other applications.

CHAPTER XXXVIII.

COMMERCIAL PREPARATION OF THE ELEMENTS.

THE preparation of many of those elements which are of commercial importance has already been indicated; the reactions by which some are obtained are, however, somewhat complicated, and their manufacture is best described here.

1. **Sodium.**—The process which has now superseded all others is that of Castner.* It consists in heating to bright-redness a mixture of iron, carbon, and caustic soda. Its advantages over the older method, in which a mixture of lime, sodium carbonate, and coke was heated, is that the carbon, being weighted by the iron, sinks, and is thus brought in contact with the fused caustic soda; whereas, by the older process, contact between the reacting materials was by no means so perfect. The iron in a spongy, finely-divided state, reduced from its oxide without fusion by carbon monoxide or hydrogen, is impregnated with tar and heated to redness. The hydrocarbon is decomposed, and a mixture of 70 per cent. of iron with 30 per cent. of carbon is left. This mixture is ground and mixed with such a quantity of caustic soda as to correspond with the equation

$$6NaOH + 2(Fe,C_2) = 6Na + 2Fe + 2CO + 2CO_2 + 3H_2;$$

this corresponds to 22 lbs. of carbon to every 100 lbs. of caustic soda. The mixture is placed in large cast-iron crucibles, each of which stands on a circular platform, which, when raised, is flush with the hearth of the furnace, but which can be lowered by aid of hydraulic power, the platform and crucible then sinking into a chamber below the furnace. The crucibles, when raised, fit a retort-head, also of iron, an asbestos collar being interposed to secure better junction. The heat is derived from a Siemens gas furnace. When the crucibles reach 1000° C., sodium distils freely, and passes through the tube projecting from the crucible cover, whence it falls into heavy oil. As soon as an operation is finished,

* *Chem. News*, **54**, 218.

the crucible is lowered, seized with travelling tongs, emptied, refilled, and before its temperature has had time to fall, replaced in position. The residue consists of a little sodium carbonate and the iron of the so-called "carbide." It is treated with warm water, and the soluble carbonate is afterwards causticised with lime; while the iron is dried, mixed with tar, and re-coked, to serve for another operation. A crucible is charged every two hours, which is the length of time occupied by a distillation.

2. **Magnesium.**—The process is sufficiently described on p. 35. The equation is $KCl.MgCl_2 + 2Na = KCl + 2NaCl + Mg$.

3. **Zinc.**—The ore, consisting of a mixture of *blende* (*black-jack*), ZnS, *calamine*, $ZnCO_3$, *calamine-stone*, $ZnSiO_3.H_2O$, and *gahnite*, $ZnO.Al_2O_3$, is roasted on a flat hearth at a dull red heat, to expel sulphur as sulphur dioxide, and also water; the resulting oxide and silicate of zinc is then mixed with coke and distilled from tubular retorts of fire-clay, and condensed in tubes of sheet-iron, secured to the mouths of the retorts by fire-clay; or it is distilled downwards, as with magnesium (see p. 34, Fig. 4). The reaction is:—$ZnO + C = CO + Zn$.

4. **Aluminium.**—There are three processes in operation for the commercial preparation of aluminium—(*a.*) Reduction of the chloride by means of sodium. The double chloride of aluminium and sodium, prepared by passing chlorine over a bright red-hot mixture of clay, finely ground charcoal, lamp-black, oil, and salt, is volatile, and sublimes in crystals. It always contains a little chloride of iron, which even in small quantity impairs the quality of the aluminium obtained. The iron is removed by fusing the double chloride, and introducing a little metallic aluminium, which displaces metallic iron from its chloride. The metallic iron sinks, leaving perfectly white double chloride, which is much less deliquescent than the impure substance. This double chloride, of the formula $3NaCl.AlCl_3$, is mixed with finely cut sodium, in a wooden agitator, and placed on the hearth of a Siemens' regenerative furnace; a brisk action takes place at once, and the aluminium is run off into moulds. The residue is treated with water to recover the salt, together with any undecomposed chloride, from which the alumina is recovered by precipitation as hydrate.

(*b.*) Reduction of *cryolite* (sodium aluminium fluoride, $3NaF.AlF_3$) by means of sodium.—The furnace is a flat chamber, in the upper surface of which there are holes, through which crucibles containing melted cryolite and salt may be reached. An iron rod, with a hollow cylinder at one end, is made use of for the purpose of conveying the sodium into the melted mass. The hollow cylinder is fillėd with sodium, and plunged by a workman into the crucible; at the high temperature of the fused cryolite, the sodium gasifies, and its vapour passing upwards through the molten cryolite, deprives it of its fluorine, metallic aluminium being produced. The metal sinks to the bottom of the crucible, and after a sufficient quantity has been reduced, it is poured out of the crucible into moulds.

(c.) By means of the electric furnace.—This process is not successful in producing metallic aluminium, for it remains mostly disseminated through the carbon; but it is well adapted for the manufacture of alloys. The furnace consists of an oblong fireclay box, into which project at each end thick rods of gas-carbon, connected by means of copper cables with a powerful dynamo-electric machine. The furnace is charged with a mixture of *corundum*, (Al_2O_3), metallic copper, and fine particles of charcoal coated with lime to render it non-conducting. On passing the current, an enormously high temperature is produced; the carbon poles, which at first are almost in contact, are gradually drawn apart, and the electric arc leaps between them. The alumina is deprived by the carbon of its oxygen, and the copper boils; the separated aluminium is washed down by the liquid copper and the aluminium bronze, as the alloy is termed, which contains over 15 per cent. of aluminium, collects on the bottom of the furnace, and is removed by tapping a hole. That the process is a true reduction, and not dependent on the electrolysis of alumina, is proved by the fact that an alternating current may be employed; and such a current is incapable of electrolysing a compound.

Magnesium and aluminium are obtained by the use of sodium. The remainder of the metals industrially prepared are reduced by aid of carbon.

5. **Iron.**—A list of the ores of iron is given on pp. 244, 248, 251, and 288. The ores are roasted to remove water, carbon dioxide, and carbonaceous matter; to render them more dense; and to oxidise any ferrous iron (as in *spathic ore*, *black band*, and *clay-band*) into ferric oxide. The roasted ore is then stamped or crushed into fragments as large as a fist. It is then introduced into a blast furnace along with alternate layers of coal and limestone. The blast furnace, which is often 80 feet in height, consists of an outer wall of brick, an inner space, fitted with loose scoriæ, or refractory sand, to allow for expansion; and an inner wall of firebrick. The upper portion of the furnace is termed the "shaft;" the cup-shaped part of the furnace is named the "boshes," and the lower cylindrical part is named the "throat," or "tunnel-hole," terminating in the "crucible," or

"hearth." The combustion of the fuel is furthered by a blast of hot air (at 200—400° C.) through the "tuyères," or "twyers," forced in under a pressure of 3 or 4, or even 10, lbs. per square inch. It is usual now, instead of allowing the furnace gases to escape, to cause a conical cover, which can be depressed into the mouth of the furnace, to force the products of combustion through "scrubbers," or iron absorbing vessels, containing water, whereby potassium cyanide and other products are condensed. Although pure iron has a very high fusing-point, iron containing 3 to 4 per cent. of carbon melts at about 1100°, and hence it is possible to smelt it in such a furnace.

The reactions occurring are very complicated. The coal burns to carbon monoxide and dioxide; it also contains nitrogen and salts of potassium, and yields potassium cyanide, which has a reducing action; the reduced iron acts on the oxides of carbon, reproducing carbon, and re-forming oxides of iron. The following equations express the changes which occur:—

1. $Fe_2O_3 + CO = 2FeO + CO_2$.
2. $FeO + CO = Fe + CO_2$.
3. $Fe + CO_2 = FeO + CO$.
4. $2FeO + CO_2 = Fe_2O_3 + CO$.
5. $2FeO + CO = Fe_2O_3 + C$.
6. $Fe + CO = FeO + C$.
7. $2CO = C + CO_2$.
8. $Fe_2O_3 + C = 2FeO + CO$.
9. $2Fe_2O_3 + C = 4FeO + CO_2$.
10. $FeO + C = Fe + CO$.
11. $C + CO_2 = 2CO$.

Many of these equations, it will be noticed, are the converse of others; the reactions take place in an inverse sense in different parts of the furnace. Besides these changes, others take place in which potassium cyanide plays a part:—

12. $2KCN + 3FeO = K_2O + 2CO + N_2 + 3Fe$.
13. $2CO = C + CO_2$; and 14. $K_2O + CO_2 = K_2CO_3$.

The carbonate is carried down and reconverted into cyanide in the throat of the furnace. The actual reduction takes place near the tuyères; only one half or one quarter of the carbon monoxide is utilised; carbon deposits, however, in the middle of the furnace, about 25 feet above the hearth.

When a sufficient amount of iron has collected in the crucible, a workman makes a hole in the clay plug which confines the iron, and allows it to flow into wide channels termed "sows," whence it diverges into narrower moulds, named "pigs;" hence the name "pig iron." The slag which is formed by the combination of the silica existing as an impurity in the ore with the lime, and with some of the iron oxide, is lighter than iron, and floats on its surface. It fuses, and runs off after the iron has flowed away. This slag is sometimes made use off for coarse glass.

There are two main varieties of pig iron, grey and white; and there are intermediate varieties known as "mottled." These all contain carbon, often as much as 6 per cent.; but while the carbon in the white pig is in combination with the iron forming a carbide, of which the formula, however, has not been determined, that in the grey pig is partly present in the free state as graphite. On treatment with acids, the combined carbon escapes in combination with hydrogen, chiefly in the form of hydrocarbons of the ethylene series; while the free carbon remains unaffected by acids. But both varieties are left if the iron is treated with a solution of copper sulphate, which dissolves the iron as sulphate, leaving copper in its place. The specific gravity of the white iron is the higher, varying between 7·58 and 7·68; that of grey pig has a specific gravity of about 7. The production of one or other variety depends on the temperature of the furnace. The white cast iron is produced at the lower temperature, while the grey pig is formed as the temperature rises. If the grey pig be melted and suddenly cooled, it solidifies as white pig, the carbon being retained; but if heated strongly and cooled slowly, the carbon has time to separate. For castings, a mixed pig is best, being more fluid, and, when it solidifies, closer-grained. The iron is remelted in a cupola, a cylindrical furnace about 9 or 12 feet high, cased in iron-plate, and the iron is run into moulds made of a mixture of sand and powdered charcoal with a little clay, or of loam, or of iron. If iron moulds are used, the casting is rapidly cooled on the exterior, and is said to be case-hardened.

The operation of removing carbon and other impurities from the iron is termed "refining." This is accomplished either by heating the iron on a hearth, a blast of air being directed on to the melted iron. The carbon is oxidised to carbon monoxide; the silicon in the oxide iron is converted into silica, which combines with ferrous oxide resulting from the oxidation of the iron to form ferrous silicate; this forms a slag and protects the iron. This slag is subsequently mixed with forge-scales (oxides of iron), and made use of in refining a further charge of crude iron; the oxygen of the iron oxides unites with the carbon of the crude iron forming carbon monoxide, which escapes, while malleable iron is left. The "bloom" of iron, a spongy semifused mass, is placed under a steam hammer, and the enclosed slag removed by repeated blows. Another plan of refining, which has much similarity with the one described, is termed "puddling." The flame of a reverberatory furnace plays on the white cast iron, placed on a hearth of slag containing iron-scales; when the iron has melted it is spread over the hearth by means of a rake, and continually stirred about; this is the operation termed "puddling." Flames of burning carbon monoxide appear on the surface of the iron, due to the action of the oxide of iron in the slag on the carbon of the crude iron; the mass becomes pasty, and is finally scraped together into lumps (blooms); these are placed under the steam-hammer, and forged into bars.

The Bessemer process—The third plan of producing soft iron (wrought iron) is by the Bessemer process. This consists in running the molten cast iron from the blast-furnace into pear-shaped vessels of iron-plate, termed converters; the lining of such converters used to be of "ganister," a variety of very siliceous clay; but of late years the magnesia bricks introduced by Messrs. Thomas and Gilchrist ("basic lining") have supplanted ganister. Fire-clay tubes lead to the bottom of the converter, through which a blast of air can be forced into the molten iron. When the lining is of ganister the phosphorus and sulphur are not wholly removed, for the free oxides are reduced by molten iron, producing phosphide and sulphide of iron. But with a magnesia lining, lime may be thrown on to the surface of the molten iron; it combines with the oxides of phosphorus and sulphur, forming phosphate and sulphate of calcium, which then escape reduction. With a siliceous lining, it is impossible to make use of lime, for an easily fusible slag is at once produced, and the lining of the furnace is destroyed, owing to the formation of an easily fusible silicate of calcium and iron.

Flames issue from the mouth of the converter; the carbon, silicon, sulphur, phosphorus, and manganese burn to oxides; and when the flames cease, the iron, approximately pure, may be run out into moulds. Such iron is the purest form of commercial iron.

Steel.—The difference between steel, wrought iron, and cast iron consists in the amount of carbon which they contain. To convert wrought iron into steel, it is necessary to add carbon. This is done in the Bessemer converter by throwing into the fused wrought iron a known quantity of a variety of iron containing a known amount of manganese and carbon, named spiegel-iron (*i.e.*, mirror-iron), owing to the bright crystalline facets which it shows when broken. The spiegel-iron mixes with the decarbonized iron, and the mixture is completed by turning on the blast for a few seconds. The converter is then tilted, and the steel is poured out into moulds.

It is also possible to prepare steel by adding wrought bar iron nearly free from carbon, to pig-iron kept melted in the hearth of a Siemens furnace. This is the principle of Martin's process. Steel produced by the Bessemer or by the Martin process is useful for rails, ship-plates, and ordnance.

For cutting instruments steel is chiefly made by the "*cementation-process.*" The best qualities of bar iron are employed. They are placed in fire-clay boxes, and packed in charcoal; the boxes are then kept at a red heat for six or seven days. A sample bar is withdrawn from time to time and tested; when a sufficient amount of carbon has been absorbed, the boxes are allowed to cool, and when cold the bars are removed and forged under a steam hammer. It has long been a matter of speculation as to the manner in which the iron absorbs the carbon. In view of the recent discovery of the compound of nickel with carbon monoxide, $Ni(CO)_4$, a compound which is decomposed into nickel, carbon, and a mixture of carbon monoxide and dioxide at a low temperature, it may be conjectured that iron also possesses some tendency to form a similar compound, which, however, is too dissociable to be isolated, but which is formed and decomposed during the process of the conversion of bar iron into steel, and which serves to convey carbon into the interior of the iron.

The steel, thus made, is refined by rolling or hammering out the bars, placing a number of the rods together, and welding them into a compact whole. Such steel goes by the name of *shear steel*, owing to its imployment for cutting instruments.

Cast steel is made by fusing such bars in crucibles made of a mixture of graphite and fire-clay, and then casting the steel in moulds.

It is also possible to convert the surface of objects made of soft iron into steel, by heating them to redness and then sprinkling them with powdered ferrocyanide of potassium. This process is termed *surface-hardening*.

Steel has a fine granular fracture, and does not exhibit the coarse crystalline structure of cast iron, nor the fibrous appearance of wrought iron. It contains amounts of carbon varying from 0·6 to 1·9 per cent., according to the use for which it is intended; the hardness, toughness, and tenacity increase with the amount of carbon.

Tempering, hardening, and annealing of steel.—The result of rapidly cooling strongly heated steel is to *harden* it; by raising it to a much lower temperature, and cooling it quickly, it is *tempered;* and it is *annealed* by heating it to a temperature higher than that required to temper it, and cooling it slowly.

It appears that there is some evidence of the existence in steel and white cast iron of a carbide of the formula Fe_2C; and also that at a temperature of 850° a change takes place in metallic iron, whereby it is changed into an allotropic modification. The specific heat of iron undergoes at that temperature a sudden change; and it also alters its electromotive force at that temperature; and these changes imply some change in molecular aggregation or structure. And it is suggested by Osmond, who has investigated this change, that the molecular arrangement or aggregation which exists at a high temperature may remain permanent if the iron contain carbon, and if it be rapidly cooled.

This capacity of retaining the molecular structure, which it possesses at temperatures above 850°, is much influenced by the presence of foreign ingredients. If manganese be present to the amount of 7 per cent., no sudden change of specific heat, &c., occurs; and if present in smaller proportions, it exerts a like influence, though to a less degree. Tungsten has an even greater effect. The effect of suddenly cooling steel containing these elements is to harden it.

In annealing, the carbide of iron becomes diffused through the iron in small crystals, and the iron itself develops a finely granular crystalline structure. It thereby becomes tougher, and at the same time softer.

The hardness of steel varies also with the temperature to which it is heated before being cooled, as well as on the suddenness of the cooling. If cooled rapidly from a high temperature, it is harder than glass, and brittle; if cooled from a comparatively low temperature, it is elastic; and at intermediate temperatures, it displays hardness and elasticity in various degrees.

6. **Nickel.**—The chief ore of nickel is the double silicate of nickel and magnesium, named *garnierite*, which contains from 8 to 10 per cent. of the metal, and is exported in large quantity to France from New Caledonia. Almost all the nickel in the market is now produced from this ore, and the process of extraction is a metallurgical one.

The finely ground ore is mixed with about half its weight of alkali-waste (calcium sulphide) or of gypsum, and about 5 per cent. of ground coal, moistened, and made into bricks. These are then smelted in a reverberatory furnace, or in a small cupola, the reduction of iron being as much as possible prevented. A "matt" is obtained, containing about 60—70 per cent. of nickel and 12 per cent. of iron, together with sulphur and graphite. As iron has a greater affinity for oxygen than nickel, while nickel combines more readily with sulphur than iron does, the iron in the ground regulus, which is roasted at a dull red heat, is converted into oxide. To convert this oxide into silicate or iron slag, the roasted mass is then thoroughly mixed with fine sand, and fused in a small reverberatory furnace. The sulphide of nickel forms a fused layer under the molten slag. The process is repeated, sometimes as often as five times, to remove iron as thoroughly as possible. The slags all contain nickel; they are ground and re-smelted with sand and gypsum, when they yield a regulus poor in nickel and a slag practically free from that metal. The poor regulus is then crushed, mixed with gypsum and sand, and again fused. The calcium sulphate is attacked by the silica, giving oxygen, which oxidises the iron in the regulus, and the resulting oxide forms an easily fusible slag with the lime and silica.

The sulphide of nickel, freed as described from iron, is crushed and exposed to a dull-red heat on the hearth of a reverberatory furnace. This oxidises both the sulphur and the nickel; a little nitre is sometimes added towards the end of the operation. The resulting oxide is finally reduced by making it into cakes with powdered wood charcoal, and heating it in crucibles to a bright-red heat.

7. **Tin.**—The only available ore is *tin-stone*, SnO_2. The ore is purified by "dressing" and washing, in which it is to some extent freed from gangue. It is then roasted in reverberatory furnaces to expel arsenic and sulphur, present as arsenical pyrites; it is again washed to remove copper sulphate and, after drying, it is mixed with slag from former operations and with anthracite coal, and smelted; occasionally fluor-spar is added to flux the silica still present; the tin collects in a compartment analogous to the crucible of a blast furnace, from which it overflows into a second receptacle. To free it from iron and arsenic, its usual impurities, it is "liquated," *i.e.*, heated to its fusing point on a sloping bed; the pure tin melts and runs down, leaving a less fusible alloy with iron and arsenic behind. It then undergoes a process known as "boiling": it is stirred with a log of wet wood, and the steam rising to the surface carries with it impurities. The metal thus prepared is known as refined tin, and is very nearly pure.

For a sketch of the methods of tinning iron and copper, see pp. 583 and 586.

8. **Lead.**—The chief source of lead is *galena*, PbS. There are two chief systems of extraction; the first, by use of the "Scotch hearth," consisting in effecting the reactions between lead oxide,

sulphate, and sulphide (see pp. 296 and 429) at as low a temperature as possible, while in certain recent processes the temperature is raised by means of a blast.

The hearth of the furnace used in the first method is constructed of slag, and slopes towards one side. The powdered galena is spread on the hearth and heated by a charge of coal. Lime is raked in small quantity into the ore to separate silica, with which it combines. The galena is partly oxidised to oxide and to sulphate, sulphur dioxide escaping along with fumes of lead sulphate and oxide. To condense and trap such fumes is very difficult, and long flues are often employed, debouching here and there into chambers, and provided with *bafflers*, *i.e.*, wooden spars kept wet by an intermittent flow of water. In spite of all precautions, much fume escapes.

When the lead sulphide is partially oxidised, the temperature is raised and the reactions occur:—

$$PbS + 2PbO = 3Pb + SO_2; \text{ and } PbS + PbSO_4 = 2Pb + 2SO_2.$$

The lead runs off, and is cast into pigs.

If the second method be employed, the ore is spread on a hearth with coal and lime, and exposed to a high temperature by means of a hot blast. The reactions already mentioned occur, but much fume escapes; it is drawn off through a hood above the furnace by means of a fan, through wide iron tubes, in which it is cooled. After passing the fan, it is forced into woollen bags stretched from top to bottom of large chambers. The solid matter is trapped in the flannel, while the gaseous products of combustion escape through the pores. The fume is shaken out of the bags and again heated with coke, a blast being again employed. A further yield of metallic lead is obtained, and the fume, which is quite white, finds some market as a paint.

Lead usually contains silver, and sometimes a trace of gold. The silver is extracted by Pattinson's process, or by Parkes's process (see p. 579, 587, and 663).

9. **Antimony.**—The sulphide, Sb_2S_3, or *grey antimony ore*, is the principal ore. It is easily fused, and is liquated from the gangue with which it is associated by heating it in a hearth provided with an opening below, through which the fused sulphide is run off. The metal is obtained from the sulphide by roasting it in a reverberatory furnace until it is converted into the oxide, Sb_2O_4. This oxide, still mixed with some sulphide, is transferred to crucibles, mixed with a little crude tartar or argol (hydrogen potassium tar-

trate, $HKC_4H_4O_6$), or with charcoal and sodium carbonate, and heated. Reduction to metal takes place, partly owing to the mutual action of oxide on sulphide, and partly to the reduction of the oxide by the carbon.

It is also possible to prepare antimony directly from the sulphide by heating it with scrap iron and a little sodium carbonate or sulphate to promote fusion. The antimony settles to the bottom of the crucible.

To purify antimony from arsenic, it is fused under a layer of potassium nitrate. A considerable loss of antimony occurs.

10. **Bismuth.**—Bismuth is generally found native, and is freed from gangue by liquation. When obtained as a bye-product from cobalt and other ores, it is precipitated as bismuthyl chloride, and reduced by fusion with charcoal and sodium carbonate.

11. **Copper.**—The extraction of copper from its ores varies according to the nature of the ore. If it is an oxide, simple reduction with coal is sufficient; but as the ore almost always contains sulphide, other processes have to be employed. The sulphide may be treated in the "dry way," *i.e.*, in a furnace, or in the "wet way," *i.e.*, by precipitation on iron.

1. The ores are calcined in order to volatilise a portion of the arsenic, antimony, and sulphur; some sulphate of copper is thereby formed. The calcined ore is then heated with a flux in a reverberatory furnace. The effect of this is to reduce any oxide present to metallic copper; to reduce sulphate to oxide, which reacts with sulphides of copper and iron, giving metallic copper, oxide of iron, and sulphur dioxide, in the same manner as the similar compounds of lead react; and to separate some of the iron as a silicate in combination with the constituents of the flux. But metallic copper is miscible with copper sulphide, and the resulting regulus is re-smelted in a similar manner. The product consist of metallic copper mixed with cuprous oxide, Cu_2O. To finally convert the remaining oxide into metallic copper, the "rose-copper" is rapidly melted under a layer of charcoal, and stirred with a birch-wood pole. It is then cast into cakes.

2. The "wet" method of extracting copper from its sulphide consists in allowing the ore to lie in heaps exposed to the air and rain. The sulphide is oxidised to sulphate, which dissolves, and its solution runs into ponds or tanks. Scrap iron is then added, and the copper precipitates as a mud on the surface of the iron, mixed with basic ferric sulphate. The copper, being specifically

heavier, is freed from the basic sulphate by washing, and is then smelted.

Large quantities of iron pyrites, containing about 4 per cent. of metallic copper, are now imported into this country. It is delivered, first to the sulphuric acid manufacturer, where the sulphur is burned in "pyrites-kilns." The ore, then consisting of ferric oxide and oxide and sulphide of copper, is then transferred to the copper works. It is roasted, at a low red heat, with salt; the copper is thus converted, first into sulphate, and then, by means of the salt, into chloride, and on treatment with water it passes into solution. The residue of iron oxide, after drying, is passed on to the iron-smelters. As such ores usually contain some silver, it, too, is converted into chloride, and as silver chloride forms a soluble double chloride with sodium chloride, it is dissolved along with the copper chloride. The silver is removed by careful addition of solution of potassium iodide, which converts it into the insoluble iodide, which is allowed to settle, and removed. The solution of cupric chloride is treated, as already described, with scrap iron, and the copper precipitated in the metallic state. It is separated, mechanically, from the precipitated basic sulphate of iron, and is then smelted.

12. **Silver.**—Silver occurs chiefly as sulphide and as thio-antimonate. If the ores are rich, they are ground to a very fine powder, moistened with salt-water, and treated with roasted iron and copper pyrites, *i.e.*, with a mixture of ferrous and cupric sulphate and sulphide, and with mercury. The sulphates are converted partly into chlorides, the silver also becoming changed into chloride, and dissolving in the excess of salt as double chloride of silver and sodium. The silver chloride reacts with the mercury, forming calomel, HgCl, and an amalgam of silver, which is pressed in canvas bags. As silver amalgam is solid, and only sparingly soluble in mercury, it remains behind for the most part; it is then distilled, and the recovered mercury is used in a subsequent operation. This process is very wasteful, inasmuch as mercury equivalent to the silver is lost as calomel.

By another process, the sulphides are reduced to powder and roasted; the sulphides are converted into sulphates. As silver sulphate withstands a higher temperature than the other sulphates, it remains unchanged, while the other sulphates are decomposed. On treatment with a solution of salt, the silver passes into solution as double chloride, and is precipitated on metallic copper. It is also possible to omit treatment with salt, and to dissolve the silver sulphate and precipitate silver from the solution by metallic copper.

The ores are also sometimes treated with melted lead, which decomposes the sulphide, forming lead sulphide and metallic silver, which dissolves in the excess of lead, and is extracted therefrom by cupellation.

If silver is contained in metallic lead, it is separated by fractional crystallisation, a process invented in 1833 by Mr. H. L. Pattinson; this method depends on the fact that when a dilute solution is cooled, the pure solvent crystallises out at a temperature below the usual freezing-point, while the remaining solution has a lower freezing-point, and remains liquid. By a series of fractional crystallisations, the lead is divided into two portions, one pure and free from silver, and a quantity of lead very rich in silver is obtained, which is then cupelled.

Another process, due to Mr. Parkes, is to add to the molten lead about one-twentieth of its weight of zinc, and to mix the two, as well as possible, by stirring. The zinc dissolves the silver, and as zinc and lead do not appreciably mix to form an alloy, the zinc floats to the surface, bringing the silver with it. The zinc solidifies at a higher temperature than the lead, and the solid cake is removed. The zinc is separated from the silver by distillation from an iron crucible, similar to that employed in producing sodium, which may be raised to fit its lid; the condensing-tube issues from the lid. The silver is left, along with a little lead, which is dissolved by the zinc.

To refine the silver, it is cupelled. The lead containing silver is fused in a good draught on a hearth of bone-ash, pressed tightly into an iron grating of an oval shape, depressed in the middle. The lead and other metals oxidise; the oxides soak into the porous bone-ash, and at a certain point the dull appearance of the metal changes; a brilliant display of iridescent colours appears on its surface, due to diffraction colours, caused by a thin film of lead oxide; and suddenly the brilliant metallic lustre of the pure molten silver is seen. The metal is cooled by water, and removed. The resulting litharge is reconverted into lead.

13. **Gold.**—Gold usually occurs native. It may be associated with quartz, in which case its extraction is, for the most part, mechanical; or with sulphides and tellurides of zinc, bismuth, lead, and other metals, from which it must be separated chemically.

1. If associated with quartz, the rock is stamped to powder, and washed, by allowing a stream of water to carry away the

gangue. But the smaller particles of gold are apt to be carried away; hence the washings are generally caused to traverse an amalgamated copper runnel, small bridges being placed at intervals across the channel, to act as traps. Much of the gold sticks to the mercury; and it is found of advantage to add a small percentage of sodium to the mercury, the effect of which is to keep its surface clean.

The mercury is scraped from the copper runnel with india-rubber scrapers, and squeezed through leather; the solid amalgam is distilled when a sufficient quantity has been collected.

2. Many processes have been proposed for the extraction of gold from ores containing sulphides. The ores are, in every case, roasted to oxidise the sulphides; they may then be treated with chlorine water, or with a solution of bleaching-powder, best under pressure, which dissolves the gold as chloride; the oxides are not thereby attacked. The gold is precipitated from its solution by boiling it with a solution of ferrous sulphate or oxalic acid.

A recent process for extracting gold or silver from very poor ores consisted in treating the crushed ore with a 5 per cent. solution of potassium cyanide. The gold and silver dissolve with evolution of hydrogen, while the sulphides, &c., are unattacked. To separate the noble metals from the cyanide solution, which contains them as double cyanides (see p. 571), the liquid is filtered through zinc turnings. The metals deposit in a film on the surface of the zinc, and are easily removed by washing. The cyanide is available for a second extraction.

14. **Mercury.**—The only available ore of mercury is cinnabar, HgS.

Two methods of extraction are practised. The first consists of calcining the ore in a shaft-furnace, and condensing the mercurial vapours in vessels of boiler-plate, or in brick chambers; or in the old form of condensers, named "aludels," which consist of earthenware bottles open at both ends, and fitted together like drain pipes. The second method is to distil the ore with lime, or with forge-scales.

The first method depends on the equation:—

$$HgS + O_2 = Hg + SO_2.$$

The second involves the formation of calcium sulphide and thiosulphate, from the action of the lime on the sulphur (see p 444), or of iron sulphide, while the mercury distils off, and is condensed by passing the vapour through water.

Mercury is sold in iron bottles containing about 80 lbs.

15. **Phosphorus.**—The present process for the commercial preparation of phosphorus is to prepare phosphoric acid from *apatite*, *phosphorite*, or *bone-ash*, all of which mainly consist of calcium orthophosphate. This is carried out by adding "chamber-acid" (*i.e.*, aqueous sulphuric acid, as it comes from the chambers) to the ground mineral, in such proportion as to correspond to the equation :—

$$Ca_3(PO_4)_2 + 3H_2SO_4 = 3CaSO_4 + 2H_3PO_4.$$

The solution of phosphoric acid is decanted from the precipitated calcium sulphate; the precipitate is washed; and the solution of orthophosphoric acid is evaporated. During evaporation, coke, or charcoal, in coarse powder, is added, and the mixture is dried and heated to dull redness. The product is a mixture of carbon with metaphosphoric acid, the insoluble variety (probably mono-metaphosphoric acid) being produced. Cylinders of Stourbridge clay are charged with the mixture; they are placed in tiers in a furnace, preferably heated by regenerator gases. To the mouths of the cylinders are attached copper tubes through which the phosphorus gas, carbonic oxide, and hydrogen issue. These tubes dip below the surface of warm water in pots provided with lids. The temperature is raised to bright redness, and the phosphorus distils over. The equation representing the change is :—

$$4HPO_3 + 12C = 2H_2 + 6CO + P_4.$$

The phosphorus is then in greyish-coloured lumps; to purify it, it is usually redistilled; it is subsequently melted, and moulded into sticks by causing it to flow into horizontal tubes, cooled by cold water.

It is possible to prepare phosphorus by distilling calcium orthophosphate with charcoal or coke, but the temperature required is a very high one, and it appears to be impossible to construct retorts of a sufficiently infusible material. Even the so-called "graphite" crucibles liquefy at the necessary temperature. It is generally stated that the calcium orthophosphate is converted into dihydrogen calcium orthophosphate, $Ca(H_2PO_4)_2$, and that a solution of this substance is evaporated in contact with carbon, producing calcium metaphosphate; and that, on distilling the metaphosphate with carbon, orthophosphate remains in the retort, while phosphorus distils. Such a process is certainly possible, but it is not practised, owing to the exceedingly high temperature required, and the destructive action on the retorts.

A process has recently been patented whereby phosphorus is produced directly from apatite, or from any substance containing calcium orthophosphate, by distilling it with carbon. To produce the enormously high temperature required, an electric furnace, somewhat similar to that employed in the manufacture of aluminium alloys by the Cowles process, is made use of. The floor of the furnace consists of a bed of cast iron, which serves as one of the electrodes; the other electrodes, as in the Cowles' furnace, consist of rods of gas-carbon. The cast iron becomes charged up to 8 or 9 per cent. with phosphorus, forming a phosphide; but its conductivity is not thereby impaired. The phosphorus distils over, and is condensed and purified as usual. This ingenious method has not as yet been carried out on a scale sufficiently large to render its success certain, but it is at present in operation.

The preparation of chlorine, bromine, iodine, and sulphur is intimately connected with that of various compounds; hence a description of the methods employed is reserved to the next chapter.

CHAPTER XXXIX.

PROCESSES OF MANUFACTURE.

1. **Utilisation of sulphur occurring as sulphur dioxide in furnace gases, &c.**—If more than 4 per cent. by volume of the furnace gas consists of sulphur dioxide, it is most profitable to pass it into sulphuric acid chambers. But below that amount the operation is unprofitable. Hence various expedients have been suggested for absorbing the dioxide by some substance which will permit of its recovery in the form of sulphur, or as dioxide free from admixed gases. As sulphur dioxide is produced in large amount by the calcination of sulphides, which is the usual preliminary to the extraction of metals from their ores, the problem of utilising it, or of preventing nuisance, is one of great importance.

Two methods are in practical operation. One of these depends on the absorption of the gas by water. The furnace gases, having been cooled, are passed into a tower filled with coke, down which cold water flows. If the gases contain 1 per cent. of sulphur dioxide, 1 cubic metre of water dissolves 3 or 4 kilograms; if $2\frac{1}{2}$ per cent., 8 to 10 kilograms. The aqueous solution is boiled to expel the dioxide (an operation which must obviously be performed by heat which would otherwise be wasted), and the gas is either utilised for the manufacture of sulphuric acid, or condensed by cold and pressure, or it may be converted into sulphites.

The other process consists in passing the furnace gases, cooled to 100°, through milk of magnesia (magnesium oxide mixed with water). A sparingly soluble sulphite is formed, which is collected. On heating it to 200°, magnesia is regenerated, the sulphite evolving from 30 to 33 per cent. of its weight of dioxide. The magnesia serves for a further operation.

2. **Manufacture of sulphuric acid.**—The sources of sulphur for sulphuric acid are: (*a*) Sicilian sulphur; (*b*) sulphur recovered from alkali-waste (see below); (*c*) hydrogen sulphide, also from alkali-waste; or (*d*) copper pyrites. If the last source be employed,

the pyrites, after being burned for the sulphur which it contains, is passed on to the copper works, where the copper is extracted by the wet process, described on p. 661.

The sulphur or the pyrites is burnt in kilns especially constructed for the purpose, usually rectangular boxes of fire-brick, into which the necessary amount of air may be admitted by dampers. Too little air causes sublimation of sulphur; too much causes an increase in the consumption of nitre, and lowers the yield of sulphuric acid. Iron pots containing sodium nitrate and sulphuric acid stand on the floor, or, better, in the flue, of the furnace; nitric acid is thereby generated, and its products of reaction, along with sulphur dioxide and a little trioxide, pass up flues leading from the kilns. These flues enter a dust chamber, where dust is deposited if pyrites be burned, which contains sand, arsenious and lead and iron oxides, sulphuric acid, and sometimes a little thallium oxide.

The mixture of sulphur dioxide, sulphur trioxide (3—10 per cent.), nitrous fumes, some excess of oxygen, and nitrogen then passes into a tower, named from its inventor the "Glover's tower," where it meets a descending current of sulphuric acid and water from which nitrous fumes have been liberated. The source of this acid will afterwards be described; the hot gases evaporate some of the water, and are themselves cooled thereby; the mixture of gases is now richer in nitrous fumes, and contains water-vapour in addition.

The gases now enter the "chambers." These consist of large rectangular boxes, made of lead, the joints in which are fused together, not soldered. As lead is a soft metal, the leaden chambers are supported on frameworks of wood at some distance from the ground. A number of such chambers (from three to five) are placed in a double row; they communicate by means of wide leaden pipes. The bottoms of the chambers are covered with about 2 inches of water, and are provided with valves, through which the weak acid is drawn off from time to time. Sometimes, when nitre-pots are not used for generating nitric acid in the burners, nitric acid is introduced into the first chamber, falling on an erection of earthenware pots, over which it flows, and is exposed to the action of sulphur dioxide. Steam is passed from a boiler into each chamber by a jet at one end, and a draught is produced through the whole set of chambers, usually by connecting the Gay-Lussac tower (afterwards to be described) with a chimney; the issuing gas should contain 5 to 6 per cent. of free oxygen, together with the nitrogen equivalent to the oxygen in the air admitted into the burners. Before passing into the chimney, however, these gases, which should always contain excess of oxides of nitrogen, are made to pass up a tall tower constructed of lead and packed with hard coke. Down this tower, which is called after its inventor, Gay-Lussac (1827), a regular supply of strong sulphuric acid trickles; it absorbs the nitrous fumes, forming hydrogen nitrosyl sulphate, $H.(NO)SO_4$, which dissolves in the excess of sulphuric acid. The reaction is

$$2NO + O + 2H_2SO_4 = 2H(NO)SO_4.$$

The "$2NO + O$" may consist of $NO_2 + NO$; the gas is pale orange. The issuing "nitrous vitriol" runs into an egg-shaped iron vessel, from which it is forced up a stout leaden tube to a tank at the top of the Glover's tower. Generally about half of the whole of the sulphuric acid made is passed down the Gay-Lussac tower.

The object of the Glover tower is the opposite of the Gay-Lussac tower, viz., to denitrate the sulphuric acid, and to return the nitrous fumes to the chambers. The Glover tower is also constructed of lead, but it is lined with brick and packed with flints, for coke soon becomes disintegrated by the hot gases. The gases from the pyrites-burners enter the tower at its base, and meet on their ascent with a stream of nitrous vitriol diluted with ordinary chamber acid. The degree of dilution depends on the temperature of the gases from the kilns, and on the amount of nitrous compounds in the nitrous vitriol. The gases are themselves cooled by their passage through the tower, and at the same time the diluted nitrous vitriol is concentrated, and passes out below in a concentrated form at a temperature of 120° to 130°. The steam evaporated from the diluted acid passes, along with the sulphur dioxide and nitrous fumes, into the chambers. The reaction which takes place in the Glover tower is

$$2H(NO)SO_4 + SO_2 + 2H_2O = 3H_2SO_4 + 2NO.$$

As nitrogen trioxide, N_2O_3, cannot exist in the gaseous state, it cannot be present in the chambers as such; hence the active gases must consist of nitric peroxide, nitric oxide, sulphur dioxide, and steam. There can be little doubt that the sulphur dioxide reacts with the peroxide and some of the water-gas to form hydrogen nitrosyl sulphate, thus:—

$$2SO_2 + 3NO_2 + H_2O = 2H(NO)SO_4 + NO.$$

The hydrogen nitrosyl sulphate, however, has only an ephemeral existence, being decomposed by the steam into sulphuric acid and nitric peroxide and nitric oxide, the latter of which is reoxidised by the oxygen present to peroxide, again to react with a further quantity of sulphur dioxide and steam. There is, however, always a certain loss of oxides of nitrogen, partly caused by some nitric oxide escaping through the Gay-Lussac tower, and partly, as some suppose, owing to a further reduction to nitrous oxide, which is not recoverable.

The acid from pyrites usually contains arsenic, from which it is purified, if desired, by precipitating the arsenic as sulphide, either with sulphuretted hydrogen or with a sulphide of sodium or barium. Nitrous compounds are generally removed by the addition of a little ammonium sulphate during the concentration of the acid. The escaping gas is nitrogen. To prepare pure acid, the concentrated acid must be distilled from glass retorts.

Should the Glover tower not be employed, the chamber acid must be concentrated by heating it in leaden pans, best from above. It may thus be concentrated until it has the specific gravity 1·72, containing about 79 per cent. of acid, and boiling at 200°. Up to that point, only water evaporates, and the acid does not appreciably attack the lead.

Further concentration is carried out in vessels of platinum or glass, in the form of stills; it may even be boiled down in iron basins, provided the top portion of the iron is protected from the hot weak acid. It is then filled into carboys, and brought to market. The strong acid is known as "oil of vitriol."

The method of manufacturing **sulphur trioxide,** or sulphuric anhydride, is described on p. 411. **Chlorosulphonic acid,** $Cl—SO_2OH$, is produced by passing gaseous hydrochloric acid over the hot sulphur trioxide; and anhydrosulphuric acid by mixing trioxide with oil of vitriol.

Alkali manufacture.—A large number of processes are connected with the manufacture of sodium carbonate, Na_2CO_3, among the more important of which are the preparation of caustic soda; of chlorine, with its concomitants bleaching powder and potassium chlorate; of sulphuric acid; of hydrochloric acid; of pure sulphur from pyrites; and of sodium thiosulphate.

As the object of the manufacturer is to make use of the cheapest materials, the choice of a starting point is limited. Two compounds of sodium occur in enormous quantity on the earth's surface, viz., salt, or sodium chloride, and *caliche*, or sodium nitrate from Peru. The latter is made use of chiefly as a manure; but it also serves as a source of nitric acid, which is employed in the manufacture of sulphuric acid, and, consequently, of sodium sulphate.

Carbon dioxide is a product of combustion; but, as it is thereby largely diluted with nitrogen and with unburned oxygen, it is not generally available when obtained from fuel.

The chief available source is limestone, or calcium carbonate.

If it were possible to cause salt to react quantitatively with limestone, so as to realise the equation $CaCO_3 + 2NaCl = CaCl_2 + Na_2CO_3$, the alkali maker's business would be a simple one. It is true that limestone moistened with a solution of salt does yield calcium chloride and sodium carbonate after some weeks, but only a small proportion of the whole mass reacts; hence it is necessary to introduce secondary reactions, so as to obtain a reasonable yield of the desired product from the raw materials.

There are two methods of achieving this object which are practically successful. These are:—

1. **The Leblanc soda-process**; and
2. **The ammonia soda-process.**

We shall consider these in their order.

1. **The Leblanc soda-process.**

The equation to be realised is, as already mentioned,

$$CaCO_3 + 2NaCl = CaCl_2 + Na_2CO_3.$$

In fact, an attempt is made to realise the still simpler equation,

$$2NaCl + H_2O + CO_2 = Na_2CO_3 + 2HCl;$$

or, if caustic soda is required, the corresponding equations,

$$Ca(OH)_2 + 2NaCl = CaCl_2 + 2NaOH, \text{ or}$$
$$H_2O + NaCl = HCl + NaOH.$$

The operations in the Leblanc process consist—

A. In preparing sodium sulphate from salt.

B. In converting the sulphate into sulphide of calcium and sodium carbonate by heating with lime and coal.

C. In crystallising out the decahydrated sodium carbonate, $Na_2CO_3.10H_2O$ (soda-crystals), or in producing dry sodium carbonate or soda-ash.

To these are added:—

D. The causticising of the sodium carbonate, producing sodium hydroxide; and

E. The recovery of the sulphur from the calcium sulphide.

A. **The preparation of sodium sulphate.**—This is achieved

either (*a*) by exposing salt to the action of sulphur dioxide, steam, and air (Hargreaves' process); or (*b*) by treating salt with sulphuric acid.

(*a*.) **The Hargreaves' process** for manufacturing sodium sulphate and hydrochloric acid.—The gases obtained by the combustion of sulphur, or more usually of pyrites, are passed *downwards* through salt contained in cast-iron cylinders enclosed in a fire-brick casing, and provided with fire-places and flues. Much depends on the physical state of the salt. It should be porous, and yet not too closely packed. It is moistened and moulded by pressure into cakes, and they are broken up and packed in the cylinders, which are furnished with perforated shelves, or grids, so that the pressure of the salt in the cylinder may not consolidate the lower layers. The sulphur dioxide produced in pyrites-kilns should contain the requisite excess of oxygen to convert it into trioxide, and is mixed with steam, which is blown into the pipes leading from the pyrites-burners. The temperature of the cylinders should be maintained as close as possible to 500—550°. At first external heat is required, but the heat developed by the reaction is afterwards sufficient. Each cylinder is capable of holding 40 tons of salt, and eight cylinders form a set. The reaction is a slow one, and takes several weeks.

The issuing gases are drawn off by means of a fan; they consist of hydrogen chloride and nitrogen, along with the excess of oxygen. The hydrogen chloride is condensed in coke towers, as will afterwards be described.

The reaction is of the simplest kind, and is shown by the equation:—

$$2SO_2 + O_2 + 2H_2O + 4NaCl = 2Na_2SO_4 + 4HCl.$$

(*b*.) **Sulphate from salt and sulphuric acid.**—This is the original process for manufacturing sodium sulphate. Coarse-grained salt is mixed with sulphuric acid of 70—80 per cent. (140° Tw.), in spoon-shaped iron pans, covered by a close arch of brick, through which a stoneware pipe passes. The sulphuric acid is mixed hot. When action has ceased, the salt is converted into hydrogen sodium sulphate, thus:—$NaCl + H_2SO_4 = HCl + HNaSO_4$. The hydrogen chloride passes off through the stoneware pipe in the arch of the furnace to a tower filled with coke, down which water flows. It is dissolved, and runs out at the foot of the tower in a saturated condition. The hydrogen sodium sulphate, containing excess of salt (for the total quantity of salt corresponding to the equation $2NaCl + H_2SO_4 = Na_2SO_4 + 2HCl$

is added at the commencement), is raked from the pan into the "roaster," a trap being lifted to permit of the transfer. Sometimes a "blind roaster," *i.e.*, a furnace heated from outside, is made use of; but such furnaces are difficult to keep tight; it is more usual to employ an open roaster, where the products of the combustion of coke play directly on the mixture of salt and acid sulphate. The gases are then condensed in a separate coke-tower, and furnish a weaker acid, which is made use of instead of water for the coke-tower connected with the "pan." When all action has ceased, the "salt-cake" is raked out of the furnace.

It is now common to use a mechanical furnace, consisting of a rotating disc of brick-work on an iron frame, covered by a stationary hood, and provided with mechanical stirrers. The heat is supplied from a furnace at the side, the products of combustion playing directly on the mixture of salt and sulphuric acid. This mixture enters from a hopper at the top of the hood, and is distributed slowly towards the side of the disc by the mechanical stirrers; it is from time to time dropped through traps into trucks placed to receive it. The gases pass out through the top of the hood, entering an iron pipe, for iron is not attacked by gaseous hydrogen chloride above a certain temperature. As the gas cools, it enters pipes of stoneware or glass, which lead it to the condensing tower.

Manufacture of sodium carbonate by the Leblanc process.—The materials are, *salt-cake*, which should be porous and spongy; *limestone*, or roughly crushed chalk, as free as possible from magnesia or silica; and *small-coal*, or "slack," as free from ash as possible, so as to avoid formation of calcium silicate. These materials are crushed and mixed together, usually in the proportion of 100 parts of sulphate, 80 of limestone, and 40 of coal.

When hand-furnaces are used, the materials are heated directly by the gases of combustion and stirred by means of rakes; the mixture sinters, but does not quite fuse. It is moved about, and finally gathered into balls of "black-ash." These are withdrawn from the furnace and allowed to cool.

Mechanical revolving furnaces are now common; a cylinder of iron plate, lined with fire-brick, lies horizontally on rollers, so that it can be made to revolve on its axis. One end of the cylinder abuts on a furnace, of which the combustion-products pass through the cylinder, escaping through a flue, similarly abutting on its other end. The cylinder is charged through an opening in the middle, through which it can be filled when the opening is above, or emptied when the cylinder is turned round. This hole is closed with a door, luted on with clay, after filling the cylinder with its charge, which consists usually of 100 parts of sulphate, 72 of limestone, and 40 of coal. The cylinder is made slowly to rotate, and the charge is thereby mixed and tossed, while the flames from the furnace play through. After $2\frac{1}{4}$ to $2\frac{1}{2}$ hours the operation is ended. The opening is brought to the top, and 10 parts of quicklime mixed with 12 to 16 parts of cinders are thrown in. The door is replaced, and for a few minutes the cylinder is rapidly rotated, so as to mix the black-ash with these additions. The object of adding them is that, on subsequently lixiviating the ash, the lime may slake and burst up the lumps, and thus allow the water quickly to dissolve the carbonate. It is also customary to add some fresh sodium sulphate at the end of the operation, in order to decompose cyanide, thus:—$Na_2SO_4 + 4NaCN = Na_2S + 4NaCNO$. The rotation of the cylinder is finally stopped, the door removed, and the charge emptied into iron trucks, brought successively below the opening. The charge of black-ash in the trucks sends off from all parts of its surface flames of carbon monoxide, coloured yellow, of course, by sodium.

The action which takes place in hand-furnaces is believed to consist of the reduction of the sulphate in the upper portion of the mixture to sulphide, and the expulsion of carbon dioxide from the limestone. But the lime is recarbonated by the carbon dioxide produced by the reduction of sulphate in the lower layers by the coal. When all the sulphate is reduced, the temperature rises, and calcium carbonate is decomposed, reacting with the sodium sulphide. The escape of carbon monoxide takes place only at the end of the operation, and renders the mass porous. The reactions in a hand-furnace are probably represented by the equations:—

$$5Na_2SO_4 + 10C = 5Na_2S + 10CO_2;$$
$$5Na_2S + 5CaCO_3 = 5Na_2CO_3 + 5CaS;$$ and, at the end,
$$2CaCO_3 + 2C = 2CaO + 4CO.$$

The black-ash, when nearly cold, is broken into lumps, and placed in the lixiviating tanks, so arranged that the strongest liquor comes in contact with the fresh black-ash, and is drawn off saturated, while the ash already partially extracted is exposed to the action of the weakest liquors. The ash must not remain too long in contact with the water, nor must the temperature be high, else back action commences, and the calcium sulphide and sodium carbonate react to form calcium carbonate and sodium sulphide. The liquor, when drawn off, is turbid, and has a green colour, due to a trace of sodium ferrous sulphide. It is allowed to settle, and transferred to pans heated by the waste heat from the black-ash furnace, preferably from above. On evaporation, salts separate, which are "fished" out with perforated ladles. They consist at first of sodium chloride and sulphate. The liquor contains carbonate of soda, and an equal quantity, or even more, caustic. On further concentration, $Na_2CO_3.H_2O$ separates and is removed. The red liquor, as the mother liquor is termed, is often made into caustic, but often it is evaporated to dryness and calcined with sawdust; or, better, treated before boiling down with carbon dioxide, either from furnace gases or from lime-kilns, in order to carbonate the caustic. A special advantage of the last process is that sulphide of sodium is thereby converted into carbonate, whereas by ignition with sawdust it remains unaffected; it is sometimes converted into sulphate by ignition in air after carbonation.

The carbonate is sometimes purified by redissolving, settling, evaporating, fishing, and igniting. It is then ground and packed.

Crystal soda, $Na_2CO_3.10H_2O$, is made from the impure yellow carbonate. It is dissolved, and after settling, it is run into tanks, and allowed to cool. The mother liquor is boiled down; it contains sulphate, but is useful for glass-making.

The "**bicarbonate,**" or hydrogen sodium carbonate, $HNaCO_3$, is produced by treating the crystals with carbon dioxide made from limestone and acid. The water of the hydrated carbonate drops away, and masses of bicarbonate are left, which retain the form of the original crystals.

Caustic soda.—To prepare "caustic," the tank liquors, which already contain much caustic, are run into iron tanks and mixed with lime. The calcium carbonate is filtered off through cloth filters, and returned to the black-ash furnace. The liquors are concentrated in a "boat-pan," *i.e.*, a pan shaped like a boat, of which the sides alone are heated from below, and the salts deposit on the bottom. They are fished, and returned to the black-ash furnace. A little sodium nitrate is then added to oxidise the sulphide, thus:—

$$2Na_2S + NaNO_3 + 3H_2O = Na_2S_2O_3 + 3NaOH + NH_3.$$
$$2Na_2S_2O_3 + 3NaOH + NaNO_3 = 4Na_2SO_3 + NH_3.$$
$$4Na_2SO_3 + NaNO_3 + 2H_2O = 4Na_2SO_4 + NaOH + NH_3.$$

The sulphide is thus ultimately converted into sulphate and caustic, with escape of ammonia.

The liquors, after concentration, are then transferred to the "finishing pans," thick hemispherical iron pots, where the final water is removed. At a certain stage graphite separates, owing to the decomposition of the cyanide, and even here a little sodium nitrate is added to remove the last trace of sulphide. The fused caustic is then poured into iron drums, in which it is brought to market.

Utilisation of tank-waste.—The recovery of sulphur from the calcium sulphide in the tank-waste would, if complete, cause the manufacture of sodium carbonate ultimately to resolve itself into a reaction between calcium carbonate and sodium chloride, if calcium carbonate is the final waste product; or between water and sodium chloride, if the calcium be recovered as carbonate and returned to the soda-ash furnace. The latter is the more perfect theoretically.

As a sample of the first, *Schaffner's process* may be cited. In it the muddy deposit, after lixiviation of the black-ash and removal of the sodium carbonate, was allowed to lie exposed to air; Mond introduced a blast of air for oxidation; and the oxidised waste, consisting largely of calcium thiosulphate mixed with a portion of unoxidised material, was treated with hydrochloric acid. A reaction similar to the following took place:—

$$CaS_2O_3.Aq + 2CaS_x + 6HCl.Aq = 3CaCl_2.Aq + 3H_2O + (2x + 2)S.$$

The sulphur sludge was allowed to deposit, or, better, the sulphur was melted by heating under pressure, and recovered; the calcium chloride was run to waste. Were the Leblanc process theoretically perfect, such recovery of sulphur would dispose of all the hydrochloric acid made, leaving no surplus for the manufacture of bleaching powder.

Schaffner and Helwig are the inventors of another process, which, however ingenious, has not succeeded commercially. In it the waste was treated with magnesium chloride, calcium chloride and magnesium sulphide (or hydrosulphide) being obtained, thus:—

$$CaS + MgCl_2.Aq = CaCl_2.Aq + MgS.Aq.$$

The solution of magnesium sulphide (or hydrosulphide) when heated to 80° yielded oxide and hydrogen sulphide, thus:—

$$MgS + H_2O = MgO + H_2S.$$

The hydrogen sulphide was stored in gas-holders, leakage,

owing to diffusion through water, being prevented by sealing the gasometers with heavy petroleum oil. The sludge of calcium chloride and magnesium oxide was treated with carbon dioxide under pressure, and yielded calcium carbonate and magnesium chloride, the latter being utilised for further treatment of waste, thus:—

$$CaCl_2.Aq + MgO + CO_2 = CaCO_3 + MgCl_2.Aq.$$

The precipitated calcium carbonate was returned in a dry state to the black-ash furnace. The hydrogen sulphide was utilised in the manufacture of sulphuric acid by direct burning, or, by mixing with sulphurous acid, made to yield up its sulphur.

Chance's recovery process promises better results. It consists in saturating with carbon dioxide the waste, mixed with water, and contained in a series of vertical cylinders. In the first cylinder into which the carbon dioxide enters, a portion of the calcium sulphide is converted into carbonate, while the hydrogen sulphide converts the remainder into calcium hydrosulphide thus:—

$$2CaS + H_2CO_3.Aq = CaCO_3 + Ca(SH)_2.Aq.$$

The further action of the carbon dioxide is to decompose the hydrosulphide, the sulphuretted hydrogen passing into the second cylinder, thus:—

$$Ca(SH)_2.Aq + H_2CO_3.Aq = CaCO_3 + 2H_2S + Aq.$$

The waste in the second cylinder is thus converted into soluble hydrosulphide of calcium, in its turn to be decomposed by the carbon dioxide. The hydrogen sulphide is thus driven from cylinder to cylinder, until finally it is expelled. When the first cylinder is exhausted of hydrogen sulphide, it is thrown out of circuit, to be recharged with fresh waste; it then becomes the end cylinder of the circuit. The resulting calcium carbonate may, when dried, be returned to the black-ash furnace.

The hydrogen sulphide may be utilised in the manufacture of sulphuric acid, but it pays better to recover it in the form of sulphur. This is done by *Claus's process*. The sulphuretted hydrogen is burned below a layer of ferric oxide, air being carefully regulated, so that one-third of the total gas is converted into dioxide. The dioxide reacts with the hydrogen sulphide, in contact with the hot and porous ferric oxide, yielding sulphur and water, thus:—

$$2H_2S + SO_2 = 2H_2O + 3S.$$

The sulphur distils over, and is condensed in suitable brick chambers, the nitrogen of the air passing on. The sulphur is melted under hot water at a high pressure, and brought to market.

Manufacture of chlorine.—The manufacture of chlorine is intimately connected with the Leblanc soda process, for hydrogen chloride is thereby produced in large amount. There are two remunerative processes for the manufacture of chlorine: the usual one, viz., the treatment of manganese dioxide with hydrochloric acid; and the mutual action of air and hydrogen chloride at a high temperature, in presence of some material (best, copper chlorides) capable of inducing their action. The last process is generally called the "Deacon chlorine process," for Mr. Deacon was the first to make it a commercial success. The first process is always worked so as to recover the manganese; this improvement is due to the late Mr. Weldon.

1. **The manganese chlorine process.**—The chief source of the manganese ore is the Spanish province of Huelva; the ore consists essentially of dioxide. It is broken into fragments smaller than a hen's egg, and placed in "chlorine stills," built of sandstone boiled in tar; such stills are sometimes cut from a single block of sandstone. They are circular or octagonal troughs, covered with a block of sandstone, the junction between the cover and the trough being made tight by a circular band of india-rubber, on which the cover rests. The acid is admitted and the chlorine evolved through holes cut in the cover. The ore (6 to 10 cwt.) is placed on a grating or table standing in the still, and acid is run in till the still is three-quarters full. The first reaction takes place quickly; it consists probably in the formation of $MnCl_3$, thus:—$2MnO_2 + 8HCl = 2MnCl_3 + 4H_2O + Cl_2$. When the first action has ceased, steam is blown in for about 10 minutes, so as to complete the reaction—$2MnCl_3 = MnCl_2 + Cl_2$; fresh steam is introduced at successive periods of an hour. Only a portion of the hydrochloric acid is used; 6 to 10 per cent. remains free after the action is completed, and the chlorine produced amounts to only about one-third of that contained in the acid used.

Weldon's manganese-recovery process.—The manganese chloride from the stills is run off and mixed in a covered tank with calcium carbonate to neutralise the free acid, and to precipitate the iron from the ore as ferric hydroxide. The precipitate is allowed to settle, and the clear liquor is mixed with milk of lime free from magnesia. This precipitates manganese hydroxide, $Mn(OH)_2$, and the object of the process is to convert this hydroxide into hydrated dioxide by means of a blast of air. If air were blown through such moist hydroxide at a high temperature, only one-third of the manganese would be oxidised, the product being hydrated Mn_3O_4, which may be viewed as $MnO_2.2MnO$; at the ordinary temperature the action would be very slow, and would lead to the formation of $Mn_2O_3 = MnO_2.MnO$. But if *excess* of lime be added in addition to that required to precipitate manganous hydroxide, and if the mud of hydroxides of manganese and calcium be exposed to air in a hot state, a mixture of manganites of calcium, of the formulæ $CaO.2MnO_2$ and $CaO.MnO_2$, is formed, in which all manganese is in the state of peroxide. The presence of calcium chloride in the liquid is desirable, inasmuch as it is then better able to dissolve lime, and to bring about its combination with the peroxide.

The mud is placed in tall iron cylinders and heated by blowing in steam; air is then forced in. The temperature should not exceed 65°, else Mn_3O_4 and Mn_2O_3 are formed. The manganese rapidly oxidises, and the colour of the mud changes to black. When the amount of dioxide no longer increases, manganous chloride is run in, when the reaction occurs:—

$$2(CaO.MnO_2) + MnCl_2 + H_2O = CaO.2MnO_2 + Mn(OH)_2 + CaCl_2.$$

The mud is then run into tanks and the calcium chloride drawn off from the precipitated sludge.

A different form of chlorine still, taller in proportion to its diameter, and unprovided with a grating, is made use of. It is charged with hydrochloric acid, and the mud is run in as long as chlorine is evolved; the mixture grows hot, and nearly all the hydrochloric acid may be utilised, only ½ to 1 per cent. remaining free at the end of the reaction. It is advisable to employ the acid liquor from the stills charged with manganese ore along with fresh hydrochloric acid in the mud stills; the free acid is thus saved.

Another process of manganese recovery, practised on a limited scale, is due to Mr. Dunlop. It consists in converting the manganous chloride into carbonate by heating its solution under pressure with chalk; the calcium chloride is removed, and the precipitated manganous carbonate, while still moist, heated in a current of air. It is thus converted into dioxide, which is used for a subsequent operation, as in the Weldon process.

2. **The Deacon chlorine process.**—The fundamental reaction of this process is expressed by the equation $4HCl + O_2 = 2H_2O + 2Cl_2$. But the action is a very incomplete one unless some porous substance be present, and even then it would amount to only a small fraction of the theoretical product. The presence of copper chlorides causes it to take place, and on the small scale as much as 90 per cent. of the hydrogen chloride has been thus decomposed. The cuprous chloride effects the reaction $2Cu_2Cl_2 + 4HCl + O_2 = 4CuCl_2 + 2H_2O$, and the cupric chloride evolves chlorine, regenerating cuprous chloride, thus :—$2CuCl_2 = 2Cu_2Cl_2 + Cl_2$. The reaction can take place only at such a temperature that the absorption and evolution of chlorine are at a balance; it begins at 204°, and is most active between 373° and 400°. At 417°, cupric chloride volatilises. The reaction depends not on the amount, but on the surface, of the copper chlorides, and to increase surface, fragments of brick soaked in copper sulphate are employed. The sulphate is rapidly converted into chloride.

In this process the hydrogen chloride may be taken directly from the salt-cake pans or the decomposing furnace. It enters a set of pipes, in which its temperature is raised to 400°; it then passes into a cylindrical chamber, filled with prepared brick, and surrounded by a non-conducting jacket to prevent heat escaping. At the commencement, this chamber is heated by the waste heat from the fire employed to heat the gas, but it maintains its own temperature after a short time. The temperature must be carefully regulated during the whole operation. The exit from the brick-chamber leads to a series of glass or earthenware pipes, in which the mixture of escaping gases—chlorine, hydrogen chloride, nitrogen, excess of oxygen, and steam—is cooled; hydrogen chloride is removed by passing the gases upwards through a coke-tower, down which water flows; and, if required for the manufacture of bleaching powder, the gases are dried by passing through a similar tower fed with oil of vitriol. They then pass through a Root's blower; this blower ensures the entry of sufficient air from leakage in the decomposing furnace, pipes, &c., to supply oxygen for the hydrogen chloride.

The cuprous bricks become exhausted after some time, probably losing their porosity, and one of the disadvantages of the process is the necessity of frequently replacing them by freshly prepared ones. The amount of decomposed hydrogen chloride on the large scale seldom exceeds 45 per cent. of the total amount present.

Bleaching powder.—The chief use of chlorine is in the manufacture of bleaching powder, "bleach," or "chloride of lime." The upshot of many researches on the formula of bleaching powder has been to show that it undoubtedly consists of the compound $Ca{<}^{Cl}_{OCl}$, with a little free lime mechanically protected from the action of chlorine, and about 15 per cent. of water. It has been proved to contain no calcium chloride, because all chlorine is expelled by passing over it a current of carbon dioxide; calcium chloride treated thus of course remains unaltered (see also p. 463).

For the manufacture of good bleaching powder, the lime must be specially pure, well slaked, and free from lumps, and it should contain no magnesia. It is exposed to the action of chlorine, made by the manganese process, and therefore nearly pure, spread on the floors of chambers of brickwork or lead or cast-iron, six feet high, and of considerable area; these chambers are protected internally by a layer of cement and tar. The gas is introduced in the roof, and descends, owing to its weight. The progress of the operation is seen through windows in the cast-iron doors; when the chambers are seen to be green, admission of chlorine is stopped. The chlorine being forced in under a slight pressure, some fresh lime is thrown in to absorb that remaining in the chambers; workmen then enter, their mouths protected by bandages of wet cloth, and rake the powder so as to expose fresh surfaces. Chlorine is again introduced, and when absorption again ceases, the powder is removed and packed in casks. The amount of available chlorine, *i.e.*, the amount which is capable of liberating iodine, acting as an oxidiser, &c., is usually 37 to 38 per cent., but may in exceptional circumstances rise to 43 per cent. During the whole operation the temperature is kept as low as possible.

Should the chlorine be made by the Deacon process, and be consequently diluted with nitrogen and oxygen, a larger surface of lime may be exposed, because there is less danger of heating, and because the chlorine is not so greedily absorbed. The chambers are filled by a series of slate shelves, distant from each other about a foot, and so arranged that the gas zig-zags, passing over shelf after shelf on its way from the top of the chamber to the ground. The weaker chlorine is brought into contact with fresh lime, and the fresh chlorine with lime already partially charged. By this method the bleaching powder contains a less percentage of available chlorine than if purer chlorine be used; the amount seldom exceeds 36 per cent.

Potassium chlorate.—The chlorine may also be utilised in the manufacture of potassium chlorate, though for this substance there is only a limited demand. To prepare chlorate, the chlorine is led into a closed leaden tank, filled with milk of lime, continually agitated and splashed by a mechanical stirrer. It is rapidly absorbed, with great evolution of heat; the hypochlorite of calcium first formed is rapidly changed into chlorate. The reaction occurs:—

$$6Ca(OH)_2.Aq + 6Cl_2 = 5CaCl_2.Aq + Ca(ClO_3)_2.Aq + 6H_2O.$$

In actual practice it is found that the ratio of chlorine in chloride to that in chlorate is 5·5 to 1, instead of 5 to 1, as required by theory; some oxygen is said to escape (?). After the lime mud has settled, potassium chloride equivalent to the calcium chlorate is added to the liquor drawn off from the sediment, and the mixture is evaporated in iron pans. On cooling, the less soluble potassium chlorate crystallises out, leaving the very soluble calcium chloride in the mother liquor. On cooling the mother liquor, a fresh crop of crystals of chlorate separates. The potassium chlorate is rendered sufficiently pure by a second crystallisation from hot water.

Sodium chlorate is made from potassium chlorate by treating a solution of the latter with hydrosilicifluoric acid, prepared by passing silicon fluoride into water; the insoluble silicifluoride of potassium is removed, and the liquid is evaporated till crystals separate.

All these operations are often concurrently carried out in an alkali-work employing the Leblanc soda process. It is a matter of choice whether the manganese chlorine process or that of Deacon be employed, but most works prefer the former. It will be seen that with recent improvements most of the by-products are utilised. The sulphur is recovered; the carbon dioxide required to decompose the waste may be obtained from the lime-kilns; the calcium carbonate produced from the waste is returned to the black-ash furnace; but of the chlorine of the salt, two-thirds are rejected as calcium chloride by the manganese chlorine process, whereas, by Deacon's process, one half is at least utilised. The manganese dioxide employed in the manufacture of chlorine is regenerated, but here again with an expenditure of lime.

The other important process for the manufacture of alkali is theoretically more perfect, but up to the present it has not been found possible to combine it with the profitable manufacture of chlorine.

2. **The ammonia-soda process.**—The fundamental reaction involved by this process is:—

$$NaCl.Aq + NH_3.Aq + H_2O + CO_2 = HNaCO_3 + NH_4Cl.Aq.$$

It was first made successful by M. Solvay, about 1864.

Brine from a salt-mine, purified from salts of magnesium by the addition of a little milk of lime, and of the added lime by ammonium carbonate, is filtered and cooled; it is brought up to saturation by addition of solid salt, and saturated with ammonia, produced by heating ammonium chloride with calcium hydrate. It is then introduced into vertical cylinders of considerable height (36 to 63 feet), provided with perforated shelves; or, more usually, it is placed in a set of shorter cylinders, arranged in a vertical column, their united height being equal to that mentioned. Carbon dioxide from the lime-kiln in which the lime-stone is burned in order to provide lime to decompose the ammonium chloride, after being washed and cooled, is pumped in at the bottom of the lowest cylinder, in a series of jerks, by a powerful pump. This carbon dioxide is necessarily dilute, containing about 25 per cent. CO_2. Towards the end, as the liquid becomes saturated, purer carbon dioxide, obtained by heating sodium hydrogen carbonate, is made use of. Crusts of hydrogen sodium carbonate deposit on the perforated shelves; when the operation is complete, the liquid containing undecomposed salt, ammonium chloride, and excess of ammonia, is run off; it enters the stills, where it is treated with milk of lime and where the ammonia is recovered. During the passage of carbon dioxide, heat is generated in the cylinders, but they are cooled from the outside by cold water, and, moreover, the expansion of the carbon dioxide on its passage upwards absorbs heat, so that the temperature does not rise greatly. The escaping dioxide is passed through fresh brine, so as to deprive it of ammonia.

Water is then run into the cylinders and steam is blown in. The crusts of bicarbonate of soda dissolve, and, on cooling, separate in crystals. It is collected, dried at 60°, and heated in closed vessels to expel carbon dioxide, which again passes into the decomposing cylinders.

This process should theoretically realise the equation 2NaCl +

$CaCO_3 = CaCl_2 + Na_2CO_3$, if the ammonia were perfectly recovered; but in practice the loss of ammonia amounts to about 6 to 8 per cent. of the carbonate of soda formed.

The product is, of course, free from sulphides and caustic soda, but it is less dense than that obtained by the Leblanc process. A very large amount of carbonate is now made by this process; it may be causticised in the usual way if caustic soda is required, or ferrate of sodium, Na_2FeO_3, may be made by heating it with ferric oxide with free access of air, and then decomposed by water, the ferric oxide being precipitated, while caustic soda remains in the liquor.

A modification of this process which is also worked consists in the preparation of solid hydrogen ammonium carbonate by the action of carbon dioxide on ammonia solution. This solid is filtered on to cloth filters and then watered with a solution of salt; it is thereby converted into bicarbonate of soda, while liquor containing ammonium chloride, one-third of the salt used, and one-third of undecomposed ammonium hydrogen carbonate, runs through. It is, however, difficult to prevent considerable loss of ammonia, and the large mass of material on the filters is difficult to handle. It is then ignited in a reverberatory furnace, the carbon dioxide being lost.

These are among the most important chemical processes carried out on a large scale. For detailed information the reader is advised to consult Lunge's *Sulphuric Acid and Alkali Manufacture;* Richardson and Watts' *Technical Dictionary;* Thorpe's *Dictionary of Applied Chemistry;* and, above all, the *Journal of the Society of Chemical Industry.*

INDEX.

MINERALS AND ORES.

GENERAL INDEX.

NOTE.—Fluorides, chlorides, bromides, and iodides are included under the head **halides**.

Sulphides, selenides, and tellurides, under the head **sulphides**.

Nitrides, phosphides, arsenides, and antimonides, under the head **nitrides**.

2 Y

HARRISON AND SONS, PRINTERS IN ORDINARY TO HER MAJESTY, ST. MARTIN'S LANE, LONDON

Lightning Source UK Ltd.
Milton Keynes UK
UKHW010422191218
334175UK00003B/41/P

9 780656 116171